LANDSCAPES OF SURVIVAL

sidestonepress

LANDSCAPES OF SURVIVAL

THE ARCHAEOLOGY AND EPIGRAPHY OF JORDAN'S NORTH-EASTERN DESERT AND BEYOND

edited by
PETER M.M.G. AKKERMANS

Published by Sidestone Press, Leiden
www.sidestone.com

Lay-out & cover design: Sidestone Press

Photograph cover: Peter Akkermans – Jebel Qurma Archaeological Landscape Project

ISBN 978-90-8890-942-9 (softcover)
ISBN 978-90-8890-943-6 (hardcover)
ISBN 978-90-8890-944-3 (PDF e-book)

Contents

Foreword

The Black Desert, or *Al-Harrah*, extends from southern Syria into eastern Jordan and northern Saudi Arabia. It is covered by dark basalt stones and was inhabited by ancient civilisations. This region is rich and diverse, and is a promising place to carry out a range of scientific research, for both cultural and environmental topics. There are plains, lakes, and temporary streams in large valleys that cross the desert and support unique plant and animal species. There is a rich diversity of life in the desert, attractive for human settlement. Despite the harsh climatic conditions of these areas at present, perhaps these environments were different and more suitable for supporting human communities in the past. The evidence for these communities is prevalent across the Black Desert, with preserved architectural remains, burials, inscriptions, rock art, and other traces of material culture.

Many of these aspects were addressed by the scientific papers presented by scientists and researchers at the conference dedicated to the Black Desert held at Leiden University. The conference was organised by Professor Peter Akkermans, who has conducted many scientific studies related to the history of the Black Desert. Contributors to the conference included researchers working on a range of current and ongoing research projects in this field. This is evidence that the Black Desert still contains many secrets that require additional research and study.

The research papers presented at this conference are important for the region, since they highlight relevant historical and scientific aspects that will have a major impact on the state of archaeological investigation in the Black Desert. Taken together, the results included within this volume will play an important role in increasing the interest of archaeologists in conducting future research in the Black Desert. At the same time, they shed light on previously unknown aspects of the area's archaeological and environmental history. The scientific developments presented in these papers offer new points of departure for the investigation of the very rich past of the Black Desert.

It is my wish that this book and the wide range of topics it addresses will be of great interest and use to researchers.

Amman, 1 May 2020

Aktham Oweidi
Director of Excavations and Surveys
Department of Antiquities of Jordan

Introduction: landscapes of survival

Peter M.M.G. Akkermans

Introduction

The present volume presents papers arising from an international conference of the same title, held at Leiden University, The Netherlands, in spring 2017. The meeting addressed the need for dialogue among researchers involved in the amazingly rich archaeological and epigraphic record of Jordan's 'Black Desert' and its adjacent regions. It aimed to explain the prominent yet understudied achievements of the indigenous peoples through the ages and to develop new, comparative perspectives on desert cultural landscapes. Guiding themes of this meeting were regional outlooks, chronologies, population dynamics, transitions, habitations, burial practices, mobility and landscape, ecology and environment, connectivity, literacy, marginality, the role of rock art, and the constitution of local material culture.

The idea for the conference (and, hence, this publication) grew out of several research projects in the north-eastern Black Desert, initiated by the author and carried out under the auspices of Leiden University since 2012. One of these initiatives is the Jebel Qurma Archaeological Landscape Project, comprising survey and excavation in the Jebel Qurma heights, east of Azraq. Another work is the Landscapes of Survival Project, an important offshoot of the former programme, which focused on the collaborative study of sites, rock art, and (Safaitic) inscriptions in Jebel Qurma. Before introducing the papers in this volume, it seems useful to present briefly these projects and their geographical background.

The Black Desert

The 'Black Desert' or Harrat al-Sham begins just south of Damascus and comprises some 40,000 km^2 of dark basalt fields, which stretch from southern Syria across north-eastern Jordan and reach the sand sea of the Nefud in Saudi Arabia. The broken basalt cover derives from the weathering of a complex series of lava flows that emerged from volcanic vents and fissures (Bender 1968; Edgell 2006). The vast and desolate wasteland consists of gently rolling country with black basalt boulders and endless gravel plains, alternating with ranges of steep-sided, flat-topped mounds of thick basalt (Fig. 1). The rough and rocky dissected terrain is often difficult to access and travel through. Particularly in the southern half of the Harrat al-Sham, the surface roughness promotes the accumulation of wind-blown sands and the formation of dunes. Extensive mud pans form hard, flat, and glaring tracts of white silt and sand in the dry part of the year and shallow, marshy lakes in the wet season.

The local climate is harsh, with much seasonal and annual variation. The basalt area is highly arid, with average annual precipitation ranging from 200 mm in southern Syria to less than 50 mm in Jordan and north-western Saudi Arabia. Most rainfall occurs intermittently in the form of cloud bursts from November until March, which results in considerable surface runoff and the subsequent flooding of stream channels and mud flats.

In: Peter M. M. G. Akkermans (ed.) 2020: *Landscapes of Survival - The Archaeology and Epigraphy of Jordan's North-Eastern Desert and Beyond*, Sidestone Press (Leiden), pp. 9-16.

Figure 1. The desolate, harsh, basalt landscape in the Black Desert in north-eastern Jordan (photograph: Jebel Qurma Project Archive).

The summers are dry and hot, with mean temperature maxima of 35-38 °C and common outliers as high as 45 °C. Occasionally strong winds lead to dust storms. Winters may be severe, with cold air gusts and an average temperature of 2-9 °C, and minima as low as −10 °C. Such a range of temperatures adds to the harsh and inhospitable character of the basalt expanse (*e.g.* Betts 1998; Dutton *et al.* 1998). Captain Lionel Rees, who was in the Black Desert in the 1920s, described the area: "Except for a short period in the spring the whole of this country looks like a dead fire – nothing but cold ashes." (Rees 1929, 389).

The rather uninviting appearance of the Black Desert is difficult to reconcile with its very large numbers of stone-built installations of different types and sizes (enclosures, huts, hunting installations, burial cairns, *etc.*) as well as the innumerable pieces of rock art and North Arabian inscriptions: the enduring testimonies of those who – almost a century ago – were referred to as 'The Old Men of Arabia' (Maitland 1927). The use of the region may have been facilitated by wetter and greener environmental conditions during some periods in the past than today, although this is still a matter of investigation (see *e.g.* the discussions in Rollefson 2016; Akkermans 2019).

Satellite imagery, aerial photographs, and archaeological fieldwork have revealed the immense archaeological and epigraphic record of the region. This material testifies to a high degree of success achieved by the nomadic peoples in exploiting the basalt range through hunting and herding many hundreds or thousands of years ago. Dismissing these thriving desert communities on the fringes of (urban) civilisation as 'marginal' or 'insignificant' would fail to do justice to them: "From their perspective, the notion of marginality may have been of no meaning whatsoever: marginal to whom or what? Why would they consider themselves to be living on the margins, as they were able to successfully sustain a long-lived desert culture that was fully adapted to a difficult terrain and climate? Even if the rough basalt uplands were ecologically and economically peripheral in comparison to the fertile Levantine regions and trading cities to the west, they were still culturally central to the communities that continually used them for many centuries." (Akkermans 2019, 427).

Although we should not underestimate the degree of continual change and modification, to a very large extent the basalt uplands represent what has been termed 'landscapes of preservation' or 'landscapes of survival'. Such landscapes preserve settlement remains of great antiquity with remarkable clarity. These are relic landscapes where the tangible imprints of both prehistoric and historic local lifeways as well as their dynamic social constructs are often still highly intact, because eradication by later activities or incorporation into later cultural landscapes was relatively restricted. Successive phases of habitation and use have left a relatively light

Figure 2. The Jebel Qurma range: basalt-covered mounds and plateaus, with endless plains of gravel and stone in front of them (photograph: Jebel Qurma Project Archive).

imprint on the landscape (Taylor 1972; Wilkinson 2003; 2004; Lawrence 2012). One of these typical landscapes of survival is the Jebel Qurma range in north-eastern Jordan.

Jebel Qurma Archaeological Landscape Project

To date, there is a considerable (and ever-increasing) interest in the archaeology and epigraphy of the Black Desert and its adjacent arid landscapes. In particular, Jordan is home to a wide range of desert-oriented research projects, which convincingly demonstrate the archaeological affluence of the *badia* and its potential for human use throughout the ages (see the contributions in the present volume).

One of these research initiatives is the Jebel Qurma Archaeological Landscape Project, east of Azraq, with annual fieldwork since 2012. Through survey and excavation in Jordan's north-eastern basalt expanse, this project aims to address continuities and changes in local ways of life and death through the ages, as well as to explore the intimate relationship between these factors and the highly diverse landscape. The project is directed by the author and under the auspices of Leiden University (The Netherlands), in close collaboration with the Department of Antiquities of Jordan.

The current work in the Jebel Qurma area aims to reconstruct the nomadic landscape as well as its underlying strategies and social fabrics. Research focuses on the tangible remnants of regional nomadic praxis and their implications for the recognition of adaptive practices, strategic choices, and modifications in the natural and social environments (see *e.g.* Khazanov 1984; Gamble and Boismier 1991; McGlade 1995; Varien 1999; Frachetti 2008). It comprises the detailed assessment of sites and other features, in order to reconstruct the extent, nature, and duration of settlement, together with the re-use and renewal of occupation locales. Also important in this regard is the identification of natural and social routes and boundaries in the landscape. The location of Jebel Qurma at the convergence of several major caravan tracks to and from Arabia facilitated frequent contact with populations from beyond the desert, which left their material imprint in the local archaeological record. Soundings at selected sites and features helped to establish much-needed local chronologies and to provide contextual site information (Akkermans *et al.* 2014; Akkermans and Huigens 2018; Huigens 2019).

To a very large extent, current fieldwork in the Jebel Qurma area focuses on the social and funerary landscape (Fig. 3). The numerous monuments in the region for the disposal of the dead are of different types and often have prominent visibility, because of their size and location on high grounds (Akkermans and Brüning 2017; Akkermans *et al.* 2020). Evidently, the study of social landscapes

Figure 3. A typical burial cairn on top of a basalt-covered hill in the Jebel Qurma area, dated to the late first millennium BC. Inside the low mound of basalt blocks is a small corbelled chamber where the deceased was placed to rest (photograph: Jebel Qurma Project Archive).

cannot be separated from environment and land-use history. Yet, as mentioned above, local environmental constraints should not distract us from positioning the desert communities at the forefront of our analyses and to study them in their own right.

Landscapes of Survival

Indeed, putting these desert groups at the forefront was key to another Leiden University research project directed by the author, entitled: 'Landscapes of Survival: Pastoralist Societies, Rock Art and Literacy in Jordan's Black Desert, *c.* 1000 BC to 500 AD.' It was funded by the Netherlands Organisation for Scientific Research (NWO) between 2014 and 2018. As an important spin-off of the fieldwork in the Jebel Qurma area, this project strived to bring rich, new data on settlement, rock art, and inscriptions from Jebel Qurma within a single interpretive framework, something which had not been done before. The project aimed to develop an understanding of the desert's cultural landscapes and to explain the prominent achievements of its indigenous peoples between roughly the first millennium BC and the early half of the first millennium AD. What were the fundamental social, political, economic and ideological strategies which allowed the populations of the Black Desert to successfully exploit this difficult-to-inhabit region in this period? How did people cope with the fragile and often uninviting environment in which they lived? And how were local communities embedded in the supra-regional political and trade networks of their time?

In order to answer these questions, the research programme consisted of three related, complementary doctoral theses, which investigated nomadic lifeways and the treatment of the dead in the desert, the role of rock art in signing the landscape, and the implications of widespread literacy among the local groupings. In his doctoral thesis, Harmen Huigens examined patterns of mobility and the development of the ways of life and death across the desert landscape from the Hellenistic to the Early Islamic period (Huigens 2019; see also Huigens, this volume). His research allowed for the reconstruction of a dynamic social landscape, in which mobility was vitally important.

A second doctoral thesis focused on the many thousands of Safaitic petroglyphs in the Jebel Qurma region, *c.* 300 BC to 300/400 AD. Nathalie Brusgaard produced a systematic and contextual analysis of a unique visual culture in a landscape-based approach, delving into the relevance that the desert groups gave to inscribing their landscape in such fashion and abundance (Brusgaard 2019; see also Brusgaard, this volume).

Another PhD thesis, by Chiara Della Puppa, investigated the thousands and thousands of Safaitic inscriptions in the

Jebel Qurma region. This study explored the meaning and implications of the local Safaitic epigraphy vis-à-vis the cultural, nomadic landscape to which it was inextricably tied (Della Puppa, forthcoming).

Contents of this volume

The present volume (and the underlying 2017 conference) is yet another result of the Landscapes of Survival Project. It brings together a set of studies that promotes our general understanding of the archaeology and epigraphy of Jordan's *badia* and in particular focuses on its north-eastern basalt expanse. The book reflects the many research interests in the region and new developments therein. It offers a wealth of new data from the field, and synthetises previous and novel insights into the astonishingly rich history of the Black Desert.

Tobias Richter opens the contributions with an extensive review of the early prehistory of north-eastern Jordan, when the first humans arrived in the region about 400,000 years ago. Fieldwork has produced much evidence for settlement in the Epipalaeolithic and Early Neolithic, including the spectacular site of Shubayqa. People adapted to the local physical constraints in several productive ways. Richter concludes that north-eastern Jordan certainly was not always a marginal or peripheral region in deep prehistory.

Daniella Vos examines questions of transitory, ephemeral settlement and the numerous problems associated with the identification and interpretation of temporary camp sites. She goes into recent methodological developments in geoarchaeology that may help to improve the data sets, and illustrates her arguments by using evidence from the Neolithic sites of Wadi Faynan in southern Jordan and Wadi al-Jilat in north-eastern Jordan.

Yorke Rowan, Gary Rollefson, and Alexander Wasse discuss recent insights into the Late Neolithic period (*c.* 7000-5000 BC) of the *badia*, primarily on the basis of their fieldwork in Wadi al-Qattafi and Wisad Pools in the Black Desert. They conclude that conditions for living in the region in the Neolithic were undoubtedly better than today. Local hunter-pastoralists used substantial buildings for prolonged dwelling, with evidence of recurrent use and rebuilding.

Alexander Wasse and his colleagues continue their discussion on the Late Neolithic of the Black Desert, with an emphasis on the so-called 'burin sites' so typical for the region: places with shallow cultural deposits and lithic assemblages dominated by truncation burins. They argue that these burin sites were key to the region, as they probably represented the herding component of local communities that otherwise relied on hunting and gathering. The numerous burin stations demonstrate that Late Neolithic people intensively exploited the Black Desert for its ample opportunities for sheep pastoralism and secondary dairy production.

Maria Guagnin also emphasises the relevance of pastoralism in the Neolithic in her contribution, albeit in a very different ecological setting: the Nefud desert of north-western Saudi Arabia. She examines the so-called 'hearth sites', characterised by ephemeral fireplaces but no architectural remains, that formed a substantial, long-lived part of Neolithic herding economies in the interior of Arabia. Although they show evidence for contact with marginal areas in the southern Levant, these 'hearth sites' were distinctly local in terms of their lithic assemblages and distributions.

David Kennedy's contribution reviews the many ancient stone-built structures ('The Works of the Old Men') found in the lava fields of Syria, Jordan, and Saudi Arabia. Scholars in the region are reasonably familiar with the features in the Harrat al-Sham, the most northerly of several major lava fields extending in succession down the western part of the Arabia Peninsula. However, Kennedy draws specific attention to the countless installations in the other, much less explored, lava expanses, which often are notably different – or even unique – in their shapes and sizes. Kennedy emphasises that it remains crucial for fieldwork to be undertaken in conjunction with the analysis of satellite imagery and aerial photography.

Bernd Müller-Neuhof returns our attention to the later prehistory of north-eastern Jordan, with a focus on several recently discovered hillforts in the lava field east of Jawa. Dating to the Late Chalcolithic and Early Bronze Age I, evidence from these sites has changed our perception about the socio-economic complexity and potential of the volcanic belt in this period. Müller-Neuhof challenges the customary, straight-forward attribution of hillforts with fixed, hierarchical social structures; instead, he argues for more heterarchically organised communities that were able to undertake large-scale projects on a collective basis.

Stefan Smith's paper offers a further exploration of the Late Chalcolithic and Early Bronze Age periods in the basalt expanse of north-eastern Jordan. This contribution examines the appeal of these constrained, arid environments to populations of the fifth to third millennium BC, as well as the effects of the natural and anthropogenic environment on site morphologies. Based on survey work in the region south of Safawi, Smith focuses on the so-called 'wheels': roughly circular arrangements of enclosures, often surrounded by a string of round hut structures. Different types of 'wheels' can be distinguished, probably linked to functional and/or chronological use.

Peter Akkermans and Merel Brüning begin their account of settlement and burial in the Jebel Qurma region in the Late Chalcolithic period, but extend their chronological range through the Bronze Age and into the Iron Age. The new data from Jebel Qurma demonstrate considerable diversity in site layout as well as clear shifts in habitation patterns and locational preferences over time. While sites from the mid-fifth to fourth millennium BC regularly were of an impressive size, those of the later periods predominantly were small, temporary camps. Burials also show distinctive patterning through time, with most of the tombs belonging to the first millennium BC.

Harmen Huigens returns to the Jebel Qurma region of north-eastern Jordan in his contribution, and discusses several types of temporary camps (enclosures, clearings) used in the area from the Hellenistic to the Early Islamic period. Ceramic scatters and fireplaces in these often ephemeral features suggest that pastoralists used these camps for short-lived, domestic uses, in addition to the penning of animals. Differences in morphology and location may relate to the use of the installations in different parts of the year. Relevant also at these sites is the stark rise in pottery in the Byzantine and Early Islamic periods, which may reflect ever-increasing, intimate bonds between the nomadic groups in the Jebel Qurma area and more sedentary communities at its fringes.

Will Kennedy's paper deals with the hinterland of Petra in the Nabataean-Roman period, in particular with its more ephemeral sites and structures pertaining to pastoral modes of production. He provides evidence that there was a substantial pastoral component in the Nabataean way of life, and that mobility was a large component of daily life. Kennedy suggests that Nabataeans were indeed 'travellers between lifestyles': travellers between the desert and the sown.

Jørgen Christian Meyer's paper offers a further examination of the desert-and-the sown perspective, from the Palmyrene area in Syria. He makes it clear that although nomadic and sedentary ways of life were complementary economic systems, they had to be kept in check through direct state control of the territory, including forts and a complicated tax system regarding grazing rights. He suggests that the nomadic groups in the Palmyrene were not an integrated part of the local population but occasional visitors with their herds. Palmyra's control over these nomadic groups in its hinterland ensured enduring and peaceful relations between the desert and the sown.

Karin Bartl's contribution concerns Jordan's north-eastern *badia* in the Early Islamic period, presenting evidence of a settlement pattern that was partly a remodeling of earlier Roman-Byzantine land use and partly a development in its own right. She discusses the range of permanently used sites and the diversity of architecture in this period, including the 'desert castles' and the sometimes sizeable 'nomad villages'. Many of these sites resulted from state or elite intervention in the steppe-desert setting, although the significance and function of these places remain controversial. The number of permanent settlements is rather small in the *badia* from the seventh to ninth/tenth centuries. However, the increasing evidence from surveys of farmsteads, camps, and other smaller settlements from this period, indicates a more dense and complex use of the region than previously assumed.

Nathalie Brusgaard's paper turns our attention to the (Safaitic) rock art so typical of the Black Desert from the late first millennium BC to the early first millennium AD. Relying on data from the Jebel Qurma region, she focuses on the many depictions on stone of Arabia's most iconic animal: the dromedary camel. It features prominently in the local visual culture, suggesting tightly interwoven relationships between the economic (everyday) importance, prestige value, and social significance of the animal. Both stylistic analysis of the prevalent dromedary motif and study of the *chaîne opératoire* of carving indicate that the carvers depicted a truly significant theme of their desert society, and not only one aspect of their society.

Keshia Akkermans also explores the rock art of Jebel Qurma, although her focus is on the weaponry of hunting and battle scenes shown in many carvings. She agrees with Brusgaard that the local rock art is socially defined and normative, rather than an assemblage of random depictions of personal interest. Akkermans distinguishes four categories of weaponry: bows-and-arrows, spears/lances, swords, and shields. Patterns in the use of these objects vary for each category. Most notable are the close association of pole weapons with people riding animals, and the common depiction of archers on foot. The paper also considers the tangible weapon remains recovered from excavations of Bronze Age and Iron Age burial cairns in the Jebel Qurma region.

Koen Berghuijs' contribution addresses a form of rock art, which has received little comprehensive study so far: the so-called *wusūm* (singular *wasm*), or markings in the form of animal brands and petroglyphs. Bringing together a multitude of relevant primary sources and archaeological data from the Jebel Qurma region, this paper offers highly useful insights into the phenomenon of *wusūm* marking systems in Arabia. Berghuijs aims to bridge the gap between ethnographic sources and archaeological data. He makes it clear that the markings tend to derail any systematic investigation, because of their high ambiguity and the multiplicity of contexts in which they were used. However, Berghuijs argues that it was precisely because of this ambiguity that *wusūm* were able to function widely and successfully within the largely oral and tribal communities of the Middle East.

While the other rock-art papers primarily rely on carvings from Jordan, Charly Poliakoff's contribution presents a multitude of recently found rock art from the Riyadh and Najrān regions in Saudi Arabia. The petroglyphs mostly depict animals *en profile* (some of which are nearly lifesize), hunting scenes, and warriors brandishing their weapons. Comparable to Jordan's Harrat al-Sham, the petroglyph repertoire from central Saudi Arabia is selective, limited, and normative, reflecting a small but apparently highly significant portion of the nomadic social and natural world.

The final four papers of this book focus on a category of material culture that is frequently found in close association with the rock art, namely the texts in Safaitic and other scripts on basalt boulders in north-eastern Jordan. Through the lens of these often casual writings, Michael Macdonald investigates the fluidity and complexity of the relationships between the nomads of the basalt desert and the larger political powers of their time, most notably the Romans and Nabataeans. He concludes that there was a great deal of personal, commercial, and military interaction between the nomads and the sedentary communities. It is therefore not surprising, he states, that the inscriptions in the basalt desert are often remarkably well-informed about events in the wider world.

Ahmad Al-Jallad, Zeyad Al-Salameen, Yunus Shdeifat, and Rafe Harahsheh are concerned with the interaction between the nomads and the sedentary groups in Jordan's basalt expanse from a military perspective. While Macdonald had earlier suggested that the Romans raised auxiliary military units from among the nomadic tribes of the *harrah*, actual proof for such cooperation was still lacking. The paper by Al-Jallad and his colleagues provides the first solid epigraphic evidence for mixed troops, consisting of both Romans and local nomads. They suggest that the Romans could have deployed such units against incursions by nomadic groups from north Arabia, or against the Nabataeans, either before the annexation of the kingdom or against rebels after the fall of Petra.

Jérôme Norris' paper also delves into the complex interplay between nomads and sedentaries, in particular regarding the Nabataean kingdom. He re-examines a number of Safaitic and Nabataean inscriptions from north-eastern Jordan, highlighting the considerable ambivalence in this relationship. The inscriptions mention the Nabataeans either as enemies or as allies, and in one instance the carver identifies himself as being from 'the Nabataean people'. The paper devotes special attention to the mention of a probably Nabataean governor of 'Gilead' and to the so-called 'revolt of Damaṣī'. This may have been nothing else than a local event in the *harrah*, instead of the long-assumed huge rebellion against the Nabataean king.

Philip Stokes presents two Safaitic inscriptions recently discovered in north-eastern Jordan, published here for the first time. They are expressions of sorrow over the death of someone's close relative, who was buried in a cairn. The inscriptions have considerable philological importance, because they provide the first unambiguous attestations of a plural demonstrative pronoun *'ly* in the pre-Islamic epigraphic corpora.

Acknowledgements

The international conference 'The Archaeology and Epigraphy of Jordan's North-Eastern Desert' took place within the framework of the Landscapes of Survival Project and was funded by the Netherlands Organisation for Scientific Research (NWO) (Grant No. 360-63-100). The conference was held at the National Museum of Ethnology in Leiden, The Netherlands, on 17-18 March, 2017. The meeting formed the very foundation of the present volume; I am very grateful to the participants in the conference for the enthusiastic discussions and to the authors in this book for their stimulating papers.

Participants in the conference were (in alphabetical order): Peter M.M.G. Akkermans, Leiden University, The Netherlands; Ahmad Al-Jallad, Ohio State University, USA; Karin Bartl, Deutsches Archäologisches Institut, Germany; Koen Berghuijs, Leiden University, The Netherlands; Robert Bewley, EAMENA, University of Oxford, United Kingdom; Merel Brüning, Leiden University, The Netherlands; Nathalie Brusgaard, Leiden University, The Netherlands; Rémy Crassard, CNRS-Lyon, France; Chiara Della Puppa, Leiden University, The Netherlands; Wesam Esaid, Department of Antiquities, Jordan; Maria Guagnin, University of Oxford, United Kingdom; Hani Hayajneh, Yarmuk University, Jordan; Harmen Huigens, Leiden University, The Netherlands; Monther Dahash Jamhawi, Department of Antiquities, Jordan; David Kennedy, University of Western Australia, Australia; Will M. Kennedy, Humboldt-Universität Berlin, Germany; Michael Macdonald, University of Oxford, United Kingdom; Jørgen Christian Meyer, University of Bergen, Norway; Bernd Müller-Neuhof, Deutsches Archäologisches Institut, Germany; Jérôme Norris, Université de Lorraine, France; Aktham Oweidi, Department of Antiquities, Jordan; Charly Poliakoff, Université Paris 1- Sorbonne, France; Tobias Richter, University of Copenhagen, Denmark; Gary Rollefson, Whitman College, USA; Yorke Rowan, University of Chicago, USA; Stefan Smith, Ghent University, Belgium; Philip Stokes, University of Texas, USA; Daniella Vos, Bournemouth University, United Kingdom; Alan George Walmsley, Melbourne, Australia; and Alexander Wasse, University of East Anglia, United Kingdom.

The organisation of the conference was in the hands of Merel Brüning, with the help of Keshia Akkermans, Sufyan al-Kharaimeh, Koen Berghuijs, Thomas Vijgen, and Maikel van Stiphout. My thanks go to all. Special thanks go to the participants in the Landscapes of Survival Project, either as a researcher or as a supervisor: Harmen Huigens, Nathalie Brusgaard, Chiara Della Puppa, Merel Brüning, Monique Arntz, Koen Berghuijs, Ahmad al-Jallad, and Maarten Kossmann.

My sincere gratitude goes to the staff of the Department of Antiquities in Amman, Jordan, and its branch in Azraq, for their continued assistance and encouragement concerning the research in the Jebel Qurma region. Particular thanks go to Dr Yazid Elayyan (Director-General), Aktham Oweidi (Director of Excavations and Surveys), and Wesam Esaid (Head of the Department's Azraq branch) for their much-valued help.

The fieldwork in the Jebel Qurma area, and subsequent laboratory analyses, were made possible by the support of the Faculty of Archaeology of Leiden University (The Netherlands); the Netherlands Organisation for Scientific Research (NWO); the Leiden University Fund; the Netherlands Institute for the Near East (NINO); the Stichting Nederlands Museum voor Anthropologie en Praehistorie (SNMAP); the Centre for Isotope Research of Groningen University; the Jordan Oil Shale Company; Migchel Migchelsen, and some other private sponsors. I am very grateful to all for their invaluable help. Aerial Photographic Archive for Archaeology in the Middle East (APAAME) is thanked for the highly useful aerial imagery of the Jebel Qurma region.

Last but not least, the yearly surveys and excavations in the Jebel Qurma region would not have been possible without the help of a most dedicated team in the field and at home. Many thanks go to all participants in the Jebel Qurma Archaeological Landscape Project in the past years.

References

Akkermans, P.M.M.G. 2019. Living on the edge or forced into the margins? Hunter-herders in Jordan's north-eastern badlands in the Hellenistic and Roman periods. *Journal of Eastern Mediterranean Archaeology and Heritage Studies* 7, 412-431.

Akkermans, P.M.M.G. and Brüning, M.L. 2017. Nothing but cold ashes? The burial cairns of Jebel Qurma, north-eastern Jordan. *Near Eastern Archaeology* 80, 132-139.

Akkermans, P.M.M.G., Brüning, M.L., Arntz, M., Inskip, S.A. and Akkermans, K.A.N. 2020. Desert tombs: recent research into the Bronze Age and Iron Age cairn burials of Jebel Qurma, north-east Jordan. *Proceedings of the Seminar for Arabian Studies* 50, 1-17.

Akkermans, P.M.M.G. and Huigens, H.O. 2018. Long-term settlement trends in Jordan's north-eastern badia: the Jebel Qurma Archaeological Landscape Project. *Annual of the Department of Antiquities of Jordan* 59, 503-515.

Akkermans, P.M.M.G., Huigens, H.O. and Brüning, M.L. 2014. A landscape of preservation: late prehistoric settlement and sequence in the Jebel Qurma region, north-eastern Jordan. *Levant* 46, 186-205.

Bender, F. 1968. *Geologie von Jordanien*. Berlin: Gebrüder Borntraeger.

Betts, A.V.G. (ed.) 1998. *The harra and the hamad: excavations and explorations in eastern Jordan*. Sheffield: Sheffield Academic Press.

Brusgaard, N. 2019. *Carving interactions: rock art in the nomadic landscape of the Black Desert, north-eastern Jordan*. Oxford: Archaeopress.

Della Puppa, C. Forthcoming. *The Safaitic scripts – An ethno-palaeographic investigation*. Leiden: Leiden University (PhD Thesis).

Dutton, R.W., Clark, J.I. and Battikhi, A. (eds.) 1998. *Arid land resources and their management: Jordan's desert margin*. London: Routledge.

Edgell, H.S. 2006. *Arabian deserts: nature, origin, and evolution*. Dordrecht: Springer.

Frachetti, M.D. 2008. Variability and dynamic landscapes of mobile pastoralism in ethnography and prehistory. In: Barnard, H. and Wendrich, W. (eds.). *The archaeology of mobility*. Los Angeles, CA: Cotsen Institute of Archaeology at UCLA, 366-396.

Gamble, C.S. and Boismier, W.A. (eds.) 1991. *Ethnoarchaeological approaches to mobile campsites*. Ann Arbor, MI: International Monographs in Prehistory.

Huigens, H.O. 2019. *Mobile peoples – Permanent places. Nomadic landscapes and stone architecture from the Hellenistic to Early Islamic periods in north-eastern Jordan*. Oxford: Archaeopress.

Khazanov, A.M. 1984. *Nomads and the outside world*. Cambridge: Cambridge University Press.

Lawrence, D.E. 2012. *Early urbanism in the northern Fertile Crescent: a comparison of regional settlement trajectories and millennial landscape change*. Durham: University of Durham.

Maitland, P. 1927. The 'Works of the Old Men' in Arabia. *Antiquity* 1, 196-203.

McGlade, J. 1995. Archaeology and the ecodynamics of human modified landscapes. *Antiquity* 69, 113-132.

Rees, L.W.B. 1929. The Transjordan desert. *Antiquity* 3, 389-407.

Rollefson, G.O. 2016. Greener pastures: 7th and 6th millennia pastoral potentials in Jordan's eastern Badia. In: Reindel, M., Bartl, K., Lüth, F. and Benecke, N. (eds.). *Palaeoenvironment and the development of early settlements*. Rahden: Verlag Marie Leidorf, 161-170.

Taylor, C.C. 1972. The study of settlement pattern in pre-Saxon Britain. In: Ucko, P.J., Tringham, R. and Dimbleby, G.W. (eds.). *Man, settlement and urbanism*. London: Duckworth, 109-114.

Varien, M.D. 1999. *Sedentism and mobility in a social landscape*. Tucson, AZ: The University of Arizona Press.

Wilkinson, T.J. 2003. *Archaeological landscapes of the Near East*. Tucson, AZ: University of Arizona Press.

Wilkinson, T.J. 2004. The archaeology of landscape. In: Bintliff, J. (ed.). *A companion to archaeology*. Oxford: Blackwell, 334-356.

First inhabitants: the early prehistory of north-east Jordan

Tobias Richter

Abstract

The semi-arid to arid steppe and desert regions of eastern Jordan have produced a remarkable record for human occupation during the Pleistocene and early Holocene. Humans arrived in this region as early as 400,000 years ago during the Lower Palaeolithic. Fieldwork has produced particularly substantial evidence for human habitation during the Epipalaeolithic, and recent fieldwork has also demonstrated considerable human settlement during the Pre-Pottery Neolithic A. This contribution outlines the history of prehistoric research in eastern Jordan, and summarises the evidence for each of the key periods. Cumulatively, the evidence suggests that eastern Jordan was not always a marginal or peripheral region during the Palaeolithic and early Neolithic. Human groups used the steppe and desert to the best of their advantage and adapted to the physical constraints of these landscapes in different ways.

Keywords: Palaeolithic, Epipalaeolithic, Pre-Pottery Neolithic, Azraq, harrah

Introduction

With human presence attested at least 400,000 years ago, the Azraq basin and adjacent areas of north-east Jordan preserve one of the most continuous and well-known sequences of early prehistoric settlement in the Levant. Although the north-east *badia* of Jordan had for a long time been thought of as a largely inhospitable region, hostile to permanent human occupation, archaeological fieldwork over the past four decades has demonstrated that the region's early occupation spans from the later part of the Lower Palaeolithic to the end of the Pre-Pottery Neolithic. The idea that north-east Jordan was a landscape in which, due to the environmental constraints of the semi-arid steppe and arid desert, survival was difficult and life was harsh, has therefore been undergoing active revision (*e.g.* Richter 2014; Maher *et al.* 2016). Thus, it has become clear that east and north-east Jordan were not a cultural 'periphery', as has sometimes been implied (*e.g.* Bar-Yosef 1998; Bar-Yosef and Belfer-Cohen 2000; 1989; Bar-Yosef and Meadow 1995; Garrard *et al.* 1996; Goring-Morris 1987). Key cultural, economic and social advances have commonly been described as having occurred outside this region and were introduced to the *badia* only later. But at least in terms of early human settlement, this is not necessarily a given, as I hope to show in this article. Certain developments in the prehistory of north-east Jordan can actually be seen as locally-specific processes, independent of events taking place elsewhere. Furthermore, it is important not to judge the archaeological record of this semi-arid steppe and desert through the perspective of a modern, western individual. Oftentimes, we tend to judge the suitability of landscapes for human occupation based

In: Peter M. M. G. Akkermans (ed.) 2020: *Landscapes of Survival - The Archaeology and Epigraphy of Jordan's North-Eastern Desert and Beyond*, Sidestone Press (Leiden), pp. 17-36.

on the idea how suitable, or not, such an area is or was for agriculture, especially when it comes to more recent periods. While it is clear that farming was probably always opportunistic and seasonal in eastern Jordan, we ought to recognise that the inhabitants of this region would not have seen it the same way. Instead, I argue that this landscape was the centre of their social lives. Their experience of living in and off this land made them experts at constructing their very own niche, developing modes of existence and experiences that were particular to this and similar parts of south-west Asia. It is also important to recognise that the landscape that characterises north-east Jordan today has changed dramatically since the beginning of human settlement in the region. There is ample evidence to suggest that the landscape we consider today as characteristic of eastern Jordan was radically different in the deep past. I will return to these twin issues throughout this contribution.

Geography, research history and palaeoenvironment of north-east Jordan

The region under discussion encompasses the geological areas of the north-eastern part of Jordan's central plateau, the Azraq-Sirhan basin; the northern basalt plateau, part of the Harrat al-Sham volcanic field; and the north-eastern limestone plateau. It totals *c.* 40,000 km^2 in area. The northern basalt plateau is comprised of lava flows dating from the Miocene-Pliocene, which overlie older Cretaceous, Eocene and Oligocene limestone formations. The latter form the bedrock on the central plateau, the north-eastern limestone plateau and parts of the Azraq-Sirhan depression (Ames and Cordova 2014; 2015; Ames *et al.* 2014; Cordova *et al.* 2013; Betts 1998; Bender 1974; Garrard 2013; Allison *et al.* 2000; Dottridge 2009; Haaland 2009; Noble 2009). While the limestone areas are dominated by wadis and hills, the northern basalt plateau also features a series of extinct volcanic, table-top mountains (*mesas*) and *playas*.

The area comprises two major hydrological systems that are of some importance: the Azraq basin and the *hamad* basin (Cordova *et al.* 2013; Garrard 2013; Haaland 2009; Noble 2009; Betts 1998). While the *hamad* basin mostly drains towards the north, the Azraq basin drains towards the Azraq-Sirhan depression, where, at one of the lowest points in the basin, a series of springs and a very large mud flat (Arabic: *qa'*) make up the Azraq oasis. Although all of the springs in Azraq are nowadays extinct due to extraction of groundwater to supply the major urban centres further west, they were a focus for settlement throughout the human past. The Azraq wetlands provided a rich micro-habitat for a wide range of wildlife and plants, which has been severely diminished in the past twenty years, despite efforts to maintain two wildlife refuges in the area. Under current climatic conditions, rainfall declines from *c.* 200-300 mm in the north-west to 60 mm in the south-west of the *badia*, whereas mean annual rainfall on the eastern limestone plateau declines from 90 to 70 mm from west to east. The area is thus classified as a hot, semi-arid steppe and desert.

However, the catchment area of the Azraq basin extends over an area of 12,000 km^2 and reaches as far up as the Jebel al-'Arab/Jebel Druze in southern Syria, where mean annual precipitation ranges between 350 and 500 mm (Haaland 2009; Noble 2009). Almost all of the water on the southern, south-eastern and eastern slopes of the Jebel al-'Arab drain towards the Azraq oasis, transporting large quantities of surface run-off during the autumn and winter seasons. The springs of the oasis are fed by the uppermost of the three Azraq aquifers, which are replenished through this seasonal rainfall. Their discharge was low compared to recharge, creating a permanent water source even during drier climatic intervals, but this equilibrium has been disturbed by recent water extraction. Seasonal rainfall also leads to flooding of *playas*, which stay flooded for several weeks and months, usually drying up by the late spring/early summer. The largest of these is the Azraq *qa'*, which can flood up to 50-60 km^2. Outside of the oasis, there is little or no fresh water across much of the area during the summer period.

Following initial reports of archaeological structures spotted from the air by air-mail pilots Maitland (1927) and Rees (1929), Henry Field was the first archaeologist to visit the area to conduct an archaeological survey on the ground (Field 1960). Field's first visit was part of an overland journey from Amman to Baghdad, during which he found several prehistoric sites. This was followed up by further expeditions to the area between 1926 and 1934. Over the course of four survey seasons, Field and his colleagues located a significant number of sites and collected representative prehistoric artefacts. Garrod identified many Lower and Middle Palaeolithic artefacts amongst these collections, while also noting a lack of Upper Palaeolithic material (Garrod 1960).

The first archaeological excavations of a prehistoric site in north-east Jordan were carried out by John Waechter and Veronica Seton-Williams (Waechter *et al.* 1938; Waechter 1947). They excavated two sites, Wadi Dhobai B and K, which are now identified as the Late Neolithic site Wadi Jilat 13 and the Early-Middle Epipalaeolithic locality Wadi Jilat 6, respectively (Garrard 2013). Further survey work was undertaken by Zeuner and colleagues in 1955, which led to the discovery of additional Palaeolithic and Neolithic sites in eastern Jordan, including Kharaneh IV (Zeuner *et al.* 1957).

The first Lower Palaeolithic site was accidentally excavated as part of well digging at 'Ain al-Assad in the Azraq oasis (Harding 1958). Thereafter, there followed a hiatus that lasted nearly twenty years, until Andrew Garrard initiated his Azraq project in 1975: an ambitious effort to investigate Palaeolithic and Neolithic sites at multiple locations throughout the Azraq basin (Garrard *et al.* 1977). Over the course of nearly twenty years, Garrard and his team surveyed and excavated Upper Palaeolithic and Epipalaeolithic, as well as early and late Neolithic, sites in the Azraq oasis, Wadi Uwaynid and Wadi al-Jilat (Garrard 1991; 1998; Garrard *et al.* 1985; 1987; 1988; 1994; 1996; Byrd and Garrard 1989; Garrard and Byrd 1992; 2013). Over the course of this project, which explicitly aimed to gain a better understanding of environmental conditions and human adaptation, the Epipalaeolithic and Neolithic occupation of the Azraq basin was revealed in great detail for the first time and extensively published.

Shortly thereafter, other archaeologists also took a renewed interest in the Azraq basin. Gary Rollefson undertook new fieldwork at 'Ain al-Assad in 1979, following up on Harding's report of Late Acheulean artefacts found here in 1956 (Rollefson 1980a; 1980b; 1982; 1983). During the work at 'Ain al-Assad, he also located Middle Acheulean artefacts in Wadi Uwaynid (Rollefson 1984). Further work focusing on the Lower and Middle Palaeolithic occupation of the Azraq basin was undertaken by a French CNRS team between 1982 and 1986 (Copeland and Hours 1989). The team surveyed wadis in the western part of the Azraq basin, including Wadis Janab, Kharaneh, Mushash, Uwaynid, Butm, Enoqqiya and Ratam, as well as the Azraq oasis springs and mud flat. The Jordanian archaeologist Mujahed Muheisen also undertook surveys in Wadi Kharaneh during the early 1980s, and relocated the major Epipalaeolithic site Kharaneh IV, which he briefly tested in 1981 and excavated at a larger scale in 1985 (Muheisen 1983; 1988a; 1988b).

Also during the 1980s, Alison Betts initiated the first comprehensive survey project of the north-eastern part of the Harrat al-Sham basalt desert and the *hamad* limestone plateau to the east. Although this project did not produce significant evidence for settlement in the Lower, Middle or Upper Palaeolithic, it did turn up a number of Late Epipalaeolithic, Neolithic and later sites (Betts 1982; 1983; 1984; 1985; 1986; 1988a; 1988b; 1993a; 1998; Garrard *et al.* 1987; 1988; Betts *et al.* 1990; 1991; 2013). Together, these projects highlighted that north-east Jordan was occupied regularly and repeatedly, if not continuously, since the Lower Palaeolithic, and especially during the Epipalaeolithic.

Following a period of intensive survey and excavation during the 1980s and early 1990s, work in north-eastern Jordan decreased significantly. Although Betts continued to undertake work in the basalt desert up to 1996, little to no prehistoric archaeology was carried out elsewhere in north-east Jordan. However, the lowering of water tables in the southern Azraq marshland (Azraq Shishan) exposed prehistoric localities in the vicinity of 'Ain Sawda and 'Ain Qasiyah, which were investigated by Rollefson and colleagues (Rollefson *et al.* 1997). They reported Lower Palaeolithic, Middle Palaeolithic and Epipalaeolithic artefacts from 'Ain Sawda, and Middle Palaeolithic and Epipalaeolithic artefacts from 'Ain Qasiyah. As part of this work, they also located an additional Natufian, PPNB and late Neolithic site at Bawwab al-Ghazal in the Azraq Wetlands Reserve,[1] and excavated four test trenches here in 1998 (Rollefson *et al.* 1999). A subsequent detailed survey of the Azraq Wetlands Reserve located a number of additional prehistoric sites, including AWS-48, a Geometric Kebaran locality (Rollefson *et al.* 2001).

On the invitation of Wilke, Quintero and Rollefson, I initiated an excavation project at 'Ain Qasiyah to investigate the Epipalaeolithic occupation at the spring. The project, which lasted from 2005-2007, found Early Epipalaeolithic, residual Late Epipalaeolithic, and PPNB materials across four trenches, including one of the earliest human burials in Jordan (Richter and Röhl 2006; Richter *et al.* 2007; Richter 2009; 2011; Richter *et al.* 2009; 2010; Jones and Richter 2011; Maher *et al.* 2014).

Shortly thereafter, a new project was launched at Kharaneh IV, to re-investigate this outstanding Early and Middle Epipalaeolithic site (Maher 2007; Maher *et al.* 2011; 2012; 2014; 2016; Jones *et al.* 2016). Wasse and Rollefson (2005) reported the finding of additional Natufian, PPNB and Late Neolithic sites in the Jebel Tharwa area south-east of Azraq. The Lower and Middle Palaeolithic of the Azraq oasis also came back under scrutiny with the launch of a new project investigating Palaeolithic sites in north Azraq (Ames and Cordova 2014; 2015; Ames *et al.* 2014; Cordova *et al.* 2008; 2009; 2013). More recently, the same team also re-investigated the Lower Palaeolithic at 'Ain Sawda (Nowell *et al.* 2016).

In 2012, I initiated a new fieldwork project in the Qa' Shubayqa, in the *harrah*, to re-investigate a Late Epipalaeolithic Natufian site originally reported by Betts. In the course of this work, several new Late Epipalaeolithic and PPNA sites were found (Richter *et al.* 2012; 2014; 2016; Richter 2014; 2017a; 2017b). Rowan *et al.* (2015) have reported Lower and Middle Palaeolithic finds from Maitland's Mesa, and Early Epipalaeolithic and Neolithic finds from the Wisad pools area. Finally, the Jebel Qurma Archaeological Landscape Project has also turned up evidence for early prehistoric remains in the southern part of the *harrah* (Akkermans and Huigens 2018; Akkermans *et al.* 2014).

The Azraq oasis currently provides the most detailed and continuous record of geomorphological evidence available in this region, that allows us to reconstruct the

1 A reserve that encompasses the former marshland and springs of south Azraq, administrated by the Royal Society for the Conservation of Nature.

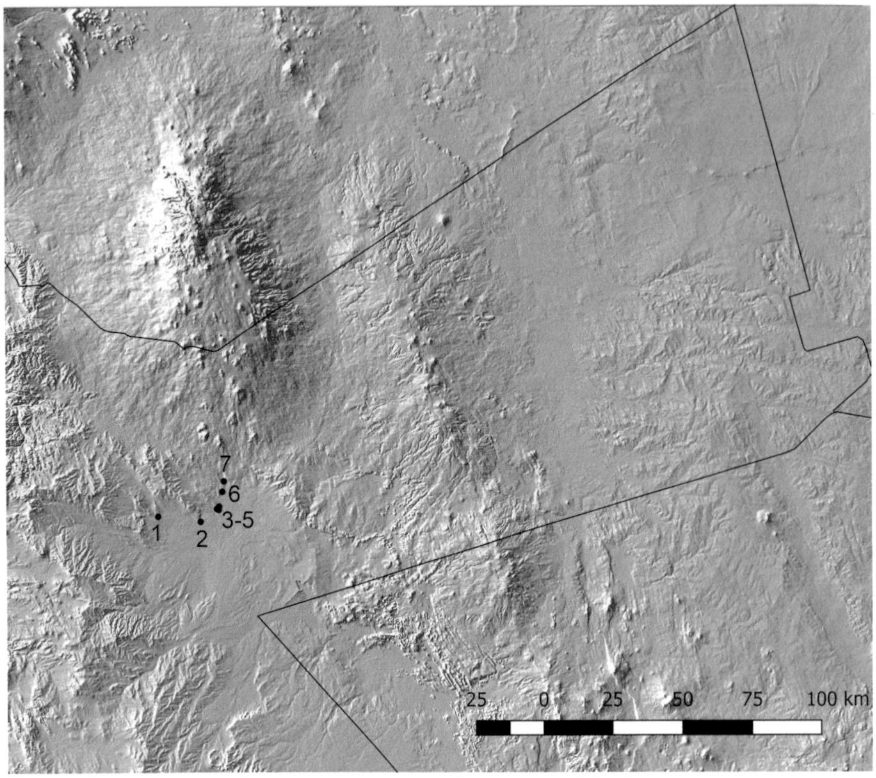

Figure 1. Locations with Lower and Middle Palaeolithic occupations mentioned in the text. 1: Wadi Butm. 2: Wadi Uwaynid. 3-5: 'Ain al-Assad, C-Spring, 'Ain Sawda. 6: Azraq ad-Druze. 7: Wadi Enoqiyya.

north-east Jordanian landscape during the Pleistocene and early Holocene. This data was recently summarised by Cordova *et al.* (2013; see also Ames *et al.* 2014; Ames and Cordova 2015; Pokines *et al.* 2019), stemming from their work at Azraq ad-Druze and other locations in the oasis. This work, in turn, draws on previous studies conducted in the wider Azraq basin, such as those of Abed (2008); Abed and Yaghan (2000); Besançon and Sanlaville (1988); Besançon *et al.* (1989); Copeland (1988); Copeland and Hours (1989); Davies (2005); Frumkin *et al.* (2008); Garrard *et al.* (1988); Garrard and Hunt (1989); Hunt (1989); Ibrahim (1996); Jones and Richter (2011); Kelso and Rollefson (1989); Noble (2009); Rollefson *et al.* (1997); and Woolfenden and Ababneh (2011).

Cordova *et al.* (2013, 105 and Fig. 12) divided the Azraq sedimentary sequences into seven phases, covering the time frame between *c.* 160,000 BP to the early Holocene. This sequence shows a succession of palaeolakes, wetlands and marshlands that periodically expanded and receded, corresponding to broader palaeoclimatic changes. Although these apparently reflect growing and shrinking bodies of water in north and south Azraq, occupations are present in both 'wet' and 'dry' periods. This suggests that water was available, although more restricted, also in drier periods. Based on their analysis of the 'Ain Qasiyah sediments, Jones and Richter (2011) also argued that, although spring activity varied, water was available almost all of the time in south Azraq from 60,000 years ago to the present. In sum, the Azraq oasis provided a more or less stable water source, even in periods when the climate was markedly drier due to the delayed discharge-recharge ratio of the Azraq aquifers. While the oasis therefore provided a stable refuge during drier periods, the connection of the Azraq basin with surrounding regions to the south, west and north may have been more limited at certain times. Hominins would have likely found it difficult to roam across an effectively drier, more sparsely vegetated landscape, whereas it would have been easier to 'hop' from one wetland or lake to another during wetter periods (Breeze *et al.* 2016). On the other hand, it is difficult to reconstruct exactly how extreme some of these conditions were; during certain seasons, water would have probably become available in short-lived streams and *playas* throughout eastern Jordan.

Nevertheless, there is compelling evidence that wetter periods enabled greater possibilities for the migration of hominins throughout the Pleistocene. We have evidence for the presence of substantial palaeolakes and wetlands outside the Azraq oasis during some parts of the Pleistocene (Cordova *et al.* 2013), particularly during the Epipalaeolithic. Evidence for soil formation related to greater moisture was reported from Wadi Jilat, occurring in association with substantial Epipalaeolithic settlements (Garrard 1998; Garrard *et al.* 1988; 1994; Hunt and Garrard 2013). The Early Epipalaeolithic site Khanareh IV is likewise associated with the presence of wetland and marshland

Figure 2. Main excavation area at the Lower Palaeolithic site of 'Ain Sawda in the Azraq Wetlands Reserve, Azraq Shishan, in 1997 (photograph by Gary Rollefson).

settings (Jones *et al.* 2016; Maher *et al.* 2014; 2016; Ramsey *et al.* 2016). Evidence for a substantial wetland and lake is now also emerging from the final Pleistocene in the Qa' Shubayqa, where it is associated with substantial Natufian and PPN settlements (Arranz-Otaegui *et al.* 2018; Richter *et al.* 2014; 2016; Richter 2017a; 2017b; Yeomans and Richter 2016). These conditions appear to have continued into the early Holocene, when conditions overall were wetter but associated with rising temperatures.

Lower and Middle Palaeolithic settlement

The first humans that arrived in north-east Jordan would have encountered a landscape quite unlike the one that we see today. Geomorphological evidence from the Azraq oasis has shown that, although spring activity and size of the pools and Azraq lake varied over the course of the past 350,000 years, water would have always been available and was one of the key factors attracting humans and animals to the Azraq oasis (Cordova *et al.* 2008; 2013; Ames *et al.* 2014; Ames and Cordova 2014; 2015; Garrard and Hunt 1989; Besançon *et al.* 1989; Copeland 1988; Garrard *et al.* 1977; Hunt 1989; Kelso and Rollefson 1989). The springs supported wetlands and marshlands that provided a rich habitat for both wildlife and humans. Even at times during which spring activity was reduced, when the Azraq pools and lakes shrunk or perhaps even dried out, water would have been available in some form. Outside the oasis, the landscape was characterised by an open parkland and steppe vegetation during wetter periods, which would have degraded to an arid steppe vegetation during drier periods. During the Lower and Middle Palaeolithic, onager, rhinoceros, bison, wild camel and elephants roamed this landscape, as is evidenced by the faunal material from C-Spring (Clutton-Brock 1970; 1989) and Shishan Marsh 1 (Pokines *et al.* 2019). These would have undoubtedly been accompanied by a range of other African and Eurasian species, such as lion, gazelle, and a hugely diverse range of avifauna that relied on the Azraq oasis as a migration resting and breeding location. The hunters and gatherers of the Lower and Middle Palaeolithic appear to have mainly exploited large mammals at this time.

The densest concentration of Lower and Middle Palaeolithic sites, as well as the only stratified sites, are found in the Azraq oasis. 'Ain al-Assad, C-spring and 'Ain Sawda are all located in south Azraq (Azraq Shishan) and feature significant lithic assemblages, as well as some faunal material (Rollefson 1980a; 1980b; 1982; 1983; 2000;

Rollefson *et al.* 1997; Garrard and Hunt 1989; Nowell *et al.* 2016) (Figs. 1 and 2). Bifaces are considerably more common at all of these three sites, as well as generally throughout the Azraq basin, than at other Lower Palaeolithic sites in the Levant (Rollefson *et al.* 2006). This seems to suggest that these were highly task-specific sites, focused on the butchering of large game. The overall characteristics of the Lower Palaeolithic lithic assemblages in the Azraq region suggest affinity with other sites located in an 'eastern wing' of the Levantine corridor (*e.g.* in the Jafr basin and the Palmyra basin to the north).

Additional Lower and Middle Palaeolithic localities have more recently been explored in the former marshland at Azraq ad-Druze (Cordova *et al.* 2009; Ames and Cordova 2014; 2015; Ames *et al.* 2014). Here, occupations seem to have occurred along the margins of the pools, spring and lakes that existed in this location at various points. Other Lower Palaeolithic sites are known from Wadi Uwaynid (Rollefson 1984), Qasr Amra (Copeland 1988; Copeland and Hours 1989), in addition to more isolated find spots throughout the region (Bartl *et al.* 2013). There is very little evidence for occupation of the basalt region to the east and north during the Lower and Middle Palaeolithic up to now (Betts 1986; 1998).

Although much of the Lower Palaeolithic material appears to date to the Late Acheulean, Rollefson (1984) dated the Wadi Uwaynid material to the Middle Acheulean. Thus, the earliest inhabitants of the Azraq basin may have arrived (based on current evidence) sometime around 750,000 BP. However, significant occupations only occurred during the Late Acheulean, from *c.* 400,000 BP onwards. Middle Palaeolithic occupations, dated to between 200,000 and 50,000 BP, have been reported from the Druze Marsh sites, Wadi Enoqiyya, 'Ain Qasiyah, 'Ain Sawda and C-spring (Fig. 2). These assemblages are generally smaller and less well stratified than the earlier sites. These lithic assemblages are often dominated by Levallois products, including oval and narrow flakes, as well as points. Some of the points are quite elongated, and *racloirs* and Mousterian points also occur. Copeland (1988) dated some of this material tentatively to the late Middle Palaeolithic.

It seems clear that the Azraq oasis provided a crucial settlement locale for hominins probably as early as 750,000 years ago. It is difficult to show how continuous occupation was during the Lower and Middle Palaeolithic, but the evidence does suggest that it was at least persistent and recurrent. Based on their observations of the geomorphology of the Azraq oasis, Cordova *et al.* (2013, 108) concluded that "the Azraq Oases seem to have played an important role in hominin population dynamics in the northern Arabian Desert because of its location at cross-roads along thoroughfares of paleolakes and springs connecting the Levant and Arabian Peninsula via the Wadi Sirhan Depression – Greater Nafud corridor facilitating the diffusion of hominin populations and technologies." This assessment is supported by the recent geohydrological modelling of 'wet corridors' through the Arabian peninsula, that connected different oases and wetland refugia in northern Arabia and the eastern Levant at certain points in time (Breeze *et al.* 2016). This recent work reinforces the argument by Rollefson *et al.* (2006) that eastern Jordan, and the eastern interior Levant, should be included in the so-called 'Levantine Corridor' through which early hominin populations migrated from Africa into Eurasia, and potentially also *vice versa*. Thus, the Lower and Middle Palaeolithic occupations of eastern Jordan are a crucial cornerstone in our understanding of human migrations in and out of Africa and Eurasia, and were not as isolated from the western Levant as has often been assumed.

North-east Jordan during the Upper Palaeolithic and Epipalaeolithic

The Upper Palaeolithic and, especially, Epipalaeolithic phases of human occupation of eastern Jordan are among the best explored in the prehistoric sequence of this region (Fig. 3). Although the period between *c.* 70,000-30,000 BP is currently poorly understood in eastern Jordan, with no sites or assemblages known to date to this interval, the period from *c.* 24,000 to 11,500 years ago stands out as one of the better investigated in the Levant. Two aspects of the Upper Palaeolithic and Epipalaeolithic settlement in eastern Jordan are particularly noteworthy: (1) the diversity of lithic assemblages that occur at different sites, and (2) the appearance of large aggregation sites that suggest the presence of large groups in certain locations.

This phase starts with a number of Upper Palaeolithic Ahmarian lithic assemblages, generally dated by lithic typology and some radiocarbon dates to between *c.* 30,000-24,000 BP. Ahmarian-style assemblages appear at Azraq 17 (Trench 2), Jilat 6 (basal phase), Jilat 9 and Uwaynid 18 (lower phase) (Garrard and Byrd 1992; Garrard *et al.* 1994; Hunt and Garrard 2013; Byrd and Garrard 2013; 2017). These assemblages are characterised by blade/bladelet reduction, and while backed and retouched microliths occur, the tool assemblage is usually dominated by scrapers, burins and simple retouched tools. Although occupations during this interval appear quite ephemeral, with all of these sites having produced only very small assemblages, it is clear that late Upper Palaeolithic hunter-gatherers had a firm foothold in the Azraq basin.

Evidence for the very earliest Epipalaeolithic occupations is known from Uwaynid 18 (Trench 2), Uwaynid 14 and Wadi Jilat 6 (lower phase). These have been assigned to the so-called Nebekian lithic tradition by Byrd and Garrard (2013; 2017). Additional Nebekian industries are known from 'Ain Qasiyah (Area D) (Richter 2011; Richter *et al.* 2009; 2013; 2014b). The radiocarbon

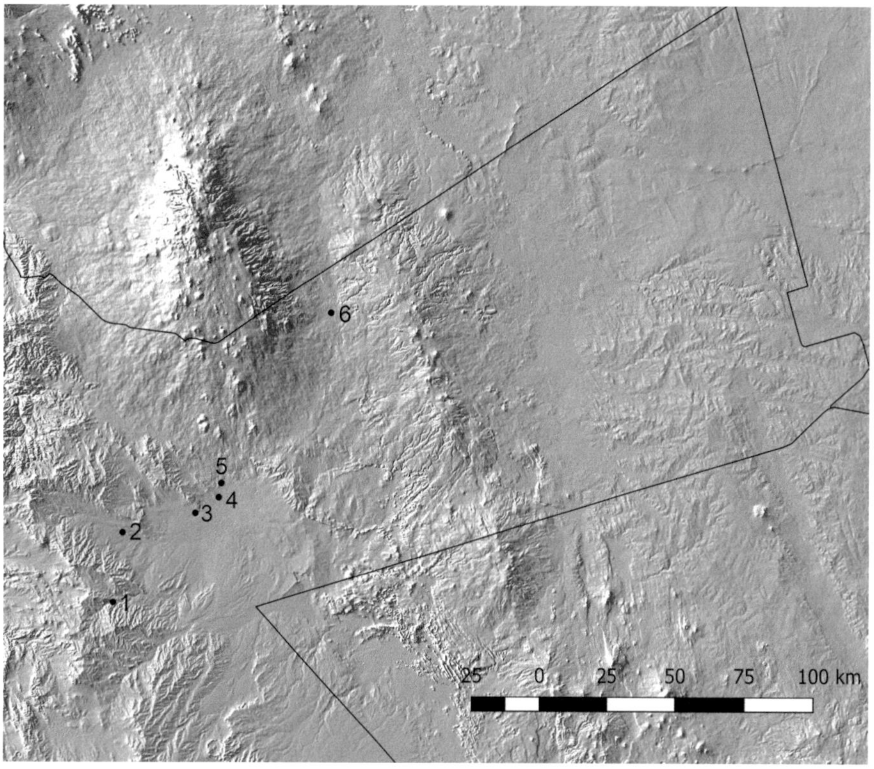

Figure 3. Upper Palaeolithic and Epipalaeolithic locations mentioned in the text. 1: Wadi al-Jilat. 2: Kharaneh IV. 3: Wadi Uwaynid. 4: Azraq Shishan, 5: Azraq ad-Druze. 6: Qa' Shubayqa.

dates of the 'Ain Qasiyah (Area D) assemblage place it slightly later than the available dates from Uwaynid 14 and 18, suggesting that the Nebekian in the Azraq basin is quite long-lived. In this phase, assemblages are characterised by abundant bladelet production and microlithic tools, especially arched-backed and pointed bladelets.

According to the chronology of Byrd and Garrard (2013; 2017), a mix of different industries is apparent during the Early Epipalaeolithic phase of occupation in the Azraq basin. While 'Ain Qasiyah Area A and B and Kharaneh IV have produced lithic assemblages that can be assigned to the Kebaran techno-complex (Richter 2011; Richter *et al.* 2011; 2013; 2014b; Maher *et al.* 2014; 2016; 2018; Macdonald *et al.* 2018), the assemblages from Azraq 32 and Jilat 6 (middle phase) have been assigned to the Qalkhan complex by Byrd and Garrard (2013; 2017), a techno-complex/industry not widely accepted by all specialists (Olszewski 2001; 2006; Maher and Richter 2011).

These assemblages are followed by a Nizzanian occupation at Jilat 6 (upper phase), and an assemblage characterised by wide, symmetric and asymmetric trapezes and rectangles with a high diversity of backing styles at Kharaneh IV Area D. Although Byrd and Garrard (2013; 2017) consider this industry a separate entity and refer to it as 'Kharanan', Maher and MacDonald (2013) have demonstrated that it should be considered part of the Geometric Kebaran complex. Their detailed analysis of the geometric microliths from the recent excavations at Kharaneh IV has shown that the wide range of different rectangles and trapezes can be found at other Middle Epipalaeolithic sites in the southern Levant. However, only at Kharaneh IV do these different microliths appear together in one assemblage. They argue that this underlines the idea of Kharaneh IV as a regional aggregation site.

The Middle Epipalaeolithic in the Azraq basin is represented by Wadi Jilat 10 (Trench 2), Wadi Jilat 22 (lower and middle phase), Wadi Jilat 8, Azraq 17 (Trench 2) and AWS-48 (Garrard *et al.* 1988; 1994; Byrd 1988; Byrd and Garrard 1989; 2013; 2017; Garrard and Byrd 1992; Hunt and Garrard 2013; Richter *et al.* 2014b). The lithic assemblages recovered from these sites are quite different from each other. AWS-48 produced a classic Geometric Kebaran assemblage dominated by trapeze-rectangles, while Jilat 8 and Jilat 22 (upper phase) have produced Mushabian assemblages with many non-geometric microliths, as well as the use of the microburin technique. But even between these two assemblages there are considerable differences: Wadi Jilat 22 (middle and lower phase) yielded yet another distinct lithic assemblage, which Byrd and Garrard termed 'Jilatian' for lack of comparison with any other assemblages elsewhere (Garrard and Byrd 1992; Byrd and Garrard 2013; 2017). This industry appears to be dominated by large blade production from single and opposed-platform cores, and the retouched assemblage is also dominated by large blade tools. Many were shaped into the distinctive so-called Jilat knives, while microliths were rare.

Figure 4. Early Epipalaeolithic 'sitting' burial from 'Ain Qasiyah. This is one of only two complete Early Epipalaeolithic burials from Jordan found to date (photograph by Tobias Richter).

Figure 5. Bird's eye view of the Early Epipalaeolithic aggregation site Kharaneh IV, which forms a low mound of dense lithic material covering an area of about two hectares (photography by The Fragmented Heritage Project, University of Bradford).

The Early and Middle Epipalaeolithic in eastern Jordan is focused on the Azraq basin, with few sites reported in the *harrah* or *hamad*, none of which are substantial (Betts 1998). This focus on the central, western and south-western Azraq basin may simply be a result of coverage bias of surveys; further work across eastern Jordan may yet produce new sites. The diversity of lithic assemblages evident in the Azraq basin is intriguing and may be related to the use of the area by a wide range of different social groups, if we accept that lithic technologies are related to cultural traditions that were confined to particular groups. It is intriguing to note that some of these 'traditions' have correlates with other Early and Middle Epipalaeolithic techno-complexes in the western Levant, while others have been seen as distinctly east Levantine entities. At the same time, exotic materials, especially marine shell beads, were imported from the Mediterranean and the Red Sea, which suggests wide-ranging contacts of groups in the Azraq basin with people to the west and south. This suggests a high degree of contact and social interaction across the Levant, which shows that the Azraq basin and eastern Jordan in general was not such a peripheral place as has often been argued (Richter *et al.* 2011).

This is also exemplified by the presence of two very large Early to Middle Epipalaeolithic aggregation sites: Wadi Jilat 6 and Kharaneh IV (Garrard and Byrd 1992; Richter *et al.* 2013; Maher *et al.* 2014; 2016; Muheisen 1988a) (Fig. 5). These two mounds, which cover 1.7 and 2 ha respectively, are characterised by very dense artefact accumulations, structures, and, in the case of Kharaneh IV, human burials. There can be little doubt that these sites represent seasonal meeting points of larger groups of hunter-gatherers, who may have taken advantage of seasonally abundant gazelle present in the area during and after the birthing seasons (Martin *et al.* 2010). Both sites were situated near reliable sources of water, which may have been permanent wetlands. Recent work by Byrd *et al.* (2016) has shown that these sites may have sat at the heart of distinct territories of different hunter-gatherer groups, since they are situated at roughly equidistant walking time from each other. These two large sites are indicative of a settlement pattern characterised by seasonal aggregations, or 'fission-fusion' as Byrd *et al.* (2016) termed it, that are unique in the context of the Epipalaeolithic of the southern Levant. The aggregations of people at these localities were highly repetitive and spatially very circumscribed. They resemble much later sites from the Late Epipalaeolithic or Early Neolithic, in more resource-rich environments. This shows that the environment in the Azraq basin was able to support large and fairly stable aggregations of people during certain periods.

Until recently, the Late Epipalaeolithic was somewhat underrepresented in eastern Jordan. Two Mushabian occupations, dated to 15,800-12,500 cal BP, have been reported from Wadi Jilat 8 and Wadi Jilat 22 (upper phase) (Byrd and Garrard 1989; 2013; 2017; Garrard 1998; Garrard *et al.* 1994). Although these lithic assemblages are quite different from one another, Garrard and Byrd nevertheless group them under the Mushabian complex.

The next phase is the Natufian. Until recently, only a few Natufian sites were known from eastern Jordan, both from the Azraq oasis and the *harrah* (Richter and Maher 2013). Garrard and his team excavated Azraq 18, a late Early Natufian site, which produced evidence for a human group burial, some of which exhibit skulls painted with ochre (Bocquentin and Garrard 2016; Garrard *et al.* 1977; 1988; 1994; Garrard 1991; 1998; 2013). Betts reported the presence of several Natufian sites in the *harrah*, and conducted excavations at Khallat Anaza and, briefly, Shubayqa 1 (Betts 1991; 1998). Other Natufian occurrences were later reported from 'Ain Qasiyah (Richter *et al.* 2007; 2009; 2014b), Bawwab al-Ghazal (Rollefson *et al.* 1999) and Jebel Tharwa 1d (Wasse and Rollefson 2005). Apart from Khallat Anaza, Shubayqa 1 and Azraq 18, all of these sites were only known from surface collections, and only Khallat Anaza and Azraq 18 were published in greater detail. None of these sites had produced material suitable for radiocarbon dating, and they also produced only small assemblages of material culture or animal remains. Thus, until fairly recently, our understanding of the Natufian in eastern Jordan had to rely on a somewhat exiguous record (see Richter and Maher 2013 for a more detailed discussion). Although Azraq 18 was dated, on the basis of its lithic artefacts, to the later part of the Early Natufian, the general consensus was that eastern Jordan only saw significant Natufian settlement during the Late Natufian, as suggested by the Khallat Anaza assemblage and collections from surface sites (Betts 1991; 1998; Henry 2013). This view corresponded with the broader idea that the Natufian originated in a 'core zone' in the Mount Carmel, Galilee and Jordan valley area, from where it spread to more 'marginal areas' after the onset of the Younger Dryas (Bar-Yosef 1998; 2002a; 2002b; 2004; 2009; Bar-Yosef and Belfer-Cohen 2000; 2002). This expansion into the semi-arid and arid periphery of the Levant was seen as reversion to a more mobile way of life dominated more by hunting than gathering, as opposed to the more sedentary way of life more dominated by plant gathering during the Early Natufian. Correspondingly, the Natufian sites in eastern Jordan were seen as temporal hunting stations.

The new excavations at Shubayqa 1, situated at the northern edge of the Azraq basin in the *harrah*, have recently contributed a great deal of new information about the Natufian occupation of eastern Jordan, challenging our previous understanding of the Late Epipalaeolithic in this region (Richter 2014; 2017a; 2017b; Richter *et al.* 2012; 2014a; 2016a; 2016b; 2017). Following on from the brief

Figure 6. The earliest stone architecture in eastern Jordan: the Early Natufian Structure 1 at Shubayqa 1, dating to *c.* 14,400-14,000 cal BP (photograph by Shubayqa Archaeological Project).

Figure 7. Example of one of the many Natufian ground-stone vessels from Shubayqa 1 (photograph by Shubayqa Archaeological Project).

and unpublished test excavations carried out by Betts, Shubayqa 1 was investigated by a team from the University of Copenhagen between 2012 and 2015. The site sits atop a low mound, situated at the northern edge of a large mud flat. Excavations across an area of 92 m^2 have revealed seven phases of occupation, including two well-preserved buildings and evidence for other structures (Fig. 6). There were also graves and rich assemblages of chipped and ground stone (Fig. 7), beads, worked bone, incised objects, and faunal and botanical remains. Shubayqa 1 was a substantial settlement, that appears to have been occupied on a multi-seasonal basis. This was likely enabled by the presence of a large lake or wetland near the site, which then occupied the mud flat. The recovery of numerous bones of waterfowl (Yeomans and Richter 2016) and charred wetland plant species suggests that this body of water was reliable and present for large parts of the year, if not permanent. This in turn attracted Natufian hunter-gatherers to the area and enabled stable settlement.

Shubayqa 1 is the first Natufian settlement in eastern Jordan that has been radiocarbon dated. In fact, the outstanding preservation conditions of charred botanical remains at the site has helped to make Shubayqa 1 one of the best dated Natufian sites in the Levant at large (Richter *et al.* 2017). The comprehensive series of dates suggests that the site was first established as early as 14,600-14,400 cal BP and occupied until around 14,200-14,000 cal BP. The site was then apparently abandoned for around 700-800 years, before people resettled it between 13,300-13,100 cal BP. Thereafter, it was abandoned again and re-occupied around 12,200-11,600 cal BP. With these interruptions in mind, the site therefore saw settlement in the Early, Late and Late/Final Natufian. The presence of such a large Natufian site, with architecture, ground-stone tools, graves, and a wide range of other material culture, not only suggests that it was not a temporal hunting camp. The dates also clearly demonstrate that Early Natufian groups were present in eastern Jordan from the start of the Natufian era. Import of marine shell and greenstone beads

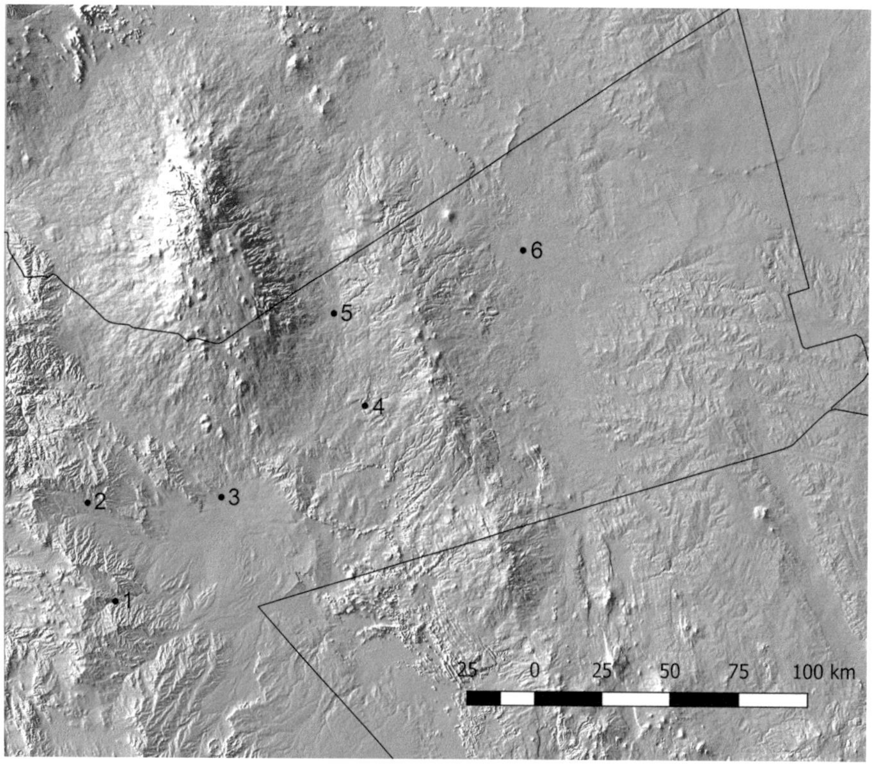

Figure 8. PPNA and PPNB localities mentioned in the text. 1: Wadi al-Jilat. 2: Mushash 163. 3: Azraq 31. 4: Dhuweila. 5: Shubayqa 6. 6: Qasr Burqu'.

in Shubayqa 1, as well as flint and chalcedony, shows that the site was linked into regional exchange networks that seem to have reached as far as the Mediterranean coast, where some of the marine gastropod shells originate. Thus, eastern Jordan was not settled only during the Late Natufian as a result of an exodus of hunter-gatherers leaving the Mediterranean core zone during the Younger Dryas. While the origins of the inhabitants of Shubayqa 1, and their affinity to other earlier and contemporary groups in the Azraq basin, are still under investigation, this new evidence shows that eastern Jordan was not a periphery in the Epipalaeolithic but rather a region that was well linked to other parts of the Levant.

The Pre-Pottery Neolithic

While our previous knowledge of the Late Epipalaeolithic period was limited, next to nothing was known about the earliest Neolithic (Pre-Pottery Neolithic A; PPNA) in eastern Jordan until recently. Garrard and his team found no PPNA sites during their work in the Azraq oasis and its surroundings (Garrard 1998; Garrard *et al.* 1994). Betts' survey of the *harrah* also turned up no substantial PPNA sites (Betts 1998). There therefore existed a gap between the end of the Epipalaeolithic Natufian and the earliest known Neolithic in the region, represented by the site of Jilat 7, where Garrard and his team uncovered evidence for an Early Pre-Pottery Neolithic B (EPPNB) occupation. This 1000-year gap seemed to suggest that eastern Jordan was sparsely, if at all, occupied during the earliest part of the Holocene, when wetter and warmer conditions returned to the region. Some have suggested that Late/Final Natufian and early PPNA populations coalesced in the Jordan valley during the final stage of the Younger Dryas and the early Holocene, since more semi-arid to arid regions may have become less inhabitable during the Younger Dryas (Kuijt and Goring-Morris 2002; Bar-Yosef 2002a; 2002b; 2009; Bar-Yosef and Belfer-Cohen 1991; 1992; 2000; 2002; Bar-Yosef and Kislev 1989; Bar-Yosef and Meadow 1995; Twiss 2007). However, the discovery of a substantial PPNA site at Shubayqa 6, north of Safawi, in 2012 has now begun to close the gap between the end of the Natufian and the EPPNB in eastern Jordan.

Shubayqa 6 is a low mound to the north of the Qa' Shubayqa, situated *c.* 800 m east of the Natufian site Shubayqa 1. Following its discovery in 2012, excavations at the site have been under way since 2014 (Richter 2017a; 2017b; Richter *et al.* 2016a; 2016b). The site has extensive architectural remains from at least two major phases of construction during the EPPNB and the PPNA (Fig. 9). Radiocarbon dates and material culture remains suggest that the occupation started during the Late Natufian and continued until the end of the PPNA. The site has produced large collections of chipped and ground-stone artefacts, faunal and botanical remains, beads and bead manufacturing waste, worked bone, and other material culture. The site's architecture is extensive and complex, with many phases of rebuilding and refurbishment. Although analyses of the faunal and botanical assemblages

Figure 9. Late PPNA oval building at Shubayqa 6 after excavation (photograph by Shubayqa Archaeological Project).

is ongoing, the site's economy appears to have been based predominantly on hunting of gazelle and other animals, as well as the exploitation of local wild plant species, including tubers and cereals. Numerous chipped-stone perforators, production waste, and unfinished and finished stone beads also show that the inhabitants of Shubayqa 6 produced beads on a regular basis. The raw material for their manufacture was imported from the south, probably from around the Azraq oasis or the Wadi al-Jilat area, although more precise sourcing studies have yet to be undertaken. Obsidian and other exotic materials, such as bitumen, also occur at Shubayqa 6, which shows that the site was linked in to a wide ranging regional exchange network. Survey elsewhere in the Qa' Shubayqa has produced evidence for additional PPNA sites, which have so far not been explored in any depth however. Shubayqa 6 appears to have been occupied throughout the entire PPNA and into the EPPNB, providing the 'missing link' between the Late Epipalaeolithic and the start of the PPNB in eastern Jordan.

Adding to the evidence from Shubayqa 6, recent work around Qasr Mushash at the western edge of the Azraq basin has led to the discovery of a new late PPNA/EPPNB site, *i.e.* Mushash 163 (Bartl *et al.* 2013; Bartl 2017; Bartl and Rokitta-Krumnow 2017; Tvetmarken 2015). Excavations at this site have revealed several round structures accompanied by a late PPNA/EPPNB chipped-stone assemblage. Mushash 163 lies adjacent to the Wadi Mushash, which drains into the Wadi Kharaneh which in its turn drains towards the Azraq oasis. This would have likely been a key route of east-west movement through the Azraq basin leading down from the Jordanian plateau into the steppe and desert.

Another EPPNB site is Jilat 7, which measures *c.* 2000 m^2 in size and has several structures visible on the surface (Garrard *et al.* 1994; 1996; Garrard 1998). Excavations revealed one complete building and parts of several other round structures. At Jilat 7, there is no evidence for the presence of domestic animals or the use of domesticated plants, although cereals are present at the site and may have been under cultivation.

Settlement density increases – if present data are to be taken for granted – during the Middle to Late PPNB. Jilat 7, 13, 26 and 32, Azraq 31, Bawwab al-Ghazal, Burqu' and Dhuweila have all produced evidence for Middle to Late PPNB occupations (Garrard *et al.* 1988; 1994; 1996; Garrard 1998; Betts 1982; 1983; 1984; 1985; 1986; 1988a; 1989; 1993a; 1993b; 1998; Betts *et al.* 1990; 1991; Finlayson and Betts 1990; Rollefson *et al.* 1999) (Fig. 10). At this point, plant cultivation using domesticated plants is evident, although this appears to have been mostly opportunistic. The key economic practice appears to have focused instead on

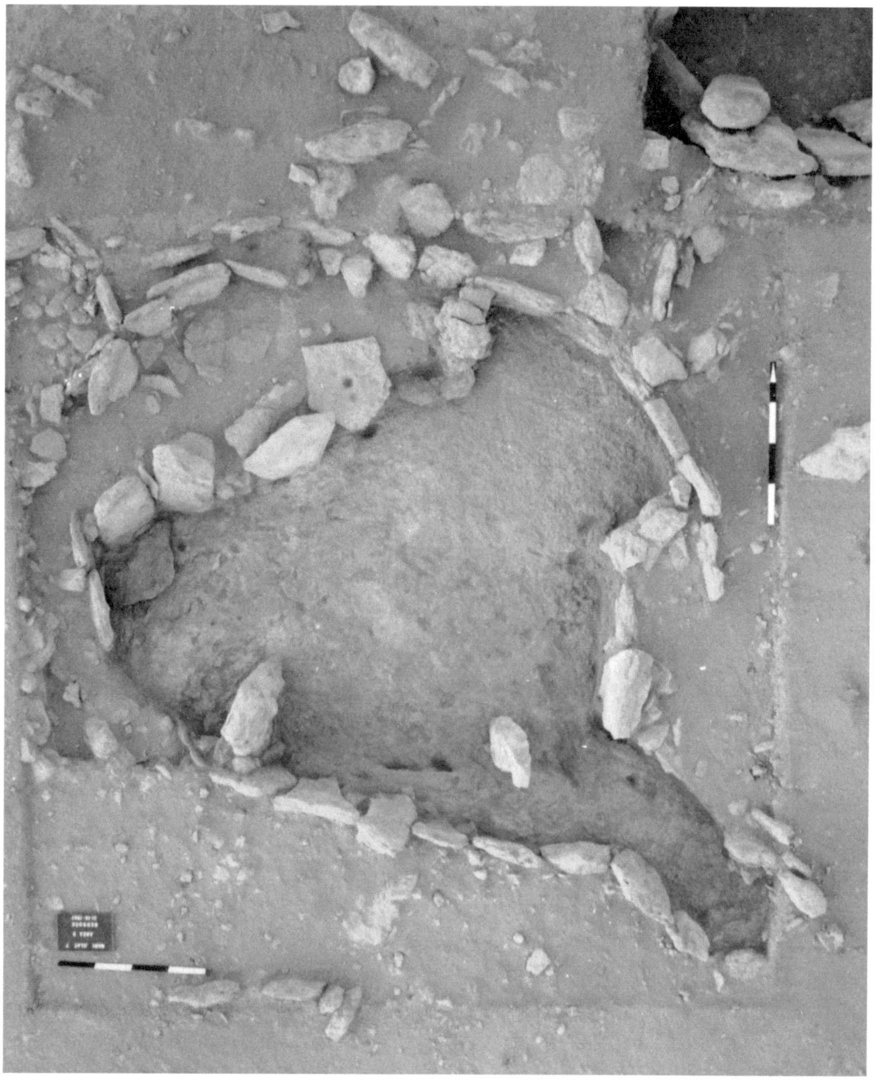

Figure 10. One of several MPPNB buildings at the Neolithic site of Jilat 7 (photograph by Andrew N. Garrard).

herding of sheep and goat, which were introduced during the LPPNB (Garrard *et al.* 1994; 1996; Martin 1994; Miller *et al.* 2018). Although hunting of gazelle and other animals likely still played an important role in the daily subsistence practices of Neolithic people in eastern Jordan at that time, the introduction of domestic livestock marked the beginning of a nomadic pastoralist way of life that would dominate the economy of this region for millennia to come, as people continued to inhabit the east Jordanian *badia*.

Although large PPNA and PPNB sites did not appear on the same scale in north-east Jordan as elsewhere in the southern Levant, the landscape nevertheless appears to have been quite well populated, particularly during the PPNB. Rather than concentrated in a few particular locales, the settlement pattern took on a more dispersed character, consisting of smaller groups that likely moved around the area more frequently. This mode of existence continued in north-east Jordan until recently and represented the best fitting solution to living off the land in this region.

Conclusion

Intensive research since the mid-1970s has produced a rich record of the near-continuous human occupation of north-east Jordan, stretching from the Lower Palaeolithic to the end of the PPNB. Although there are periods during which less material is present, or is absent entirely (notably from the later part of the Middle Palaeolithic to the earlier part of the Upper Palaeolithic), it is important to bear in mind that the research potential of north-east Jordan is not yet exhausted. Large parts of this region have not yet been intensively surveyed. Our knowledge and understanding of the early human occupation of this region is therefore not complete.

The archaeological record of the early inhabitants of north-east Jordan is of crucial importance for the wider Levant and south-west Asia as a whole. The archaeological signatures that have been documented in this region to date paint a different perspective on some of the key transitions in human prehistory that occurred in the region. The evidence for Lower Palaeolithic settlement in the Azraq oasis shows that this area was a refuge for hominin groups and that an easterly migration route, which took advantage of oases and wetland corridors during climatically favourable periods, existed in the Levantine interior. The appearance of large aggregation sites during the Early Epipalaeolithic, which are rare not just in south-west Asia but generally in the early human past, is also noteworthy. They represent a settlement type that only re-appears 5,000 or so years later, during the Late Epipalaeolithic. The presence of Natufian and Early Neolithic hunter-gatherers and hunter-gatherer-cultivators in north-east Jordan shows signs of different, yet complementary, pathways of human land-use during the transition to food production, which have previously not been recognised. The emergence of nomadic pastoralism during the LPPNB in north-east Jordan provides further evidence for the emergence of a distinctly different way of life. Thus, study of the early inhabitants of north-east Jordan shows that changes in the economy, settlement pattern and social life of south-west Asia was not synchronous or followed along the same tracks everywhere. Regionally specific social and economic practices are evident, that do not necessarily fit with models of cultural and social change developed elsewhere.

The near-continuous occupation of north-east Jordan during the Palaeolithic and Early Neolithic, combined with specific adaptations to the landscape, also shows that ideas concerning the marginality and periphery of this region are ill-conceived. People made north-east Jordan their home for millennia. They had the technological, social and cultural capacity to adapt to this landscape and use it to their advantage, developing different strategies and approaches than in other regions of south-west Asia. These repeated, habitual enactment of practices made north-east Jordan the centre of their social lives. We should focus on what the differences in human habitation of this landscape can tell us about broader patterns and models of social change, rather than trying to fit north-east Jordan into preconceived schemes of social evolutionary change.

Acknowledgements

I am very grateful to the organisers of the Landscapes of Survival conference for the invitation to Leiden to talk about the early inhabitants of north-east Jordan. Over the years, my work in Azraq and in the Qa' Shubayqa has been made possible by generous financial and logistical support from the Palestine Exploration Fund, Arts and Humanities Research Council, Council for British Research in the Levant, the Danish Council for Independent Research, Danish Institute in Damascus and the H.P. Hjerl Mindefondent for Dansk Palæstinaforskning. I am also extremely grateful to the Department of Antiquities of Jordan for its long support of this research, and particularly the archaeologists working in the Department's offices in Azraq and Mafraq. I would also like to thank the people of Azraq and Safawi for their hospitality and generosity without which none of this work would have been possible. I am grateful to Andrew Garrard, Gary Rollefson and Lisa Maher for their helpful comments on an earlier draft of this paper, although none of them should be blamed for any mistakes or omissions in this paper.

References

Abed, A.M. 2008. The paleoclimate of the eastern desert of Jordan during marine isotope stage 9. *Quarternary Research* 69, 458-468.

Abed, A.M. and Yaghan. R. 2000. On the paleoclimate of Jordan during the last glacial maximum. *Palaeogeography, Palaeoclimatology, Palaeoecology* 160, 23-33.

Akkermans, P.M.M.G. and Huigens, H.O. 2018. Long-term settlement trends in Jordan's northeastern badia: the Jabal Qurma Archaeological Landscape Project. *Annual of the Department of Antiquities of Jordan* 59, 503-515.

Akkermans, P.M.M.G., Huigens, H.O. and Brüning, M.L. 2014. A landscape of preservation: late prehistoric settlement and sequence in the Jebel Qurma region, north-eastern Jordan. *Levant* 46, 186-205.

Allison, R.J., Grove, J.R., Higgitt, D.L., Kirk, A.J., Rosser, N.J. and Warburton, J. 2000. Geomorphology of the eastern badia basalt plateau, Jordan. *The Geographical Journal* 166, 352-370.

Ames, C.J.H. and Cordova, C.E. 2014. Geoarchaeology and prehistoric transitions in the Azraq Druze basin, Jordan: preliminary data and hypothesis for a 4-dimensional model of landscape change. In: Jamhawi, M. (ed.). *Jordan's prehistory: past and future research.* Amman: Department of Antiquities, 69-80.

Ames, C.J.H. and Cordova, C.E.. 2015. Middle and Late Pleistocene landscape evolution at the Druze Marsh Site in northeast Jordan: implications for population continuity and hominin dispersal. *Geoarchaeology* 30, 307-329.

Ames, C.J.H., Nowell, A., Cordova, C.E., Pokines, J.T. and Bisson, M.S. 2014. Paleoenvironmental change and settlement dynamics in the Druze Marsh: results of recent excavation at an open-air Paleolithic site. *Quaternary International* 331, 60-73.

Arranz-Otaegui, A., González Carretero, L., Roe, J. and Richter, T. 2018. "Founder crops" v. wild plants: assessing the plant-based diet of the last hunter-gatherers in southwest Asia. *Quaternary Science Reviews* 186, 263-83.

Bar-Yosef, O. 1998. The Natufian culture in the Levant. Threshold to the origins of agriculture. *Evolutionary Anthropology* 6, 159-177.

Bar-Yosef, O. 2002a. The Natufian culture and the Early Neolithic: social and economic trends in southwest Asia. In: Bellwood, P. and Renfrew, C. (eds.). *Examining the farming/language dispersal hypothesis.* Cambridge: McDonald Institute for Archaeological Research, 113-126.

Bar-Yosef, O. 2002b. The role of the Younger Dryas in the origin of agriculture in West Asia. In: Yasuda, Y. (ed.). *The origins of pottery and agriculture*. New Delhi: Roli Books.

Bar-Yosef, O. 2004. The Natufian: a complex society of foragers. In: Fitzhugh, B. and Habu, J. (eds.). *Beyond foraging and collecting. Evolutionary change in hunter-gatherer settlement systems*. New York: Kluwer Academic Press/ Plenum Publishers, 91-152.

Bar-Yosef, O. 2009. Social changes triggered by the Younger Dryas and the early Holocene climatic fluctuations in the Near East. In: Feinman, G.M. (ed.). *The archaeology of environmental change: socionatural legacies of degradation and resilience*. Tucson: University of Arizona Press, 192-208.

Bar-Yosef, O. and Belfer-Cohen, A. 1989. The origins of sedentism and farming communities in the Levant. *Journal of World Prehistory* 3, 447-498.

Bar-Yosef, O. and Belfer-Cohen, A. 1991. From sedentary hunter-gatherers to territorial farmers in the Levant. In: Gregg, S.A. (ed.). *Between bands and states*. Carbondale, IL: Southern Illinois University, 181-202.

Bar-Yosef, O. and Belfer-Cohen, A. 1992. From foraging to farming in the Mediterranean Levant. In: Gebauer, A.B. and Price, T.D. (eds.). *Transitions to agriculture in prehistory*. Madison, WI: Prehistory Press, 21-48.

Bar-Yosef, O. and Belfer-Cohen, A.. 2000. Early sedentism in the Near East: a bumpy ride to village life. In: I. Kuijt, I. (ed.). *Life in Neolithic farming communities. Social organization, identity, and differentiation*. New York: Kluwer Academic, 19-62.

Bar-Yosef, O. and Belfer-Cohen, A. 2002. Facing environmental crisis: societal and cultural changes at the transition from the Younger Dryas to the Holocene in the Levant. In: Cappers, R.T. and Bottema, S. (eds.). *The dawn of farming in the Near East*. Berlin: ex oriente, 55-66.

Bar-Yosef, O. and Kislev, M. 1989. Early farming communities in the Jordan valley. In: Harris, D. and Hillman, G. (eds.). *Foraging and farming: the evolution of plant exploitation*. London: Unwin Hyman, 632-642.

Bar-Yosef, O. and Meadow, R.H. 1995. The origins of agriculture in the Near East. In: Price, T.D. and Gebauer, A.B. (eds.). *Last hunters – first farmers: new perspectives on the prehistoric transition to agriculture*. Santa Fe: School of American Research Press, 39-94.

Bartl, K. 2017. Mushash 163, Jordanien. Die Grabungskampagnen 2017. *e-Forschungsberichte* 2, 140-145.

Bartl, K. and Rokitta-Krumnow, D. 2017. Mushash 163, Jordanien. Die Grabungskampagnen 2015/2016//Die lithischen Kleinfunde. *e-Forschungsberichte* 1, 97-104.

Bartl, K., Bisheh, G., Blochand, F. and Richter, T. 2013. Qasr Mushah survey: first results of the archaeological fieldwork in 2011 and 2012. *Annual of the Department of Antiquities* 57, 179-193.

Beller, J.A., Ames, C.J. and Nowell, A., 2020. Exploring Mid-Late Pleistocene lithic procurement strategies at Shishan Marsh 1: preliminary geochemical characterization of chert sources around the Greater Azraq Oasis Area, Jordan. *Journal of Archaeological Science: Reports* 29, 102091.

Bender, F. 1974. *Geology of Jordan*. Berlin: Gebr. Borntraegger.

Besançon, J. and Sanlaville, P. 1988. L'évolution géomorphologique du bassin d'Azraq (Jordanie) depuis le Pléistocene Moyen. *Paléorient* 14, 23-30.

Besançon, J., Geyer, B. and Sanlaville, P. 1989. Contribution to the study of the geomorphology of the Azraq basin, Jordan. In: Copeland, L. and Hours, F. (eds.). *The hammer on the rock: studies in the Early Palaeolithic of Azraq, Jordan*. Oxford: British Archaeological Reports (International Series 540), 7-64.

Betts, A. 1982. Prehistoric sites at Qa'a Mejalla, eastern Jordan. *Levant* 14, 1-34.

Betts, A. 1983. Black Desert survey, Jordan: first preliminary report. *Levant* 15, 1-10.

Betts, A. 1984. Black Desert survey, Jordan: second preliminary report. *Levant* 16, 25-34.

Betts, A. 1985. Black Desert survey, Jordan: third preliminary report. *Levant* 17, 29-52.

Betts, A. 1986. *The prehistory of the basalt desert, Transjordan: an analysis*. London: University of London (unpublished PhD Thesis).

Betts, A. 1988a. 1986 excavations at Dhuweila, eastern Jordan: a preliminary report. *Levant* 20, 7-21.

Betts, A. 1988b. The Black Desert survey. Prehistoric sites and subsistence strategies in eastern Jordan. In: Garrard, A.N. and Gebel, H.G.K. (eds.). *The prehistory of Jordan. The state of research in 1986*. Oxford: British Archaeological Reports (International Series 396), 369-391.

Betts, A. 1989. The Pre-Pottery Neolithic B period in eastern Jordan. *Paléorient* 15, 147-153.

Betts, A. 1991. The late Epipalaeolithic in the Black Desert, eastern Jordan. In: Bar-Yosef, O. and Valla, F.R. (eds.). *The Natufian culture in the Levant*. Ann Arbor: International Monographs in Prehistory, 217-234.

Betts, A. 1993a. The Burqu'/Ruwaysid Project: preliminary report on the 1991 field season. *Levant* 25, 1-11.

Betts, A. 1993b. The Neolithic sequence in the east Jordan badia. A preliminary overview. *Paléorient* 19, 49-53.

Betts, A. (ed.) 1998. *The harra and the hamad. Excavations and surveys in eastern Jordan*. Sheffield: Sheffield Academic Press.

Betts, A., Cropper, D., Martin, L. and McCartney, C. 2013. *Later prehistory of the badia: excavation and surveys in eastern Jordan*. Oxford: Oxbow Books.

Betts, A., Helms, S.W., Lancaster, W., Jones, E., Lupton, A., Martin, L. and Matsaert, F. 1990. The Burqu'/Ruweishid project: preliminary report on the 1988 field season. *Levant* 22, 1-20.

Betts, A., Helms, S.W., Lancaster, W. and Lancaster, F. 1991. The Burqu'/Ruweishid project: preliminary report on the 1989 field season. *Levant* 23, 7-28.

Bocquentin, F. and Garrard, A. 2016. Natufian collective burial practice and cranial pigmentation: a reconstruction from Azraq 18 (Jordan). *Journal of Archaeological Science: Reports* 10, 693-702.

Breeze, P.S., Groucutt, H.S., Drake, N.A., White, T.S., Jennings, R.P. and Petraglia, M.D. 2016. Palaeohydrological corridors for hominin dispersals in the Middle East □250-70,000 years ago. *Quaternary Science Reviews* 144, 155-185.

Byrd, B.F. 1988. Late Pleistocene settlement diversity in the Azraq basin. *Paléorient* 14, 257-264.

Byrd, B.F. and Garrard, A.N. 1989. The last Glacial Maximum in the Jordanian desert. In: Soffer, O. and Gamble, C. (eds.). *The world at 18,000 BP*. London: Unwin Hyman, 78-96.

Byrd, B.F. and Garrard, A.N. 2013. Section C: The Late Palaeolithic – Chipped stone assemblages. In: Garrard, A.N. and Byrd, B.F. (eds.). *Beyond the Fertile Crescent. Late Palaeolithic and Neolithic communities of the Jordanian steppe*. Oxford: Oxbow Books, 137-394.

Byrd, B.F. and Garrard, A.N. 2017. The Upper and Epipalaeolithic of the Azraq basin, Jordan. In: Enzel, Y. and Bar-Yosef, O. (eds.). *Quaternary of the Levant. Environments, climate change and humans*. Cambridge: Cambridge University Press, 669-678.

Byrd, B.F., Garrard, A.N. and Brandy, P. 2016. Modeling foraging ranges and spatial organization of Late Pleistocene hunter-gatherers in the southern Levant – A least-cost GIS approach. *Quaternary International* 396, 62-78.

Clutton-Brock, J. 1970. The fossil fauna from an Upper Pleistocene site in Jordan. *Journal of Zoology* 162, 19-29.

Clutton-Brock, J. 1989. A re-consideration of the fossil fauna from C-spring, Azraq. In: Copeland, L. and F. Hours, F. (eds.). *The hammer on the rock: studies in the Early Palaeolithic of Azraq, Jordan*. Oxford: British Archaeological Reports (International Series 540), 391-398.

Copeland, L. 1988. Environment, chronology and Lower-Middle Paleolithic occupations of the Azraq basin, Jordan. *Paléorient* 14, 66-75.

Copeland, L. and Hours, F. 1989. *The hammer on the rock. Studies in the early Palaeolithic of Azraq, Jordan*. Oxford: British Archaeological Reports (International Series 540).

Cordova, C.E., Nowell, A., Bisson, M., Ames, C.J.H., Pokines, J., Chang, M. and Al-Nahar, M. 2013. Interglacial and glacial desert refugia and the Middle Paleolithic of the Azraq oasis, Jordan. *Quaternary International* 300, 94-110.

Cordova, C.E., Nowell, A., Bisson, M., Pokines, J., Ames, C.J.H. and Al-Nahar, M. 2009. The Druze Marsh Paleolithic Project, north Azraq, Jordan: stratigraphic sequences from the 2008 and 2009 seasons. *Annual of the Department of Antiquities of Jordan* 53, 311-320.

Cordova, C.E., Rollefson, G.O., Kalchgruber, R., Wilke, P. and Quintero, L. 2008. Natural and cultural stratigraphy of 'Ayn as-Sawda, Azraq Wetland Reserve: 2007 excavation report and discussion of finds. *Annual of the Department of Antiquities of Jordan* 52, 417-425.

Davies, C.P. 2005. Quarternary paleoenvironments and potential for human exploitation of the Jordan Plateau desert interior. *Geoarchaeology* 20, 379-400.

Dottridge, J. 2009. Water resource quality, sustainability and development. In: Dutton, R., Clarke, J. and Battikhi, A. (eds.). *Arid land resources and their management: Jordan's desert margin*. London: Routledge, 67-80.

Field, H. 1960 (ed.). *North Arabian desert archaeological survey, 1925-1950*. Cambridge, MA: Peabody Museum, Harvard University.

Finlayson, B. and Betts, A. 1990. A functional analysis of chipped stone artifacts from the late Neolithic site of Gabal Na'ja, eastern Jordan. *Paléorient* 16, 13-30.

Frumkin, A., Bar-Matthews, M. and Vaks, A. 2008. Palaeoenvironment of Jawa basalt plateau, Jordan, inferred from calcite speleotherms from a lava tube. *Quaternary Research* 70, 358-367.

Garrard, A.N. 1991. Natufian settlement in the Azraq basin, eastern Jordan. In: Bar-Yosef, O. and Valla, F.R. (eds.). *The Natufian culture in the Levant*. Ann Arbor: International Monographs in Prehistory, 235-244.

Garrard, A.N. 1998. Environment and cultural adaptations in the Azraq basin: 24,000 – 7,000 B.P. In: Henry, D.O. (ed.). *The prehistoric archaeology of Jordan*. Oxford: British Archaeological Reports (International Series 705), 139-148.

Garrard, A. 2013. Section A: project background. In: Garrard, A. and Byrd, B. (eds.). *Beyond the Fertile Crescent: Late Palaeolithic and Neolithic communities of the Jordanian steppe*. Oxford: Oxbow Books, 1-51.

Garrard, A.N. and Byrd, B.F. 1992. New dimensions to the Epipalaeolithic of the Wadi el-Jilat in central Jordan. *Paléorient* 18, 47-62.

Garrard, A.N. and Byrd, B.F. 2013. *Beyond the Fertile Crescent: Late Palaeolithic and Neolithic communities of the Jordanian steppe*. Oxford: Oxbow Books.

Garrard, A. and Hunt, C. 1989. The 1985 excavations at C Spring. In: Copeland, L. and Hours, F. (eds.). *The hammer on the rock. Studies in the Early Palaeolithic of Azraq, Jordan.* Oxford: British Archaeological Reports (International Series 540), 319-323.

Garrard, A.N., Baird, D. and Byrd, B.F. 1994. The chronological basis and significance of the late Palaeolithic and Neolithic sequence in the Azraq basin, Jordan. In: Bar-Yosef, O. and Kra, R.S. (eds.). *Late Quarternary chronology and paleoclimates of the eastern Mediterranean.* Ann Arbor: Radiocarbon, 177-199.

Garrard, A.N., Baird, D., Colledge, S., Martin, L. and Wright, K.. 1994. Prehistoric environment and settlement in the Azraq basin: an interim report on the 1987 and 1988 excavation season. *Levant* 26, 73-109.

Garrard, A.N., Betts, A., Byrd, B.F., Colledge, S. and Hunt, C. 1988. Summary of the palaeoenvironmental and prehistoric investigations in the Azraq basin. In: Garrard, A.N. and Gebel, H.G. (eds.). *The prehistory of Jordan: the state of research in 1986.* Oxford: British Archaeological Reports (International Series 396), 311-337.

Garrard, A.N., Betts, A., Byrd, B. and Hunt, C. 1987. Prehistoric environment and settlement in the Azraq basin: an interim report on the 1985 excavation season. *Levant* 19, 5-25.

Garrard, A., Byrd, B., Harvey, P. and Hivernel, F. 1985. Prehistoric environment and settlement in the Azraq basin: a report on the 1982 survey season. *Levant* 17, 1-28.

Garrard, A.N., Colledge, S. and Martin, L. 1996. The emergence of crop cultivation and caprine herding in the 'marginal zone' of the southern Levant. In: Harris, D. (ed.). *The origins and spread of agriculture and pastoralism in Eurasia.* London: University College London Press, 204-226.

Garrard, A.N., Price, S. and Copeland, L. 1977. A survey of prehistoric sites in the Azraq basin of eastern Jordan. *Paléorient* 3, 109-126.

Garrod, D.A.E. 1960. The flint implements. In: Field, H. (ed.). *North Arabian desert archaeological survey, 1925-1950.* Cambridge, MA: Peabody Museum, Harvard University, 111-135.

Goring-Morris, N.A. 1987. *At the edge: terminal Pleistocene hunter-gatherers in the Negev and Sinai.* Oxford: British Archaeological Reports (International Series 361).

Haaland, A. 2009. Hydrogeology and hydrochemistry of the Sirhan and Hammad basins. In: Dutton, R., Clarke, J. and Battikhi, A. (eds.). *Arid land resources and their management: Jordan's desert margin.* London: Routledge, 95-102.

Harding, G.L. 1958. Recent discoveries in Jordan. *Palestine Exploration Quarterly* 90, 7-18.

Henry, D.O. 2013. The Natufian and the Younger Dryas. In: Bar-Yosef, O. and Valla, F.R. (eds.). *Natufian foragers in the Levant: terminal Pleistocene social changes in Western Asia.* Ann Arbor, MI: International Monographs in Prehistory, 584-610.

Hunt, C. 1989. Notes on the sediments of some Paleolithic sites at Azraq, Jordan. In: Copeland, L. and Hours, F. (eds.). *The hammer on the rock. Studies in the Early Palaeolithic of Azraq, Jordan.* Oxford: British Archaeological Reports (International Series 540), 469-479.

Hunt, C. and Garrard, A. 2013. Section B: The Late Palaeolithic – Geological context. In: Garrard, A. and Byrd, B. (eds.). *Beyond the Fertile Crescent: Late Palaeolithic and Neolithic communities of the Jordanian steppe.* Oxford: Oxbow Books, 53-136.

Ibrahim, K.M. 1996. *The regional geology of the Al-Azraq area.* Amman: Natural Resources Authority.

Jones, M. and Richter, T. 2011. Palaeoclimatic and archaeological implications of Pleistocene and Holocene environments in Azraq, Jordan. *Quaternary International* 76, 363-372.

Jones, M.D., Maher, L.A., Macdonald, D.A., Ryan, C., Rambeau, C., Black, S. and Richter, T. 2016. The environmental setting of Epipalaeolithic aggregation site Kharaneh IV. *Quaternary International* 396, 95-104.

Kelso, G.K. and Rollefson, G.O. 1989. Two late Quaternary pollen profiles form Ain El-Assad, Azraq, Jordan. In: Copeland, L. and Hours, F. (eds.). *The hammer on the rock. Studies in the Early Palaeolithic of Azraq, Jordan.* Oxford: British Archaeological Reports (International Series 540), 259-275.

Kuijt, I. and Goring-Morris, A.N. 2002. Foraging, farming, and social complexity in the Pre-Pottery Neolithic of the southern Levant: a review and synthesis. *Journal of World Prehistory* 16, 361-440.

Macdonald, D.A., Allentuck, A. and Maher, L.A. 2018. Technological change and economy in the Epipalaeolithic: assessing the shift from early to middle Epipalaeolithic at Kharaneh IV. *Journal of Field Archaeology* 43, 437-456.

Maher, L.A. 2007. Archaeological survey at the Epipalaeolithic site of Kharaneh IV. *Annual of the Department of Antiquities of Jordan* 51, 263-272.

Maher, L, and Macdonald, M. 2020. Communities of interaction: tradition and learning in stone tool production through the lens of Epipalaeolithic Kharaneh IV, Jordan. In: Groucutt, H.S. (ed.). *Culture history and convergent evolution: can we detect populations in prehistory?* London: Springer International Publishing.

Maher, L.A. and Macdonald, D.A. 2013. Assessing typo-technological variability in Epipalaeolithic assemblages: preliminary results from two case studies from the southern Levant. In: Borrell, F., Ibáñez, J.J. and Molist, M. (eds.). *Stone tools in transition: from hunter-gatherers to farming societies in the Near East.* Barcelona: Universitat Autònoma de Barcelona, 29-44.

Maher, L.A. and Richter, T. 2011. PPN predecessors: current issues in Late Pleistocene chipped stone analyses in the southern Levant. In: Healey, E., Osamu, M. and Campbell, S. (eds.). *Proceedings of the 6th conference on PPN chipped and ground stone industries of the Fertile Crescent. Manchester, March 3rd-6th 2008*. Berlin: ex oriente, 25-31.

Maher, L., Macdonald, D., Jones, M., Stock, J., Martin, L. and Allentuck, A. 2016. Occupying wide open spaces? Late Pleistocene hunter-gatherer activities in the eastern Levant. *Quaternary International* 396, 79-94.

Maher, L., Richter, T., Jones, M. and Stock, J.T. 2011. The Epipalaeolithic foragers in Azraq project: prehistoric landscape change in the Azraq basin, eastern Jordan. *CBRL Bulletin* 6, 21-27.

Maher, L.A., Richter, T., Macdonald, D., Jones, M.D., Martin, L. and Stock, J.T. 2012. Twenty thousand-year-old huts at a hunter-gatherer settlement in eastern Jordan. *PloS One* 7, e31447.

Maher, L., Richter, T., Stock, J.T. and Jones, M. 2014. Preliminary results of recent excavations at the Epipalaeolithic site of Kharaneh IV. In: Jamhawi , M. (ed.). *Jordan's prehistory: past and future research.* Amman: Department of Antiquities of Jordan, 81-92.

Maitland, P.E.. 1927. The 'Works of the Old Men' in Arabia. *Antiquity* 1, 196-203.

Martin, L. 1994. *Hunting and herding in a semi-arid region: an archaeozoological and ethnological analysis of the faunal remains from the Epipalaeolithic and Neolithic of the eastern Jordanian steppe*. Sheffield: University of Sheffield.

Martin, L., Edwards, Y., and Garrard, A.N. 2010. Hunting practices at an eastern Jordanian Epipalaeolithic aggregation site: the case of Kharaneh IV. *Levant* 42, 107-135.

Miller, H., Baird, D., Pearson, J., Lamb, A.L., Grove, M., Martin, L. and Garrard, A. 2018. The origins of nomadic pastoralism in the eastern Jordanian steppe: a combined stable isotope and chipped stone assessment. *Levant* 50, 281-304.

Muheisen, M. 1983. *Le Paléolithicque et l'Épipaléolithique en Jordanie*. Bordeaux: University of Bordeaux.

Muheisen, M. 1988a. The Epipalaeolithic phases of Kharaneh IV. In: Garrard, A.N. and Gebel, H.G. (eds.). *The prehistory of Jordan. The state of research in 1986*. Oxford: British Archaeological Reports (International Series 396), 353-367.

Muheisen, M. 1988b. Le gisement de Kharaneh IV, note sommaire sur la phase D. *Paléorient* 14, 265-282.

Noble, P. 2009. Quantification of recharge to the Azraq Basin. In: Dutton, R., Clarke, J. and Battikhi, A. (eds.). *Arid land resources and their management: Jordan's desert margin*. London: Routledge, 103-110.

Nowell, A., Walker, C., Cordova, C.E., Ames, C.J.H., Pokines, J.T., Stueber, D., DeWitt, R. and Al-Souliman, A.S.A. 2016. Middle Pleistocene subsistence in the Azraq oasis, Jordan: protein residue and other proxies. *Journal of Archaeological Science* 73, 36-44.

Olszewski, D.I. 2001. The Palaeolithic, including the Epipalaeolithic. In: MacDonald, B. (ed.). *The archaeology of Jordan.* Sheffield: Sheffield Academic Press, 31-65.

Olszewski, D.I. 2006. Issues in the Levantine Epipalaeolithic: the Madamaghan, Nebekian and Qalkhan (Levant Epipalaeolithic). *Paléorient* 32, 19-26.

Pokines, J.T., Lister, A.M., Ames, C.J., Nowell, A. and Cordova, C.E. 2019. Faunal remains from recent excavations at Shishan Marsh 1 (SM1), a Late Lower Paleolithic open-air site in the Azraq Basin, Jordan. *Quaternary Research* 91, 768-791.

Ramsey, M.N., Maher, L.A., Macdonald, D.A. and Rosen, A. 2016. Risk, reliability and resilience: phytolith evidence for alternative 'Neolithization'pathways at Kharaneh IV in the Azraq Basin, Jordan. *PloS One*, https://doi.org/10.1371/journal.pone.0164081.

Rees, L.W.B. 1929. The Transjordan desert. *Antiquity* 3, 389-407.

Richter, T. 2009. *Marginal landscapes? The Azraq oasis and the cultural landscapes of the final Pleistocene Levant*. London: University College London, Institute of Archaeology.

Richter, T. 2011. Nebekian, Qalkhan and Kebaran: variability, classification and interaction. New insights from the Azraq oasis. In: Healey, E., Osamu, M. and Campbell, S. (eds.). *Proceedings of the 6th conference on PPN chipped and ground stone industries of the Fertile Crescent. Manchester, March 3rd-6th 2008*. Berlin: ex oriente, 33-49.

Richter, T. 2014. Margin or centre? The Epipalaeolithic in the Azraq oasis and the Qa' Shubayqa. In: Finlayson, B. and Makarewicz, C.A. (eds.). *Settlement, survey and stone: essays on Near Eastern prehistory in honour of Gary Rollefson*. Berlin: ex oriente, 27-36.

Richter, T. 2017a. The Late Epipalaeolithic and Early Neolithic in the Jordanian badia: recent fieldwork around the Qa' Shubayqa. *Near Eastern Archaeology* 80, 30-37.

Richter, T. 2017b. Natufian and Early Neolithic in the Black Desert, eastern Jordan. In: Enzel, Y. and Bar-Yosef, O. (eds.). *Quaternary of the Levant. Environments, climate change and humans*. Cambridge: Cambridge University Press, 715-722.

Richter, T. and Maher, L.A. 2013. The Late Epipalaeolithic in the Azraq basin: a reappraisal. In: Bar-Yosef, O. and Valla, F.R. (eds.). *Natufian foragers in the Levant: terminal Pleistocene social changes in Western Asia*. Ann Arbor, MI: International Monographs in Prehistory, 429-448.

Richter, T. and Röhl, C. 2006. Rescue excavations at Epipalaeolithic 'Ayn Qassiyah: report on the 2005 season. *Annual of the Department of Antiquities of Jordan* 50, 189-203.

Richter, T., Colledge, S., Luddy, S., Jones, D., Jones, M., Maher, L. and Kelly, R. 2007. Preliminary report on the 2006 season at Epipalaeolithic 'Ayn Qasiyah, Azraq es-Shishan. *Annual of the Department of Antiquities of Jordan* 51, 313-328.

Richter, T., Allcock, S., Jones, M., Maher, L.A., Martin, L., Stock, J.T. and Thorne, B. 2009. New light on Final Pleistocene settlement diversity in the Azraq basin: recent excavations at 'Ayn Qasiyyah. *Paléorient* 35, 49-68.

Richter, T., Stock, J.T., Maher, L. and Hebron, C. 2010. An Early Epipalaeolithic sitting burial from the Azraq oasis, eastern Jordan. *Antiquity* 84, 321-334.

Richter, T., Garrard, A.N., Allcock, S. and Maher, L.A. 2011. Interaction before agriculture: exchanging material and sharing knowledge in the Final Pleistocene Levant. *Cambridge Archaeological Journal* 21, 95-114.

Richter, T., Bode, L., House, M., Iversen, R., Arranz-Otaegui, A., Saehle, I., Thaarup, G., Tvede, M.L. and Yeomans, L. 2012. Excavations at the Late Epipalaeolithic site of Shubayqa 1: preliminary report on the first season. *Neo-Lithics* 2/12, 3-14.

Richter, T., Maher, L.A., Garrard, A.N., Edinborough, K., Jones, M.D. and Stock, J.T. 2013. Epipalaeolithic settlement dynamics in southwest Asia: new radiocarbon evidence from the Azraq basin. *Journal of Quaternary Science* 28, 467-479.

Richter, T., Arranz-Otaegui, A., House, M., Rafaiah, A. and Yeomans, L. 2014a. Preliminary report on the second season of excavation at Shubayqa 1. *Neo-Lithics* 1/14, 3-10.

Richter, T., Jones, M., Maher, L. and Stock, J.T. 2014b. The Early and Middle Epipalaeolithic in the Azraq oasis: fieldwork at Ayn Qasiyya and AWS-48. In: Jamhawi, M. (ed.). *Jordan's prehistory: past and future research.* Amman: Department of Antiquities of Jordan, 93-108.

Richter, T., Arranz-Otaegui, A., Boaretto, E., Bocaege, E., Estrup, E., Martinez-Gallardo, C., Pantos, G.A., Nørskov Pedersen, P., Sæhle, I. and Yeomans, L. 2016a. A Late Natufian and PPNA settlement in north-east Jordan: interim report on the 2014-2016 excavations at Shubayqa 6. *Neo-Lithics* 1/16, 13-21.

Richter, T., Arranz-Otaegui, A., Boaretto, E., Bocaege, E., Estrup, E., Martinez-Gallardo, C., Pantos, G.A., Pedersen, P., Sæhle, I. and Yeomans, L. 2016b. Shubayqa 6: a new Late Natufian and Pre-Pottery Neolithic A settlement in north-east Jordan. *Antiquity* 90 (354). doi: 10.15184/aqy.2016.182.

Richter, T., Arranz-Otaegui, A., Yeomans, L. and Boaretto, E. 2017. High resolution AMS dates from Shubayqa 1, northeast Jordan reveal complex origins of Late Epipalaeolithic Natufian in the Levant. *Scientific Reports* 7, 17025. doi: 10.1038/s41598-017-17096-5.

Rollefson, G.O. 1980a. Late Acheulian artifacts from 'Ain el-Assad ("Lion's Spring"), near Azraq, eastern Jordan. *Bulletin of the American Schools of Oriental Research* 240, 1-20.

Rollefson, G.O. 1980b. The Palaeolithic industries of Ain El-Assad (Lion's Spring) near Azraq, eastern Jordan in Gerald Lankester Harding memorial volume. *Annual of the Department of Antiquities Amman* 24, 129-144.

Rollefson, G.O. 1982. Preliminary report on the 1980 excavations at Ain El-Assad. *Annual of the Department of Antiquities* 26, 5-35.

Rollefson, G.O. 1983. Two seasons of excavation at Ain el-Assad, eastern Jordan, 1980-1981. *Bulletin of the American Schools of Oriental Research* 252, 25-34.

Rollefson, G.O. 1984. A Middle Acheulian surface site from Wadi Uweinid, eastern Jordan. *Paléorient* 10, 127-133.

Rollefson, G.O. 2000. Return to 'Ain el-Assad (Lion Spring), 1996: Azraq Acheulian occupation in situ. In: Stager, L.E., Greene, J.A. and Coogan, M.D. (eds.). *The archaeology of Jordan and beyond: essays in memory of James A. Sauer.* Winona Lake, IN: Eisenbrauns, 418-428.

Rollefson, G.O., Quintero, L. and Wilke, P. 1999. Bawwab al-Ghazal: preliminary report on the testing season 1998. *Neo-Lithics* 1/99, 2-4.

Rollefson, G.O., Quintero, L. and Wilke, P. 2001. Azraq Wetlands survey 2000: preliminary report. *Annual of the Department of Antiquities of Jordan* 45, 71-82.

Rollefson, G.O, Quintero, L.A. and Wilke, P.J. 2006. Late Acheulian variability in the southern Levant: a contrast of the western and eastern margins of the Levantine corridor. *Near Eastern Archaeology* 69, 61-72.

Rollefson, G.O., Schnurrenberger, D., Quintero, L., Watson, R.P. and Low, R. 1997. 'Ain Soda and 'Ain Qasiya: new late Pleistocene and early Holocene sites in the Azraq Shishan area, eastern Jordan. In: Gebel, H.G.K. and Rollefson, G.O. (eds.). *The prehistory of Jordan II. Perspectives from 1997.* Berlin: ex oriente, 45-58.

Rowan, Y.M., Rollefson, G.O., Wasse, A., Abu-Azizeh, W., Hill, A.C. and Kersel, M.M. 2015. The "land of conjecture": new late prehistoric discoveries at Maitland's Mesa and Wisad Pools, Jordan. *Journal of Field Archaeology* 40, 176-189.

Tvetmarken, C.L. 2015. Mushash 163, Jordanien. *e-Forschungsberichte* 2, 46-49.

Twiss, K.C. 2007. The Neolithic of the southern Levant. *Evolutionary Anthropology: Issues, News, and Reviews* 16, 24-35.

Waechter, J. 1947. Some surface implements from Transjordan. *Proceedings of the Prehistoric Society* 13, 178-180.

Waechter, J., Seton-Williams, V., Bate, D.M. and Picard, L. 1938. The excavations at Wadi Dhobai 1937-1938 and the Dhobaian industry. *Journal of the Palestine Oriental Society* 18, 172-186.

Wasse, A. and Rollefson, G.O. 2005. The Wadi Sirhan project: report on the 2002 archaeological reconnaissance of Wadi Hudruj and Jabal Thawra, Jordan. *Levant* 37, 1-20.

Woolfenden, W.B and Ababneh, L. 2011. Late Holocene vegetation in the Azraq Wetland Reserve, Jordan. *Quaternary Research* 76, 345-351.

Yeomans, L. and Richter, T. 2016. Exploitation of a seasonal resource: bird hunting during the Late Natufian at Shubayqa 1. *International Journal of Osteoarchaeology* 28, 95-108.

Zeuner, F., Kirkbride, D. and Park, B. 1957. Stone Age exploitation in Jordan 1. *Palestine Exploration Quarterly* 89, 17-44.

New techniques for tracing ephemeral occupation in arid, dynamic environments: case studies from Wadi Faynan and Wadi al-Jilat, Jordan

Daniella Vos

Abstract

Can we identify transitory, ephemeral camp sites in dynamic environments? How can we maximise the information gained from such sites, depicting mobile-pastoral subsistence, to enable a consideration of spatial patterns of activity? Ephemeral occupation is underrepresented within archaeological investigations, perhaps because short-lived sites are notoriously difficult to interpret due to the poor preservation of their remains. However, information about ancient modes of existence in peripheral areas carries much value for the interpretation of past ways of life that are currently understated within archaeological narratives. This paper will discuss recent methodological developments in geoarchaeology, which may enable us to maximise the information gained from ephemeral sites, even after a long period of abandonment. The value of reconstructing 'marginal' lifestyles for archaeological accounts will be discussed, addressing the visibility of subsistence strategies which have dominated many landscapes in the Near East since the Neolithic. The potential of the application of a dual methodology, using phytolith and geochemical soil analysis, to achieve a better understanding of the use of space at ephemeral archaeological sites will be explored by presenting two case studies from Jordan.

Keywords: geoarchaeology, phytolith analysis, soil analysis, ephemeral sites, Wadi Faynan, Wadi al-Jilat

Introduction

The completeness, and thereby representativeness, of the archaeological record is a re-occurring uncertainty within the investigation of past landscapes. Schiffer's influential consideration of processes leading to the preservation, state and location of artefacts (Schiffer 1988; 1995) might offer a way to address the effects of formation processes on the material record, but it does not model their influence on the visibility of entire sites and past activities. In order to assess how well past human activity is detectable across entire landscapes we must consider the durability of anthropogenic sites. While more substantial settlement forms may leave a clear mark in the landscape for thousands of years, ephemeral occupation is underrepresented in the landscape.

Though understated, ephemeral sites carry much value for the reconstruction of past lifestyles. Transient occupation is characteristic of many pastoral and hunter-gatherer societies, whose settlements reflects the demands of their highly mobile

In: Peter M. M. G. Akkermans (ed.) 2020: *Landscapes of Survival - The Archaeology and Epigraphy of Jordan's North-Eastern Desert and Beyond*, Sidestone Press (Leiden), pp. 37-58.

lifestyles. Entire landscapes and periods characterised by ephemeral occupation can be difficult to interpret due to the low intensity of occupation characterising them. The lack of durable structures and poor preservation of organic remains at these sites pose challenges for their identification and interpretation (Gifford 1977; Banning and Köhler-Rollefson 1983; Cribb 1991). Without being able to estimate what has been lost over time, it is difficult to distinguish between evidence of absence and absence of evidence in the archaeological record. And in order to properly consider pastoral nomadic ways of existence in the past, certain issues must be addressed: how durable are ephemeral traces of human activity? How fast do short-lived sites disappear in arid, dynamic environments? And how can archaeologists make the most of what is left for them to study?

In addition to their visibility, understanding the use of space in ephemeral structures is vital for their interpretation. This can shed light on past ways of life that are currently underrepresented within archaeological narratives. The division of space within human built environments can inform us about subsistence and daily activities, and can also reveal a great deal about notions of cleanliness, sacrality or gender, and relationships with animals or the natural environment (Douglas 1966; Bourdieu 1990; Parker Pearson and Richards 1994).

Until recently, most archaeological studies of spatial patterns have focused on a reconstruction of the location of activities based on the distribution of artefacts (Whallon 1973; Hodder and Orton 1979; Simek 1987; Hardy-Smith and Edwards 2004; Kuijt and Goodale 2009). There is, however, another level of evidence for the spatial patterning of activities which is more direct than the location of artefacts in abandoned sites: their sediments. These are often overlooked in spatial reconstructions, perhaps because they do not visually appear to contain evidence of activities, or perhaps because floors in modern western societies are not associated with soil but with hard surfaces of wood, stone and concrete. These are easily kept clean and are, in most cases at least, devoid of evidence of activities. Soils in archaeological sites, on the other hand, are central to the interpretation of past activities. They are both the carpet on which life takes place and the product of human endeavours.

Soils were often considered to be a product of natural processes but are increasingly seen as cultural products that should be studied as part of an investigation of social processes (Wagstaff 1987). As part of a shift in archaeology towards understanding past landscapes and environments as a whole rather than focusing on a single site, Wells (2006) offers the concept of cultural 'soilscape' as including a magnitude of materials reflecting both the use of resources and social frameworks by humans within their physical surroundings. Through the study of cultural soilscape the ways in which humans interact with their environment, both on the site level and beyond, can be understood within a framework of spatial activities. This is important because human environments are the physical manifestations of palimpsests of a range of behaviours and ideas. Although these records of human presence may be altered through time, they are tied to space.

Making sense of human space

The dimension of space is a fundamental aspect of cultural soilscapes, yet it has often been neglected in favour of a focus on time and history in western social sciences throughout most of the previous century (Soja 1989). When offered, discussions of the role that the material environment had on human well-being and consciousness mostly focused on two types of modern structures: dwellings and monuments. The majority of these, however, are characteristically different to the spaces that represent a wide range of functions and meanings at archaeological sites. Nevertheless, some approaches to space within the social sciences have provided important perspectives on the role of buildings, among others things: their part in allowing people to dwell in the metaphysical, spiritual and corporeal senses (Heidegger 1971); the agency of constructed space within a human belief system (Durkheim 1915); the instrumentality of the built space in the communication of power (Foucault 1982); the role of the material environment in articulating human consciousness (Husserl 1990); and the notion of *habitus* in regard to the built environment as a means to establish, express and sustain identities and social relationships (Bourdieu 1990).

The notion of correlations between spatial activity patterns and social structure has been put to the test in ethnographic studies of modern traditional societies. Yellen's (1977) study of the !Kung is one of the most well-known ethnoarchaeological recordings of the use of space in hunter-gatherer societies. In an examination of the use of household and communal areas, he links the location of objects within the domestic unit of a nuclear family to social context rather than function. Social space, as well as considerations such as messiness, or the time of day dictating the location of shade, were the main factors determining the location of activities and in turn that of the distribution of related artefacts in space. Yellen argues that straightforward, functional reconstruction of activities at the !Kung camp sites would be of no more use in the interpretation of the spatial trends at these sites than abstract speculations (*ibid.*).

A different emphasis on the cause of spatial patterning is presented by Binford (1978), whose account of a Nunamiut hunting stand in Alaska focused on the use of non-residential, ephemeral sites located away from main settlements, and the type of objects left behind there.

He argued that by studying a structure and the spatial organisation of activity areas within it, such as hearths and 'drop and toss zones', one can derive information about the number of participants and their activities. Relying on his own work on hunter-gatherer communities in Alaska, backed up by additional comparative studies, Binford developed influential models for understanding how activity areas in archaeological sites are shaped by the basic mechanics of the human body. His studies have been applied widely to the study of activity areas at various Palaeolithic sites (Audouze 1988; Guan *et al.* 2011; Koetje 1994; Simek 1987; Sørensen 2008).

Yet another consideration for the interpretation of the distribution of activity areas is provided by O'Connell (1987), who studied the occupation and abandonment of Alyawara camp sites in Australia. There he noticed that past a certain duration of occupation, the living areas would be swept, and large objects were removed to a secondary place of deposition, while small artefacts mostly remained *in situ*. This created a blurred spread of indicators of activity, according to which the location of activity areas would be difficult to discern. The outcomes of this case study have consequences for the interpretation of the spatial distribution of activity areas within sites, which could depend to a large degree on the duration and frequency of occupation. A site which has been revisited or cleaned, or in which the location of activities frequently changed, will be difficult to interpret (*ibid.*).

The different approaches to the correlation between the use of space and social and cultural domains provided by the ethnographic works outlined above demonstrate the power of such studies in shaping ideas about human societies. They suggest that spatial patterning at anthropogenic sites can reveal a lot about human lifestyles, from subsistence and daily routines to social structures, ceremonial events and cultural preferences. At the same time, they advise caution when interpreting archaeological remains. Ethnographic analogy ought to open up avenues of interpretation rather than limit these to universal models.

The work of Karl Heider (1967), who confronted archaeologists with their inability to truly conceptualise the rich variety of human cultures, revealed how misleading our common sense and imprinted assumptions can be. Other ethnographers enabled archaeologists to consider 'real life' scenarios for different archaeological patterns for the first time, such as what happens during the abandonment of structures (Cameron and Tomka 1993), the relationship between technology and social interaction (Gosselain 1998), or between material culture and inter-group relations (Hodder 1979). These studies opened room for discussion about the connection between the social and the material spheres of human cultures.

Spatial archaeology

It is up to the archaeologist to use all that remains of ancient occupation to reach a better understanding of the past use of the built environment and the role it played in different aspects of human life. This is not an easy task at the best of times. Even when studying ethnographic cases, where activities can be observed as they take place, the ambiguity and intricacy of human behaviour complicate interpretation. This task becomes more difficult when the material record of a site is very limited, whether because of poor preservation or the limited deposition of remains in the first place. In these instances the importance of a site's soilscape becomes clearer, as it enables us to reconstruct past behaviour *in situ*. The testing and application of methods of soil analysis to these sites is therefore vital if we want to understand their spatial use, which in turn can provide important insights into past behaviour. By establishing the value of soil analysis to the interpretation of ephemeral sites one also ascertains the potential to further explore periods characterised by ephemeral occupation, which are, as a result, poorly understood, such as the Neolithic of the Near East.

Theories of behavioural archaeology (Schiffer 1988) and spatial archaeology (Clarke 1977) have been used over the past four decades to link the spatial distribution of artefacts in archaeological sites with perceived past activities and behaviours of the groups that occupied them. To do this, the spatial patterns of artefact dispersal must be considered in relation to the cause of past human behaviour rather than a random scattering of objects. Spatial archaeology offers an approach that legitimises this idea by proposing that the spatial patterning of the remains of a site reflect behavioural patterns of the society that created them. Both social and functional interpretations are suggested based on the spatial distributions of artefacts, structures or activities (*ibid.*). Behavioural archaeology, as expanded by Schiffer (1988, 1995), extends the notion of spatial archaeology and provides a framework for culturally meaningful distribution patterns by describing the relationship between human action and the material record.

With the rise of post-processual archaeology came other changes in approaches to, and notions of, space. Earlier functional interpretations were accompanied by phenomenological ones, seeing space an as active force both structured by and structuring human life and behaviour. Space became a social construct, a concept, perceived and determined by individual agents (Tilley 1994). The study of space within archaeology began to extend across multiple scales, from entire landscapes and regions to individual houses or areas (Salisbury 2007).

Following these theoretical changes came advances in methods and techniques, and space started to gain a cultural importance within archaeology. Careful visual examinations of the locations of individual artefacts, features or sites, an analysis technique called point patterns, had already been in use for a while (Bradley and Small 1985). The use of quantitative methods to investigate spatial correlations became more widespread during the 1970s, replacing the earlier visual examinations. These included different statistical tests such as nearest-neighbour, Thiessen polygons, and more recently also more extensive GIS analysis (Hodder and Orton 1979).

Geoarchaeological methods for the analysis of space

Although archaeological studies of spatial patterning cover a range of techniques to analyse spatial relationships, previous attempts concentrated on the distribution of artefacts rather than soils (Hardy-Smith and Edwards 2004; Hodder and Orton 1979; Kuijt and Goodale 2009; Simek 1987; Whallon 1973). These reconstructions of activity areas carry limitations in the form of both pre- and post-depositional taphonomic processes influencing the location of artefacts, and often portray problematic links between the location of artefacts and other contextual, functional or chronological evidence (Manzanilla and Barba 1990; Ullah *et al.* 2015).

The need for geoarchaeological approaches for the study of spatial activity patterns at archaeological sites has driven several research projects in the past two decades seeking to test and apply various microscopic techniques to the study of activity areas, such as micromorphology, geochemistry, phytolith analysis and mineralogy (Banerjea *et al.* 2015; Manzanilla and Barba 1990; Middleton and Price 1996; Shahack-Gross *et al.* 2004; Tsartsidou *et al.* 2009). Canti and Huisman (2015) provide an overview of the developments in the use of geoarchaeological techniques in archaeology during this time and emphasise the need for continued validation through experimentation and performing multi-proxy studies. While such studies were previously rare, they have now gained popularity to a degree that the term 'geo-ethnoarchaeology' has recently been coined (Friesem 2016). It is important to keep in mind however, that whether spatial analysis of archaeological sites relies on the distribution of artefacts, micro-refuse or soil analysis, it is always based on the premise that human occupation results in a non-random distribution of the remains of past activities.

This paper will focus on the use of phytolith analysis and geochemistry for spatial analysis in particular, though other geoarchaeological techniques should not be considered less or more valuable. Each particular situation, research question or site will call for the use of a specific geoarchaeological method or a combination of these.

The advantages of the use of phytolith analysis are that phytoliths often represent *in situ* deposition, usually preserve better than organic remains (especially in arid conditions), and enable us to distinguish between different plant parts. Nevertheless, phytoliths too may suffer from chemical dissolution depending on their depositional environment, and may not always be identifiable to the species or even genus level. Geochemical analysis benefits from a long history of use within archaeology, and the simultaneous identification of geochemical elements in archaeological sites is currently easily achieved with modern analytical tools such as Inductively Coupled Plasma (ICP) or X-ray Fluorescence Spectrometer (XRF) instruments. On the other hand, certain unresolved issues regarding the correlation of geochemical signatures to anthropogenic activities, understanding of the baseline geochemistry of the parent material and processes affecting elements in this (Matschullat *et al.* 2000), difficulties distinguishing the archaeological input from modern or geological ones (Oonk *et al.* 2009), and problems of equifinality must be considered prior to analysis.

In order to tackle some of these issues, recent geochemical studies of anthropogenic sites aimed at identifying activity areas use combinations of several geochemical elements, which can often be correlated to specific types of activities (Middleton and Price 1996; Oonk *et al.* 2009; Parnell and Terry 2002; Vyncke *et al.* 2011). During the past two decades, multi-elemental examinations of archaeological, historical and modern houses revealed that activity areas and different features can be correlated to certain (combinations of) elements, and that household, production and even ceremonial practices can be distinguished. Another approach for improving archaeological interpretations of geochemical signals is the testing of processes that influence the creation of anthropogenic soil signatures by studying ethnographic or experimental cases.

Many scholars stress the importance of such analogies to our understanding of geochemical signatures and the activities that produce these (Fernandez *et al.* 2002, 488; King 2008, 1225; Middleton and Price 1996; López Varela and Dore 2010; Wilson *et al.* 2008). Ethnoarchaeological observations laid the ground for better interpretations of general patterns of human input in soils, and later studies related a suite of elements to specific activities. Middleton (2004) for example, was able to distinguish activity areas in buildings at two sites, Çatalhöyük in Turkey and Ejutla in Oaxaca, Mexico. He managed to identify the chemical remains of burning (P, Na, Mn and K), food storage and preparation (P and Ca), plastered surfaces (by alkalinity), high traffic zones (lower reading of elements than off-site controls) and craft production (burning and high Fe). However, as with the case of even well informed ethnographic studies, some of the observed patterns in this analysis were left unexplained. Most of the sites examined through geochemical analysis so far were substantial buildings with a clear division of space, and some of these produced very comprehensive and convincing reconstructions (Hutson and Terry 2006; King 2008; Milek and Roberts 2013; Terry *et al.* 2004). While geochemical studies at ephemeral sites benefit from this knowledge, there is a need for additional targeted geoarchaeological studies of short-lived occupation.

In a similar way to geochemistry, phytoliths are increasingly being used to inform archaeologists about ancient activities which took place within and around ancient households, often in combination with other micro-techniques. Both quantitative and morphological studies of phytoliths are useful aids in identifying spatial activity patterns. A study of abandoned Maasai settlements by Shahack-Gross *et al.* (2004) demonstrated that ashy and trash deposits, livestock enclosures and even associated large gates could be recognised by using a suite of micromorphological, mineralogical and phytolith analyses. They suggest that together with information from features such as post holes, artefact and faunal and botanical studies, a comprehensive reconstruction of archaeological sites and ancient lifestyles can be achieved.

Following their study, other scholars started to explore the potential of phytolith analysis for spatial reconstructions. Tsartsidou *et al.* (2008; 2009) conducted phytolith analyses at both ethnographic and archaeological sites. Phytolith analysis was also used in combination with micromorphology in order to characterise outdoor activity areas at Çatalhöyük, Turkey (Shillito and Ryan 2013). The analysis was able to distinguish between episodes of construction, dumping, accumulation, exposure and trampling, demonstrating a dynamic use of these areas through time as middens, yards or traffic zones. The same techniques were able to achieve the same detailed level of interpretation at the Iron Age site of Tel Dor, Israel, revealing that deposits which were first considered to be plaster floors were in fact compressed layers of grasses and animal dung (Shahack-Gross *et al.* 2005). A study of phytoliths and faecal spherulites by Portillo *et al.* (2009) demonstrated that certain areas of the PPNB site Ayn Abu Nukhayla, Jordan, contained evidence of the processing of cereals, while others were used as animal pens. The combination of phytoliths and spherulites allowed the researchers to differentiate between plant material that was introduced into the building from dung sources and other origins.

Although these studies illustrate the usefulness of phytolith analysis for identifying activity areas in anthropogenic site, the nature of this type of information carries limitations which must be addressed. Since the use of plants varies across sites due to local availability of vegetation and human preferences, phytolith signatures from specific activities are not uniform across sites. When it comes to fire installation for example, Shahack-Gross *et al.* (2004) identified elevations in two types of phytoliths in hearth contexts from the Maasai compound in relation to other localities (one characteristic of grasses and the other of wood/bark), but no higher concentrations of other phytolith forms. They reported that the fuel type used in the settlement was wood. Portillo *et al.* (2014) found large amounts of grass phytoliths in the Neolithic fireplaces, which they associated with an abundance of faecal spherulites suggesting the use of dung for fuel. Tsartsidou *et al.* (2008) reported a high concentration of irregular phytoliths (comprising a high percentage of variable morphology phytoliths) in the hearth deposits of an ethnographic village in Greece, which they interpreted as the presence of wood ash. The same is true for phytolith evidence of dung deposits. Although high concentrations of phytoliths are a frequent characteristic of animal enclosures, the associated morphologies will vary according to fodder and the local availability of plant species grazed, and evidence of dung can be missing if it is removed for secondary use (Tsartsidou *et al.* 2008, 611). Phytolith evidence of specific activities is therefore site dependent and frequently ambiguous, it is often combined with other sources of information in order to cope with issues of equifinality.

Tracing pastoral lifestyles

The use of ethnoarchaeology to gain insights into ancient habitation is not new. Towards the end of the twentieth century, a growing interest in ephemeral and pastoral archaeological sites coincided with a revival of ethnoarchaeological studies in the Near East, within Bedouin groups in Jordan in particular. By establishing the nature of pastoral occupation during the recent past, and assessing the potential for identifying ancient pastoral activity following abandonment, they addressed our ability to interpret the archaeological pastoral landscape. What type of evidence of pastoral habitation is left in the landscape? Can we speak of evidence of absence, or merely absence of evidence? Although pastoral life would have undoubtedly changed through time, these studies recognise the need to establish a better understanding of different aspects of pastoral and nomadic activities across a varied landscape today (Palmer *et al.* 2007; Saidel 2009, 179).

Banning and Köhler-Rollefson (1983; 1986; 1992) were two of the pioneers of ethnoarchaeological studies in Jordan, who applied ideas about the relationship between spatial deposition patterns and the material record explored by earlier ethnoarchaeologists (Binford 1978; Gifford 1977; Yellen 1977) to the study of Bedouin camp sites in Jordan. They documented the remains of numerous abandoned pastoralist sites in the vicinity of Petra with the aim of contributing to the finding of archaeological pastoral sites and distinguishing them from those of settled agriculturalists. Their research focused on the material remains left behind after abandonment of such sites, and the identification of typical features indicating pastoral-nomadic occupation.

Figure 1. Map of Jordan showing the location of Wadi al-Jilat and Wadi Faynan.

Around the same time, Simms (1988) studied one of the camp sites of the Bedul Bedouin of Petra, Jordan, in order to compare the site's structure to those of hunter-gatherer sites that had been the subject of earlier ethnoarchaeological studies. The findings from this research represent a focus on functional explanations to the spatial distribution of activity remains, which can be used to understand cross-cultural patterns of the use of space at pastoral sites, and advise future excavation strategies. Findings made in this investigation include the location of refuse which was different from the location of activities, the cleaning of hearths which meant that their contents only represent their terminal use, and an indicator of animal domestication in the form of '*laban*' platforms for the processing of dairy products. The background to this study was the need for a better understanding of the processes leading to spatial distribution patterns in the archaeological record, especially after previous ethnoarchaeological studies questioned contemporary assumptions about the relationship between refuse and activities (Simms 1988; Yellen 1977; Kent 1984).

Later studies set out to expand both the methodologies used to study Bedouin camp sites, which focused on the identification and layout of the sites, and the area of Jordan where ethnoarchaeology took place – which at the time was limited to the Petra region. The Bedouin Ethnoarchaeological Survey Project, led by Saidel (2001), set out to position the studied Bedouin sites within a microenvironment with the aim of discovering correlations between local conditions and the size and spatial organisation of camp sites. Additional goals included establishing the patterns of artefact deposition within the camp sites, and the collection of soil samples for geoarchaeological analysis. The collection of geoarchaeological samples was likely inspired by an earlier micromorphological study of a Bedouin tent floor, which illustrated the potential of this technique to identify formation processes and evidence of human activities at nomadic-pastoral sites (Goldberg and Whitbread 1993).

The aims of ethnoarchaeological investigations of Bedouin camp sites in the 1990s and the beginning of the twenty-first century were not very different to those guiding research during the 1980s, including establishing cross-cultural functional explanations for the use of space at pastoral sites. However, the methodology for achieving them had changed to include more detailed studies of artefact distributions and the application of geoarchaeological analyses.

Case studies: Wadi Faynan and Wadi al-Jilat

The study described in this section sought to explore the potential of a dual phytolith-geochemical methodology for spatial analysis at ephemeral sites, particularly those located in the dynamic environments of the Near East. Analysing the data using two sources of information could potentially help combat issues of equifinality (*i.e.*, a state can be reached by multiple potential means) and equivocality (*i.e.*, a single process may result in several outcomes) that occur with the use of one technique. By verifying or contradicting the identification given by one method through additional information from the other, a more reliable and comprehensive account of the social use of space at a site can be reached. In addition, the combination of geochemical and phytolith analysis has the potential to capture signals from different types of activities, the phytoliths representing exploitation of plant material and the geochemistry reflecting other types of anthropogenic enrichment such as burning or craft production.

By applying this methodology to sites that are difficult to interpret because of their short-lived nature, information can be gained about the use of space that was previously unavailable because of the poor preservation of structures, artefacts and the limited incidence of organic remains. The dual methodology was first tested through an ethnoarchaeological study of Bedouin camp sites at Wadi Faynan in Jordan (Fig. 1). The Bedouin sites provide an excellent subject for the testing of the dual phytolith-geochemical methodology; the use of space by Bedouins at Wadi Faynan has been thoroughly documented so that known activities can be correlated to the analysis results. The sites reflect a seasonal, ephemeral occupation in a dynamic, arid environment, and they represent a range of abandonment periods. The same methodology was then applied to the excavated Neolithic sites in Wadi al-Jilat, Jordan, in order to test its efficacy on archaeological material (Fig. 1). The sites of Wadi al-Jilat provide an ideal case study to test the applicability of a dual phytolith-geochemical methodology for distinguishing activity areas in ephemeral occupation deposits as they represent seasonal occupation in an arid, dynamic environment and were completely excavated.

Methods

Laboratory methods

The geochemical analysis in this study focused on the following chemical elements, measured in PPM: magnesium (Mg), silicon (Si), potassium (K), calcium (Ca), phosphorus (P), iron (Fe), titanium (Ti), manganese (Mn), aluminium (Al), strontium (Sr), sulphur (S), chlorine (Cl), zinc (Zn), chromium (Cr) and zirconium (Zr). The analysis was performed using a Thermo Scientific Niton XL 3t Goldd+ (geometrically optimised large area drift detector) handheld XRF analyser (pXRF), with an Ag anode 50 kV, 200 μA tube. A helium purge was used to lower the detection limits for light elements. The samples were placed in 9 mm plastic cups, covered with a thin polypropylene film, and analysed using a mobile test stand. The pXRF machine was set to the 'mining Cu/Zn mode' and the exposure time for each of the ranges was adjusted to achieve the following settings: the main range was run for 40 seconds, the high and low ranges for 30 seconds each, and the light element range for 80 seconds to allow for reliable readings for elements on the edge of the detection limits of pXRF such as Mg and P. In total each reading took 180 seconds.

One silica (blank) standard and three National Institute of Standards and Technology (NIST) standards were analysed using the same setting as the soil samples during each analysis session; SRM 2711a (Montana II soil), SRM 2709 (San Joaquin Soil), and SRM 1646a (Estuarine Sediment). The measurements of the NIST standards confirmed the precision of the pXRF instruments (for details, see Vos *et al.* 2018).

Phytolith extraction was performed using the dry ashing method, where the soil sample is burnt in a muffle furnace in order to remove organic matter and isolate phytoliths (Rosen 1992). Slides containing the phytolith material were counted using a Meiji infinity polarising microscope at a magnification of x400, using a modern Jordanian phytolith reference collection prepared from plants collected in Jordan (housed at Bournemouth University and the CBRL British Institute in Amman). At least 250 phytoliths were counted per slide, and the entire slide was counted if this amount was not reached. The counted quantities of different phytolith types and (when relevant) taxa were documented on a tally recording sheet. The names of the phytolith types followed the International Code for Phytolith Nomenclature (Madella *et al.* 2005).

Statistical analysis

Separate databases for geochemical and phytolith data were created for each site, and a combination of sites, using IBM SPSS statistics version 23. The geochemical database included the readings of the chosen elements (see previous section) for each sample, which contained error readings of ≤3%. Other elements containing error readings (two-sigma precision) of ≥10% were excluded from the analysis. An exception to this rule was made for Mg, Mn and Zn, which contained error readings of 20%, 23% and 13% (respectively) but were kept in the analysis as they are valuable indicators of anthropogenic activities. The phytolith database included the morphological categories used in the counting sheets and additional variables calculated from the raw data: dicotyledon (dicot – here we use the term according to the pre-1990s definition to mean non-monocotyledon), monocotyledon (monocot), single-cell, multi-cell, Panicoideae, Pooideae, Chloridoideae, Arundinoideae, Palmaceae, Hordeum sp., Triticum sp., leaf, leaf/husk, leaf/stem, husk, awn, weight percent of extracted phytoliths (weight of phytoliths exctracted after processing divided by weight of the initial dried sample × 100), and number of phytoliths per gram of original sediment processed. As the total amount of counted phytoliths varied per slide, the data were transformed to percentages by dividing the number for each counted category by the number of phytoliths counted for the relevant slide, and then multiplied by 100. The number of phytoliths per gram of sediment was calculated using the following formula:

- no. per slide = (phytolith count / no. of counted fields) × total no. of fields on slide
- no. per gram = (no. per slide / mass of phytoliths mounted in mg) × (mass of phytoliths extracted in mg / total sediment weight in mg) ×1000

The data was explored using box plots and bar charts that were created for every variable and for related variables (such as plant parts or genus categories). When analysing the results, it became clear that several categories plotted very similarly, in most cases these were variations of floor surfaces. For example, samples collected from the edges of hearths did not differ from the general floor samples, and so were grouped under the floor category.

Principal component analysis (PCA) was run in SPSS using the correlation matrix, a method which standardises the variables. No rotation was applied to the analysis, and the components were extracted based on eigenvalues greater than 1, and saved as variables based on regression. Discriminant function analysis was carried out with the independents entered together and the prior probabilities computed from group size, including leave-one-out classification in the display option. A two-tailed Pearson correlation test was run with variables from both the geochemical and phytolith analyses in order to identify patterns that could influence the results of the PCA analysis.

Ethnographic case study: Wadi Faynan

The majority of ethnographic samples discussed in this research were collected as part of an extensive ethnoarchaeological survey of abandoned Bedouin camp sites at Wadi Faynan during 1999 and 2000, led by Carol Palmer and Helen Smith as part of the Wadi Faynan Landscape Survey (WFLS) (Barker 2000). The aims of this survey were to explore the nature of pastoral activity in Wadi Faynan during the recent past and assess the potential for identifying ancient pastoral activity following abandonment. By doing so, the project intended to address our ability to interpret the archaeological pastoral landscape – what type of evidence of pastoral habitation is left in the landscape? And is there evidence of absence, or merely absence of evidence? Furthermore, the survey helped reveal practical and social aspects of Bedouin life, including use of space, and the changes in this through time and across seasonal and tribal variations (Palmer *et al.* 2007).

The research questions stated above were addressed by recording the material culture left behind during abandonment of modern Bedouin camp sites at Wadi Faynan. The study focused on sites that had been abandoned for various durations of time in order to evaluate the influence of taphonomic processes on the presence of material remains during different stages of abandonment.

An initial survey during April 1999 documented the locations and main architectural characteristics (both durable and perishable) of Bedouin tents in the landscape; in total 83 sites were visited. During the visits several physical attributes were recorded, including tent size, orientation, position, spatial arrangement and both common and supplementary features such as storage facilities or outdoor hearths. These data were accompanied by the accounts of the occupants of the area, who provided information about the abandoned camp sites and the activities that took place at these. The team conversed with the tent inhabitants in order to get a better understanding of the use of space at these camp sites and where possible, about the individuals that were living there and the animals owned by them. An accompanying local informant, Jouma' 'Aly of the 'Azazma tribe, enabled a good flow of conversation with the interviewees and a deeper understanding of local lifestyles and use of space to be achieved (Palmer *et al.* 2007).

During 2000, the same camp sites were revisited and studied in greater detail, an artefact distribution study was undertaken, and the soil samples used for the research presented in this paper were collected from chosen sites (Palmer and Daly 2006). In addition to the sites that were

Figure 2 (right). Plans of the Bedouin camp sites at Wadi Faynan (plans of WF953, WF940 and WF982 after Palmer *et al.* 2007, 381-387. Plans of WF916, JTS and JTW created by Daniella Vos, based on schematic drawings made in the field).

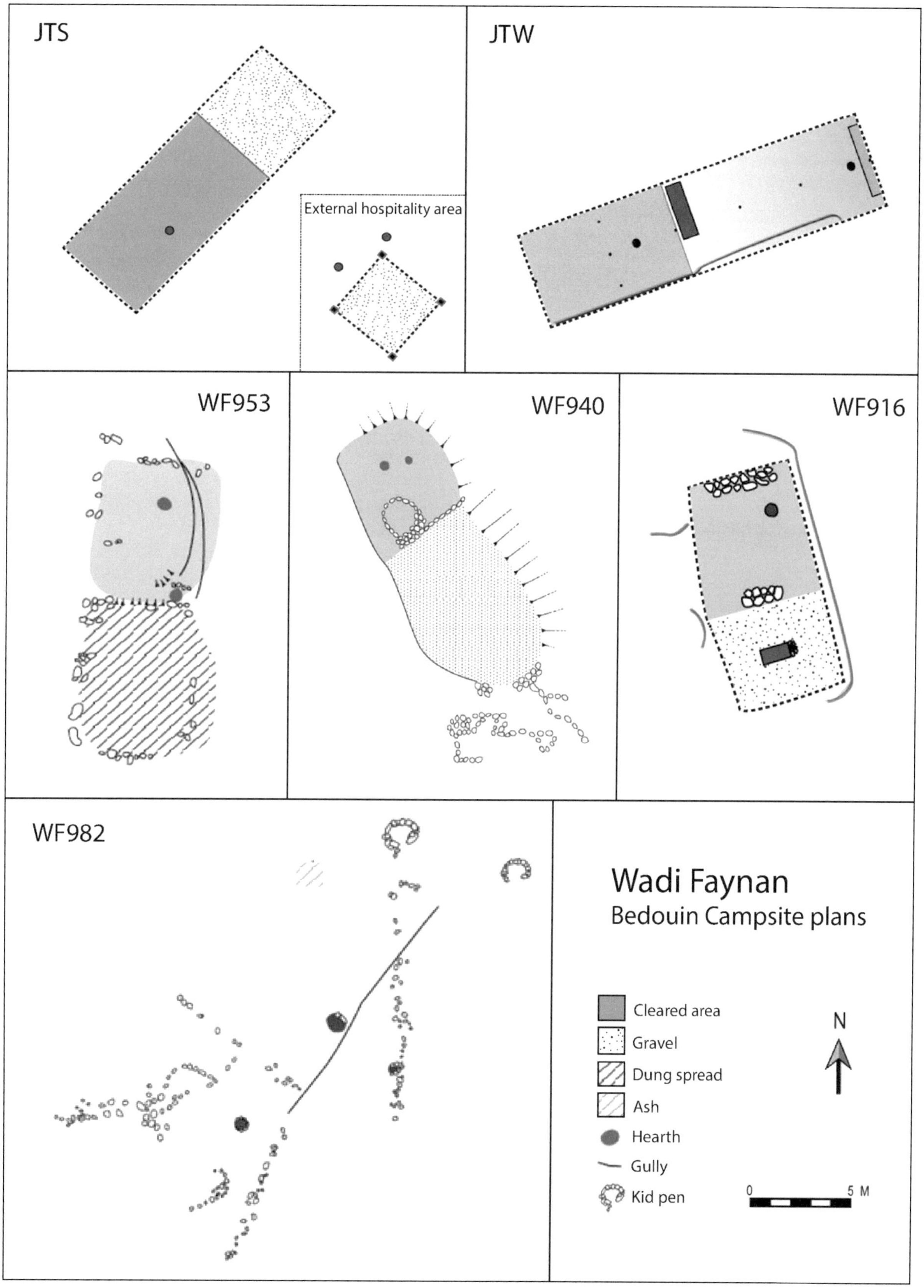
JTS
External hospitality area
JTW
WF953
WF940
WF916
WF982
Wadi Faynan
Bedouin Campsite plans
Cleared area
Gravel
Dung spread
Ash
Hearth
Gully
Kid pen
N
0
5 M

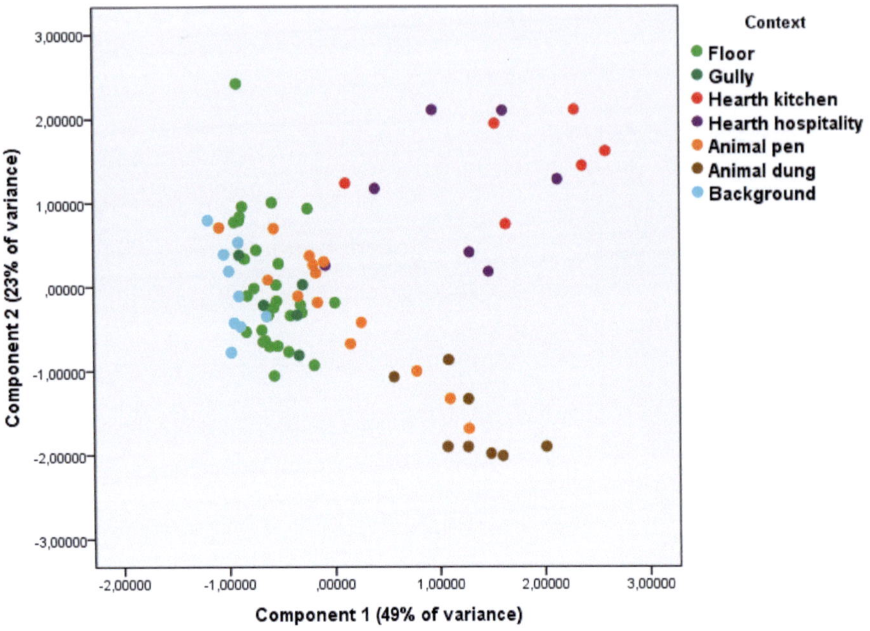

Figure 3. PCA biplot for all Wadi Faynan sites. The first component is driven by P, K, Zn and negatively by Si, Al, Ti and Zr. The second component is driven by Ca, Mn and Mg.

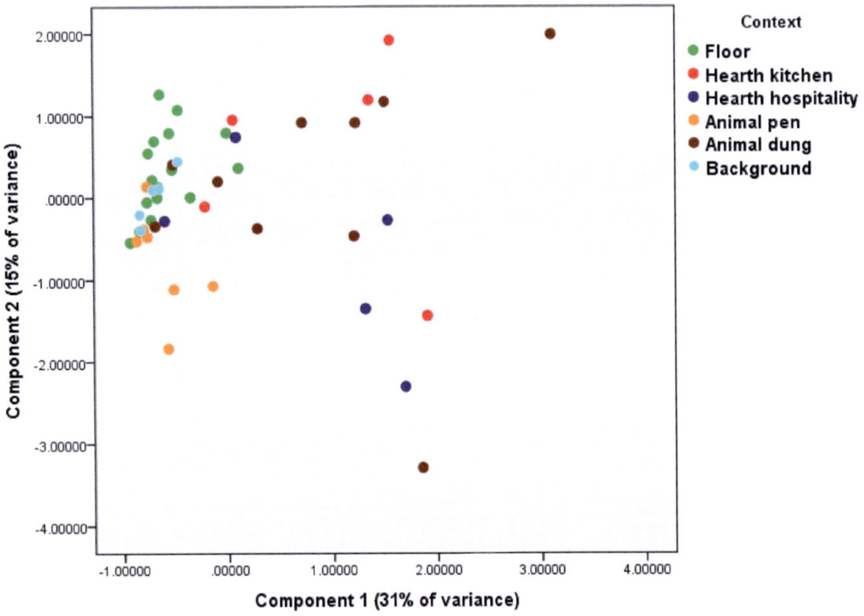

Figure 4. Combined PCA biplot for the sites JTS, JTW, WF916 and WF953. The first component is driven by monocots vs. dicots, multi-celled vs. single-celled phytoliths, husk material and Pooideae. The second component is driven by unidentified phytoliths, leaf, negatively by no. per gram, weight percent and *Triticum* sp.

sampled in 1999 and 2000, two additional camp sites were later sampled by Carol Palmer, Jouma' 'Aly and the author at Wadi Faynan in 2014. The ethnographic soil samples discussed here were collected from one occupied and five abandoned sites at Wadi Faynan (Fig. 2) (Vos *et al.* 2018).

The seasonally occupied ephemeral Bedouin camp sites at Wadi Faynan were chosen as a case study for testing the efficacy of geochemistry and phytolith analysis to identify activity areas because much is known about the use of space in Bedouin tents. Generally, the tents are divided into public-male areas (*shigg*) and private-female (*mahram*) areas. The public area is used for hospitality; coffee, tea or food are served to honoured guests here, and the central hearth can be used for preparing coffee and sometimes tea, though tea is usually prepared in the *mahram* and brought to guests. Various household activities take place within the private area, which includes a kitchen with a hearth which is used for cooking. The use of space within Bedouin households at Wadi Faynan has both static and dynamic aspects. While activities take place within designated areas, each section of the tent can change its function throughout the day. For example, the private area can be used for activities such as weaving, churning butter or entertaining female guests, and will be used for sleeping at night. And when guests are not present, the public area is used by all members of the household.

The use of space at the Bedouin camp sites of Wadi Faynan is in many ways fixed and guided by cultural principles, but some flexibility is maintained through the dynamic use of spaces for different purposes at various points in time throughout the day. The types of camp sites analysed in this research all include a private area which contains a kitchen, and all but one include a hospitality area, animal pens, and in some cases internal animal sleeping areas. All but one camp sites (WF940) contain two hearths, one used for food preparation in the kitchen and another for coffee making in the hospitality area.

Results

The traditional use of space at the Wadi Faynan Bedouin camp sites resulted in relative fixed locations of activity. These leave clear traces of burning and animal husbandry, even after post-abandonment exposure to the elements in this dynamic environment. Activity areas with a strong anthropogenic input were clearly distinguishable from the background and floor related samples through both means of analysis: the hearths, dung sediments, and, to a lesser degree, the animal pen floors (Vos *et al.* 2018).

The hearths are clearly visible within the ethnographic data. They have the largest enrichment of Mg, Ca, Sr, and in some of the sites also S and Zn. The evidence from the phytolith analysis is less straightforward. Elevations of monocots and multi-celled phytoliths, and in some cases Panicoideae grasses, were found in most hearths. An increase in phytoliths that were indicative of various plant parts is correlated to the large amount of monocots identified within the hearth context. The kitchen hearth samples at some of the sites contained higher levels of husk material, but so did many of the dung samples. This might reflect the preference for dung cake fuel in the kitchen hearth (Vos *et al.* 2018). While the PCA biplot created for the geochemistry results shows that the two hearth types form a cluster (Fig. 3), the PCA biplot based on the phytolith analysis displays less clustering and differentiation between the hearths and the animal dung samples (Fig. 4). Interestingly, the two groups of kitchen and hospitality hearths plot separately in the PCA biplot based on the phytolith results. This suggests that the difference between the two types of hearths is better observed through the phytolith data (*ibid.*).

Dung deposits at Wadi Faynan were rich in grass phytoliths, and contained high proportions of conjoined phytolith material. However, the dung samples did not contain higher phytolith concentrations with the exception of the samples from WF916. This could be due to the use of dung cakes in the other sites, which might have caused a reduction of dung within the animal enclosures (Vos *et al.* 2018). The same trend can be seen within some of the elements chosen for the geochemical analysis. P levels are elevated in all dung samples, but are higher still within the hearths of all of the sites apart from WF916. In addition to these, concentrations of K and Cl are highest in dung samples, and S and Zn are slightly elevated in relation to the background samples (*ibid.*).

Floors and gullies display similar patterns to each other and to the background samples in all of the Wadi Faynan sites, and plotted similarly to these in the PCA scatterplot (Figs. 3-4). They contain no elevations in the anthropogenic chemical markers mentioned above, such as Mg, P, K, Mn, Sr, Ca, or the phytolith categories related to anthropogenic input such as high levels of monocots and multi-cells, although slight Cl enrichments can be seen in floor and gully samples from the majority of sites. Unlike floor areas that have been described as high traffic zones (Middleton 2004, 56), the floors and gullies at Wadi Faynan do not show signs of a depletion in concentrations of chemical elements. They plot similarly to the background samples, which suggests that signatures of activity remained local and did not spread out across the floor surfaces.

Archaeological case study: Wadi al-Jilat

The Neolithic of the Levant is characterised by very gradual changes in lifestyle, leading to a transition from hunter-gatherer societies to early sedentary farming communities. This transition, however, is not a linear and inclusive change that affected all human societies in the Levant. Rather, a mosaic of human cultures and modes of subsistence would be a more suitable description of the situation during the Neolithic. Alongside the so-called mega-sites of the Pre-Pottery Neolithic B (PPNB) period, which consisted of permanent architecture, other sites such as Wadi al-Jilat show a more ephemeral occupation during the Neolithic (Goring-Morris *et al.* 2009; Goring-Morris and Belfer-Cohen 2008; 2011). At these ephemeral sites, a mixture of subsistence activities seems to have taken place, and the occupation of the Wadi al-Jilat structures appears to have been seasonal. Ephemeral habitation has been studied at less depth than more substantial settlements during the Neolithic, and the difficulty of interpreting the use of space at these sites limits our view of lifestyles during the Neolithic.

Is was therefore important to explore new ways to study the use of space at such sites. To this end, 36 soil samples from the Neolithic site of Wadi al-Jilat 13 (WJ13) and 17 from Wadi al-Jilat 7 (WJ7) were analysed in this study. Fieldwork at Wadi al-Jilat was part of a series of excavations at the Azraq Basin during the 1980s under direction of Dr Andrew Garrard. The project aimed to provide new insights into settlement and subsistence in the steppe and desert regions of the Levant during the

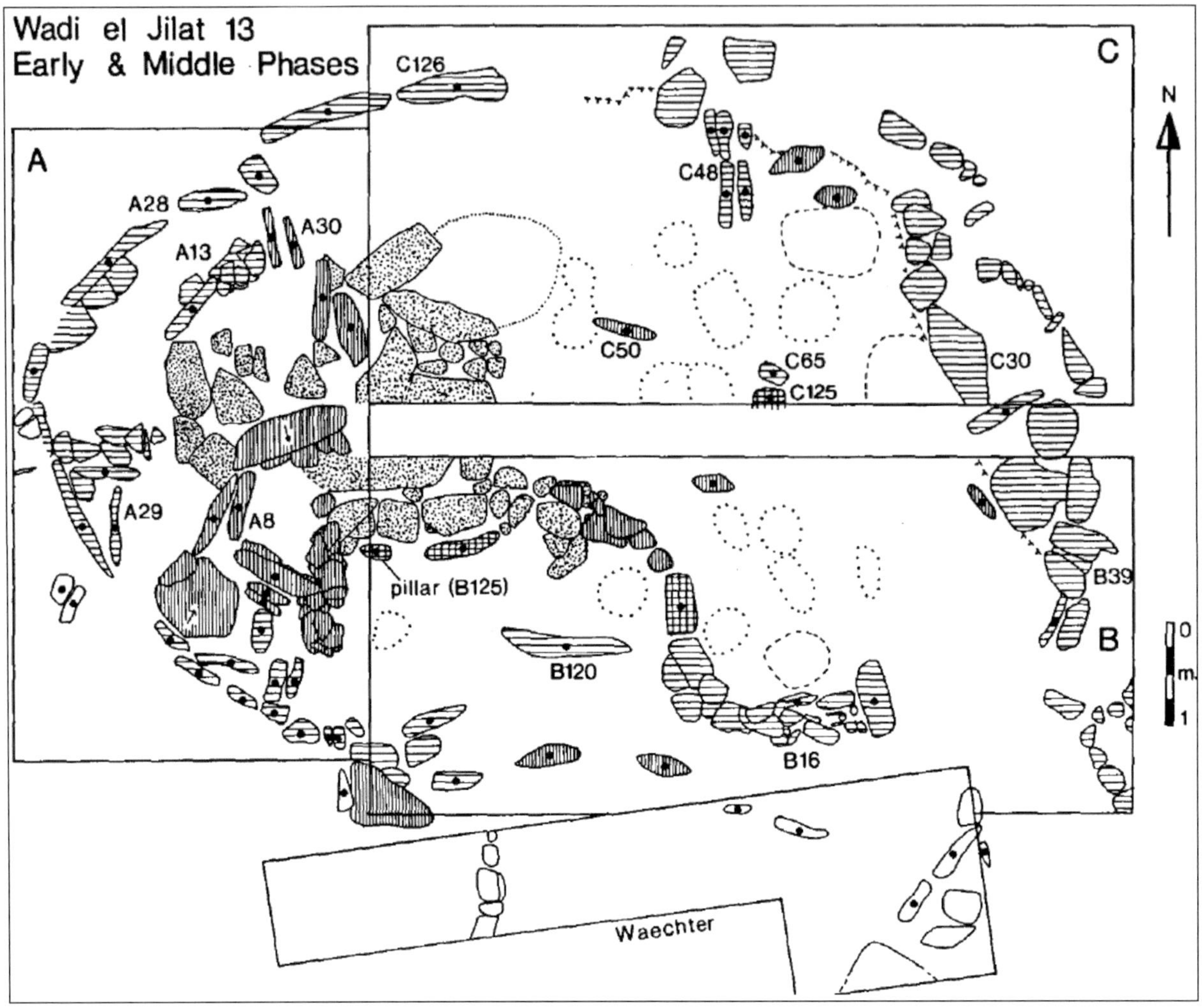

Figure 5. A plan of early and middle phases at WJ13 (from Garrard *et al.* 1994, 80).

early stages of sedentism, agriculture and pastoralism (Garrard *et al.* 1988). The great advantage of using the Neolithic sites at Wadi al-Jilat is that complete structures have been excavated and a soil sample from each context (including hearths and other internal features) was collected. This meant that a full sequence of occupation at these sites was available to choose from, and the detailed records for each context make a reconstruction of the occupation history a straightforward task.

Wadi al-Jilat is situated on the banks of the Jilat gorge, a tributary of the Wadi al-Dabi in the south-west of the Azraq basin and located approximately 55 km south-west of the modern town of Azraq in Jordan. The site lies in a transition area between steppe and desert, receiving approximately 100 mm precipitation yearly, and cuts into late Cretaceous and early Tertiary limestones, chalks and marls which contain a large concentration of flint beds (Garrard *et al.* 1994). The dynamic environment which Wadi al-Jilat makes part of is not unlike that of Wadi Faynan (for an overview, see Vos 2017). The availability of a nearby seasonal water source and presence of diverse ecological zones formed by the topography of the region, together with the restraints set by the arid and variable climatic conditions, could have been exploited by the Neolithic inhabitants of Wadi al-Jilat using a range of subsistence strategies. Each of these strategies might have been preferred under different circumstances. It is in this aspect that the two types of data analysed in this research, ethnographic and archaeological, may show the most similarity. If patterns of mobility during the Neolithic reflect communities' negotiation with frequently changing environmental, socio-economic and internal factors in the same way that mobility patterns at Wadi Faynan did in the recent past, it is not surprising that we find ephemeral patterns of settlement at both. These would allow for the flexibility needed when interacting with a highly dynamic, arid environment.

Description of the sites

The vast majority of Neolithic buildings at Wadi al-Jilat are circular or oval semi-subterranean constructions, with upright slabs forming the fragile external walls, which often enclosed shallow deposits. Many of these structures had internal divisions, hearths and other features such as benches or storage bins (Garrard *et al.* 1988, 40-41). Nevertheless, unlike contemporary sites in moister regions of the Levant, which present substantial architectural remains, the Neolithic settlement at Wadi al-Jilat left traces of somewhat flimsy structures. These, according to the excavators, hint towards a seasonal occupation, as is the case with many ephemeral structures used today by modern nomadic populations (Garrard 1994; Köhler-Rollefson 1992).

WJ13 is comprised of one (relatively large) oval structure measuring 10 x 6.5 m that has been fully excavated, with the exception of a single baulk. The structure takes advantage of a natural crescent shaped gully in the bedrock and follows this natural line, along which the western and north-western walls were erected from upright stone slabs. No clear wall was found bordering its southern end, but some features and stone slabs along the southern boundary could have been part of a wall in the past. Several bedrock post holes in the centre of the gully could have provided support for a superstructure. The excavation surface was divided into three areas, A, B and C (Fig. 5). The building was dated to the final PPNB according to four radiocarbon dates, ranging between 6840 ± 150 and 6739 ± 152 cal BC.[1] The four dates are similar to each other, which might suggest that this site was in use for only a short duration of time. Nevertheless, during its occupation history the sites was prone to substantial remodelling, resulting in a complex stratigraphic sequence and probably significant changes in the use of space.

Three phases of occupation were recognised, during each of these the interior of the structure had been divided up by platforms and partition walls (in the form of lying or upright stone slabs). During the initial phase, following the construction of the building, a series of occupation fills was deposited within the structure, and a pavement of stone slabs was laid on top of these at the western end. Within the primary deposits in the southern and eastern sections several stone-lined hearths were used. The middle phase of occupation included the construction of a partition wall separating the western part of the structure, above the previous pavement. A niche or sub-compartment was added as part of this wall, and in the eastern sector two pits and a number of stone-lined hearths were created. Isolated upright slabs were erected within the structure, the function of which is unclear. The last phase of occupation at WJ13 saw the placement of a stone-slab pavement on top of a rubble foundation, extending from the entrance in the south-east to the partition wall at the western end.

The occupation of WJ7 took place during the Early and Middle PPNB period, and two radiocarbon samples from the building provided the dates of 7942 ± 197 and 7571 ± 106 cal BC. The site was divided into areas A, B and C (Figs. 6-7). The initial deposit on the bedrock in areas A and C was a layer of compact ashy material dated to the Early PPNB, which covered most of the excavated surface. Several sub-structures and walls were set into or overlay this primary deposit. In area B a silty layer covered the bedrock, not including much archaeological material, and above it a series of ashy midden deposits and two unlined hearths were found. During the later phases, dated to the Middle or Late PPNB, a number of stone alignments were built in the centre of area A, and a pit was cut through earlier deposits and the bedrock in its south-west corner. In area B, a pavement and upright slabs were added to a sub-compartment in the north-west area. Above the pavement a compact occupational deposit was excavated. After this phase, the building seems to have fallen into disuse (Garrard *et al.* 1994).

The faunal assemblages found at these sites show a reliance on wild populations of gazelle and hare during the PPNB, and the introduction of caprines into the area by humans during the early Late Neolithic (LN), when hunting seems to have decreased but was still significant. While 78% of the faunal assemblage at PPNB WJ7 consisted of hare and gazelle, within the faunal remains at LN WJ13 hare and gazelle represent 42% of the assemblage and caprines make up 20% of the assemblage (Garrard *et al.* 1994; Baird *et al.* 1992). The faunal remains at the sites have been interpreted as representing a range of subsistence strategies, including hunting, trapping and, from the early LN onwards, also sheep and goat herding (Martin 1999).

The results of the faunal analysis tie in well with those of the botanical examination, which likewise suggests a broad use of subsistence strategies including foraging and crop cultivation. Colledge (2001) found domestic glume wheats and barley in early PPNB levels at WJ7, and tentatively identified einkorn. It is unclear if these were cultivated nearby the site or imported. While only opportunistic cultivation takes place in the Jilat area today, cereals could have been grown there in the past if rainfall was sufficient during the Neolithic. Legumes, chenopods, fruits and seeds were also identified (Garrard *et al.* 1988, 47; 1994, 104-105). The botanical assemblages at WJ13 and WJ7 are similar, with large amounts of carbonised plant remains and poor preservation of the specimens.

Interestingly, Colledge mentions that species diversity was larger at WJ7 and WJ13 compared to

1 All dates in this section were taken from Garrard *et al.* 1994, and calibrated through www.calpal-online.de.

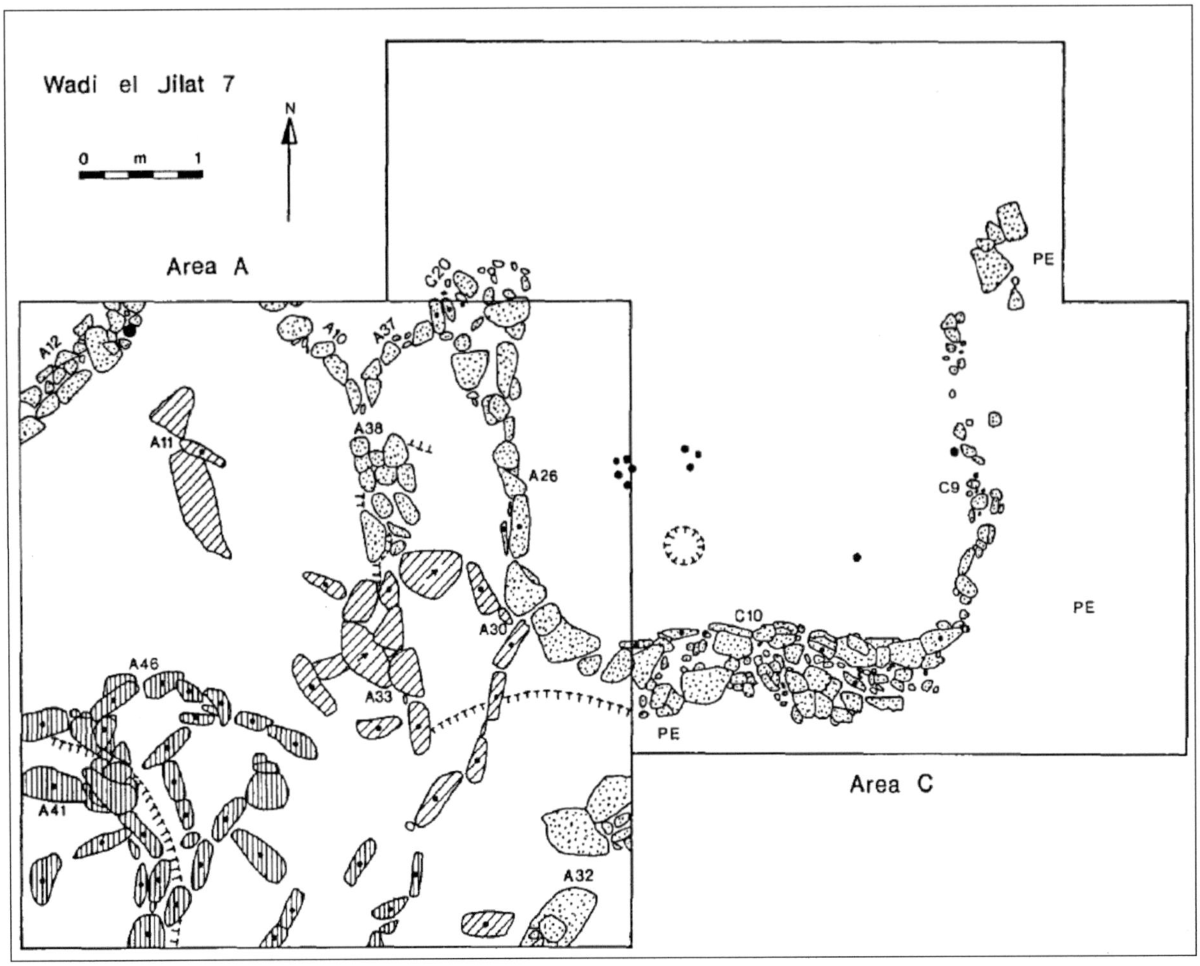

Figure 6. Plan of areas A and C at WJ7 (from Garrard *et al*. 1994, 74).

Wadi Fidan and Beidha, which are located in the Mediterranean woodland region and seem to have relied more heavily on cereals. The latter sites also contained higher levels of charcoal residue then the Wadi al-Jilat sites (Colledge 2001). Although this could be the result of excavation or collection biases, this observation could also reflect a reliance on a wider range of plant species at Wadi al-Jilat than the perhaps more specialised cultivation taking place during the Neolithic at Wadi Fidan and Beidha. Charcoal concentrations were higher in WJ7 and WJ13 than the other Wadi al-Jilat sites, these are also the two sites with the deepest stratigraphies. This trend could either relate directly to the extent of burning activities at the sites, or reflect taphonomic processes. It is worth noting that hearth features at the Wadi al-Jilat sites contained relatively low amounts of charcoal in comparison to the occupation fills.

The two sites of Wadi al-Jilat encompass various structures that were occupied, probably seasonally, between around 8000 and 6000 cal BC. It is likely that the extensive time span separating between the occupation of the various areas at this site encompassed differences in subsistence strategies, cultural practices and other aspects of life. On the other hand, the inhabitants of Wadi al-Jilat across the Neolithic are connected by sharing the same terrain, and probably similar environmental conditions. In this respect, they share similarities with the ephemeral sites at Wadi Faynan, where patterns of mobility and subsistence changed through time in relation to varying circumstances (Vos 2017). The use of the ephemeral architecture at these sites corresponded with these. These changes might be better understood through the incorporation of new techniques for gaining information about the spatial use of such structures. At the same time, the range of purposes and uses which might be represented at the Wadi al-Jilat sites must be kept in mind when analysing the phytolith and geochemical soil signature at these sites, as they affect the ability to juxtapose the results of such analysis.

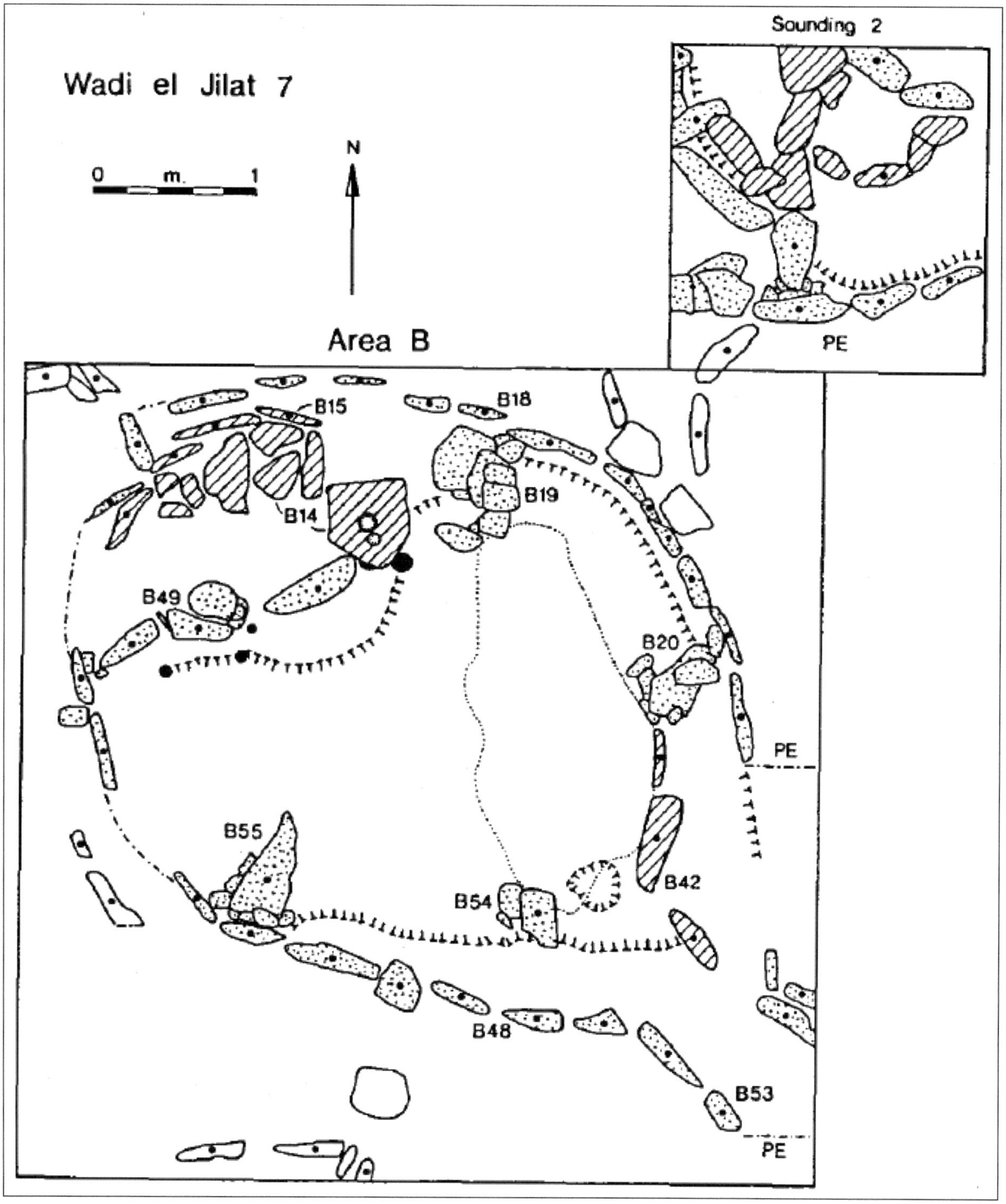

Figure 7. Plan of area B at WJ7 (from Garrard *et al*. 1994, 74).

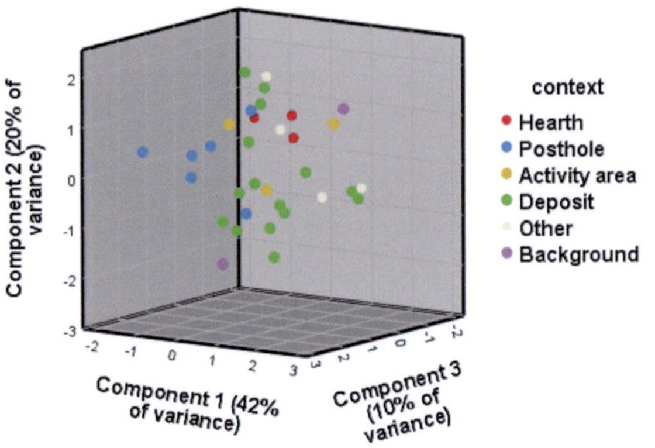

Figure 8. 3D PCA biplot, WJ13. The first component is driven by Ti, Si, Fe, K, Al, Zr and Nb. The second component is driven by Mg, Ba, Sr and Ca. The third component is driven by Cr, P, Rb, Cl and negatively by V.

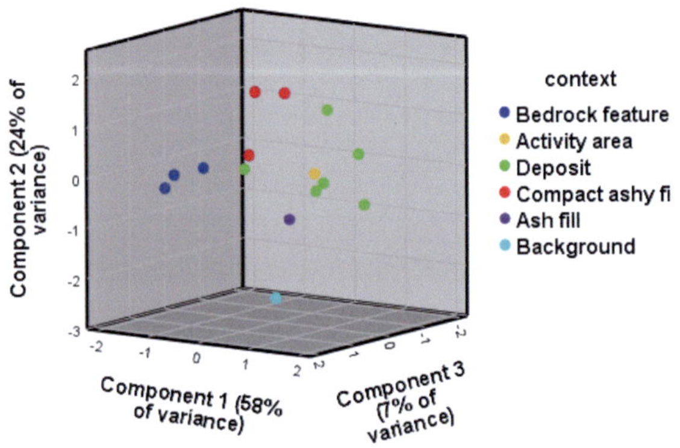

Figure 9. 3D PCA biplot, WJ7. The first component is driven by Mg, Si, Ti, Fe, S, Zr, K and P, and the second component by Ca, Sr and Rb.

Results

At first sight, the results of the geochemical analysis at Wadi al-Jilat do not appear to show clear trends of anthropogenic anomalies when it comes to individual chemical elements, and are not as straightforward as those obtained for Wadi Faynan. The geochemical variables that drive most of the variance within the PCA analysis of the Neolithic sites do not correspond well with the elements that were found to indicate anthropogenic input in the analysis of the Wadi Faynan sites or in earlier studies. In addition, WJ7 and WJ13 portray more differences then the Wadi Faynan sites, where most trends were representative of all sites.

The largest variance within the geochemical results of WJ13 is driven by background elements such as Ti, Fe, Al and Si, represented in the first component (Fig. 8). However, the anthropogenic input is better represented by the second, third and fourth components. Scatterplots combining the first three factors show a clustering of the bedrock features, hearths, and to a certain degree also the deposits and activity areas (Figs. 9-10). The main elements that drive the second, third and fourth components are P, Mg, Cl, Mn, Zn, Ca, Ba, Cr, Sr and S negatively. The first six elements are also important indicators of anthropogenic activity in the Wadi Faynan sites, which might indicate that the signal of human activity might still be similar to other sites after all.

The PCA scatterplot created for the first, second and third components of the geochemical results of WJ7 provided a better result than the one representing the first two components for WJ13, explaining 82% of variance (Fig. 9). These were driven by both chemical elements associated with anthropogenic activity such as Mg and Sr, and those related to the natural background such as Si and Ti. However, although the overall trends at this site enable us to distinguish between context categories based on geochemical variables considered to reflect anthropogenic activity, the individual elements do not appear to show remarkable trends or share similarities with findings in previous studies. This, however, was the case with the site of WJ13, where trends of specific elements provide interesting insights. P levels are increased in all anthropogenic contexts in comparison to the background samples, noticeably mostly in the posthole samples (Fig. 10). This could be explained by leaching of P downwards, but then one would expect to see a similar pattern in the other Wadi al-Jilat sites, which is not the case. Interestingly, there is a slight elevation of K and Mg in the hearths, and of Mn in activity areas (Fig. 10). These trends are similar to the observations at Wadi Faynan.

Generally, the context category that stands out in relation to the rest is postholes, as was the case in WJ13. However, it varies from the other contexts for different reasons, and seems similar to the background sample in some respects. Bedrock features at WJ7 had the lowest levels of Mg, K and P, yet the highest amount of S. Deposits generally contained high levels of most elements, but low levels of S, which was higher in the background and compact ashy deposits in addition to the bedrock features. Nevertheless, the PCA scatterplot above (Fig. 9) reveals that overall, samples in the same context category do cluster and that all categories vary significantly from the background sample.

The results of the phytolith analysis at Wadi al-Jilat revealed only very subtle patterns of differentiation between activity areas within the sites, while the background samples were clearly different to the on-site material. A high monocot to dicot ratio, the abundance of grass husks and the high weight percent and number of phytoliths per gram all appear to be associated with anthropogenic activity at the Neolithic sites. The bedrock features at WJ13 contained very low counts of phytoliths

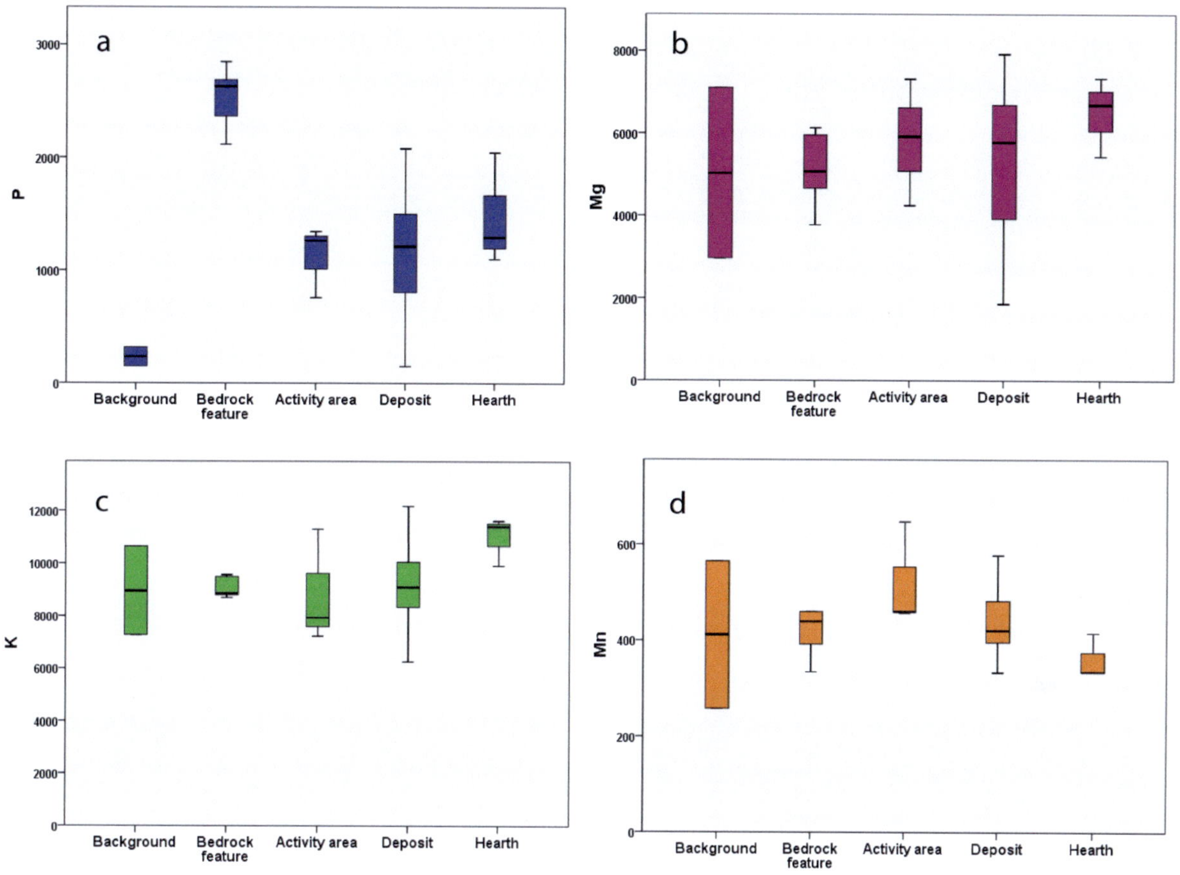

Figure 10. Average measurements in PPM for WJ13 per context category for the following chemical elements: (a) P, (b) Mg, (c) K, (d) Mn.

and most of them were associated with large amounts of silica aggregate material (which is considered to be an indicator of woody material by Schiegl *et al.* 1994). In addition, the weight percent of this context category was much higher than the other activity areas (calculations of phytolith number per gram would not suffice as silica aggregate does not fall within the phytolith counts). The background samples clearly vary from all the on-site ones, having lower amounts of weight percent and number of phytoliths per gram, and a lower monocot to dicot ratio. The phytolith analysis results at WJ7, which provided the best results for the geochemical analysis, demonstrate the most variability in context categories. While all contexts show an increase of monocots in relation to the background samples, the categories 'activity area' and 'compact ashy fill' (which probably reflect hearths) contained the highest concentrations of these. These two categories show resemblance when it comes to plant parts, containing the largest amounts of husk material in relation to the other context categories.

The anthropogenic enrichment within these two context categories at WJ7 appears to reflect high activity, strengthening the association between the mentioned variables and human occupation. In addition, enrichment of silica aggregate material in combination with low phytolith counts at the bedrock features of WJ13 might indicate a high anthropogenic input, albeit of a different kind. Interestingly, the background sample is devoid of husks, but contains larger amounts of silica aggregates.

The PCA scatterplots created for the phytolith results at these sites portray some clustering. As with the results of the geochemical analysis, the second and third components represent less of the overall variance but demonstrate better clustering of context categories than the scatterplots created for the first two components (Figs. 11-12). All in all, a high monocot to dicot ratio, the abundance of grass husks and the high weight percent and number of phytoliths per gram all appear to be associated with anthropogenic activity at the Neolithic sites.

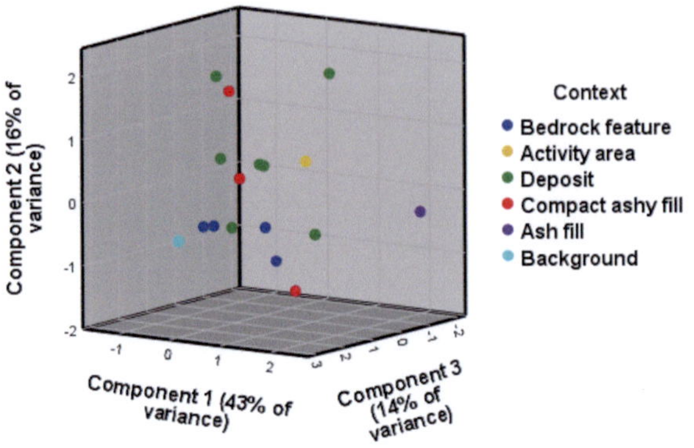

Figure 11. 3D PCA biplot, WJ7. The first component is driven by monocots, unidentified and degraded phytoliths, leaf, leaf/stem, Pooideae and single-cell phytoliths. The second component is driven by weight percent, Chloridoideae and negatively by burnt phytoliths. The third component is driven by Panicoideae, leaf/husk and weight percent.

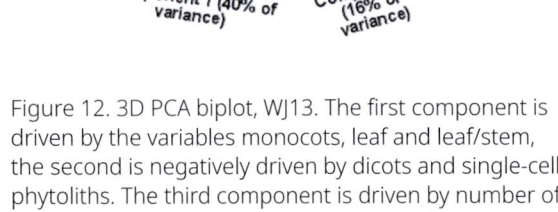

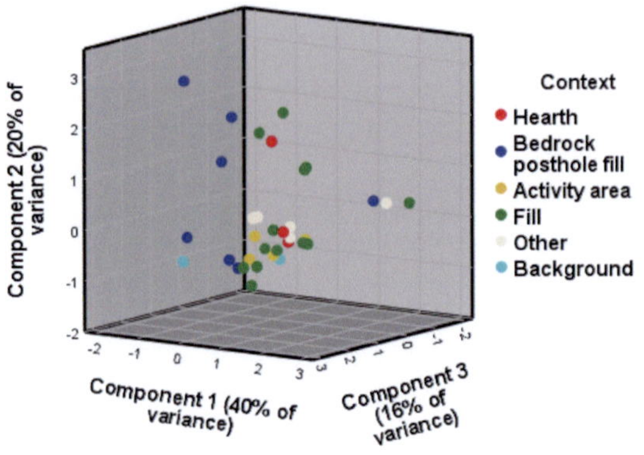

Figure 12. 3D PCA biplot, WJ13. The first component is driven by the variables monocots, leaf and leaf/stem, the second is negatively driven by dicots and single-cell phytoliths. The third component is driven by number of phytoliths per gram and multi-cell phytoliths.

Discussion

The results of the geo-ethnoarchaeological analysis suggest that the geochemical and phytolith analyses provide useful methods for studying activity areas at the Bedouin camp sites at Wadi Faynan. Activity areas with a strong anthropogenic input were clearly distinguishable from the background and floor related samples through both means of analysis. Individual trends within the geochemical and phytolith analysis were found to correspond with the known context categories within the areas of high anthropogenic activity. These findings support observations made in previous geo-ethnoarchaeological studies, indicating that specific (groups of) chemical elements are correlated to certain human activities and that anthropogenic anomalies can also be observed through phytolith analysis, though the latter trends will be more site specific.

The analysis of the Neolithic sites suggests that there is great potential in identifying, or at least distinguishing between categories of activity areas at ephemeral archaeological sites. Although WJ13 and WJ7 share the same environmental and historical setting and are adjacent to one another, the dual geochemical-phytolith approach worked differently with each site. WJ7 exhibits distinguishable context categories when examined through PCA scatterplots (mainly due to the geochemical input), while the geochemical and phytolith analysis of WJ13 demonstrate subtle trends within individual variables.

The geochemical variables that represent most of the variance between context categories in the Neolithic sites are not always the same ones that were found in the analysis of the Wadi Faynan sites or earlier studies. The PCA scatterplots exhibited far better clustering of context categories when plotted according to the second and third components, which did include variables more similar to the ones found to represent anthropogenic input. This might indicate that the signal of human activity might be comparable to other sites after all but has been diluted, and could be found at such sites once anthropogenic traces are filtered from other chemical 'background noise'. Additional studies are needed to establish the effects of long term abandonment of ephemeral sites on the presence (and relative abundance) of specific chemical elements in more detail.

Within the phytolith analysis results, it appears that the same variables indicate a strong anthropogenic input at the Wadi al-Jilat sites as the ones identified for Wadi Faynan, although the signals of activity within the archaeological data are weaker than for the ethnographic data. These results are encouraging especially considering the general sampling strategy (soil samples were collected from the general area of each context rather than targeting smaller, specific zones), the ephemeral and shallow nature of the Neolithic sites and the long duration since abandonment, which made the deposits prone to mixing, dissolution, and various other taphonomic disturbances.

It is therefore likely that the length of time since abandonment and perhaps the remodelling activities that took place within the buildings have affected the ability to identify activity specific soil signatures in these samples. One of the issues that complicates the interpretation of activity areas at Wadi al-Jilat is the difference in period of occupation and perhaps also in use between the two sites, which is responsible for some of the variation between the context categories. While the Bedouin camp sites were used contemporarily and in the same manner (domestic occupation) and therefore portray similar soil signatures, WJ7 and WJ13 could have had been used for different purposes which would have affected their geochemical and phytolith characteristics. After all, even within the ethnographic sites at Wadi Faynan, there were differences between the results of WF916 and the other camp sites due to the limited use of dung cakes at this site (Vos *et al.* 2018).

The geochemical and phytolith analyses at WJ13 portray a less straightforward clustering into the predefined context categories than is the case with WJ7, even though it is a more substantial site. This, however, might have contributed to the complexity of its interpretation. WJ13 had a long sequence of occupation and re-use, which could have caused mixing of material within the building. In addition, it was excavated in three parts, and a baulk was left between areas B and C which might have added difficulty to the systematic excavation of its three areas. WJ7 enjoyed a less extensive occupation then WJ13 and contained shallow deposits, and although it was also excavated in three parts it portrayed a simpler stratigraphic sequence than WJ13. It could be that the short-lived nature and relative simplicity of the occupation sequence at WJ7 actually contributed to the ease of its interpretation.

These findings support the observations made by O'Connell (1987) during his study of occupation and abandonment patterns at the Alyawara camp sites. The longer a site is in use, the more prone it is to cleaning activities which can affect the distribution of signals of activity. In addition, a long sequence of occupation including episodes of reconstruction can cause a shift in activity areas and evidence of these within the site, making the spatial patterns more difficult to interpret. In this respect one could propose that ephemeral archaeological sites with a straightforward stratigraphic sequence and a fixed, structured, spatial use of activity areas can benefit from geoarchaeological analysis techniques to a greater degree than sites with a complex stratigraphy which have been regularly modified.

Conclusions

This article discussed the value of (ethno-) geoarchaeological studies to aid the interpretation of ephemeral anthropogenic sites. The results of the two case studies support earlier reports and suggest that geochemistry and phytolith analysis carry much potential for the spatial reconstruction of activity areas within ephemeral sites situated in the dynamic, arid environments of the Near East. Such sites may retain signs of anthropogenic enrichment over thousands of years, though the anthropogenic signals might be diluted and thus more difficult to identify.

The successful application of the dual phytolith-geochemical methodology was more site dependent at Wadi al-Jilat than at Wadi Faynan. While the identification of activity areas at WJ13 was fruitful to a limited degree, the application of the dual methodology to WJ7 provided clear differentiation of activity signals and a profound clustering of context categories within the PCA scatterplots. It is likely that the individual buildings at Wadi al-Jilat were used in a different way, or for different purposes, which did not always comply with the predefined context categories. It was therefore not possible to study the sites together, or in comparison to each other. An investigation into spatial patterning is therefore best restricted to an individual, contemporary site context, which contains a large enough sample size to establish general trends for each context category.

The interpretation of ephemeral archaeological sites can greatly benefit from the use of geoarchaeological techniques. These have so far mostly been applied to substantial sites, where repetitive activities in fixed locations often leave clearer traces in the soil. However, shallow and straightforward sequences of occupation in ephemeral sites used in a 'habitual' manner, where activities had fixed locations, may prove to be as promising candidates for spatial analysis as more substantial ones. This might be a consequence of the limited cleaning and change in location of activity areas that take place at a site which is only occupied for a short time. Additional studies should be encouraged to fully explore the potential of geoarchaeology in contributing to our understanding of the use of space at sites which are less visible, and therefore underrepresented in the archaeological record.

Acknowledgments

The author wishes to thank the conference organiser, Peter Akkermans, for inviting this contribution to the proceedings following a fruitful meeting in Leiden. I would also like to thank Andrew Garrard, Carol Palmer and Emma Jenkins for their involvement in this project, who kindly provided the study material in addition to invaluable guidance and advice. This research was supported by an AHRC/Bournemouth University PhD studentship and by Arts and Humanities Research Council (AHRC) grant no. AH/K002902/1.

References

Audouze, F. 1988. Des modèles et des faits: les modèles de A. Leroi-Gourhan et de L. Binford confrontés aux résultats récents. *Bulletin de la Société Préhistorique Française* 84, 343-352.

Baird, D., Garrard, A., Martin, L. and Wright, K. 1992. Prehistoric environment and settlement in the Azraq basin: an interim report on the 1989 excavation season. *Levant* 24, 1-31.

Banerjea, R.Y., Bell, M., Matthews, W. and Brown, A. 2015. Applications of micromorphology to understanding activity areas and site formation processes in experimental hut floors. *Archaeological and Anthropological Sciences* 7, 89-112.

Banning, E.B. and Köhler-Rollefson, I. 1983. Ethnoarchaeological survey in the Beidha area, southern Jordan. *Annals of the Department of Antiquities of Jordan* 27, 375-384.
Banning, E.B. and Köhler-Rollefson, I. 1986. Ethnoarchaeological survey in the Bedā area, southern Jordan. *Zeitschrift des Deutschen Palästina-Vereins* 102, 152-170.
Banning, E.B. and Köhler-Rollefson, I. 1992. Ethnographic lessons for the pastoral past: camp locations and material remains near Beidha, southern Jordan. In: Bar-Yosef, O. and Khazanov, A.M. (eds.). *Pastoralism in the Levant: archaeological materials in anthropological perspective.* Madison: Prehistory Press, 181-204.
Barker, G. 2000. Farmers, herders and miners in the Wadi Faynan, southern Jordan: a 10,000-year landscape archaeology. In: Barker, G. and Gilbertson, D. (eds.). *The archaeology of drylands.* London: Routledge, 63-85.
Binford, L.R. 1978. Dimensional analysis of behavior and site structure: learning from an Eskimo hunting stand. *American Antiquity* 45, 255-273.
Bourdieu, P. 1990. Appendix: the Kabyle house or the world reversed. In: Bourdieu, P. *The logic of practice.* Cambridge: Polity Press, 271-283.
Bradley, B. and Small, C. 1985. Looking for circular structures in post hole distributions: quantitative analysis of two settlements from Bronze Age England. *Journal of Archaeological Science* 12, 285-98.
Cameron, C.M. and Tomka, S.A. 1993. *Abandonment of settlements and regions: ethnoarchaeological and archaeological approaches.* Cambridge: Cambridge University Press.
Canti, M. and Huisman, D.J. 2015. Scientific advances in geoarchaeology during the last twenty years. *Journal of Archaeological Science* 56, 96-108.
Clarke, D.L. 1977. *Spatial archaeology.* London: Academic Press.
Colledge, S. 2001. *Plant exploitation on Epipalaeolithic and early Neolithic sites in the Levant.* Oxford: British Archaeological Reports (International Series 986).
Cribb, R.L.D. 1991. *Nomads in archaeology.* Cambridge: Cambridge University Press.
Douglas, M. 1966. *Purity and danger: an analysis of the concepts of pollution and taboo.* London: Routledge.
Durkheim, E. 1915. *The elementary forms of the religious life.* New York: Macmillan.
Fernandez, F.G., Terry, R.E., Inomata, T. and Eberl, M. 2002. An ethnoarchaeological study of chemical residues in the floors and soils of Q'eqchi' Maya houses at Las Pozas, Guatemala. *Geoarchaeology* 17, 487-519.
Foucault, M. 1982. Afterword: the subject and power. In: Dreyfus, H. and Rabinow, P. *Michel Foucault: beyond structuralism and hermeneutics.* Chicago, IL: University of Chicago Press, 208-226.
Friesem, D.E. 2016. Geo-ethnoarchaeology in action. *Journal of Archaeological Science* 70, 145-157.
Garrard, A.N., Baird, D., Colledge, S., Martin, L. and Wright, K. 1994. Prehistoric environment and settlement in the Azraq basin: an interim report on the 1987 and 1988 excavation season. *Levant* 26, 73-109.
Garrard, A.N., Colledge, S., Hunt, C. and Montague, R. 1988. Environment and subsistence during the late Pleistocene and early Holocene in the Azraq basin. *Paléorient* 14, 40-49.
Gifford, D. 1977. *Observations of contemporary human settlements as an aid to archaeological interpretation.* Berkeley, CA: University of California at Berkeley (unpublished PhD thesis).
Goldberg, P. and Whitbread, I. 1993. Micromorphological study of a Bedouin tent floor. In: Goldberg, P., Nash, T.D. and Petraglia, M.D. (eds.). *Formation processes in archaeological context.* Madison, WI: Prehistory Press, 165-188.
Goring-Morris, A.N. and Belfer-Cohen, A. 2008. A roof over one's head: developments in Near Eastern residential architecture across the Epipalaeolithic-Neolithic transition. In: Bocquet-Appel, J.P. and Bar-Yosef, O. (eds.). *The Neolithic demographic transition and its consequences.* New York: Springer, 239-286.
Goring-Morris, A.N. and Belfer-Cohen, A. 2011. Neolithization processes in the Levant. *Current Anthropology* 52, 195-208.
Goring-Morris, A.N., Hovers, E. and Belfer-Cohen, A. 2009. The dynamics of Pleistocene settlement patterns and human adaptations in the Levant: an overview. In: Shea, J.J. and Liebermann, D. (eds.). *Transitions in prehistory. Papers in honor of Ofer Bar-Yosef.* Oakville, CT: Oxbow Books, 187-254.
Gosselain, O.P. 1998. Social and technical identity in a clay crystal ball. In: Stark, M.T. (ed.). *The archaeology of social boundaries.* Washington, DC: Smithsonian Institution Press, 78-106.
Guan, Y., Gao, X., Wang, H., Chen, F., Pei, S., Zhang, X. and Zhou, Z. 2011. Spatial analysis of intra-site use at a Late Palaeolithic site at Shuidonggou, northwest China. *Chinese Science Bulletin* 56, 3457-3463.
Hardy-Smith, T. and Edwards, P.C. 2004. The garbage crisis in prehistory: artefact discard patterns at the early Natufian site of Wadi Hammeh 27 and the origins of household refuse disposal strategies. *Journal of Anthropological Archaeology* 23, 253-289.
Heidegger, M. 1971. *Poetry, language, thought.* New York: Harper and Row.
Heider, K. 1967. Archaeological assumptions and ethnographical facts: a cautionary tale from New Guinea. *Southwestern Journal of Anthropology* 23, 52-64.
Hodder, I. 1979. Economic and social stress and material culture patterning. *American Antiquity* 44, 446-454.

Hodder, I. and Orton, C. 1979. *Spatial analysis in archaeology*. Cambridge: Cambridge University Press.
Husserl, E. 1990. *On the phenomenology of the consciousness of internal time (1893-1917)*. Trans. J.B. Brough. Dordrecht: Kluwer.
Hutson, S.R. and Terry, R.R. 2006. Recovering social and cultural dynamics from plaster floors: chemical analyses at ancient Chunchucmil, Yucatan, Mexico. *Journal of Archaeological Science* 33, 391-404.
Kent, S. 1984. *Analyzing activity areas: an ethnoarchaeological study of the use of space*. Albuquerque, NM: University of New Mexico Press.
King, S.M. 2008. The spatial organization of food sharing in Early Postclassic households: an application of soil chemistry in ancient Oaxaca, Mexico. *Journal of Archaeological Science* 35, 1224-1239.
Koetje, T.A. 1994. Intrasite spatial structure in the European Upper Paleolithic: evidence and patterning from the SW of France. *Journal of Anthropological Archaeology* 13, 161-169.
Köhler-Rollefson, I. 1992. A model for the development of nomadic pastoralism on the Transjordanian plateau. In: Bar-Yosef, O. and Khazanov, A.M. (eds.). *Pastoralism in the Levant. Archaeological materials in anthropological perspectives*. Madison, WI: Prehistory Press, 11-18.
Kuijt, I. and Goodale, N. 2009. Daily practice and the organization of space at the dawn of agriculture: a case study from the Near East. *American Antiquity* 74, 403-422.
López Varela, S. and Dore, C. 2010. Social spaces of daily life: a reflexive approach to the analysis of chemical residues by multivariate spatial analysis. *Journal of Archaeological Method and Theory* 17, 249-278.
Madella, M., Alexandre, A. and Ball, T. 2005. International code for phytolith nomenclature 1.0. *Annals of Botany* 96, 253-260.
Manzanilla, L. and Barba, L. 1990. The study of activities in classic households: two case studies from Coba and Teotihuacan. *Ancient Mesoamerica* 1, 41-49.
Martin, L. 1999. Mammal remains from the eastern Jordanian Neolithic, and the nature of caprine herding in the steppe. *Paléorient* 25, 87-104.
Matschullat, J., Maenhaut, W., Zimmermann, F. and Fiebig, J. 2000. Aerosol and bulk deposition trends in the 1990's, eastern Erzgebirge, Central Europe. *Atmospheric Environment* 34, 3213-3221.
Middleton, W.D. 2004. Identifying chemical activity residues on prehistoric house floors: a methodology and rationale for multi-elemental characterisation of a mild acid extract of anthropogenic sediments. *Archaeometry* 46, 47-65.
Middleton, W.D. and Price, T.D. 1996. Identification of activity areas by multi-element characterization of sediments from modern and archaeological house floors using inductively coupled plasma-atomic emission spectroscopy. *Journal of Archaeological Science* 23, 673-687.
Milek, K.B. and Roberts, H.M. 2013. Integrated geoarchaeological methods for the determination of site activity areas: a study of a Viking Age house in Reykjavik, Iceland. *Journal of Archaeological Science* 40, 1845-1865.
O'Connell, J.F. 1987. Alyawara site structure and its archaeological implications. *American Antiquity* 52, 74-108.
Oonk, S., Slomp, C.P. and Huisman, D.J., 2009. Geochemistry as an aid in archaeological prospection and site interpretation: current issues and research directions. *Archaeological Prospection* 16, 35-51.
Palmer, C. and Daly, P. 2006. Jouma's tent. Bedouin and digital archaeology. In: Evans, T.L. and Daly, P. (eds.). *Digital archaeology. Bridging method and theory*. London: Routledge, 97-127.
Palmer, C., Gilbertson, D., El-Rishi, H., Hunt, C., Grattan, J., McLaren, S. and Pyatt, B. 2007. The Wadi Faynan today: landscape, environment, people. In: Barker, G., Gilbertson, D. and Mattingly, D. (eds.). *Archaeology and desertification. The Wadi Faynan landscape survey, southern Jordan*. Oxford: Oxbow Books, 25-57.
Palmer, C., Smith, H. and Daly, P. 2007. Ethnoarchaeology. In: Barker, G., Gilbertson, D. and Mattingly, D. (eds.). *Archaeology and desertification. The Wadi Faynan landscape survey, southern Jordan*. Oxford: Oxbow Books, 369-395.
Parker Pearson, M. and Richards, C. 1994. *Architecture and order: approaches to social space*. London: Routledge.
Parnell, J.J. and Terry, R.E. 2002. Soil chemical analysis applied as an interpretive tool for ancient human activities in Piedras Negras, Guatemala. *Journal of Archaeological Science* 29, 379-404.
Portillo, M., Albert, R.M. and Henry, D.O. 2009. Domestic activities and spatial distribution in Ain Abū Nukhayla (Wadi Rum, southern Jordan): the use of phytoliths and spherulites studies. *Quaternary International* 193, 174-183.
Portillo, M., Kadowaki, S., Nishiaki, Y. and Albert, R.M. 2014. Early Neolithic household behavior at Tell Seker al-Aheimar (Upper Khabur, Syria): a comparison to ethnoarchaeological study of phytoliths and dung spherulites. *Journal of Archaeological Science* 42, 107-118.
Rosen, A.M. 1992. Preliminary identification of silica skeletons from Near Eastern archaeological sites: an anatomical approach. In: Rapp, G. and Mulholland, S.C. (eds.). *Phytolith systematics: emerging issues*. New York: Plenum Press, 129-147.
Saidel, B.A. 2001. Abandoned tent camps in southern Jordan. *Near Eastern Archaeology* 64, 150-157.
Saidel, B.A. 2009. Coffee, gender and tobacco. Observations on the history of the Bedouin tent. *Anthropos* 104, 179-186.
Salisbury, R.B. 2007. Introduction: examining archaeology's spatial fabric. In: Salisbury, R. B. and Keeler, D. (eds.). *Space: archaeology's final frontier? An intercontinental approach*. Cambridge: Cambridge Scholars Publishing, 2-17.

Schiffer, M.B. 1988. The structure of archaeological theory. *American Antiquity* 53, 461-485.

Schiffer, M.B. 1995. *Behavioral archaeology: first principles*. Salt Lake City, UT: University of Utah Press.

Schiegl, S., Lev-Yadun, S., Bar-Yosef, O., El Goresy, A. and Weiner, S. 1994. Siliceous aggregates from prehistoric wood ash: a major component of sediments in Kebara and Hayonim caves (Israel). *Israel Journal of Earth Science* 43, 267-278.

Shahack-Gross, R., Albert, R.-M., Gilboa, A., Nagar-Hillman, O., Sharon, I. and Weiner, S. 2005. Geoarchaeology in an urban context: the uses of space in a Phoenician monumental building at Tel Dor (Israel). *Journal of Archaeological Science* 32, 1417-1431.

Shahack-Gross, R., Marshall, F., Ryan, K. and Weiner, S. 2004. Reconstruction of spatial organization in abandoned Maasai settlements: implications for site structure in the pastoral Neolithic of East Africa. *Journal of Archaeological Science* 31, 1395-1411.

Shillito, L.-M. and Ryan, P. 2013. Surfaces and streets: phytoliths, micromorphology and changing use of space at Neolithic Çatalhöyük (Turkey). *Antiquity* 87, 684-700.

Simek, J.F. 1987. Spatial order and behavioural change in the French Palaeolithic. *Antiquity* 61, 25-40.

Simms, S.R. 1988. The archaeological structure of a Bedouin camp. *Journal of Archaeological Science* 15, 197-211.

Soja, E.W. 1989. *Postmodern geographies: the reassertion of space in critical social theory.* New York: Verso.

Sørensen, M. 2008. Spatial analysis by dynamic technological classification: a case study from the Palaeolithic-Mesolithic transition in Scandinavia. In: Sørensen, M. and Desrosiers, P. (eds.). *Technology in archaeology*. Copenhagen: The Danish National Museum, 107-125.

Terry, R.E., Fabian, G., Fernández, F.G., Parnella, J.J. and Inomata, T. 2004. The story in the floors: chemical signatures of ancient and modern Maya activities at Aguateca, Guatemala. *Journal of Archaeological Science* 31, 1237-1250.

Tilley, C.Y. 1994. *A phenomenology of landscape: places, paths, and monuments*. Oxford: Berg.

Tsartsidou, G., Lev-Yadun, S., Efstratiou, N. and Weiner, S. 2008. Ethnoarchaeological study of phytolith assemblages from an agro-pastoral village in northern Greece (Sarakini): development and application of a Phytolith Difference Index. *Journal of Archaeological Science* 35, 600-613.

Tsartsidou, G., Lev-Yadun, S., Efstratiou, N. and Weiner, S. 2009. Use of space in a Neolithic village in Greece (Makri): phytolith analysis and comparison of phytolith assemblages from an ethnographic setting in the same area. *Journal of Archaeological Science* 36, 2342-2352.

Ullah, I.I., Duffy, P.R. and Banning, E.B. 2015. Modernizing spatial micro-refuse analysis: new methods for collecting, analyzing, and interpreting the spatial patterning of micro-refuse from house-floor contexts. *Journal of Archaeological Method and Theory* 22, 1238-1262.

Wagstaff, J.M. 1987. *Landscape and culture: geographical and archaeological perspectives*. London: Basil Blackwell.

Wells, E.C. 2006. Cultural soilscapes. In: Frossard, E., Blum, W.E.H. and Warkentin, B.P. (eds.). *Function of soils for human societies and the environment*. London: Geological Society Special Publications 266, 125-132.

Whallon, R. 1973. Spatial analysis of occupation floors I: application of dimensional analysis of variance. *American Antiquity* 38, 266-278.

Wilson, C., Davidson, D. and Cresser, M. 2008. Multi-element soil analysis: an assessment of its potential as an aid to archaeological interpretation. *Journal of Archaeological Science* 35, 412-424.

Vos, D. 2017. *Out of sight, but not out of mind: exploring how phytolith and geochemical analysis can contribute to understanding social use of space during the Neolithic in the Levant through ethnographic comparison*. Bournemouth: Bournemouth University (unpublished PhD thesis).

Vos, D., Jenkins, E. and Palmer, C. 2018. A dual geochemical-phytolith methodology for studying activity areas in ephemeral sites: insights from an ethnographic case study from Jordan. *Geoarchaeology* 33, 680-694.

Vyncke, K., Degryse, P., Vassilieva, E. and Waelkens, M. 2011. Identifying domestic functional areas. Chemical analysis of floor sediments at the Classical-Hellenistic settlement at Düzen Tepe (SW Turkey). *Journal of Archaeological Science* 38, 2274-2292.

Yellen, J.E. 1977. *Archaeological approaches to the present*. New York: Academic Press.

Populating the Black Desert: the Late Neolithic presence

Yorke M. Rowan, Gary O. Rollefson
and Alexander Wasse

Abstract

Our perception of the Late Neolithic period (*c.* 7000-5000 cal BC) is changing dramatically, particularly in the arid regions of eastern Jordan. Investigations by the Eastern Badia Archaeological Project at Wisad Pools and Wadi al-Qattafi in the Black Desert demonstrate that hunter-pastoralists occupied substantial dwellings in hamlets that enjoyed significantly more salubrious conditions than the current harsh environment allows. In this paper we summarise evidence that indicates denizens of the Black Desert established solid structures for long term occupation with recurrent use and rebuilding.

Keywords: Late Neolithic, desert environment, Yarmoukian, Jordan, pastoralism, hunting

Introduction

Until relatively recently, in the southern Levant – as elsewhere (*e.g.* Cyprus, northern Levant and upper Mesopotamia; Akkermans and Schwartz 2003, 99; Clarke 2007, 5) – few prehistoric periods were more neglected, or at least treated with less interest, than the Late Neolithic. Implicitly, the period seems to be regarded largely as an uninteresting interlude following the major transformations of the Pre-Pottery Neolithic and prior to the dramatic changes attributed to the Chalcolithic period. Our understanding of the socio-economic dimension is relatively poor, and there is little consensus on the basic chronology for the period. Nonetheless, in recent years there have been important insights that provide counter-weight to the common misconception that the Late Neolithic witnessed a collapse or hiatus of cultural progress. Recent discoveries in the Black Desert of eastern Jordan contribute to these insights. In this paper we highlight the existence of Late Neolithic occupation more extensive and intensive than originally considered possible, and which may extend beyond our study region to the wider dry steppe and sub-desert (*badia*) regions to the south and north, underscoring the flexibility of the Neolithic population to exploit a diverse array of environmental niches.

Scholars of the mid-twentieth century regarded the Late Neolithic period as an interruption, a period of local abandonment (Kenyon 1957; Perrot 1968; Vaux 1966), even a Dark Age. There are several reasons why this perspective persisted for decades, leading scholars to be disinclined to focus on the Late Neolithic. In part, this reflects a period considered less valuable for addressing 'big' questions, such as the emergence of agriculture or the rise of hierarchical societies. Moreover, as Gibbs and Banning (2013) point out, Late Neolithic sites can be difficult to identify, with few diagnostic sites on the

In: Peter M. M. G. Akkermans (ed.) 2020: *Landscapes of Survival - The Archaeology and Epigraphy of Jordan's North-Eastern Desert and Beyond*, Sidestone Press (Leiden), pp. 59-78.

Period (Culture)	Starts (cal BC)	Ends (cal BC)	Primary Sites	References
Late Chalcolithic – Ghassulian, Golanian	4600/4500	3700/3600	Shiqmim, Gilat, Tulaylut al-Ghassul	Rowan and Golden 2009; Burton and Levy 2001
Early Chalcolithic – Besorian	5100/4900	4600/4500	Tel Tsaf	Garfinkel 1999; Banning 2007
Late Neolithic – Wadi Rabah	5800/5600	5100/4900	Munhata, N. Zehora, Ein el-Jarba	Banning 2007
Late Neolithic – Jericho IX	6000-5500	5700/5400	Jericho, Lod	Banning 2007
Late Neolithic – Yarmoukian	6500-6400	5800/5700	Sha'ar Hagolan, Munhata	Banning 2007; Rollefson and Kafafi 2013
Pre-Pottery Neolithic C	7000-6900	6500/6400	'Ain Ghazal	Rollefson and Kafafi 2013
Late Pre-Pottery Neolithic B	7500	7000-6900	'Ain Ghazal, Beisamoun	Rollefson and Kafafi 2013

Table 1. Periodisation outline for the southern Levant during late prehistory.

ground surface, or buried under colluvium, a problem compounded by the limited database for comparison and interpretation. Our understanding of Late Neolithic society remains fairly rudimentary, but this has improved as scholars publish major sites and additional survey results become available.

Much of this new information derives from sites located in the southern Levantine Mediterranean zones, with arable land and significant water resources steering the discussion. Within this area, typology and chronology remain debated but there is general agreement that the end of the Pre-Pottery Neolithic falls in the mid-seventh millennium BC (Banning 2018; Rollefson and Kafafi 2013) (see Table 1). At some sites, the Pre-Pottery Neolithic C, a final transitional phase of the Late Pre-Pottery Neolithic B, is followed by a direct outgrowth, the Yarmoukian, dating from about 6500/6400 to 5900/5800 BC. There may be overlap with Jericho IX (or Lodian) but this is less clearly defined, has few dates, and is probably more regionally restricted (*c.* 6000-5700/5500 BC). Finally, we find Wadi Rabah (*c.* 5800/5600-5100/4900 BC) sites, which like the Yarmoukian are restricted more to northern Palestine (Rollefson 2008, 92) and may also overlap with the Yarmoukian (Banning 2007). This chronology focuses on the southern Levantine Neolithic cultural complexes, which cannot be directly imposed upon the *badia*, where equid hunting, Badia points, and an early adoption of transverse arrowheads define different cultural trajectories in the Black Desert (outlined in greater detail below).

There is little consensus on the internal chronology of the Late Neolithic, or the transition to the Chalcolithic. Although some would consider Wadi Rabah to be early Chalcolithic (Garfinkel 1999), following Banning (2007) we will consider the Late Neolithic to include the Yarmoukian and the Wadi Rabah, or roughly 6500/6400 to approximately 5100/5000 BC. To date, much of the material culture dated by this rough chronology has been restricted to sites from along the Jordan valley to the Mediterranean. It may also be noted that during this period, the greater Syrian desert – of which the Black Desert may be considered part – was bounded to the north and east by the very different Late Neolithic entities of the northern Levant and Mesopotamia (cf. Akkermans and Schwartz 2003, 99ff.). This cautions against the imposition of a unidirectional, southern Levantine cultural affiliation upon the desert regions.

The Black Desert Neolithic

Rather than a gap, variability characterises the southern Levantine Late Neolithic, and this may reflect an increased fluidity and flexibility beyond that of people living during the earlier aceramic Neolithic phases. Although probably linked in part to adaptive responses to the 8.2 kya climate event (Alley *et al.* 1997; Alley and Agustsdottir 2005; Migowski *et al.* 2006), in the Black Desert this notable shift to greater variability in local adaptation began earlier. One aspect of this Late Neolithic variability now includes the steppe and desert region of Jordan, where multiple research projects are discovering much richer occupations than previously realised, particularly dating to the late prehistoric period (Müller-Neuhof 2012; 2013; 2014; Akkermans *et al.* 2014). Until recently, most of our information for the eastern steppe and desert derived from research conducted by Garrard (Garrard *et al.* 1994; 1996; Garrard and Byrd 2013) and Betts (Betts 1982; 1983; Betts 1998; Betts *et al.* 2013; Betts and Helms 1987), which primarily uncovered shallow, temporary camps of small groups (*e.g.* Jilat 13, Jebel Naja). As we have argued (Rollefson *et al.* 2014), based on our recent research at the sites of Wadi al-Qattafi and Wisad Pools in the Black Desert, a substantial population increase in the *badia* seems to date to the Late Neolithic. In this paper we wish to build upon and update that research, which highlights the emergent herder-hunter groups harnessing the potential of the eastern *badia*, by comparing the data from excavation of the structures at Wisad Pools and Wadi al-Qattafi. Through this detailed study, we can begin to build a typology and chronology of attributes found in the Black Desert Neolithic, an archaeological complex contextualised elsewhere in this volume (see Wasse *et al.*, this volume). Here, we concentrate on comparing attributes of Late Neolithic features and their immediate context.

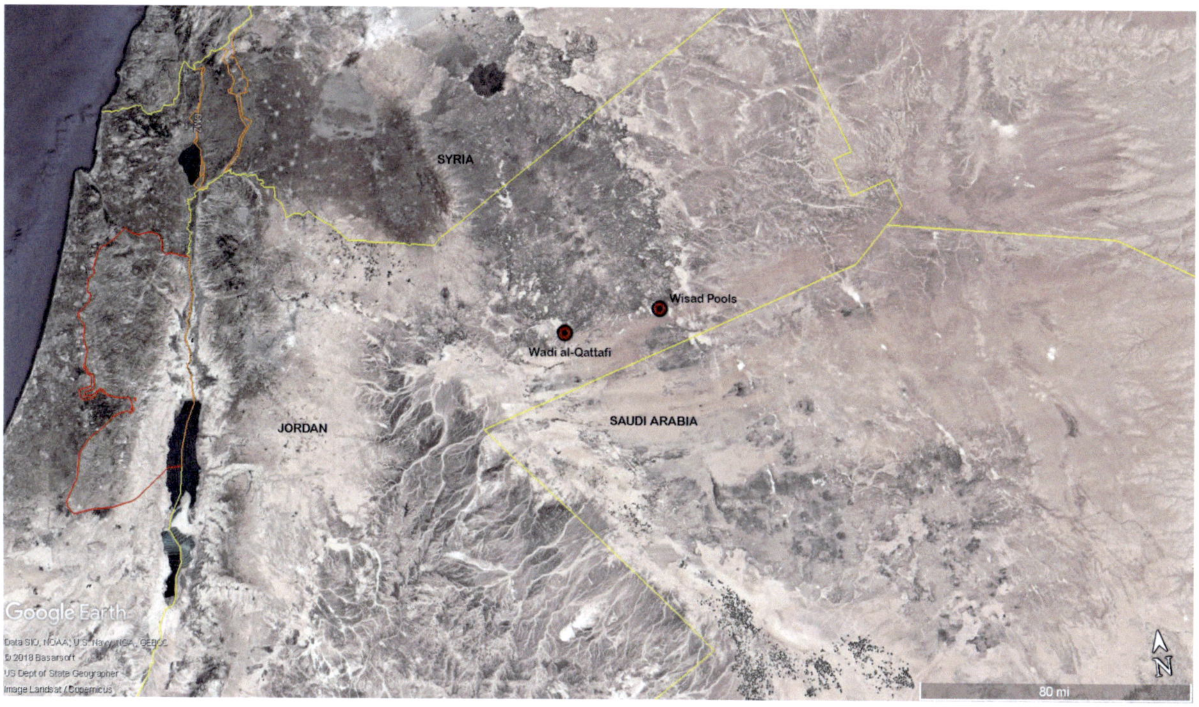

Figure 1. Map of the study area.

The Eastern Badia Archaeological Project (EBAP) study area comprises a west-east transect across the southern part of the Jordanian 'panhandle', selected to include a variety of ecological zones and to provide opportunities to assess the evidence for links with the 'Levantine corridor', the Hauran, northern Arabia and Mesopotamia. Our broader goal is to record and study the architecture, artefacts, and petroglyphs, integrating that data with biological and palaeoclimatic data in order to understand human occupation and exploitation of the region. We are particularly interested in preliminary evidence suggesting that a florescence of human activity in the Black Desert was possible during later prehistory because of environmental conditions significantly better than the modern situation.

Since 2008, our surveys and excavations have concentrated on two areas, Wisad Pools and Wadi al-Qattafi, located on the eastern and western margins of the *harrah* respectively (Fig. 1). This research sought to explore the hypothesis that a few cultural innovations (*e.g.* herding, pottery use) dispersed into the area, possibly from west to east. At the same time, the possibility that connections with people to the north were also significant cannot be discounted, although this is more difficult to examine given the slender evidence from field research on the Late Neolithic of the Hauran immediately to the north.

Wadi al-Qattafi, a major north-south drainage approximately 60 km east of Azraq, is dominated by more than 20 basalt capped *mesas* that rise 40-60 m above the wadi; more than 600 structures have been identified among these prominences. Wisad Pools, 60 km further to the east, includes over 300 structures and hundreds of petroglyphs concentrated around a series of pools. Both areas attest to a striking presence of Late Neolithic communities that evidently invested a great amount of time and energy into building sturdy structures of basalt. In each area, excavations indicate domestic structures of Late Neolithic construction, occupation and re-use. However, there are important differences between the two locations in the longevity of the buildings (although they overlap), their associated artefacts and, perhaps, the faunal assemblages. Our goal is to highlight some preliminary observations on the similarities and differences between these two areas and the distinct environmental contexts.

The *mesas* of Wadi al-Qattafi

The *mesas* (or *ghura* in local Arabic) along Wadi al-Qattafi are remnants of an extensive cap of the flood basalts. Architecture atop each *mesa* varies greatly, although virtually all have at least a tower tomb on top, invariably looted. Maitland's Mesa (M-4) is named after the RAF pilot who first spotted the many structures which he believed to be an ancient hill fort (Maitland 1927). Atop Maitland's, there are numerous small buildings including both single and double-roomed '*ghura* huts' on the summit, typically outside the larger enclosures (Rowan *et al.* 2015b, 179). Excavation of two huts, one single-cell and one double-cell, produced nothing diagnostic (*ibid.*, 179-180 and Fig. 4). At Tulul

Figure 2. South Slope structure SS-11 at Maitland's Mesa, with hearth and courtyard walls.

al-Ghusayn, however, Bernd Müller-Neuhof excavated very similar *ghura* huts, where his team found Early Bronze Age pottery and charcoal supporting a late fourth millennium date (Müller-Neuhof 2014, Fig. 11; see also Akkermans and Brüning, this volume, on *ghura* huts in the Jebel Qurma region).

As we began recording the structures, other clusters of collapsed buildings on the lower slopes became apparent, most of them circular in outline with substantial amounts of large basalt slabs. On the southern slope, there are approximately 26 structures ranging from 2-10 m in diameter. A preserved doorway in one of the buildings, South Slope no. 11 (SS-11), was reminiscent of *nawamis* (Rowan *et al.* 2014), the fourth millennium BC standing burial chambers of dry masonry best known from the Sinai (Bar-Yosef *et al.* 1977; 1986).

Excavation of this structure (SS-11) quickly revealed that it was not a funerary chamber, but a 4.7 m^2 domestic structure with two doorways and very low, corbelled walls (Wasse *et al.* 2012). Each doorway had vertical slabs as door jambs, capped with a lintel (*ibid.*, 19). Space on the slope was cleared and levelled, and on the upslope wall, the exterior apparently included large upright boulders, perhaps fortifying this side of the building against run-off from the slope of the *mesa*. A later interior use of the building included well-placed paving stones that would have lowered the ceiling further. Both the low doorways and minimal height of the walls suggest that people would only be able to crouch once they had crawled inside. An elongated basalt slab found near the centre of the interior was almost certainly a toppled central support pillar, as documented in the other Black Desert Neolithic buildings (discussed below). On the exterior of the building, at least one adjacent storeroom still had a central standing stone supporting the roof (Fig. 2). Two low walls attached to the exterior of the structure, one on the north-eastern face, and the other on the south-eastern face, may represent an enclosure wall (cf. Timnian structures in the Negev; Rosen 2017, 134-138). Inside this possible enclosure a small hearth was built immediately outside the eastern entrance. Another hearth, sealed by the paved floor within the structure, provided the sole radiocarbon date of 5480-5320 cal BC

Figure 3. Looking north over Mesa 7 (M-7), with collapsed structures visible on the lower slopes (photograph by A.C. Hill).

(2σ). This date conforms to the rather poorly dated Wadi Rabah of the western agricultural zones as well as the appearance of Timnian huts and enclosures in more arid regions (*ibid.*, 138), but we should not necessarily expect the Black Desert Neolithic to precisely match that chronology. The Yarmoukian and Haparsa points recovered during excavation fit with this date, although relatively archaic in the context of the wider southern Levant (Gopher 1994, Fig. 6.10).

Just over one kilometre to the north of Maitland's Mesa, another cluster of collapsed structures was recognised along the lower slopes of M-7 and the larger M-5. Over 350 structures group in small clusters along the lower slopes of these two *mesas* (Fig. 3). Two structures were selected for excavation, South Slope nos. 1 and 2 (SS-1 and SS-2). Immediately to the south-west of SS-1, SS-2 consisted of a curvilinear wall built of horizontally stacked flat basalt slabs still standing approximately 1 m. This wall defined one cell, while another cell was apparently part of the same structure. A well-constructed paving of large flat basalt slabs extended between these two cells (Rollefson *et al.* 2017, Fig. 17). The lack of walls on the southern aspects and dearth of finds indicates that this building suffered some destruction or was robbed of stones for other buildings, yet proximity suggests it may also date to the Late Neolithic.

South Slope no. 1 (SS-1), a roughly circular building, was excavated over two seasons (Rollefson *et al.* 2016; 2017). Although superficially similar to SS-11 at M-4, this building varies in some important ways (Fig. 4). External dimensions of roughly 6.3 x 5.4 m, and interior dimensions of 4.6 x 4 m (14.5 m^2), are three times the area of M-4 SS-11. Yet this building did not seem corbelled. In fact, the south-western half of the 'interior' may have been open to the elements given the much higher frequency of heavily patinated chipped stone. Also unlike SS-11, a number of constructed features were set into the floor, including a large well-built stone-lined hearth on the north-eastern interior (Fig. 5), a plastered cylindrical pit near the central pillar (Fig. 6), and a shallow plastered basin set into the south-west interior floor, renewed with another layer of plaster. A large hearth was also found outside of the structure, approximately 2 m to the south-east (Fig. 7).

Figure 4. Overhead view of structure SS-1 (at M-7) after excavation.

Figure 5. Interior hearth at SS-1 prior to excavation.

Figure 6. Plastered cylindrical plaster pit in the centre of SS-1, before (left) and after (right) excavation.

Figure 7. Exterior hearth south of SS-1.

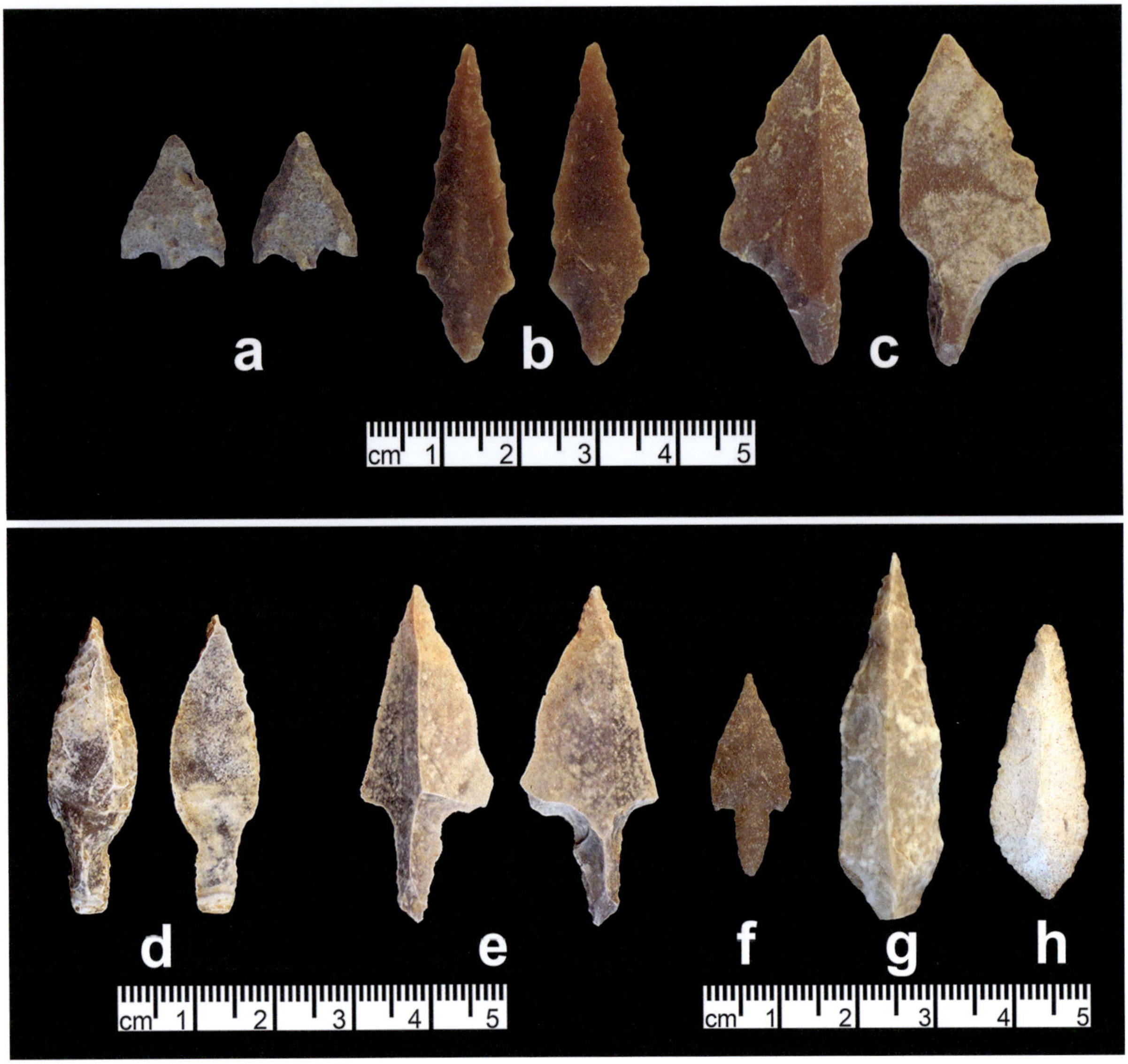

Figure 8. Projectile points recovered from excavations at M-7. Haparsa points (a, f), Nizzanim points (b, d, g, h) and Badia points (c, e).

The builders of SS-1, after levelling an area of the steep limestone slope, constructed a thick wall laid in horizontal courses externally but with basal large vertical slabs internally. In similar fashion to SS-11, large blocks were then placed upslope, possibly to protect the exterior of the building from slope wash. Unlike M-4 SS-11, M-7 SS-1 does not appear to have had a corbelled roof. One section of the wall was apparently rebuilt, visible as a straight section on the northern aspect; it is possible an earlier entrance existed here. Standing pillars (*c.* 1.25-1.3 m in height), including one in the centre of the room and others built into the walls, probably supported some form of organic roof over the eastern half, although a pillar is also visible in the western section. The well-built, very narrow entrance may have been a later reconfiguration of the exterior walls. On the exterior of this narrow entrance, a sort of portico was apparently built on the virgin sediment and bedrock, probably a later addition to the building.

Material culture was much richer in this building in comparison to SS-11, suggesting more intensive use, and possibly over a longer period. A variety of Badia, Haparsa and Nizzanim points, comprising 15% of the formal tools, contrasts to the two points found at SS-11 (Fig. 8). In addition to the points, beads and bead blanks, a small obsidian biface, and a high number of burins further underscore the differences between these two Wadi al-Qattafi structures (Table 2). Two particularly significant features of the chipped stone assemblage are

	Wadi al-Qattafi				*Wisad Pools*			
Architecture & features	SS-11 (M4)		SS-1 (M7)		W-80		W-66	
Length (m, interior)	3.0		4.6		6.5		4.3	
Width (m, interior)	2.0		4.0		5.5		4.0	
Square meters	4.7		14.5		28.1		13.4	
Hearths	1		2		4+		-	
Pillars	1?		3		3-4		1	
Plaster	-		+		-		+	
Bench	-		-		2		-	
Exterior, storage rm.	1		-		-		-	
Courtyard	1		1		-		-	
Attached structures	-		1		3		1	
Artefacts	n=	%	n=	%	n=	%	n=	%
Projectile points	4		147 (98%)		154 (19.5%)		100 (45.7%)	
Transverse forms	-		3 (2%)		635 (80.5%)		119 (54.3%)	
	4 (100%)	4.8	*150 (100%)*	5.1	789 (100%)	26.7	219 (100)	11.2
Burins	25	30.1	906	30.8	95	3.2	40	2.0
Scrapers	13	15.7	142	4.8	265	9.0	138	7.0
Knives	13	15.7	127	4.3	214	7.2	107	5.5
Drills/borers/awls	4	4.8	307	10.4	277	9.3	99	5.0
Notches	7	8.4	197	6.7	284	9.6	177	9.0
Denticulates	10	12.1	143	4.9	343	11.6	251	12.8
Other tools	7	8.4	973	33.0	691	23.4	934	47.5
Totals	*83*	*100*	*2945*	*100*	*2958*	*100*	*1965*	*100*
Microflake cores	(-)		(25 of 844)	3.0	(139 of 842)	16.5	(114 of 449)	25.4
Ground stone	2		74		114		12+	
Pottery fragments	(0)		(0)		12		15	
^{14}C								
14 dates (cal BC)	5475-5325		6383-6236		5710-5610		6600-6460	
			6432-6336		5765-5670			
			6455-6390		5890-5740			
			6490-6430		6000-5840			
					6590-6580			

Table 2. Summarised data from excavated structures at Wadi al-Qattafi and Wisad Pools.

the relatively high number of burins (n= 220, 25% of formal tools) and the rarity of transverse arrowheads. Field observations suggest that gazelle and sheep/goat are the main taxa represented.

Based on radiocarbon dates, the earliest phases at SS-1 predate SS-11 by roughly a millennium. Four samples from within the structure yielded radiocarbon dates spanning approximately from as early as 6490-6430 to 6383-6236 cal BC (Table 2). This overlaps with dates from structures at Wisad Pools (discussed below) and tentatively supports the suggestion that structures with attached courtyards or pens, as is the case at SS-11, come later in the local Late Neolithic sequence (Betts *et al.* 2013, 189; Rosen 2017, 138). In addition, the virtual absence of transverse arrowheads during contemporaneous periods of W-80 building use at Wisad Pools may hint at an important functional difference.

Wisad Pools

Located on a broad, relatively level basalt covered limestone plateau, Wisad Pools includes hundreds of structures concentrated around approximately nine pools. The pools, perhaps enhanced by barrages, were created by a small wadi (*c.* 1.5 km long) dropping in elevation about 10 m from north to south. Concentrated around the pools are over 400 petroglyphs, primarily representing horned animals (ibex, kudu, cattle) and the animal traps known as kites (Hill *et al.* 2020; Rowan and Hill 2014). Although there are tower tombs close

Figure 9. Interior of Wisad structure W-66 with its central pillar and patches of plaster.

Figure 10. Handle of Yarmoukian vessel from W-66.

to the pools, virtually all looted to some degree, our initial assumption that most of the collapsed structures are mortuary-related is probably not accurate. We investigated several other structures, but for the present discussion two buildings excavated over four seasons so far provide the majority of information (Rollefson *et al.* 2012; 2013; Rowan *et al.* 2015a). Similar in general outline, the two structures differ in size, construction techniques, and duration of occupation and re-use. Material culture is also broadly similar but there are several significant differences.

The first, more basic construction, W-66 was a corbelled Late Neolithic house with maximum internal dimension of roughly 4.25 m (about 13.4 m^2). In the centre of the room a pillar stood with patches of floor plaster still *in situ* (Fig. 9). Wall construction was not consistent around the building. On the southern, south-eastern and western walls, the interior construction consisted of stacked basalt slabs, with each short segment angled to create a roughly curving line (more accurately, polygonal). These stacks were low, with corbelling towards the central pillar above that point (Rollefson *et al.* 2011). The exterior wall of the building was difficult to identify, covered by piles of stone apparently cantilevering the weight of the slabs for the corbelling of the huge stones (*c.* 1 m long). On the northern side of the room, an alcove had four identifiable re-plastered levels (Rowan *et al.* 2015a, Fig. 10), from which *Anabasis*

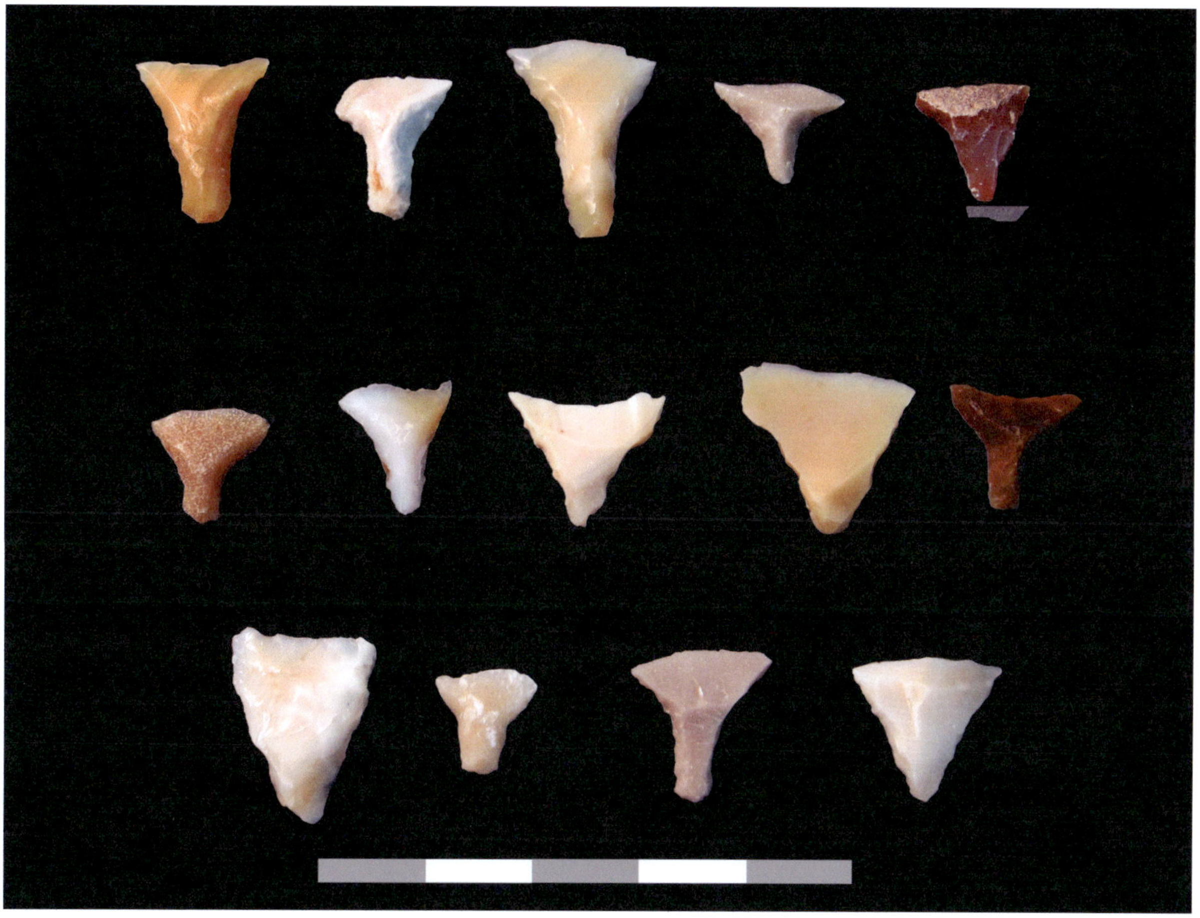

Figure 11. Transverse arrowheads from W-80.

charcoal in the plaster yielded our single radiocarbon date of 6600-6460 cal BC. Below the plaster alcove, an elliptical plastered basin or shallow pit was set into the floor, and up against exposed bedrock. Relatively little ash was identified within the structure, although some ash was found on cobbles to the north of the pillar.

Finds were richer than SS-11, and a painted Yarmoukian vessel handle with herring-bone incisions was recovered (Fig. 10). Chipped-stone finds were plentiful, including 147 arrowheads, of which 65% were transverse (Table 2 and Fig. 11), in striking contrast to the arrowhead frequencies at SS-11 and more similar to those of W-80 (see below). Other arrowheads were Haparsa and Nizzanim forms. An obsidian bladelet fragment is bilaterally notched, similar to a Helwan point (Fig. 12). Burins, on the other hand, were relatively few, similar to the relative frequency at W-80. Although ground-stone artefacts were not abundant, a cache of eleven pestles was found against an interior wall, as if curated for a return visit. Other ground-stone pieces may be figurines, one of sandstone that may represent a dog or goat, and another of fine basalt that could be human in form.

Figure 12. Obsidian notched arrowhead, similar to Helwan point, from W-80.

Figure 13. Initial exposure of W-80, probably an Iron Age tomb atop of a Neolithic complex.

Approximately 100 m south of W-66 is W-80, a larger (*c.* 8-9 m exterior diameter, and 6.5 x 5.5 interior, or a floor area of *c.* 28.1 m^2), collapsed structure selected for excavation in 2013. Massive basalt slabs and large boulders left a mound approximately 2 m above the surrounding surface. Excavation is nearly complete after three seasons, and thus the following discussion summarises our current understanding, with details to be found in earlier publications (Rollefson *et al.* 2013; Rowan *et al.* 2015a; 2015b). From the beginning, it was evident that this was a larger collapsed structure with exterior features visible beyond the central area. Initial clearance revealed a later tomb built atop the Neolithic complex below (Fig. 13). Although the human remains were basically unsalvageable, after two seasons (2013-2014) we were able to determine the tomb dated to the terminal Late Bronze II or more likely, the early Iron Age (Rowan *et al.* 2015a). A copper ring, silver (?) earring, bronze arrowhead and two glass beads originated with the tomb (Dussubieux *et al.* 2018), with a few carnelian and cowrie shell beads probably also originally associated with the burial. We are uncertain whether a copper bead (Rollefson *et al.* 2013, Fig. 22) found below the tomb context should be associated with the burial. In terms of technology and typology, it seems superficially comparable with Pre-Pottery Neolithic examples of native copper (*e.g.* Molist *et al.* 2009). Additional discussion of these finds will be forthcoming.

The Neolithic structure below included the largest slabs of any structure excavated by EBAP to date, part of the reason multiple seasons have been required. Deposits are also much deeper than originally recognised, with a wealth of artefacts, animal bones, and botanical remains. The building was clearly rebuilt and modified on numerous occasions, with periods of disuse and re-occupation extending over nearly one thousand years. Five radiocarbon dates range from 6590-6580 to 5710-5610 cal BC (Table 2), which complements the rare fragments of Late Neolithic Yarmoukian pottery. We anticipate additional radiocarbon dates which may push back the chronology for the earliest deposits in the building.

Figure 14. Partially paved exterior area of W-80 with central worked basalt slab.

Figure 15. Small built feature with an orthostat built against the exterior uprights.

Figure 16. Two caches of gazelle/caprine astragali.

Figure 17. Grinding slabs with cup marks set in the northern floor of W-80.

Two curvilinear external areas demarcated by low upright basalt slabs may have been later additions and served as exterior work areas, one of which was apparently roughly paved (Fig. 14). The paved exterior 'porch' seemed to lead to a south-western entrance, associated with the creation of the doorway. Immediately inside of this entrance, a curvilinear alcove may also be a bit later than the original building's walls. Other exterior features include a possible ritual feature with a small orthostat (Fig. 15), and a small external built hearth to the south-west, similar to that found in the courtyard of SS-11 at Wadi al-Qattafi. The doorway on the north-eastern side was apparently rebuilt and narrowed at some point. Directly on the interior of the doorway, gazelle/caprine astragali were found in a cached deposit close to a vertically-placed, large red ochre 'core' and pierced mother-of-pearl pendant. A similar cache of astragali was found near the central standing pillar (Fig. 16). During a middle phase of occupation, very large basalt slabs with mortar cup marks were set into the interior floor north of the central pillar (Fig. 17) and, in places, small hand stones were resting on low benches built along the inner wall. Multiple hearths exposed inside the structure provide additional carbonised material for radiocarbon dating.

Ground-stone artefacts, particularly hand stones, are common, although lower grinding slabs of vesicular basalt typical of agricultural villages are rare. Other ground-stone items, such as stone 'bracelet' fragments, smashed mace heads, shaft straighteners (incised or grooved stones), sandstone palettes and beads were also found. The chipped-stone tool assemblage, which includes scrapers, knives, borers and drills, notches, and denticulates, clearly underscores the long period of use, but there are important distinctions in functional terms between this structure and those of Wadi al-Qattafi as well. The flint projectile points recovered from the later phases at W-80 diverge significantly from those recovered from Wadi al-Qattafi. Of 789 identifiable point types, including Nizzanim, Haparsa, and Herzliya types, 80.5% (n= 635) are transverse arrowheads (Table 2). The large number of animal bones found inside of the structures suggests that many of these points fell out of animal carcasses as they were butchered. Preliminary results from analysis of the W-80 faunal assemblage confirm that hunted taxa (gazelle, hare, fox, onager, bird) were predominant, with only a minority of goat and sheep bones. Also of interest are a number of large felid specimens, including a lion distal phalanx. A single cattle tooth was also recovered.

We have not completed the excavation of the basal deposits of this building, so we are not yet certain of the earliest dates. Nevertheless, we detect at least four significant phases of use. The earliest phase that requires completion includes a red, gritty sediment and greenish basalt slabs. The subsequent early phase includes the main, wide doorway, smaller Late Neolithic arrowheads, few pottery sherds and an apparent absence of the massive basalt slabs with cup marks. A subsequent phase may see an increase of transverse arrowheads, followed by a phase with the re-configured, narrow doorway, massive slabs with cup marks and a very high proportion of transverse arrowheads.

Comparing the structures and assemblages

Although these four Late Neolithic structures represent a small sample size, we are beginning to understand that many of the collapsed structures visible along Wadi al-Qattafi and around Wisad Pools may date to the Late Neolithic. Given the long span of time represented by the Late Neolithic, these hundreds of structures need not, of course, represent a large resident population, since we cannot demonstrate contemporaneity. Nonetheless, the investment in constructing these substantial structures, and their frequent re-building and re-occupation, argues strongly for a substantial uptick in the presence of hunter-herders after the Pre-Pottery Neolithic B (Rollefson 2014, 299). The scale of some of the Late Neolithic construction at these sites is such that it must have required at least a degree of communal labour.

Comparison of the assemblages is difficult, given the very different quantities, but some general trends are becoming clear. Perhaps one of the foremost distinctions is the highly variable quantities of burins. The very low number of burins (n= 95, or 3.2% of the formal tools) recovered from W-80 at Wisad Pools contrasts dramatically with the hundreds of burins recovered from SS-1 at M-7 (n= 906, 30.8% of formal tools). We discuss the significance of burin sites in the larger context elsewhere (Wasse *et al.*, this volume) but should emphasise that the abundance of these sites hints at the potential significance of herding within the Black Desert Neolithic economy.

Equally striking is the difference of arrowheads between structures at the *mesas* and Wisad Pools (Table 2). Although the material culture is limited at SS-11, the artefact assemblage at SS-1 is a large sample size, yet the clumsily made transverse arrowheads are few (n= 3, or 2% of arrowheads). In contrast, transverse forms dominate the arrowheads at both Wisad Pool structures: 635 transverse arrowheads at W-80 constitute 80.5% of that arrowhead assemblage, with the 119 transverse forms at W-66 representing 54.3% there. Also notable is that transverse arrowheads occur throughout the sequence of W-66 and much of W-80, with radiocarbon samples that date them through much of the Late Neolithic sequence. Ongoing analysis of W-80, however, suggests the possibility that transverse arrowheads may be rare or absent in the earliest centuries of the Late Neolithic. Even so, there is some evidence to suggest that transverse arrowheads had appeared by the third quarter of the seventh millennium, and indisputable evidence that these were predominant by the start of the sixth millennium BC. This preceded their predominance in the Negev and Sinai by about 1500 years (cf. Rosen 2011, 75-86 and Fig. 6.9). This could support an association of the Black Desert Neolithic with northern influences as much as with the adjacent southern Levantine Mediterranean zone. Nevertheless, and as we have noted previously, the presence of occasional Yarmoukian pottery does suggest at least some exchange or movement of people from the west. In the other direction, the exchange of cortical flake scrapers and knives from *badia* sources to the Mediterranean zone seems very likely, if difficult to pinpoint geochemically (Müller-Neuhof 2012; 2013; Quintero *et al.* 2002). In addition, the desert kites, acting as large 'machines' for gazelle hunting, may have created a useful exchange item in the form of meat and skins. At the same time, we should not look solely to east-west movement of exchange. The few pieces of obsidian sourced to Anatolian outcrops indicates north-south movement, either of people or exchange routes, that certainly makes sense in view of the lie of the land.

Final remarks

The increasing quantity and resolution of archaeological data emerging from the EBAP study area are enabling us to explore more sophisticated interpretative models than has hitherto been the case. It has long been known that the *harrah* was occupied by hunter-foragers during the late Epipalaeolithic (Arranz-Otaegui *et al.* 2018; Richter 2014; Yeomans *et al.* 2017; see also Richter, this volume). There is emerging evidence for a human presence in the area during the Pre-Pottery Neolithic A but currently this is restricted to a few particularly advantageous locations. At the start of the southern Levantine Pre-Pottery Neolithic B, hunter-foragers started to move back out into the western steppe and then, during the Late Pre-Pottery Neolithic B, the *harrah*. Evidence for Pre-Pottery Neolithic B occupation of the *hamad*, to the east of the *harrah*, remains extremely limited (see Betts and Cropper 2013, Figs. 7:1-3).

Major upheaval is suggested by the changes in the so-called 'mega-sites' in the Jordanian highlands around the end of the eighth millennium BC. Ranging in size from 7 ha (*e.g.* 'Ain Jammam) to 14 ha ('Ain Ghazal, Basta), these mega-sites may have included populations up to 4000 people (Rollefson and Pine 2009, Table 1). These apparently decreased in size, dramatically depopulating to the point where some were apparently abandoned, such as Basta (Nissen *et al.* 1987) and As-Sifiya (Mahasneh and Bienert 2000), or dwindled by as much as 90% at 'Ain Ghazal (Rollefson 2015) and Wadi Shu'eib (Simmons *et al.* 2001). Rollefson (2011) argues that at 'Ain Ghazal, the competition between pasturage for caprines and arable soil for crops necessitated taking herds tens of kilometres away from the settlement, requiring long periods of time away from the main settlement, what Köhler-Rollefson (1992) terms 'tethered pastoralism'.

The Black Desert supplies evidence for where at least some of the population from these mega-sites might have turned for additional subsistence options, although there is a substantial lag time between their contraction and the expansion of the *badia* population. Although some Middle and Late Pre-Pottery Neolithic B hunting camp sites are known, they are rare, and sparsely distributed (Wasse and Rollefson 2005). Architecture, in particular, is very uncommon, although Betts (1998, 37-55, 191) records a number of Late Pre-Pottery Neolithic B structures at and around Dhuweila. By the middle of the seventh millennium BC, however, structures appear more frequently, with significant concentrations being indicative of larger groups in the landscape. The possibility that these populations derived from others to the north and north-east must also be considered, given the wider burin site presence across the wider region. With the emergence of Google Earth and the work of APAAME, additional clusters of sturdy structures ('permanent') similar to those documented at Wisad Pools and the *mesas* along Wadi al-Qattafi are recognisable. Whether or not these were continually occupied structures (at least by part of the community) or repeatedly visited on a seasonal basis remains to be determined.

Sheep and goat, presumably domestic, appear in small numbers in the zooarchaeological record as early as the Late Pre-Pottery Neolithic B (*e.g.* Azraq 31, Ibn el-Ghazi, Dhuweila 1, Bawwab al-Ghazal) but seem not to have been present in significant numbers until the Pre-Pottery Neolithic C/early Late Neolithic. Even by the Late Neolithic their representation is subject to significant variation, with hunted taxa remaining predominant in most assemblages. However, this may not be an accurate reflection of potentially larger numbers of domesticates that were not killed on-site because they were kept for milk and possibly fleece (see Wasse *et al.*, this volume).

A key issue is clearly to investigate whether the variation seen in the data represents functional variation within a single community of Late Neolithic *badia* hunter-herders, or whether separate communities of migrant pastoralists from the west (perhaps associated with the concentration of burin sites along the western margins of the *harrah*) and indigenous hunter-forager-herder communities occupying well-watered niches further to the east can be identified. Either way, the mechanisms by which cultural innovations and exchange items were dispersed among and adopted by communities in the *harrah* will be of particular interest. The western edge of the *harrah* and seasonal/perennial water sources throughout the *badia* may have had an important role to play in this regard.

From the many substantial, well-constructed Late Neolithic buildings and extensive systems of kites (perhaps starting in the Early Neolithic and continuing for millennia) to the botanical evidence for trees and marshy plants, various lines of evidence paint a very different picture of the desert that we see today. Rather than a virtually empty territory with only brief, temporary visits, we seem to have small hamlets or extended families spending a great part of the year hunting, herding, and exploiting local plants from semi-permanent basecamps or 'stations' and seasonal camping places. Nor is this limited to the Neolithic period in the EBAP study area; the evidence from the Jebel Qurma project and the Jawa Hinterland project spatially and chronologically broadens this understanding. We are quickly realising that rather than a marginal environment of little utility, this landscape was once rich in animals, plants, and people.

These thriving Late Neolithic communities seem to have disappeared from the eastern *badia* towards the end of the sixth millennium BC. In contrast to the southern deserts of Jordan, where there is evidence for Chalcolithic occupation, in the *harrah* there is an apparent gap in the archaeological record (presumably reflecting reduced occupation intensity/increased mobility, rather than actual abandonment) of up to a millennium until the middle of the second half of the fifth millennium BC. The reasons for the disappearance of these Late Neolithic communities remain unclear, but the impact of climatic change on a landscape potentially heavily degraded by well over a millennium of intensive Neolithic activity must surely be considered as a possibility. The apparent gap in the *harrah* but not in the more marginal southern *badia* may indicate that the already mobile pastoral economies of the south were better able to adapt to mid-Holocene climatic change than those of the more architecturally-invested Black Desert Neolithic.

A final point concerns the subsequent, likely terminal Chalcolithic/Early Bronze Age, emergence of fortified (whether naturally or deliberately) sites with gardens/terraces in the study area (*e.g.* Maitland's Mesa and M-7). This probably occurred in response to the same processes that led to the appearance of the walled settlement of Jawa, *c.* 70 km to the north-west, in Early Bronze IB (Müller-Neuhof and Abu-Azizeh 2016). Although the immediate (*i.e.* Chalcolithic) antecedents of this phenomenon have yet to be elucidated, the evidence accumulated by the EBAP and others for the presence of significant, likely semi-sedentary communities in the *harrah* during the Late Neolithic, inhabiting substantial stone-built dwellings and capable of organising at least a degree of communal labour, suggests that the 'Jawa phenomenon' may have evolved over a considerably greater period of time than has hitherto been assumed.

Acknowledgments

We would like to thank Peter Akkermans and Ahmad al-Jallad, and their conference organisers, for the invitation to participate in the Landscapes of Survival conference, and for their wonderful hospitality in Leiden. In addition, we would like to thank the American Center for Oriental Research in Amman, the Council for British Research in the Levant, the Oriental Institute of the University of Chicago, and Whitman College for logistical and funding support through the years. Part of the lithic analysis cited in this paper was supported by a research fellowship of the National Endowment for the Humanities. We also wish to thank the Department of Antiquities of Jordan for their support and assistance in our field research. Special thanks to the many students and colleagues who have participated or otherwise supported the efforts of the Eastern Badia Archaeological Project through the seasons.

References

Akkermans, P.M.M.G. and Schwartz, G.M. 2003. *The archaeology of Syria: from complex hunter-gatherers to urban societies (ca. 16,000 – 300 BC).* Cambridge: Cambridge University Press.

Akkermans, P.M.M.G., Huigens, H.O. and Brüning, M.L. 2014. A landscape of preservation: late prehistoric settlement and sequence in the Jebel Qurma region, north-eastern Jordan. *Levant* 46, 186-205.

Alley, R.B. and Agustsdottir, A.M. 2005. The 8k event: cause and consequences of a major Holocene abrupt climate change. *Quaternary Science Reviews* 24, 1123-1149.

Alley, R.B., Mayewski, P.A., Sowers, T., Stuiver, M., Taylor, K.C. and Clark, P.U. 1997. Holocene climatic instability: a prominent widespread event 8200 yr ago. *Geology* 25, 483-486.

Arranz-Otaegui, A., Gonzalez Carretero, L., Ramsey, M.N., Fuller, D.Q. and Richter, T. 2018. Archaeobotanical evidence reveals the origins of bread 14,400 years ago in northeastern Jordan. *Proceedings of the National Academy of Sciences* 115, 7925-7930.

Banning, E. 2007. Wadi Rabah and related assemblages in the southern Levant: interpreting the radiocarbon evidence. *Paléorient* 33, 77-101.

Banning, E.B. 2018. It's a small world. Work, family life, and community in the Late Neolithic. In: Yasur-Landau, A., Cline, E. and Rowan, Y.M. (eds.). *The social archaeology of the Levant.* Cambridge: Cambridge University Press, 98-121.

Bar-Yosef, O., Belfer, A., Goren, A. and Smith, P. 1977. The nawamis near 'Ein Huderah (eastern Sinai). *Israel Exploration Journal* 27, 65-88.

Bar-Yosef, O., Belfer-Cohen, A., Goren, A., Hershkovitz, I., Ilan, O., Mienis, H.K. and Sass, B.. 1986. Nawamis and habitation sites near Gebel Gunna, southern Sinai. *Israel Exploration Journal* 36, 121-167.

Betts, A.V.G. 1982. "Jellyfish": prehistoric desert shelters. *Annual of the Department of Antiquities of Jordan* 26, 183-188.

Betts, A.V.G. 1983. Black Desert survey, Jordan: first preliminary report. *Levant* 15, 1-10.

Betts, A.V.G. and Helms, S. 1987. A preliminary survey of Late Neolithic settlements at el-Ghirqa, eastern Jordan. *Proceedings of the Prehistoric Society* 53, 327-336.

Betts, A.V.G. (ed.) 1998. *The harra and the hamad. Excavations and surveys in eastern Jordan.* Vol. 1. Sheffield: Sheffield Academic Press.

Betts, A.V.G. and Cropper, D. 2013. The eastern badia. In: Betts, A.V.G., Cropper, D., Martin, L. and McCartney, C. *Later prehistory of the badia: excavation and surveys in eastern Jordan.* Vol. 2. Oxford: Oxbow Books, 179-191.

Betts, A.V.G., Cropper, D., Martin, L. and McCartney, C. 2013. *Later prehistory of the badia: excavation and surveys in eastern Jordan.* Vol. 2. Oxford: Oxbow Books.

Clarke, J.T. 2007. *On the margins of southwest Asia: Cyprus during the 6th to 4th millennia BC.* Oxford: Oxbow Books.

Dussubieux, L., Schmidt, K., Rowan, Y.M., Wasse, A.M.R. and Rollefson, G.O. 2018. Two glass beads from Wisad Pools in the Jordanian Black Desert. *Journal of Bead Studies* 60, 303-306.

Garfinkel, Y. 1999. *Neolithic and Chalcolithic pottery of the southern Levant.* Jerusalem: The Hebrew University of Jerusalem.

Garrard, A.N., Colledge, S. and Martin, L. 1996. The emergence of crop cultivation and caprine herding in the "marginal zone" of the southern Levant. In: Harris, D. (ed.). *The origins and spread of agriculture and pastoralism in Eurasia.* London: UCL Press, 204-226.

Garrard, A.N., Baird, D., Colledge, S., Martin, L. and Wright, K.I. 1994. Prehistoric environment and settlement in the Azraq basin: an interim report on the 1987 and 1988 excavation seasons. *Levant* 26, 73-109.

Garrard, A.N. and Byrd, B.F. 2013. *Beyond the Fertile Crescent: Late Palaeolithic and Neolithic communities of the Jordanian steppe.* Oxford: Council for British Research/ Oxbow Books.

Gibbs, K. and Banning, E.B. 2013. Late Neolithic society and village life: the view from the southern Levant. In: Nieuwenhuyse, O., Bernbeck, R., Akkermans, P.M.M.G. and Rogasch, J. (eds.), *Interpreting the Late Neolithic of Upper Mesopotamia.* Turnhout: Brepols, 355-366.

Gopher, A. 1994. *Arrowheads of the Neolithic of the Levant.* Winona Lake, IN: Eisenbrauns.

Hill, A.C., Rowan, Y.M., Rollefson, G.O. and Wasse, A.M.R. 2020. Inscribed landscapes in the Black Desert: petroglyphs and kites at Wisad Pools, Jordan. *Arabian Archaeology and Epigraphy* 00, 1-18 https://doi org/10.1111/aae.12158.

Kenyon, K.M. 1957. *Digging up Jericho.* London: Benn.

Köhler-Rollefson, I. 1992. A model for the development of nomadic pastoralism on the Transjordanian plateau. In: Bar-Yosef, O. and Khazanov, A. (eds.). *Pastoralism in the Levant. Archaeological materials in anthropological perspectives.* Madison, WI: Prehistory Press, 11-18.

Maitland, P.E. 1927. The "Works of the Old Men" in Arabia. *Antiquity* 1, 197-203.

Mahasneh, H. and Bienert, H.-D. 2000. Unfolding the earliest pages of sedentism: the Pre-Pottery Neolithic Settlement of es-Sifiya in southern Jordan. In: Bienert, H.-D. and Müller-Neuhof, B. (eds.). *At the crossroads. Essays on the archaeology, history and current affairs of the Middle East.* Amman: German Protestant Institute of Archaeology in Amman, 1-13.

Migowski, C., Stein, M., Prasad, S., Negendank, J.F.W. and Agnon, A. 2006. Holocene climate variability and cultural evolution in the Near East from the Dead Sea sedimentary record. *Quaternary Research* 66, 421-431.

Molist, M., Montero-Ruiz, I., Clop, X., Rovira, S., Guerrero, E. and Anfruns, J. 2009. New metallurgic findings from the Pre-Pottery Neolithic: Tell Halula (Euphrates Valley, Syria). *Paléorient* 35, 33-48.

Müller-Neuhof, B. 2012. The Wadi ar-Ruwayshid mining complex: Chalcolithic/Early Bronze Age cortical tool production in northeast Jordan. *Annual of the Department of Antiquities of Jordan* 56, 351-362.

Müller-Neuhof, B. 2013. SW-Asian Late Chalcolithic/EB demand for 'big tools': specialised flint exploitation beyond the fringes of settled regions. *Lithic Technology* 38, 220-236.

Müller-Neuhof, B. 2014. A 'marginal' region with many options: the diversity of Chalcolithic/Early Bronze Age socio-economic activities in the hinterland of Jawa. *Levant* 46, 230-248.

Müller-Neuhof, B. and Abu-Azizeh, W. 2016. Milestones for a tentative chronological framework for the late prehistoric colonisation of the basalt desert (north-eastern Jordan). *Levant* 48, 220-235.

Nissen, H.J., Muheisen, M. and Gebel, H.G. 1987. Report on the first two seasons of excavation at Basta (1986-1987). *Annual of the Department of Antiquities of Jordan* 31, 79-119.

Perrot, J. 1968. La préhistoire palestinienne. In: *Supplément au dictionnaire du la bible.* Vol. 8. Paris: Letouzey et Ané, 286-446.

Quintero, L., Wilke, P. and Rollefson, G. 2002. From flint mine to fan scraper: the late prehistoric Jafr industrial complex. *Bulletin of the American Schools of Oriental Research* 327, 17-48.

Richter, T. 2014. Margin or centre? The Epipalaeolithic in the Azraq Oasis and the Qa' Shubayqa. In: Finlayson, B. and Makarewicz, C. (eds.). *Settlement, survey and stone. Essays on Near Eastern prehistory in honour of Gary Rollefson.* Berlin: ex oriente, 27-36.

Rollefson, G.O. 2008. The Neolithic period. In: Adams, R.B. (ed.). *Jordan: an archaeological reader*. London: Equinox, 71-108.

Rollefson, G.O. 2011. The greening of the badlands: pastoral nomads and the "conclusion" of neolithization in the southern Levant. *Paléorient* 37, 101-109.

Rollefson, G.O. 2014. The fat of the land: Neolithic origins of "wealth" in the southern Levant. In: Swinnen, M. and Gubel, E. (eds.). *From Gilead to Edom. Studies in the archaeology and history of Jordan*. Leuven: Peeters, 19-26.

Rollefson, G.O. 2015. 'Ain Ghazal (Jordan). In: Barker, G. and Goucher, C. (eds.). *Cambridge History of the World*. Vol. 2. Cambridge: Cambridge University Press, 243-260.

Rollefson, G.O. and Kafafi, Z. 2013. The town of 'Ain Ghazal. In: Schmandt-Besserat, D. (ed.). *Symbols at 'Ain Ghazal*. Berlin: ex oriente, 1-29.

Rollefson, G.O. and Pine, K. 2009. Measuring the impact of LPPNB immigration into highland Jordan. *Studies in the History and Archaeology of Jordan* 10, 473-481.

Rollefson, G.O., Rowan, Y.M. and Perry, M. 2011. A Late Neolithic dwelling at Wisad Pools, Black Desert. *Neo-Lithics* 11/1, 35-43.

Rollefson, G.O., Rowan, Y.M., Perry, M. and Abu-Azizeh, W. 2012. The 2011 season at Wisad Pools, Black Desert: preliminary report. *Annual of the Department of Antiquities of Jordan* 56, 29-44.

Rollefson, G.O., Rowan, Y.M. and Wasse, A. 2013. Neolithic settlement at Wisad Pools, Black Desert. *Neo-Lithics* 13/1, 11-23.

Rollefson, G.O., Rowan, Y.M. and Wasse, A.. 2014. The Late Neolithic colonization of the Eastern Badia of Jordan. *Levant* 46, 285-301.

Rollefson, G.O., Rowan, Y.M., Wasse, A.M.R., Hill, A.C., Kersel, M.M., Lorentzen, B., Al-Bashaireh, K. and Ramsay, J. 2016. Investigations of a Late Neolithic structure at Mesa 7, Wadi al-Qattafi, Black Desert, 2015. *Neo-Lithics* 1/16, 3-12.

Rollefson, G., Wasse, A., Rowan, Y.M., Kersel, M.M., Jones, M., Lorentzen, B., Hill, A.C. and Ramsay, J. 2017. The 2016 excavation season at the Late Neolithic structure SS-1 on Mesa 7, Black Desert. *Neo-Lithics* 2/17, 19-29.

Rosen, S.A. 2011. Desert chronologies and periodization systems. In: Lovell, J.L. and Rowan, Y.M. (eds.). *Culture, chronology and the Chalcolithic. Theory and transition*. Oxford: Oxbow Books, 71-83.

Rosen, S.A. 2017. *Revolutions in the desert*. New York and London: Routledge.

Rowan, Y.M. and Hill, A.C. 2014. Pecking at basalt: photogrammetric documentation of petroglyphs in the Black Desert, Jordan. In: Finlayson, B. and Makarewicz, C. (eds.). *Settlement, survey, and stone: essays on Near Eastern prehistory in honour of Gary Rollefson*. Berlin: ex oriente, 209-217.

Rowan, Y.M., Rollefson, G.O. and Wasse, A.M.R. 2014. Survey of Maitland's Mesa: a basalt prominence in the eastern badia of Jordan. In: Jamhawi, M. (ed.). *Jordan's prehistory: past and future research*. Amman: Department of Antiquities of Jordan, 277-284.

Rowan, Y.M., Wasse, A., Rollefson, G., Kersel, M., Jones, M. and Lorentzen, B. 2015a. Late Neolithic architectural complexity at Wisad Pools, Black Desert. *Neo-Lithics* 1/15, 3-10.

Rowan, Y.M., Rollefson, G.O., Wasse, A., Abu-Azizeh, W., Hill, A.C. and Kersel, M.M. 2015b. The 'land of conjecture': new late prehistoric discoveries at Maitland's Mesa and Wisad Pools. *Journal of Field Archaeology* 40, 175-188.

Simmons, A.H., Al-Nahar, M., Rollefson, G.O., Cooper, J., Kafafi, Z., Köhler-Rollefson, I., Mandel, R. D. and Durand, K.R. 2001. Wadi Shu'eib, a large Neolithic community in central Jordan: final report of test investigations. *Bulletin of the American Schools of Oriental Research* 321, 1-39.

Vaux, R. de. 1966. Palestine during the Neolithic and Chalcolithic periods. *Cambridge Ancient History* 1, 499-538.

Wasse, A. and Rollefson, G.O. 2005. The Wadi Sirhan project: report on the 2002 archaeological reconnaissance of Wadi Hudruj and Jabal Tharwa, Jordan. *Levant* 37, 1-20.

Wasse, A., Rowan, Y.M. and Rollefson, G.O. 2012. A 7th millennium BC Late Neolithic village at Mesa 4 in the Wadi al-Qattafi, eastern Jordan. *Neo-Lithics* 1/12, 15-24.

Yeomans, L, Martin L. and Richter, T. 2017. Environment, seasonality and hunting strategies as influences on Natufian food procurement: the faunal remains from Shubayqa 1. *Levant* 49, 85-104.

Flamingos in the desert: how a chance encounter shed light on the 'Burin Neolithic' of eastern Jordan

Alexander Wasse, Gary Rollefson and Yorke Rowan

Abstract

This article proposes that so-called burin sites represent the herding element of what is here defined as the Black Desert Neolithic cultural complex, that being the material manifestation of the hunter-gatherer-herder society that utilised the *harrah* and limestone steppe as far north as the Palmyra range during the Late Neolithic. An association with sheep herding in particular is suggested by an observed increase in the frequency of truncation burins and sheep bones at 'Ain Ghazal from the LPPNB onwards. The possibility that milking may have been the mainstay of the Black Desert Neolithic herding economy, with gazelle being hunted for their meat and possibly hides by means of 'desert kite' hunting traps, is explored. It is argued that the '*domus*' of the Black Desert Neolithic is to be found in the 'Late Neolithic stations' of the *harrah*, seasonally sedentary base camps associated with intensive plant-food economies and episodes of macro-banding. Sustained by 1500 years of relatively high annual rainfall, productive microenvironments in the *harrah* appear to have supported a process of cultural evolution not unlike that which characterised the Early Natufian some six millennia earlier. Once again, this resulted in rapid population growth, generally higher levels of occupational intensity, intensifying plant food economies and increasing levels of social, if not quite physical sedentism.

Keywords: Black Desert, burin sites, desert kites, Jordan, milking, Neolithic, pastoralism, Saudi Arabia, sedentism, sheep, Syria, transhumance

Prologue

The well-used Toyota pick-up picked its way over the black-blue basalt paving of the TAP line, eight days out from Amman and now *en route* from Al-Ghirqa to the pools at Wisad. Somewhat east of Dhuweila, after threading through a herd of creamy white camels grazing at the side of the track, the ground fell away in front of the truck in a long tilt at the east. A few hundred yards away was a small mud pan that had flooded during recent rain. Standing ankle-deep in the shallow water, a flash of white and pink against the reflected blue of the sky, was a small flock of flamingos in the desert.

Introduction

The genesis of this article lay in the encounter described above, which happened in late October 2007. That trip to the *harrah* was a brief follow-up to the short-lived Wadi Sirhan project (Wasse and Rollefson 2005), which aimed to examine "whether the relatively well-

In: Peter M. M. G. Akkermans (ed.) 2020: *Landscapes of Survival - The Archaeology and Epigraphy of Jordan's North-Eastern Desert and Beyond*, Sidestone Press (Leiden), pp. 79-102.

watered Wadi Sirhan might have been one of the routes by which sheep and goat husbandry entered the Arabian peninsula, following the first appearance of domestic sheep and goats at eastern Jordanian Neolithic sites" (*ibid.*, 1). Two locations in the greater Wadi Sirhan basin, Jebel al-Dharwa in the *harrah* and Wadi Hudruj in the southern limestone steppe,[1] were selected for investigation by that project, during which time it became abundantly clear, as others have observed before (*e.g.* Rolston and Rollefson 1982) and since (*e.g.* Betts and Tarawneh 2010, 69; Fujii 2013, 110-111), that the archaeology and landscape of these two parts of the Jordanian *badia* (and, for that matter, the Negev; see Rosen 2017) are very different.

In contrast to the waterless gravel plains of the south, which have to date yielded little evidence of intensive Neolithic utilisation (Gebel and Mahasneh 2013; Tarawneh and Abudanah 2013; Abu-Azizeh 2014), the *harrah* is a mosaic of productive seasonal microhabitats, many fed by phenomenal quantities of winter rainfall draining off the southern slopes of Jebel Druze ("a veritable water tower", Lancaster and Lancaster 1999, 99; see also Al-Homoud *et al.* 1995; Dutton *et al.* 1998; Allison *et al.* 2000). The abundant and obvious traces (Maitland 1927; Rees 1929) of once-thriving Neolithic communities are concentrated in many of these locations.

It had already become clear during regular visits to the *harrah* by the first author between 1999 and 2002 that, even under the highly degraded conditions of today, sufficient winter and early spring rainfall would transform the region for a few weeks into a magical landscape of seemingly endless grassland and flowers, dotted with flooded mud pans and pools which in turn attracted a multitude of animals and birds. That chance encounter with flamingos in the desert was, in many ways, the inspiration to set in train a quest (in the guise of the 'Eastern Badia Archaeological Project', co-directed by the authors) to discover whether, under the very different environmental conditions of the Late Neolithic (LN), that magic might once have extended throughout the year and in so doing helped to shape the prehistoric archaeology of the Black Desert.

The significance of burin sites

For almost 80 years, burin sites, that is to say sites with little or no cultural deposit and chipped stone assemblages heavily dominated by truncation burins, have been the subject of intense debate amongst prehistorians working in the deserts of Badiyat al-Sham and northern Arabia. Over that time, having frustrated efforts to arrive at a satisfactory explanation for their nature and presence, they have assumed a near-iconic status in the archaeology of that region. Their proposed but until now undemonstrated association with the sheep and goat herding that underpins the spectacular Bedouin culture of today, coupled with the grimly impressive nature of the black basalt desert in which they are concentrated, has detracted little from their mystique.

History of research to the mid-1990s

1920s and 1930s: discovery

The history of the discovery of the burin sites is well known (Betts 1986, 201; Baird 1993, 10-12; Betts *et al.* 2013, 4-5; Garrard and Byrd 2013, 23-24), so only a brief summary is provided here. Sites with high frequencies of truncation burins and an absence of arrowheads were first observed in the *badia* by Field between 1925 and 1934, initially along the then new Mandate-era routes across the north Arabian desert but subsequently more widely, especially around the basalt hill of Jebel Um Wual in northern Saudi Arabia (Field 1960; see also Garrard and Byrd 2013, 23). High frequencies of truncation burins were subsequently also found by Waechter in Wadi Dhobai, now known as Wadi al-Jilat, just before the Second World War, this time in association with Byblos points (Waechter *et al.* 1938; Waechter 1947). While the former industry was dubbed Wualian by Garrod (1960), who analysed the material, the presence of arrowheads in the latter led Waechter to differentiate it with the name Dhobaian, whilst at the same time drawing a parallel between these arrowheads and those by then emerging from what are now known to be Pre-Pottery Neolithic (PPN) B levels at Jericho (Garstang 1935; Gopher 1994, 6-7). The Dhobaian/Wualian split has clouded the debate around the nature of the so-called 'burin Neolithic' (Garrard *et al.* 1988) to this day.

1980s: insights from the Black Desert

The debate was picked up by Betts (1986) in her doctoral research on the prehistory of the Black Desert. In addition to identifying more than 80 new burin sites in the western *harrah*, she defined the industry in more detail on the basis of excavations at what was to become the 'type site' of Jebel Naja: "an overwhelming proportion of burins on concave truncations..., some crude flake scrapers, bifacial pieces, borers and drills, the former on flakes or thick blades, the latter on burin spalls, and a very few small pressure-flaked arrowheads." (Betts 1986, 205). This was clearly Garrod's Wualian, which could now be dated by association to the LN on the basis of a single radiocarbon date from Jebel Naja of 6455-6080 cal BC (OxA-375; Betts *et al.* 2013, Table 1.2) and the presence on western *harrah* burin sites more generally of small numbers of bifacial knives and the occasional small, LN-type arrowhead.

1 For the purpose of this article, the Jordanian limestone steppe is divided into northern and southern sectors by a line projected arbitrarily eastwards from the modern settlement of Al-Hasa on the Desert Highway.

Although the relationship of this industry with the apparently PPNB Dhobaian was left unresolved ("evidence from Wadi Dhobai B on which the tentative dating for the industry is based must be considered anomalous", Betts 1986, 261), Betts made three observations that would prove prescient: (1) an association with bead manufacturing at Jebel Naja and Site 2331 in the Al-Ghirqa area (Betts 1986, 203, 242); (2) a proposed link with herding on the basis of the recurring discovery of burin sites at locales favoured by modern Bedouin (*ibid.*, 264); and (3) tight clustering of burin-site debitage at Site 1601, which had structures, albeit undated (*ibid.*, 223).

1990s: a challenge from the northern limestone steppe

By the late 1980s, high-resolution data were beginning to emerge from renewed excavations at Waechter's sites in Wadi al-Jilat (Garrard *et al.* 1986; 1987; 1994; Baird *et al.* 1992). On the basis of doctoral research on the chipped stone assemblages from these sites, Baird (1993) entered the burin site debate from the perspective of the Dhobaian. Betts' assertion that burin sites should be dated to the LN was challenged on the basis of an observed "quantum leap" (Baird 1993, 520) in the frequency of truncation burins in association with Byblos points from the second quarter of the eighth millennium cal BC onwards. This is seen most clearly at Jilat 26, which remains the earliest well-dated evidence for raised frequencies of truncation burins in either the steppe or Mediterranean zone. Additionally, as the faunal data then available suggested that caprine herding did not appear in the *badia* until the PPNC, Baird disputed Betts' proposed association of burin sites with herding, though it was accepted that it was likely a "by-product of the settlement system of mobile societies" (*ibid.*, 523).

More significantly, and related to Betts' observation that burin site debitage was clustered at Site 1601, Baird independently observed that the frequency of burins was subject to a high degree of contextual variability at PPNB and PPNC sites in the northern limestone steppe (Baird 1993, 520). Even at sites where the frequency of burins as a proportion of tools did not exceed 40% of whole assemblages, in some specific contexts their frequencies were much higher. Significantly, many of these contexts were from sites at which quite substantial domestic structures were also present, *e.g.* Jilat 26 Areas B and C, Jilat 13 and Azraq 31 (*ibid.*, 520). Baird thus proposed a functional, if not then a chronological, association between the open Wualian burin sites in the *harrah* and more substantial Dhobaian sites in the limestone steppe via specific contexts at the latter that shared the same high frequencies of truncation burins. Both types of site were interpreted as foci of the same burin-associated activity or activities, with the former representing locations at which "limited activities were pursued (…) The frequency of such localities is likely to be highest in mobile settlement systems" (Garrard *et al.* 1994, 91) and the latter displaying the "more diverse range of activities that might be expected in longer term, or frequently reoccupied settlements" (*ibid.*). In so doing, Baird narrowed the troubling gap between the Wualian and Dhobaian that had persisted since before the Second World War, whilst at the same time contextualising the 'burin Neolithic' within a site hierarchy of multi-activity base camps and functionally specific temporary camps.

Definitions

Having by this stage identified LN base camps in the *harrah*, Betts neatly formalised the emerging site hierarchy thus: "It is proposed … to refer to sites with few or no structures and a tool assemblage consisting almost exclusively of concave truncation burins as *"burin Neolithic" camps* …. Sites with substantial structures and toolkits with truncation burins in a mixed chipped stone assemblage including diagnostic artefacts will be referred to as *Late Neolithic stations*" (Betts 1992, 112 [own emphases]). A third site category was the *hunting camp*, typified by Dhuweila. Betts' (1992) terminology is used throughout this article.

Interpretation in the mid-1990s

By the mid-1990s, the various elements of the burin-site debate had coalesced around two alternative, though not necessarily mutually exclusive, interpretations that were closely bound up with their respective adherents' views of the beginnings of pastoralism in the *badia*.

Pastoralism

By the mid-1990s, driven largely by the discovery and definition of the PPNC phase at 'Ain Ghazal, a school of thought emerged which argued that caprines were first introduced to the limestone steppe in the PPNC in response to environmental degradation around substantial agricultural villages ("towns", Quintero *et al.* 2004) in the Mediterranean zone. Noting an increase in the frequency of caprines in steppic faunal assemblages from the PPNC onwards, the model argued that herders from these villages would have moved their flocks into the steppe between autumn and late spring to protect crops from caprine depredations, returning after harvest so that flocks could graze on crop by-products (Köhler-Rollefson 1988; 1992). In this model, 'burin Neolithic' camps in the steppe were explicitly associated with the seasonal migration of transhumant herders from agricultural villages to the west (*e.g.*, Quintero *et al.* 2004, 210).

Bead production

Rejecting the notion that caprines were introduced to the steppe by transhumant herders in favour of an 'indigenous adoption' model (Byrd 1992; Martin 1994; 1999) and mindful of Baird's (1993, 520) observation that increased frequencies of truncation burins at Wadi al-Jilat predated the widespread appearance of caprines in the *badia* by approximately a millennium, others felt compelled to look beyond pastoralism to explain the 'burin Neolithic'. An earlier suggestion that was picked up was that burins might have served primarily as cores for the production of spall drills used in stone-bead manufacturing (Betts 1987; Finlayson and Betts 1990). However, this was not universally accepted, not least because of a frequently observed discrepancy between the number of burins and number of spall drills actually recovered. Nevertheless, a contextual association between bead-manufacturing debris, burins and spall drills at some sites is undeniable (Wright *et al.* 2008, 144-145).

Developments since the mid-1990s

Over the past two decades, significant new data have emerged regarding the context, if not yet the interpretation, of burin sites. These include: (1) substantial new Neolithic-focused fieldwork in the *badia* (Quintero *et al.* 2004; Rollefson *et al.* 2014; Richter *et al.* 2016; Rowan *et al.* 2017), especially the hitherto neglected southern limestone steppe (Fujii 2013; Abu-Azizeh 2014; Abu-Azizeh and Tarawneh 2015; Fujii *et al.* 2017a; 2017b); (2) definitive publications of earlier fieldwork (Betts *et al.* 2013; Garrard and Byrd 2013); (3) primary studies of relevant archaeological material (Cropper 2006; Wright *et al.* 2008; Makarewicz 2014; Yeomans *et al.* 2017); (4) the increasing availability and consequent use of high-resolution, open-source satellite imagery (Kennedy 2009, 2011; Kennedy *et al.* 2014); (5) lipid residue analyses that have pushed the beginnings of milking in the wider region back to the late seventh millennium cal BC (Nieuwenhuyse *et al.* 2015; see also Vigne and Helmer 2007); and (6) a number of valuable syntheses (Betts 2008; Betts and Tarawneh 2010; Cropper 2011; Rosen 2017). It is therefore an apposite moment to review our understanding of burin sites and assess whether extant models and interpretations might usefully be updated.

A 'low chronology': the case for caprine herding in the *badia* during the PPNB

By the mid-1990s, a consensus had emerged that although domestic cereals may have been cultivated in the limestone steppe as early as the MPPNB, there was a time lag before the widespread adoption of herding, which was not thought to have happened before the PPNC (Garrard *et al.* 1994; 1996). This seemed reasonable at the time, as there was widely believed to be a gap of more than a millennium between plant and animal domestication in the Levant more generally (*e.g.* Bar-Yosef and Meadow 1995).

Over subsequent decades, new data and AMS dating of old samples have demonstrated that both developments are actually likely to have occurred more or less synchronously, "with signs of initial management of morphologically wild future plant and animal domesticates reaching back to at least 11,500 cal BP, if not earlier" (Zeder 2011, S230; see also Vigne 2011; Weiss and Zohary 2011; Martin and Edwards 2013). In light of this, the view that crop cultivation, the more rainfall-dependent of the two adaptations, was practiced – however opportunistically (Rosen 2017, 98-102 and 125) – on the arid margins of the Mediterranean zone *without* its more drought-tolerant herding counterpart for a thousand years becomes harder to sustain. It would have meant that the first PPNB communities in the *badia* chose not to adopt the herding component of the mixed farming package available to them, at a time when they were culturally speaking most closely aligned with the Mediterranean zone, only to adopt it later after centuries of cultural divergence.

It is worth recalling the example of Cyprus, where the arrival of the PPNB cultural complex in the mid-ninth millennium cal BC (Manning 2013, Table A2) was accompanied by the appearance of the full mixed-agricultural package. Subsequently, elements of the package that were found to be superfluous in that low-intensity environment seem either have been jettisoned (*e.g.*, cattle herding) or marginalised (*e.g.*, caprine herding) as local adaptations focused primarily on hunting subsequently evolved (Croft 1991, 2002; Vigne *et al.* 2003). "This left Cypriot economies "frozen" in the far more typically transitional phase of the mainland Early PPNB, in which agriculture was practiced as a minor adjunct to hunting and gathering" (Wasse 2007, 60).

The arrival of the PPNB cultural complex in the limestone steppe, demonstrably another low-intensity area at that time, may have followed a comparable path. In this interpretation, the paucity of caprine remains in the *badia* prior to the PPNC would be attributed to an absence of the selective pressures driving intensification in the Mediterranean zone. As a result, there would have been little incentive for herding to have been undertaken in the *badia* as anything more than an adjunct practice, possibly non-economic in the earliest stages (cf. Martin 1999, 101), until selective pressures such as population growth made themselves felt from the PPNC onwards (Garrard *et al.* 1994, 106). It should also be borne in mind that, in the highly attenuated faunal assemblages so characteristic of the *badia*, all but the most prevalent human/animal relationships are likely to slip below the threshold of zooarchaeological visibility. On theoretical grounds alone, then, there is strong reason to suspect that caprine herding was practiced in the steppe as a minor adjunct to hunting following the appearance of PPNB cultural traits there at the start of the eighth millennium cal BC (Edwards 2016, 60) and that its importance subsequently increased.

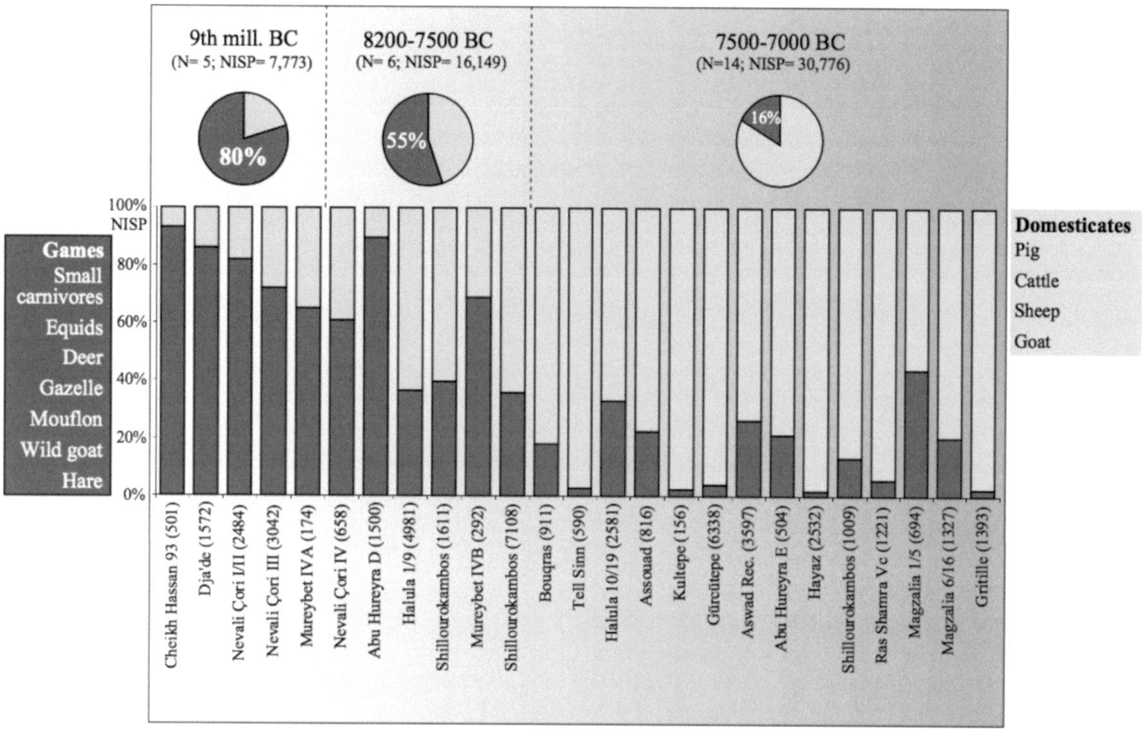

Figure 1. Relative proportions of hunted and domestic taxa in selected Early, Middle and Late PPNB faunal assemblages from the northern Levant (reproduced with permission from Vigne and Helmer 2007, Fig. 13).

Since the mid-1990s, empirical evidence to support this view has started to emerge. Hongo *et al.* (2013, 11-16) argued that caprines were herded as an adjunct to gazelle hunting in the southern limestone steppe at M-LPPNB Wadi Abu Tulayha. This interpretation is supported by spherulite evidence for herd-animal penning at the contemporary late MPPNB site of Ayn Abu Nukhayla in Wadi Rum (Albert and Henry 2004, 90) and the identification of a reasonably substantial assemblage of apparently domestic caprines at LPPNB Bawwabah al-Ghazal in the Azraq basin (Quintero *et al.* 2004, 205; Rollefson 2011; G. Rollefson, pers. comm.). These data augment the handful of previously reported LPPNB caprine remains from the *harrah* sites of Azraq 31 (Baird *et al.* 1992, 28-29 and Table 3), Ibn al-Ghazi (Martin 1999, Table 3), Dhuweila I (Betts *et al.* 1998, 171-173) and perhaps Burqu' 35000 (Betts *et al.* 2013, 80). All things considered, a sufficient body of evidence has perhaps now emerged to suggest that although herding did not become a mainstream economic practice in the *badia* until the PPNC, it may nevertheless have been practiced as a minor adjunct to gazelle hunting (Martin and Edwards 2013, Table 4.3) for at least a millennium beforehand.

It is worth emphasising that such a situation would by no means have been unusual at this time. Drawing on zooarchaeological data from the northern Levant, Vigne and Helmer (2007, 34) have argued that for much of the MPPNB, even in the precocious PPNB core area of northern Syria, "hunting continued to provide meat as it had for millennia, while domestic animals were at least partly exploited for their "secondary products", especially milk, which hunting could not provide. This would explain how these early Neolithic societies could have had a dairy economy in spite of low milk yield, since hunting supplied important animal protein" (see Fig. 1). This model is particularly germane to the burin-site debate and deserves the most serious consideration. In the low-intensity *badia*, just as in Cyprus, it is highly likely that earlier PPNB subsistence strategies would have endured for centuries longer than in the 'Levantine corridor', certainly until the PPNC and possibly beyond the end of the Neolithic period altogether.

In sum, multiple theoretical and empirical considerations now counsel reconsideration of Baird's (1993, 523) assertion that the MPPNB roots of the 'burin Neolithic' rule out a primary association with herding. Indeed, it is not so much that caprine remains are consistently absent in pre-PPNC Neolithic faunal assemblages from the *badia* as it is that they are consistently present, albeit in small numbers, from at least the late MPPNB onwards. This reflects, almost certainly, a continued ease of acquiring 'free food' by hunting until well into the Neolithic period and,

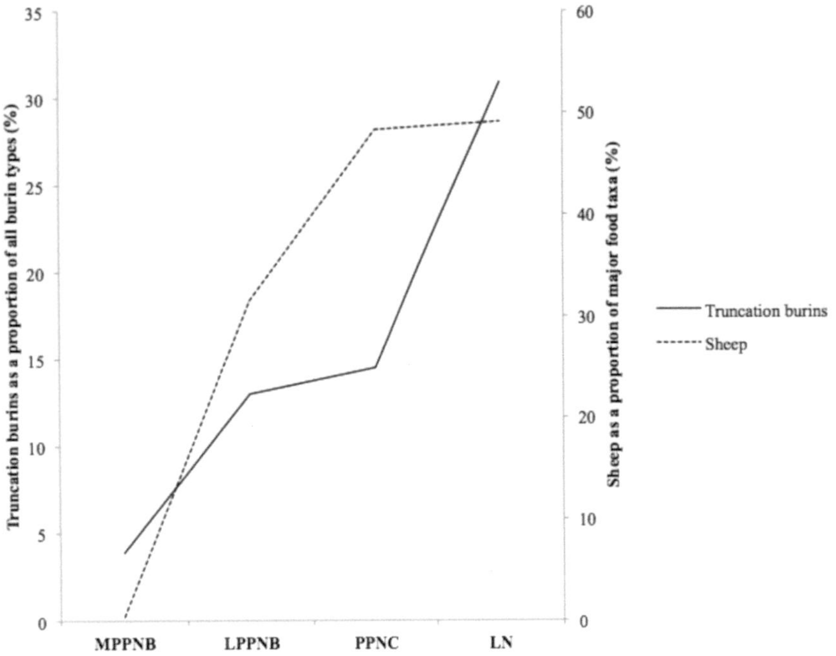

Figure 2. Proportions of truncation burins and sheep by occupational phase at 'Ain Ghazal (Rollefson *et al.* 1992, Table 5; Wasse 2002, Table 3).

perhaps, the retention of milk-producing caprines (cf. Vigne and Helmer 2007; for an alternate view, see Rosen 2017, 98-102 and 122-129).

From flints to sheep: the case for an association between the 'burin Neolithic' and herding

The observations of Rollefson *et al.* (1992, 459 and Table 5) regarding the frequency of truncation burins at the agricultural village of 'Ain Ghazal on the current eastern margin of the Mediterranean zone, where caprine herding was almost certainly practiced from the early MPPNB onwards (von den Driesch and Wodtke 1997; Wasse 2000; 2002; Martin and Edwards 2013; Makarewicz 2014), are of critical importance to the burin site debate. Although published absolute counts are quite small, a marked increase in the frequency of truncation burins has nevertheless been documented there. However, this started not during the late MPPNB, as in the northern limestone steppe, but several hundred years later, in the LPPNB. "From the PPNC onwards at 'Ain Ghazal a burin assemblage is found that has some relationship with those found in the steppe/desert to the east." (Baird 1993, 526).

This suggests, first, that the 'burin Neolithic' could have dispersed from the steppe to the Mediterranean zone rather than the other way round and, second, that it might have been associated with the well-documented (von den Driesch and Wodtke 1997; Wasse 2000; 2002) appearance at 'Ain Ghazal of large numbers of sheep from the LPPNB onwards (see Fig. 2). Although the proposal of a link between raised frequencies of truncation burins and sheep herding in no way implies a functional association between the two, it is worth recalling the suggestion that burin spalls may have been "set into pieces of wood or other material and used to harvest wool by combing" (Quintero *et al.* 2004, 209).

That sheep are likely to have been an important element in early pastoral economies in the steppe is not a new idea (Lancaster and Lancaster 1991; Ducos 1993; Perrot 1993; Quintero *et al.* 2004; Wasse 2007, 63). However, most models have hitherto argued that sheep dispersed into the southern Levantine Mediterranean zone from Syria, specifically the Damascus basin, via the 'Levantine corridor' (Ducos 1993; Stordeur *et al.* 2010; Martin and Edwards 2013, 61). As now seems clear, the frequencies of both truncation burins and sheep increased at 'Ain Ghazal in near-lockstep from the LPPNB onwards. This raises the possibility that domestic sheep were introduced to the Mediterranean zone from the *badia* as part of a wider package of steppic cultural traits that included a focus on truncation burins. Green-blue Dabba marble (Wright *et al.* 2008, 137-138), which started to appear at 'Ain Ghazal in the LPPNB where it replaced similarly coloured but no doubt harder-to-source copper ores from southern Jordan (Rollefson 2011, 106), may have been a third element in such a package. It is tempting to speculate, in view of the recent identification of a PPNA population of wild sheep at Shubayqa (Yeomans *et al.* 2017), whether the north-western *badia* might have been a focus of autochthonous sheep domestication following the as yet poorly understood appearance of PPNB cultural traits in that region.

If confirmed by future research, this scenario would be evidence for a more dynamic process of multi-directional cultural interchange between the Mediterranean zone and *badia* than previous models have allowed for. Under such circumstances, Makarewicz's (2014, 128) proposal that at 'Ain Ghazal the "pronounced intensification of sheep husbandry by *c.* 6600 cal BC (PPNC) coincides with the appearance of a new, but small, population of large-bodied sheep that probably originated either from the northern Levant or *herders inhabiting the eastern badia*" (own emphasis) should not be considered as much a possibility as a probability. More detailed work on the chronology, dispersal and interchange of cultural traits between the Mediterranean zone and *badia* in the eighth and seventh millennia BC is sorely needed.

'Population', 'ethnic group' or 'economic specialisation': what exactly is chipped stone technology telling us?

To date, at least three doctoral theses have focused on chipped stone assemblages from the *badia* (Baird 1993; McCartney 1996; Cropper 2006). All concluded that, although Mediterranean-zone typologies and reduction strategies persisted in the *badia* for a few hundred years after the first appearance there of PPNB cultural traits at around the start of the eighth millennium cal BC, local adaptations subsequently evolved and remained distinct from those of the Mediterranean zone for the remainder of the Neolithic. What is much less clear is how this observation should be interpreted.

A technological distinction of this type ('industry'; 'strategy') is all too frequently used, whether implicitly or explicitly, to impose a social or ethnic distinction on the makers ('group'; 'population'). These distinctions tend then to be appended to descriptors of economic specialisation ('hunter-gatherer'; 'herding'; 'farming') to create wobbly and ill-defined compounds of archaeological analysis. Resolution of this issue insofar as it affects the debate on the 'burin Neolithic' is beyond the scope of this article, other than to acknowledge, as a first step, that it exists, both here and elsewhere.

If the chipped stone evidence from Wadi al-Jilat is accepted at face value (Baird 1993, 652-653; Garrard *et al.* 1994, 91), it might be concluded that specific behavioural patterns, perhaps representing a discrete *badia* population, had emerged in the northern limestone steppe by as early as the late MPPNB. The technological evidence from Dhuweila I suggests that elements of this population had dispersed deep in to the *harrah* by the LPPNB (Baird 1993, 438; Betts *et al.* 2013, 187). That subsequent evolution occurred within these *badia* lithic traditions, especially across the PPNB-PPNC transition, is a given. However, it's worth emphasising that these changes display a high degree of continuity with previous practice: "If these continuing and closely related techniques do not reflect use of the same locales by related groups over extended time periods it would be surprising" (Baird 1993, 651).

Moving forward in time to the LN, Cropper's (2006; 2011) doctoral research is a valuable contribution to the debate on the nature of relationships between the Mediterranean zone and *badia* during that period as evidenced by chipped stone technology. Her conclusions are worth quoting at length: "When considering the two regions, the primary differences in the reduction strategies include the frequencies of the core types, the preparation of facetted striking platforms, and the consistent modification of the overhang prior to the detachment of the flake… I have come to the conclusion that while there was contact between these groups as attested by the diffusion of ideas and goods, the knapping methods indicate that they were disparate populations. Because these populations were largely isolated from each other, the cultural norms that governed their daily life slowly diverged, allowing for differences in the lithic technology to develop." (Cropper 2011, 136).

Crucially, the chipped stone assemblages from sites with raised frequencies of truncation burins, *viz.* Jebel Naja, Burqu' 03000 and Burqu' 35000 (Betts *et al.* 2013, Tables 2.12, 3.13 and 3.27), were all assessed as being more closely affiliated with *badia* as opposed to Mediterranean-zone reduction strategies (Cropper 2011, Figs. 5.1-5.3). As noted by Betts *et al.* (2013, 187), this deeply embedded (cf. Ingold 2000, Ch. 19) cultural trait "would certainly not change simply because a population moved from the verdant to the steppic zones". Although recent analyses of chipped stone assemblages from the Late Neolithic stations at Wisad Pools and Wadi al-Qattafi (G. Rollefson, pers. comm.) suggest that there may have been more variation in *badia* reduction strategies than that described by Cropper (2006; 2011), current evidence still offers little support to the notion that the 'burin Neolithic' was directly associated with the seasonal migration of transhumant herders from agricultural villages to the west. Instead, it tends to reaffirm Baird's (1993, 509-511) belief that the roots of the phenomenon lay in the late MPPNB of the northern limestone steppe.

Garden of Eden and a Howling Wilderness: divergent trajectories in the northern and southern *badia*

PPNA

Until recently, an apparent hiatus in the archaeological record of the *badia* during the PPNA led scholars to speculate that much of the region might have been abandoned (Betts *et al.* 2013, 5; Rosen 2017, 96), perhaps from as early as the Late Natufian in response to the onset of colder and dryer (Bar-Yosef 2011) conditions that characterised the

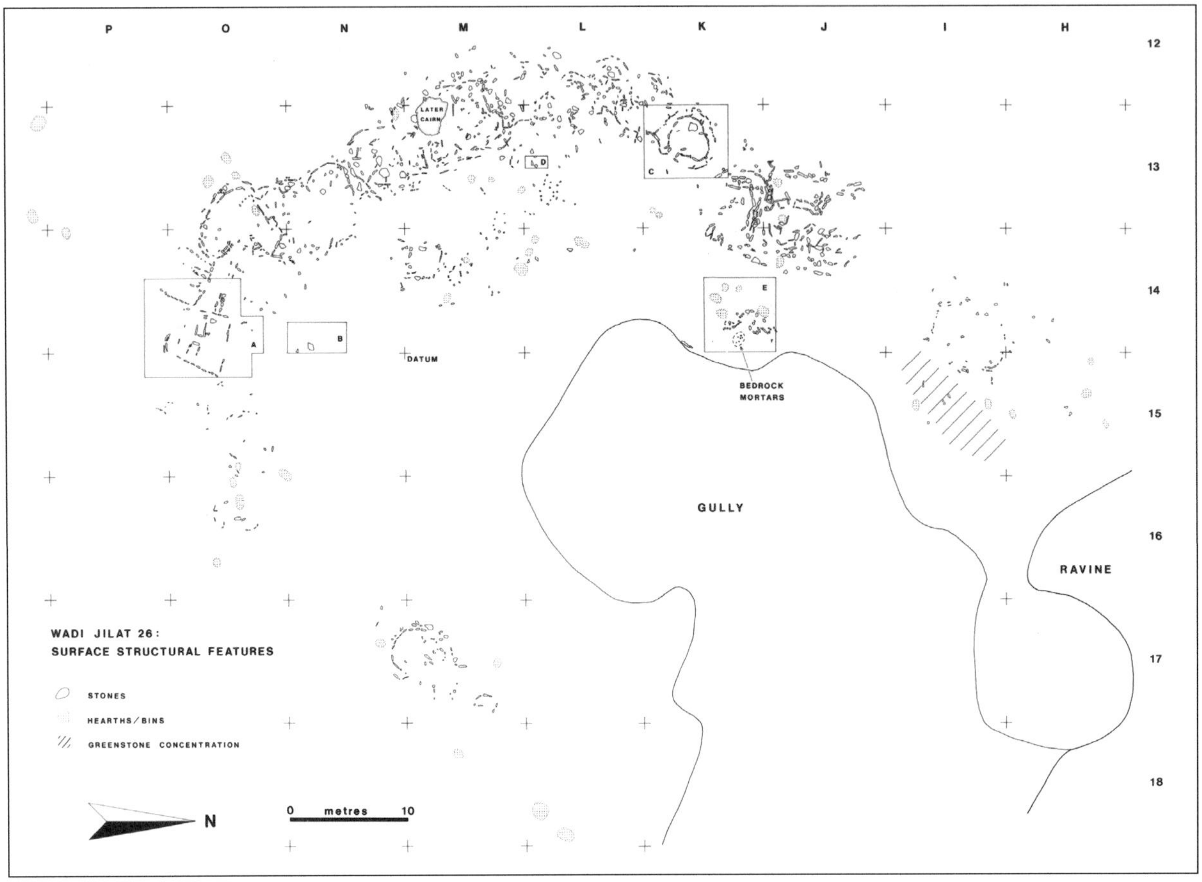

Figure 3. Structural features at Wadi Jilat Site 26 (courtesy of Dr Andrew Garrard, reproduced with permission).

Younger Dryas. Recent work in the Shubayqa area has however exposed several structures radiometrically dated to the PPNA, with the possibility of re-occupation in the EPPNB (Richter *et al.* 2016; Yeomans *et al.* 2017). There are also traces of PPNA-EPPNB occupation in the Azraq basin (A. Betts, pers. comm.). Nevertheless, the extent to which this emerging PPNA horizon may have been restricted to a small number of better-watered refugia in the north and west of the region remains uncertain, as current evidence still suggests that the eastern *harrah* remained sparsely utilised until the LPPNB (Betts 2008, 35).

PPNB

Regardless of the extent to which the *badia* as a whole was or was not utilised during the PPNA, it is generally accepted that western parts of the limestone steppe saw increasing occupation intensity (Munro 2004, S7) from the start of the eighth millennium cal BC onwards (Baird 1993; Garrard *et al.* 1994; Betts 2008; Fujii 2013; Garrard and Byrd 2013, 7; Edwards 2016). Whether this was associated with an influx of communities from the Mediterranean zone or reflects a process of acculturation on the part of existing communities who remain, at present, below the threshold of archaeological visibility is still uncertain, as is whether it occurred within EPPNB or MPPNB chronological and cultural horizons (Edwards 2016 *contra* Baird 1993). However, there is little doubt that by around 8000 cal BC (Edwards 2016, 60) the northern limestone steppe was occupied by communities who used similar chipped stone reduction strategies to those documented at contemporary agricultural villages in the Mediterranean zone to the west, with the southern limestone steppe apparently following suit a few hundred years later (Fujii 2014, Table 1). As such, it is highly probable that these events were associated with the wider dispersal of the PPNB cultural complex across the Levant and beyond from a core area in north Syria (Edwards 2016).

With regard to M-LPPNB Jilat and Azraq, Baird (1993, 461) observed that "there are no contrasts in the employment of reduction strategies between the sedentary communities in the moister areas and the mobile groups exploiting the arid zone". Similarly, at contemporary Wadi Abu Tulayha in the Jafr basin, "flint artifacts were produced by means of the naviform core-and-blade technique, a landmark of the Levantine PPNB flint industry" (Fujii 2013, 56). Some cultural traits remained specific to each area of course, but this would have been inevitable given that "the PPNB

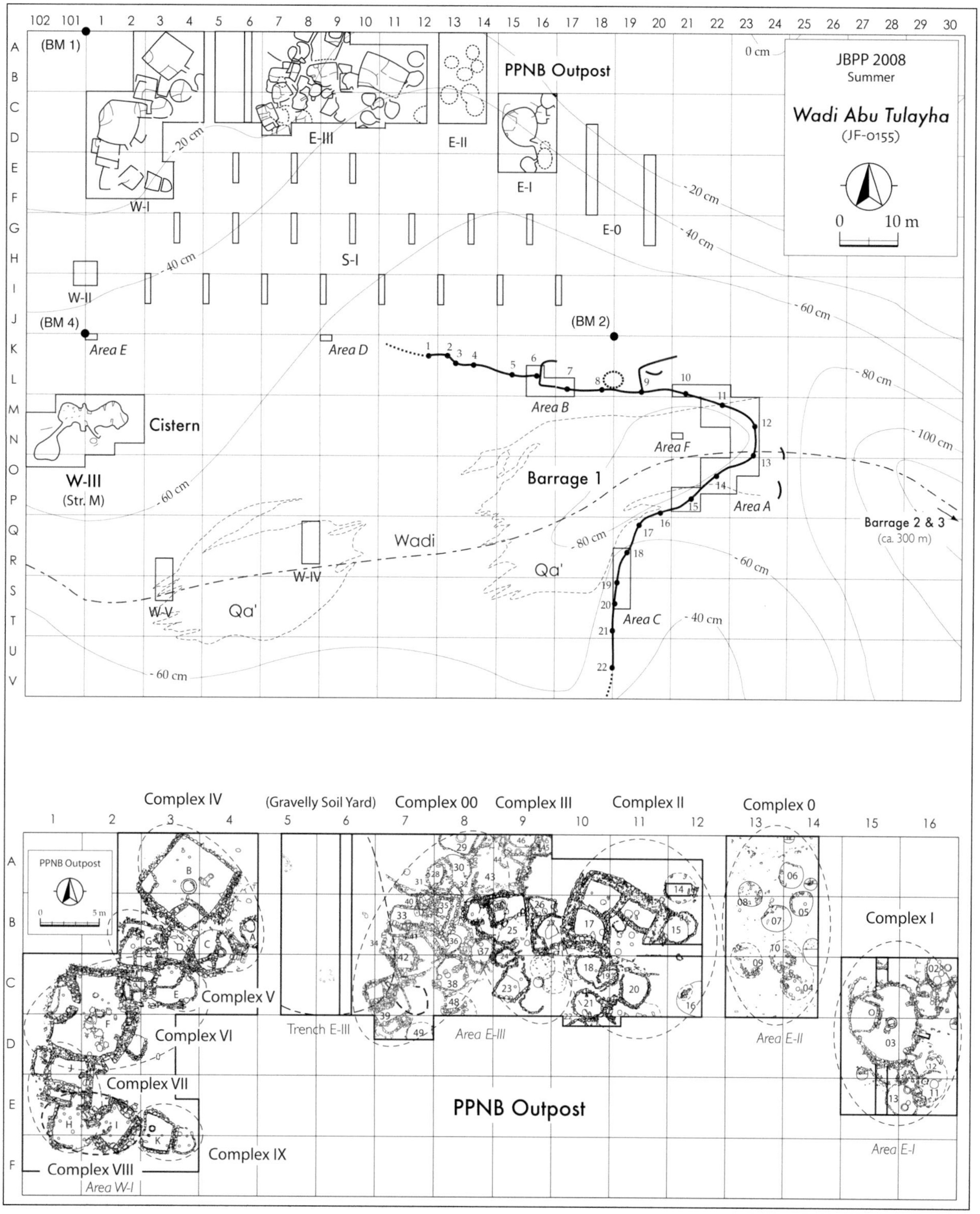

Figure 4. Structural features at Wadi Abu Tulayha (courtesy of Prof. Sumio Fujii, reproduced with permission).

should best be considered as a polythetic cluster of material culture traits and behavioural practices, some of which were autochthonous to the various regions north and south, and some of which are ultimately derived from North Syria. Each region and its community(-ies) developed idiosyncratic cultural repertoires" (Edwards 2016, 54). Nevertheless, it remains the case that cultural similarities between northern and southern sectors of the limestone steppe outweighed the differences between them at this stage.

Further evidence that these areas shared elements of a common cultural heritage can be seen in the architecture of Wadi Jilat 26 (Garrard *et al.* 1994, Fig. 3) and Wadi Abu Tulayha (Fujii 2013, Fig. 4). This is not a new observation; that "the Jilat structures share much… with contemporary dwellings in other arid areas of the southern Levant" was noted by Garrard *et al.* (1994, 106) more than two decades ago. Nevertheless, the subsequent discovery of Wadi Abu Tulayha has brought these similarities into sharper focus, particularly with regard to the nature and strength of the relationship with the Mediterranean zone during the PPNB. Both sites consist of a substantial (*c.* 80-100 m long) arc of adjoining or adjacent structures overlooking a gully (see Figs. 3-4). In the case of Wadi Abu Tulayha, the arc is thought to have developed in linear fashion by a process of successively building on to or immediately adjacent to an existing structure, with only one or a few structures being utilised at any one time (Fujii 2013, 56). Wadi Abu Tulayha, dated to the M-LPPNB transition, displays an *in-situ* transition from curvilinear to rectilinear architecture in the later portions of the arc. The possibility that MPPNB Jilat 26 developed and was utilised in a similar way is supported by the presence of a remarkably similar transition to rectilinear architecture at the extreme south-eastern end of its arc.

The above example illustrates how the current limestone steppe mirrored at least some Mediterranean-zone cultural developments during the PPNB, albeit with a not-inconsiderable time lag discussed at length by Edwards (2016). It also begs the question of why, just a few hundred years later (Fujii 2013, Fig. 38), rectilinear domestic architecture disappeared from the *badia* for millennia, despite being the prevalent architectural form in the Mediterranean zone for much of that time? The most economical explanation is that *badia* communities, whether consciously or unconsciously, began to assert a greater degree of cultural maturity and independence over the course of the second half of the eighth millennium cal BC, noting that this seems not to have come with any reduction in contact with the Mediterranean zone. It is, perhaps, no coincidence that it is at precisely this time that overt expressions of a more self-referencing cultural identity start to appear (*e.g.* 'desert kite' hunting traps) or increase (*e.g.* green stone beads) in the archaeological record of *badia* (Wright *et al.* 2008, 134; see also below).

PPNB structures in the northern and southern limestone steppe were probably occupied seasonally, with Fujii (2013, Fig. 43) going so far as to position Wadi Abu Tulayha within a system of pastoral transhumance tethered to agricultural villages to the west. No such claim was made for the Wadi al-Jilat sites, not least because of the belief of the excavators that herding was not widely adopted in the *badia* until the PPNC. It is however interesting to note that the quantity of bone at Jilat 26 was very much less than at other Jilat sites (Garrard *et al.* 1994, 97-98), raising the possibility that it too may have been utilised in a transhumant context.

As noted above, there is currently little evidence for intensive utilisation of the *harrah* during the first half of the eighth millennium cal BC (but see Wasse and Rollefson 2005, 14; Rowan *et al.* 2014, 281-283). Nevertheless, the much wider processes behind the dispersal of the PPNB cultural complex across the Levant did not fizzle out on reaching the western limestone steppe, but continued unabated – though painfully slowly (about 20 km per century) – eastwards into the second half of the eighth millennium. Over the course of that time, the *harrah* found itself being increasingly utilised by communities who, in an idiosyncratic twist, may also have introduced desert kites to the region (Betts *et al.* 1998; Betts and Tarawneh 2010, 69; Betts and Burke 2015; see also Abu-Azizeh *et al.* 2016).

PPNC – LN

The nature of PPNB utilisation of the limestone steppe has been discussed at length because it was, as currently understood, the start point for the emergence of the 'burin Neolithic'. In this regard, understanding where the 'burin Neolithic' did not develop is of as much importance as understanding where it did. From the PPNC onwards, the hitherto shared culture of the northern and southern limestone steppe developed along increasingly divergent paths. This may have been linked to the presence at that time of a north-south rainfall gradient (cf. Rosen 2017, 97 and 121), with the southern sector finding itself beyond the limits of marginal dry farming. That this was the case is supported by the introduction of a basin-irrigation barrage system at Wadi Abu Tulayha early in the LPPNB (Fujii 2013, 96). These remarkable features seem not to have been present at Wadi al-Jilat (Garrard *et al.* 1994; 1996), located 150 km further up the gradient to the north, presumably because annual rainfall there was adequate for the task at hand.

Southern limestone steppe

Over the course of more than 25 seasons of fieldwork in the wider Jafr basin since 1997, Fujii and his colleagues have forensically documented what they interpret as a process of pastoral nomadisation spanning more than four and a half millennia, from the MPPNB to at least EB III (Fujii 2013, Fig. 38). In this regard, it is significant that, following the

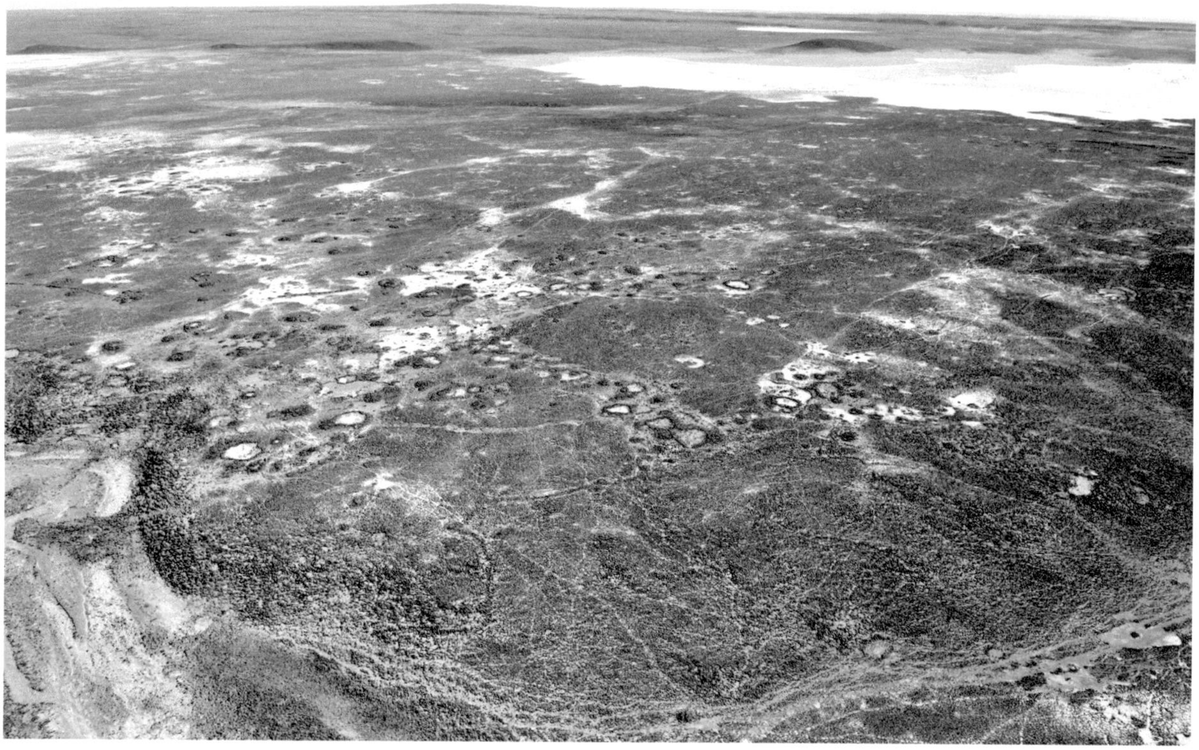

Figure 5. Aerial view of the Late Neolithic core area at Wisad, showing the concentration of structures around the eponymous pools (APAAME_20080909_DLK-0347. Photograph by David Kennedy. Courtesy of APAAME).

abandonment of the small encampment at Khashm al-'Arfa in the eastern Jafr basin some time around the LPPNB-PPNC transition (Fujii 2017a), few substantial Neolithic domestic structures are currently known from south-eastern Jordan (but see Gebel and Mahasneh 2013; Abu-Azizeh 2014; Abu-Azizeh *et al.* 2016). Notwithstanding the consequent absence of primary economic evidence in the form of faunal and botanical samples, Fujii (2013) has convincingly argued that, from the PPNC onwards, the region was utilised by increasingly nomadic pastoralists. In consequence, it is proposed, stone-built domestic architecture was replaced in the archaeological record by elaborate symbolic and funerary monuments that served the twin purposes of reinforcing territorial claims and consolidating lineage groups. Such activity is frequently associated with pastoral groups (*e.g.* Di Lernia 2006, 60-61; Gebel and Mahasneh 2013; Rosen 2017, 162-163), to the extent that it is frequently (as in this case) used as a proxy for the presence of otherwise invisible nomadic pastoralists in the archaeological record, especially where settlement evidence is absent.

Northern limestone steppe and harrah

In marked contrast to the reduction in occupation intensity and increase in mobility documented in the southern limestone steppe during the PPNC and LN, cultural evolution in the northern limestone steppe and *harrah* appears to have proceeded in a near-opposite direction. In Wadi al-Jilat, Garrard *et al.* (1994, 106) noted that the structures "from the Early LN are substantially larger than the structures from the PPNB, suggesting use by larger co-resident groups or by individuals with livestock." An even more pronounced trend in the direction of occupation intensity has recently been described for the Late Neolithic in the *harrah* by Rollefson *et al.* (2014, 299): "the sheer number of buildings, and their density close to seasonal water sources, suggests a much larger population in the Badia, than, perhaps, previously imagined." (see Fig. 5). Although PPNC and LN structures in this part of the *badia* are typically interpreted as seasonally occupied (Garrard *et al.* 1994, 106; Betts *et al.* 2013, 188; Rollefson *et al.* 2014, 285), the evidence for this has not yet been systematically tested against criteria commonly associated with the appearance (or not) of sedentism in, for example, the Early Natufian period (*e.g.* Hardy-Smith and Edwards 2004; Boyd 2006). This should be regarded as a priority area for future research.

Although raised frequencies of truncation burins have been documented in the northern limestone steppe as early as the MPPNB (see above), the 'burin Neolithic'

Figure 6. Aerial view of structure M-7 SS-1 at Wadi al-Qattafi, illustrating both the substantial nature of many structures at 'Late Neolithic' stations and the investment in their internal features.

in its developed (cf. Betts 1992) manifestation cannot be said to have emerged until the widespread appearance of 'burin Neolithic' camps in the LN (noting that Rowan *et al.* [2014, 281-282] have documented a possible LPPNB 'burin Neolithic' camp [MF-1] on the southern slope of Maitland's 'hill-fort' in Wadi al-Qattafi). The patchy survey data currently available suggest that, by the LN, the centre of gravity of the 'burin Neolithic' had shifted east into the *harrah* from its apparent area of origin, as especially dense distributions of 'burin Neolithic' camps have been recorded along the western *and* eastern margins of the *harrah* and into the *hamad* beyond the latter (Betts 2008, 35; Betts *et al.* 2013, 31 and 184-186). This process is poorly understood at present, but was doubtless associated with the cultural wake left by the slow, eastwards dispersal of the PPNB cultural complex across the *badia* over the course of the course the eighth millennium cal BC.

Just as the origins of the 'burin Neolithic' appear to have lain within MPPNB domestic structures in the northern limestone steppe, so its LN manifestation seems to close in on the plethora of 'Late Neolithic' stations (see Fig. 6) that actually emerged in the *harrah* from at least the late PPNC onwards (Rollefson *et al.* 2014; Rowan *et al.* 2017). As the widespread 'burin Neolithic' camps have been convincingly identified as a mobile and functionally specific manifestation of *the same* cultural complex as the stations (Baird 1993, 521; Garrard *et al.* 1994, 91), the most likely interpretation of the former is that they represent the seasonally dispersed sheep, or at least mixed, herding subcomponent of much broader hunter-gatherer-herder economies (cf. Garrard *et al.* 1996; Martin 1999) centred on 'Late Neolithic' stations such as those at Wisad Pools (Rollefson *et al.* 2014; Rowan *et al.* 2017), Wadi al-Qattafi (*ibid.*), Al-Ghirqa (Betts and Helms 1987) and Burqu' (Betts *et al.* 2013). As the 'Late Neolithic' stations are themselves thought (but see above) to have been seasonally occupied, what we are left with is a system of pastoral transhumance tethered more to a social ('group') than a physical ('site') entity.

Sadly, a direct association of 'burin Neolithic' camps with sheep herding has yet to be demonstrated, as such sites seldom yield bone debris. "This is not merely a result of prevailing conditions of preservation, as bone can be seen in surface collections from sites with faunal remains" (Betts *et al.* 2013, 180). Nevertheless, the frequent location of such sites on sheltered lee slopes along wadis, many with seasonal pools and extensive areas of grazing nearby, strongly hints at an association with herding at some level, as does the virtual absence of arrowheads (Betts 1992, 112).

If 'burin Neolithic' camps were indeed associated with herding, their largely anosseous nature indicates that animals were not slaughtered on-site and most likely had some value or significance beyond that as a source of meat. "If animals are being used for purposes other than meat, they are too valuable a resource to slaughter simply when required for food" (Sherratt 1981, 275). Milking (see above) is perhaps the most credible option (Lancaster and Lancaster 1991, 135; Köhler-Rollefson 1997; Vigne and Helmer 2007), especially as recent lipid residue analyses have conclusively demonstrated the presence of milk fats on pottery from Tell Sabi Abyad in upper Mesopotamia at a time coeval with the *badia* LN (Nieuwenhuyse *et al.* 2015). However, the possibility of exploitation for fleece (cf. Quintero *et al.* 2004; see also Helmer *et al.* 2007; Saña and Tornero 2014) or the maintenance of flocks as sources of social (Hodder 1982) or economic (Rollefson 2011, 106-107) capital are all equally feasible, and not in any way mutually exclusive.

Distribution of 'burin Neolithic' camps and 'Late Neolithic' stations

More than three decades ago, Betts (1986, 1987) mapped the distribution of burin sites as then understood, describing them as having been "recorded from Rutbah in Iraq up to the very rim of the Jordan Rift Valley and from Palmyra down to Jawf and Sakaka south of the basalt in Saudi Arabia" (Betts 1986, 259). This description has proved remarkably resilient over the intervening period, with subsequent fieldwork adding little to challenge its basic essentials, although refinements have inevitably been made.

Absence of the 'burin Neolithic' from the southern limestone steppe?

One potentially significant development has been the documentation for the first time of somewhat raised frequencies of truncation burins at sites south of Wadi al-Hasa, at LPPNB Jafr-17 (Wilke and Quintero 1998), LPPNB-PPNC Khashm al-'Arfa (Fujii *et al.* 2017a, Table 2) and LN Jafr-144 (G. Rollefson, pers. comm.), all located on the northern and eastern margins of the Jafr basin. However, there is currently little to suggest that the 'burin Neolithic' thrived here following the cessation of basin-irrigation agriculture and shift from pastoral transhumance to incipient pastoral nomadism at around the time of the LPPNB-PPNC transition (Fujii 2013, Fig. 43; Fujii 2017b, 589 and Fig. 15; see also Abu-Azizeh *et al.* 2016). Certainly, LN 'burin Neolithic' camps still appear to be virtually absent from the southern limestone steppe. The apparent association of this early variant of the 'burin Neolithic' with pastoral transhumance and marginal agriculture, but *not* incipient pastoral nomadism, is highly significant and will be returned to below.

Spatial correlates of the 'burin Neolithic'

If raised frequencies of truncation burins were not associated with incipient nomadic pastoralism, it is pertinent to consider what the phenomenon was associated with.

Rainfall, plant foods and fodder. On current understanding, one variable with which the distribution of the 'burin Neolithic' seems closely correlated is rainfall. Allowing for the possibility of sampling bias, especially with regard to northern Saudi Arabia, few burin sites are currently known beyond the area in which marginal agriculture or relatively intensive gathered plant food economies are known to have been practiced (see below). This suggests, first, that the 'burin Neolithic' was positioned within the context of what was, at least seasonally, a mixed economy and, second, that its area of distribution would have provided abundant seasonal grazing for flocks. It is important to emphasise that environmental conditions in the *badia* during the LN are likely to have been more benign than today: "The time interval between 8.5 and 7 ky is characterized by a ... deluge period when annual precipitation was extremely high." (Bar-Matthews *et al.* 1999, 91; see also Bar-Matthews and Ayalon 2004, 385; Rosen 2017, 83-84). By way of example, this is manifested in the *harrah* by the documented presence of oak stands and marsh plants in their respective environmental niches (Rowan *et al.* 2017, 109-110).

Although archaeobotanical evidence for domestic cereals in the PPNC and LN is currently restricted to Wadi al-Jilat (Garrard *et al.* 1994, 100-105; 1996, 214; Betts *et al.* 2013, 5), most sites have yielded evidence for exploitation of wild plant foods of one sort or another (*ibid.*; Betts *et al.* 1998, 185-189; 2013, 186; Rollefson *et al.* 2016, 6). Additionally, the recently excavated 'Late Neolithic' stations at Wisad Pools have yielded abundant evidence (see Figs. 7 and 8) for intensive plant food processing in the form of domestic grinding and pounding tools (Rollefson *et al.* 2014, 291). If 'burin Neolithic' camps were indeed associated with some sort of sheep herding, there is little need to look as far as agricultural villages far to the west to find thriving plant-food economies with which they might have been associated, most likely in short-range, or at least intra-*badia*, systems of pastoral transhumance.

Stone-built domestic structures. Another correlate of the 'burin Neolithic' is the presence within its core area of distribution of reasonably substantial stone-built domestic structures (Garrard *et al.* 1994, 75-85; Betts *et al.* 1998, 2013; Rowan *et al.* 2017), or at least the basal components thereof. This is essentially the point made by Baird when he argued that "developments in Jilat/Azraq clearly relate

Figure 7. Cache of basalt pestles from structure W-66 at Wisad Pools.

Figure 8. Grinding slabs with a small central depression set into the floor of structure W-80 at Wisad Pools (cf. LN Dhuweila II; Betts *et al*. 1998, Pls. 5-6).

to this 'burin site' phenomenon" (Garrard *et al.* 1994, 90), owing to the discovery of high frequencies of truncation burins in some structural contexts there.

Clearly, "permanence of structure does not necessarily reflect permanence of occupation" (Boyd 2006, 170). Nevertheless, regardless of whether or not PPNC and LN structures in the *badia* were permanently occupied, it remains the case that "construction in stone (as opposed to wood or brush) indicates the elaboration or embellishment of particular places – locales – in the landscape" (*ibid.*, 171), which almost certainly reflects an intention to return to that same location on future occasions. (As an aside, the construction of desert kites might be viewed in much the same light). The 'burin Neolithic' cannot therefore be understood solely as the material manifestation of seasonal wanderings by herders, guided primarily by the needs of their flocks. Instead, it is much better understood as being socially grounded in those regions in which structures containing high frequencies of truncation burins are located. By the LN, foremost amongst these regions was the *harrah*, which might therefore reasonably be considered the '*domus*' (cf. Hodder 1990) of the 'burin Neolithic'.

'Desert kite' hunting traps. One of the most intriguing spatial correlates of 'burin Neolithic' camps concerns their apparent collocation with some types of desert kite. It should be emphasised that kites have an extremely wide spatial and chronological distribution within the Levant and beyond (Zeder *et al.* 2013; Abu-Azizeh and Tarawneh 2015; Barge *et al.* 2015; Betts and Burke 2015). What concerns us here are the G1- and G2-type kites of Barge *et al.* (2015, Table 1), which are the main types found in the Jordanian *badia*. The G1 type is restricted to the south-east part of the *harrah*, including northern Saudi Arabia, while the G2 type is concentrated in the north-west part of the *harrah* with an extension north as far as the Palmyra range (Barge *et al.* 2015, Fig. 10; see also Helms and Betts 1987; Echallier and Braemer 1995). Despite some inevitable blurring around the edges, the combined distribution of these two kite types (see Fig. 9) is a remarkably good fit with the distribution of 'burin Neolithic' camps as described by Betts (1986, 259).

The question, then, is whether a cultural and chronological association can be inferred from the apparent collocation of 'burin Neolithic' camps and G1 and G2-type kites and, if so, whether camps and kites were constructed and used at the same time by a single population? More than three decades ago, Helms and Betts (1987, 54-55) argued for the construction and use of these kites in the Jordanian *badia* as early as the PPNB (see also Betts *et al.* 1998; Betts and Burke 2015). This was by no means universally accepted, with some subsequently going on to argue for a peak in their use and construction in the Negev and upper Mesopotamia during later millennia in the context of emergent and early urbanism (*e.g.* Holzer *et al.* 2010; Zeder *et al.* 2013; Rosen 2017, 142-143). The wide geographical distribution of kites, multiplicity of types, difficulty of dating them and enormously lengthy period of use, up to and including the nineteenth century (Betts 1989; Simpson 1994), has doubtless contributed to this lack of agreement. Recent data (*e.g.* Akkermans *et al.* 2014, 189-190; Morandi Bonacossi 2014, 37-38; Richter 2014, 23) and analyses (*e.g.* Kempe and Al-Malabeh 2013; Betts and Burke 2015) have however tended to bolster the argument that at least some, and possibly very many, kites were constructed in the *harrah* prior to the LN. This is tentatively supported by new radiometric dates from subtly different (Abu-Azizeh and Tarawneh 2015, 115-116) kite types in the southern limestone steppe that hint at a peak in the construction and use of these structures between the MPPNB and PPNC (but less so during the LN; Abu-Azizeh *et al.* 2016).

As "herding and hunter-gatherer lifestyles require contrasting choices in terms of site location and movement" (Betts *et al.* 2013, 188), the overlapping distribution of 'burin Neolithic' camps and G1- and G2-type kites is unlikely to be coincidental. Evidence from Dhuweila suggests that this particular kite system and its associated structures were constructed during the LPPNB, but were subsequently renovated and reutilised in the LN (Betts *et al.* 1998, 176 and 195-200). A strong case can therefore be made for a chronological association between 'burin Neolithic' camps and the secondary use of at least some kites. If the argument that the former were associated with sheep herding is accepted, it must be concluded that hunting and herding were both practiced in the *harrah* during the LN (cf. Martin 1994; 1999; Garrard *et al.* 1996; Betts *et al.* 2013). This is supported by the fact that most LN faunal assemblages, if not sites, in the region have yielded at least some caprine bones (Martin 1999). These are generally thought to represent herded animals, although Betts and Tarawneh (2010, 78) rightly caution against the assumption that these were necessarily close herded throughout the year. There is also a good case for a cultural association between 'burin Neolithic' camps and some kites. The chipped stone assemblage from the LN hunting camp of Dhuweila II, which has a stratified physical association with a kite system (Betts *et al.* 1998, 41 and Fig. 3.3) as well as a faunal assemblage consisting almost entirely of gazelle (*ibid.*, Table 8.2b), was assessed by Cropper (2011, 122-123 and Figs. 5.1-5.3) as being more closely affiliated with the *badia* reduction strategies that characterised the burin sites of Jebel Naja, Burqu' 03000 and Burqu' 35000 than with those of the Mediterranean zone.

The question of whether or not 'burin Neolithic' camps and hunting camps / kites were constructed and used at the same time by a single population is harder to answer. Whilst it has been demonstrated (see above) that both were likely to have been used by a single LN population, insofar as that is implied by a shared lithic technology and typology, the extent to which LN practice was to re-use already ancient kites or

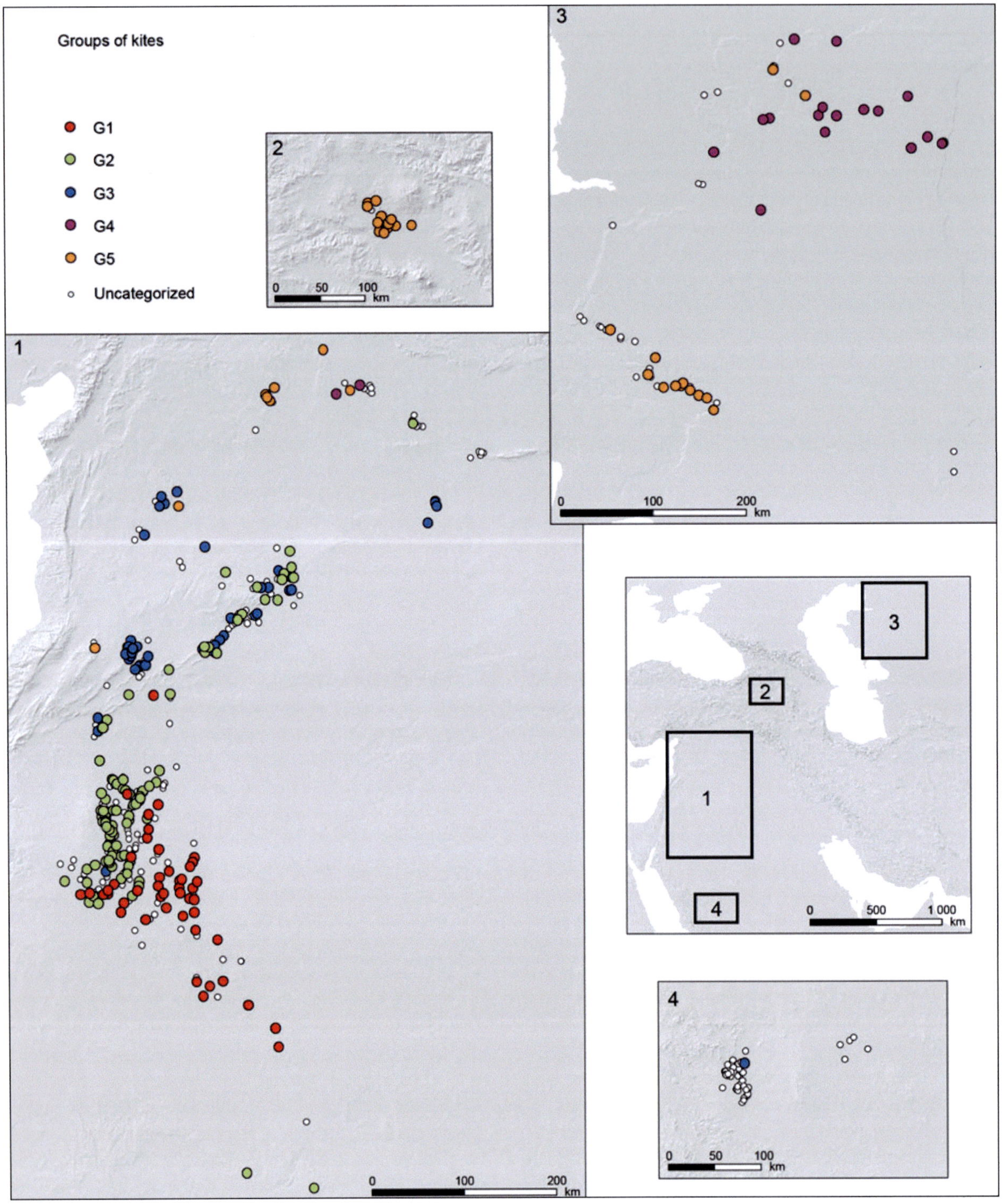

Figure 9. Distribution of G1- (red) and G2-type (green) kites (reproduced with permission from Barge *et al*. 2015, Fig. 10. © 2015, John Wiley and Sons).

construct new ones remains unknown at present, as does the frequency and intensity of that use. Further examination of this issue would be a valuable avenue of enquiry.

Discussion

An integrative approach: the Black Desert Neolithic

This article has, it is hoped, made the case that 'burin Neolithic' camps, 'Late Neolithic' stations, hunting camps and at least some desert kites represent an idiosyncratic combination of functionally specific elements within a single LN hunter-gatherer-herder cultural complex. This had its roots in traditions and practices that evolved in the northern limestone steppe and *harrah* following the cultural divergence of the former from the Mediterranean zone which started in the second quarter of the eighth millennium, some 1500 years earlier, and went on to become more rather than less apparent. Acknowledging the fact that the centre of gravity of this cultural complex, though not the ultimate extent of its dispersed elements, seems to have lain within the *harrah*, it is proposed to designate it, without prejudice to modern borders, the Black Desert Neolithic.

Professional shepherds, social hunters?

The designation of a single Black Desert Neolithic cultural complex in no way implies that it did not have multiple constituent parts. Betts *et al.* (2013, 188) noted that the inhabitants of the *badia* "quite likely represented distinct groups who may have had varying balances of economic strategies", although the possibility that such distinctions may have been based as much on social factors, such as lineage, should also be borne in mind. One objection to the argument for economic segmentation, *i.e.* that the users of kites and burin sites were separate groups within the same Black Desert Neolithic cultural complex, is that it is hard to reconcile with the fact that current evidence provides little support for the presence of a pre-LPPNB eighth-millennium hunter-gatherer population in the *harrah*. Unless the material remains of a substantive MPPNB hunter-gatherer population lie below the current threshold of archaeological visibility (a possibility, as surveys in the *harrah* have picked up occasional Jericho points; see *e.g.* Rowan *et al.* 2014, 283), a 'two-group' scenario would imply that 'hunters' (*e.g.* Dhuweila I) and 'herders' (*e.g.* Maitland's hill-fort MF-1; perhaps Burqu' 35000) both moved into the *harrah* at approximately the same time, during the LPPNB, occupied and elaborated overlapping territories in which they practiced competing economic strategies, and generated lithic assemblages in which the technological similarities outweigh the differences.

If the LN users of kites and burin sites were indeed one and the same group of people, it is pertinent to consider why they chose to invest considerable resources of time, energy and materiel in at least renovating and possibly constructing kites and their associated features when they already possessed an economic strategy, herding, with greater potential for intensification than hunting. Indeed, it is equally pertinent to consider why kites were constructed in the first place when evidence from late Epipalaeolithic, PPNA and E-MPPNB sites in the *badia* (Garrard *et al.* 1996, Fig. 11.3; Yeomans *et al.* 2017, 4) demonstrates that adequate technology with which to hunt gazelle already existed.

It may of course have been that earlier methods of gazelle hunting proved impossible to upscale to meet the demands of what appears by then to have been a rapidly expanding population (Rollefson *et al.* 2014, 298-299), particularly if (cf. Vigne and Helmer 2007) flocks were kept mainly for milk and hunting continued to be the primary source of meat (and possibly hides; Zeder *et al.* 2013, 118; Bar-Yosef 2016). Indeed, the addition of kite technology to the repertoire of earlier hunting techniques is much easier to understand if we assume that milking *was* practiced in the *harrah* during the LPPNB. The fact that kites would have facilitated the killing of large numbers of prey in a short period of time may have been of real benefit to herders whose primary focus would necessarily have been the welfare and productivity of their flocks, to say nothing of the temporal demands of milking and the processing of milk products (cf. Nieuwenhuyse *et al.* 2015, 62).

There are good grounds, also, for looking beyond purely economic motives for the construction and use of kites in the *harrah*. One possibility is that kites were constructed as much with social as with economic goals in mind (cf. Zeder *et al.* 2013, 122). Whilst the Black Desert Neolithic can in no way be considered an emergent urban society (but see Helms 1982), it would undoubtedly have faced its own challenges, social as well as economic, as a result of its rapid expansion. If the interpretation of 'burin Neolithic' camps as foci of seasonally dispersed sheep herding tethered to seasonally sedentary 'Late Neolithic' stations is accepted, maintenance of group identity and social cohesion under conditions of such fluidity may well have presented a problem.

One way of dealing with this challenge may have been through the communal construction of substantial stone structures, such as kites, and their use during seasonal episodes of macro-banding (Rosen 2008, 120). Indeed, the attraction of adding kites to the repertoire of earlier hunting techniques at this time of social intensification may have lain precisely in its communal nature. Fujii (2013) has documented the quite remarkable lengths that early pastoral communities in the southern limestone steppe went to over millennial time spans in order to maintain a cultural connection, however faint, with a dimly remembered pre-nomadic past. This was achieved through the medium of symbolic and funerary architecture that recalled elements of their last domestic structures. Within the very different milieu of the Black Desert Neolithic, oft-dispersed groups of herders may have sought to reinforce a cultural identity under stress through their own medium of

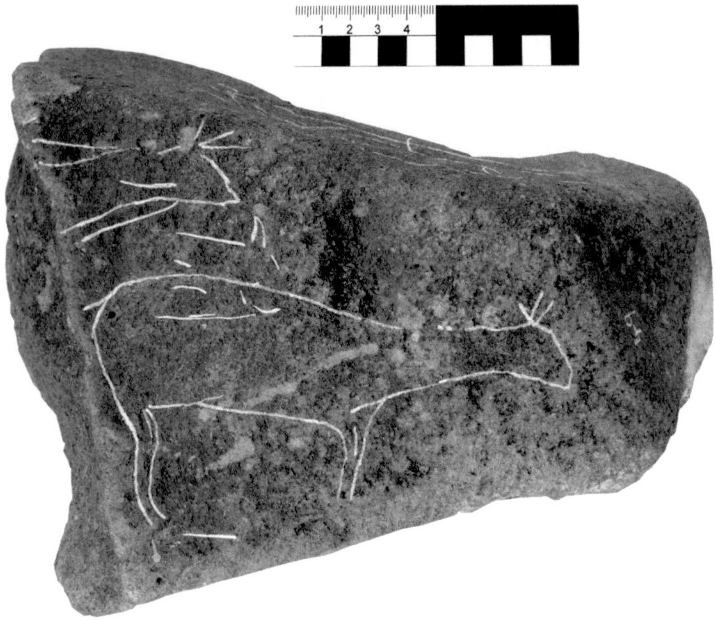

Figure 10. Incised rock carving depicting horned animals, probably gazelle, from Abu Masiad al-Sharqi near Dhuweila (courtesy of Prof. Alison Betts, reproduced with permission. Marks enhanced with water-soluble white ink for publication).

what may (though not necessarily; see Svizzero 2016, 65) have been spectacular mass kill events and subsequent communal feasting that hearkened back to and possibly even amplified their shared hunting heritage.

Of the three main domestic site types identified to date within the Black Desert Neolithic, *viz.* 'burin Neolithic' camps, 'Late Neolithic' stations and hunting camps (see Betts *et al.* 2013, 179-182 for a more detailed breakdown), 'burin Neolithic' camps appear to be by some margin the most numerous and widely distributed. That they are widely distributed is unsurprising in view of their proposed identification as dispersed herding camps. However, their very abundance also hints that herding may have played a much larger role in the economy of the Black Desert Neolithic than the rather limited representation of caprine bones in faunal assemblages might suggest (Martin 1994; 1999).

It is tempting to speculate whether herding may have been the steady pulse of these economies, punctuated by intensive episodes of gazelle hunting (cf. Simpson 1994, 79) whose social significance may have been just as important as their calorific contribution, if not more so. Under such conditions it may have been hunting rather than herding that referenced the social, if not the economic, identity of the Black Desert Neolithic. This is supported by the fact that the iconic rock carvings from Dhuweila (see Fig. 10), which probably date to the second half of the eighth millennium cal BC, appear to depict hunters and their prey rather than herders and their flocks (Betts *et al.* 1998, Ch. 7). Hodder and Meskell's (2011, 251) observation that, in Neolithic Turkey, "domestication occurred in the context of societies thoroughly engaged in building histories around their manipulation of non-domestic species" might equally apply to the exploitation of herd animals in the Black Desert Neolithic. In this sense, not only would kites have served as large and highly visible territorial statements of social identity in their own right ("elaboration of place"; Boyd 2006, 171), their permanent presence in stone may also have served the social function of 'fixing' (*ibid.*) the Black Desert Neolithic cultural complex within its landscape.

Conclusions

This article has suggested that 'burin Neolithic' camps might represent the herding element of the hunter-gatherer-herder society thought to have utilised the *harrah* and limestone steppe as far north as the Palmyra range (Akazawa 1979; Cauvin and Cauvin 1993, 26; Stordeur 1993; 2000) during the LN. On current evidence the lands of southern Jordan were not part of its core territory, although an extension to the south-east into the basalt country along the eastern side of the Wadi Sirhan basin seems likely (see Abu-Azizeh *et al.* 2016 for the southern limestone steppe). An integrated view of the archaeology is taken, in which 'burin Neolithic' camps, 'Late Neolithic' stations, hunting camps and desert kites are regarded as functionally specific elements of what is referred to here as the Black Desert Neolithic cultural complex. Chipped stone evidence suggests that this cultural complex was rooted in *badia* adaptations and practices that had diverged from those of the Mediterranean zone by the second quarter of the eighth millennium cal BC and remained distinct for the remainder of the Neolithic.

The *'domus'* of the Black Desert Neolithic is thought to be found in the 'Late Neolithic' stations of the *harrah*, seasonally sedentary base camps associated with intensive plant food economies and episodes of macro-banding. The latter are likely to have been of great importance in reinforcing notions of a shared social identity amongst an otherwise dispersed population. Participation in possibly intensive episodes of communal gazelle hunting utilising desert kites and hunting camps may have played a disproportionately significant role in referencing the social identity of the Black Desert Neolithic.

The absolute abundance of 'burin Neolithic' camps vis-à-vis other site types suggests that herding played a more significant role in the economy of the Black Desert Neolithic than the relative paucity of caprine remains in LN faunal assemblages might suggest. This strongly suggests (cf. Lancaster and Lancaster 1991, 135; Vigne and Helmer 2007) that milking may already have been a mainstay of the herding economy. An association with sheep herding in particular is suggested by an observed increase in the frequency of truncation burins *and* sheep bones at 'Ain Ghazal from the LPPNB onwards, although this in no way implies that sheep and goats were not herded together. This model implies that a proportion of the Black Desert Neolithic population would have been dispersed on transhumant 'burin Neolithic' camps that pushed out into the limestone steppe to the west, east and north (Helms' [1984] "land behind Damascus") of the *harrah* for some part(s) of the year, doubtless in order to exploit seasonally abundant grazing. As a result, establishing whether the Late Neolithic stations were abandoned at these times or whether they were occupied by another part of the population year-round will be a critical avenue of future research.

The encounter with flamingos in the desert in 2007 symbolised the environmental opportunities afforded by the presence of remarkably productive microenvironments in the broken landscape of the *harrah*. Sustained by 1500 years of relatively high annual rainfall, these appear to have supported a process of cultural evolution not unlike that which characterised the Early Natufian some six millennia earlier. Once again, this resulted in rapid population growth, generally higher levels of occupational intensity, intensifying plant food economies and increasing levels of social, if not quite physical sedentism. The presence of abundant and accessible building material in the *harrah* was doubtless also a powerful draw, serving to 'fix' the Black Desert Neolithic in its landscape. Indeed, in some instances (*e.g.*, the Wadi al-Qattafi *mesas*) it appears to have been the determining factor in influencing site location. This stands in sharp contrast to the situation in the southern limestone steppe, where absolutely lower rainfall seems to have encouraged a move in the direction of greater mobility, smaller group size and lower occupation intensity that was likely instrumental in the emergence of incipient pastoral nomadic economies (cf. Di Lernia 2006, 59-60).

In the sense that strong ties seem still to have bound them to 'Late Neolithic' stations, the occupants of burin sites *strictu senso* were therefore more transhumant herders than incipient nomads: very much in the world of pastoralism, but not of it.

Acknowledgments

Grateful thanks are due to Prof. Peter Akkermans and the Landscapes of Survival conference team, both for inviting us to Leiden and for their gracious hospitality whilst there. Prof. Alison Betts, Dr Joanne Clarke, Dr Andrew Garrard and Dr Louise Martin kindly commented on an inchoate draft of this article, which is greatly improved as a result of their counsel. Any remaining errors and omissions are however the sole responsibility of the authors. Thanks are also due to the many colleagues, students and friends who have made the long journey out to the *harrah* during Eastern Badia Archaeological Project field seasons and have contributed, through their observations and discussions, to the gestation of many of the ideas discussed in this article.

References

Abu-Azizeh, W. 2014. Stone enclosures and late prehistoric pastoral nomadic campsites: a methodological review of al-Thulaythuwat case study, southern Jordan. In: Finlayson, B. and Makarewicz, C. (eds). *Settlement, survey, and stone. Essays on Near Eastern prehistory in honour of Gary Rollefson*. Berlin: ex oriente, 187-208.

Abu-Azizeh, W. and Tarawneh, M. 2015. Out of the harra: desert kites in south-eastern Jordan. New results from the South Eastern Badia Archaeological Project. *Arabian Archaeology and Epigraphy* 26, 95-119.

Abu-Azizeh, W., Tarawneh, M., Sanchez-Priego, J., Crassard, R., Chambrade, M.-L. and Abu-Danah, F. 2016. *Desert kites and Neolithic hunter's campsites in Jibal al-Khashabiyeh area. New results of the South Eastern Badia Archaeological Project.* Paper presented at the 13th International Conference on the History and Archaeology of Jordan. Amman, Jordan.

Akazawa, T. 1979. Prehistoric occurrences and chronology in Palmyra basin, Syria. In: Hanihara, K. and Akazawa, T. (eds). *Paleolithic sites of Douara Cave and paleogeography of Palmyra basin in Syria*. Tokyo: University of Tokyo Press, 201-220.

Akkermans, P.M.M.G., Huigens, H.O. and Brüning, M. 2014. A landscape of preservation: late prehistoric settlement and sequence in the Jebel Qurma region, north-eastern Jordan. *Levant* 46, 186-205.

Albert, R.M. and Henry, D.O. 2004. Herding and agricultural activities at the early Neolithic site of Ayn Abu Nukhayla (Wadi Rum, Jordan). The results of phytolith and spherulite analyses. *Paléorient* 30, 81-92.

Al-Homoud, A., Allison, R., Sunna, B. and White, K. 1995. Geology, geomorphology, hydrology, groundwater and physical resources of the desertified Badia environment in Jordan. *GeoJournal* 37, 51-67.

Allison, R., Grove, J., Higgitt, D., Kirk, A., Rosser, N. and Warburton, J. 2000. Geomorphology of the eastern Badia basalt plateau, Jordan. *Geographical Journal* 166, 352-370.

Baird, D. 1993. *Neolithic chipped stone assemblages from the Azraq basin, Jordan and the significance of the Neolithic of the arid zones of the Levant*. Edinburgh: University of Edinburgh (unpublished PhD thesis).

Baird, D., Garrard, A.N., Martin, L. and Wright, K.I. 1992. Prehistoric environment and settlement in the Azraq basin: an interim report on the 1989 excavation season. *Levant* 24, 1-31.

Bar-Matthews, M. and Ayalon, A. 2004. Speleothems as paleoclimate indicators, a case study from Soreq cave located in the eastern Mediterranean region, Israel. In: Battarbee, R.W., Gasse, F. and Stickley, C.E. (eds). *Past climate variability through Europe and Africa*. Dordrecht: Kluwer Academic Publishers, 363-391.

Bar-Matthews, M., Ayalon, A., Kaufman, A. and Wasserburg, G. 1999. The eastern Mediterranean paleoclimate as a reflection of regional events: Soreq cave, Israel. *Earth and Planetary Science Letters* 166, 85-95.

Bar-Yosef, O. 2011. Climatic fluctuations and early farming in west and east Asia. *Current Anthropology* 52, Supplement 4, S175-S193.

Bar-Yosef, O. 2016. Changes in 'demand and supply' for mass killings of gazelles during the Holocene. In: Marom, N., Yeshuran, R., Weissbrod, L. and Bar-Oz, G. (eds). *Bones and identity: zooarchaeological approaches to reconstructing social and cultural landscapes in southwest Asia*. Oxford: Oxbow Books, 113-124.

Bar-Yosef, O. and Meadow, R.H. 1995. The origins of agriculture in the Near East. In: Price, T.D. and Gebauer, A.-B. (eds). *Last hunters, first farmers: new perspectives on the prehistoric transition to agriculture*. Santa Fe, NM: School of American Research Press, 39-94.

Barge, O., Brochier, J.E. and Crassard, R. 2015. Morphological diversity and regionalisation of kites in the Middle East and Central Asia. *Arabian Archaeology and Epigraphy* 26, 162-176.

Betts, A.V.G. 1986. *The prehistory of the basalt desert, Transjordan: an analysis*. London: University of London (unpublished PhD thesis).

Betts, A.V.G. 1987. Recent discoveries relating to the Neolithic periods in eastern Jordan. *Studies in the History and Archaeology of Jordan* 3, 225-230.

Betts, A.V.G. 1989. The Solubba: nonpastoral nomads in Arabia. *Bulletin of the American Schools of Oriental Research* 274, 61-69.

Betts, A.V.G. 1992. Eastern Jordan: economic choices and site location in the Neolithic periods. *Studies in the history and archaeology of Jordan* 4, 111-114.

Betts, A.V.G. 2008. Things to do with sheep and goats: Neolithic hunter-forager-herders in north Arabia. In: Barnard, H. and Wendrich, W. (eds). *The archaeology of mobility: Old World and New World nomadism*. Los Angeles, CA: Cotsen Institute of Archaeology, UCLA, 25-42.

Betts, A.V.G. and Burke, D. 2015. Desert kites in Jordan: a new appraisal. *Arabian Archaeology and Epigraphy* 26, 74-94.

Betts, A.V.G. and Helms, S.W. 1987. A preliminary survey of Late Neolithic settlements at el-Ghirqa, eastern Jordan. *Proceedings of the Prehistoric Society* 53, 327-336.

Betts, A.V.G. and Tarawneh, M. 2010. Changing patterns of land use and subsistence in the Badiyat al-Sham in the Late Neolithic and Chalcolithic periods: new data from Burqu and Bayir. In: Dentzer-Feydy, J. and Vallerin, M. (eds). *Hauran V: La Syrie du Sud du Néolithique à l'Antiquite tardive*. Beirut: Presses de l'IFPO, 69-80.

Betts, A.V.G., Colledge, S., Martin, L., McCartney, C., Wright, K.I., Yagodin, V., Cooke, L., Garrard, A.N., Hather, J., McClintock, C., Lancaster, W. and Reese, D. 1998. *The harra and the hamad: excavations and surveys in eastern Jordan*. Vol. 1. Sheffield: Sheffield Academic Press.

Betts, A.V.G., Cropper, D., Martin, L., McCartney, C. 2013. *Later prehistory of the badia: excavation and surveys in eastern Jordan*. Vol. 2. Oxford: Oxbow Books.

Boyd, B. 2006. On 'sedentism' in the later Epipalaeolithic (Natufian) Levant. *World Archaeology* 38, 164-178.

Byrd, B.F. 1992. The dispersal of food production across the Levant. In: Gebauer, A.-B. and Price, T.D. (eds). *Transitions to agriculture in prehistory*. Madison, WI: Prehistory Press, 49-61.

Cauvin, M.-C. and Cauvin, J. 1993. La séquence néolithique PPNB au Levant nord. *Paléorient* 19, 23-28.

Croft, P. 1991. Man and beast in Chalcolithic Cyprus. *Bulletin of the American Schools of Oriental Research* 282-283, 63-79.

Croft, P. 2002. Game management in early prehistoric Cyprus. *Zeitschrift für Jagdwissenschaft* 48 (Supplement), 172-179.

Cropper, D. 2006. *Bridging the gap between the Mediterranean and the steppe: lithic technology in Late Neolithic Jordan*. Sydney: University of Sydney (unpublished PhD thesis).

Cropper, D. 2011. *Lithic technology and regional variation in Late Neolithic Jordan*. Oxford: Archaeopress.

Di Lernia, S. 2006. Building monuments, creating identity: cattle cult as a social response to rapid environmental changes in the Holocene Sahara. *Quaternary International* 151, 50-62.

Driesch, A. von den and Wodtke, U. 1997. The fauna of 'Ain Ghazal, a major PPN and Early PN settlement in central Jordan. In: Gebel, H.G.K., Kafafi, Z. and Rollefson, G.O. (eds). *The prehistory of Jordan, II. Perspectives from 1997*. Berlin: ex oriente, 511-556.

Ducos, P. 1993. Proto-élevage et élevage au Levant sud au VIIe millénaire b.c.: les données de la Damascène. *Paléorient* 19, 153-173.

Dutton, R., Clarke, J. and Battikhi, A. (eds). 1998. *Arid land resources and their management: Jordan's desert margin*. London: Kegan Paul International.

Echallier, J.C. and Braemer, F. 1995. Nature et fonctions des 'desert kites': données et hypothèses nouvelles. *Paléorient* 21, 35-63.

Edwards, P.C. 2016. The chronology and dispersal of the Pre-Pottery Neolithic B cultural complex in the Levant. *Paléorient* 42, 53-72.

Field, H. 1960. *North Arabian desert archaeological survey, 1925-50*. Papers of the Peabody Museum of Archaeology and Ethnology, Harvard University 45/2. Cambridge, MA: Peabody Museum, Harvard University.

Finlayson, B. and Betts, A.V.G. 1990. Functional analysis of chipped stone artefacts from the Late Neolithic site of Gabal Na'ja, eastern Jordan. *Paléorient* 16, 13-20.

Fujii, S. 2013. Chronology of the Jafr prehistory and protohistory: a key to the process of pastoral nomadization in the southern Levant. *Syria* 90, 49-125.

Fujii, S. 2014. A half-buried cistern at Wadi Abu Tulayha: a key to tracing the pastoral nomadization in the Jafr basin, southern Jordan. In: Jamhawi, M. (ed.). *Jordan's prehistory: past and future research*. Amman: Department of Antiquities, 159-167.

Fujii, S., Adachi, T., Yamafuji, M. and Nagaya, K. 2017a. Khashm al-'Arfa: an Early Neolithic encampment in the eastern Jafr basin, southern Jordan. *Annual of the Department of Antiquities of Jordan* 58, 567-584.

Fujii, S., Adachi, T., Yamafuji, M. and Nagaya, K. 2017b. A preliminary report on the Neolithic barrage surveys in the eastern Jafr basin, 2013-2014. *Annual of the Department of Antiquities of Jordan* 58, 585-599.

Garrard, A.N. and Byrd, B.F. 2013. *Beyond the Fertile Crescent: Late Palaeolithic and Neolithic communities of the Jordanian steppe*. Oxford: Oxbow Books.

Garrard, A.N., Byrd, B.F. and Betts, A.V.G. 1986. Prehistoric environment and settlement in the Azraq basin: an interim report on the 1984 excavation season. *Levant* 18, 5-24.

Garrard, A.N., Colledge, S. and Martin, L. 1996. The emergence of crop cultivation and caprine herding in the "marginal zone" of the southern Levant. In: Harris, D. (ed.). *The origins and spread of agriculture and pastoralism in Eurasia*. London: UCL Press, 204-226.

Garrard, A.N., Betts, A.V.G., Byrd, B.F. and Hunt, C. 1987. Prehistoric environment and settlement in the Azraq basin: an interim report on the 1985 excavation season. *Levant* 19, 5-25.

Garrard, A.N., Baird, D., Colledge, S., Martin, L. and Wright, K.I. 1994. Prehistoric environment and settlement in the Azraq basin: an interim report on the 1987 and 1988 excavation seasons. *Levant* 26, 73-109.

Garrard, A.N., Betts, A.V.G., Byrd, B.F., Colledge, S. and Hunt, C. 1988. Summary of palaeoenvironmental and prehistoric investigations in the Azraq basin. In: Garrard, A.N. and Gebel, H.G.K. (eds). *The prehistory of Jordan*. Oxford: British Archaeological Reports (International Series 396).

Garrod, D. 1960. The flint implements. In: Field, H. *North Arabian desert archaeological survey, 1925-50*. Papers of the Peabody Museum of Archaeology and Ethnology, Harvard University 45/2. Cambridge, MA: Peabody Museum, Harvard University, 111-131.

Garstang, J. 1935. Jericho: city and necropolis (fifth report). *Liverpool Annals of Archaeology and Anthropology* 22, 143-184.

Gebel, H.G.K. and Mahasneh, H. 2013. Disappeared by climate change: the shepherd cultures of Qulban Beni Murra (2nd half of the 5th millennium BC) and their aftermath. *Syria* 90, 127-158.

Gopher, A. 1994. *Arrowheads of the Neolithic Levant*. Winona Lake, IN: Eisenbrauns.

Hardy-Smith, T. and Edwards, P.C. 2004. The garbage crisis in prehistory: artefact discard patterns at the Early Natufian site of Wadi Hammeh 27 and the origins of household refuse disposal strategies. *Journal of Anthropological Archaeology* 23, 253-289.

Helmer, D., Gourichon, L. and Vila, E. 2007. The development of the exploitation of products from Capra and Ovis (meat, milk and fleece) from the PPNB to the Early Bronze in the northern Near East (8700 to 2000 BC cal.). *Anthropozoologica* 42, 41-69.

Helms, S.W. 1982. Paleo-beduin and transmigrant urbanism. *Studies in the History and Archaeology of Jordan* 1, 97-113.

Helms, S.W. 1984. The land behind Damascus: urbanism during the 4th millennium in Syria/Palestine. In: Khalidi, T. (ed.). *Land tenure and social transformation in the Near East*. Beirut: American University of Beirut, 15-31.

Helms, S.W. and Betts, A.V.G. 1987. The desert 'kites' of the Badiyat esh-Sham and north Arabia. *Paléorient* 13, 41-67.

Hodder, I. 1982. *Symbols in action: ethnographic studies of material culture*. Cambridge: Cambridge University Press.

Hodder, I. 1990. *The domestication of Europe. Structure and contingency in Neolithic societies*. Oxford: Blackwell.

Hodder, I. and Meskell, L. 2011. A "curious and sometimes a trifle macabre artistry." Some aspects of symbolism in Neolithic Turkey. *Current Anthropology* 52, 235-263.

Holzer, A., Avner, U., Porat, N. and Horwitz, L.K. 2010. Desert kites in the Negev desert and northeast Sinai:

their function, chronology and ecology. *Journal of Arid Environments* 74, 806-817.

Hongo, H., Omar, L., Nasu, H. and Fujii, S. 2013. Faunal remains from Wadi Abu Tulayha: a PPNB outpost in steppe desert of southern Jordan. In: De Cupere B., Linseele, V. and Hamilton-Dyer, S. (eds). *Archaeozoology of the Near East X: Proceedings of the Tenth International Symposium on the Archaeozoology of South-Western Asia and Adjacent Areas.* Leuven: Peeters Publishers, 1-25.

Ingold, T. 2000. *The perception of the environment.* London: Routledge.

Kempe, S. and Al-Malabeh, A. 2013. Desert kites in Jordan and Saudi Arabia: structure, statistics and function, a Google Earth study. *Quaternary International* 297, 126-146.

Kennedy, D. 2009. Desktop archaeology. *Saudi Aramco World* 60/4, 3-9.

Kennedy, D. 2011. The 'Works of the Old Men' in Arabia: remote sensing in interior Arabia. *Journal of Archaeological Science* 38, 3185-3203.

Kennedy, D., Banks, R. and Houghton, P. 2014. *Kites in Arabia.* Apple iBook.

Köhler-Rollefson, I. 1988. The aftermath of the Levantine Neolithic Revolution in the light of ecological and ethnographic evidence. *Paléorient* 14, 87-93.

Köhler-Rollefson, I. 1992. A model for the development of nomadic pastoralism on the Transjordanian plateau. In: Bar-Yosef, O. and Khazanov, A. (eds). *Pastoralism in the Levant. Archaeological materials in anthropological perspectives.* Madison, WI: Prehistory Press, 11-18.

Köhler-Rollefson, I. 1997. Proto-élevage, pathologies, and pastoralism: a post-mortem of the process of goat domestication. In: Gebel, H.G.K., Kafafi, Z. and Rollefson, G.O. (eds). *The prehistory of Jordan, II. Perspectives from 1997.* Berlin: ex oriente, 557-565.

Lancaster, W. and Lancaster, F. 1991. Limitations on sheep and goat herding in the eastern badia of Jordan: an ethno-archaeological enquiry. *Levant* 23, 125-138.

Lancaster, W. and Lancaster, F. 1999. *People, land and water in the Arab Middle East: environments and landscapes in the Bilâd ash-Shâm.* Amsterdam: Harwood Academic Publishers.

Makarewicz, C. 2014. Bridgehead to the badia: a biometric and isotopic perspective of caprine husbandry at Pre-Pottery Neolithic 'Ain Ghazal. In: Finlayson, B. and Makarewicz, C. (eds). *Survey, settlement, and stone: essays in honour of Gary Rollefson.* Berlin: ex oriente, 117-131.

Maitland, R.A. 1927. The 'Works of the Old Men' in Arabia. *Antiquity* 1, 197-203.

Manning, S.W. 2013. Appendix: a new radiocarbon chronology for prehistoric and protohistoric Cyprus, ca. 11,000-1050 cal BC. In: Knapp, A.B. *The archaeology of Cyprus: from earliest prehistory through the Bronze Age.* Cambridge: Cambridge University Press, 485-533.

Martin, L. 1994. *Hunting and herding in a semi-arid region: an archaeozoological and ethological analysis of the faunal remains from the Epipalaeolithic and Neolithic of the eastern Jordanian steppe.* Sheffield: University of Sheffield (unpublished PhD thesis).

Martin, L. 1999. Mammal remains from the eastern Jordanian Neolithic, and the nature of caprine herding in the steppe. *Paléorient* 25, 87-104.

Martin, L. and Edwards, Y. 2013. Diverse strategies: evaluating the appearance and spread of domestic caprines in the southern Levant. In: Colledge, S., Conolly, J., Dobney, K., Manning, K. and Shennan, S. (eds). *The origins and spread of domestic animals in southwest Asia and Europe.* Walnut Creek, CA: Left Coast Press, 49-82.

McCartney, C. 1996. *Analysis of variability in simple core technologies: case studies of chipped stone technology in post-PPN assemblages from the Levant.* Edinburgh: University of Edinburgh (unpublished PhD thesis).

Morandi Bonacossi, D. 2014. Desert-kites in an aridifying environment. Specialised hunter communities in the Palmyra steppe during the Middle and Late Holocene. In: Morandi Bonacossi, D. (ed.). *Settlement dynamics and human-landscape interaction in the dry steppes of Syria.* Wiesbaden: Harrassowitz Verlag, 33-47.

Munro, N.D. 2004. Zooarchaeological measures of hunting pressure and occupation intensity in the Natufian: implications for agricultural origins. *Current Anthropology* 45, Supplement, S5-S33.

Nieuwenhuyse, O.P., Roffet-Salque, M., Evershed, R., Akkermans, P.M.M.G. and Russell, A. 2015. Tracing pottery use and the emergence of secondary product exploitation through lipid residue analysis at Late Neolithic Tell Sabi Abyad (Syria). *Journal of Archaeological Science* 64, 54-66.

Perrot, J. 1993. Rémarques introductives. *Paléorient* 19, 9-21.

Quintero, L., Wilke, P. and Rollefson, G.O. 2004. Highland towns and desert settlements: origins of nomadic pastoralism in the Jordanian Neolithic. In: Bienert, H., Gebel, H.G.K. and Neef, R. (eds). *Central settlements in Neolithic Jordan.* Berlin: ex oriente, 201-213.

Rees, L.W.B. 1929. The Transjordan desert. *Antiquity* 3, 389-407.

Richter, T. 2014. Rescue excavations at a Late Neolithic burial cairn in the east Jordanian badya. *Neo-Lithics* 1/14, 18-24.

Richter, T., Arranz-Otaegui, A., Boaretto, E., Bocaege, E., Estrup, E., Martinez-Gallardo, C., Pantos, G.A., Pedersen, P., Sæhle, I. and Yeomans, L. 2016. Shubayqa 6: a new Late Natufian and Pre-Pottery Neolithic A settlement in north-east Jordan. *Antiquity* 90, 1-5.

Rollefson G.O. 2011. The greening of the badlands: pastoral nomads and the "conclusion" of Neolithization in the southern Levant. *Paléorient* 37, 101-109.

Rollefson, G.O., Rowan, Y. and Wasse, A.M.R. 2014. The Late Neolithic colonization of the eastern badia of Jordan. *Levant* 46, 285-301.

Rollefson, G.O., Simmons, A. and Kafafi, Z. 1992. Neolithic cultures at 'Ain Ghazal, Jordan. *Journal of Field Archaeology* 19, 443-470.

Rollefson, G., Rowan, Y., Wasse, A., Hill, A., Kersel, M., Lorentzen, B., Al-Bashaireh, K. and Ramsay, J. 2016. Investigations of a Late Neolithic structure at Mesa 7, Wadi al-Qattafi, Black Desert, 2015. *Neo-Lithics* 1/16, 3-12.

Rolston, S. and Rollefson, G.O. 1982. The Wadi Bayir paleoanthropological survey. *Annual of the Department of Antiquities of Jordan* 26, 211-219.

Rosen, S. 2008. Desert pastoral nomadism in the longue durée: a case study from the Negev and the southern Levantine deserts. In: Barnard, H. and Wendrich, W. (eds). *The archaeology of mobility: Old World and New World nomadism*. Los Angeles, CA: Cotsen Institute of Archaeology, UCLA, 115-140.

Rosen, S. 2017. *Revolutions in the desert: the rise of mobile pastoralism in the southern Levant*. London: Routledge.

Rowan, Y., Wasse, A.M.R. and Rollefson, G.O. 2014. Survey of Maitland's Mesa: a basalt prominence in the eastern badia of Jordan. In: Jamhawi, M. (ed.). *Jordan's prehistory: past and future research*. Amman: Department of Antiquities, 277-284.

Rowan, Y., Rollefson, G.O., Wasse, A.M.R., Hill, A. and Kersel, M. 2017. The Late Neolithic presence in the Black Desert. *Near Eastern Archaeology* 80, 102-113.

Saña, M. and Tornero, C. 2012. Use of animal fibres during the Neolithisation in the middle Euphrates valley (Syria): an archaeozoological approach. *Paléorient* 38, 79-91.

Sherratt, A. 1981. Plough and pastoralism: aspects of the secondary products revolution. In: Hodder, I., Isaac, G. and Hammond, N. (eds). *Pattern of the past: studies in honour of David Clarke*. Cambridge: Cambridge University Press, 261-305.

Simpson, St J. 1994. Gazelle-hunters and salt-collectors; a further note on the Solubba. *Bulletin of the American Schools of Oriental Research* 293, 79-81.

Stordeur, D. 1993. Sédentaires et nomades du PPNB final dans le désert de Palmyre (Syrie). *Paléorient* 19, 187-204.

Stordeur, D. 2000. *El Kowm 2. Une île dans le désert. La fin du Néolithique précéramique dans la steppe syrienne*. Paris: CNRS.

Stordeur, D., Helmer. D., Jamous, B., Khawam, R., Molist, M. and Willcox, G. 2010. Le PPNB de Syrie du sud à travers les découvertes récentes à Tell Aswad. In: Dentzer-Feydy, J. and Vallerin, M. (eds.). *Hauran V: La Syrie du Sud du Néolithique à l'Antiquite tardive*. Beirut: Presses de l'IFPO, 41-67.

Svizzero, S. 2016. Hunting strategies with cultivated plants as bait and the prey pathway to animal domestication. *International Journal of Research in Sociology and Anthropology* 2, 53-68.

Tarawneh, M. and Abudanah, F. 2013. Subsistence of early pastoral nomadism in the southern Levant: new data from eastern Bayir. *Syria* 90, 231-252.

Vigne, J.-D. 2011. The origins of animal domestication and husbandry: a major change in the history of humanity and the biosphere. *Comptes Rendus Biologies* 334, 171-181.

Vigne, J.-D. and Helmer, D. 2007. Was milk a "secondary product" in the Old World Neolithisation process? Its role in the domestication of cattle, sheep and goats. *Anthropozoologica* 42, 9-40.

Vigne, J.-D., Carrère, I. and Guilaine, J. 2003 Unstable status of early domestic ungulates in the Near East: the example of Shillourokambos (Cyprus, IX-VIIIth millennia cal. B.C.). In: Guilaine, J. and Le Brun, A. (eds.). *Le Néolithique de Chypre*. Paris: Bulletin de Correspondance Hellénique, Supplément 43, 239-251.

Waechter, J. 1947. Some surface implements from Transjordan. *Proceedings of the Prehistoric Society* 13, 178-180.

Waechter, J., Seton-Williams, V.M., Bate, D.A. and Picard, L. 1938. The excavations at Wadi Dhobai, 1937-38 and the Dhobaian industry. *Journal of the Palestine Oriental Society* 18, 172-186, 292-298.

Wasse, A.M.R. 2000. *The development of goat and sheep herding during the Levantine Neolithic*. London: University of London (unpublished PhD thesis).

Wasse, A.M.R. 2002. Final results of an analysis of the sheep and goat bones from 'Ain Ghazal, Jordan. *Levant* 34, 59-82.

Wasse, A.M.R. 2007. Climate, economy and change. In: Clarke, J., McCartney, C. and Wasse, A.M.R. *On the margins of south-west Asia: Cyprus during the 6th to 4th millennia BC*. Oxford: Oxbow Books, 43-63.

Wasse, A.M.R. and Rollefson, G.O. 2005 The Wadi Sirhan Project: report on the 2002 archaeological reconnaissance of Wadi Hudruj and Jabal Tharwa, Jordan. *Levant* 37, 1-20.

Weiss, E. and Zohary, D. 2011. The Neolithic southwest Asian founder crops: their biology and archaeobotany. *Current Anthropology* 52, Supplement 4, S237-S254.

Wilke, P. and Quintero, L. 1998. New Late Pre-Pottery Neolithic B sites in the Jordanian desert. *Neo-Lithics* 1/98, 2-4.

Wright, K.I., Critchley, P., Garrard, A.N., Baird, D., Bains, R. and Groom, S. 2008. Stone bead technologies and early craft specialization. Insights from two Neolithic sites in eastern Jordan. *Levant* 40, 131-165.

Yeomans, L., Martin, L. and Richter, T. 2017. Expansion of the known distribution of Asiatic mouflon (Ovis orientalis) in the Late Pleistocene of the southern Levant. *Royal Society Open Science* 4, 1-15.

Zeder, M.A., 2011. The origins of agriculture in the Near East. *Current Anthropology* 52, Supplement 4, S221-S235.

Zeder, M., Bar-Oz, G., Rufolo, S.J. and Hole, F. 2013. New perspectives on the use of kites in mass-kills of Levantine gazelle: a view from northeastern Syria. *Quaternary International* 297, 110-125.

Pastoralists of the southern Nefud desert: inter-regional contact and local identity

Maria Guagnin

Abstract

The recent excavation of three Neolithic hearth sites in the Nefud desert of north-western Saudi Arabia allows a first comparison with settlement and subsistence strategies in the marginal areas of the southern Levant. The latter are characterised by clusters of curvilinear dwellings and animal corrals, and the complete absence of architectural remains at Arabian hearth sites is therefore striking. Faunal remains and ephemeral fireplaces suggest repeated short-term occupation by Neolithic pastoralists, which may often only have lasted mere hours and occasionally stretched to days and perhaps even weeks. Hearth sites are common across the sand seas of the Arabian interior and may represent an adaptation to these marginal environments. Their frequency and distribution suggest that hearth sites were not a short-lived phenomenon but formed a substantial part of Neolithic settlement patterns in Arabia. All known Neolithic sites in the Nefud desert show evidence for contact with the Levant. However, lithic assemblages show distinct local traditions, and the distribution of 'Jubbah style' rock art and of 'gate' stone structures allows a tentative identification of local Neolithic traits. The distribution of this local Neolithic may relate to the rainfall patterns of the Holocene humid period, when the northern extent of the African summer monsoon provided an ecological connection with the south-west. The Neolithic herders of north-western Saudi Arabia may thus have been part of a mobility pattern that was driven by the availability of water and pastures and extended between the Jubbah oasis and the eastern Khaybar.

Keywords: pastoralism, Neolithic, settlement, hearth, rock art, gates, desert kites

Introduction

The Neolithic of northern Arabia remains poorly known. Few sites have been excavated on the Arabian Peninsula, and sites with faunal remains are predominantly known from the Arabian east coast and from Yemen (Fig. 1). Evidence from these coastal sites suggests that between 6800 and 6200 BC, domesticated cattle, sheep and goat were introduced to the Arabian Peninsula from the Levant (Drechsler 2007). However, known sites are located at a distance of over 1000 km from comparable sites in the Levant. The recent discovery of three Neolithic occupation sites in the Nefud desert now, for the first time, provides some information on subsistence strategies and settlement patterns in an area that links the Levant with the rest of the Arabian Peninsula. This allows a first assessment of the extent to which we can identify local adaptations to the environment and landscapes of the Nefud desert, and contact or exchange of ideas with the Levant. Neolithic sites in the eastern *badia* and

In: Peter M. M. G. Akkermans (ed.) 2020: *Landscapes of Survival - The Archaeology and Epigraphy of Jordan's North-Eastern Desert and Beyond*, Sidestone Press (Leiden), pp. 103-116.

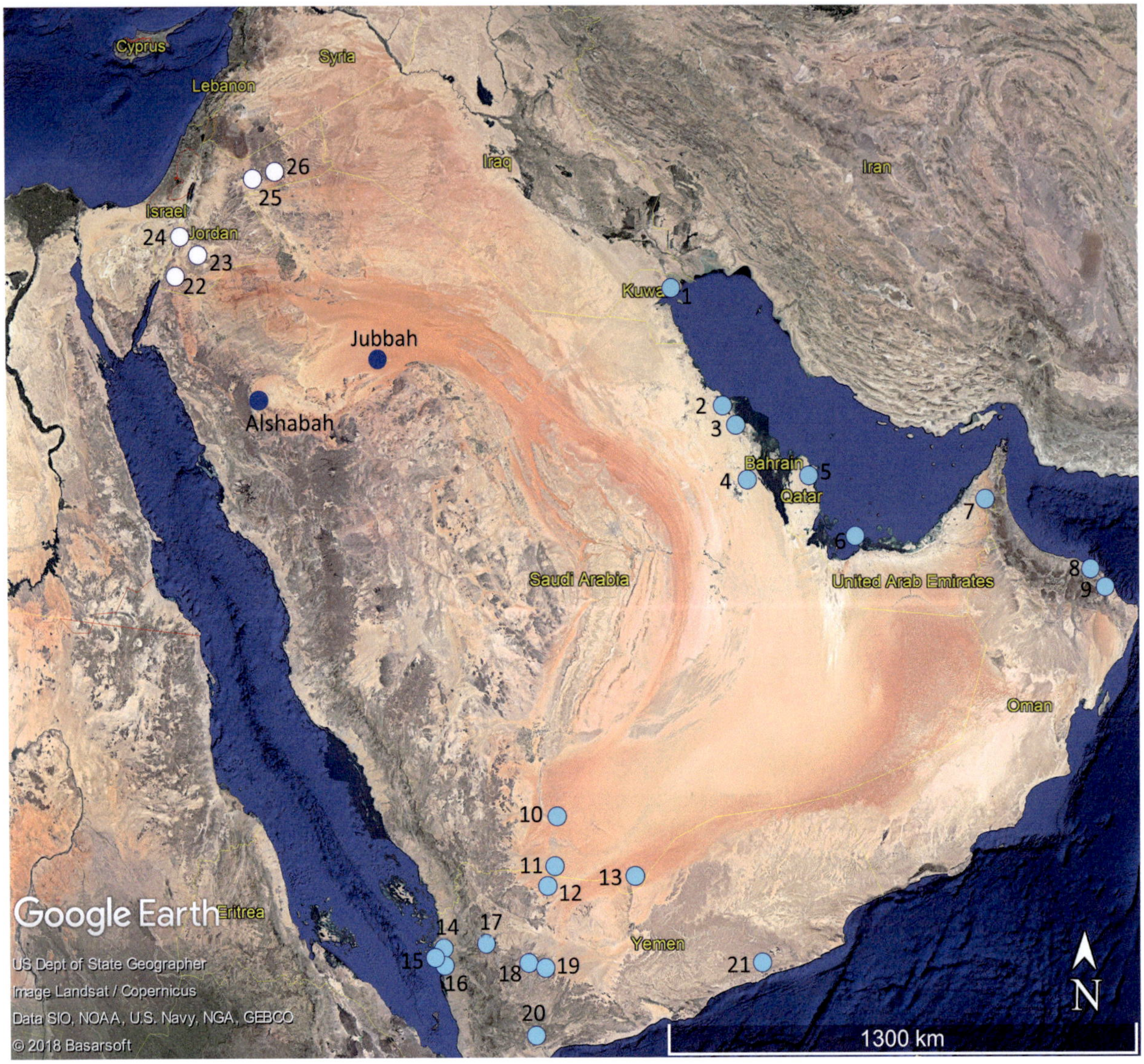

Figure 1. White dots: key sites in southern and eastern Jordan discussed in the text. Turquoise dots: Neolithic sites with faunal remains on the Arabian Peninsula (as listed by Drechsler 2009, 32). Blue dots: excavated Neolithic occupation sites with faunal remains in northern Arabia. 1: H3. 2: Abu Khamis. 3: Dosariyah. 4: Ain Qannas. 5: Khor 2. 6: Dalma 11. 7: Buhais 18. 8: Ras al-Hamra. 9: Khor Milkh. 10: Jiledah. 11: Janub al-Mutabthat. 12: Shaqqat el-Khariyta. 13: Sharorah. 14: Surud1. 15: ash-Shumah. 16: JHB1. 17: Ak5. 18: WTHiii. 19: HARiii. 20: MK 2. 21: Khuzma as-Shumlya. 22: Ayn Abu Nukhayla. 23: Wadi Abu Tulayha. 24: Faynan. 25: Azraq. 26: Wadi al-Qattafi.

in southern Jordan can be used as case studies to compare settlement and subsistence strategies adapted to marginal environments in the Levant with sites in the Nefud desert. In addition, rock art, and particularly the depiction of human figures can be used to explore local identities. Moreover, recent large-scale documentation of stone structures such as desert kites and gates from satellite imagery (Barge *et al.* 2015; Kennedy 2017) now makes it possible to compare their distribution with the location of rock art sites and allows us to link cultural markers with broader patterns of landscape use in the Neolithic of northern Arabia.

Settlements and subsistence north of the Nefud desert

Eastern badia

In the Fertile Crescent, sedentism, resource abundance, and population density are thought to have been some of the most important factors in the transition from foraging to farming (Price and Bar-Yosef 2011). In the ecologically more marginal areas, considerable temporal and spatial variation in the uptake of plant cultivation and herding can thus be observed in the archaeological record. During the

Pre-Pottery Neolithic B (PPNB) the steppe of Jordan's eastern *badia* was mostly populated by hunter-gatherers, although first steps towards the herding of sheep and goat may have been taken during this period. Pastoralism was probably not wide spread until the Late Neolithic (6250-5300 BC) and, coupled with a diversification of subsistence strategies, allowed a more intensive exploitation of the *badia*. Neolithic communities followed a relatively mobile lifestyle, where village-like clusters of substantial dwellings were occupied seasonally, and herding was supplemented by hunting and opportunistic agriculture (Betts *et al.* 2013; Henry *et al.* 2003; Martin and Edwards 2013; Rollefson *et al.* 2014; 2016b). In the Bronze Age, there is a drastic reduction in the number of sites, although Betts and colleagues (2013) suggest that this lack of visibility may partly be caused by a lack of characteristic artefacts.

Throughout the Neolithic, sites are characterised by structural remains such as dwellings, animal pens, storage facilities, and funerary features. Residential structures are generally built in drystone walls, sometimes supported with upright slabs at the base, and form sub-circular, irregularly shaped small rooms that are often arranged in a honeycomb pattern (Betts 1993). Only a single site was reported not to have traces of structures. At Azraq 31, a series of superimposed hearths was covered by a crude pavement of angular pebbles, dated to around 7300 cal BC. Faunal remains from this site show a high proportion of gazelle, in addition to caprines and remains of cultivated wheat and barley (Betts 1989). Recent research at Wadi al-Qattafi in the Black Desert of eastern Jordan also showed that many stone structures that were thought to be mortuary features are in fact corbelled Neolithic houses, many with animal corrals attached to them (Rollefson *et al.* 2016b; Rowan *et al.* 2015; see also Rowan *et al.*, this volume). In general, Neolithic residential structures in eastern Jordan seem to have been built as permanent structures, and were probably re-occupied in cyclical transhumance movements (Betts 1989; Rollefson *et al.* 2016b). Spatial analysis of over 9000 clustered enclosures, dating from the Late Neolithic to the present, has recently confirmed this pattern of seasonal movement. Meister and colleagues were able to show that the distribution of clustered enclosures appears to reflect a movement east/south towards the open landscapes of the *hamad* in late autumn, returning to the Azraq basin in spring. This pattern may have persisted since the Late Neolithic and is thought to have been driven by the seasonal availability of water and pastures (Meister *et al.* 2019).

Southern Jordan

Evidence from Ayn Abu Nukhayla and Wadi Abu Tulayha in southern Jordan indicates that Desert Neolithic groups began to herd sheep and goats as early as 7500 BC (Henry *et al.* 2017; Fujii 2010). Moister conditions also enabled the cultivation of cereals, which together with caprine herding, foraging and trade provided a broad economic base that was coupled with seasonal movements to the wetter uplands of the Ma'an plateau during the dry season (Henry *et al.* 2017). Sites are characterised by curvilinear structures, often arranged in a honeycomb pattern, and in some cases paving and even plaster are preserved on the floor. At Wadi Abu Tulayha, the excavation of a possible cistern suggests that early Neolithic groups may have already employed advanced methods of water management (Fujii 2010; Henry *et al.* 2003).

In southern Jordan, pastoral populations of the Chalcolithic are linked with the so-called 'Timnian complex', which stretches across the Negev desert and into Sinai (Abu-Azizeh 2013). From the late sixth millennium BC, pastoral groups are thought to have followed a central-based transhumance pattern where groups of several families aggregated into longer-term camps during the winter, and dispersed into smaller, ephemeral camps during the dry season in the summer. Both long-term and ephemeral sites consist of the familiar curvilinear structures clustered in honeycomb layouts. Clusters include houses, pens, and storage facilities, although dwellings tend to be larger (*ibid.*; Henry *et al.* 2017). This period coincided with an increase in sites and population, although stocking rates remained low, below the carrying capacity of the local landscape, until the Early Bronze Age (Henry *et al.* 2017). Characteristic for this period is the large-scale production and trade of cortical flakes for the production of tabular scrapers, which may have been used in the shearing of sheep (Abu-Azizeh 2013; Henry *et al.* 2017). Further east, on the northern margins of the Nefud desert, Chalcolithic sites at Rajajil, dating to the fifth millennium BC, are also characterised by curvilinear animal pens and domestic structures, and remains of wells, troughs and canals attest to the use of advanced water management strategies (Gebel 2016).

Settlements and subsistence in the Jubbah oasis and the Nefud desert

The Jubbah oasis is located in north-western Saudi Arabia, on the southern edge of the Nefud desert, *c.* 60 km inside the sand sea (Fig. 1). A number of sandstone hills form barriers for wind-blown sand, opening up shallow depressions on their eastern side, in which palaeolake deposits are visible (Fig. 2). The Jubbah oasis is rich in rock art, and cliffs and boulders at the base of hills are often densely covered in engravings in a rock art tradition that spans most of the Holocene period (Guagnin *et al.* 2017a; Khan 2007; 2011). In 2015, the rock art of Jubbah and the nearby site of Shuwaymis was inscribed on the UNESCO World Heritage list (UNESCO 2016).

Palaeolake sediments in the Jubbah basin attest to the presence of surface water in the past. At Al-Rabyah,

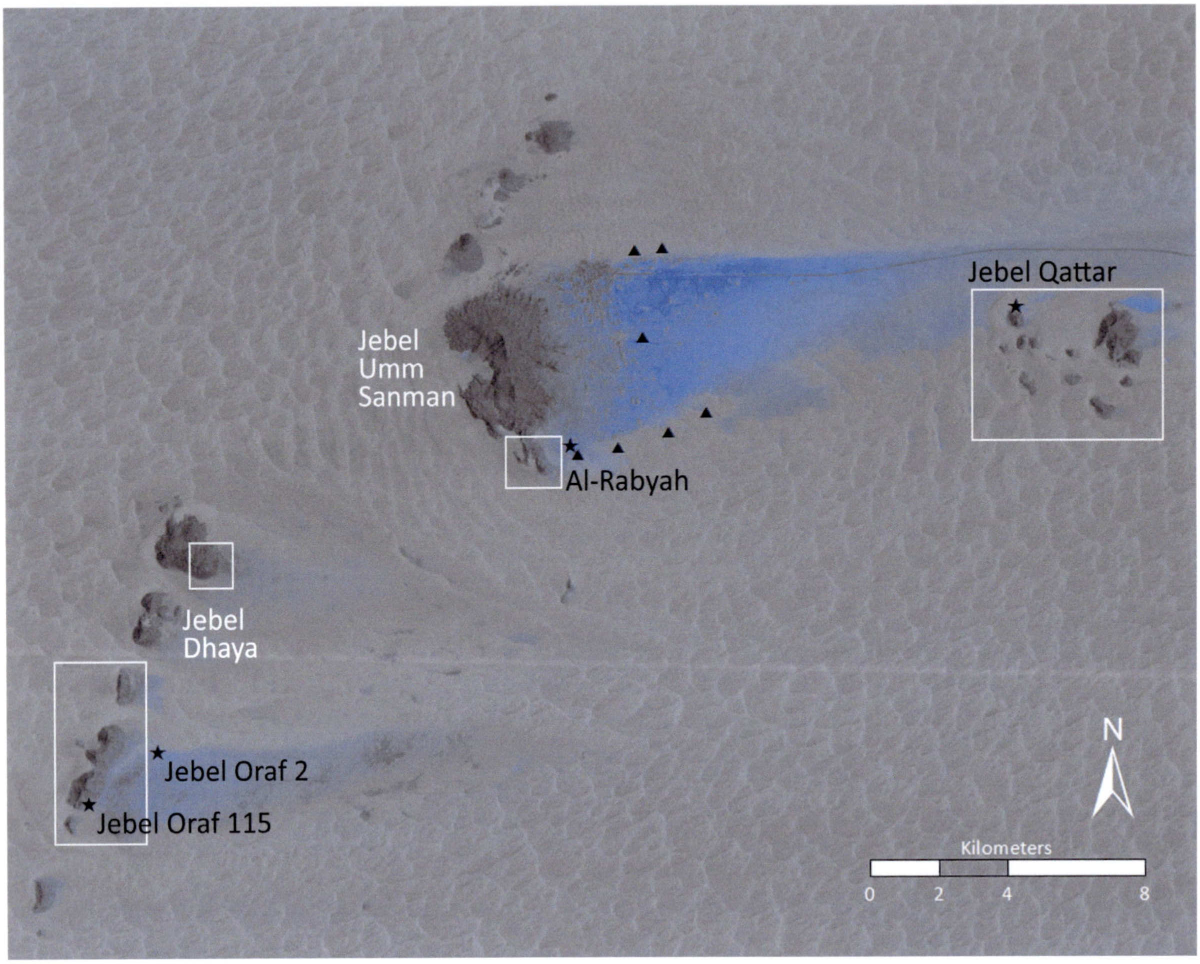

Figure 2. False colour Landsat TM satellite image of the Jubbah oasis and surrounding hills (bands 1 and 4; band 4 modified in blue for better visibility of lake deposits). Palaeolake deposits are visible in blue. Sand seas surround the hills and palaeolake deposits. Neolithic/Chalcolithic sites mapped by Garrard *et al.* (1981) are indicated with triangles. Early Holocene sites excavated by the Palaeodeserts project are indicated with stars (Crassard *et al.* 2013; Guagnin *et al.* 2017b; Hilbert *et al.* 2014). Other hills mentioned in the text are indicated in white font. Areas that have been surveyed for rock art are marked with white boxes.

periods of Holocene lake expansion have been dated to around 10,000 BC and 4600 BC; molluscs and ostracods recovered from these deposits are indicative of a shallow freshwater lake, surrounded by lush vegetation (Hilbert *et al.* 2014). An additional, smaller lake at Jebel Qattar was dated to between 6700 and 6000 BC (Crassard *et al.* 2013). Evidence from rock art suggests that lakes and vegetation of the Holocene humid period supported a range of animal species including lesser kudu, wild ass, onager, leopard and lion (Guagnin *et al.* 2016; 2017a; 2018a; 2018b).

In the late 1970s, the Northern Province survey identified twelve Neolithic and Chalcolithic sites in Jubbah (Fig. 2). Sites were dated based on lithic materials, which included arrowheads of the Rub' al-Khali type and tabular scrapers. No structural remains were found associated with these sites (Parr *et al.* 1978; see also Garrard *et al.* 1981). Reports suggest that in the landscapes of northern Saudi Arabia sites dating to the fourth millennium BC are more common, with lithic materials and occasionally pottery recovered from scatters, stone circles, kites and cairns (*ibid.*).

More recently, excavations at Al-Rabyah identified a lithic assemblage that includes bladelets and geometric microliths similar to Levantine Epipalaeolithic assemblages. Sediments associated with the assemblage yielded a minimum age of 9200 years (Clark-Balzan *et al.* 2018; Hilbert *et al.* 2014). At Jebel Qattar 101, a number of arrowheads with similarities to El-Khiam and Helwan points of the Pre-Pottery Neolithic in the Levant were associated with a palaeolake dated to 6700-6000 BC (Crassard *et al.* 2013). Faunal or botanic remains were not recovered at either site, preventing a reconstruction of

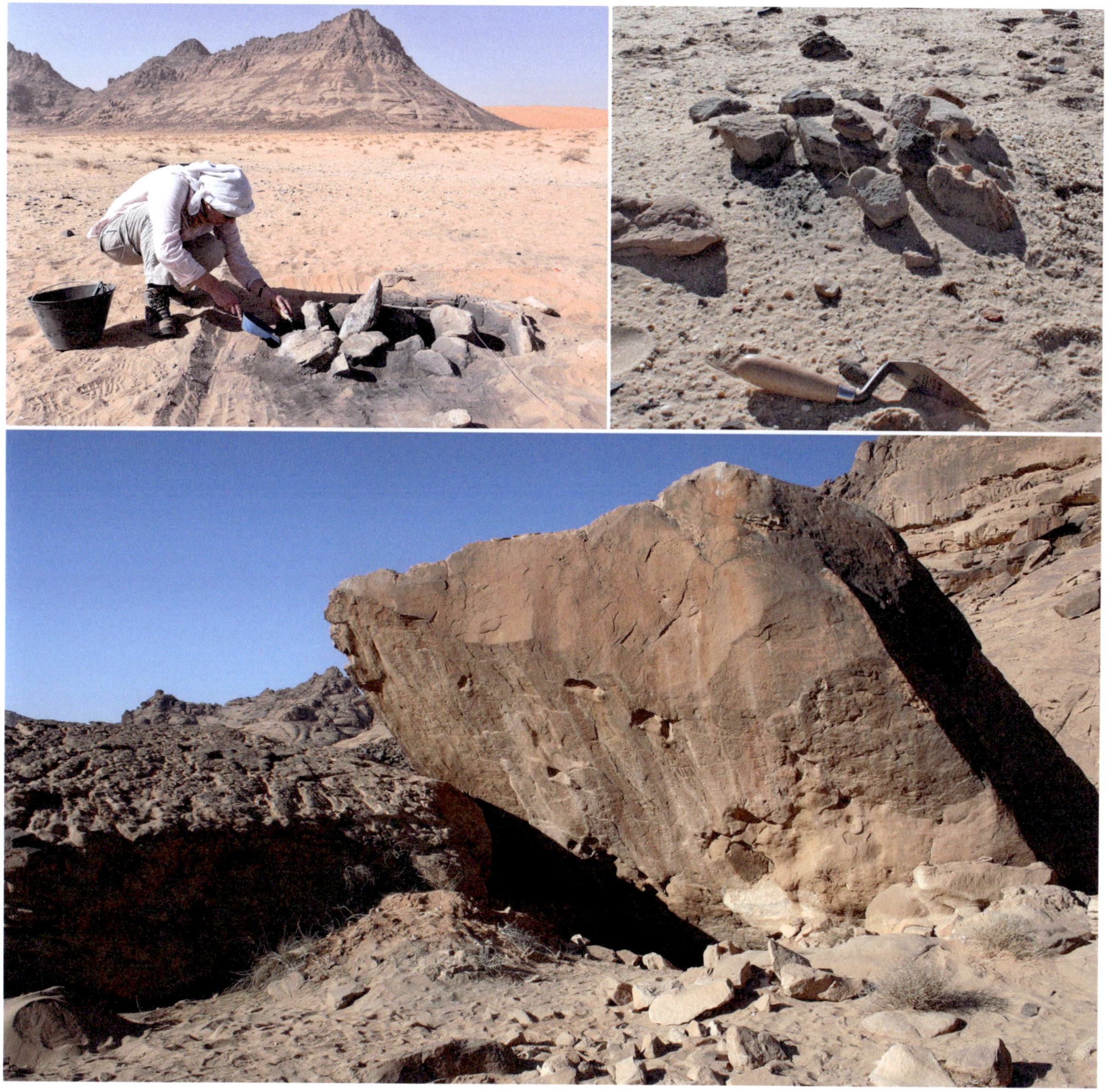

Figure 3. Neolithic occupation sites at Jebel Oraf. Top left: a stone-lined hearth at ORF2. Top right: hearth 1 before excavation, which revealed a simple, shallow pit filled with charcoal (Guagnin *et al.* 2017b, Fig. 3). Bottom: two large boulders form a shelter at ORF115.

subsistence patterns. However, lithic assemblages provide a link to pre-pastoral occupations of the Jordanian *badia* and were thus likely used by hunters rather than herders.

At Jebel Oraf (ORF), in the south of the Jubbah basin, two Neolithic sites were recently excavated, ORF2 and ORF115. The site of ORF2 is located on the edge of a palaeolake. Small clusters of stones indicate the position of individual hearths and numerous lithics, grinding stone fragments and faunal remains are visible on the surface. In total, 170 hearths were mapped at this site (Guagnin *et al.* 2017b). Initial test excavations of two hearths showed that hearths were constructed as simple pits in the sand, *c.* 45-50 cm in diameter, and are filled with charcoal rich sand. These ephemeral hearths are probably the result of a single, small fire. Radiocarbon dates obtained from charcoal fragments date both hearths between 5300-5000 cal BC. At hearth 2, tooth fragments of an adult *Bos* were recovered, making up a minimum of two teeth. Based on their morphology and size, and given their age these teeth are likely remains of domestic cattle (*ibid.*). Based on an initial assessment of the site, ORF2 appears to have been seasonally occupied by Neolithic pastoralists. The discovery of stone-lined hearths and the presence of grinding stones indicates that not all occupation was as short lived as the construction of hearth

1 suggests (Fig. 3), and some visits may have stretched over several days and perhaps extended to weeks (*ibid.*). However, neither structural remains nor clearance areas that could indicate the use of tents (see for example Palmer *et al.* 2007) were found associated with the site. Where these pastoralist groups spent the rest of the year remains an open question.

An archaeological survey along the base of Jebel Oraf identified a second Neolithic occupation site, ORF115. Here, two boulders form a small shelter (Fig. 3). Although the site had partly been destroyed, the remaining deposits contain a stratigraphic sequence of Neolithic hearths, similar to those recorded at ORF2. While the shelter would have provided some degree of protection from the elements, the absence of dwellings, animal pens, and storage facilities suggests that settlement patterns and landscape use at Jebel Oraf differed noticeably from those proposed for contemporary sites in the eastern *badia* and in southern Jordan.

The hearth sites of the Jubbah oasis are by no means unique in Arabia. Recent fieldwork at Alshabah, in the western Nefud desert, identified a cluster of 125 hearths on the shores of an interdune lake (Scerri *et al.* 2018). The hearths are ephemeral structures of *c.* 50 cm in diameter, topped with small clusters of rocks that were collected from the immediate landscape. A number of highly weathered and fragmented animal bones were recovered from this site. Only a single fragment could be identified as a fragment of goat or sheep. Radiocarbon and OSL dates obtained from three excavated hearths indicate that the site was occupied between 5300 and 4500 cal BC. The dates place occupation of the site towards the end of the Holocene humid period, when pastoralists exploited seasonal pastures in interdunal basins. Lithic artefacts recovered from this site include a tabular scraper, which may have been used for the shearing of sheep, and provides a link to the Chalcolithic/Early Bronze Age of southern Jordan. Grinding stones are common at Alshabah, although it remains unclear whether they were used for the manufacture of ground stone tools, or for the processing of plants or cereals. Differences with lithic assemblages from the Levant, and an absence of southern Arabian forms suggests that the Neolithic occupants of Alshabah had formed their own traditions (*ibid.*).

The site of Alshabah was probably used in repeated short-term occupation by small communities or family groups, related to the seasonal availability of water and pasture. Again, no structural remains were found near the site (Scerri *et al.* 2018.) and it remains unclear where the pastoral groups of Alshabah spent the rest of the year, particularly the dry season. The discovery of eleven further hearth sites in the Nefud desert (Breeze *et al.* 2017) suggests that hearth sites were not merely a short lived and seasonal phenomenon, but formed a substantial component of Neolithic settlement patterns in Arabia.

Reeler and Al-Shaikh (2015) report similar hearth sites from central Arabia, at Thumamah, and from the Rub' al-Khali. Lithics assemblages associated with these hearths include classic Arabian Neolithic types such as barbed and tanged arrowheads, blades, spearheads and scrapers. The use of charcoal in the hearths of the Rub' al-Khali suggests that they were in use during a time when wood and vegetation were available in this area. Although these sites remain unexcavated, no structural remains were visible on the surface. Similar sites have also been reported from the Tuwayq escarpment west of Riyadh, where photos of Neolithic 'structures' show features of a similar size and shape to hearths recorded in Jubbah and Alshabah (Zarins *et al.* 1979, Pl. 6A). Reeler and Al-Shaikh (2015) suggest that hearth sites may be common across the dune systems of the Arabian interior. The phenomenon of hearth sites may therefore represent an adaptation to the marginal environments of sand seas.

Rock art

The rock art of north-western Saudi Arabia is very distinctive, and the so-called 'Jubbah-style' is generally attributed to the Neolithic period (Khan 2007; 2011). This rock art is typically engraved in deep, semi-naturalistic outlines and is dominated by the depiction of human figures, dogs and cattle (Guagnin *et al.* 2017a). Cattle are generally shown with very large horns that often curve slightly outwards at the tips, while the outline of the head is stylised and merged with the neck. Eyes and mouth are rarely shown, but ears are generally depicted in outline, behind the horns (Figs. 4-5). Human figures are elongated, with slim upper bodies and long legs that are usually bent at the knee. Arms are very thin and are always held out to one side, in a twist of perspective from the frontal view of the shoulders to a profile view of the arms. The majority of human figures is male, and depicted wearing a penis sheath. Headdresses are also a common feature (Guagnin 2018; Guagnin *et al.* 2017a) (Figs. 4-5).

Systematic analysis of superimpositions on rock art panels has identified a period of hunting imagery that was later superimposed by engravings of cattle, which were sometimes integrated into earlier hunting scenes (Guagnin *et al.* 2015; 2017a). The depiction of human figures shows marked similarities across both periods, cultural markers such as headdress and penis sheaths continue to be depicted and presumably to be worn (Guagnin *et al.* 2015). The engraving of Jubbah style human figures is thus a marker of continuity across the transition from hunting to herding. The recent discovery of a panel near ORF115 that shares many features with the Jubbah style, but shows human figures with metal daggers (Guagnin *et al.* 2018b) indicates that this type of human representation was long lived and likely ended in the Early Bronze Age.

Figure 4. Large panel at Jebel Qattar, Jubbah, showing a large cattle, three goats and five human figures superimposed/re-engraved over a hunting scene with ibex and other wild animals.

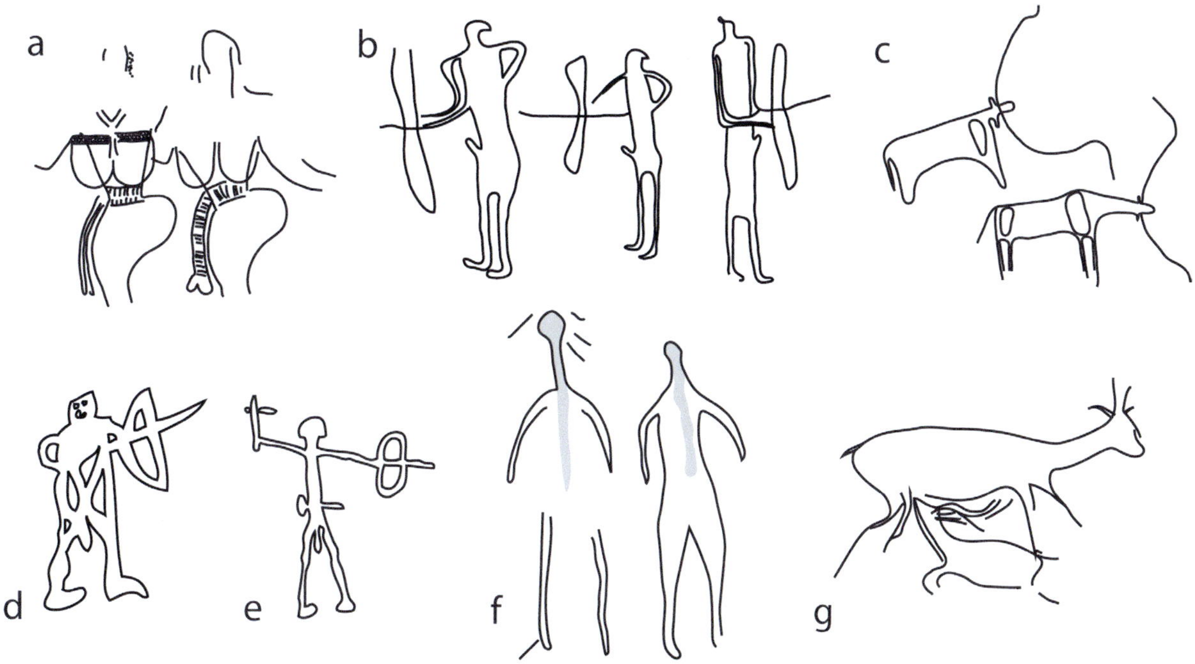

Figure 5. Comparison of rock art from Jubbah, Shuwaymis, the Negev and Dhuweila. a: curvaceous women, Jubbah (Guagnin *et al*. 2017a). b: 'Jubbah style' human figures, Shuwaymis (Guagnin 2018). c: 'Jubbah style' cattle, Shuwaymis. d: figure from a hunting scene, Wadi Ramlije, central Negev, dated to Style III by Anati (1981). e: figure from a large chariot scene, Timna, Wadi Araba, dated to Style IV (dispersal of herders) by Anati (1981). f: 'dancing men with spears and bow', Dhuweila, deeper grooves within the engraving are shown shaded in grey. g: gazelle engraving, Dhuweila (d and e are traced from Anati 1981; f and g are traced from photographs published in Betts 1987. All images traced by author to allow better comparison).

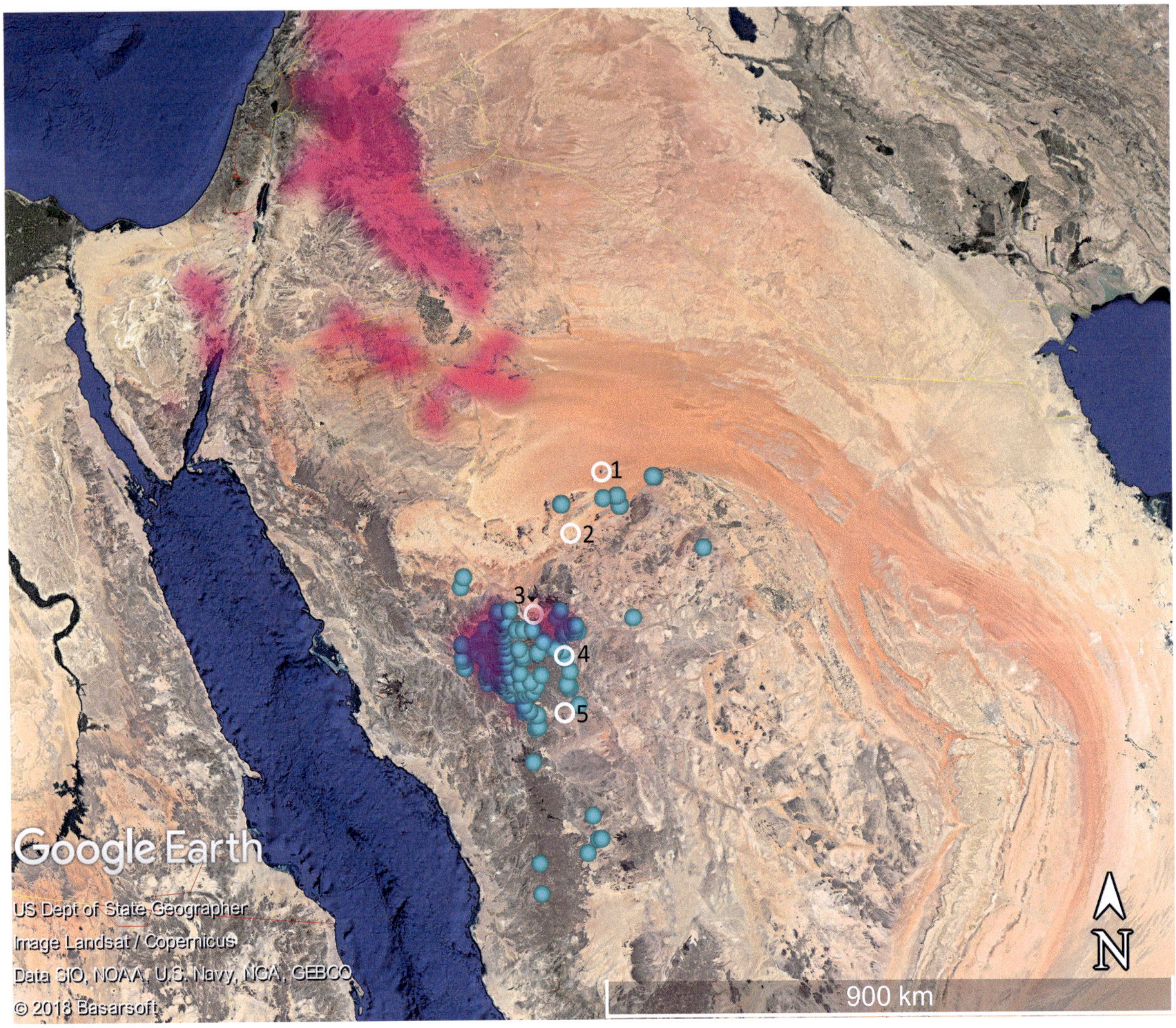

Figure 6. Distribution of desert kites (pink), gates (turquoise) and rock art (white circles). Distribution of kites traced from https://www.globalkites.fr, distribution of gates based on Kennedy (2017). 1: Jubbah. 2: Jabal al-Mismā. 3: Shuwaymis. 4: east Khaybar. 5: Hanakiyah. Location of Jabal al-Mismā is estimated based on an overview map published by Bednarik and Khan (2017).

In Jubbah, an earlier phase of rock art production can be identified. These show almost life-sized, curvaceous women, with exaggerated hips and breasts, and often with braided hair. Where the arms of these female figures are preserved, they are naturalistic, held out to each side, and the hands show all five fingers (Guagnin *et al.* 2017a; Fig. 5). While these early engravings appear to be unique to the rock art of Jubbah, human and animal representations of the 'Jubbah style' are also known from other sites in this area, including Shuwaymis, Hanakiyah (Nayeem 2000), Jabal al-Misma (Bednarik and Khan 2017), and the east Khaybar (Vic Camp, pers. comm. 2017).

The rock art of the eastern *badia*, southern Jordan and the Negev desert is distinctly different to that of north-western Saudi Arabia. At Dhuweila, in eastern Jordan, rock art attributed to the PPN shows lightly incised naturalistic outlines of gazelles (Betts 1987) (Fig. 5g). Human figures from the same site have elongated necks, engraved in a groove that continues across the chest, and their heads appear be formed by small cupules (*ibid.*). Compared to human figures of the 'Jubbah style' there is a similar degree of stylisation and lack of detail, but the Dhuweila figures are facing forward, their bodies are relatively naturalistic and lack elongation and angular shoulders. Particularly the proportion and size of the head and neck is distinctly different. Human figures in the Negev desert are depicted in a range of different forms, but are generally small and simplified to the point where they are reduced to stick figures. Neither the human figures in Dhuweila, nor the figures in the Negev are shown with penis sheaths, and although one of the figures may be depicted with a headdress (Fig. 5f, left), they are very different to those of the 'Jubbah style'. Rock art recorded as part of the Wadi Faynan project is dominated by stick figure-like engravings of ibex and shows some similarity to engravings recorded in the Negev desert (Pinkett and Mithen 2007; Eisenberg-Degen and Rosen 2013).

However, 'Jubbah style' cattle engravings and human figures are also well known from Shuwaymis and Hanakiyah (Guagnin *et al.* 2015; 2016; Nayeem 2000). More recently similar sites have also been reported from Jebel al-Misma (Bednarik and Khan 2017), and from the eastern Khaybar (Vic Camp, pers. comm.). This suggests that during the Neolithic the north-west of Saudi Arabia was occupied by a population with a shared tradition of wearing headdresses and penis sheaths, and a shared understanding of how human figures are to be depicted in rock art. The distance between these sites is such that they could feasibly have been part of a transhumance cycle, possibly within a population where multiple groups aggregate and disperse across the year. Climatic data simulated by the Community Earth System Models (COSMOS) climate model suggests that the area around Shuwaymis may have been on the northern edge of the African summer monsoon rainfall regime. The north-west of Saudi Arabia was therefore ecologically connected with the south-west of the Arabian Peninsula. Data from climate modelling and rock art also indicates that the northward extent of the monsoon fluctuated, and that droughts may have been common during the Holocene humid period (Guagnin *et al.* 2016). Pastoralists and wild animal populations likely responded by retreating into oases or by moving south, where the rains were more reliable. It is possible that the distribution of the 'Jubbah style' was linked to changes in the availability of water and pastures during the Holocene humid period.

Stone structures

The increasing availability of satellite imagery has led to a more comprehensive mapping of stone structures across the Levant and the Arabian Peninsula. Desert kites, which are characterised by long guiding walls that converge in an enclosed space, are thought to have been used for hunting. Kites have been recorded across the Levant, on the northern margins of the Nefud desert, and in the Khaybar region of western Saudi Arabia. While all kites share distinctive features such as guiding walls and enclosed spaces, regional variations can be observed (Barge *et al.* 2015; Kennedy *et al.* 2015). Regional differences have also been observed in other stone structures, such as 'pendants', which tend to have longer 'tails' in the Khaybar region, and in 'wheels' (Kennedy 2011). Although many stone structures such as cairns, pendants or keyhole tombs are associated with later periods *(ibid.)*, OSL dates obtained from a small number of kites and wheels suggest that construction of these features was well established by the Late Neolithic (Rollefson *et al.* 2016a). Despite the presence of regional differences, the distribution of these stone structures implies some form of contact between the Khaybar area and the Levant (Fig. 6).

However, one type of stone structure has only been identified in the area of the Harrat Khaybar and surrounding areas of northern Saudi Arabia. 'Gates' are rectangular stone structures that are formed by two shorter, thicker lines of stones set in parallel, linked by two or more longer and thinner walls. A recent survey of satellite imagery available on Google Earth identified over 400 gates in western Saudi Arabia. In all instances where superimpositions of different types of stone structures were recorded, gates were found beneath older structures and thus appear to be the oldest man-made structures in the landscape (Kennedy 2017). Although kites and gates are partly found in the same region, they are not generally found in the same location. Kennedy (2017, 171) only notes one instance where a kite appears so overlie a gate. Parr and colleagues also report a kite near Hail that was probably superimposed by a gate (Parr *et al.* 1978, 40), although the site can no longer be identified on satellite imagery and may have been destroyed in the recent expansion of settlements. Kites and gates therefore appear to have some overlap in age.

The distribution of gates in north-western Saudi Arabia largely overlaps with rock art sites of the 'Jubbah style', and given their timing it is possible that both were linked to the same (pre-)Neolithic tradition. Two gates were found in the area around ORF2 and ORF115. One gate is located on the slopes of Jebel Oraf, a second gate was found further north, on a low-lying promontory at Jebel Dhaya (Fig. 7). A third, much smaller gate is located on a low-lying promontory on the eastern side of Jebel Umm Sanman. Given the fact that highest rock art concentrations occur on the slopes of smaller outcrops on the north-eastern and south-eastern side of Jebel Umm Sanman, where no gates were recorded, and that rock art concentrations are low in the vicinity of the gates at Jebel Oraf and Jebel Dhaya, there appears to be no direct spatial link between both site types, and perhaps they are deliberately set apart. In the Jubbah oasis, numerous later cairns are placed around and onto the gates, some of which re-use stones from the gate, confirming Kennedy's observation (2017) that gates are the oldest man-made structures in the landscape.

Discussion

The hearth sites of the Nefud desert and the Arabian interior show considerable differences to contemporary sites in the eastern *badia* and in southern Jordan, which are characterised by substantial curvilinear dwellings, often clustered in honeycomb patterns. In the marginal areas of the Levant, subsistence was based on a combination of herding, hunting, and agriculture, and seasonal mobility

Figure 7. 'Gate' recorded at Jebel Dhaya in the Jubbah oasis. The short sides of this gate form shallow platforms, a member of the team is standing on the northern end, which is aligned with a small cairn. The longer, thinner side of the gate in the background of the photograph is only partially preserved and has been re-used for the construction of a number of cairns (visible between the gate and the edge of the escarpment).

allowed Neolithic groups to maximise the exploitation of the *badia*. Hearth sites in north-western Arabia appear to have also been used by mobile Neolithic herders, and faunal remains attest the herding of caprines and cattle. However, the absence of structural remains and even clearing areas is striking. Whether this difference is merely architectural, caused by the use of small, lightweight shelters, or indicative of a higher degree of mobility remains an open question. While some of the hearths appear to contain the remains of a single fire, the presence of grinding stones, evidence for stone tool production, and the discovery of stone-lined hearths all suggest that on some occasions occupation may have been longer lived, perhaps spanning weeks rather than days.

Neolithic hearth sites have been recorded in the Jubbah oasis, the western Nefud desert, central Saudi Arabia, and the Rub' al-Khali, and may be common across the sand seas of Arabia. Comparable sites have also been recorded in the interdune corridors of the central Sahara, where fireplaces and grinding stones dating to the Middle Pastoral Period (*c.* 5000-3800 BC) are found on the shores or ephemeral lakes (Di Lernia 1999). The number and distribution of these sites suggests that in Arabia hearth sites formed a substantial part of the Neolithic settlement pattern and may have been a specific adaptation to the marginal areas of sand seas.

The rock art of the Neolithic in north-western Saudi Arabia shows distinct local characteristics. Human figures and cattle engravings of the so-called 'Jubbah style' are found at numerous sites between the Jubbah oasis and the central Khaybar. Rock art north of the Nefud desert, on the other hand, shows very different rock art traditions, both in the attributes associated with human figures, and in the way the shape and proportion of the human body is represented on rock surfaces. The known distribution of the 'Jubbah style' rock art appears to overlap with the distribution of gates (Fig. 6). Although uncertainties remain in the dating of gates and rock art, the superimpositions of other stone structures such as cairns and pendants over gates suggest that the latter are the oldest stone structures in the landscape, and were likely built in the Neolithic period. The continued representation of key characteristics of the 'Jubbah style' (such as proportions, twisted perspective, headdress, penis sheath) from the pre-Neolithic until the onset of the Bronze Age suggests that this particular type of rock art was long lived. The building of gates and the creation of 'Jubbah style' rock art are therefore likely to have had at least some temporal overlap, giving the Neolithic of north-western Saudi Arabia a distinctly local character. Modelling of rainfall patterns shows that the area around Jubbah may have been on the northern extent of the African summer monsoon and thus ecologically connected with the south-west. The distribution of cultural traits in the Neolithic of north-western Arabia may therefore relate to transhumance patterns driven by the availability of water and pastures.

The distribution of desert kites (Fig. 6) and wheels, the presence of exotic raw materials and lithic types with similarities to Levantine assemblages, and the introduction of livestock overlay these local characteristics with

Levantine influences. However, the absence of architectural remains, and the relative scarcity of Levantine lithic types in Arabian assemblages show that the adoption of Levantine technologies was very selective, and was adapted to local environments and landscapes. Contact with the Levant is evident at all known Neolithic sites in the Nefud desert. Given the location of the Jubbah oasis, and evidence for the exploitation of interdune depressions at Alshabah, it is possible that contact was maintained across the interdune corridors of the Nefud desert. Increased precipitation during the Holocene humid period would have improved conditions along a route that was still used by caravans in the nineteenth century and provided a vital trade link between the Azraq basin and the Nejd (Euting 1896).

Although archaeological research is beginning to identify distinct local traits in the Neolithic of north-western Saudi Arabia, many questions remain. Further fieldwork is now needed to establish when herding was first introduced into northern Arabia, and to identify the subsistence and mobility patterns that underpinned the occupation of hearth sites that can only be described as extremely ephemeral.

Acknowledgements

The Palaeodeserts project is grateful to His Royal Highness Prince Sultan bin Salman, President of the Saudi Commission for Tourism and National Heritage, and Prof. Ali Ghabban, Vice President, for permission to carry out fieldwork at Shuwaymis and Jubbah. The fieldwork in Jubbah could not have been done without the support of our Saudi colleagues, especially Abdullah Alsharekh, Abdulaziz al-Omari and Habeeb Turki. Financial support was provided by the European Research Council, in a grant to Michael Petraglia (no. 295719), the Saudi Commission for Tourism and National Heritage, the Max Planck Society, and the Dahlem Research School (Freie Universität Berlin). Finally, I wish to thank Huw Groucutt for constructive comments on the manuscript.

References

Abu-Azizeh, W. 2013. The south-eastern Jordan's Chalcolithic-Early Bronze Age pastoral nomadic complex: patterns of mobility and interaction. *Paléorient* 39, 149-176.

Anati, E. 1981. *Felskunst im Negev und auf Sinai*. Bergisch-Gladbach: Gustav Lūbbe Verlag.

Barge, O., Brochier, J.E. and Crassard, R. 2015. Morphological diversity and regionalisation of kites in the Middle East and Central Asia. *Arabian Archaeology and Epigraphy* 26, 162-176.

Bednarik, R.G. and Khan, M. 2017. New rock art complex in Saudi Arabia. *Rock Art Research* 34, 179-188.

Betts, A.V.G. 1987. The hunter's perspective: 7th millennium BC rock carvings from eastern Jordan. *World Archaeology* 19, 214-225.

Betts, A.V.G. 1989. The Pre-Pottery Neolithic B period in eastern Jordan. *Paléorient* 15, 147-153.

Betts, A.V.G. 1993. The Neolithic sequence in the eastern Jordan badia. A preliminary overview. *Paléorient* 19, 43-53.

Betts, A.V.G., Martin, L. and McCartney, C. 2013. Background and methodology. In: Betts, A.V.G., Cropper, D., Martin, L. and Mc Cartney, C. (eds.). *The later prehistory of the badia. Excavations and surveys in eastern Jordan*. Oxford: Oxbow Books, 1-12.

Breeze, P.S., Groucutt, H.S., Drake, N.A., Louys, J., Scerri, E.M.L., Armitage, S.J., Zalmout, I.S.A., Memesh, A.M., Haptari, M.A., Soubhi, S.A., Matari, A.H., Zahir, M., al-Omari, A., Alsharekh, A. and Petraglia, M.D. 2017. Prehistory and palaeoenvironments of the western Nefud desert, Saudi Arabia. *Archaeological Research in Asia* 10, 1-16.

Clark-Balzan, L., Parton, A., Breeze, P.S., Groucutt, H.S. and Petraglia, M.D. 2018. Resolving problematic luminescence chronologies for carbonate- and evaporite-rich sediments spanning multiple humid periods in the Jubbah Basin, Saudi Arabia. *Quaternary Geochronology* 45, 50-73.

Crassard, R., Petraglia, M.D., Parker, A.G., Parton, A., Roberts, R.G., Jacobs, Z., Alsharekh, A. Al-Omari, A., Breeze, P., Drake, N.A., Groucutt, H.S., Jennings, R., Régagnon, E. and Shipton, C. 2013. Beyond the Levant: first evidence of a Pre-Pottery Neolithic incursion into the Nefud desert, Saudi Arabia. *PLoS ONE* 8 (7), e68061. Doi: 10.1371/journal.pone.0068061

Di Lernia, S. 1999. Discussing pastoralism. The case of the Acacus and surroundings (Libyan Sahara). *Sahara* 11, 7-20.

Drechsler, P. 2007. The Neolithic dispersal into Arabia. *Proceedings of the Seminar for Arabian Studies* 37, 93-109.

Drechsler, P. 2009. *The dispersal of the Neolithic over the Arabian Peninsula*. Oxford: British Archaeological Reports (International Series 1969).

Eisenberg-Degen, D. and Rosen, S.A. 2013. Chronological trends in Negev rock art: the Har Michia petroglyphs as a test case. *Arts* 2, 225-252.

Euting, J. 1896. *Tagebuch einer Reise in Inner-Arabien*. Leiden: Brill.

Fujii, S. 2010. Domestication of runoff surface water: current evidence and new perspectives from the Jafr pastoral Neolithic. *Neo-Lithics* 2/10, 14-32.

Garrard, A., Harvey, C.P.D. and Switsur, V.R. 1981. Environment and settlement during the Upper Pleistocene and Holocene at Jubba in the Great Nefud, northern Arabia. *Atlal* 5, 137-148.

Gebel, H.G.K. 2016. The socio-hydraulic foundations of oasis life in northwest Arabia: the 5th Millennium BCE shepherd environs of Rajajil, Rasif and Qulban Beni Murra. In: Luciani, M. (ed.). *The archaeology of north Arabia. Oases and landscapes*. Vienna: Austrian Academy of Sciences Press, 79-113.

Globalkites. 2016. Globalkites project website: https://www.globalkites.fr/Interactive-Map (accessed 30 January 2018).

Guagnin, M. 2018. The hunters and herders of Shuwaymis: new evidence for the population dynamics of the Neolithic transition in Saudi Arabia. In: Huyge, D. and Van Noten, F. (eds.). *What ever happened to the people? Humans and anthropomorphs in the rock art of northern Africa*. Brussels: Royal Academy of Overseas Sciences, 231-241.

Guagnin, M., Jennings, R., Eager, H., Parton, A., Stimpson, C., Stepanek, C., Pfeiffer, M., Groucutt, H.S., Drake, N.A., Alsharekh, A. and Petraglia, M.D. 2016. Rock art imagery as a proxy for Holocene environmental change: a view from Shuwaymis, NW Saudi Arabia. *The Holocene* 26, 1822-1834.

Guagnin, M., Jennings, R., Clark-Balzan, L., Groucutt, H.S., Parton, A. and Petraglia, M.D. 2015. Hunters and herders: exploring the Neolithic transition in the rock art of Shuwaymis, Saudi Arabia. *Archaeological Research in Asia* 4, 3-16.

Guagnin, M., Perri, A.R. and Petraglia, M.D. 2018a. Pre-Neolithic evidence for dog-assisted hunting strategies in Arabia. *Journal of Anthropological Archaeology* 49, 225-236.

Guagnin, M., Shipton, C., El-Dossary, S., Al-Rashid, M., Moussa, F., Stewart, M., Ott, F., Alsharekh, A. and Petraglia, M.D. 2018b. Rock art provides new evidence on the biogeography of kudu (Tragelaphus imberbis), wild dromedary, aurochs (Bos primigenius) and African wild ass (Equus africanus) in the early and middle Holocene of north-western Arabia. *Journal of Biogeography* 45, 727-740.

Guagnin, M., Shipton, C., Al-Rashid, M., Moussa, F., El-Dossary, S., Bin Sleima, M., Alsharekh, A. and Petraglia, M. 2017a. An illustrated prehistory of the Jubbah oasis: reconstructing Holocene occupation patterns in northwestern Saudi Arabia from rock art and inscriptions. *Arabian Archaeology and Epigraphy* 28, 138-152.

Guagnin, M., Shipton, C., Martin, L. and Petraglia, M. 2017b. The Neolithic site of Jebel Oraf 2, northern Saudi Arabia: first report of a directly dated site with faunal remains. *Archaeological Research in Asia* 9, 63-67.

Henry, D.O., Cordova, C., White, J.J., Dean, R.M., Beaver, J.E., Ekstrom, H., Kadowaki, S., McCorriston, J., Nowell, A. and Scott-Cummings, L. 2003. The Early Neolithic site of Ayn Abū Nukhayla, southern Jordan. *Bulletin of the American Schools of Oriental Research* 330, 1-30.

Henry, D.O., Cordova, C.E, Portillo, M., Albert, R.M., Dewitt, R. and Emery-Barbier, A. 2017. Blame it on the goats? Desertification in the Near East during the Holocene. *The Holocene* 27, 625-637.

Hilbert, Y.H., White, T.S., Parton, A., Clark-Balzan, L., Crassard, R., Groucutt, H.S., Jennings, R.P., Breeze, P., Parker, A., Shipton, C., Al-Omari, A., Alsharekh, A. and Petraglia, M.D. 2014. Epipalaeolithic occupation and palaeoenvironments of the southern Nefud desert, Saudi Arabia, during the Terminal Pleistocene and Early Holocene. *Journal of Archaeological Science* 50, 460-474.

Kennedy, D. 2017. 'Gates': a new archaeological site type in Saudi Arabia. *Arabian Archaeology and Epigraphy* 28, 153-174.

Kennedy, D. 2011. The "Works of the Old Men" in Arabia: remote sensing in interior Arabia. *Journal of Archaeological Science* 38, 3185-3203.

Kennedy, D., Banks, R. and Dalton, M. 2015. Kites in Saudi Arabia. *Arabian Archaeology and Epigraphy* 26, 177-195.

Khan, M. 2007. *Rock art of Saudi Arabia across twelve thousand years*. Riyadh: Deputy Ministry of Antiquities and Museums.

Khan, M. 2011. *Jubbah: the land of golden sands and the lost civilization of Arabia*. Riyadh: Saudi Commission for Tourism and Development.

Martin, L. and Edwards, Y. 2013. Diverse strategies: evaluating the appearance and spread of domestic caprines in the southern Levant. In: Colledge, S., Conolly, J., Dobney, K., Manning, K. and Shennan, S. (eds.). *The origins and spread of domestic animals in southwest Asia and Europe*. Walnut Creek, CA: Left Coast Press, 49-82.

Meister, J., Knitter, D., Krause, J., Müller-Neuhof, B. and Schütt, B. 2019. A pastoral landscape for millennia: investigating pastoral mobility in northeastern Jordan using quantitative spatial analyses. *Quaternary International* 501, 364-378.

Nayeem, M.A. 2000. *The rock art of Arabia. Saudi Arabia, Oman, Qatar, The Emirates and Yemen*. Hyderabad: Hyderabad Publishers.

Palmer, C., Smith, H. and Daly, P. 2007. Ethnoarchaeology. In: Barker, G., Gilbertson, D. and Mattingly, D. (eds.). *Archaeology and desertification. The Wadi Faynan landscape survey, southern Jordan*. Oxford: Oxbow Books & Council for British Research in the Levant, 369-395.

Parr, P.J., Zarins, J., Ibrahim, M., Waechter, J., Garrard, A., Clarke, C., Bidmead, M. and Al Badr, H. 1978. Preliminary report on the second phase of the Northern Province survey 1397/1977. *Atlal* 2, 29-50.

Pinkett, S. and Mithen, S. 2007. The rock art of WF400, Wadi Ghuwayr. In: Finlayson, B. and Mithen, S. (eds.). *The early prehistory of Wadi Faynan, southern Jordan. Archaeological survey of Wadis Faynan, Ghuwayr and al-Bustan and evaluation of the Pre-Pottery Neolithic A site of WF16*. Oxford: Oxbow Books & Council for British Research in the Levant, 115-133.

Price, D.T. and Bar-Yosef, O. 2011. The origins of agriculture: new data, new ideas. *Current Anthropology* 52, S163-174.
Reeler, C. and Al-Shaikh, N. 2015. A discussion of Neolithic settlement patterns in Saudi Arabia and Bahrain during the Holocene Pluvial Period. *Proceedings of the Seminar for Arabian Studies* 45, 1-16.
Rollefson, G., Rowan, Y. and Wasse, A. 2014. The Late Neolithic colonization of the eastern badia of Jordan. *Levant* 46, 1-17.
Rollefson, G.O. Althanassas, C.D., Rowan, Y.M. and Wasse, A.M.R. 2016a. First chronometric results for 'works of the old men': late prehistoric 'wheels' near Wisad Pools, Black Desert, Jordan. *Antiquity* 90, 939-952.
Rollefson, G.O., Rowan, Y., Wasse, A., Hill, A.C., Kersel, M., Lorentzen, B., Al-Bashaireh, K. and Ramsay, J. 2016b. Investigations of a Late Neolithic structure at Mesa 7, Wadi al-Qattafi, Black Desert, 2015. *Neo-Lithics* 1/16, 3-12.
Rowan, Y.M., Rollefson, G.O., Wasse, A., Abu-Azizeh, W., Hill, A.C. and Kersel, M.M. 2015. The "land of conjecture:" new late prehistoric discoveries at Maitland's Mesa and Wisad Pools, Jordan. *Journal of Field Archaeology* 40, 176-189.
Scerri, E.M.L., Guagnin, M., Groucutt, H.S., Armitage, S.J., Parker, L.E., Drake, N., Louys, J., Breeze, P.S., Zahir, M., Alsharekh, A. and Petraglia, M.D. 2018. Neolithic pastoralism in marginal environments during the Holocene humid period, northern Saudi Arabia. *Antiquity* 2018, 1180-1194.
UNESCO. 2016. World Heritage List. Rock art in the Hail region of Saudi Arabia. Available at: http://whc.unesco.org/en/list/1472_(accessed 29 January 2018).
Zarins, J., Ibrahim, M., Potts, D.T. and Edens, C. 1979. Saudi Arabian archaeological reconnaissance 1978. The preliminary report on the third phase of the comprehensive archaeological survey program – The central province. *Atlal* 3, 9-42.

The works of the old men in Arabia: a comparative analysis

David Kennedy

Abstract

The ancient stone-built structures found in the Harrat al-Sham and known to Bedouins as the 'Works of the Old Men' are familiar and increasingly well-documented, especially for the part in Jordan. Several site types (Kites, Wheels, Pendants, Cairns, Chain Walls, *etc.*), totaling many thousands, have been recorded and plotted as well as hundreds of kilometres of low, meandering walls. Similar structures are found more widely in 'Arabia'. Kites, the best-known, have been recorded as far away as Armenia and Yemen (and even in central Asia). There are also parallels elsewhere in Arabia for some of the other Harrat al-Sham structures in a variety of landscapes. However, just as it is precisely on this lava field of Syria-Jordan-Saudi Arabia that overwhelmingly hosts these structures in huge numbers, so, too, one should look to the succession of other lava fields (*harrat*; singular *harrah*) found along the west coast of the Arabian peninsula. There has been little fieldwork on these latter and they are far less well-explored on the satellite imagery of Google Earth and Bing. Nevertheless, a growing number of high-resolution 'windows' in this imagery have been interpreted systematically and with varying degrees of completeness. Tens of thousands of sites have now been recorded on these other *harrat* revealing not just some expected parallels but also surprising differences. Kites have been found in very large numbers in the Harrat Khaybar, but seldom in the other *harrat* and many are of a design notably different from elsewhere; indeed, unique to this lava field. Likewise, Pendants and Wheels are abundant but the designs are different. Even more notable are site types found in the Harrat al-Sham but not elsewhere, and others found elsewhere but not in the Harrat al-Sham. For example, Chain Walls are found just once outside the Harrat al-Sham; conversely, Gates, Barred Rectangles, Triangles and two types of Pendants (Trumpet and Keyhole) are found only in some of the *harrat* of Saudi Arabia. Variants in design, differences in site types and the distribution patterns and associations between site types may be of significance in interpreting and explaining human activity and at least the relative chronology of the sites. It remains crucial for fieldwork to be undertaken in parallel with the interpretation and analysis of the imagery.

Keywords: aerial archaeology, Saudi Arabia, Google Earth, prehistory, funerary monuments

In: Peter M. M. G. Akkermans (ed.) 2020: *Landscapes of Survival - The Archaeology and Epigraphy of Jordan's North-Eastern Desert and Beyond*, Sidestone Press (Leiden), pp. 117-144.

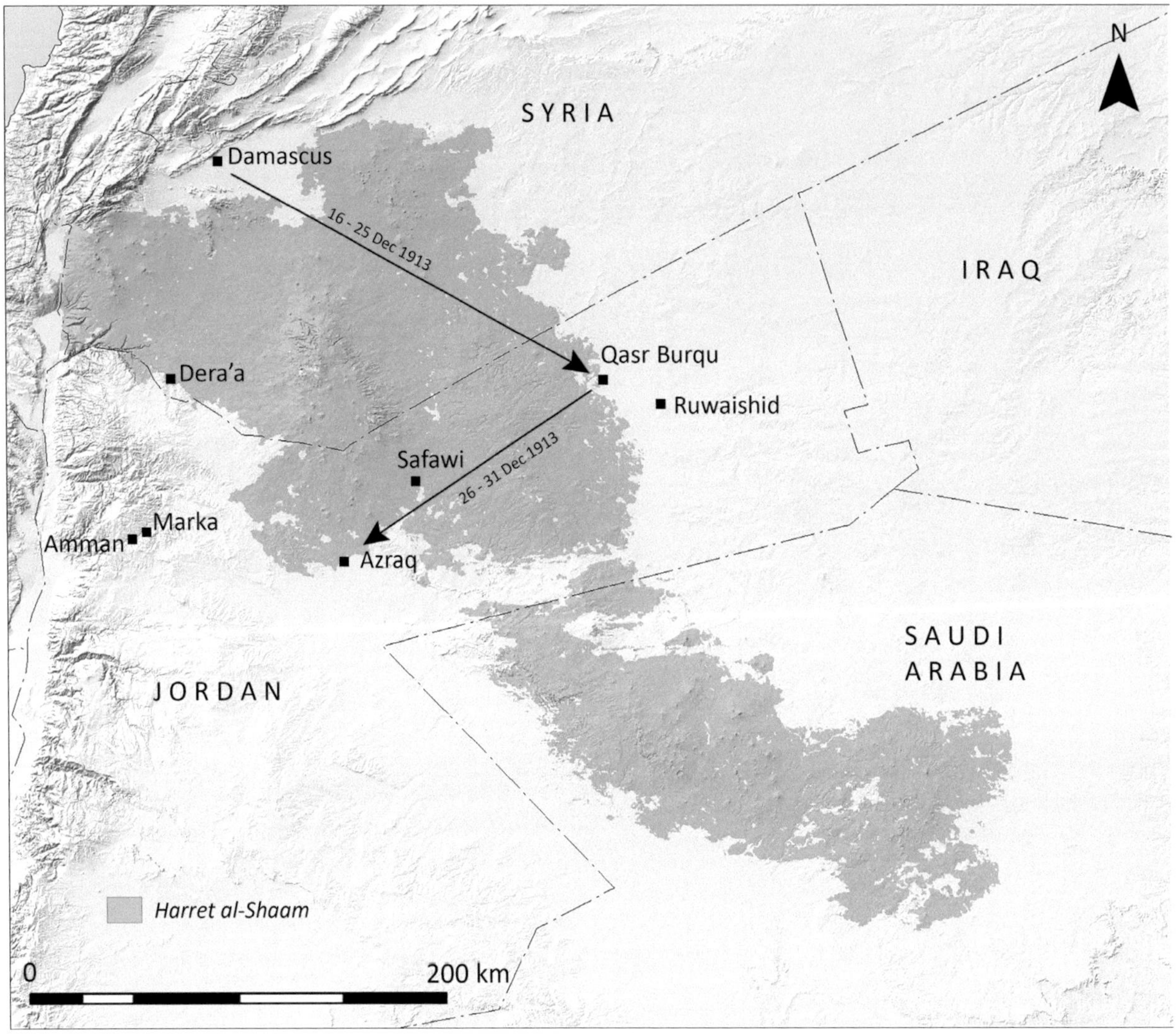

Figure 1. The journey of Gertrude Bell across the Harrat al-Sham in December 1913 (drawn by M. Dalton, T. Hearn).

Introduction

Despite the increasing numbers of western travellers 'east of Jordan' in the century before the First World War, few entered the great lava field, the Harrat al-Sham, stretching from southern Syria through the present Jordanian panhandle and into northern Saudi Arabia.[1] The north-western fringes, *i.e.* the Hauran, where rainfall was higher and the underlying soils more fertile, was extensively settled in the past, not least in the Nabataean to Early Islamic periods, and extensive remains of towns, villages and farms have been reported and often well-recorded (*e.g.* in Jordan, Umm el-Jimal; De Vries 1998). A little further east, however, it presents a bleak landscape. A basalt boulder-covering overlies thin sandy soils with little natural vegetation and scant scope for cultivation: few water sources, minimal rainfall and fierce summer temperatures (with little shade) and cold winter days. The landscape presents significant impediments to travel.

A few western travellers did penetrate beyond the Hauran part of the lava field passing through the western edge and to the major water source of the Azraq oasis and down the Wadi Sirhan, *e.g.* Huber in 1878 and 1883, Lady Anne and Mr Wilfrid Blunt in 1879, and Captains Aylmer and Butler came north from Jawf to the Hauran in 1908 (Huber 1885, 2ff.; 1891, 21ff.; Blunt 1881, Chapters III-IV; Butler 1909). More notable was Gertrude Bell. On 16 December 1913 she rode and walked from Damascus for *c.* 200 km across the northern *harrah* to Qasr Burqu', then she turned south-west and made for the Azraq oasis *c.* 135 km away, arriving on 31 December (Bell 1927, 311-20; cf. unpublished diaries at the University of Newcastle)

1 Lava field, in Arabic *harrah* (singular), *harrat* (plural). In Jordan, it is simply thought of as *the* lava field and commonly referred to as *al-harrah*.

Figure 2. A typical landscape in the Harrat al-Sham near Safawi, Jordan. Note the kite tail along right and stone heaps on the horizon (APAAME G_20091014_DLK-2).

(Fig. 1). Bell was an observant traveller and explorer and experienced archaeologist but her contemporary diary and letters have little to report about the remains encountered. Indeed, what is striking is that she so seldom did so, except at a handful of major sites where there were substantial structures: an early Islamic village (Jabal Seys), a Byzantine tower (Qasr Burqu'), Roman forts (Qasr Aseikhin and Qasr Azraq). In between she recorded a few Safaitic inscriptions but otherwise had virtually nothing to say. That might seem natural in view of the bleak character of the region.

Bell's second stage, from Burqu' to Azraq, extended over five days. As we now know, that took her close to and probably often *over* hundreds of ancient structures. She would certainly have been aware of heaps of rocks or crude lines of boulders marking something man-made but making no sense. At ground level and largely on a great natural carpet of boulders they merited no comment by Bell (or other western travellers before her), as indeed they still do today for most visitors (Fig. 2).

That was all about to change: the same month had seen the first appearance of aircraft in the region (competing in a race from Paris to Cairo) and the first known aerial photographs of archaeological sites in the Near East were published (Anonymous 1914). Within five years, the belligerents in the First World War brought hundreds of aircraft to the Middle East and established reconnaissance and aerial photography as outstanding tools for exploring and recording landscapes. In Syria, German pilots certainly overflew the lava field around Dera'a but the surviving aerial photographs are all of the airfield and vicinity. In Jordan, flying by both German and British and Australian pilots was largely confined to the region west of the lava field. It was not till 1921 and the start of the British Mandate that aircraft began any serious overflights of the great expanse of the lava field. The scene was set for unexpected discoveries.

PART A

'The Works of the Old Men'

First German and Ottoman, then British and Australian pilots flew in 1917-1918 from airfields at Marka (Amman) and Dera'a, then the great mud flat south-west of the Azraq pools was used as an airfield (now the RJAF base) at the time of the northwards thrust of the Arab Revolt forces in 1917-1918. One of those flown there at that time was T.E. Lawrence, a Near Eastern archaeologist in peacetime. Further flights for reconnaissance and map-making of north-western Jordan were undertaken from airfields in Palestine (Thomas 1918; 1920). Although most of the flying in Jordan would have been in the triangle encompassed by those airfields (Fig. 1) rather than the interior of the lava field, there were many sites visible and unavoidable, especially around Azraq itself and around Dera'a. No report survives, however, and none of the surviving aerial

	Syria	Jordan	KSA	TOTAL	Rating
Kites	821	1595	174	2590	9
Wheels	394	3035	5	3434	6
Pendants	53	1016	200	1269	6
Walls	26	211	15	252	4
Chain Walls	21	148	-	169	5
Cairns	4	249	5	288	1
Cairn Rings	1	34	1	36	3
Cairn Fields	5	7	-	12	7
Enclosure	-	32	-	32	3
Circular Paths	-	378	-	378	3
Bullseyes	3	19	2	24	5
Camps	135	128	37	300	1
Corrals	-	64	1	65	1
Total	1463	6916	440	8849	

Table 1. Principal site types in the three components of the Harrat al-Sham. Rating out of 10 reflects a rough assessment of the degree of completeness of the count in each case, with 10 as high.

photographs include examples of the stone-built structures soon to be 'discovered'.

The first indication that there was much to be seen, came that year from a senior RAF officer being flown across part of the lava field: "A peculiar thing about this lava country is that there are traces of some ancient civilisation on it. There are several circles of lava blocks which may very likely be the remains of houses, but in addition there are straight walls, which from the air look like the boundaries between fields, ..." (Brooke-Popham 1921, 573).

That same year (1921) saw the initiation of the Amman to Baghdad section of the Airmail Route and regular flights by RAF pilots across the Jordanian part of the Harrat al-Sham. Six years later, the first volume of the new periodical, *Antiquity*, published an article which revealed these "traces of some ancient civilization" for the first time in some detail, including four aerial photographs. This publication marked the discovery of a vast prehistoric landscape in the Harrat al-Sham. It was a revelation (Maitland 1927; cf. Rees 1929; Insall 1929; Kennedy 2012a for discussion and references). A region so obviously inhospitable for life was in fact thickly strewn with thousands of stone-built structures, vastly outnumbering the mere handful of historic structures (mainly Roman), such as those at the Azraq oasis, Qasr Burqu', Deir al-Kahf, Deir al-Qinn, Qasr Uweinid and (Bronze Age) Jawa (cf. Kennedy and Bewley 2004, *passim*). Most famous amongst these new site types were those the pilots called Kites, but even the earliest records by these pilots included references to, and even aerial photographs of, other site types: Pendants, Wheels, Cairns and Walls. Today, that list can be extended further to include Cairn Rings, Cairn Fields, Chain Walls, Enclosures, Camps, Enclosure Paths and Corrals. In addition, there are a small number of quite significant prehistoric settlement sites (see Müller-Neuhof 2015). Since the 1920s these stone-built structures have been known collectively (adopting the description of the Bedouins) as 'The Works of the Old Men' (Maitland 1927).

The numbers of sites of each type is very large (Table 1). As far as imagery allows, only Kites have been counted systematically across the entire lava field. Large parts of the Hauran have been extensively developed but vertical survey aerial photographs of 1953 provide evidence of Kites now lost in at least the southern part. The total may be close to complete. Pendants, Wheels and Chain Walls have been counted less systematically but still extensively and the numbers provide a guide to orders of magnitude. The Rating column reflects the conjectured degree of completeness of the count. The tabulated total is nearly 9000; the likely total perhaps as much as 50,000. Although the walls of the structures are often a metre or less in height and span a period of several thousand years, collectively they represent an enormous output of activity.

Kites are the best-known and most characteristic of the site types in the Harrat al-Sham. They are also found more widely and in quite large numbers in other parts of north Arabia including extensively beyond the lava fields (cf. Kennedy 2012b; 2014). As the map shows (Fig. 3), Kites are found in central and northern Syria, into adjacent parts of north-western Iraq and south-eastern Turkey (and further north still in Armenia). A few have been recorded west of the Jordanian part of the lava field, just east of the airfield at Marka, a dozen more in south-eastern Jordan (Abu-Azizeh and Tarawneh 2015), and at least two on the Jordanian side of the Wadi Araba.

Considerable numbers of Bullseye Cairns and Pendants are also found elsewhere in north Arabia but mainly in the hill chain extending from south-west to north-east of Palmyra and interspersed with dozens of Kites. Wheels are seldom encountered elsewhere in northern Arabia.

Lava fields in the Arabian Peninsula

The Harrat al-Sham is the most northerly of several major lava fields extending in succession down the western coast of the Arabia Peninsula as far as south-western Yemen (Fig. 4). (Indeed, the volcanic belt continues across the Red Sea and Gulf of Aden into an immense area of Djibouti, Eritrea, Ethiopia and Somalia). They vary considerably in size. After the Harrat al-Sham (*c.* 40,000 km^2), the two largest are in Saudi Arabia and adjacent to one another: the Harrat Khaybar (including the attached Harrat Ithnayn and small Harrat Kura) (20,564 km^2) and Harrat Rahat (19,830 km^2). Some have witnessed eruptions in relatively recent times. For example, there were eruptions in the Harrat Rahat

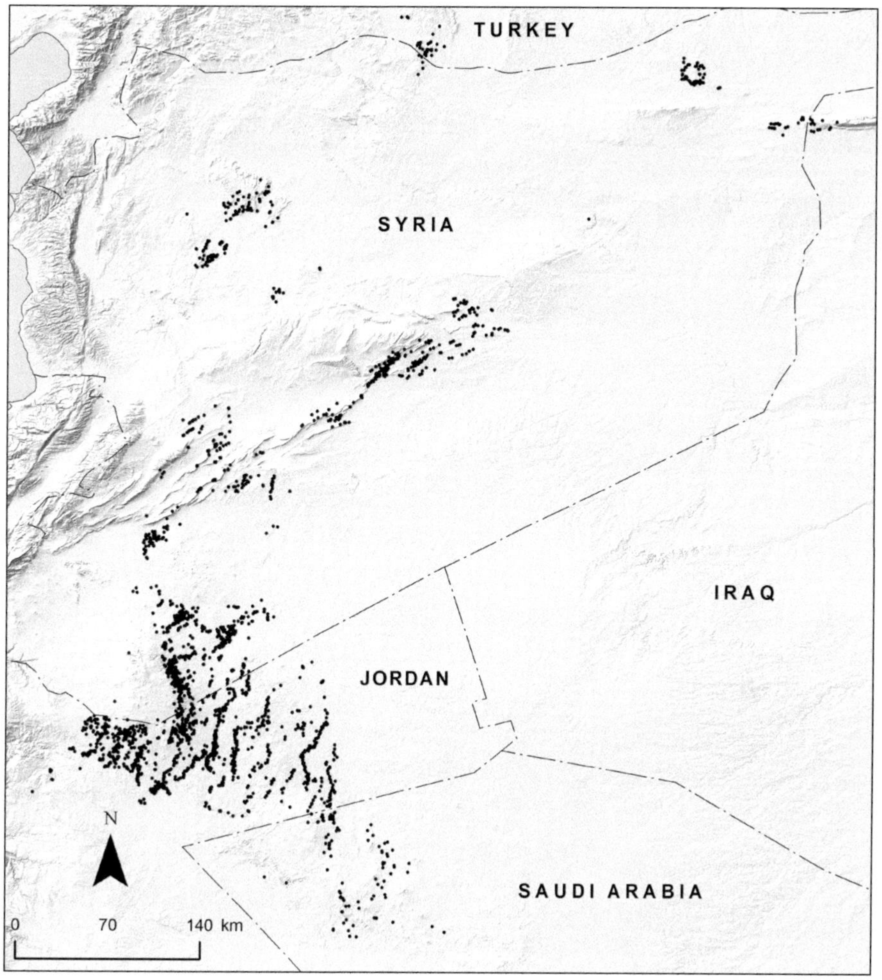

Figure 3. Distribution map of Kites in north Arabia (drawn by M. Dalton).

in AD 641 and then again in AD 1256, when a flow of 23 km from cones in the north of the *harrah* threatened Medina (Moufti *et al.* 2013). Also in the seventh century AD is a reported eruption in Harrat Khaybar, possible eruptions in Harrat Uwayrid in AD 640, and in Harrat Lunayyir in the tenth century AD (Global Volcanism Program 2011).[2]

In light of the central place of Harrat al-Sham in the mass of evidence for the 'Works' and their overwhelming concentration in that lava field itself, it is natural to explore and compare these other large lava fields. We should note first, however, that just as Kites are also found beyond and well away from the lava field, so too many Pendants and Bullseye Cairns are found in the Arabian Peninsula beyond its lava fields. Especially notable are those recorded in large numbers in the Hadramaut of central Yemen (Cleuziou *et al.* 1992; De Maigret 1996, 333; 2009, 337; McCorriston *et al.* 2002, 62; Steimer-Herbet, 2001, 223; 2004, 26, 74, 80, 96). As we shall see, however, the striking concentrations are again in some of these south-Arabian lava fields.

2 See also http://www.sgs.org.sa/English/NaturalHazards/Pages/Volcanoes.aspx.

Although high-resolution imagery is not available for coverage of all the *harrat* in the Arabian peninsula, it is now extensive and permits generally reliable evidence for at least the presence or absence of 'Works'. Google Earth is the most user-friendly platform but it can often be supplemented by Bing, which sometimes has a different and superior coverage. Systematic analysis of the imagery and recording of sites is time-consuming and inevitably uneven. In some instances, specific 'windows' of high-resolution imagery have been explored systematically; elsewhere such windows have only been explored briefly to assess the presence/absence of structures and form impressions of types and numbers. Together they permit generalisations to be made and in some instances, such as Harrat Khaybar and large parts of other *harrat*, detailed interpretation and comparison both of site types and of individual sites. Harrat Khaybar has been especially interesting as it has been explored systematically at different times over several years with additional high-resolution windows, allowing interpretation to be progressively extended until we have now near-complete coverage and very extensive cataloguing of sites (Kennedy *et al.* 2015, Figs. 8-9).

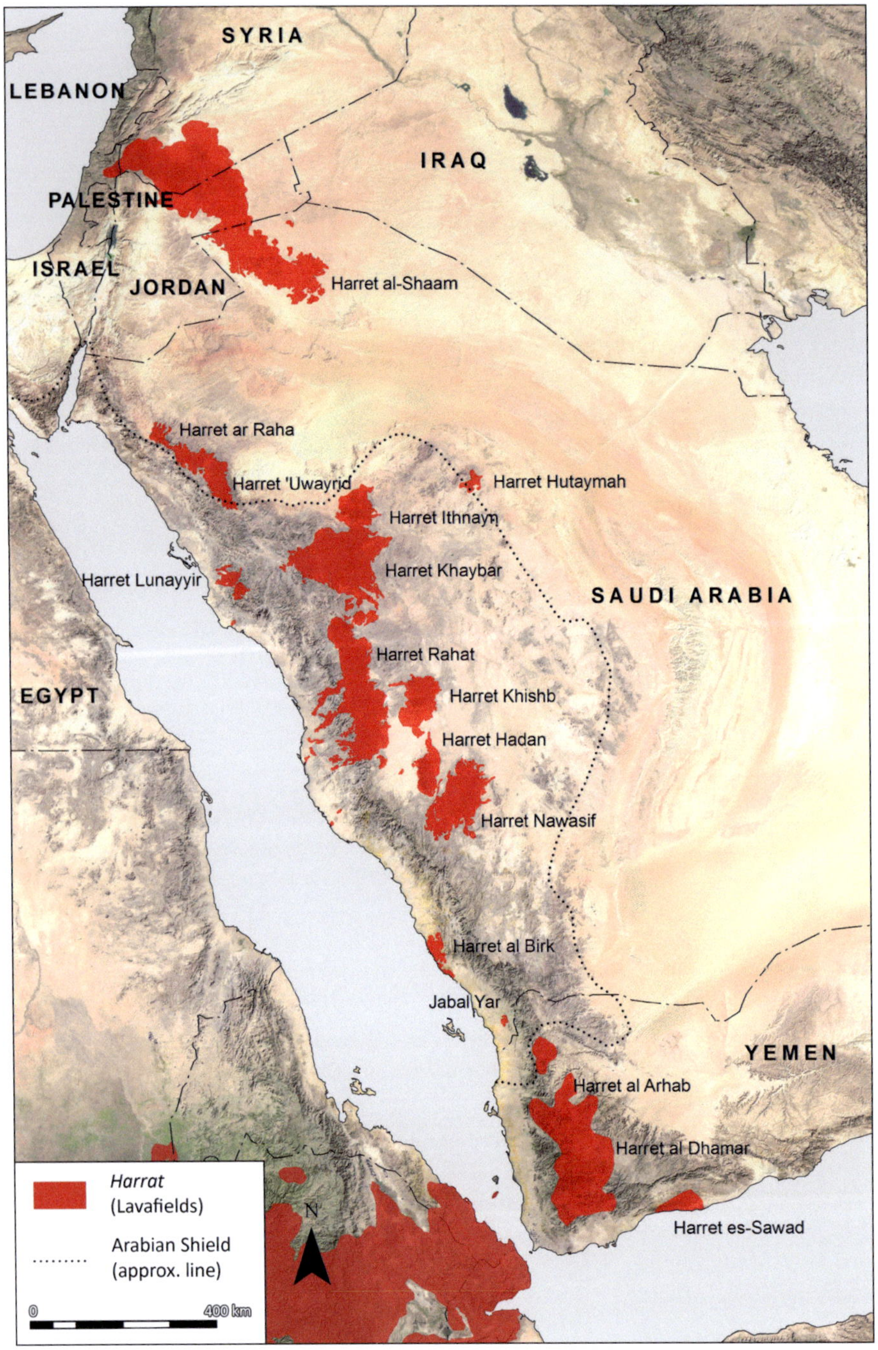

Figure 4. Map of lava fields in Arabia (drawn by S. Smith and T. Hearn).

The lava fields of Ar-Rahah, Ithnayn, Hutaymah and Lunayyir have relatively few sites visible on the satellite imagery of Google Earth and Bing – generally only simple Pendants or Bullseye Cairns. Al-Birk shows no sites at all. In contrast, Uwayrid, Khaybar, Rahat and Haddan have many thousands of sites, comparable in density to what one encounters in the Harrat al-Sham. The picture emerging for Harrat Nawasif suggests dense concentrations in at least some parts and Kishb certainly has hundreds of sites beyond the area of most recent lava flow. What is striking, too, is that with the exception of the Harrat Khaybar, the range of site types in all of these lava fields is restricted to Pendants, Bullseye Cairns, Barred Rectangles, Triangles and Cairns.

PART B

Novelty in site types and/or forms

Table 2 shows graphically the range of principal site types in the lava fields of south Arabia. It does not adequately reveal the variations which often

	Harrat al-Sham	Ar-Rahah	Uwayrid	Khaybar	Ithnayn (= part of Khaybar)	Lunayyir	Rahat	Kishb	Hadan	Nawasif	Al-Birk	Yemen	Totals
Kites	2590	-	-	804	-	-	19	-	-	-	-	20	**3433**
Wheels	3434	-	-	75	-	-	159	23	-	-	-	-	**3691**
Pendants	1269	2	44	3919	24	18	2076	324	68	476	1	115	**8336**
Walls	252	-	-	64	-	5	17	-	-	-	4	-	**342**
Chain Walls	169	-	-	1	-	-	-	-	-	-	-	-	**170**
Cairns	288	-	12	648	17	1	929	42	14	12	-	1	**1964**
Cairn Rings	36	-	-	6	-	-	-	-	-	-	-	-	**42**
Cairn Fields	12	-	2	1	-	-	1	-	-	-	-	-	**16**
Corrals	65	-	-	1239	36	-	-	-	-	-	-	-	**1341**
Enclosures	32	-	-	21	-	10	4	-	-	2	16	-	**85**
Enclosure Paths	378	-	-	-	-	-	-	-	-	-	-	-	**378**
Camps	300	3	19	530	-	-	12	-	-	-	-	-	**864**
Bullseyes	24	5	390	501	7	12	60	23	1	20	-	-	**1043**
Triangles	1	-	158	89	-	1	10	1	-	1	-	-	**261**
Gates	-	-	-	256	-	-	5	3	-	-	-	-	**264**
Rings	-	2	19	-	-	-	35	1	1	10	-	-	**68**
Circles	13	-	-	3	-	-	33	1	-	-	-	-	**50**
Barred Circles	1	-	-	4	-	-	40	1	-	-	-	-	**46**
Rectangles	-	-	-	2	-	-	29	3	-	-	1	-	**35**
Barred Rectangles	-	-	-	13	-	-	140	6	-	-	-	-	**159**
TOTAL	8890			8176	84								**22514**

Table 2. Principal site-types in the Arabian lava fields. The numbers are of varying degrees of completeness. The figure for Kites is probably close to the number still surviving; those even for Wheels and Pendants for which there are large numbers are less complete, because so much of the Saudi lava fields remain unexamined. Cairns are found in large numbers and all sizes almost everywhere and the final count is likely to be tens of thousands. 'Corrals' is a term taken over from the American-produced K737 1:50,000 maps of Jordan. It is applied to usually small curvilinear enclosures, often in clusters, and, as the name implies, used as animal pens. Most are likely to be the work of Bedouins in the last century but some are certainly older and some are built over and from the stones of earlier structures. Their ubiquity can be gauged from the one part of the Harrat Khaybar (Al-Hiat), where all were recorded systematically.

distinguish a type found in one lava field from those in others. That was illustrated in the case of Kites where the large numbers in the Harrat Khaybar included a Barbed form and a Y form not found in the Harrat al-Sham or, indeed, elsewhere (Kennedy *et al.* 2015; Kennedy, in press; cf. Kennedy 2012b). Few Kites have been found anywhere else in Saudi Arabia; several dozen have been recorded in Yemen but once again of a type not just different in form but wholly unparalleled anywhere else (Brunner 2008; 2015). Details also vary; the numerous very complex, re-modelled and multi-tailed Kites of the Harrat al-Sham are completely absent in Harrat Khaybar.

Variations can be seen in other site-types known from the Harrat al-Sham and there are site types found in or around other lava fields on the Arabian Shield which are unique to those areas (below; cf. Kennedy 2017).

Cairns

Cairns are ubiquitous, most consisting of simple circular heaps of boulders very varied in size, which could be of any and every date. Some are demonstrably funerary but yet others have revealed no burial (cf. Kennedy 2012d for Jordan). Few have been individually counted but there are certainly tens of thousands in Arabia as a whole. Indeed, in Jordan and now occasionally elsewhere, there are 'Cairn fields' consisting of large numbers of small heaps of stone, very close together, seemingly not the water-harvesting technique seen in the Negev Desert (Evenari *et al.* 1982) but of no obvious purpose (cf. Kennedy 2012d, 497-499 and Figs. 14-15, where n= *c.* 1300 small cairns). Recent work, however, reported elsewhere in this volume has revealed that some such as those on the slopes of the Jebel Aseikhin and around Maitland's Hill-Fort are in fact simple huts.

Figure 5. Cairns in Harrat al-Sham. (a) Two Cairns, one a Bullseye (Burqu' Kite 6, Burqu' Bullseye Cairn 1 (APAAME_20111024_DLK-0096C) inside the head of Burqu' Kite 6 (and presumably therefore more recent than it). (b) Low-level oblique photograph of a chambered cairn which shows from vertically above as a Bullseye Cairn (Qattafi Pendant 4; APAAME_20100601_SES-0103C). (c) A Bullseye Pendant as sketched by a RAF pilot (Rees 1929, 3).

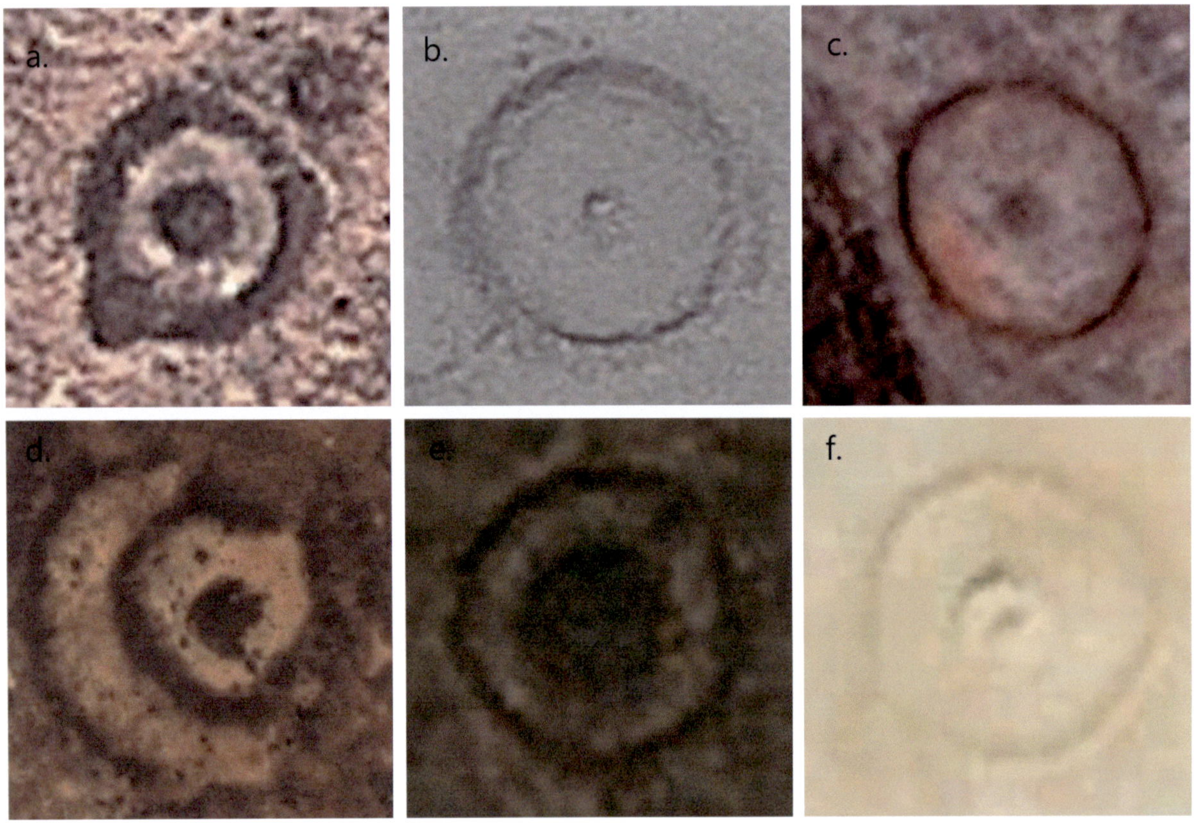

Figure 6. Bullseye Cairns in Saudi Arabia. (a) 3627-44 Az-Zawiyah: Bullseye 2 (18 and 6 m in diameter). (b) 3725-24: Bi'r ash-Shu'aybiyah Bullseye 1 (30 and 2.5 m). (c) 3627-21: Bi'ar al-Khalas Bullseye 15 (25+ m; 4.5 m). (d) 3921-11: Madrakah Bullseye 1 (80, 55 and 17 m). (e) 4023-44: Sha'ibal Batra' Bullseye 2 (20 and 11 m). (f) 4823-21: Yabrin (North) Bullseye 7 (24 and 8 m) (source: Google Earth).

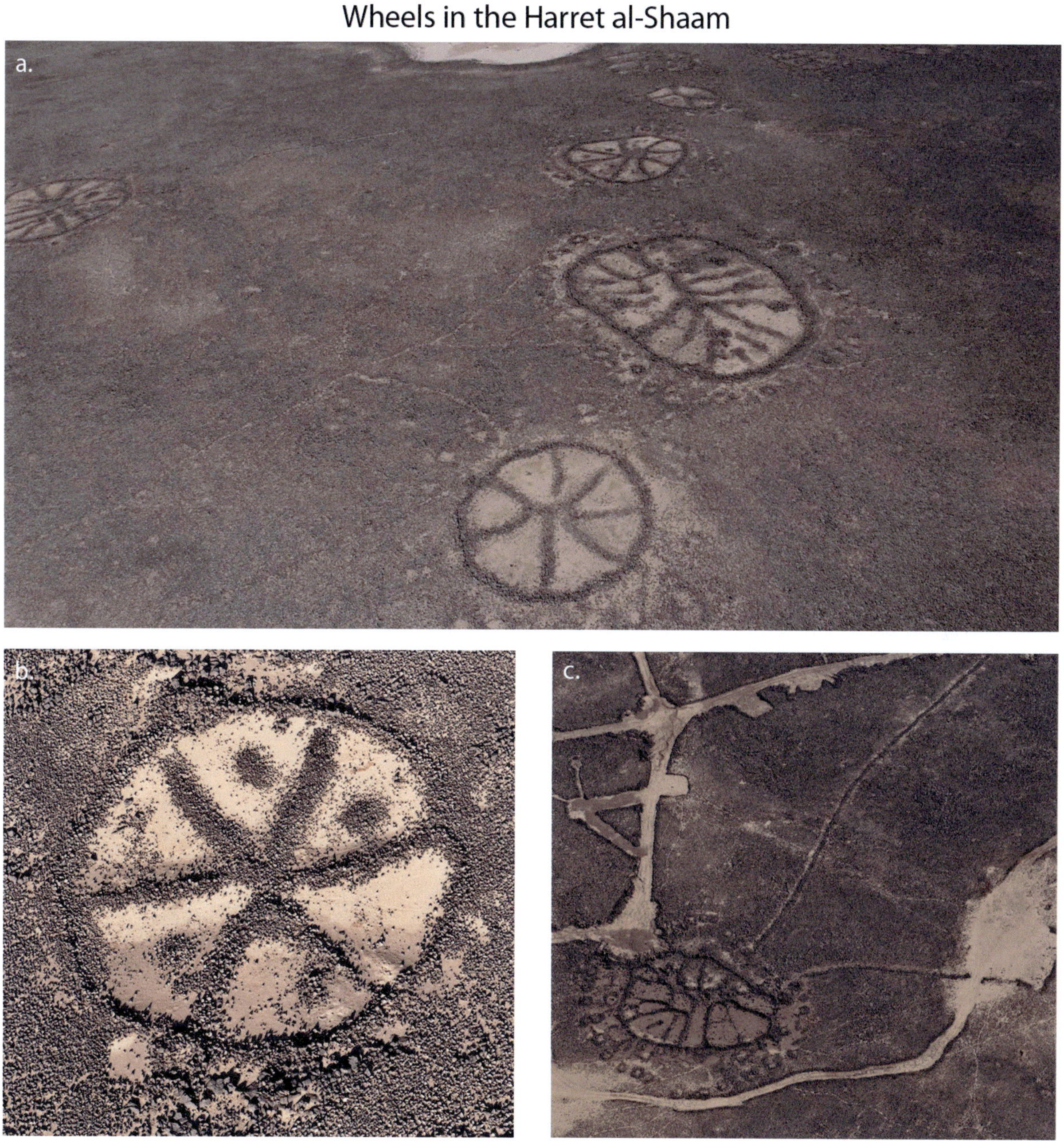

Figure 7. Wheels in the Harrat al-Sham (Kennedy 2012c, Fig. 2).

One type that stands out in the lava fields of Saudi Arabia is the Bullseye Cairn, a ring in the centre of which is a cairn. The type is found in the Harrat al-Sham (and in Palmyrena) but is rare (n= 24). In some cases, at least the inner cairn appears to be the remains of a collapsed circular chamber: a tower tomb, set inside a low ring wall (Fig. 5a). The type is common as a component of a Bullseye Pendant (Figs. 5b-c).

Bullseye Cairns survive in immense numbers in many other lava fields (Table 2), most notably the adjacent Harrat Uwayrid and Harrat Khaybar, where more systematic counts were made (n= 390 and 508, respectively). The form in the Saudi lava fields is visually different from its Jordanian counterpart: as viewed from above, the inner cairn looks to be no more than a tiny heap of stones, too small to be more than a marker (Figs. 6a and c). When measured, however, it is immediately apparent that it is a matter of scale: in contrast to the Bullseye Cairns in Jordan where the outer ring is no more than 4-5 m in diameter, those in the Saudi lava fields are much greater. Fig. 6 illustrates examples from or around different Saudi lava fields and gives the approximate dimensions. As may be seen, the ring is often 20-30 m in diameter and the inner cairn 2-10 m across (or more). In one example (Fig. 6f) we see a type (rare, but there are other examples) consisting of *two* concentric rings, the outer in this

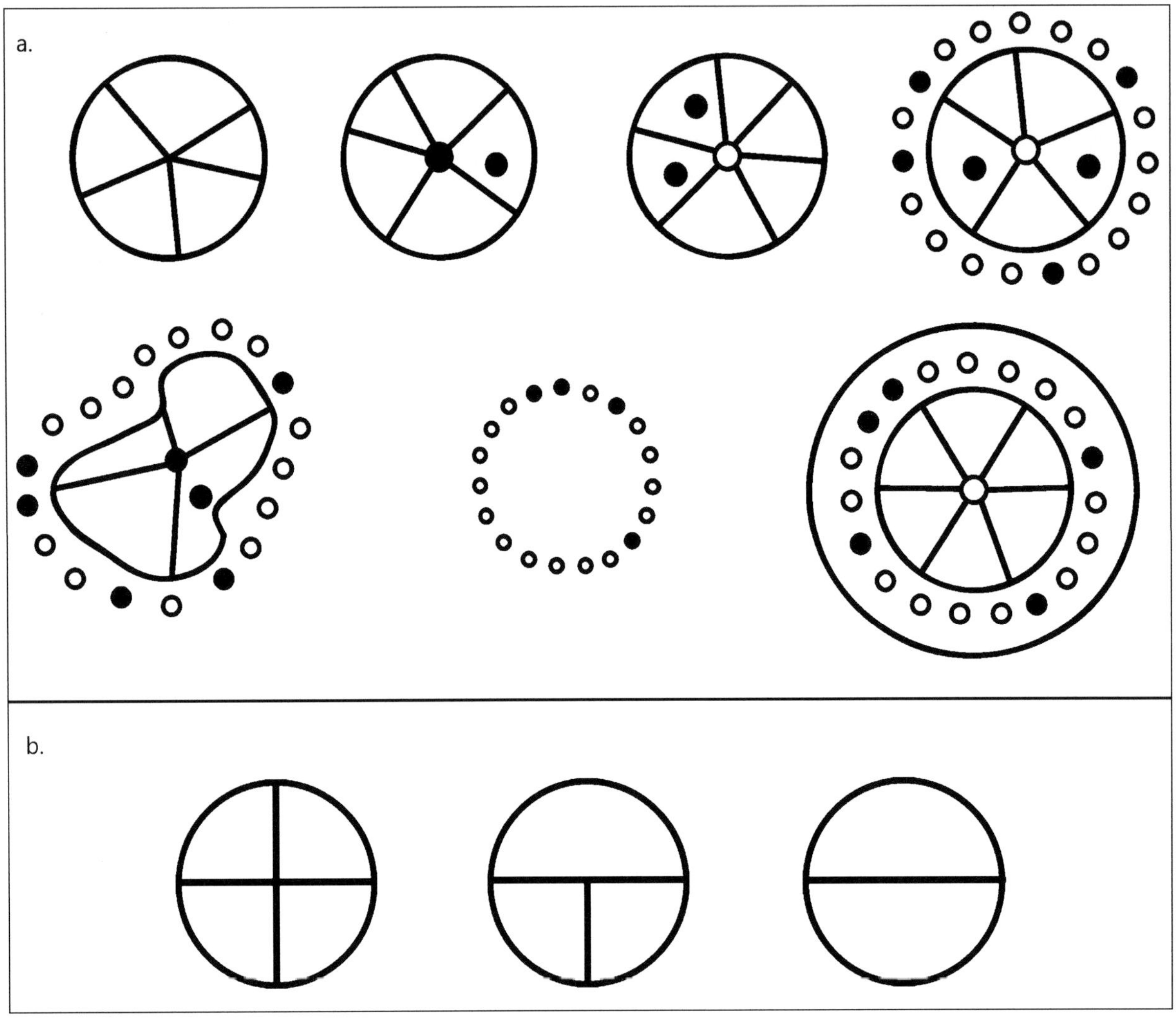

Figure 8. (a) Schematic drawings of a variety of Wheels forms in the Harrat al-Sham (Drawn by S. Smith). (b) Wheel types in south Arabia lava fields (drawn by T. Hearn).

instance being a huge 80 m in diameter. The size of the inner cairn leaves open the possibility that in at least some cases these Bullseye Cairns are funerary.

The distribution of Bullseyes within areas intensively and systematically interpreted is instructive though still in its infancy. In the map square 'SAUDI-3726-14-THARBAH' there are 154 Bullseyes but just 12 Pendants, half of which are Bullseye Pendants. It is possible that Bullseye Pendants are simply those Bullseye Cairns for which a tail has been added at some stage (cf. Fig. 21, below).

Wheels

Wheels of the type and in the numbers recorded (n= 3434) in the Harrat al-Sham (Kennedy 2012c) are rare in all other lava fields (n= 257) (Fig. 7). 'Rings' are found in large numbers but they are usually relatively small, simply divided by a single horizontal bar, a pair of bars crossing in the middle, or occasionally a T-shape. There are a few much larger rings which have multiple 'spokes' as in the Harrat al-Sham examples, but they are commonly found with tails and categorised as Pendants. There are none of the complex forms found in the northern lava field (Fig. 8) and none of the Wheel Groups found in Jordan (Kennedy 2012c).

Pendants

In the Harrat al-Sham, Pendants are common (n= 1269) and are usually a simple Cairn with a short 'tail' of small cairns (Fig. 9; cf. Fig. 5c). The Cairn can be a circular chamber collapsed to form a cairn with a surrounding ring wall, giving the appearance from above of a Bullseye Cairn, in which the central cairn is very large; the tail often has the impression of a continuous wall but many appear in fact to consist of a succession of small cairns which have collapsed and fused.

Figure 9. Safawi Pendant 52 and Wheel 290 (APAAME_20120522_DLK-98).

In a few cases the 'tail' may consist of constructed box-cairns rather than stone heaps.

Pendants are very common in other *harrat*. Although interpretation of the imagery is still only partial, there are already 50% more recorded for Harrat Rahat (n= 2076) and there are 300% more in Harrat Khaybar, where interpretation is fairly complete (n= 3943) (Table 2), than for the Harrat al-Sham (n= 1269). They are often quite distinctive and variants of the simpler types may be noted and illustrated (Fig. 10):

- Bullseye Pendants are common, with several dozen in the Harrat Rahat alone.
- In contrast to Jordan, those in other lava fields can have immensely long tails (Fig. 10). *E.g.* '4023-44 – Sha'ibal Batra Pendant 146' has a tail 9.75 km albeit not continuous; '4022-44 – Abar al-Mayayn Pendant 18' is 2.2 km long, very straight and seemingly continuous, then becoming cairns and, like some others, ends at a prominent outcrop/hill.

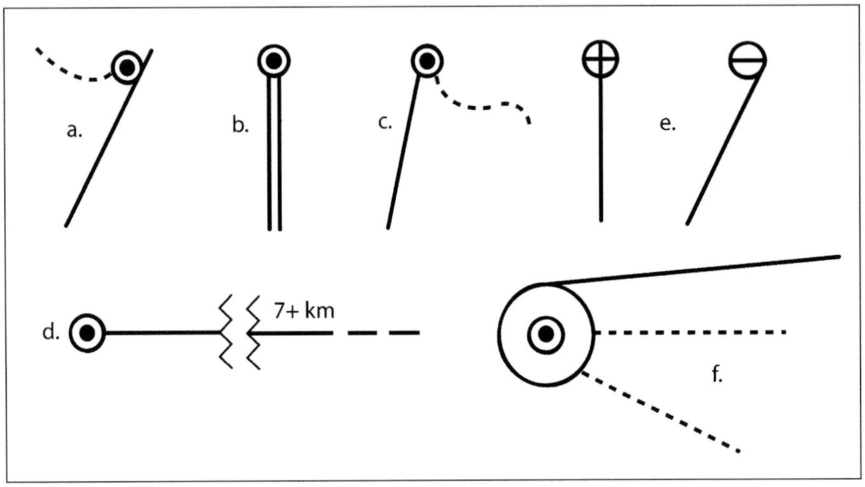

Figure 10. Pendant types in the Harrat Rahat (Kennedy 2011, Fig. 20) (drawn by S. Smith).

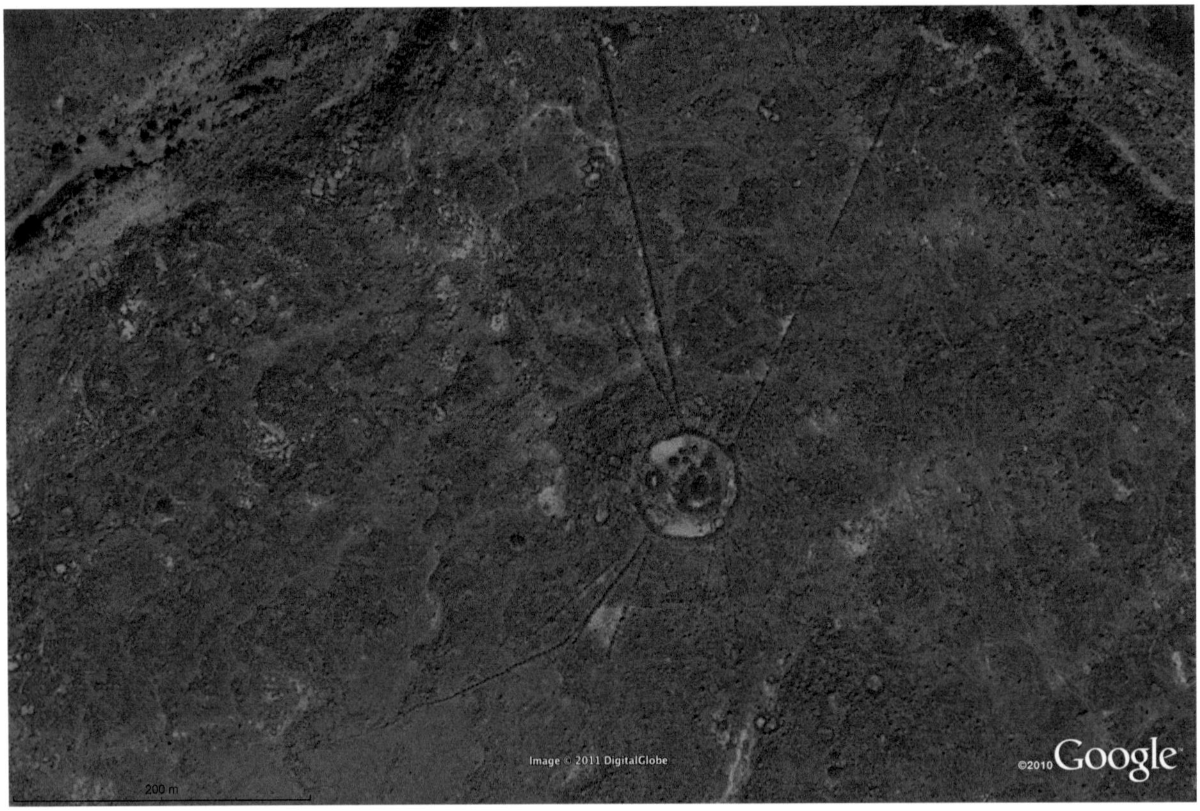

Figure 11. A complex Pendant in the Harrat Rahat with multiple tails (Kennedy 2011, Fig. 19) (source: Google Earth).

- The tails themselves can be a meandering line of small cairns or a thick and very straight continuous line.
- In some cases, the tails, especially those which are very straight and long, seem never to have been continuous but rather to consist of a series of short lengths, but on exactly the same alignment and often running over hills and through depressions.
- Some heads have two or more 'tails', sometimes of different forms: a meandering line of small cairns and a continuous straight tail (Fig. 11).
- Sometimes the tail touches the Cairn at a tangent.

Figure 12. Al-Hiat Funerary Avenue 4 in the Harrat Khaybar packed with Trumpet and Keyhole Pendants (source: Google Earth).

Trumpet and Keyhole Pendants

Several types found only in Saudi lava fields and overwhelmingly in Harrat Khaybar are so different as to cast doubt on whether the term 'Pendant' is legitimate. For present purposes, only two types are considered: Trumpet and Keyhole.

In Harrat Khaybar radiating out from each of the two major settlements, Khaybar and Al-Hiat, are several tracks, each flanked by scores of Cairns, Pendants, Trumpet Pendants and Keyholes (Figs. 12-13). They appear to be funerary avenues, some running for great lengths – around Al-Hiat these avenues total at least 13 km.

Trumpet Pendants, as the name suggests, consist of a circular head: a solid cairn or, more commonly, an open ring, sometimes with a small central cairn (a Bullseye). The form of the 'tail' is a triangular funnel, sometimes very long and slender; sometimes squat and looking increasingly like the Keyhole type (below) (Fig. 14).

Although the type is very common, one sub-category is especially notable: Segmented Trumpets. Figure 15 illustrates an outstanding example near Al-Hiat. The head consists of a Bullseye *c.* 25 m in diameter with a central cairn of *c.* 4 m. The 'trumpet' is *c.* 195 m long and *c.* 12 m wide at its furthest end. It is divided by cross-walls into at least ten segments. Not visible in the satellite imagery but revealed by ground photographs is that, in contrast to the simple use of uncut boulders in the structures of all kinds in the Harrat al-Sham, parts at least are built from roughly rectilinear blocks (Fig. 15b).

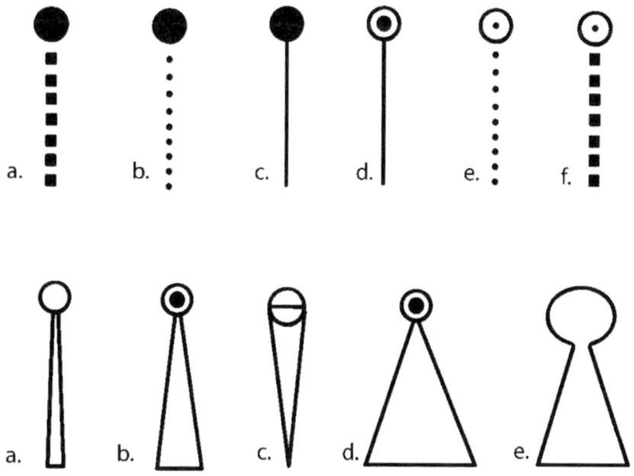

Figure 13. Schematic drawings of Simple, Trumpet and Keyhole Pendants in the Harrat Khaybar (Kennedy 2011, Fig. 15) (drawn by S. Smith).

Even more remarkable are those Pendants which have a keyhole shape. Although they are found widely in the Harrat Khaybar, one group is outstanding (Fig. 16). The seven Keyholes in the South Avenue are *c.* 54 m long. Those in the West Avenue are *c.* 40 m long but, uniquely, are laid out overlapping one another, five in total, on a very clear common base line. Alongside the Pendants, inside some Keyholes are small cairns and there are dozens of individual Cairns on the plateau to the west (Fig. 16b).

Figure 14. Pendants on an avenue near Al-Hiat. Above: looking over the Bullseye head and down the 'trumpet' (SAUDI-4025-14-Hulayfah Pendant 649 (DSC00148_GS)). Below: a wall bounded by two rows of kerb stones (SAUDI-4025-14-Hulayfah Pendant 645 (DSC00147_GS)) (photographs by Grant Scroggie).

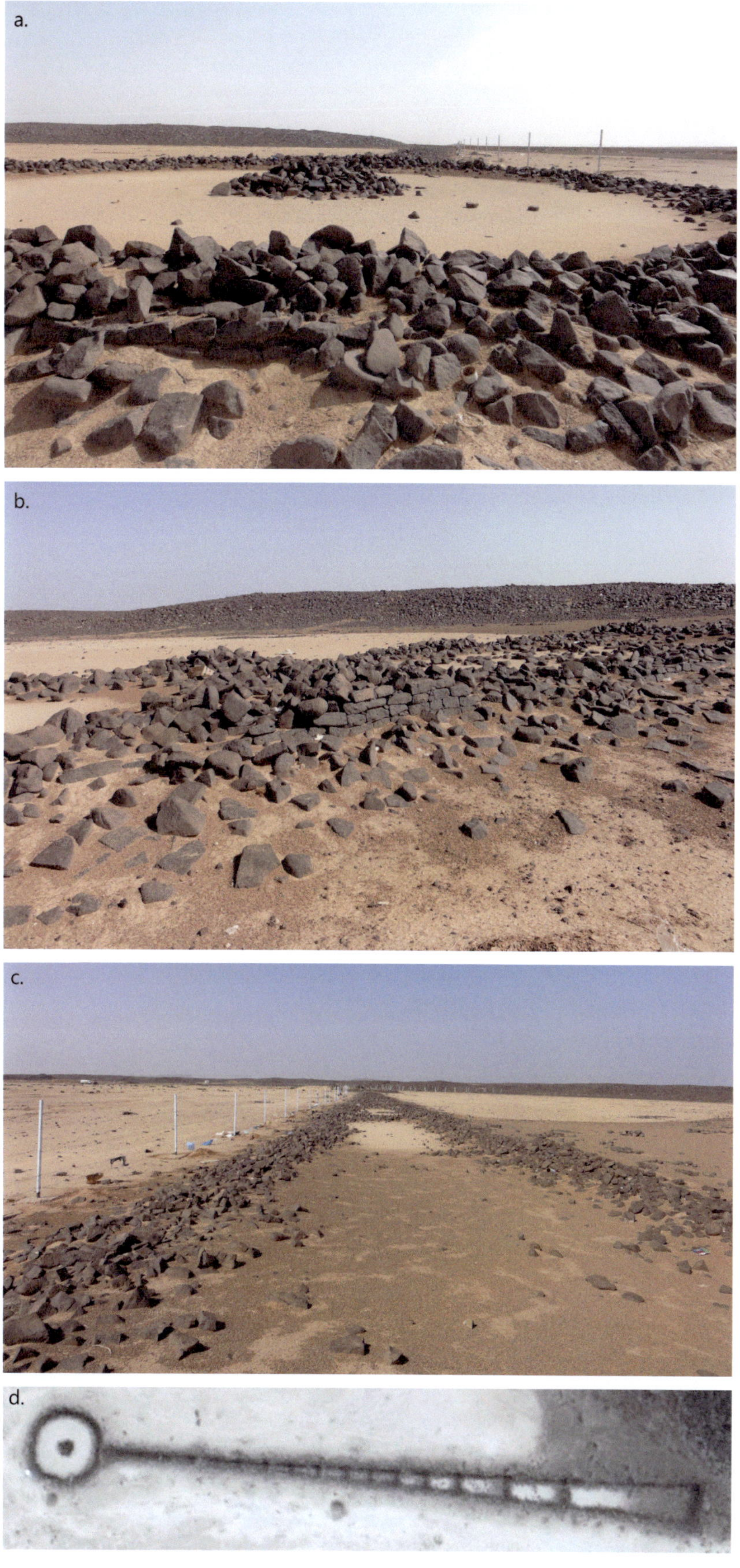

Figure 15. One of a group of Segmented Trumpet Pendants west of Al-Hiat (4025-14 -Hulayfah Segmented Pendant 929). (a) Ground view of showing the Bullseye 'head' and central cairn, with Trumpet stretching into distance (DSC00027_GS). (b) Ground view of showing the coursed and shaped masonry blocks in the walls (DSC00028_GS). (c) Ground view looking along the 'trumpet' towards the 'head' in distance (DSC00034_GS). (d) On Google Earth (ground photos by Grant Scroggie).

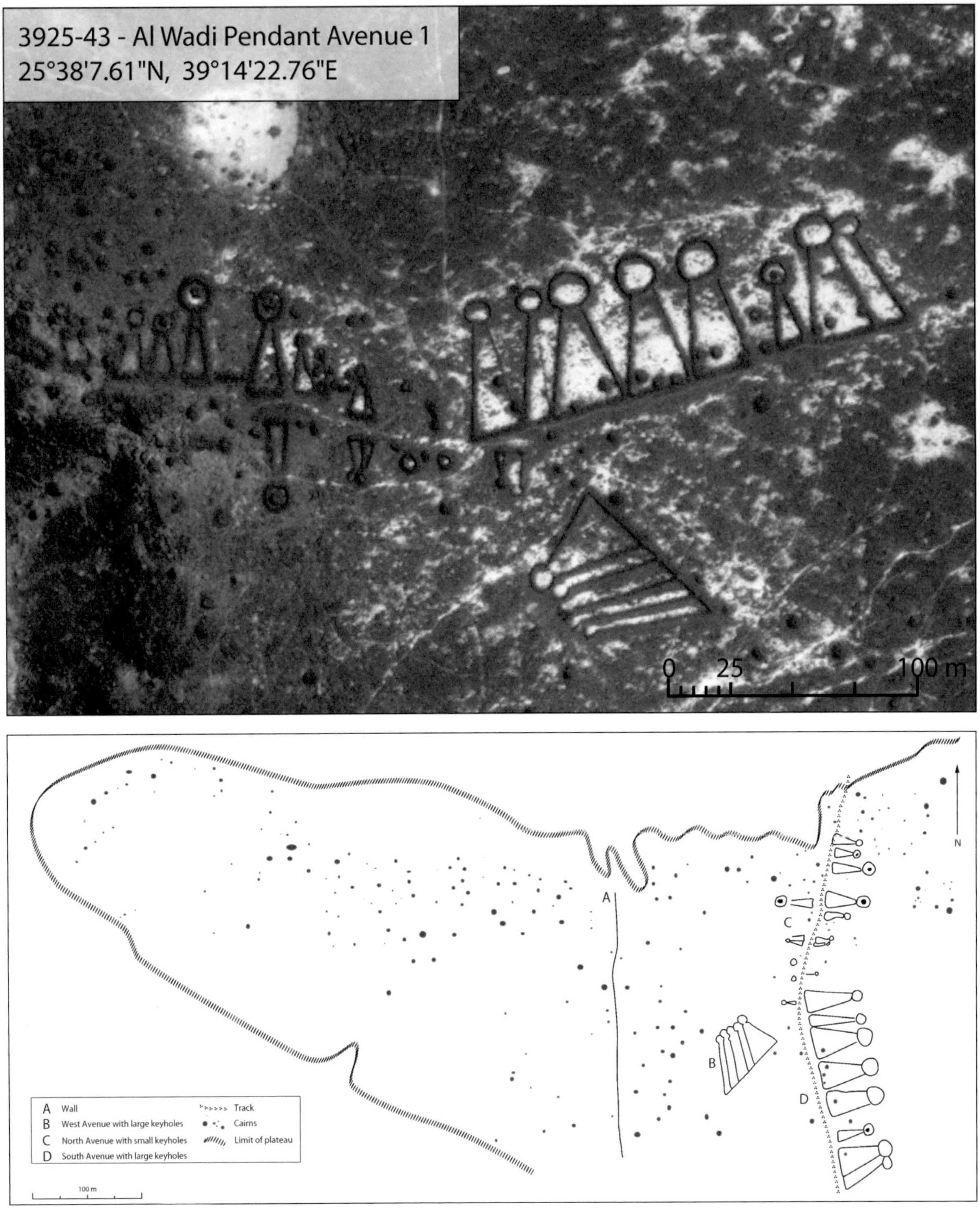

Figure 16. Keyhole Tombs: Al-Wadi Pendant Avenue 1. Above: on Google Earth. Below: sketch plan (drawn Rebecca Repper).

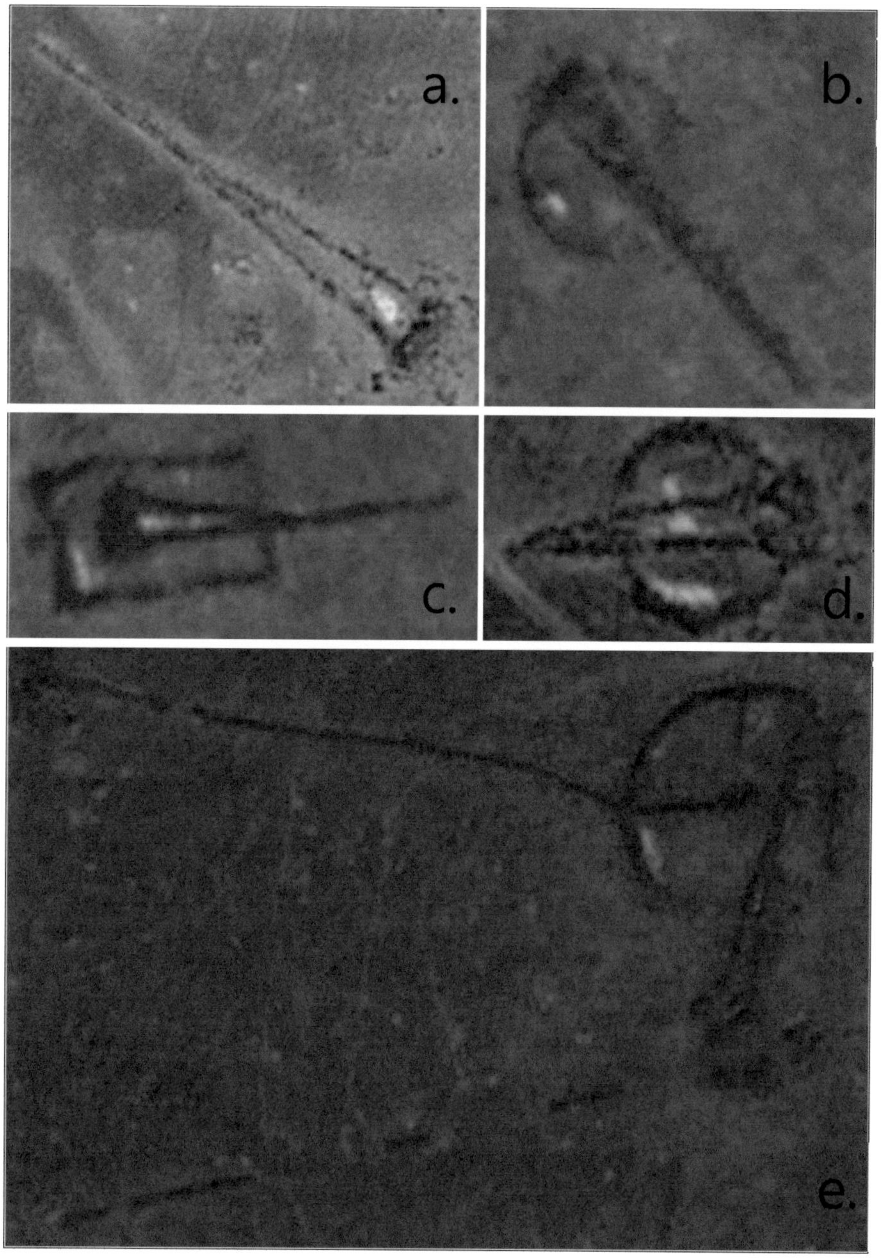

Figure 17. Inverted Trumpet Pendants. (a) 4023-11-Al 'Umaq (Jibal Sayid) Inverted Trumpet Pendant 1. (b) 4023-34-As Sidrah Inverted Trumpet Pendant and Ring 2. (c) 4123-31-Abar ar Raghiyah Inverted Trumpet Pendant and Rectangle 2. (d) 4122-44-Hafir Kishb Inverted Trumpet Pendant and Ring 1. (e) 4023-43-Jabal as Sahiliyah Inverted Trumpet Pendant and Ring 1 (source: Google Earth).

Inverted Trumpet Pendants

A small number of Pendants have been recorded in which the 'trumpet' is inverted: the broad end begins at a small cairn and then tapers to a point (Figs. 13c and 17a). In some cases the Pendant overlies a Ring (Fig. 17b, d and e), and in one case a Rectangle (Fig. 17c).

Triangles

The form of this last type of Pendant involves 'trumpets' which are triangular. A site type defined as Triangles is found in several lava fields in Saudi Arabia (Table 2). At least 261 have been catalogued so far. In a few instances the Triangle is free-standing (Fig. 18, where the lengths range from *c.* 15 to 30 m).

Overwhelmingly Triangles are found in close association with a Bullseye Cairn, sometimes clearly separated by up to 100 m, sometimes quite close together, and sometimes actually touching or even slightly overlapping. These last examples then have the appearance of stubby Trumpet Pendants. The common forms of this pairing are set out in Fig. 19.

In the map 'windows' of the Harrat Uwayrid and adjacent areas which have been systematically interpreted, over 250 Triangles have been catalogued and over 600 Bullseyes. What is notable is that both Triangles and Bullseyes, alone and in association, are found off as well as on the lava field (Fig. 20).

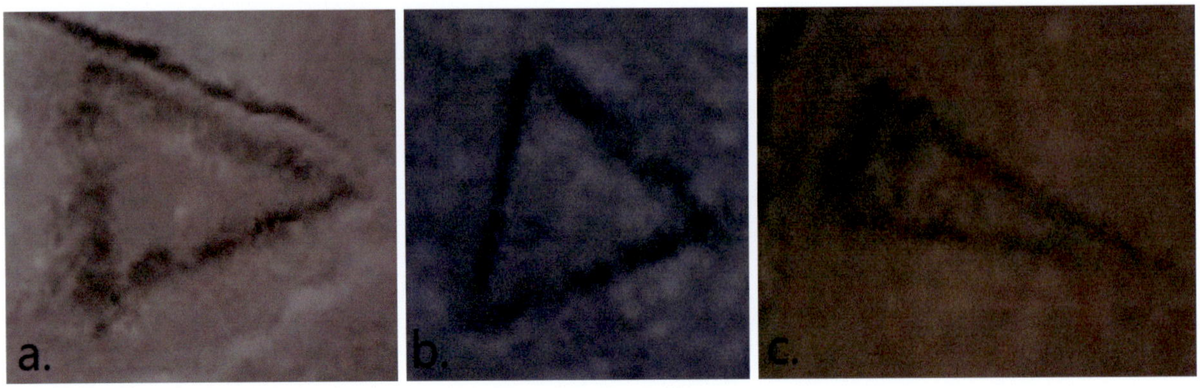

Figure 18: Basic Triangles. (a) 3726-14 – Tharbah Triangle 10 (*c.* 20 m). (b) 3727-23-Wadi Wirqan Triangle 11 (*c.* 15 m). (c) 4023-43-Jabal as Sahiliyah Triangle 1 (*c.* 30 m) (source: Google Earth).

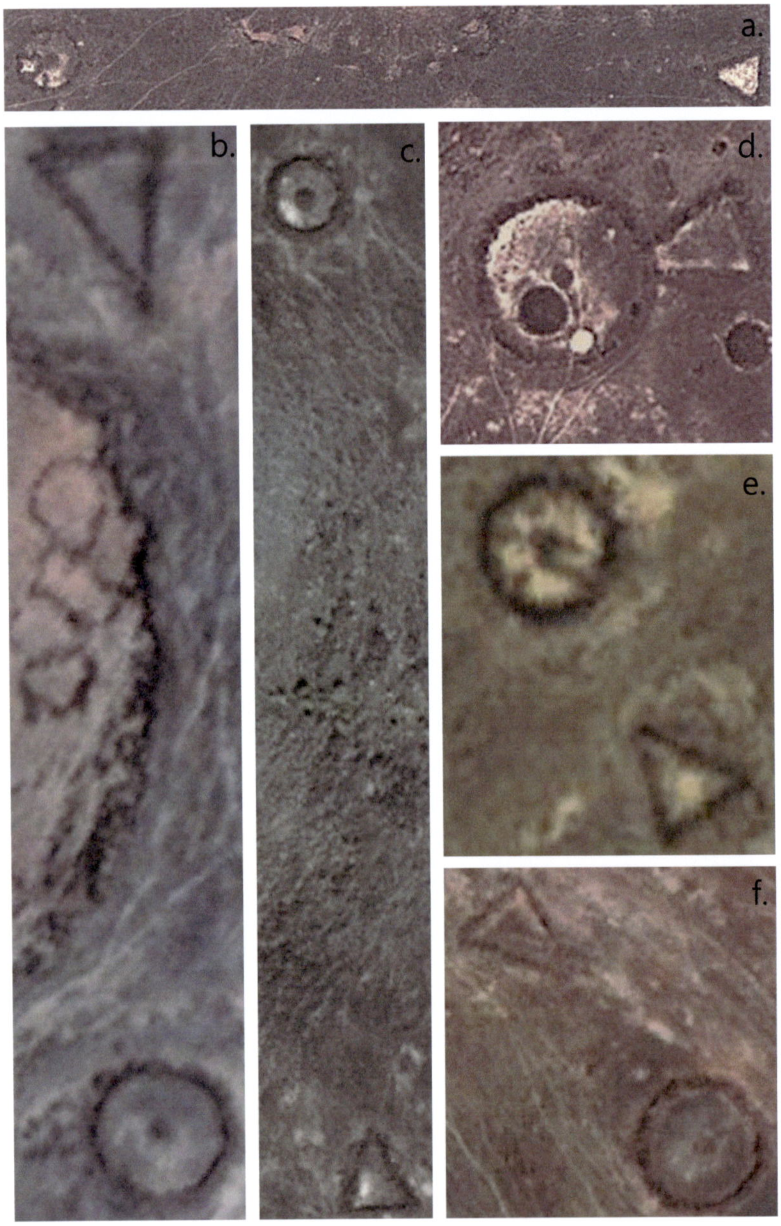

Figure 19. Triangles and Bullseye Cairns. (a) 3727-23-Wadi Wirqan Bullseye 30 and Triangle 10 (*c.* 270 m). (b) 3727-23-Wadi Wirqan Bullseye 50 and Triangle 16 (*c.* 140 m). (c) 3726-14-Tharbah Bullseye 62 and Triangle 33 (*c.* 190 m). (d) 3727-23-Wadi Wirqan Bullseye 5 and Triangle 1 (*c.* 58m). (e) 3627-21-Bi'ar al Khalas Triangle 1 and Bullseye 1 (*c.* 47 m). (f) 3727-33-Abu Arakah Triangle 14 and Bullseye 60 (*c.* 65 m) (source: Google Earth).

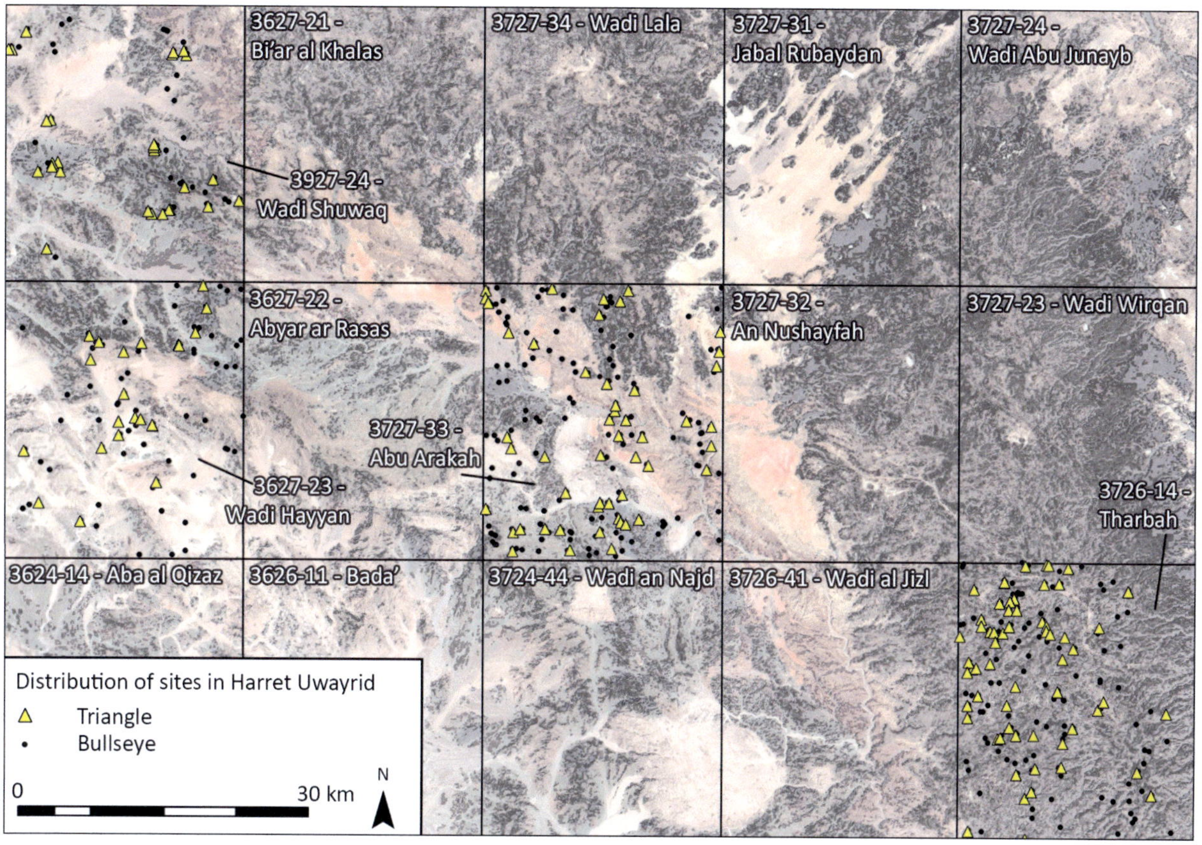

Figure 20. Distribution of Triangles in four map 'windows' of the Harrat Uwayrid in the context of Bulleye Cairns (drawn by Travis Hearn).

In the case of one map (Fig. 21), 154 Bullseyes were catalogued and 68 Triangles. Although almost half the Bullseyes had no associated Triangles, all but three Triangles had an associated Bullseye. As the map shows, there is no obvious pattern to the distribution of the Bullseye-Triangle pairs.

Although most Triangles are short and well-defined, a few are quite elongated. Most are isosceles triangles with the effect that the apex seems to 'point' to the associated Bullseye even if at some distance. The length of the Triangles varies but 10-20 m is common.

A few Triangles and Bullseye Cairns have been recorded in very unusual combinations. Fig. 22a consists of a Bullseye Pendant partly overlapped by a Triangle. There are two small cairns inside and one outside the Pendant head. The tail of the latter seems to consist of small elongated triangles. A Bullseye Cairn partly overlies the Pendant head. '3726-14-Tharbah Bullseye 130 and Triangle 61' offer a similar combination. Fig. 22b is a Triangle linked by a tail to two Bullseye Cairns. Fig. 22c has a Bullseye flanked by two Triangles side by side. Fig. 22d is one of several examples of a Bullseye Cairn and Triangle linked by a tail of small cairns. The most unusual combination is Fig. 22e, where three Triangles on three separate ridges are linked by tails to a Bullseye Cairn in the centre, a unique layout (cf. Fig. 23a).

There is no agreed explanation for these structures. We may note, however, the discussion by Avner and Avner (1999) drawing attention to a triangle as a female symbol of fertility. Likewise a circle. Is each Triangle pointing to a Bullseye burial and identifying it as that of a female?

Barred Rectangles

The name is given to a site type consisting of a rectilinear enclosure, sometimes almost square in appearance. Most are bisected horizontally by a single wall, a few have other small sections of wall, and a number have a small central cairn (Fig. 24).

Although there is not a single example in the Harrat al-Sham, Barred Rectangles are quite numerous in Saudi Arabia (n= 159). Because of the relatively large number involved and the multitude of other site types in close proximity, and the limits in some key areas of high-resolution imagery, a systematic analysis of their distribution is currently impossible. Nevertheless preliminary statements can be made.

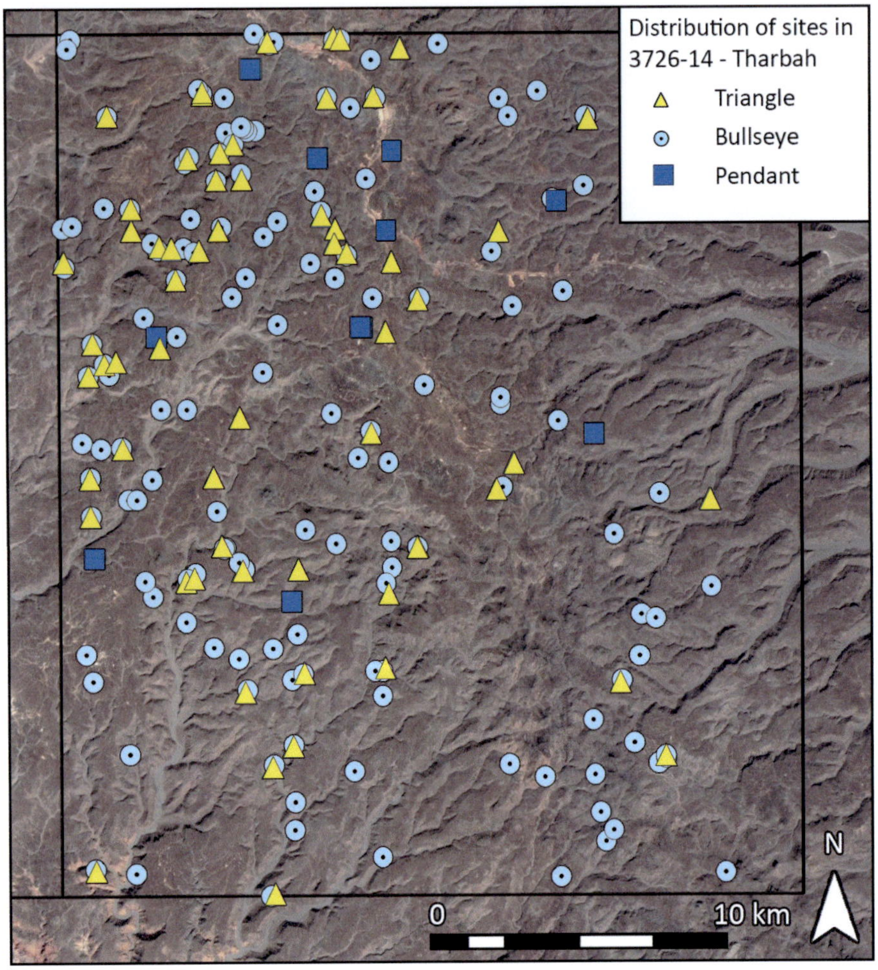

Figure 21. Distribution of Triangles, Bullseye Cairns and Pendants in the Tharbah area of the Harrat Uwayrid (drawn by Travis Hearn).

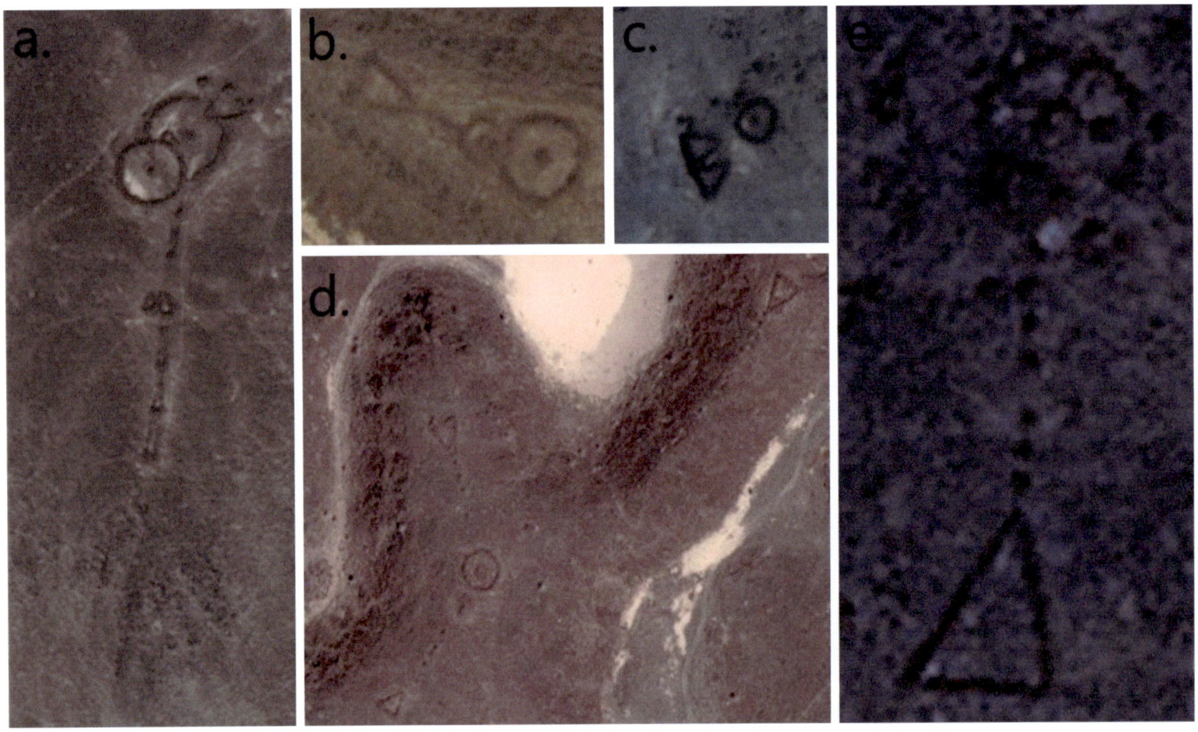

Figure 22. Triangles in combination with other structures. (a) 3726-14-Tharbah Bullseye 45 and 46 and Triangle 22. (b) 3727-31-Jabal Rubaydan Bullseyes 5 and 6 and Triangle 3. (c) 3727-33-Abu Arakah Bullseye 92 and Triangles 24 and 25. (d) 3925-41-Samhah Triangle 27 and Bullseye 77. (e) 3925-41-Samhah Pendant (Triangle Bullseye) 592 (source: Google Earth).

Figure 23 (right). Schematic drawings of Triangles set up as Pendants (drawn by Stafford Smith).

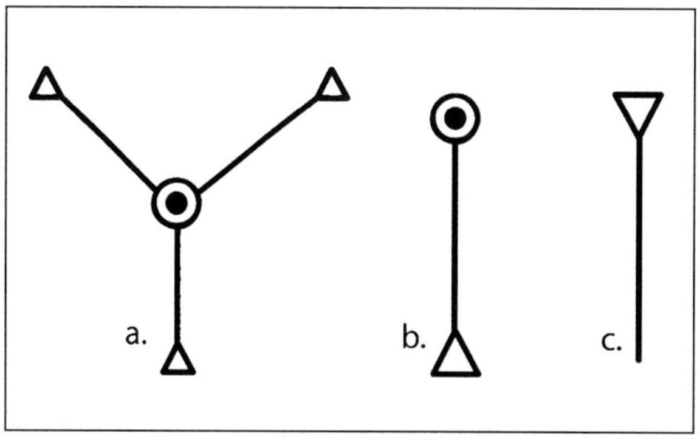

Figure 24 (below). Barred Rectangles. Left: 4023-31-As Suwayriqiyah Barred Rectangle 5 (*c.* 23 m). Right: 4023-31-As Suwayriqiyah Barred Rectangle 8 (*c.* 25 m) (source: Google Earth).

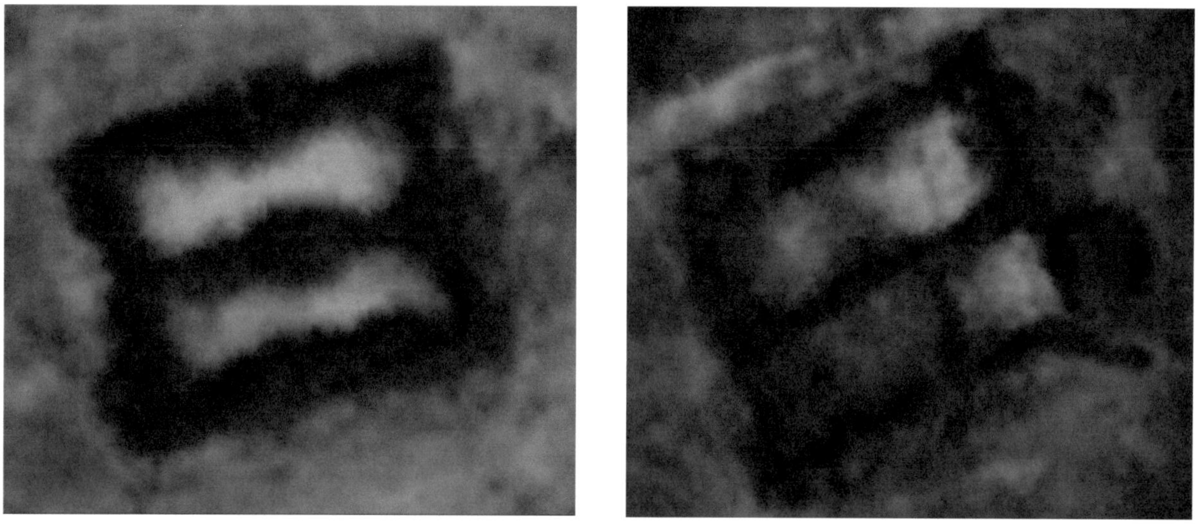

At the time of writing (February 2017), Barred Rectangles have been identified in just three adjacent lava fields in west central Saudi Arabia: Khaybar, Rahat and Kishb. Although there has been extensive interpretation of imagery for other *harrat*, and several thousand sites recorded, none, so far, includes a single Barred Rectangle (Fig. 25). Distribution between the three lava fields is uneven, in part because of unevenness in interpretation (Table 3).

As the Harrat Khaybar is covered almost entirely by high-resolution imagery and has been intensively and systematically interpreted, the small number of Barred Rectangles found there is a reflection of their rarity. We may note, too, that the thirteen examples are spread across no less than ten map squares. The Harrat Kishb has been less extensively explored but it appears to have few Barred Rectangles.

The Harrat Rahat is different. Although extensive areas of this very large lava field remain either unexamined or merely examined superficially, there have been intensive interpretations of several parts. For example, the Map 'SAUDI-4023-31-As Suwayriqiyah' covers the usual 700 km^2 but *c.* 230 km^2 are too pixelated to interpret (Fig. 26). In the remaining *c.* 470 km^2 are at least 58 Barred Rectangles, *i.e.* over one-third of the total known for this entire *harrah*. Even that figure is misleading as an indicator of frequency: an area of over 200 km^2 consists of a sandy surface devoid of sites of any kind and a further *c.* 70 km^2 is covered by a modern town and its fields. In short, 58 Barred Rectangles in an area of *c.* 200 km^2; *i.e.* they occur about one per 4 km^2, and as the map shows, they are fairly evenly distributed. Analysis of this group is revealing:

- All are at least 10 m long;
- 26 of the 57 are 10-15 m long;
- 9 are 25 m or longer;
- One is 35 m; two are 45 m long;
- 20 are 'square';
- 13 seem to have a small cairn in the centre;
- All are oriented east-west (19), or tilted a few degrees up (30) or down (8).

This is a landscape in which the other common site-type is the Barred Ring. In the case of this same map square (n= 61 Barred Rings) the two sites are intermingled (Fig. 26).

We may look further afield at the evidence of the other Barred Rectangles recorded less intensively or systematically. A survey of those (n= 101) showed exactly the same pattern. As for orientation, some are aligned east-west but more are tilted slightly either to the north

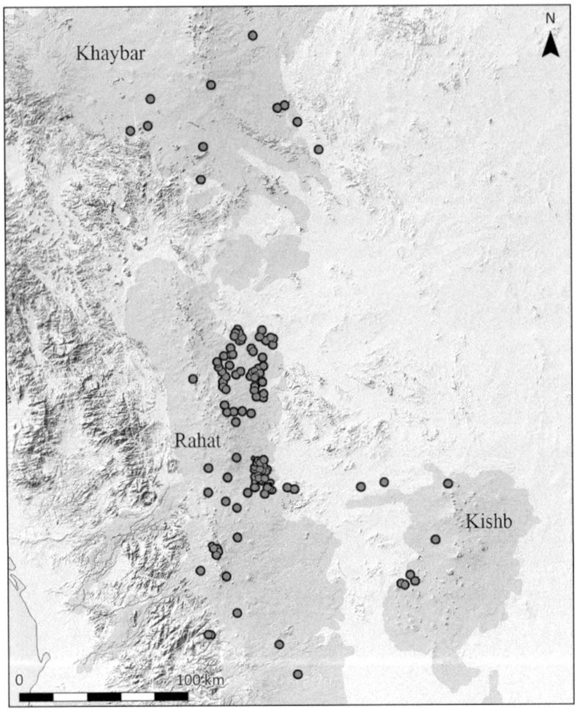

Figure 25 (right). Distribution of Barred Rectangles in Saudi Arabia. The data is incomplete, reflecting only current interpretation of some of the available high-resolution imagery (drawn by Travis Hearn).

Harrat	Area km2	Barred Rectangles
Khaybar	*c.* 20,564	13
Rahat	*c.* 20,000	140
Kishb	*c.* 5900	6
Total		159

Table 3. Distribution of Barred Rectangles in Saudi Arabia.

or south. Unlike some site types in a cluttered landscape, only three of the 159 Barred Rectangles intersect or are intersected by any other structure (Fig. 27). One Barred Rectangle has been visited on the ground and shows walls of heaped boulders without any cut stone or attempts at coursing (Fig. 28).

Gates

These are undoubtedly the most unusual and enigmatic site types found in anywhere in Arabia. None has been found in the Harrat al-Sham and, as the map shows, most have been recorded in just a limited group of lava fields in west-central Saudi Arabia (Kennedy 2017) (Fig. 29). In total 389 Gates have now been recorded, overwhelmingly in the Harrat Khaybar and mainly in the vicinity of the town of Khaybar itself. Although many exist in isolation from one another, many others survive in groups. A particularly striking group is that *c.* 40 km north-east of Khaybar town with 16 within a two km diameter circle; eight of them are illustrated in Fig. 30.

With overall dimensions of *c.* 373 x 80 m, Gate 31 is the fifth largest of the entire corpus. The thick 'posts' at either end are *c.* 10 m wide and the four 'bars' connecting them make this a relative rarity: only 13 of the total have so many 'bars' (cf. Gate 32). A ground photograph shows the core of a bar of Gate 31 to consist of carefully placed flat field stones set on edge. More common are the 2-Bar Gates in this same figure (Gates 28-30). There is also a relatively rare 3-Bar example (Gate 27) (n= 36). Notable are two of the so-called I-Type Gates (26 and 73) lying side-by-side (n= 56).

The purpose of these structures is totally unknown. They often lie set aside from other structures. In the case of one group, the landscape covered by that particular map (3925-42-Khaybar) is thickly strewn with ancient sites of various types: 327 Kites, 829 Pendants, 172 Bullseye Cairns, as well as 71 Gates. When mapped, it is immediately apparent that Gates cluster and are not intermingled with either Kites or Pendants (Fig. 31). It is evident, too, that Gates often lie immediately south or east of areas of seasonal water pools. However, one group of Gates recorded on the north-east fringe of the Harrat Khaybar are in sandy soil or rocky outcrops, far from any obvious water, even seasonal.

No fieldwork has been devoted to any Gates and we are dependent on their association with other site types to determine at least a relative chronology. Fortunately, many Gates intersect with other ancient structures. Examples show that Pendants, Triangles, Bullseye Cairns and even a Kite, all overlie the Gates with which they intersect (Fig. 32), suggesting they may be the oldest of the 'Works' in this landscape.

Discussion

Ancient stone-built structures, 'The Works of the Old Men', survive in immense numbers in interior Arabia, not least in the succession of lava fields. The latter are especially notable as the bleak and forbidding volcanic landscapes seem profoundly inimical to human activity, yet it is there these sites often survive in greatest abundance. The Works in the Harrat al-Sham are likely to remain the best-known and most intensively studied for many years to come, both because of several decades of fieldwork and the availability of thousands of aerial photographs and an on-going programme of aerial reconnaissance. However, the increasing availability of high-resolution satellite imagery for 'Arabia' as a whole and for the largely unexplored interior in particular has opened up an exciting and fruitful new avenue for exploration, recording, mapping, and at least preliminary analysis of the kind undertaken by aerial archaeologists elsewhere.

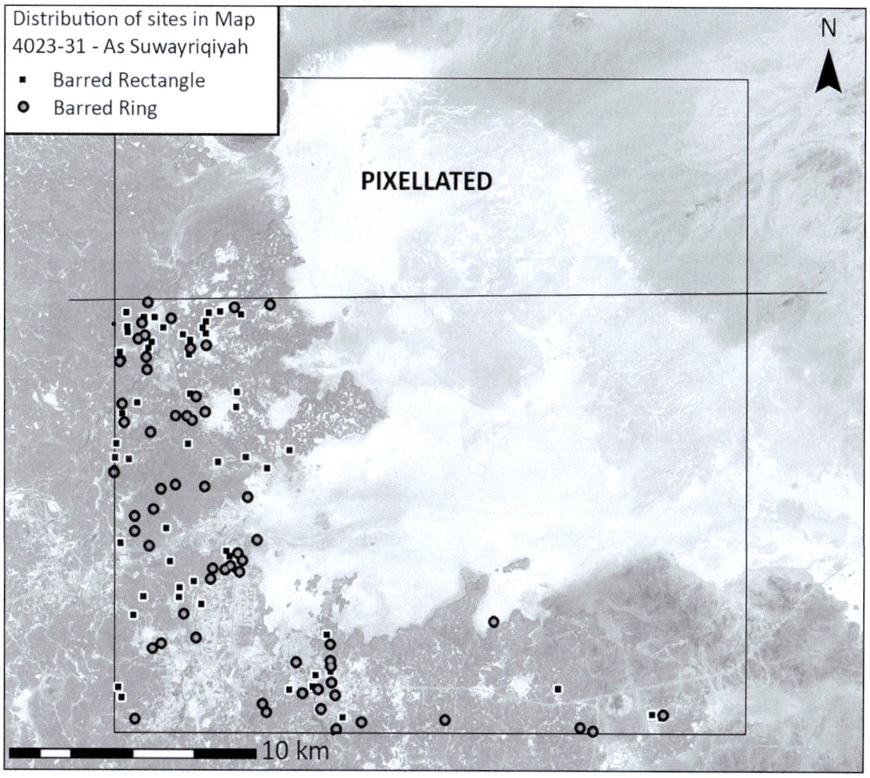

Figure 26. Distribution of Barred Rectangles and Rings in Map 4023-31-As Suwayriqiyah (drawn by Travis Hearn).

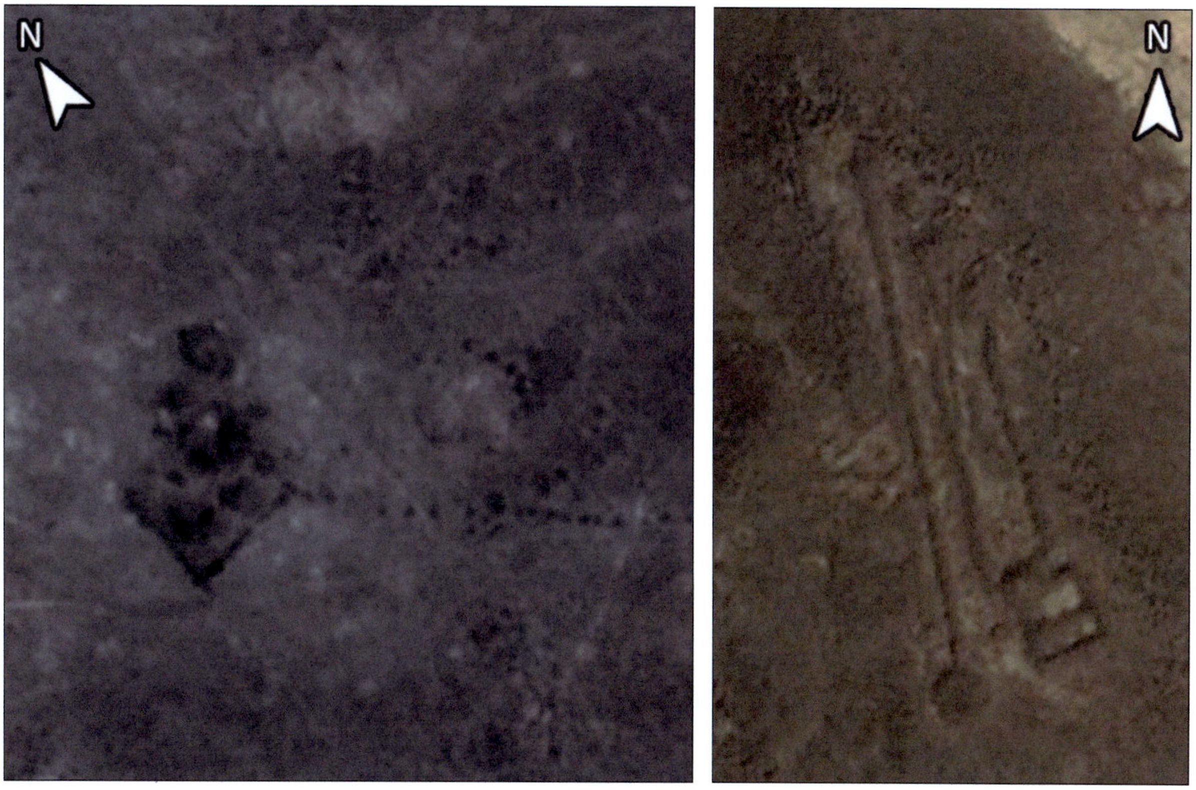

Figure 27. Barred Rectangles intersected by other site-types. Left: 4023-34-As Sidrah Barred Rectangle 3 is clearly overlain by Pendant 29 and part of its tail of small cairns. Right: 4024-32-Hazrah Barred Rectangle 1 lies very closely alongside Pendants 17 and 24, while 25 seems to overlie the Rectangle (source: Google Earth).

Figure 28. 4022-23-Jabal Hidan Barred Rectangle (GS_0841) (photograph by Grant Scroggie).

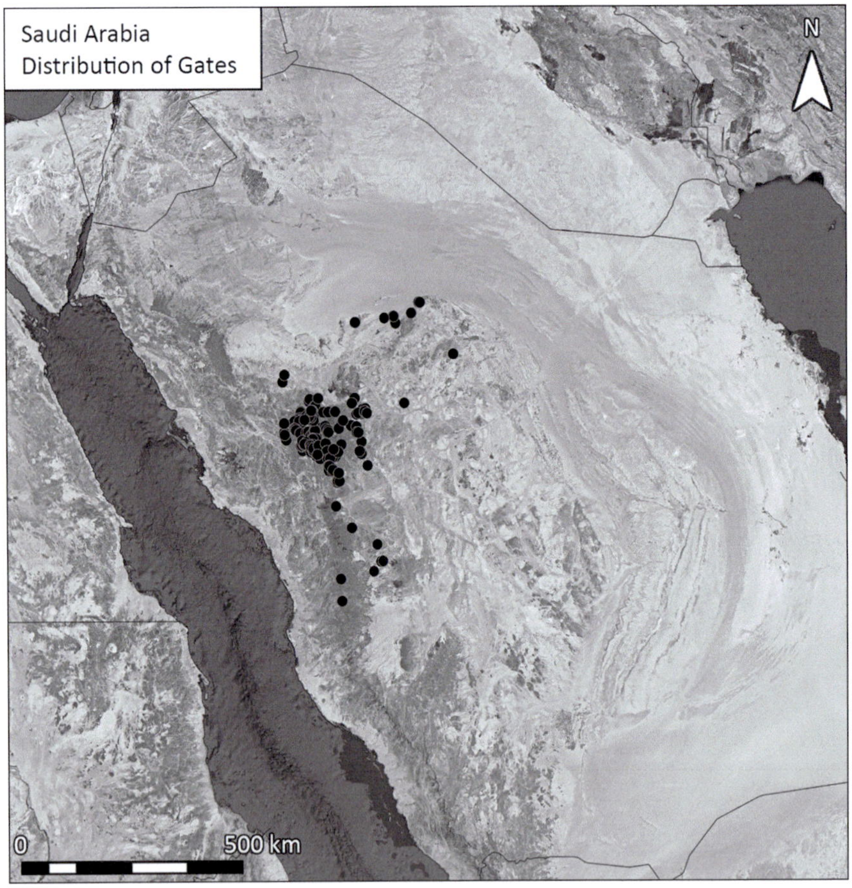

Figure 29. Distribution of Gates in 'Arabia'. The major concentration is of those in the Harrat Khaybar (drawn by T. Hearn).

Figure 30. Eight Gates of various types and sizes in the Samhah area (source: Google Earth).

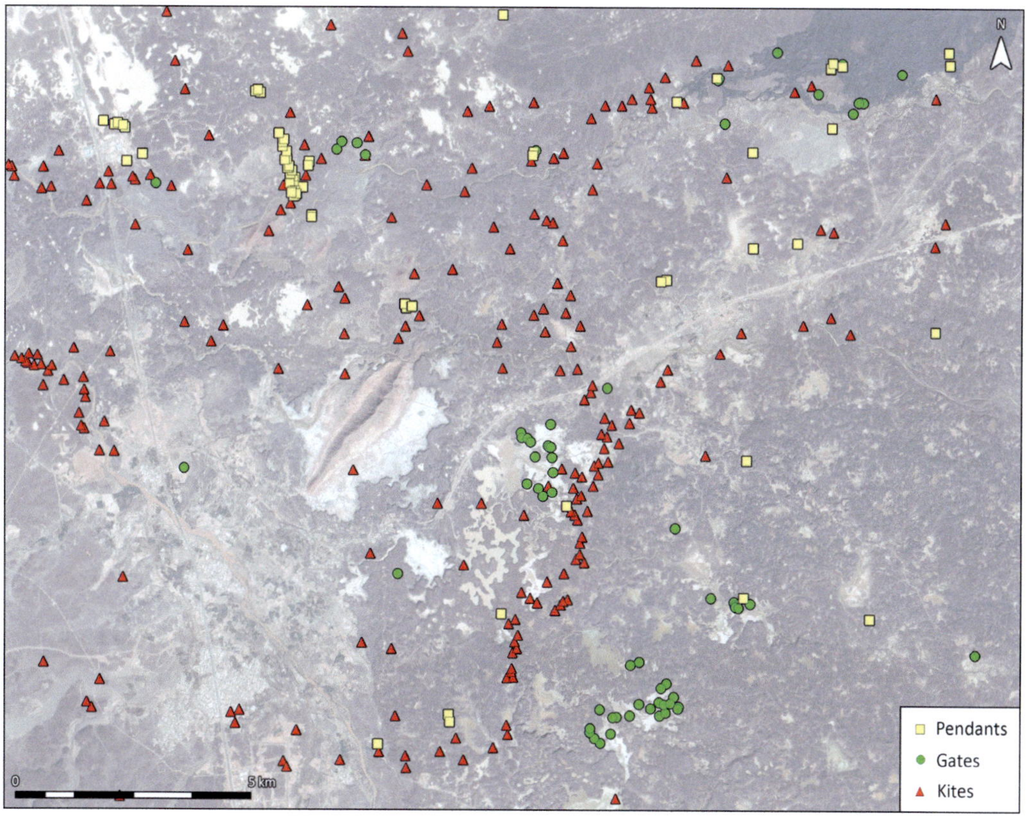

Figure 31. Gates, Kites and Pendants distribution map (drawn by Travis Hearn).

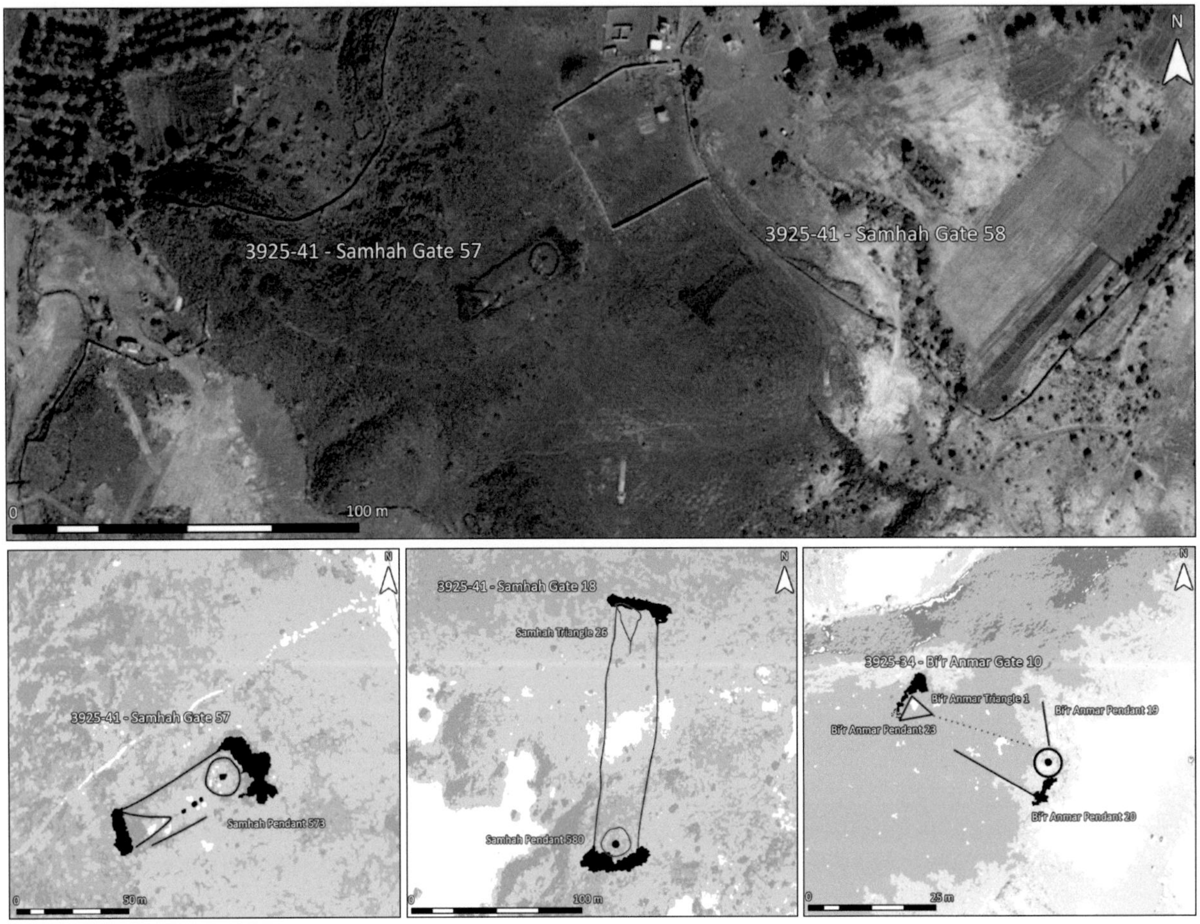

Figure 32. Gates overlain by other ancient structures (drawn by Travis Hearn).

Tens of thousands of sites have now been 'pinned' in Google Earth and given preliminary catalogue numbers. More are being added by the EAMENA team at Oxford. It seems likely the total sites in and around the lava fields, number in the hundreds of thousands.

What is already very clear is that site types identified in one lava field are not necessarily to be found in others. In some instances, site types found in the Harrat al-Sham are found more widely but in different forms – as with Kites and basic Pendants. In other instances, sites are found in Saudi lava fields that are unknown in this northern lava field, *e.g.* Triangles, Barred Rectangles and Gates, as well as the much wider range of structures labelled as Pendants.

The requirement now is for much more systematic interpretation of high-resolution windows on Google Earth for Saudi Arabia. There is a need, too, to refine the labelling of site types, not least those currently defined as 'Pendants'. Seldom included in interpretation of imagery are those sites for which there is no agreed name: a narrow cleared track usually enclosing an area of varying shapes. Perhaps 'Circular Paths' could be used as a convenient shorthand, as suggested by Kempe and Al-Malabeh who first drew attention to them (2010a, 209-210; 2010b, 60; 2012, 64; 2013, *passim*). Then there are those referred to as 'Corrals' and 'Camps'. Many are plainly quite recent in date but equally many others appear ancient and presumably the settlement sites of the people who built the other site-types.

In parallel it is vital there should be ground work: at the least examination of specific site types may well reveal more details of the kind brought to light with evidence of cut-stone and coursing in some structures. Beyond that it is still unclear what the function was of some site types. Kites are surely for trapping animals and many Cairns and Pendants are funerary. But what of Gates, Keyhole Pendants and Triangles?

Acknowledgements

None of the research on this article would have been possible without the continued generous support of the Packard Humanities Institute which provided me with time, research support and greatly increased fieldwork opportunities over an extended period. Several people have assisted at various times with the identification, cataloguing and mapping of sites in 'Arabia': Mat Dalton, Rebecca Repper (née Banks) and Grant Scroggie who provided invaluable ground photographs from his touristic visits. Fundamental to this present article has been Travis Hearn who undertook the thankless task of preparing tables of site-types and preparing illustrations for publication. None is responsible for the results set out here.

Sigla

APAAME Aerial Photographic Archive for Archaeology in the Middle East (http://www.apaame.org/)

EAMENA Endangered Archaeology in the Middle East and North Africa (http://eamena.arch.ox.ac.uk/)

References

Abu-Azizeh, W. and Tarawneh, M.B. 2015. Out of the harra: desert kites in south-eastern Jordan. New results from the South Eastern Badia Archaeological Project. *Arabian Archaeology and Epigraphy* 26, 95-119.

Anonymous. 1914. Paris – Le Caire en aéroplane. *L'Illustration* 3698 (10 Janvier), 28-29.

Avner, U. and Avner, R. 1999. Circles, triangles and lines in desert archaeological remains and rock engravings, and their tnterpretations. In: Bahn, P. and Fossati, A. (eds). *Rock art studies: NWES of the World 1. Proceedings of the International Rock Art Congress, Turin 1995.* Pinerolo (CD).

Bell, G.L. 1927. *The letters of Gertrude Bell, edited by Lady Bell.* London: Benn, Ernest Ltd.

Blunt, Lady A. 1881. *A pilgrimage to Nejd.* London: John Murray.

Brooke-Popham, R. 1921. Aeroplanes in tropical countries. *Journal of the Royal Aeronautical Society* 25, 563-58.

Brunner, U. 2008. Les pièges de chasse antiques au Yémen. *Chroniques Yéménites* 15, 29-34.

Brunner, U. 2015. The south Arabian form and its implications for the interpretation of desert kites. *Arabian Archaeology and Epigraphy* 26, 196-207.

Butler, S.S. 1909. Baghdad to Damascus via el Jauf, northern Arabia. *The Geographical Journal*, 517-535.

Cleuziou, S., lnizan, M.-L. and Marcolongo, B. 1992. Le peuplement pré- et protohistorique du système fluviatile fossile du Jawf-Hadramawt au Yemen (d'après l'interprétation d'images satellites, de photographies aériennes et de prospections). *Paléorient* 18, 5-29.

De Maigret, A. 1996. New evidence from the Yemenite 'turret graves' for the problem of the emergence of the south Arabian states. In: Reade, J. (ed.). *The Indian Ocean in antiquity.* London and New York: Taylor and Francis, 321-337.

De Vries, B. (ed.) 1998. *Umm el-Jimal. I: a frontier town and its landscape in northern Jordan.* Journal of Roman Archaeology Supplementary Series 26. Portsmouth, RI: Journal of Roman Archaeology.

Evenari, M., Shannon, L. and Tadmor, N. 1982. *The Negev. The challenge of a desert.* Cambridge, MA: Harvard University Press.

Global Volcanism Program, 2011. Report on Harrat Lunayyir (Saudi Arabia). In: R. Wunderman, R. (ed.). *Bulletin of the Global Volcanism Network* 36, 3. (https://doi.org/10.5479/si.GVP.BGVN201103-231040)

Huber, C. 1885. *Voyage dans l'Arabie centrale (1878-82).* Paris: Société de Géographie.

Insall, G.S.M. 1929. The aeroplane in archaeology. *Journal of the RAF College, Cranwell* 9.2, 174-175.

Kempe, S. and Al-Malabeh, A. 2010a. Hunting kites ('desert kites') and associated structures along the eastern rim of the Jordanian harrat. A geo-archaeological Google Earth images survey. *Zeitschrift für Orient-Archäologie* 3, 46-86.

Kempe, S. and Al-Malabeh, A. 2010b. Kites and other archaeological structures along the eastern rim of the harrat (lava plain) of Jordan, signs of intensive usage in prehistoric time, a Google Earth images study. *Proceedings 14th International Symposium of Vulcanospeleology*, 199-215.

Kempe, S. and Al-Malabeh, A. 2012. Distribution, sizes, function and heritage importance of the Harrat Al Shaam desert kites: the largest prehistoric stoneworks of mankind? *Proceedings 15th International Symposium on Vulcanospeleology*, 57-66.

Kempe, S. and Al-Malabeh, A. 2013. Desert kites in Jordan and Saudi Arabia: structure, statistics and function, a Google Earth study. *Quaternary International* 297, 126-146.

Kennedy, D.L. 2011. The 'Works of the Old Men' in Arabia: remote sensing in interior Arabia. *Journal of Archaeological Science* 38, 3185-3203.

Kennedy, D.L. 2012a. Pioneers above Jordan. Revealing a prehistoric landscape. *Antiquity* 86, 474-491.

Kennedy, D.L. 2012b. Kites: new discoveries and a new type. *Arabian Archaeology and Epigraphy* 23, 145-155.

Kennedy, D.L. 2012c. Editorial: Wheels in the Harret al-Shaam. *Palestine Exploration Quarterly* 144, 77-81.

Kennedy, D.L. 2012d. The Cairn of Hani: significance, present condition and context. *Annual of the Department of Antiquities of Jordan* 56, 483-505.

Kennedy, D.L. 2014. *Kites in 'Arabia'.* With contributions by R. Banks and P. Houghton. iBook (Apple).

Kennedy, D.L. 2017. 'Gates': a new archaeological site type in Saudi Arabia. *Arabian Archaeology and Epigraphy* 28, 153-174.

Kennedy, D.L. In press. Kites in Saudi Arabia. In: Betts, A. and van Pelt, W.V. (eds.). *The gazelle's dream: game drives of the Old and New Worlds.* Sydney: University of Sydney Press.

Kennedy, D.L. and Bewley, R. 2004. *Ancient Jordan from the air.* London :Council for British Research in the Levant.

Kennedy, D.L., Banks, R.E. and Dalton, M. 2015. Kites in Saudi Arabia. *Arabian Archaeology and Epigraphy* 26, 177-195.

Maitland, P. 1927. The 'Works of the Old Men' in Arabia. *Antiquity* 1, 196-203.

McCorriston, J., Oches, E.A., Walter, D. and Cole, K. 2002. Holocene paleoecology and prehistory in highland southern Arabia. *Paléorient* 28, 61-88.

Müller-Neuhof, B. 2015. Evidence for an Early Bronze Age (EBA) colonization of the Jawa hinterland: preliminary results of the 2015 fieldwork season in Tulul al-Ghusayn. *CBRL Bulletin* 10, 74-75.

Moufti, M.R., Németh, K., Murcia, H., Al-Gorry, S.F. and Shawali, J. 2013. Scientific basis of the geoheritage and geotouristic values of the 641 AD Al Madinah eruption site in the Al Madinah volcanic field, Kingdom of Saudi Arabia. *The Open Geology Journal* 7, 31-44.

Rees, L.W.B. 1929. The Transjordan desert. *Antiquity* 3, 389-406.

Steimer-Herbet, T. 2001. Result from the excavation of one high and circular tomb in Jabal Jidran 99. *Proceedings of the Seminar for Arabian Studies* 31, 221-226.

Steimer-Herbert, T. 2004. *Classification des sépultures à superstructure lithique dans le Levant et l'Arabie Occidentale (IVe et IIIe millénaires avant J.-C.)*. Oxford: British Archaeological Reports (International Series 1246).

Thomas, H.H. 1918. Photographic work of the R.F.C. in Sinai and Palestine during 1917. *P.R.O. AIR 1/2415/303/28*.

Thomas, H.H. 1920. Geographical reconnaissance by aeroplane photography, with special reference to the work done on the Palestine front. *The Geographical Journal* 55, 349-376.

Defending the 'land of the devil': prehistoric hillforts in the Jawa hinterland

Bernd Müller-Neuhof

Abstract

Archaeological research activities in the northern *badia* in north-eastern Jordan have revealed previously unexpected indications of late prehistoric socio-economic activities, both in the basalt desert (*harrah*) and the adjacent eastern limestone desert (*hamad*). Very important in this regard was the identification of several prehistoric hillforts in the *harrah* east of Jawa, whose occupations can be dated to the Late Chalcolithic/Early Bronze Age I (LC/EBA I; between the second half of the fifth millennium and the end of the fourth millennium BC). These discoveries changed a long-existing perception regarding human activities in this region in antiquity, which did not believe that such a bleak and inhospitable landscape could support the extended residence of humans. Whereas the climate conditions in this period are still not fully elucidated, it can be assumed that the environmental conditions in this time were unstable, which is exemplified by innovative solutions in reaction to these changing conditions. In this regard, it is especially important to mention the artificially irrigated and terraced garden systems that used surface run-off. Even though the research on the Chalcolithic/Early Bronze Age (C/EBA) occupation of this region is still at its beginning, this contribution presents a preliminary and hypothetical model of the colonisation and subsequent occupation of this region. Particularly, it will discuss the possible reasons for the existence and abandonment of these sites as well as their socio-economic background.

Keywords: hillforts, Late Chalcolithic, Early Bronze Age I, fortification, sedentism, irrigation agriculture

Introduction

"Bilad esh-Shaytan", or the "Land of the Devil" (Helms 1981, 17), is one of the many epithets for the *harrah*, the basalt desert which covers an area of about 11,000 km^2 in north-eastern Jordan. Indeed, this bleak landscape can be interpreted as the vestibule of the netherworld. A vestibule that is characterised by large and almost inaccessible, barren plains, densely covered with basalt boulders; by stumps of extinct volcanos, strung along chains indicating fissure eruption zones; and by the source of this infernal landscape: the massive volcanic mountain of Jebel Druze (or Jebel al-Arab) in the north, in modern Syrian territory. Jebel Druze can be regarded as the origin of the area's volcanic emissions, which mostly happened in geologic ages (*ibid.*, 20-22).

To make things worse, the region is known for high temperatures in summer, arid conditions throughout the year (despite some short but massive rainfall events), and cold winds in winter. In short, the *harrah* is a very inhospitable place. Even today, in

In: Peter M. M. G. Akkermans (ed.) 2020: *Landscapes of Survival - The Archaeology and Epigraphy of Jordan's North-Eastern Desert and Beyond*, Sidestone Press (Leiden), pp. 145-164.

spite of some infrastructural development in this region with villages, power lines, roads and tracks, the area is not regarded as being pleasant by many travellers. Although Bedouins still seasonally traverse the *harrah* with their flocks, as they have done for hundreds of years, for a long time it was unimaginable that humans could have survived here for much longer than a few weeks.

It was therefore a great surprise when pilots in the early twentieth century discovered abundant remains of anthropogenic structures – the "works of the old men" (see D. Kennedy, this volume) – on their flights *en route* from Cairo to Baghdad. Subsequently, archaeological surveys and excavations, especially in the 1970s and 1980s (*e.g.* Betts 1998; 2013), provided extensive evidence for prehistoric human activities in the region. These activities left substantial structural remains (*e.g.* kites and camp sites), which are mostly interpreted as being frequently but temporarily used.

While such an interpretation may be correct for the hunting kites and camp sites, it is probably not true for other structures. With new data from Late Neolithic sites at the southern fringe of the *harrah*, G. Rollefson, Y. Rowan and A. Wasse recently have challenged this assumption (see Rowan *et al.*, this volume). Based on pollen, charcoal, and sedimentological analyses, they contend that the dense architectural clusterings at Wisad Pools and Wadi al-Qattafi represent semi-sedentary habitations for months at a time during and after the rainy season.

Besides the structures that were encountered by the early explorers from the air, which referred to a possible temporary utilisation of the *harrah*, it was the impressive settlement of Jawa which struck the archaeological community when it was discovered by the French pilot Antoine Poidebard in 1931. The almost 10-ha fortified site is located on a volcanic hillock on the south-eastern foot of Jebel Druze and overlooks the adjacent Wadi Rajil and the lowlands towards the east. While the existence of such a large settlement in this barren landscape was astonishing in itself, its date was a further surprise. Lankester Harding, who visited Jawa in 1950, proposed a date in the Early Bronze Age. This was later proven through comparative pottery typologies by Svend Helms during his excavations at Jawa between 1972 and 1976. He was even able to narrow down the first occupational period to the Early Bronze Age I (EBA I),[1] by comparing the pottery from Jawa with contemporary pottery finds from his excavations at Tell Um Hammad in the Jordan Valley. Many scholars did not accept this early date for such an elaborate fortification, which instead provoked much criticism.[2] However, new radiocarbon dates from charcoal samples taken from the old excavations at Jawa have proven that Helms' chronological classification of the first occupation phase at Jawa was correct (Müller-Neuhof *et al.* 2015).

The existence of a large town with a massive fortification in an isolated location, far from the core areas of EBA Levantine settlement, made Jawa a peculiar place. This peculiarity continued for several decades until the early 2010s, when the origin of the scientific perspective focussing on Jawa changed from a western, Levantine approach towards an eastern, *harrah*-oriented outlook. The reason for this change in perspective towards Jawa was an archaeological project that started in 2010, which concentrated on the eastern hinterland of Jawa. The first four years of the 'Jawa Hinterland Project' (2010-2014)[3] aimed to shed light on this archaeologically almost unknown and – as it turned out – underestimated region. This was accomplished by looking for remains of possible Chalcolithic/Early Bronze Age (hereafter abbreviated as C/EBA) socio-economic activities in this region that might have been linked with Jawa. The archaeological surveys not only revealed abundant evidence for diverse C/EBA economic activities, such as flint mining with affiliated tool production and nomadic pastoralism, but they also provided evidence for terraced gardening with artificial irrigation by means of an intensive exploitation of surface run-off.

One of the most impressive discoveries, however, was the identification of two hillforts (Khirbet Abu al-Husayn and Tulul al-Ghusayn), which are located on volcanoes to the east of Jawa in the barren landscape of the *harrah*. This discovery induced the initiation of a follow-up project within the Jawa Hinterland Project for another three years (2015 2018).[4] This project focused exclusively on the exploration of these two hillforts and two further possible C/EBA hillfort sites, Khirbet al-Ja'bariya and Qasr Usseikhim. The latter two sites were identified in the *harrah* during the first two years of this follow-up project, as were a number of possibly contemporaneous unfortified settlements.

1 A later but much smaller re-occupation of the site dates to the beginning of the Middle Bronze Age I (*e.g.* Helms 1989).

2 See Müller-Neuhof *et al.* 2015, 126 for a summary of the critique.

3 The first project, entitled 'Aride Lebensräume im 5. bis frühen 3. Jahrtausend v.Chr.: Mobile Subsistenz, Kommunikation und Ressourcennutzung in der Nördlichen Badia (Nordostjordanien)' (Arid habitats in the 5th to the early 3rd millennium B.C.: mobile subsistence, communication and key resource use in the Northern Badia (NE-Jordan), was funded by the Deutsche Forschungsgemeinschaft (DFG) (German Research Foundation) (DFG-MU3075/1-1, DFG-MU3075/1-2).

4 The second project is called:'Die Besiedlung der nördlichen Badia (Nordostjordanien) im Spätchalkolithikum und der Frühbronzezeit (4.-3. Jt. v.Chr.). Ein Beitrag zur archäologischen Siedlungsgeographie in ariden Regionen Vorderasiens.' (The colonization of the Northern Badia (NE-Jordan) in the Late Chalcolithic and Early Bronze Age (4th to 3rd millennium BC): a contribution to archaeological settlement geography in the arid regions of Southwest Asia). It is also funded by the Deutsche Forschungsgemeinschaft (DFG) (German Research Foundation) (DFG-MU3075/3-1).

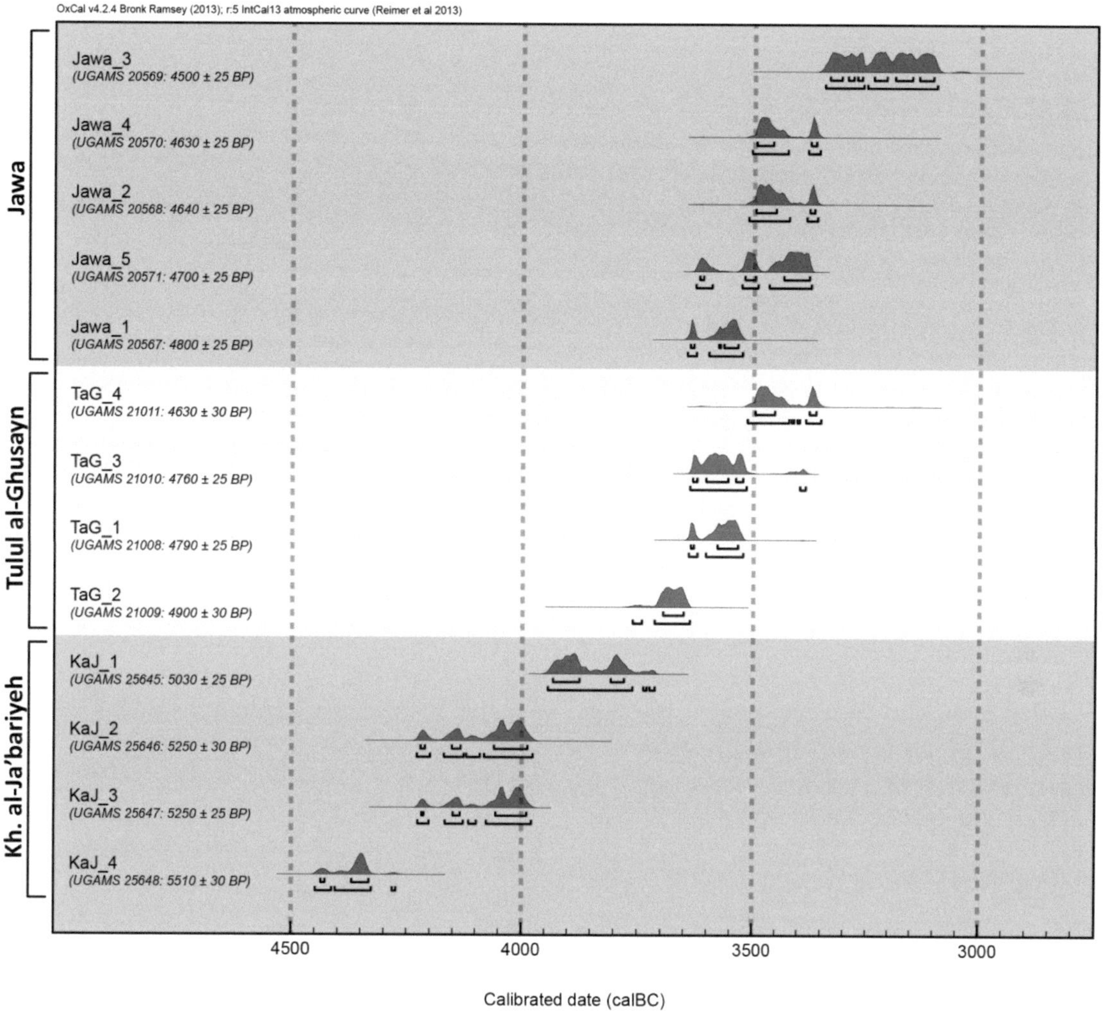

Figure 1. Plot of calibrated dates from Jawa, Tulul al-Ghusayn, and Khirbet al-Ja'bariya.

This brief introduction has shown that the research on the C/EBA colonisation of the *harrah* is still in its infancy, and that there is a minimal amount of available information and data. However, it can be stated that the "Bilad esh-Shaytan" was probably not considered to be an unhospitable countryside in the C/EBA, as it is today and has been in the recent past. It seems that the *harrah* was a densely and permanently occupied region to some degree, characterised by the fortified settlements located on defensible elevations (*i.e.* volcanoes).

A preliminary chronology of the hillfort site phenomenon in the *harrah*

In terms of the five known hillforts in the *harrah*, their surface finds tend to date these structures within the C/EBA period. Thus, it is possible to consider a C/EBA hillfort site phenomenon in the basalt desert (Müller-Neuhof 2017a). The long-known site of Jawa also belongs to the group of hillforts due to its location on a basalt hillock and its fortification. The other four newly-discovered hillforts are Khirbet Abu al-Husayn, Tulul al-Ghusayn, Khirbet al-Ja'bariya, and Qasr Usseikhim. With the exception of Jawa, which has a smaller Middle Bronze Age I re-occupation, and Qasr Usseikhim, which has a small Roman re-occupation, all the hillfort sites most probably were occupied for a single period only. Surface finds indicate a chronological range of the occupations within the Late Chalcolithic/Early Bronze Age I (LC/EBA I). So far, radiocarbon dates are available from Jawa, Tulul al-Ghusayn and Khirbet al-Ja'bariya. They cover a period from the second half of the fifth millennium to the beginning of the fourth millennium BC (Khirbet al-Ja'bariya), the first half of the fourth millennium to the beginning of the second half of the fourth millennium (Tulul al-Ghusayn), and the end of the first half of the fourth millennium until the end of the fourth millennium (Jawa) (Müller-Neuhof and Abu-Azizeh 2016, Fig. 20) (Fig. 1). Radiocarbon dates and OSL dates from Khirbet Abu al-Husayn and Qasr Usseikhim are pending.

Figure 2. Aerial view of Jawa (photograph by B. Müller-Neuhof, courtesy of APAAME).

That these hillfort sites emerged in the late fifth and fourth millennium BC is of particular interest, especially in light of the importance of this period for the cultural history of south-west Asia. Major developments in this period occurred in urbanisation, supra-regional contacts by trade and socio-economic developments, including the beginning of administration (beginning of writing and bookkeeping at the end of the fourth millennium), as well as innovations in agriculture and animal husbandry. The latter are commonly labelled with the term 'secondary product revolution'; this encompasses the emergence of horticulture by planting domesticated perennial crop plants (*e.g.* wine, date, pomegranate, fig), refinements in animal domestication (selective breeding for woolly sheep), the introduction of newly domesticated animals (donkey), as well as the modified utilisation of domesticated animals (cattle for dairy production and traction). New economic strategies and industries developed in the context of these secondary product innovations (horticulture and pastoralism, dairy and textile production, transport and trade). In turn, these led to technological innovations not only in agriculture and transport (plough and wheel), but also in other industries (potter's wheel, loom, metallurgy, *etc.*).

Focussing on Mesopotamia and the southern Levant, including the adjacent Transjordanian plateau, some of the developments of the late fifth and fourth millennium were confined to Mesopotamia (*e.g.* urbanisation, administration, textile industry, pottery industry, writing), others to the southern Levant (*e.g.* metallurgy, mineral resource exploitation). Further developments occurred in both regions (*e.g.* pastoralism, trade and transport, horticulture). These developments had major and as of yet incompletely explored impacts on these ancient societies that encompass, among other things, changes in individual and collective affiliations to territories, the growth of local societies beyond face-to-face relations, new conflict resolution strategies, as well as alterations in the social organisation of the society (including religion and cult). Additional impacts can be expected by the increasing and frequent contacts with distant regions and people due to trade relations and intensified large-scale pastoral nomadism. This is the context for the intensified exploitation of the *badia* and the involvement of the hillfort sites in one way or another, as will be discussed below.

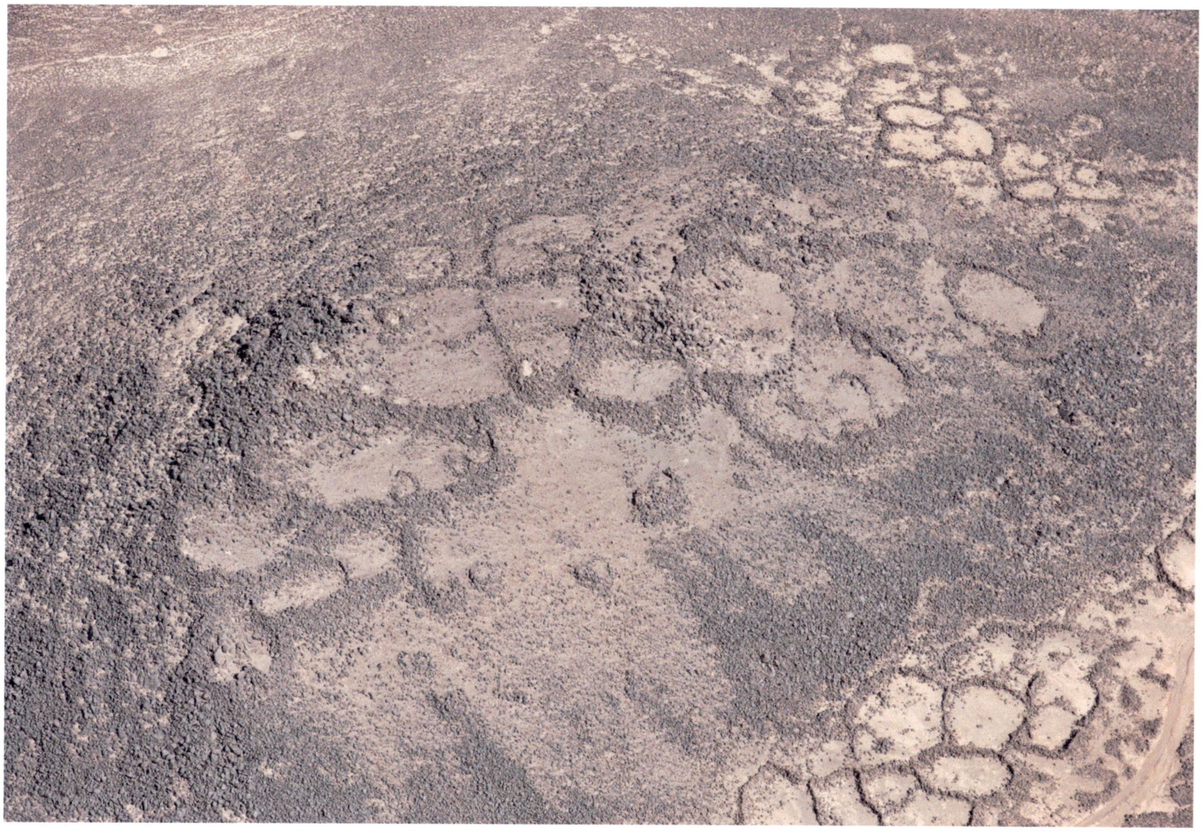

Figure 3. Aerial view of Khirbet Abu al-Husayn (photograph by R. Banks, courtesy of APAAME).

A brief portrait of the C/EBA hillforts in the *harrah*

Jawa

Jawa is located in the north-west of the *harrah*, close to the Jordanian-Syrian border, and is the longest-known hillfort in the region. Excavations were carried out here by S. Helms between 1972 and 1976 (Helms 1981; Betts 1991b). The EBA I settlement can be differentiated into a fortified 'upper town' and attached to it is a fortified 'lower town' (Fig. 2). Besides the well-known fortifications, Jawa is characterised by a water retention dam, which is currently the most ancient dam in the world and dates in its earlier phases to *c.* 3520-3350 cal BC (Müller-Neuhof *et al.* 2015, 129). Additionally, Jawa is well-known for its pool and channel system in the adjacent Wadi Rajil (Helms 1981). In 2010 and 2011, the author identified an extensive area covered with terraced gardens in three clusters in the south-west, south, and south-east vicinity of Jawa (Müller-Neuhof 2012; 2014a). According to OSL samples, the gardens date to the mid-fourth millennium BC (Meister *et al.* 2017).

Khirbet Abu al-Husayn

Khirbet Abu al-Husayn is located on the eastern fringe of the *harrah*, close to the Qa' Abu al-Husayn, beside a fissure eruption zone. The latter enabled access into the basalt desert along wadis and mud flats located beside a chain of small volcanoes. The site is located on one of these volcanoes (Fig. 3). The author identified the site in 2010 within the first transect survey of the Jawa Hinterland Project. In 2013 and 2017, Khirbet Abu al-Husayn was investigated in two one-week field seasons (Müller-Neuhof 2013a; 2016). The fortification is characterised by massive walls enclosing singular terraces that are attached to each other. Terraced gardens were identified on the southern and eastern feet of the elevation as well as just to the north of Khirbet Abu al-Husayn. Dwelling structures are primarily located outside of the fortification on the eastern, northern, and western slopes and foot of the volcano.

Tulul al-Ghusayn

Tulul al-Ghusayn is located in the north-eastern part of the *harrah*. The site was discovered by D. Kennedy and R. Bewley during one of the APAAME aerial archaeology flights in 2011, and they kindly provided the author with coordinates and photos. The author paid a first visit to

Figure 4. Aerial view of Tulul al-Ghusayn (photograph by D.L. Kennedy, courtesy of APAAME).

Figure 5. Aerial view of Khirbet al-Ja'bariya (photograph by R.H. Bewley, courtesy of APAAME)

the site in 2013 and a two-week field season took place in 2015 (Müller-Neuhof 2015; 2016; Müller-Neuhof and Abu-Azizeh 2016; 2018a). Tulul al-Ghusayn is located on a volcano that is characterised by a blown-out crater (Fig. 4). The southern rim of the crater is fortified and can be regarded as the 'upper town'. Unfortified dwelling areas are located in the crater, on the southern and the eastern slopes, and at the foot of the volcano. Terraced gardens are located inside the crater and on its southern and eastern slopes.

Khirbet al-Ja'bariya

Khirbet al-Ja'bariya is located in the western half of the *harrah*, close to the section of the Wadi Rajil south of the mud flat Qa' Shubayqa, where the wadi flows southwards. The author identified Khirbet al-Ja'bariya by examining Google Earth satellite images in 2015. In October 2015, the author participated in an aerial reconnaissance flight with David Kennedy of the APAAME project, producing the first photographs of the site (Fig. 5). In spring 2016, a two-week field season was carried out at Khirbet al-Ja'bariya (Müller-Neuhof 2017b; Müller-Neuhof and Abu-Azizeh 2018b). Khirbet al-Ja'bariya is a low, elongated volcanic elevation, of which the summit plateau is more or less entirely fortified. Dwellings were identified outside of the fortification on the southern and northern slopes. Terraced gardens were identified on the southern slope as well as on the eastern and northern foot of the volcano.

Qasr Usseikhim

Qasr Usseikhim is located on the western edge of the *harrah*, east of the northern part of the Azraq oasis. The site is located on a high and steep limestone rise that is capped by basalt. Qasr Usseikhim is well-known because of the small Roman outpost on the summit, identified by Gertrude Bell in 1913. Sir Aurel Stein took some aerial photos here in 1938. A Nabataean building probably preceded the Roman outpost (see Kennedy 2004, 65ff.). Additionally, a clearly pre-Roman and pre-Nabataean wall encloses the summit of the site and the Roman fort. A Jordanian-Italian team of both archaeologists and restorers restored this wall in the early 2000s (Al-Khouri and Infranca 2005).

In 2015, the author identified C/EBA double-cell dwelling structures on the site while reviewing Qasr Usseikhim on satellite images. Aerial photos, taken by the APAAME project in previous years and again in 2017, confirmed the existence of these double-cell dwelling structures on the site. In 2017, a first reconnaissance field season took place here. Qasr Usseikhim can be differentiated into an 'upper town', enclosed by the restored fortification wall (Fig. 6). North of

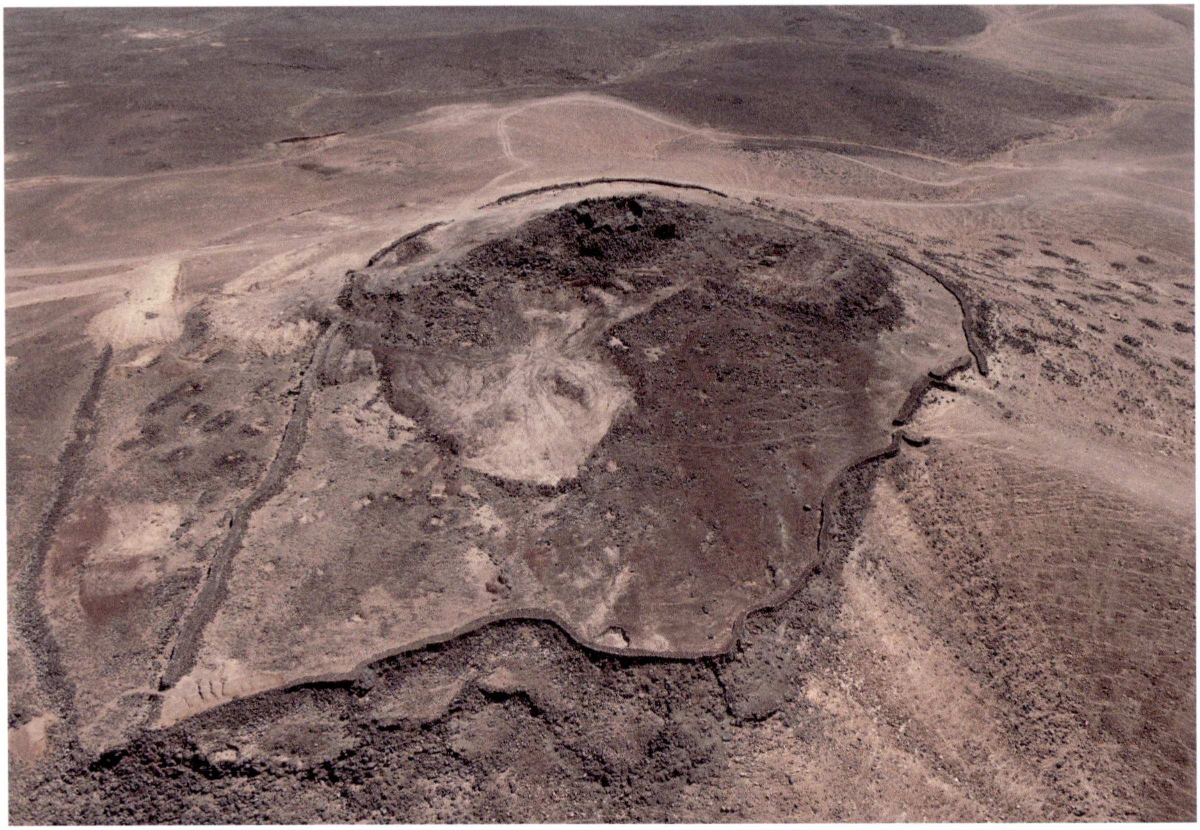

Figure 6. Aerial view of Qasr Usseikhim (photograph by D.L. Kennedy, courtesy of APAAME).

Site	Approximate enclosed area in hectare	Approximate total area in hectare
Jawa	4.5* / 8.4**	9.5
Qasr Usseikhim (QU)	2.2* / 3**	8.7
Tulul al-Ghusayn (TaG)	1.5	9.4
Khirbet Abu al-Husayn (KAH)	0.6	3.4
Khirbet al-Ja'bariya	0.4	1.8

Table 1. Settlement sizes of the LC/EBA hillfort sites in the *harrah* (*Upper 'town' fortification. **Total of upper and lower 'town' fortification).

the 'upper town' and attached to it, is a 'lower town', which consists of several double-cell dwellings, a large water storage pool with supply canals, and an unrestored fortification wall, which seems to be unfinished. Extensive dwelling areas are located on the southern slope and to the east of the elevation, close to a small (restored) ancient dam. A small and possibly terraced garden area is located *c.* 500 m south of the site.

All five hillfort sites show different sizes of their fortified ('upper town') and unfortified ('lower town') areas (see Table 1). At Khirbet Abu al-Husayn, Tulul al-Ghusayn and Khirbet al-Ja'bariya most if not all (Khirbet al-Ja'bariya and probably Khirbet Abu al-Husayn) of the dwelling structures are located outside of the fortified 'upper town'. However, Jawa and Qasr Usseikhim appear to have a fortified 'lower town' as well; at the latter, it is most likely that this 'lower town' was not completed entirely. Furthermore, a large number of dwellings at Qasr Usseikhim are located outside of both fortified areas.

Summary of the hillfort characteristics

Since the *harrah* is generally only traversable along topographic features such as wadis and mud pans, it is no surprise that the hillfort sites all are located on or close to some of the most important wadis, which can be regarded as part of a communication network. This is the case for Jawa on Wadi Rajil, Khirbet al-Ja'bariya close to Wadi Rajil, Tulul al-Ghusayn close to Wadi Ghusayn, Qasr Usseikhim on Wadi Usseikhim and within view of the Azraq oasis; also Khirbet Abu al-Husayn is on a mud flat/wadi route via the mud flats at Qa' Bakhita and Qa' al-Aza'im from the eastern entrances into to the *harrah* along the south-east/north-western fissure eruption zone (Fig. 7).

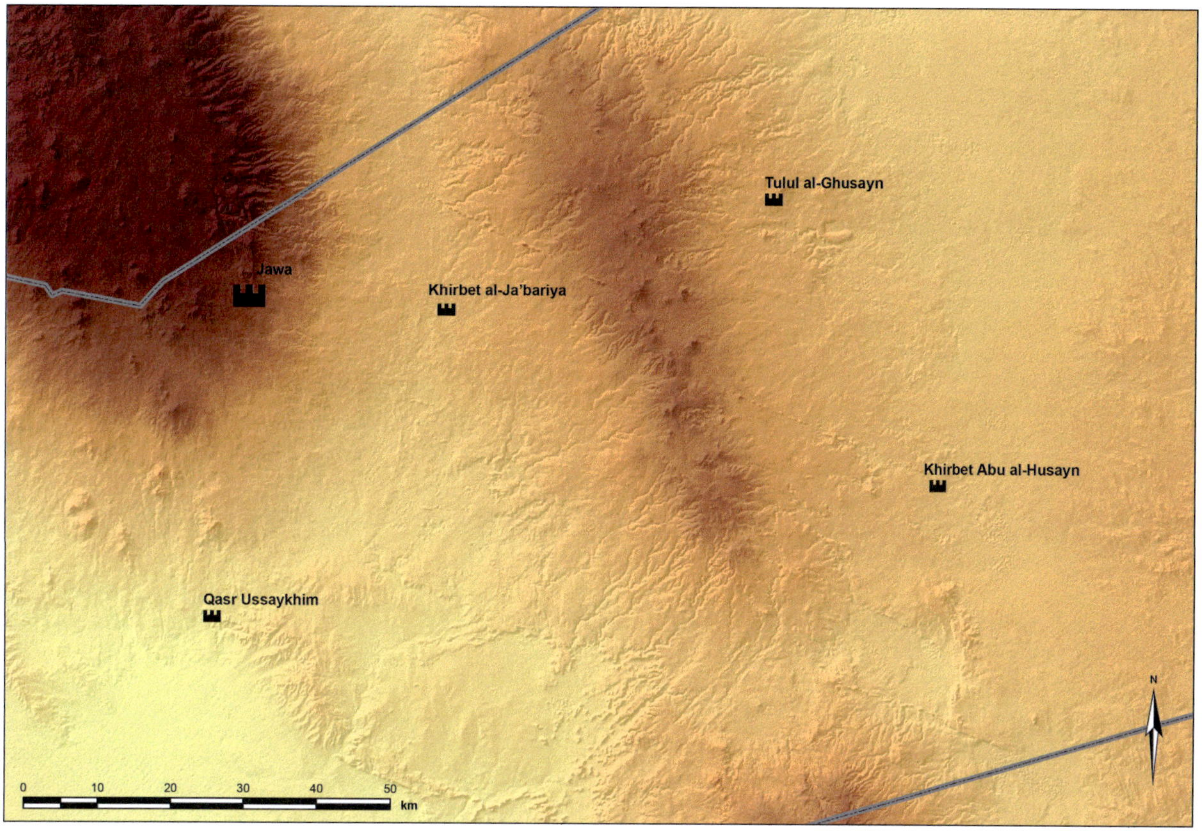

Figure 7. Map of the *harrah* with the location of the hillforts (W. Abu-Azizeh, DAI-Orientabteilung).

All sites are located on volcanic elevations with summit plateaus fortified by defensive walls. The fortification walls differ in size at each site. In terms of the size of the stones, Jawa clearly has the most advanced fortification wall of almost cyclopean dimensions. The height of the wall is partly still preserved up to *c.* 4 m, and it has a width of up to *c.* 5 m. All the preserved fortification walls at the other sites are much smaller, measuring: at Khirbet al-Ja'bariya up to *c.* 2 m high and 1.1-1.8 m wide; at Khirbet Abu al-Husayn up to *c.*1 m high and *c.* 1 m wide; at Tulul al-Ghusayn *c.* 1 m high and 0.75-1 m wide; and at Qasr Usseikhim up to *c.* 1.6 m high and 1.5 m wide. Regarding Qasr Usseikhim, as has already been mentioned, it has to be considered that the fortification wall of the 'upper town' was almost entirely restored in the early 2000s. Therefore, it is unclear if the restored height and thickness approximate the original height and thickness of the EBA I wall.

A common feature for all sites is the double shell masonry in the majority of the fortification wall sections. In addition, it is highly likely that the preserved remains of the fortification walls were just the foundation walls, which supported a superstructure made of perishable material. The close vicinity of the sites to large mud flats implies the possibility that the walls had a *pisé* or mud-brick superstructure, which also can be assumed for the dwelling structures (see below). However, the position of these walls on top of the elevations exposed them to wind and rain for some 5000-6000 years, leaving no traces of this material in place.

Several entrances/gates facilitated access to the fortified summits which were characterised by different lay-outs. These range from simple gates (Tulul al-Ghusayn and Khirbet al-Ja'bariya), simple gates flanked by tower-like structures or bastions (Khirbet Abu al-Husayn), baffled gates (Qasr Usseikhim), pincer gates (Khirbet al-Ja'bariya), to chambered gates (Jawa) (Fig. 8). It seems that the number of gates and especially their construction (layout) was related, among other factors, to the height and size of the natural elevation, the gradient of the access routes, and the size and thickness of the fortification walls. It is also possible that the stage of the development of fortification technology, the weapon technology, the kind of threat, as well as the conceivable kind of warfare (single assaults, frequent raids or siege) caused modifications in construction (see Müller-Neuhof 2005, 124-125). A re-enforcement of the fortifications by additional structures, such as towers or bastions, is possibly demonstrated at Khirbet Abu al-Husayn (Müller-Neuhof 2013a, 130-131), Khirbet al-Ja'bariya, and Jawa (Helms 1981, Figs. 43-45).

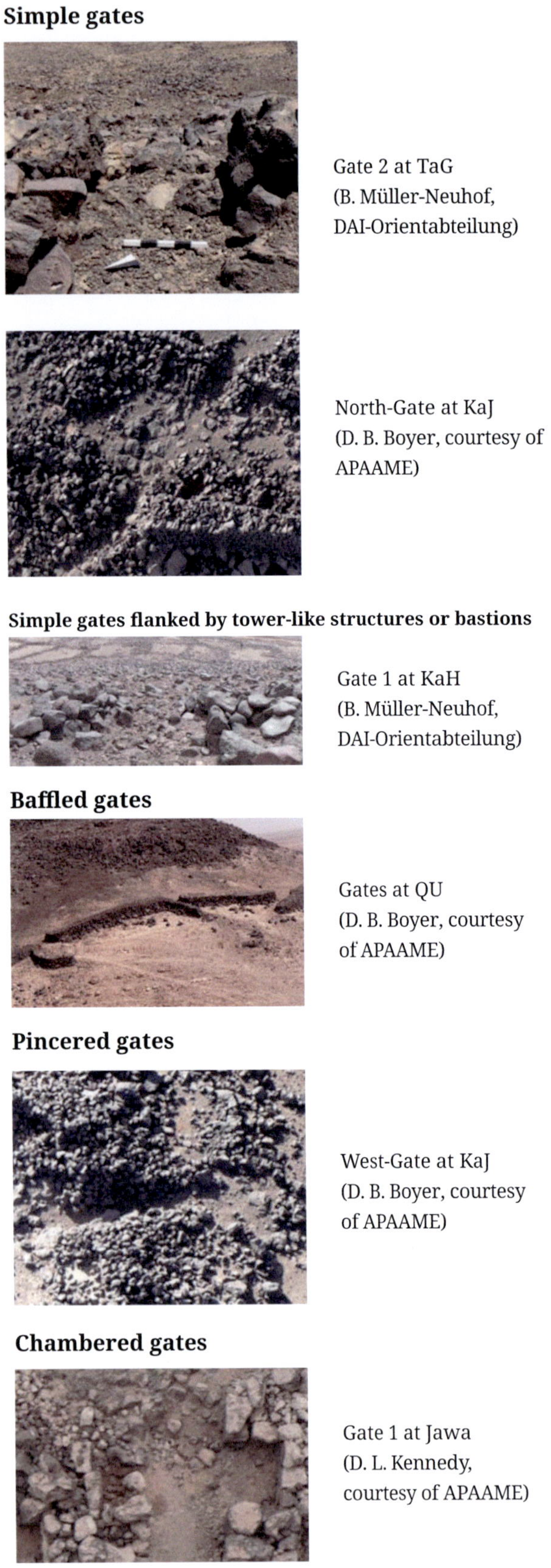

Figure 8. Typology of hillfort gates.

All of the hillfort sites possessed dwellings. The majority of the sites have a small number of dwelling structures inside the fortification in relation to the much larger number of dwellings outside the fortified area. Khirbet al-Ja'bariya and probably Khirbet Abu al-Husayn[5] are exceptions to this trend, as dwelling structures have not been encountered within the enclosed summit plateau. The fortified areas at both of these sites are characterised by internal divisions by walls and, in the case of Khirbet Abu al-Husayn, additionally by at least one non-domestic building (see below). Another exception is Jawa, where almost all the dwellings are located inside the fortified areas.

Dwelling structures are usually small in size. At Tulul al-Ghusayn and Qasr Usseikhim, the majority of the dwellings consist of double-cell structures with an interior length between 2.5 and 5.5 m and interior widths between 1.8 and 2 m. At Khirbet al-Ja'bariya, the circular dwellings have an interior diameter of *c.* 2 m. The limited dimensions suggest that the structures had flat roofs, which probably consisted of supportive beam structures; their length was dictated by the available tree heights or the length of reed bundles,[6] which restricted the width of these structures. The walls of the dwelling structures are usually double-faced walls and are preserved up to a height of 0.5 m. However, due to the presence of additional stones lying around and inside the structures, the original height of the stone walls was probably some centimetres higher. The large amount of loose loam inside the buildings suggests an original superstructure of *pisé* or even mud bricks, as was observed at Jawa (Helms 1981, 120ff.); the same is also assumed for the fortification walls. Similar to the assumed superstructure of the fortification wall, the mud superstructure of the houses has meanwhile blown and washed away.

An important observation regards the general uniformity of the dwelling structures at all sites. No building stands out in terms of a significantly larger size or different layout. Additionally, no proper special-purpose buildings or structures associated with administrative or religious purposes have been identified. However, there is some evidence for architectural features that did not serve domestic, defensive, or hydraulic purposes, at least at Tulul al-Ghusayn and Khirbet Abu al-Husayn. On the southern edge of the settlement terrace at Tulul al-Ghusayn, there are three rows of standing stones that run north-south and are oriented towards the west. These rows of standing stones show strong similarities to contemporaneous sites

5 If fireplaces are regarded as indicators for dwellings, then none of the four excavated structures at Khirbet Abu al-Husayn served as dwellings.

6 The use of reed bundles for roofing is suggested by Gary Rollefson for the Late Neolithic dwellings in Wadi al-Qattafi and Wisad Pools on the southern edge of the *harrah* (G. Rollefson, pers. comm.).

Figure 9. View of the pillar building with flagstone floor at Khirbet Abu al-Husayn (photograph by B. Müller-Neuhof, DAI-Orientabteilung).

in the Sinai, which may have had calendrical and/or ritual purposes (see Avner 2002; Müller-Neuhof 2013a, 135-136; 2014b, 244; Müller-Neuhof and Abu-Azizeh 2018a).

At Khirbet Abu al-Husayn, a small circular building with a flagstone floor and a massive central pillar was excavated in the 2017 fieldwork season (Fig. 9). This building is located close to the highest accessible point of the summit. A dwelling function can be excluded here due to the limited interior space of *c.* 1.6 m in diameter, which is furthermore reduced by the massive pillar in the centre of the room. The entrance of the building is oriented towards the east and a large open area extends in front of the building. The excavations in the building revealed no evidence for a fireplace or any small finds, which may further hint to the non-domestic function of this building. Comparable pillar buildings are known from the Sinai (*e.g.* in Wadi Radadi; see Avner 2002, 100, Fig. 4:36).

Other structures which most probably did not have a dwelling function are the 'enclosed open spaces' that have been identified at Tulul al-Ghusayn and Khirbet Abu al-Husayn. Located on prominent points of the settlements, outside of the fortifications, they show annex structures at some cardinal points (Müller-Neuhof 2014b, 244). These structures also have equivalents with some in the Sinai (Avner 2002, Table 14: nos. 50, 51 and 63), and the Judean desert (Avner 2002, fig. 5:5.55, quoting Bar-Adon 1972, 106-120). Both in the Sinai and in the Judean desert, these structures are interpreted as ritual places. However, since no excavations have been carried out yet at these structures in Khirbet Abu al-Husayn and Tulul al-Ghusayn, it is difficult to state that they actually served ritual purposes.

Another characteristic feature for the C/EBA hillfort phenomenon in the *harrah*, found at all sites, is the terraced gardens, which are located in the direct vicinity of the settlements and partly even inside the dwelling areas. A possible exception is Qasr Usseikhim, where some smaller terraced garden-like features are located *c.* 500 m away from the settlement to the south.[7]

Such gardens inside the dwelling areas have been observed at Khirbet al-Ja'bariya, where one horizontal row of terraced gardens is located on the southern slope of the volcano, directly below the fortress wall. Dwelling structures are built around this terrace row and are partly incorporated into the garden walls. An additional horizontal row of gardens is located at the northern base of the elevation. The gardens on the eastern base of the volcano, which were later converted into animal pens and camp-site structures, show only minimal evidence for dwellings inside the garden area. However, this can be simply a result of the later alterations.

A close spatial relationship between the gardens and the dwellings also has been observed at Tulul al-Ghusayn, where the dwellings are located close to the garden walls in the crater of the volcano and on its southern outer flank and base. At Khirbet Abu al-Husayn, the dwellings are located close to the cluster of terrace gardens on the eastern and north-eastern slopes, whereas the rows of gardens at the southern foot of the volcano show no evidence of dwellings.

A clear separation between dwelling area and gardens has been observed only at Jawa. Here, three terraced garden complexes were identified south of the site on the opposite side of an ancient wadi, in which Jawa's dam and reservoirs were located. Dwelling structures were not encountered in these garden complexes.

Interpretation

Climate, environmental conditions and innovative strategies to cope with these conditions

At first sight, permanent settlement in the *harrah* in the C/EBA can be explained only by much better climatic and, consequently, ecologic conditions than at present. But what were the environmental conditions in the fifth and fourth millennium? Unfortunately, local climate proxies and environmental data are not available (yet), but there is some indirect evidence to hypothesise about the possible environmental situation. Common climate proxies used for reconstructing the precipitation in the southern Levant are the Dead Sea-level fluctuations (Migowski *et al.* 2006) and the speleothem analyses of Soreq cave (Bar-Matthews and Ayalon 2011). These were taken into consideration together with marine climate proxies and terrestrial proxies from further away by Clarke *et al.* (2016). Their analyses show that the periods under consideration witnessed several distinctive and partly rapid changes in precipitation from one century to another or even from one decade to another (*ibid.*, 106-109).

Another hint of probably different environmental conditions comes from the research on the Late Neolithic occupations in the southern part of the *harrah* by Rowan, Rollefson and Wasse (this volume). At Wisad Pools, remains of red, gritty soils preserved below Late Neolithic dwelling structures and below silt layers in the mud pans were dated to the seventh millennium BC; they contained the pollen remains of plant species that reflect much wetter conditions than today (Rowan *et al.* 2017, 109). This does not necessarily mean that there was much more precipitation. However, it is evidence for environmental conditions in which water was available for a longer period of the year than today. The preserved red soils with their intact plant cover considerably increased the water absorption ability of these soils. With the introduction of sheep and goats in the Late Neolithic, these soils were increasingly degraded by overgrazing and especially by the fracturing of the soil surface by the hard and spatulate hooves of sheep and goats.[8] Therefore, sheep and goat husbandry had a negative impact on the preservation of these red soils over time and led to an increased erosion of the fertile soil cover. This process might have lasted for several millennia and was probably even accelerated and increased by the intensified sheep pastoralism in the fourth millennium BC in the region, together with the introduction of the wool sheep and the advent of a textile industry based on sheep wool in south-west Asia (see Sherratt 1983, 99).

Probably around the end of the fourth millennium BC, this degradation already had a noticeably negative effect, when environmental conditions turned for the worse. This caused the population of the *harrah* to withdraw voluntarily from this region. This may explain the absence of visible evidence for destruction at the hillfort sites, probably with the exception of Jawa (see below; see Helms 1981, 201-205). It also may account for the absence of any archaeological remains dating between the end of the EBA I and the Roman-Byzantine period in the *harrah*, as was observed by the author in two transect surveys (Müller-Neuhof 2013a, 127). The *harrah* became the unhospitable region of today, even though there was abundant precipitation at least in the Jordan valley region in the third millennium and, after another dry spell, again in the beginning of the second millennium BC (documented by distinctively higher Dead Sea levels; see Weninger 2009, 7, Fig. 2). The *harrah*, however, was not an inviting place to be, until around the beginning of the Common Era. Starting with the Roman occupation and later during the Late Roman-Byzantine period, the region was visibly re-

7 At Qasr Usseikhim. it can be assumed that the banks of the neighbouring and meandering Wadi Usseikhim have been used primarily for cultivation.

8 According to G. Rollefson (pers. comm.).

occupied by mobile groups (however, see for another view the contribution by Akkermans and Brüning, this volume). Again, climatic data refer to an increase of precipitation in the region in this era. The resumption of human activities in the *harrah* is shown by the abundant finds of nomadic-pastoral camp sites, as well as by measures to increase the infrastructural development of the *harrah* for pastoral activities and caravan trade. Examples of this include the construction of pools, cisterns, and wells, all of which date to the Roman, Roman-Byzantine, and Early Islamic periods (Müller-Neuhof 2013a). However, with the exception of a Roman watchtower north of Jawa, the Roman fortress at Qasr Usseikhim and the Roman/Byzantine fortress of Qasr Burqu', clear sedentary occupations have not been identified in the region. It can therefore be assumed that the period under consideration (the late fifth and the fourth millennium BC, or C/EBA I) was climatically and environmentally characterised by a fluctuating precipitation regime, but still with more or less intact soils.

This hypothesised fluctuating rainfall regime is probably reflected by agricultural innovations, exemplified by the development and implementation of terraced gardens that were artificially irrigated by local run-off in order to cope with the irregular rainfall. This run-off was 'harvested' from the direct vicinity of the gardens and directed into them, where it was stored in the sediments by saturation and thereby increased the water quantity in the areas under cultivation.

According to the presently available data, it seems that we can observe a local development of the terraced garden technology: from single row gardens, such as at Khirbet al-Ja'bariya, to more complex reticulated terraced gardens, as at Tulul al-Ghusayn and (in its most elaborated stage) Jawa (Müller-Neuhof 2014a; Meister *et al.* 2017). At Khirbet Abu al-Husayn, a more developed form of single-row gardens and also some smaller, reticulated terraced gardens have been identified. However, due to the present lack of chronological data, the gardens at Khirbet Abu al-Husayn cannot yet be classified chronologically.

Additionally, the size of the catchment area is of interest. At Khirbet Abu al-Husayn, Khirbet al-Ja'bariya and Tulul al-Ghusayn, the catchment areas comprise only very restricted areas, namely the basalt elevations themselves on which the settlements are located. However, the gardens at Jawa are characterised by a much larger local catchment area of almost six km^2, which extends far beyond the size of the terraced garden complexes.

Reasons for settling and defending

The establishment of sedentary communities and, especially, defensible hillfort sites in the *harrah* is puzzling. The better environmental conditions mentioned above clearly facilitated such settling activities and most probably also a year-round occupancy. In general, it is assumed that the C/EBA subsistence economy in the region was based primarily on agro-pastoralism. Specifically, this consisted of the seasonal migrations of some members of the communities with their herds to distantly located pastures, as well as local agriculture supported by the utilisation of artificially irrigated garden terraces. Additionally, the exchange and trade of animal products, as well as the exploitation and exchange of raw mineral materials from the desert with neighbouring regions were important economic pillars of the local population (see Müller-Neuhof 2013b; 2014b). This was especially the case for the exploitation of flint resources in the adjacent eastern *hamad* and the related export-oriented production of cortical scrapers on an industrial scale (*e.g.* Müller-Neuhof 2014b); the latter might have been organised and realised by the communities that dwelled in the *harrah*.

Concerning the defensive measures, which are characterised by settlements located on elevations and fortified by walls, questions remain as to the reasons for such defence. Can we infer possible reasons for this from the socio-economic context of the C/EBA sedentary populations in the *harrah*? It is obvious that these defensive measures reflect the existence of hostility and violence in the C/EBA societies in the *harrah* and probably the neighbouring regions. This must have been at such a large and 'advanced' scale that settling on mountains and constructing fortification walls was the most reasonable reaction to the threats. Often-cited symbolic reasons for such defensive measures or their protective function against wild animals can be excluded. No archaeological evidence at these sites points to a clear socio-political hierarchy that would have been a pre-condition for the construction of such architectural symbols of power. Furthermore, because these are among the earliest fortifications in south-west Asia (at least in the Levant and Transjordan), their original function (defence) is clearly more conceivable than the much later developed symbolic aspect of fortifications as emblems of power (see Müller-Neuhof 2005, 125-127). That these fortifications were not used to protect against wild animals is illustrated by the fact that most of the dwellings, at some sites even all the dwellings, are located outside of the fortifications. Furthermore, unfortified settlements are located in the region, and the protection of livestock against predators also could have been realised by the construction of animal pens of stones and/or thorny brush.

Accepting hostile and violent times in the region in the C/EBA, it is necessary to consider the causes for such hostility and violence, as well as the causes for settling on mountains and constructing hillforts in the *harrah*. Several possible reasons are conceivable and are discussed below. The already-mentioned position of the hillfort sites on or close to major communication routes that followed wadis and mud pans suggests that these locations were chosen not only in order to have access to these routes but also to control them. This implies that the access to these routes had a high value, which was probably substantiated by the transport of valuable (raw) materials and/or the access to the sources of these resources via these routes. It implies furthermore the existence of competitors who sought violent solutions to reach their aims. The *harrah* and the eastern adjacent *hamad* were intensively used economically in the C/EBA. Archaeological evidence shows that economic activities such as flint mining, pastoralism, and probably also hunting[9] had an economic output that went beyond local subsistence needs. Therefore, such surplus was intended for exchange and (supra-)regional trade. Probably, these are only a small part of a larger number of resource exploitation options in the region (see Müller-Neuhof 2013b). However, neither these other exploitation options nor the structures for storing such precious raw materials have been identified in any of the hillfort sites. Furthermore, the limited archaeological soundings and excavations have not yet revealed significant amounts of such raw materials or products. Therefore, the control of resources and routes might have been a reason for the establishment of hillfort sites, but probably not the single and most important one.

The protection of the *harrah* and the control of access to the *harrah* and its resources from neighbouring regions might be a further explanation for these fortifications. Such a motivation may be valid for Jawa, Qasr Usseikhim, and Khirbet Abu al-Husayn, which are located either on the western edge (Jawa, Qasr Usseikhim) or the eastern edge of the *harrah* (Khirbet Abu al-Husayn). Yet, this does not apply to Tulul al-Ghusayn and Khirbet al-Ja'bariya, which are located more or less in the middle of the basalt area.

Competition and conflicts inside the *harrah* between the rival populations of the different hillfort sites (which therefore were the focal points of single political entities) over resources, routes, status, *etc.*, appear to be a conceivable reason for the existence of these hillforts. However, the hillfort chronology established so far indicates a chronological succession of these sites, rather than contemporaneity.

The large, enclosed, unbuilt, open areas of the hillfort sites refer to a preparedness to accommodate large groups of people and, probably, livestock from the unfortified dwelling areas of the hillfort sites, as well as inhabitants and livestock from settlements and camp sites in their close vicinity. The hillforts therefore served as refuge forts in times of conflicts. This is a probable explanation, which would include the above-mentioned reasons for conflicts, such as access to resources and pastures, and the control of communication routes. However, it does not explain why these particular volcanoes were chosen for the hillforts as opposed to other volcanoes that were more suitable for the construction of hillforts: for example, those that are located closer to wadis and routes or were easier to defend due to their topographic characteristics, size and/or height.

However, one hypothetical reason may have been the availability of fresh water. Precipitation run-off was used by directing water into the garden sediments, where it could saturate the soil and be used for agriculture. Collection of rainfall run-off in artificial pools such as at Qasr Usseikhim, traditional *ghudrun* in mud pans, and in natural wadi pools was used for drinking water for livestock. However, such run-off water might be potable for humans right after the precipitation occurred, but not for very long if it was stored in open basins or pools. Potable water was therefore a very important commodity in a region lacking any rivers, lakes, or springs. Digging holes in wadi beds to obtain potable water implies short-term methods that satisfied only the needs of relatively small groups of people for a (very) restricted period of time. More importantly, it could only be carried out outside of the fortifications.

However, another place where fresh water is available in the *harrah* are the lava tubes. The sub-surface of the *harrah* is interspersed with lava tubes or lava caves that partly serve as natural cisterns, as observed by the author during a geological reconnaissance trip with geologists from the Technical University Darmstadt and the Hashemite University of Jordan in 2009. Surface run-off naturally penetrates through cracks into these tubes, where the water stays for a long time in cool and dark conditions. Since the water is not exposed to sunlight, it is protected from evaporation as well as from the formation of phytoplankton and other microorganisms that otherwise would turn it into brackish water. The utilisation of such natural cisterns is observable in the *harrah* in different places. In 2000, the author identified well heads in the Wadi Ghusayn, which probably date as early as the Byzantine period and which were, in reality, artificial openings made to access lava caves below the wadi bed (which served as natural cisterns). A similar natural cistern (Mugharet al-Jawa), the utilisation of which probably goes back to prehistory, was discovered by Alison Betts in the vicinity of Jawa (Betts 1991a, 224).

It seems logical that accessible lava tubes also existed below the surface of the enclosed parts of the hillfort sites. The plausibility of this assumption is illustrated by lava caves under the surface of Jawa that were recently exposed by illicit diggings (Fig. 10). It can therefore be posited that these caves at Jawa contained freshwater derived from surface run-off and from water that penetrated into them via cracks in the basalt rock from the adjacent Wadi

9 Even though most of the small number of hunting kites (artificial hunting traps) for gazelle hunting in the *harrah* investigated so far are dated to the Late Neolithic (*e.g.* Betts and Burke 2015), their re-utilisation as well as the construction of new kites in the C/EBA cannot be excluded.

Figure 10. Lava cave exposed by looting activities inside the fortification wall at Jawa (photograph by B. Müller-Neuhof, DAI-Orientabteilung).

Rajil in times of high-water levels. However, it has to be mentioned that the caves at Jawa have not been explored yet, and therefore we do not know if they contained water and were accessible in the past.

Such caves have not been encountered yet at Khirbet Abu al-Husayn, Khirbet al-Ja'bariya, Tulul al-Ghusayn, and Qasr Usseikhim. However, these sites are characterised by spatially limited remains of massive concentrations of stones and short walls. These structures do not belong to the fortification walls. They are also not comparable to dwelling structures, and the relationship between the massive amount of building material and their limited extension does not indicate administrative or ritual buildings. It can therefore be assumed that these ruins are the remains of structures whose primary role was to protect something very valuable, *i.e.* the lava caves and, hence, access to freshwater sources.

These natural cisterns were naturally or artificially accessible from the surface at these defensible locations and were such a vitally important resource for the people in the C/EBA in the *harrah* that their possession and control was crucial and therefore required protection and defence. The need to protect these sources probably initiated the establishment of fully sedentary settlements in the region, because a temporary abandonment of these sites, such as in the context of seasonal pastoral migration, would have led to their being taken over by competitive groups.

The possible competitors, against whom the defensive measures were directed, were probably local groups. While the access to, and the possession of, freshwater dictated the location of the hillfort sites, the fortifications themselves served multiple purposes. Besides the protection of these freshwater sources, they also served to protect livestock, to store resources (harvests and raw materials), and to house people. Therefore, the possible economic reasons for these early fortifications cannot be fully excluded. Rather they have to be regarded as complementary to the major aim: defending the freshwater resources.

The end

It is too early to develop clear arguments for characterising the end of the LC/EBA I hillfort phenomenon in the *harrah*. However, several observations have been made at Khirbet Abu al-Husayn, Tulul al-Ghusay, Khirbet al-Ja'bariya and Qasr Usseikhim, which give a hint about the end of the occupation of these sites. Contrary to the excavations at Jawa, where a large number of artefacts were excavated, the hitherto few and spatially restricted soundings and excavations at the other hillfort sites have yielded only very few artefacts. The few exceptions are large grinding slabs and *manos* in a building at Tulul al-Ghusayn (TAG 181) and at Khirbet al-Ja'bariya (KAJ 8), and an intentionally placed jar and a cortical scraper at Tulul al-Ghusayn (TAG 208). Most of the excavations in other dwelling structures at Tulul al-Ghusayn, and especially at Khirbet Abu al-Husayn and Qasr Usseikhim, revealed no finds at all. The surface surveys at these sites yielded a few pottery sherds and some lithic artefacts. On the whole, the small-find situation is poor. Additionally, the fortifications at these sites, as was stated above, show no evidence for destructions that might have been caused by attacks. Therefore, based on these preliminary observations, it seems that the hillfort sites most likely were abandoned voluntarily and according to a plan (probably with the exception of Jawa; see Helms 1981, 204-205).

However, such a planned abandonment of the sites does not explain the lack of evidence for fireplaces, such as ash and charcoal, in the excavated structures at Khirbet Abu al-Husayn and Qasr Usseikhim and some of the dwellings at Tulul al-Ghusayn. At the moment, this can only be explained by strong aeolian erosion of these structures, due to their location on top of the elevations and their exposure to wind and water over almost six millennia.

Climate change, which caused a reduction of the annual precipitation, in connection with environmental degradation, might have led to the voluntary abandonment of the region. Terraced farming and pastoralism could no longer be relied on year-round, so the population left the region. According to the survey data, it even seems that the *harrah* was so heavily exploited that it was not used anymore for seasonal grazing activities of nomadic pastoralists from the beginning of the third millennium on until the Roman-Byzantine era (but see Akkermans and Brüning, this volume).

Conclusion

It has to be emphasised here that the research aims of the project were to identify additional C/EBA hillfort sites in the *harrah* in order to establish an inventory of C/EBA hillforts there, to map these sites and their architectural remains, and to generate radiocarbon dates for an approximate chronological classification of the occupation periods at these sites. Since these aims have not yet been fully reached, and since more detailed studies must necessarily be carried out, the discussions above and especially the following conclusion have to be regarded as provisional and hypothetical.

With the exception of the Early and Late Neolithic fortifications at Pre-Pottery Neolithic Tell Maghzaliya in northern Iraq (Bader 1993a; 1993b) and Late Neolithic Tell as-Sawwan I in northern Iraq (Youkana 1997), it seems that the beginning of fortifications can be dated to the LC/EBA I and that they originated in the southern Levant and Transjordan. Whereas the majority of LC/EBA I fortifications hitherto have been found in the *harrah* of the northern *badia*, additional fortified sites dating to the LC/EBA I have been identified and partly excavated near Aqaba (Tell Hujayrat al-Ghuzlan; Becker 2013; Khalil and Schmidt 2009), in the Samaria region (Zertal 1993; see also Paz 2002) and, although unexcavated, in the Leja north of the Hauran (Sharaya; Nicolle and al-Maqdissi 2006; Nicolle and Braemer 2012). In north-western Mesopotamia, early LC fortifications have been encountered especially in the area of the upper Euphrates, where some fortified Uruk sites have been excavated (*e.g.* Habuba Kabira and Tell Sheikh Hassan; Strommenger 1980; Boese 1995).

It is still difficult to hypothesise about a possible origin of the fortifications in the LC/EBA I period in south-west Asia, and about the possible carriers of this technology. Tulul al-Ghusayn and Khirbet al-Ja'bariya clearly predate Jawa and the Uruk sites, so one of Helm's hypotheses – that the 'Jawaites' immigrated from the north or from the Jordan valley (Helms 1981, 60-68) – has to be questioned in this regard. However, Helms (1981, 62) also suggested the possibility of a local autochthonous origin, which currently seems to be the most plausible hypothesis.

Nevertheless, it is interesting to note that the C/EBA settlement pattern of the northern *badia* was not only characterised by fortified settlements. A Google Earth survey by the author revealed a large number of unfortified sites on wadi banks and on the shores of mud pans. They are characterised by concentrations of large numbers of double-cell houses similar to those dwellings encountered at Tulul al-Ghusayn and Qasr Usseikhim. During the last two field seasons, some of these sites were visited and the existence of these double-cell dwellings was confirmed. It can be hypothesised that these settlements are contemporaneous, assuming that these structures date to the fourth millennium (as has been shown at least for Tulul al-Ghusayn).

Besides these double-cell dwelling sites, a large number of unfortified sites with simple circular dwellings also were identified in the Google Earth survey. However, it is more difficult to date them to the C/EBA without ground inspection and excavation, since circular dwelling structures also were common in much earlier periods, such as the Late Neolithic. Nevertheless, the large number of such unfortified sites shows that the *harrah* was a proper settlement region in the C/EBA.

The identification of the hillfort sites and the existence of – at least partly contemporaneous – unfortified settlements raise the question of whether we can identify a settlement pattern in the *harrah*. However, it has to be stated that a clear settlement pattern, characterised by concentrations of unfortified settlements around hillforts, could not be established at present. This can be explained by a number of possible reasons related to research but also may be due to technical issues. First of all, the unfortified settlements, especially the double-cell dwelling sites, have not been investigated by excavations and soundings so far, and chronological data are not available. Therefore, a possible contemporaneity of these sites and (some) of the hillfort sites must remain speculative at present. Second, the topography of the *harrah*, with its large, almost inaccessible parts, and the concentration of settlements and camp sites on wadis and mud pans, affected the distribution of these settlements. This situation may hinder the characterisation of possible settlement patterns in some areas, which at this point relies solely on a visual interpretation of satellite images. Third, the resolution of some areas of the satellite imagery, such as Google Earth, is not always very high. Furthermore, aeolian sedimentation and modern human activities (quarrying, road building, *etc.*) have caused the coverage as well as the destruction of sites and therefore falsify such a settlement pattern analysis.

However, the automatic assumption that hillforts indicate a hierarchical social organisation deserves reconsideration. In the present case, it seems that such an expectation is outdated and not applicable to societies who dwelled in the *harrah* in the C/EBA I. As their economy primarily focused on agro-pastoralism, their societal structure therefore also consisted of semi-mobile elements, such as has been assumed for the EBA IA settlements in the Lejja in south-west Syria (Nicolle and Braemer 2012, 11).

There is a lack of clearly recognisable special-purpose buildings (*e.g.* for ritual, administration, or storage) as well as a lack of a clear differentiation in the lay-out and sizes of the dwellings at the hillfort sites, which would be an indication for a differentiation in the distribution of wealth and, probably, power among the settlement residents. These are other arguments that we cannot use with conventional models of societal differentiation. The question therefore arises whether these characteristics mirror egalitarian organisations. Comparable observations have been made in the EBA IA settlements in the Lejja by Nicolle and Braemer (2012, 15), who emphasise the systematic architectural reproduction of near-identical households as an argument for egalitarian group organisation.

However, the construction of the settlement fortifications, the terraced gardens, and especially the maintenance of the gardens (as at Jawa) and the irrigation management, require a specific degree of specialisation, work distribution, and management, which are far from being characterised as egalitarian. It seems that EBA settlements of the southern Levant were more heterarchically organised communities that were able to undertake large-scale projects on a communal basis (Chesson and Philip 2003, 11). It also seems that the societies of the EBA walled settlements in the southern Levant had not yet witnessed a distinctive separation into urban and rural communities (*ibid.*, 12), which would have been a precondition for urbanisation and consequently hierarchical stratification.

The research on the C/EBA in the *harrah* is still in its embryonic stage; the curtain has been lifted a little and now offers a glimpse of the dimensions and character of the C/EBA occupation of the *harrah*. The region has been underestimated in its scientific potential for contributing to discussions about the beginnings of the formation of socio-economic characteristics of south-west Asian cultural history, such as complex societies, urbanisation, long-distant relations, and technological and economic specialisation.

Renewed excavations at Jawa, which are currently in their planning stage, and forthcoming additional soundings at hillfort sites and especially unfortified sites, will hopefully contribute to a more comprehensive picture of the C/EBA society in the *harrah*.

Acknowledgements

Even though I was not able to participate at the conference in Leiden, I am thankful to its organiser, Peter Akkermans, for allowing me to contribute to the conference volume with a summarising overview on the state of research of the Jawa Hinterland Project. Additionally, I want to thank all participants of the last fieldwork seasons in which the research focussed on the exploration of the hillfort sites. My sincere thanks also go to the Department of Antiquities of Jordan and its former Director-General Dr. Monther al-Jamhawi and his staff for their excellent support. Furthermore, I wish to thank the Badia Research Program (Higher Council of Science and Technology of Jordan) with its director Mr. Nawras al-Jazi and the staff at the BRP field station in Safawi. Finally, I wish to thank Gary Rollefson for commenting on this contribution. Of course, any remaining mistakes are mine.

References

Al-Khouri, M. and Infranca, G.C. 2005. The archaeological site of Qaṣr al-Uṣeikhim. *Annual of the Department of Antiquities of Jordan* 49, 351-364.

Avner, U. 2002. *Studies in the material and spiritual culture of the Negev and the Sinai populations, during the 6th-3rd millennium B.C.* Jerusalem: Hebrew University (unpublished PhD thesis).

Bader, N.O. 1993a. Tell Maghzaliyah: an Early Neolithic site in northern Iraq. 1977-1978 results. In: Yoffee, N. and Clark, J.J. (eds.). *Early stages in the evolution of Mesopotamian civilization. Soviet excavations in Northern Iraq.* Tucson, AZ: The University of Arizona Press, 7-25.

Bader, N.O. 1993b. Tell Maghzaliyah: an Early Neolithic site in northern Iraq. 1979-1980 results. In: Yoffee, N. and Clark, J.J. (eds.). *Early stages in the evolution of Mesopotamian civilization. Soviet excavations in northern Iraq*. Tucson, AZ: The University of Arizona Press, 26-40.

Bar-Adon, P. 1972. *The Judean desert and plain of Jericho*. In: Kochavi, M. (ed.). *Judea, Samaria and the Golan – Archaeological survey 1967-1968*. Jerusalem: The Archaeological Survey of Israel and Carta, 95-152 (in Hebrew).

Bar-Mathews, M. and Ayalon, A. 2011. Mid-Holocene climate variations revealed by high-resolution speleothem records from Soreq cave, Israel and their correlation with cultural changes. *The Holocene* 21, 163-171.

Becker, N. 2013. *Die Umfassungsmauer der chalkolithischen Siedlung Hujayrat al-Ghuzlan und seine Bedeutung für die südliche Levante*. Bochum: Ruhr-Universität Bochum (unpublished MA thesis).

Betts, A.V.G. 1991a. The Late Epipalaeolithic in the Black Desert, eastern Jordan. In: Bar-Yosef, O. and Valla, F. (eds.). *The Natufian culture in the Levant*. Ann Arbor, MI: International Monographs in Prehistory, 217-234.

Betts, A.V.G. (ed.) 1991b. *Excavations at Jawa 1972-1986. Stratigraphy, pottery and other finds*. Edinburgh: Edinburgh University Press.

Betts, A.V.G. (ed.) 1998. *The harra and the hamad. Excavations and surveys in eastern Jordan*. Vol. 1. Sheffield: Sheffield Academic Press.

Betts, A.V.G. (ed.) 2013. *The later prehistory of the badia. Excavations and surveys in eastern Jordan*. Vol. 2. Oxford: Oxbow Books.

Betts, A.V.G. and Burke, D. 2015. Desert kites in Jordan: a new appraisal. *Arabian Archaeology and Epigraphy* 26, 74-94.

Boese, J. 1995. *Ausgrabungen in Tell Sheikh Hassan I. Vorläufige Berichte über die Grabungskampagnen 1984-1990 und 1992-1994*. Saarbrücken: Saarbrücker Druckerei und Verlag.

Chesson, M.S. and Philip, G. 2003. Tales of the city? 'Urbanism' in the Early Bronze Age Levant from Mediterranean and Levantine perspectives. *Journal of Mediterranean Archaeology* 16, 3-16.

Clarke, J., Brooks, N., Banning E.C., Bar-Matthews, M., Campbell, S., Clare, L., Cremaschi, M., di Lernia, S., Drake, N., Gallinaro, M., Manning, S., Nicoll, K., Philip, G., Rosen, S., Schoop, U.-D., Tafuri, M.A., Weninger, B. and Zerboni, A. 2016. Climate changes and social transformations in the Near East and North Africa during the 'long' 4th millennium BC: a compratative study of environmental and archaeological evidence. *Quarternary Science Reviews* 136, 96-121.

Helms, S. 1981. *Jawa. Lost city of the Black Desert*. London: Methuen.

Helms, S. 1989. Jawa at the beginning of the Middle Bronze Age. *Levant* 21, 141-168.

Kennedy, D. 2004. *The Roman army in Jordan*. London: Council for British Research in the Levant.

Khalil, L. and Schmidt, K. (eds.) 2009. *Prehistoric 'Aqaba I*. Rahden: Verlag Marie Leidorf.

Meister, J., Krause, J., Müller-Neuhof, B., Portillo, M., Reimann, T. and Schütt, B. 2017. Desert agricultural systems at EBA Jawa (Jordan): integrating archaeological and paleoenvironmental records. *Quarternary International* 434, 33-50.

Migowski, C., Stein, M., Prasad, S., Negendank, J. and Agnon, A. 2006. Holocene climate variability and cultural evolution in the Near East from the Dead Sea sedimentary record. *Quaternary Research* 66, 421-431.

Müller-Neuhof, B. 2005. *Zum Aussagepotenzial archäologischer Quellen in der Konfliktforschung: eine Untersuchung zu Konflikten im vorderasiatischen Neolithikum*. Berlin: Freie Universität Berlin (unpublished PhD thesis).

Müller-Neuhof, B. 2012. The gardens of Jawa: early evidence for rainwater harvesting irrigation. *Bulletin of the Council for British Research in the Levant* 7, 62-64.

Müller-Neuhof, B. 2013a. East of Jawa: Chalcolithic/Early Bronze Age settling activities in the al-harra (NE-Jordan). *Annual of the Department of Antiquities of Jordan* 57, 125-139.

Müller-Neuhof, B. 2013b. Nomadische Ressourcennutzungen in den ariden Regionen Jordaniens und der südlichen Levante im 5. bis frühen 3. Jahrtausend v. Chr. *Zeitschrift für Orient-Archäologie* 6, 64-80.

Müller-Neuhof, B. 2014a. Desert irrigation agriculture – Evidence for Early Bronze Age rainwater-harvesting irrigation agriculture at Jawa (NE-Jordan). In: Morandi Bonacossi, D. (ed.). *Settlement dynamics and human landscape interaction in the steppes and deserts of Syria*. Wiesbaden: Harrassowitz Verlag, 187-197.

Müller-Neuhof, B. 2014b. A 'marginal region' with many options: the diversity of Chalcolithic/Early Bronze Age socioeconomic activities in the hinterland of Jawa. *Levant* 46, 230-248.

Müller-Neuhof, B.2015. Evidence for an Early Bronze Age (EBA) colonization of the Jawa hinterland: preliminary results of the 2015 fieldwork season in Tulul al-Ghusayn. *Bulletin of the Council for British Research in the Levant* 10, 74-77.

Müller-Neuhof, B. 2016. Prehistoric settlements in the northern badia (Jordan). In: Bartl, K., Lüth, F. and Reindl, M. (eds.). *Palaeoenvironment and the development of early settlements*. Rahden: Verlag Marie Leidorf, 149-160.

Müller-Neuhof, B. 2017a. The Chalcolithic/Early Bronze Age hillfort phenomenon in the northern badia. *Near Eastern Archaeology* 80, 124-131.

Müller-Neuhof, B. 2017b. Khirbet al-Ja'bariya: a recently discovered fortified Early Bronze Age settlement in the Jordanian basalt desert. *Bulletin of the Council for British Research in the Levant* 12, 66-71.

Müller-Neuhof, B. and Abu-Azizeh, W. 2016. Milestones for a tentative chronological framework of the late prehistoric colonization of the basalt desert (north-eastern Jordan). *Levant* 48, 220-235.

Müller-Neuhof, B. and Abu-Azizeh, W. 2018a. Chalcolithic/ Early Bronze Age settlements in the northern badia – Part I: Tulul al-Ghusayn. *Annual of the Department of Antiquities of Jordan* 57, 411-423.

Müller-Neuhof, B. and Abu-Azizeh, W. 2018b. Chalcolithic/ Early Bronze Age settlements in the northern badia – Part II: Khirbet al-Ja'bariya. *Annual of the Department of Antiquities of Jordan* 57, 425-434.

Müller-Neuhof, B., Betts, A. and Wilcox, G. 2015. Jawa, eastern Jordan: the first 14C dates from the early occupation phase. *Zeitschrift für Orient-Archäologie* 8, 124-131.

Nicolle, C. and Al-Maqdissi, M. 2006. Sharaya: un village du Bronze Ancien IA en Syrie du Sud. *Paléorient* 23, 125-136.

Nicolle, C. and Braemer, F. 2012. Settlement networks in the southern Levant in the mid 4th millennium BC: sites with double-apsed houses in the Leja area of southern Syria during the EBA IA. *Levant* 44, 1-16.

Paz, Y. 2002. Fortified settlements of the EB IB and the emergence of the first urban system. *Tel Aviv* 29, 238-261.

Rowan, Y. M., Rollefson, G.O., Wasse, A., Hill, A.C. and Kersel, M.M. 2017. The Late Neolithic presence in the Black Desert. *Near Eastern Archaeology* 80, 102-113.

Sherratt, A. 1983. The secondary exploitation of animals in the old world. *World Archaeology* 15, 90-104.

Strommenger, E. 1980. *Habuba Kabira. Eine Stadt vor 5000 Jahren. Ausgrabungen der Deutschen Orient-Gesellschaft am Euphrat in Habuba Kabira*. Mainz: Phillip von Zabern.

Weninger, B. 2009. Yarmoukian rubble slides. Evidence for Early Holocene rapid climate change in southern Jordan. *Neo-Lithics* 1/09, 5-11.

Youkana, D.G. 1997. *Tell es-Sawwan. The architecture of the sixth millennium B.C.* London: NABU Publications.

Zertal, A. 2003. Fortified enclosures of the Early Bronze Age in the Samaria region and the beginning of urbanization. *Levant* 25, 113-125.

The Late Chalcolithic and Early Bronze Age of the *badia* and beyond: implications of the results of the first season of the 'Western Harra Survey'

Stefan L. Smith

Abstract

The climatically varied Syro-Levantine steppes feature complex dynamics of past human occupation that vary greatly across the region in terms of scale, time periods, and archaeological remains. In particular, the Late Chalcolithic (LC) and Early Bronze Age (EBA) (*c.* 4400-2100 BC) saw urbanism in north-eastern Syria, smaller-scale sedentism in central Syria, and the decline of longstanding occupation in north-eastern Jordan. Despite this, the challenges faced by prehistoric populations in these uncertain environments would have been very similar; thus it is reasonable to propose that some of their solutions were also. The region-wide project 'Human Adaptation in Climatically Marginal Environments of late-fifth to third millennium BC Syria and Jordan' takes a holistic approach to investigating these arid and semi-arid regions to determine their appeal to past populations, and the effects of the natural and anthropogenic environment on settlement morphologies and societies. It uses a variety of past and present remote sensing and ground truth data, a vital part of which is the author's 'Western Harra Survey', south of Jawa in the northern *badia* of Jordan. The first fieldwork season, conducted October-November 2015, identified large quantities of lithic material at numerous sites, a handful of which were likely occupied during the LC/EBA, as well as potential links to raw chert material sources, adding another facet to the appeal of the *harrah* to past populations, on top of the well-established arguments for the exploitation of pasture land resources. Additionally, a typological seriation of the morphology of sites known as 'wheels' was commenced, which appears to be linked to different site uses and/or periods of occupation. Establishing these connections is crucial to allow mapping occupation dynamics across the greater region and comparisons with areas in Syria and beyond.

Keywords: human adaptation, subsistence strategies, semi-arid steppes, remote sensing, surface survey, basalt desert, raw chert material, morphological site typologies

Introduction

The arid and semi-arid steppes of the Syro-Levantine region consist of a varied climatic geography that has in common intermittent and uncertain precipitation. With modern-day values ranging from nearly 350 mm per annum in north-eastern Syria to less than 100 mm in north-eastern Jordan, the potential for human use of the landscape is far from uniform (Fig. 1). However, the environmental uncertainty of the entire region, with

In: Peter M. M. G. Akkermans (ed.) 2020: *Landscapes of Survival - The Archaeology and Epigraphy of Jordan's North-Eastern Desert and Beyond*, Sidestone Press (Leiden), pp. 165-184.

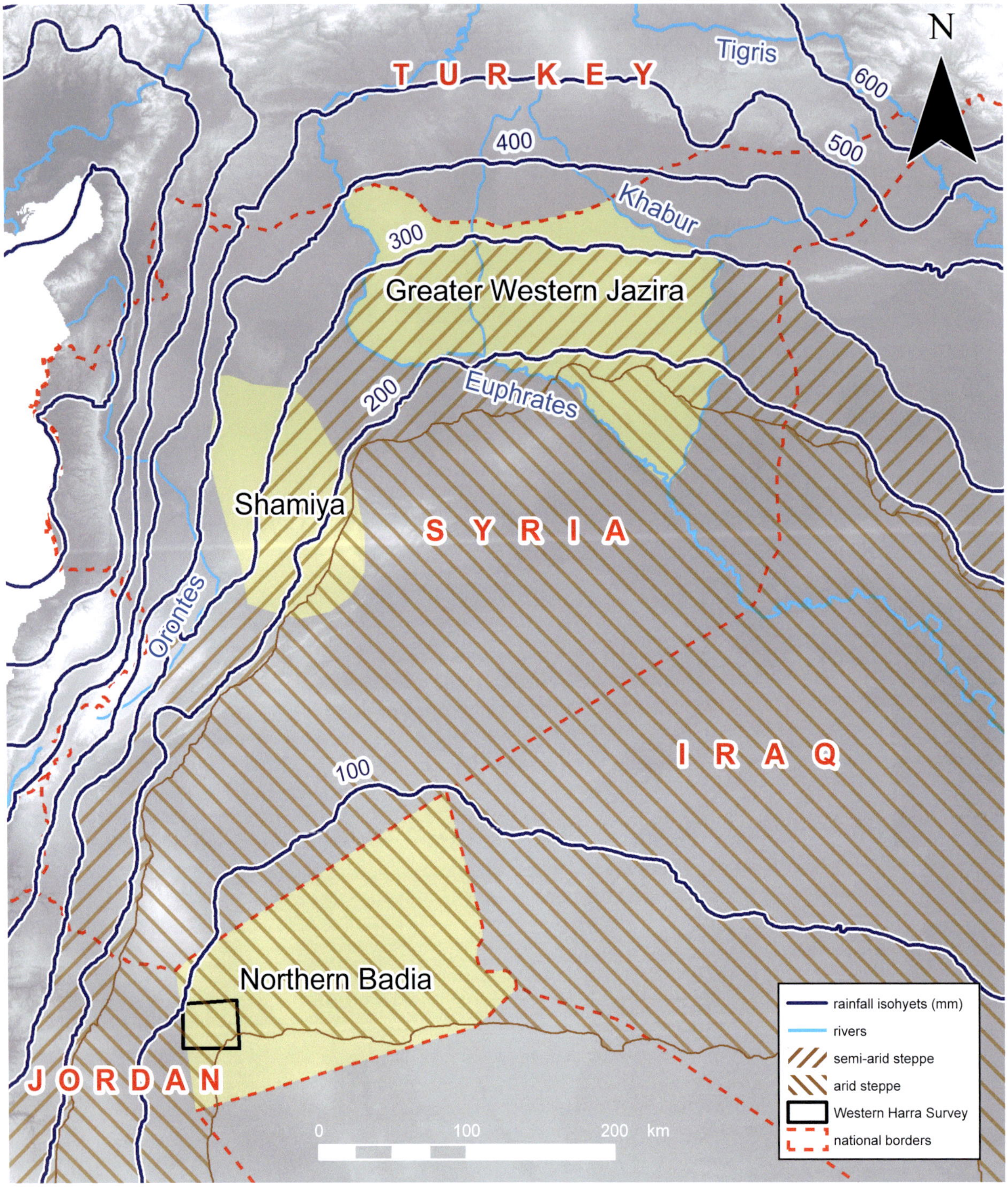

Figure 1. ASTER topographical map showing the climatic landscape of the Syro-Jordanian steppes; the three case-study regions of this project (yellow shading); and the Western Harra Survey. ASTER GDEM is a product of METI and NASA. Isohyets represent average annual precipitation from 1980 to 2010 (from Global Precipitation Climatology Centre (GPCC) data, processed by Louise Rayne, University of Leicester).

severe rainfall fluctuations being common and where any regional climate variations are hardest felt (Sanlaville 2000), is a factor likely to have remained unchanged despite possible palaeoclimatic changes, and unifies the challenges faced by its past and present populations. This, in turn, potentially means that unified solutions could be and were employed by prehistoric nomads and sedentarists, including agro-pastoralism, conservation strategies such as extensification, and a general ability to rapidly adjust any subsistence strategies to match the unpredictable environment (Smith and Wilkinson 2020; Wilkinson 1997; 2000; Wilkinson *et al.* 2014). Additionally, as suggested by McClellan and Porter (1995), the core reasons for human occupation of these steppes may themselves be broadly uniform, such as the exploitation of pastoral resources, mineral sources, and profiting from trade routes. Long neglected by supra-regional archaeological studies, these areas are now being included in wide-ranging discourses due to the dissemination of past fieldwork in Syria and continued fieldwork in Jordan. In particular, the Late Chalcolithic (LC) to Early Bronze Age (EBA) (*c.* 4400-3000 BC) is a period that saw complex settlement dynamics in north-eastern and central Syria, with low levels of settlement giving way to rapid urbanisation processes, which just as quickly collapsed again by the end of the EBA (Castel *et al.* 2008; Geyer *et al.* 2007; Hempelmann 2013, 271-276; Smith *et al.* 2014; Wilkinson *et al.* 2014).

In north-eastern Jordan, on the other hand, this period saw the eventual decline of a likely long sequence of occupation from at least the Epipalaeolithic onwards, though this may have been intermittent (Betts 1998a; Müller-Neuhof 2014a; Rollefson *et al.* 2014). Nevertheless, much evidence has been gathered by recent fieldwork that human occupation of this area during the early EBA at least was perhaps more prevalent than once thought, including at the major site of Jawa (Betts 1991; Müller-Neuhof 2014b). However, no holistic, unified study has ever been conducted that brings together the archaeological landscape of the Syro-Levantine steppes during this time period. This is the remit of a supra-regional project I commenced in 2017 which includes, as one of its most important components, fieldwork in the basalt *harrah* region of north-eastern Jordan.

Overview of the supra-regional project

The region-wide project, entitled 'Human Adaptation in Climatically Marginal Environments of late-5th to 3rd Millennium BC Syria and Jordan', aims to create a holistic overview of the origins and transformations of nomadic and sedentary settlement in the arid and semi-arid steppes of Syria and Jordan during the LC and EBA. A detailed examination of remote sensing data, most notably satellite and aerial photography, across the entire Syro-Jordanian region is being combined with available ground data from past and present site visits, excavation and survey reports, as well as from the Western Harra Survey, a co-directed fieldwork project between myself and Dr Marie-Laure Chambrade.[1] This process, already successfully implemented by the Fragile Crescent Project of Durham University (see below), allows for the analysis of such diverse aspects as settlement dynamics over time, population migrations, and links between settlement morphologies and periods of occupation (which can in turn inform remote sensing-based investigations of regions that are not accessible on the ground; Galiatsatos *et al.* 2009; Lawrence *et al.* 2012). By processing this data together with regional interpretations, this project seeks to determine the effects of the natural and anthropogenic environment on settlement, settlement morphologies, and subsequently on societies. Furthermore, the identification of coping strategies employed by varying societies in comparable settlements on a regional scale is being analysed to determine whether overarching unifying factors drove human endeavours in uncertain environmental conditions.

Greater Western Jazira

To achieve this, three case study areas were selected that provide a representative sample of the different climatic and topographic landscapes that exist across the region, and which have seen past or present archaeological investigations (Fig. 1). The first of these is the 'Greater Western Jazira', bordered by the Euphrates and Khabur rivers in north-eastern Syria – a 27,000 km^2 region of undulating steppe broken by the mountain chain of Jebel Abd al-Aziz and traversed by numerous highly seasonal wadis. This area was the subject of my doctoral research, which investigated the region by remote sensing and the collating of existing fieldwork and site visit reports, focussing on creating a holistic overview of LC and EBA settlement dynamics (Smith 2015). Through a few archaeological investigations, most notably over half a century of excavations at Tell Chuera (see Meyer 2010 for an overview), this area has long been known to contain large and complex tell settlements broadly dating to the Early Bronze Age, which appear to have formed rapidly *ex nihilo*. No integrated study of the entire landscape had formerly been undertaken, however.

The results of the investigation showed that the area exhibits a complex system of steppe habitation not limited to the known large tells, with over 300 sites of varying sizes and morphologies identified as definitely or probably dating to the periods in question (Fig. 2). Analyses carried out on site densities, settlement sizes, grain production, supporting settlements for centres, and site alignments allowed several

1 Marie-Laure Chambrade, Archéorient Research Unit, CNRS, Lyon, France

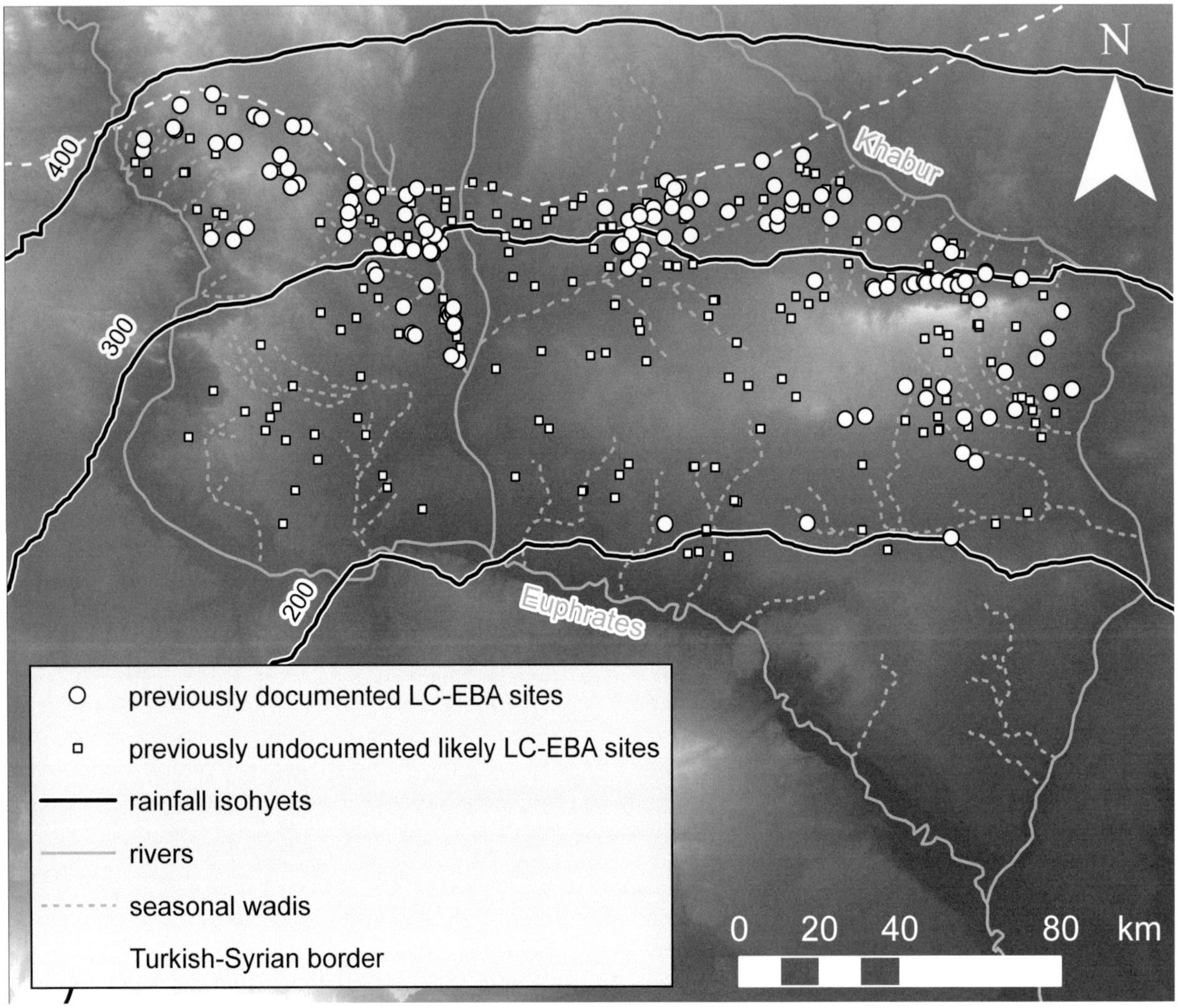

Figure 2. ASTER topographical map of the Greater Western Jazira, showing the sites that likely date to the LC and/or EBA.

economic systems to be proposed, indicating that multiple sedentarisation and possible nomadisation processes occurred at different times during the fourth and third millennia BC (Smith 2020; Smith *et al.* 2014). Specifically, two independent trajectories of EBA settlement were identified. The first is in the northern part of the Greater Western Jazira where, following sparse occupation during the early LC and little to none during the late LC, large, apparently planned urban centres emerged around 3100 BC, likely as the result of migration into the region from the north and north-west, followed by smaller settlements and farmsteads in their hinterlands (Hempelmann 2013, 271-276; Meyer and Hempelmann 2006). Including many of the large walled two-tiered tells referred to as *Kranzhügel* in the literature,[2] this represented a continuation of the development of large-scale cities during the Uruk period, which in other parts of northern Mesopotamia collapsed into a decentralised system of smaller towns (Meyer 2010; Ur 2010). The impetus for such a migration would have been a combination of the 'push' factor of the collapse of the Uruk expansion, which removed the foundations of a regionally integrated economy on which local centres had perhaps become reliant, and the 'pull' factor of a fertile steppe with over 300 mm annual precipitation (Kalayci 2013). It is probable that these settlers were not the only occupants of the steppe at this time, as nomadic pastoralists may have existed in these steppes prior to the third millennium BC (Wilkinson *et al.* 2014). Some of these mobile groups, at least, would therefore have interacted with the new settlers, perhaps for economic profit, perhaps out of coercion, but doubtless with a profound effect on their societies. This could have manifested in an acquired

2 This term, applied indiscriminately to many large fortified tells in the region, in fact refers to a number of disparate site morphologies which are far from homogenous (Smith *et al.* 2014; Smith 2020).

sedentism for some, but likely not all, nomadic tribes or kinship groups (Porter 2004).

The second trajectory, focussed on the central and southern parts of the area, did not commence until *c.* 2500 BC, and resulted in the establishment of smaller fortified tell sites and other settlements. These sites, which significantly differ in terms of morphology, internal structure, and material culture from the northern *Kranzhügel*, more likely came about as a result of the growing regional polities of Mari and Ebla, which drove pressure on their hinterlands to supplement the grain supplies of centres in unfavourable years (Ur 2010). This led to an intensification of agriculture, which in turn made the opportunities for extensification in a large empty landscape with no settlement clustering very attractive, as did the chance to exploit pastoral produce. Wool was a well-established commodity by this time, with philological evidence that shepherds were ranked highly in the labour system of Tell Beydar (McCorriston 1997; Sallaberger and Pruß 2015, 94-98; Smith *et al.* 2014). Another likely factor that drove settlement in these more arid parts of the Greater Western Jazira was trade routes, perhaps brought about by similar desires to exploit pastoral commodities, evidenced by the alignment of several large sites near the 200 mm isohyet (Fig. 2).

Thus, overall it can be said that this area was an integral part of the northern Mesopotamian economic and political landscape, belying its reputation as a 'margin'. Not only does this statement apply to the complex urban processes that occurred within the steppe, but also to their interactions with and effects upon the surrounding 'core' regions of long-term settlement. Rather than being a side-venture entered into by a few large polities, the exploitation of this region was a major component of the regional and inter-regional economic and political landscape (Smith 2020).

Nevertheless, both the northern and southern manifestations of sedentism (and urbanism) in the Greater Western Jazira came to an end beginning *c.* 2300 BC and concluding *c.* 2100 BC. This likely had multiple causes, including the expansion of the Akkadian empire, which sought to directly control trade routes and access to commodities (Liverani 2014, 141-143); the waning in power of both Ebla and Mari; and environmental degradation, both anthropogenic (overgrazing, cf. Danti 2000, 308-311; deforestation, cf. Deckers and Pessin 2011) and natural (aridification; see Kalayci 2013, 13-14; Riehl 2009; Wossink 2009, 24-25). This led to the abandonment of the settlements, more gradual at some than others, whose inhabitants either transitioned to nomadism or resettled in socio-climatically more stable regions such as the Khabur or Euphrates valleys.

Shamiya

Located east of the fertile Orontes river valley, and south-west of the Euphrates, the Shamiya region occupies *c.* 7000 km^2 in a similar climatic and topographic landscape as the Greater Western Jazira (Fig. 1), comprising an undulating steppe traversed by some major seasonal wadis, though with the additional presence of flat valley bottoms with fertile silty soils known locally as *fayda*, akin to the *qa'a* features of the northern *badia* (see below). Over a decade of investigation by the Mission des Marges Arides has identified settlement dynamics, morphologies, and regional influences that are different to those of the Greater Western Jazira, however (Castel *et al.* 2008; Geyer *et al.* 2007). Firstly, there seems to have been a complete absence of sites during the LC, as well as the preceding 'Ubaid and Halaf periods, which implies the region was at most solely occupied by mobile pastoralists during this time. Secondly, the EBA sees the emergence of a settlement network only in the second half of the third millennium BC, more than 500 years after the first emergence of urbanism in the Greater Western Jazira. These settlements include fortified tell sites such as Tell al-Rawda, which with successive lines of defence ramparts yet no clear upper or lower town do not resemble the morphologies of the *Kranzhügel*. Such tells and other smaller settlements appeared rapidly around 2500-2450 BC, indicating a migration from the west, as do material culture connections to the Orontes valley, Ebla, and Qatna (Castel and Peltenburg 2007). Furthermore, unique features such as the *Très Long Mur* speak for a planned incursion into the semi-arid steppe. This low but extensive 200 km long wall was not fortified enough to have had a defensive function, and thus most likely served as a boundary marker to nomads beyond, potentially of pasture land desirable to a large regional centre (Geyer *et al.* 2010, 67-69). This may have been due to the need for vast pasture lands by Ebla during the late third millennium BC, a model that also partially explains patterns in the central and southern parts of the Greater Western Jazira (see above). However, the absence in the Shamiya of even small sites prior to this period, and the condensed timeframe of its mid-EBA occupation, which came to an end by the late third millennium BC, are major differences. This does not preclude contact of one area with the other, though. Indeed, with Tell al-Rawda having been constructed "as if from a blueprint (...) derived from the earliest (examples ...) of this ideal city type with radial and concentric streets" (Castel and Peltenburg 2007, 611-612), it is possible that the early EBA settlement morphologies and therefore subsistence strategies of the Greater Western Jazira may have been an influence, paving the way for the general exploitation of steppe pasturelands by sedentary populations.

Northern badia

The nearly 26,000 km^2 north-eastern 'panhandle' of Jordan comprises a substantially different environment and landscape from either of the other two case study areas. Climatically, it is much drier, located on the arid south-eastern side of the 100 mm isohyet (Frumkin *et al.* 2008, 360-361) (Fig. 1). Some of the northern *badia* features a similar undulating steppe with seasonal wadis (although these are less frequently flowing) and large flat valley bottoms covered in deep silt sediments (mud flats or *qe'an*; singular: *qa'a*) in the *hamad* landscape. However, the majority is covered by a dense layer of basalt rocks in the *harrah*, part of the 40,000 km^2 Harrat al-Sham basaltic plateau that stretches from southern Syria to north-western Saudi Arabia, formed by lava flows dating from the Oligocene to the Quaternary (most recently *c.* 400,000 BP; Kempe and Al-Malabeh 2010). These stone blocks make traversing the *harrah* extremely difficult compared to the Syrian steppes, often being impossible except along wadis or across the mud flats, but also provide ample readily available construction materials for human occupants. Thus it is unsurprising that the region's dense network of visible prehistoric sites are comprised almost entirely of structures made of basalt boulders, ranging from single to multiple courses in height. These are very clearly visible on aerial photographs, which allow numerous morphological types to be identified, much clearer than for any sites in the Greater Western Jazira or Shamiya. Amongst these are the features known as wheels, pendants, kites, cairns, and meandering walls (Kennedy 2011; see also D. Kennedy, this volume).

Evidence based on lithic material and, more recently, Optically-Stimulated Luminescence (OSL; Athanassas *et al.* 2015) and Accelerator Mass Spectrometry (AMS; Müller-Neuhof and Abu-Azizeh 2016; Richter *et al.* 2017) dating indicates a settlement chronology that is very different from the other two case study areas. Evidence for some occupation exists from at least the Middle Palaeolithic, while intensive human activity is evidenced from the Epipalaeolithic (Late Natufian) onwards (*c.* 12,650 cal BC at Shubayqa, cf. Richter *et al.* 2017; *c.* 9000 BC at Dhuweila, see Betts 1998a), and is subsequently attested to for all periods up to and including the EBA (Akkermans *et al.* 2014; Müller-Neuhof 2014a; Rollefson *et al.* 2014; 2016). While this does not necessarily indicate permanent occupation, it speaks against a prolonged period of site abandonment such as during the entire LC in the Shamiya and the southern Greater Western Jazira. Settlement in the northern *badia* does not appear to have continued long into the EBA, however. By the very early third millennium BC, occupation of the *harrah* seems to cease until the Middle Bronze Age. This is another major departure of this region from those in Syria; the *badia* did not see a lengthy third millennium BC urban expansion across multiple sites. The closest example is the relatively urbanised occupation at Jawa during the late LC and early EBA, which however lasted no more than a couple of centuries (possibly until the end of the Levantine EBA IB), and appears to have been unique in terms of its large size (Betts 1991; Müller-Neuhof and Abu-Azizeh 2016).

Despite these significant differences from the Syrian steppes, some of the challenges faced by the occupants of the northern *badia*, as well as possible opportunities this landscape provided, are likely to have been similar. Most clearly, the climate, though doubtless more arid regardless of any palaeoenvironmental variances, would have been equally as uncertain (Frumkin *et al.* 2008). According to Sanlaville (2000), a 45-50% yearly fluctuation in precipitation is common across the entire Syro-Jordanian steppe, and perhaps higher in areas where the average annual rainfall is lower than 150 mm. Therefore, similar coping mechanisms, like agricultural extensification focussed on fodder crops (Müller-Neuhof 2013a), water catchment systems such as those at Jawa (Müller-Neuhof 2014b), adaptive hunting strategies as evidenced by faunal remains at Shubayqa 1 (Yeomans *et al.* 2017), or the ability to rapidly switch between nomadic and (semi-)sedentary lifestyles, may well have been employed to enable subsistence (McClellan and Porter 1995).

The reasons for the attraction of the *harrah* for widespread human occupation are still unclear, although several hypotheses can be reasonably proposed, such as safety due to the inaccessibility of the landscape, population pressures in the fertile Jordanian uplands, or the semi-sedentarisation of nomadic groups for reasons of improved economic exploitation of the landscape or cultural shifts (or a combination of both) (see *e.g.* Porter 2012, 8-64). However, there is good evidence that once the human presence had become established in the region, the mining of, and trade routes for, raw flint material and perhaps salt became important, providing a further impetus for the occupation of these regions, localised in specific areas (Müller-Neuhof 2013a; 2013b). This bears similarities to the southern part of the Greater Western Jazira, although the trade routes here would have likely been locally controlled by mobile groups which exploited the land for its pastoral potential, while at the same time engaging in long-distance trade, probably with the Jordanian uplands far to the west. Furthermore, the difficulty of traversing the landscape doubtless played a more significant role in dictating the trajectories of these routes. Lastly, interactions between nomadic tribes and (semi-)sedentary populations are likely to have been key to subsistence, with local knowledge of this hostile environment particularly important for long-term occupation. This, as in the Greater Western Jazira and possibly the Shamiya, may have led to the (temporary) sedentarisation of some mobile groups.

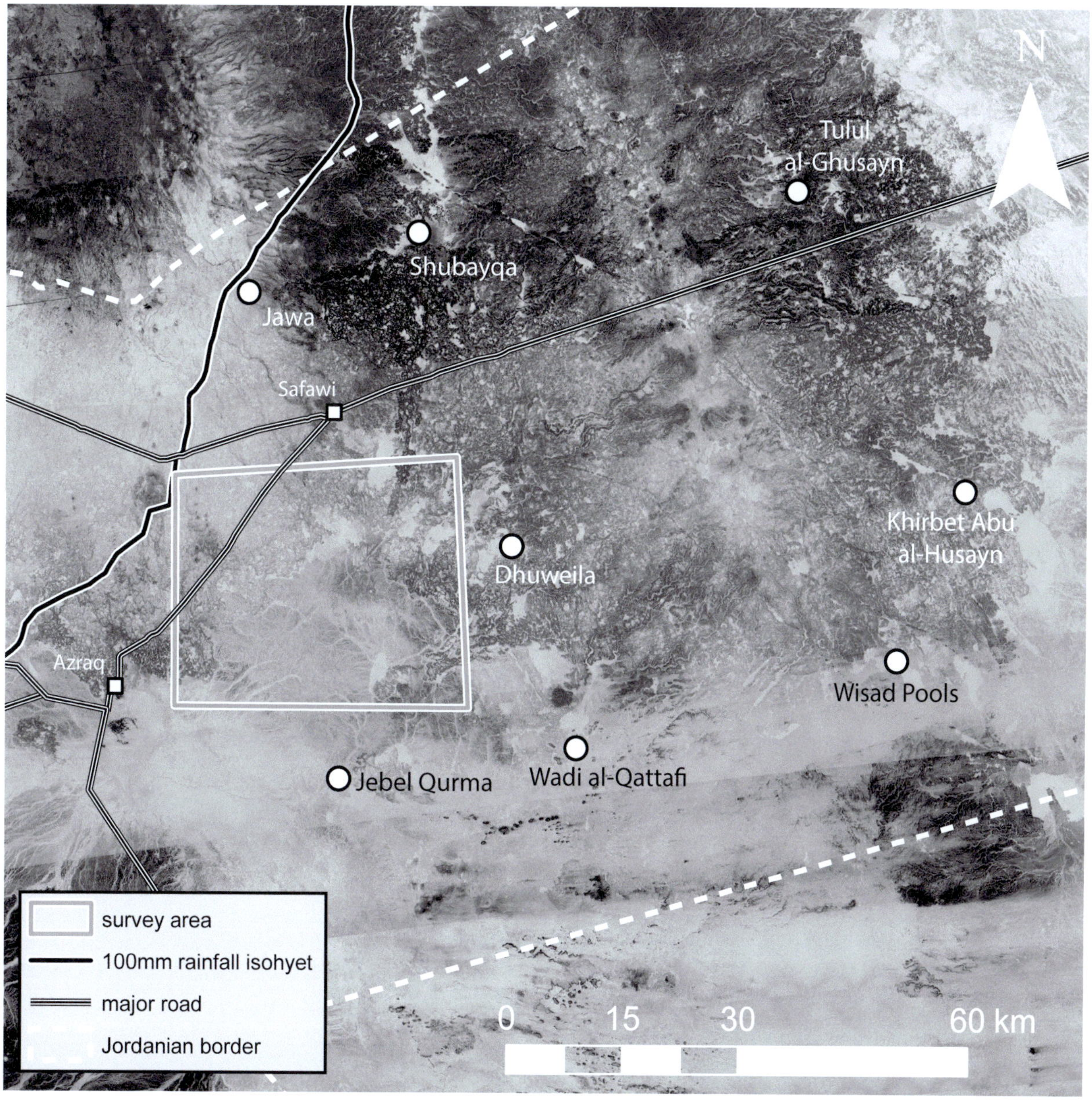

Figure 3. CORONA satellite map of the location of the Western Harra Survey in the context of other investigated areas in the region.

The Western Harra Survey

When planning the project 'Human Adaptation in Climatically Marginal Environments of late-5th to 3rd Millennium BC Syria and Jordan' in early 2015, it was clear that self-conducted fieldwork, in a region and with a data collection methodology tailored to answer the research questions it posed, would be essential. As Jordan was the only country in the project's geographical remit where this was possible, I searched for a suitable area of the northern *badia* for a survey project with the potential for future excavations. To get an initial idea, a stratified random sampling exercise was carried out, where the entire Jordanian *harrah* was divided into 15 by 15 km squares, and a randomly-selected 3 by 3 km area within each was analysed in detail on satellite imagery. The largest numbers of sites likely to date to the time periods under study were yielded on the western side of the basalt fields, just east of the Azraq-Safawi road (see Fig. 3). Though not enough research had been conducted in this region to be able to conclusively link any site morphologies with specific periods of occupation, some preliminary data could inform this planning. Amongst the most conclusive of these are the following. Desert kites, which feature many examples of other sites constructed on top of them, appear to be some of the oldest features, likely initially constructed during the

Figure 4. Photographs illustrating the typical landscapes represented in the Western Harra Survey area, corresponding to study regions A-D (see also Fig. 5) (photographs by S.L. Smith and L. Haddad).

early Neolithic (seventh millennium BC or earlier), although often subsequently re-used (Akkermans *et al.* 2014; Betts 1998b). So-called 'wheels' have been found to possibly contain material dating from the Neolithic to Chalcolithic and perhaps EBA periods, with the latter also indicated by OSL dates (Rollefson *et al.* 2016; see also Akkermans and Brüning, this volume). Meanwhile, more precise evidence of EBA occupation exists in the form of so-called '*ghura* huts', small double-celled enclosure structures identified across the *harrah*, in particular east of Azraq, and preliminarily dated to this period based on lithic typologies (with more precise data from ^{14}C samples forthcoming; see Müller-Neuhof and Abu-Azizeh 2016, 229). Therefore, the high numbers of wheels and *ghura* huts in the region immediately east of the Azraq-Safawi road indicated this to be a promising area.

Being thus drawn to the western edge of the Jordanian *harrah*, it was clear that a major area that had not been formerly investigated was the large region south of the immediate hinterland of Jawa, north of Jebel Qurma, and west of Dhuweila (Fig. 3). As LC and EBA occupation had been confirmed in multiple locations east and west of Jawa by the 'Jawa Hinterland Survey Project' (Müller-Neuhof 2014a), the question of whether the same could be found to the south remained an important one.[3]

Furthermore, this trajectory from Jawa roughly follows the modern-day 100 mm isohyet, suggesting similar climatic conditions, not to mention the same geographic situation vis-à-vis the basalt terrain, being close to its western edge. Additionally, this area encompasses four types of landscape that are representative, in microcosm, of the entire *harrah* (Fig. 4): (A) undulating steppe carpeted by a dense layer of basalt blocks, extremely difficult to traverse; (B) large traversable wadi systems surrounded by pockets of dense to medium-dense basalt outcrops; (C) large mud flats often adjoining each other over many tens of kilometres, providing access into areas otherwise similar to (A); and (D) hilly areas crossed by small wadis that are not easy to travel along, with medium-dense basalt coverage right up to the wadi edges.

There were several other, more practical, reasons to choose this location also. As accessibility is a major issue in the *harrah*, now just as much as in the past, the area east of the Azraq-Safawi road is ideal. This asphalt highway allows quick

3 Since 2015, the continuation of the Jawa Hinterland Project has identified LC/EBA sites closer to and potentially within the Western Harra Survey area also (Müller-Neuhof and Abu-Azizeh 2016; see below).

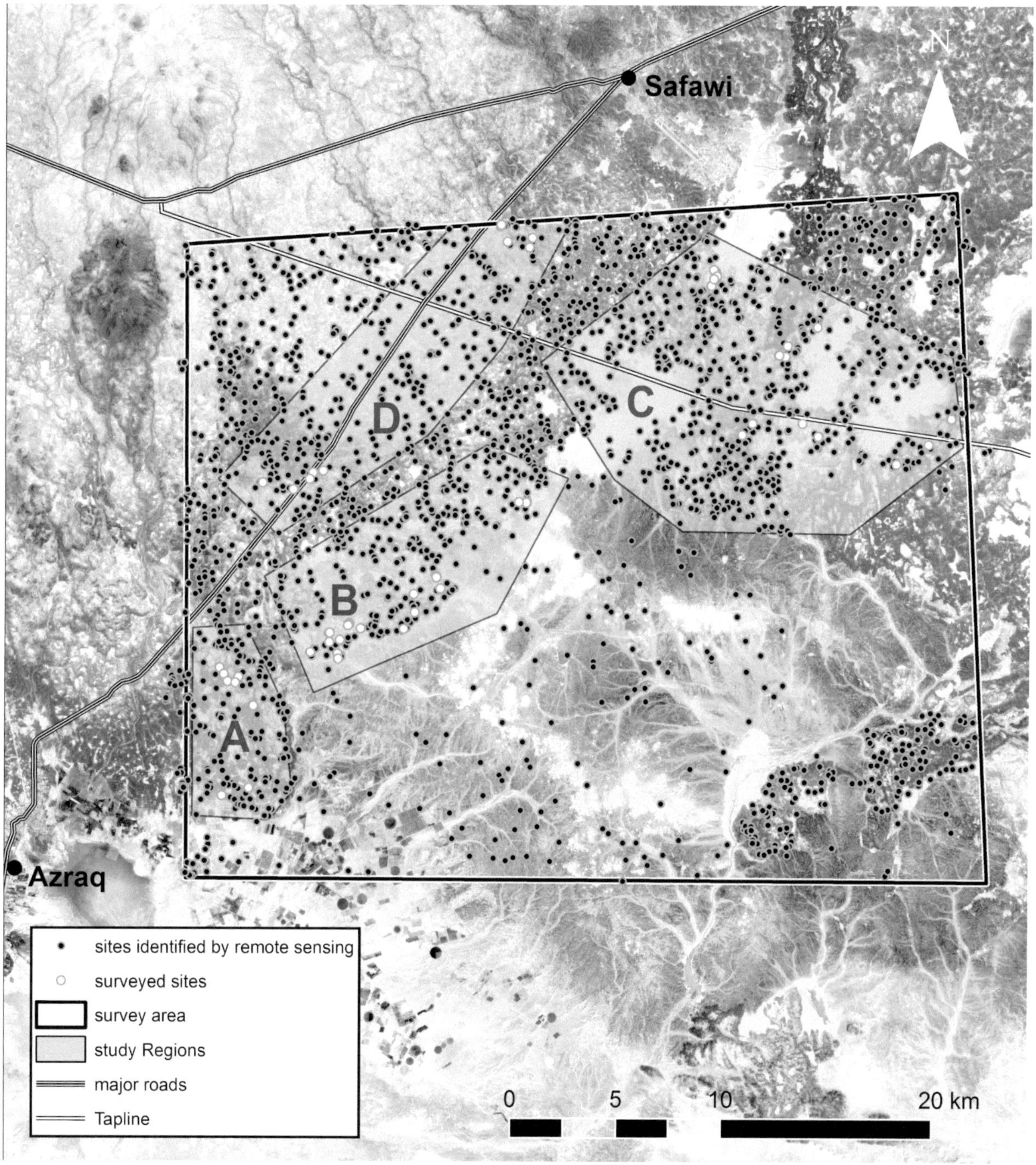

Figure 5. GeoEye satellite map of the Western Harra Survey, showing the study regions A-D, the sites identified by the remote sensing survey, and the sites visited on the ground in 2015.

access to the edge of the survey area, while a few routes lead into its interior. One of these is known as the Tapline, a road constructed in the 1940s to follow the course of the Trans-Arabian oil pipeline. Though now crumbling in a poor state of repair, it provides relatively easy access to areas otherwise inaccessible (see Fig. 5). Others include tracks constructed for quarry vehicles, and oil prospection routes from the 1980s. Though none of these methods of access are easily traversed, compared to the complete vehicle inaccessibility of much of the *harrah*, they provide vital links to its largely unexplored interior (Smith and Chambrade 2018, 16-17). Additionally, the proximity of the basalt desert to Azraq in this location makes logistics for fieldwork as easy as is possible for this part of Jordan. Finally, the size of the survey area (30 by 36 km;

Figure 6. Path leading up to the large wheel Site 1745, with visible 'steps' (photograph by M.-L. Chambrade).

c. 1100 km²) is a manageable one, large enough to provide a representative overview of the landscape, but small enough to analyse in its entirety on remote sensing, and cover well within a few fieldwork seasons of surveying.

With the survey area defined, the region was systematically analysed using GeoEye satellite imagery, accessed via the Google Earth platform, at its native resolution. This process allowed for the identification of nearly 3000 individual sites or features of all morphologies, assigned numbers in the order in which they were recorded; this then became their site number in the field also. These were pared down to those sites likely to be of interest to the research question, and that are accessible by vehicle to within one km. A selection was then chosen divided by study regions A-D that correspond to the four main landscape types represented (Figs. 4-5), and were targeted on the ground by their GPS coordinates. At each site, natural and anthropogenic landscape settings were noted (*e.g.*, proximity to wadis or vegetation, intervisibility of sites), morphologies recorded, noteworthy features sketched, and surface material collected by systematic site-walking and processing *in situ*. Accurate GPS points were also taken to allow for more detailed future analyses of site shapes than is possible from satellite imagery.

Summary of results

In the first fieldwork season, carried out October-November 2015, the goal was both to gather preliminary data pertaining to the research project, but also to explore the archaeological landscape in general. Thus a total of 38 wheels were visited, as well as 22 enclosures, 5 pendants, and 1 cairn field. These sites were largely chosen based upon their good level of preservation and distinctiveness of morphology as determined by the satellite imagery, as well as representing a roughly equal number from each study region. The majority are located near the boundary of the basalt desert with wadis or mud flats, partially due to practicalities of accessibility, partially due to these being the areas of largest site concentration (Fig. 5). Most also feature good views in at least two cardinal directions, and many have intervisibility with other sites (although this is to be expected with a site density of over two per square kilometre, and may not always have been by design, especially as definite site chronologies have not yet been established). The sites visited ranged in size from pendants of *c.* 20 m in length to large wheels over 70 m in diameter (*c.* 4000 m²).

Numerous paths formed by the linear clearing of basalt rocks were also identified in the vicinities of several sites, in particular the wheels. These are wider than sheep tracks formed by modern-day herds moving across the landscape, yet much too narrow to allow passage for vehicles, indicating a prehistoric origin as previously noted by Akkermans *et al.* (2014), and likely contemporaneous with the associated sites. Many exist directly adjacent to sites and emanate a few hundred metres into their hinterlands, while others appear to connect multiple sites with each other. One particularly interesting instance of a path in region B appears to lead from a wadi bed to the large wheel Site 1745 on the hillside, with lines of small basalt rocks arranged laterally to the course of the path at semi-regular intervals, indicating a purpose to prevent slope wash by creating 'steps' (Fig. 6) (for more detail, see Smith and Chambrade 2018). However, some visited sites, in particular several of the wheels in the south-west of the survey area, are located deep within the basalt terrain, in contradiction to the assertion that they are usually close to a wadi or mud flat.

As expected for sites in the *harrah*, the surface material overwhelmingly consisted of lithics, with a total of 19,476 pieces recorded compared to 576 ceramic pieces (all undiagnostic body-sherd fragments) and a handful of other objects such as grinding stones or Islamic-era clay pipes. The lithics comprised a range of points, scrapers, knives, cores, and flakes, with the first two types clearly in the majority. The amount of surface lithics visible at each site varied greatly, even when taking into account site size and natural landscape transformation factors, such as alluvial layers of silt from seasonal rains. For example, 42 lithics were recorded at the 2500 m² Site 1975 in region A, while only 6 km away in the same study region, 1028 lithics were counted at the 3200 m² Site 2233. The raw lithic material consists of a range of chert, including flint and porcellanite, though in some instances attempts at creating tools from basalt were registered. Certain chert types appeared to have been sourced locally, while some site morphologies appeared to contain greater or fewer proportions of particular lithic types, as is discussed below.

Preliminary typological dating of the lithics suggests that the wide range of occupation of the northern *badia* is fully represented within the survey area, from possibly Palaeolithic, but definitely Early Neolithic through to LC/EBA material. The clearest examples of the latter come from a large wadi that runs south-west to north-east within Region B, which has also been found to contain a number of *ghura* huts near the third century AD fortress Qasr Usaykhim (Müller-Neuhof and Abu-Azizeh 2016). This location is interesting regarding the spread of fourth millennium BC occupation from Jawa, since region B is located almost directly south of that site, and closest of the entire survey area to a comparable precipitation isohyet.

Ramifications of the results

Lithic assemblages, their provenances, and the region B wadi

Overall, the amount of surface material present at the sites visited was unexpectedly high. The count of materials based upon samples collected by evenly-paced site-walking documented at least 100 lithic artefacts at 75%

Figure 7. Photograph showing the large quantities of raw chert strewn across the landscape of the wadi in region B (photograph by S.L. Smith).

of sites visited, while at four individual sites, over 1000 lithics were documented. This factor alone suggests an intensity of occupation, longevity of occupation, or at least frequent re-occupation of sites that contrasts the apparent environmental hostility of the region. These findings of course do not preclude that many sites may have been seasonal camps rather than long-term settlements; however, they do suggest that they were used by relatively large groups of peoples and over many successive generations, as also interpreted for sites in the Hazimah plains south of Jebel Qurma (Akkermans *et al.* 2014, 200-202; Huigens 2015). The latter is emphasised by the fact that the large majority of sites which were confidently dated from the material were multi-period, often spanning the Early Neolithic to the Early Chalcolithic, and sometimes into the LC/EBA. A further indication of the longevity of site use is the presence of basalt grind stones at some sites. These suggest agricultural practices that would not have been feasible without long-term occupation (or at least semi-seasonal occupation) of individual sites.

Lithics from the latter time period were clearly in the minority, represented at only seven sites and even at those accounting for a minority percentage of the assemblages. Nevertheless, LC/EBA material is clearly not absent from the survey area, considering it was identified at all in a single three-week fieldwork season that sampled only 2% of total sites. It is possible, however, that these sites were occupied less frequently or by smaller populations during the fourth millennium BC. Still, this indicates that contemporary occupation with Jawa certainly existed on the western side of the *harrah*, and that it may have been more prevalent here than further east.

The raw material from which the lithic tools were fashioned varies greatly from site to site. Most notably, a lot of local material appears to have been used along the major wadi of region B. Raw chert material is strewn across the length of this valley in abundance, in many cases completely paving the surface (Fig. 7). While it is known that this can occur in the *hamad*, for example in the Hazimah plains (Huigens 2015), such a volume has not before been documented in the interior of the *harrah*. Though bedrock outcrops of chert were identified at several locations along the wadi, it is likely that its alluvial processes contributed to this density. The analysis of worked lithics at sites along this wadi show that this material was used frequently, as the locally-available chert types account for the majority of the artefacts. In fact, the presence of large numbers of half- or nearly-completed tools indicates a 'wasteful' production in these locations, no doubt precipitated by the abundance of raw material available. This contrasts sharply with the assemblages at sites deeper in the basalt desert, yet not geographically distant (*e.g.* Sites 1982, 1984, and 1985 in the north-east of region A, all under 6 km away), where not only is nearly every lithic artefact a completed object, but clear attempts have also been made to fashion tools from basalt, a material very unsuited to that purpose. However, not all material analysed at sites along the region B wadi was locally sourced; some consisted of higher-quality flint that was not found as a raw material anywhere in the surveyed portions of the area. This same flint is also found at the sites in the north-east of region A.

These findings, when combined, pose a conundrum regarding the sites in the vicinity of, but not immediately adjacent to, the raw material sources. On the one hand, sites only 6 km from a large source of chert showing little to no signs of using this as a resource would seem to indicate an extreme localisation. To be sure, the chert material located in the region B wadi is visibly not of the highest quality, and circumstantial evidence that this was recognised by its prehistoric inhabitants can be inferred from the fact that at least some imported high-quality flint was found at every site within the wadi. However, the simultaneous abundant use of local chert indicates that it cannot have been worthless either. On the other hand, while it is true that the dense basalt landscape could account for a lack of regional contact between its inhabitants, and therefore a lack of trade, the presence of clearly imported high-quality flint at sites deep within the *harrah* negates this, as does circumstantial evidence from other sites in the region, such as Shubayqa 1, where long-distance trade is documented from the Late Epipalaeolithic onwards (Richter *et al.* 2012).

Another explanation could be that, despite having long-distance connections, sites within the basalt desert may not have had frequent contact with other sites relatively nearby. This might be feasible, for example, if two sites were geographically close but separated by dense basalt terrain, whilst each being located on separate wadis leading in different directions. In this hypothetical situation, the extreme difficulty of traversing the basalt might negate the theoretical proximity of the sites. In the case in question, however, the closest natural route to the sites in the north-east of region A is indeed the region B wadi, which moreover is a particularly large and easy route into the *harrah* unlikely to have been missed as an opportunity for the movement of people and goods.

A more feasible explanation might be that the sites in the region B wadi exercised a tight control over the local raw material, and that their very purpose of being there was to exploit its resources. In this case, rather than inferring trade routes emanating to other parts of the *harrah*, as one can from the export-oriented mining at Wadi Ruwayshid (Müller-Neuhof 2013b), the material here would have been exclusively for local use, and indeed only desired as such by the population. Such a practice does not make much economic sense, however, as great benefit could doubtless have been gained from trading with the clearly material-starved occupants of sites in the nearby *harrah*.

Perhaps the most fitting explanation based on the current evidence, however, is that the sites examined in region A served a fundamentally different purpose to those in region B. As is discussed in detail below, there are clear morphological discrepancies between the structure types analysed in each region, and these may well indicate different uses. For example, if the sites in region A were ritual structures of some kind, the practices associated with these might have required high-quality flint only (*e.g.* for ceremonial tools), while the region B sites might be habitation areas that required both high-quality flint for specialised tools and low-quality chert for everyday use. This further makes sense of the isolated location of region A sites, deep within very dense basalt, since the construction of ritual sites in remote and/or hard-to-access locations is a well-known practice across almost all cultures.

Site morphologies: true wheels and encircled enclosure clusters

In the remote sensing analysis that preceded the fieldwork of the Western Harra Survey, it became clear that the site types known as 'wheels' or 'jellyfish' (Betts 1982; Kennedy 2011) encompass a range of discrete morphological characteristics visible on satellite imagery. The need for a typological seriation of these sites has been recognised before by Rollefson *et al.* (2016), who in their OSL sampling of wheels in the Wisad Pools region dated one site to the Late Neolithic and one to the LC/EBA. A similarly large range of dates for this site type has been suggested by Akkermans *et al.* (2014), who further postulate a re-use of some during the period of the Safaitic inscriptions (roughly second century BC to fourth century AD) in the form of associated cairns. In the Western Harra Survey area, two distinct forms could be typologically defined by remote sensing data, which during the course of the later fieldwork were found to have impacts on their locations and material remains.

The first form is, true to its name, indeed wheel-like in shape (Fig. 9a). Its main features comprise a roughly circular or elliptical outline, inside which enclosures are divided by mostly straight walls, arranged like the spokes of a wheel. Though these 'spokes' sometimes come to a central point, they often converge around one to three sub-circular central enclosures, from which the other arc-shaped enclosures emanate. They often include internal cairns (which, as mentioned above, may be later additions), and are always singular sites, though not always isolated. Occasionally, such sites are encircled by a series of very

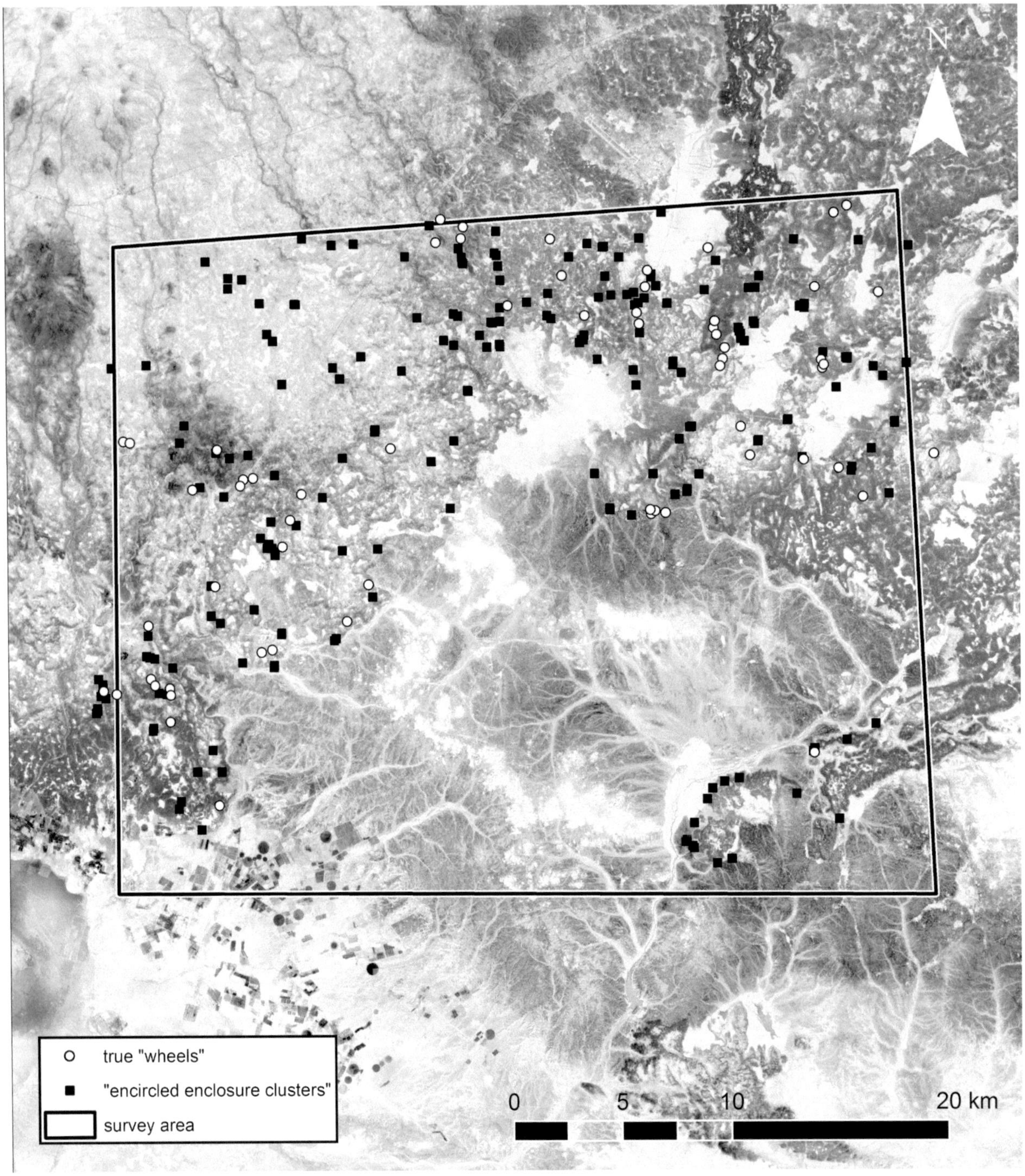

Figure 8. GeoEye satellite map showing the distribution of true wheels and encircled enclosure clusters.

small enclosures, no more than 2 m across. In the survey area, 70 of these true wheels were identified, 43% of which are located within basalt terrain, over 500 m from the edge of the *harrah*, mud flats, or major wadis (Fig. 8).

The second form, which I have termed 'encircled enclosure clusters', are each comprised of a randomly clustered set of at least four sub-circular or sub-elliptical enclosures (Fig. 9b). This creates an irregular external outline, sometimes with one or two additional protruding enclosures. Few, if any, of the internal walls are straight, and there is no clear central enclosure. Internal cairns are very rarely present at such sites. As their name suggests, they are always encircled by a series of very small enclosures, which, however, vary in clarity on remote-sensing images

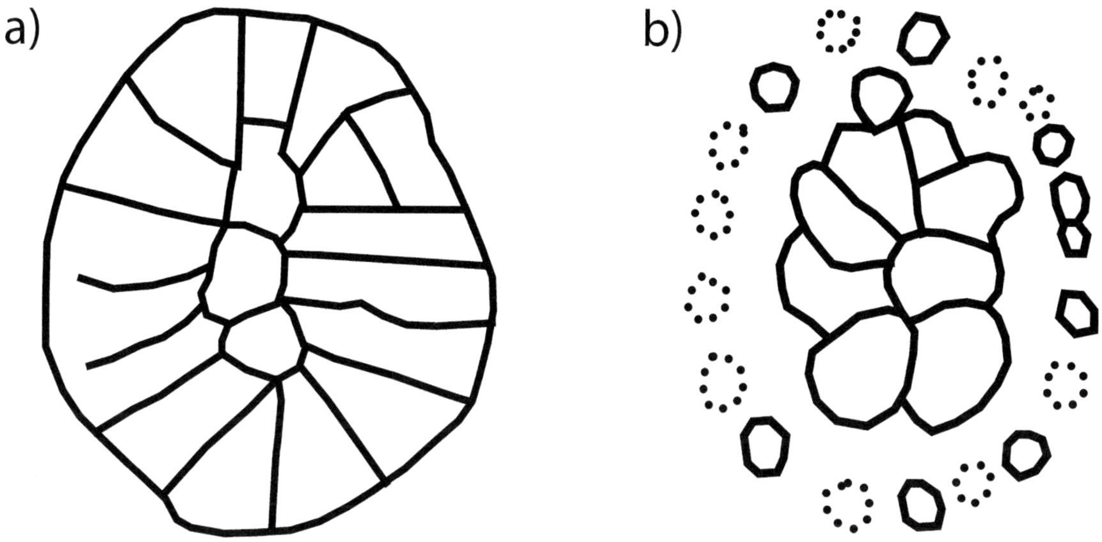

Figure 9. Line drawings highlighting the differences between (a) true wheels, and (b) encircled enclosure clusters.

and on the ground. Occasionally, these sites come in pairs, sharing some of their encircling enclosures (Fig. 10). Over three times as many encircled enclosure clusters as wheels were identified in the survey area: a total of 226 sites. In further contrast, only 30% of these sites are located within the basalt desert, the majority being on the edge of that terrain, or even somewhat within the *hamad* (Fig. 8).

Although these sites' morphologies differ significantly, they do occasionally share the property of being surrounded by a series of small enclosures, which leads me to believe that these may have been later additions to the main structures. As has been suggested for internal cairns, the encircling enclosures may date to the Safaitic period. Safaitic inscriptions were identified by the Western Harra Survey at several locations directly adjacent to both true wheels and encircled enclosure clusters, and always external to the main structures. Furthermore, upon close examination on the ground, the encircling enclosures were found to be square or rectangular in outline, again suggesting a much later date for their construction (Smith and Chambrade 2018, 11). Although this evidence remains very circumstantial, it would also explain why the material remains of encircled enclosure cluster sites match those of regular enclosures, with occasional LC/EBA material and ceramic remains, neither of which were identified at any true wheel site. Thus I consider encircled enclosure clusters to be fundamentally akin to enclosures rather than wheels. Indeed, there are several sites formed of groupings of enclosures which can be considered morphologically identical to encircled enclosure clusters, save for the presence of the encircling enclosures.

Furthermore, the material remains of true wheels and encircled enclosure clusters show some interesting discrepancies. Over half of the latter sites surveyed contained a greater density of lithics than any of the former. However, the surveyed wheels contained on average more lithic scrapers (as a percentage of the total lithics counted at each site), associated with the processing of animal hides, supporting part of the hypothesis put forward by Betts (1982) that these were corrals. They also contained a significantly greater percentage of lithic cores, while, conversely, flakes were found to be proportionally more numerous at encircled enclosure clusters. These seemingly conflicting data may indicate that wheels were used for storage purposes of materials as well as livestock, and that cores were transported elsewhere to be worked, perhaps within or close to habitation areas, which the encircled enclosure clusters may have been. An alternative explanation is that these different site types are the result of chronological variations. The presence of LC/EBA material at encircled enclosure clusters, as well as some enclosure groups which, as discussed above, are likely the same site type, would seem to suggest these as later sites. True wheels, on the other hand, feature an abundance of likely Late Neolithic material, with some potentially dating to the Early Chalcolithic also.

A third explanation is that the discrepancies in morphology stem from more significantly variant site purposes, such as ritual or cultic sites and production or habitation areas. This interpretation would explain both the significant morphological discrepancies and the differences in distribution vis-à-vis ease of access, as well as tying in to the question of the sites in the region B

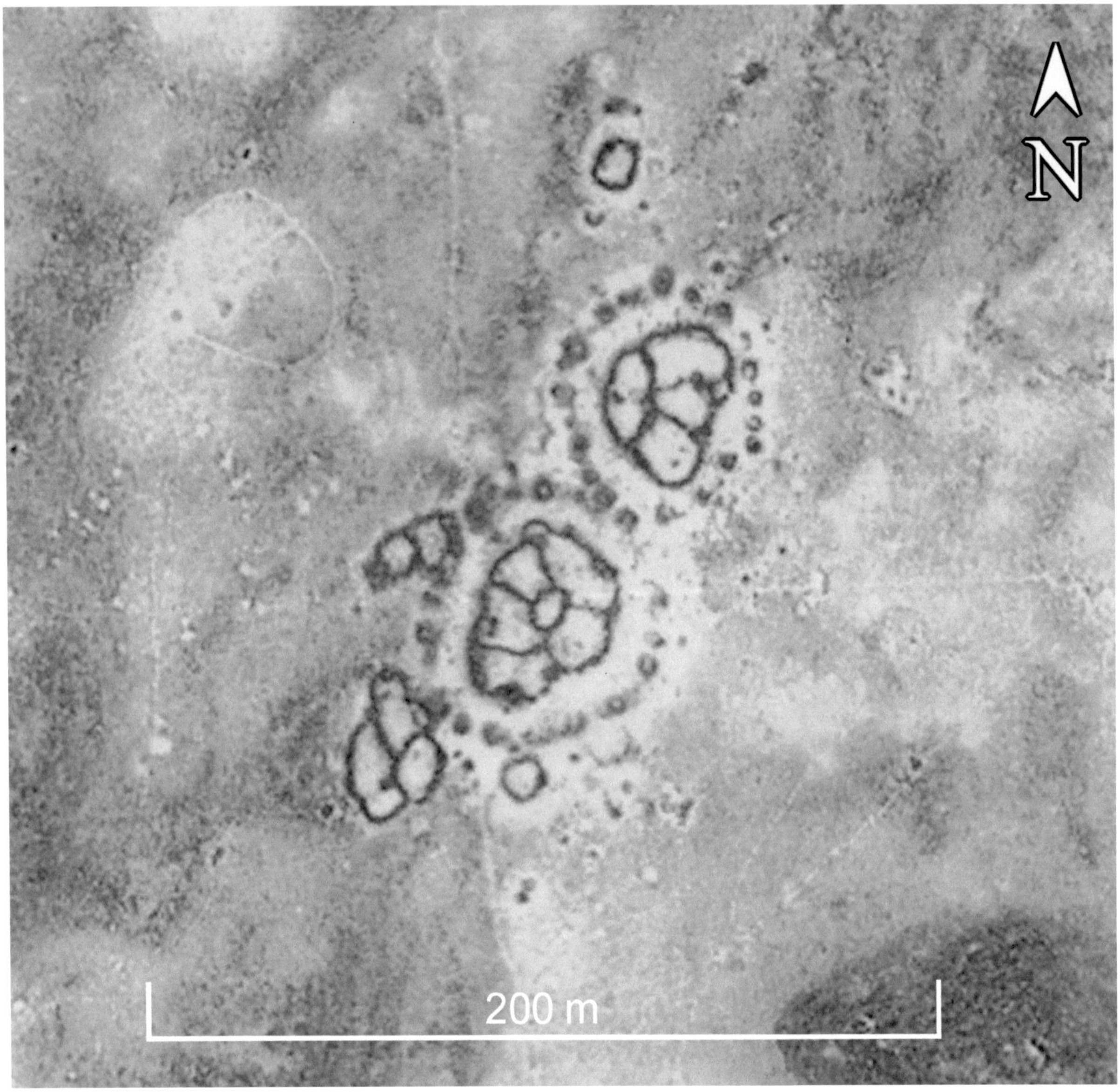

Figure 10. GeoEye satellite image of the interconnected encircled enclosure cluster Sites 1118 and 1119.

wadi, and those nearby in region A. Those surveyed in the former region were mostly encircled enclosure clusters or simple enclosure groups, while those in the latter were exclusively true wheels. This would further support the idea of region A structures (true wheels) being ritual sites, as an alternative explanation for the high numbers of lithic scrapers, and hence hide processing, is the carrying out of sacrificial practices. At the same time, the hypothesis that the region B structures (encircled enclosure clusters) were predominantly habitation areas is supported by both the greater relative numbers of lithics at this site type and by the higher presence of flakes, as production would have likely occurred in or near living spaces.

Conclusions and further work

The first season of the Western Harra Survey already started to achieve some of the overall goals of the project. Primarily, the existence of LC/EBA occupation on the western edges of the *harrah* directly south of Jawa was confirmed; it now remains to be seen how widespread and intensive this occupation was. However, the data already contributes to the growing body of evidence that numerous LC/EBA sites exist in several locations south and east of Jawa (*e.g.* Jebel Qurma; Akkermans and Brüning, this volume), and that in general post-Neolithic occupation of the basalt desert was relatively common and long-term (see Müller-Neuhof and Abu-Azizeh 2016, with further references). Furthermore, a general intensity of occupation that was long-term and/or consisted of large populations can be inferred from the large volumes of lithic materials documented at a variety of sites, although not enough data has been collected to say so for the LC/EBA specifically. Consistent with other investigations in the region, these results belie the description of the northern *badia* as 'marginal' in any sense other than for traditional agricultural practices.

As for suggestions for the impetuses of human occupations of these regions, circumstantial evidence points to the exploitation of mineral resources (raw chert material) to have been a factor for the western *harrah* at least, which along with widespread opportunities for pastureland would have contributed to the economic potential of the region (see Müller-Neuhof 2013a). Thus though different in the specific, the general model of the exploitation of semi-arid and arid regions for their economic potential appears to apply to the *harrah* as to the Greater Western Jazira and the Shamiya. Not enough data has been gathered to explore the equally important social landscape and its impact upon the patterns observed. The variations between true wheels and encircled enclosure clusters could be a manifestation of this, however, representing potential cultural and/or spiritual concepts that separated ritual areas within the basalt from habitation and working areas on the boundaries between the basalt and wadis, mud flats, or the *hamad*.

Lastly, the definition and separation of the wheel and encircled enclosure cluster site types has begun the process of developing methodologies for holistically investigating the landscape by remote sensing. A crucial requirement for mapping settlement dynamics across a large area, especially one with necessarily little uniform survey coverage on the ground such as the northern *badia*, is the creation of precise morphological classifications. Once definite correlations (with acceptable statistical variances) between certain site types and specific occupation dates or site uses can be made, accuracy and precision of remote sensing analyses and interpretations can be greatly improved. While such analyses are in no way as reliable as ground-truth data, they can significantly develop at least a broad understanding of the archaeology of the region by a method that is rapid, cost-effective, and encompasses a large scale. As mentioned earlier, this has proven successful in other parts of the Syro-Levantine steppe for both large-scale analyses and background data upon which to plan targeted fieldwork (see *e.g.* Galiatsatos *et al.* 2009; Smith 2020; and more recently for the Levant, Ansart *et al.* 2016). This is an important first step in bringing the analysis potential of the northern *badia* towards the level of the semi-arid and arid steppes of Syria, which have been investigated in several magnitudes of greater breadth and depth, both in terms of ground and remote sensing analyses. For ground investigations in north-eastern Jordan to reach the levels of over a century of fieldwork in Syria will take decades, but the targeted use of fieldwork to increase the potential of remote sensing investigations of the northern *badia* can already commence in the present. This also represents a major initial contribution of the Western Harra Survey to my supra-regional project, which relies heavily on this methodology.

It is clear that much further work needs to be done to accurately address the research questions of the regionwide project, and to improve our basic understanding of the prehistory of the northern *badia* in general. For the Western Harra Survey, this means more fieldwork seasons, as the area's potential for contributing to answering these issues has already been clearly demonstrated. As well as further similar investigations to the first season, a focus on the accurate dating of sites and components of sites, of different morphologies including wheels and encircled enclosure clusters, is vital. This process has already started with three further fieldwork seasons, carried out 2017-2019, the analysis of which is currently underway. As well as visiting further sites, material technologies were focussed on, with targeted lithic types recorded in detail and certain artefacts modelled in three dimensions for later, precise analysis. Additionally, we began collecting soil samples for analysis by OSL in order to date the initial construction of wheel and encircled enclosure cluster sites, using the methodology of Athanassas *et al.* (2015). At the latter, samples were taken both from the main structures and from several of the small rectangular encircling enclosures. Thus when the results from these are obtained, they should begin to confirm or deny many of the hypotheses discussed above. Lastly, a selection of sites, with a focus on wheels and encircled enclosure clusters, were aerially documented in detail using drone photography, which produced accurate and precise two-dimensional and three-dimensional models of the structures. These are in the process of being used to quantify morphological variations using statistical analyses, to more objectively specify the visual differences identified.

Several further fieldwork seasons are planned, which will henceforth focus on excavations of selected representative site morphologies to sure up interpretations based on survey data. Notably, no wheel or encircled enclosure cluster site has yet been excavated in its entirety. Furthermore, evidence from other parts of the *harrah* shows the existence of sites that when excavated were found to have intensive prehistoric occupation, yet showed little to no trace of surface material that could be identified by a survey. This, in microcosm, highlights also the further work needed in the northern *badia* in general in order to allow for accurate comparisons to other environmentally comparable regions, paving the way for a holistic understanding of the prehistory of the Syro-Levantine steppe and beyond.

Acknowledgements

The Western Harra Survey is co-directed by myself and Dr Marie-Laure Chambrade, whom I greatly thank for all her hard work in taking this project beyond its first season. The fieldwork would not have been possible without funding from the National Geographic Society and the Council for British Research in the Levant (2015 season), the Curtiss T. Brennan & Mary G. Brennan Foundation, managed by the Archéorient Research Unit and CNRS (2017 season), and the Institut Français du Proche Orient (2019 seasons). Thank you to all additional team members: Anne Binder, Lanah Haddad, S. Nina Mann (2015 season); David Burke (2017 season); Dr Imad Alhussain (2017 and 2019 seasons). Thanks also go to the Department of Antiquities of the Hashemite Kingdom of Jordan, in particular HE Yazid Elayan, Dr Monther Jamhawi, Aktham Oweidi, Ahmad Lash, Aisar Radaydeh, and Bilal Alboreni. The supra-regional project 'Human Adaptation in Climatically Marginal Environments of late-fifth to third millennium BC Syria and Jordan' has received funding from the Research Foundation-Flanders (FWO) and the European Union's Horizon 2020 research and innovation programme under the Marie Skłodowska-Curie grant agreement No. 665501. I thank Veronika Kudlek for providing me with unpublished fieldwork data from the Greater Western Jazira, and Dr Louise Rayne for the processing of GPCC precipitation data.

References

Akkermans, P.M.M.G., Huigens, H.O. and Brüning, M.L. 2014. A landscape of preservation: late prehistoric settlement and sequence in the Jebel Qurma region, north-eastern Jordan. *Levant* 46, 186-205.

Ansart, A., Braemer, F. and Davtian, G. 2016. Preparing an archaeological field survey: remote sensing interpretation for herding structures in the southern Levant. *Journal of Field Archaeology* 41, 699-712.

Athanassas, C.D., Rollefson, G.O., Kadereit, A., Kennedy, D., Theodorakopoulou, K., Rowan, Y.M. and Wasse, A. 2015. Optically stimulated luminescence (OSL) dating and spatial analysis of geometric lines in the northern Arabian desert. *Journal of Archaeological Science* 64, 1-11.

Betts, A.V.G. 1982. "Jellyfish": prehistoric desert shelters. *Annual of the Department of Antiquities of Jordan* 26, 183-188.

Betts, A.V.G. 1991 (ed.). *Excavations at Jawa 1972-1986*. Edinburgh: Edinburgh University Press.

Betts, A.V.G. 1998a. The Epipalaeolithic Periods. In: Betts, A.V.G. (ed.). *The harra and the hamad: excavations and survey in eastern Jordan*. Sheffield: Sheffield Academic Press, 11-35.

Betts, A.V.G. 1998b. Dhuweila: area survey. In: Betts, A.V.G. (ed.). *The harra and the hamad: excavations and survey in eastern Jordan*. Sheffield: Sheffield Academic Press,, 191-205.

Castel, C. and Peltenburg, E. 2007. Urbanism on the margins: third millennium BC Al-Rawda in the arid zone of Syria. *Antiquity* 81, 601-616.

Castel, C., Archambault, D., Awad, N., Barge, O., Boudier, T., Brochier, J., Cuny, A., Gondet, S., Herveux, L., Isnard, F., Martin, L., Quenet, P., Sanz, S. and Vila, E. 2008. Rapport préliminaire sur les activités de la mission archéologique franco-syrienne dans la micro-région d'Al-Rawda (Shamiyeh): quatrième et cinquième campagnes (2005 et 2006). *Akkadica* 129, 5-54.

Danti, M. 2000. *Early Bronze Age settlement and land use in the Tell es-Sweyhat region, Syria*. Philadelphia, PA: University of Pennsylvania (unpublished PhD thesis).

Deckers, K. and Pessin, H. 2011. Vegetation development in relation to human occupation and climatic change in the Middle Euphrates and Upper Jazirah (Syria/Turkey) during the Bronze Age. In: Deckers, K. (ed.). *Holocene landscapes through time in the Fertile Crescent*. Turnhout: Brepols, 33-48.

Frumkin, A., Bar-Matthews, M. and Vaks, A. 2008. Paleoenvironment of Jawa basalt plateau, Jordan, inferred from calcite speleothems from a lava tube. *Quaternary Research* 70, 358-367.

Galiatsatos, N., Wilkinson, T.J., Donoghue, D., Philip, G. 2009. The Fragile Crescent Project (FCP): analysis of settlement landscapes using satellite imagery. CAA 2009: Making history interactive, Williamsburg, VA, 22-26 March. Online: http://dro.dur.ac.uk/6909/ (Accessed 20 November 2017).

Geyer, B., Calvet, Y., Awad, N., al-Dbiyat, M. and Rousset, M.-O. 2010. Un 'Très Long Mur' dans la steppe syrienne. *Paléorient* 36, 57-72.

Geyer, B., Al-Dbiyat, M., Awad, N., Barge, O., Besançon, J., Calvet, Y. and Jaubert, R. 2007. The arid margins of northern Syria: occupation of the land and modes of exploitation in the Bronze Age. In: Morandi Bonacossi, D. (ed.). *Urban and natural landscapes of an ancient Syrian capital*. Udine: Forum, 269-281.

Hempelmann, R. 2013. *Tell Chuera, Kharab Sayyar und die Urbanisierung der westlichen Gazira*. Wiesbaden: Harrassowitz Verlag.

Huigens, H.O. 2015. Preliminary report on a survey in the Hazimah plains: a hamad landscape in north-eastern Jordan. *Palestine Exploration Quarterly* 147, 180-194.

Kalayci, T. 2013. *Agricultural production and stability of settlement systems in Upper Mesopotamia during the Early Bronze Age (third millennium BCE)*. Fayetteville, AR: University of Arkansas (unpublished PhD thesis).

Kempe, S. and Al-Malabeh, A. 2010. Hunting kites ('desert kites') and associated structures along the eastern rim of the Jordanian harrat: a geo-archaeological Google Earth images survey. *Zeitschrift für Orient-Archäologie* 3, 46-86.

Kennedy, D. 2011. The "Works of the Old Men" in Arabia: remote sensing in interior Arabia. *Journal of Archaeological Science* 38, 3185-3203.

Lawrence, D., Bradbury, J. and Dunford, R. 2012. Chronology, uncertainty and GIS: a methodology for characterising and understanding landscapes of the ancient Near East. In: Bebermeier, W., Hebenstreit, R., Kaiser, E. and Krause, J. (eds.). *Landscape archaeology.* Proceedings of the International Conference held in Berlin, 6th-8th June 2012. eTopoi special volume 3. Berlin: Excellence Cluster Topoi, 353-359.

Liverani, M. 2014. *The ancient Near East. History, society and economy.* Abingdon: Routledge.

Meyer, J.-W. 2010. Versuch einer historischen Einordnung von Tell Chuera in die politisch-historische Entwicklung Nordostsyriens im 3. Jt. v.Chr. In: Meyer, J.-W. (ed.). *Tell Chuera: Vorberichte zu den Grabungskampagnen 1998 bis 2005.* Wiesbaden: Harrassowitz Verlag, 11-34.

Meyer, J.-W. and Hempelmann, R. 2006. Bemerkungen zu Mari aus der Sicht von Tell Chuera – Ein Beitrag zur Geschichte der ersten Hälfte des 3.Jts.v.Chr. *Altorientalische Forschungen* 33, 22-41.

McClellan, T.L. and Porter, A. 1995. Jawa and north Syria. *Studies in the History and Archaeology of Jordan* 5, 49-65.

McCorriston, J. 1997. The fiber revolution. *Current Anthropology* 38, 517-535.

Müller-Neuhof, B. 2013a. Nomadische Ressourcennutzung in den ariden Regionen Jordaniens und der Südlichen Levante im 5. bis frühen 3. Jahrtausend v. Chr. *Zeitschrift für Orient-Archäologie* 6, 64-80.

Müller-Neuhof, B. 2013b. Chalcolithic/Early Bronze Age flint mines in the northern badia. *Syria* 90, 177-188.

Müller-Neuhof, B. 2014a. A 'marginal' region with many options: the diversity of Chalcolithic/Early Bronze Age socio-economic activities in the hinterland of Jawa. *Levant* 46, 230-248.

Müller-Neuhof, B. 2014b. Desert irrigation agriculture: evidences for Early Bronze Age rainwater-harvesting irrigation agriculture at Jawa (NE-Jordan). In: Morandi Bonacossi, D. (ed.). *Settlement dynamics and human-landscape interaction in the dry steppes of Syria.* Wiesbaden: Harrassowitz Verlag, 187-197.

Müller-Neuhof, B. and Abu-Azizeh, W. 2016. Milestones for a tentative chronological framework for the late prehistoric colonization of the basalt desert (north-eastern Jordan). *Levant* 48, 220-235.

Porter, A. 2004. The urban nomad: countering the old clichés. In: Nicolle, C. (ed.). *Nomades et sédentaires dans le Proche-Orient ancien.* Paris: Éditions Recherche sur les Civilisations, 69-74.

Porter, A. 2012. *Mobile pastoralism and the formation of Near Eastern civilizations. Weaving together society.* Cambridge: Cambridge University Press.

Richter, T., Arranz-Otaegui, A., Yeomans, L. and Boaretto, E. 2017. High resolution AMS dates from Shubayqa 1, northeast Jordan reveal complex origins of Late Epipalaeolithic Natufian in the Levant. *Scientific Reports* 7, 7025. DOI:10.1038/s41598-017-17096-5.

Richter, T., Bode, L., House, M., Iversen, R., Arranz-Otaegui, A., Saehle, I., Thaarup, G., Tvede, M.-L. and Yeomans, L. 2012. Excavations at the Late Epipalaeolithic site of Shubayqa 1: preliminary report on the first season. *Neo-Lithics* 2/12, 3-14.

Riehl, S. 2009. Archaeobotanical evidence for the interrelationship of agricultural decision-making and climate change in the ancient Near East. *Quaternary International* 197, 93-114.

Rollefson, G., Rowan, Y. and Wasse, A. 2014. The Late Neolithic colonization of the eastern badia of Jordan. *Levant* 46, 285-301.

Rollefson, G.O., Athanassas, C.D., Rowan, Y.M. and Wasse, A. 2016. First chronometric results for 'Works of the Old Men': late prehistoric 'wheels' near Wisad Pools, Black Desert, Jordan. *Antiquity* 90, 939-952.

Sallaberger, W. and Pruß, A. 2015. Home and work in Early Bronze Age Mesopotamia: "ration lists" and "private houses" at Tell Beydar/Nabada. In: Steinkeller, P. and Hudson, M. (eds.). *Labor in the ancient world.* Dresden: ISLET, 69-136.

Sanlaville, P. 2000. Environment and development. In: Mundy, M. and Musallam, B. (eds.). *The Transformation of nomadic society in the Arab East.* Cambridge: Cambridge University Press, 6-16.

Smith, S.L. 2015. *Late Chalcolithic to Early Bronze Age settlement patterns in the Greater Western Jazira: trajectories of sedentism in the semi-arid Syrian steppe.* Durham: Durham University, Department of Archaeology. Online: http://etheses.dur.ac.uk/11404/ (Accessed 10 December 2017).

Smith, S.L. 2020. The view from the steppe. Using remote sensing to investigate the landscape of "Kranzhügel" in its regional context. In: Lawrence, D., Altaweel, M. and Philip, G. (eds.). *New Agendas in Remote Sensing and Landscape Archaeology in the Near East. Studies in Honour of Tony J. Wilkinson.* Oxford: Archaeopress, 109-123.

Smith, S.L. and Chambrade, M.-L. 2018. The application of freely-available satellite imagery for informing and complementing archaeological fieldwork in the "Black Desert" of north-eastern Jordan. *Geosciences* 8/12, 491. DOI: 10.3390/geosciences8120491.

Smith, S.L. and Wilkinson, T.J. 2020. The circular cities of northern Syria in their environmental context. In: Castel, C., Meyer, J.-W. and Quenet, P. (eds.). *Circular Cities of Early Bronze Age Syria.* Turnhout: Brepols, 151-160.

Smith, S.L., Wilkinson, T.J. and Lawrence, D. 2014. Agro-pastoral landscapes in the zone of uncertainty. In: Morandi Bonacossi, D. (ed.). *Settlement dynamics and human-landscape interaction in the dry steppes of Syria*. Wiesbaden: Harrassowitz Verlag, 151-172.

Ur, J.A. 2010. Cycles of civilization in northern Mesopotamia, 4400-2000 BC. *Journal of Archaeological Research* 18, 387-431.

Wilkinson, T.J. 1997. Environmental fluctuations, agricultural production and collapse: a view from Bronze Age Upper Mesopotamia. In: Dalfes, H.N., Kukla, G. and Weiss, H. (eds.). *Third millennium BC climate change and Old World collapse*. NATO ASI Series I 49. Berlin: Springer Verlag, 67-106.

Wilkinson, T.J. 2000. Settlement and land use in the zone of uncertainty in Upper Mesopotamia. In: Jas, R. (ed.). *Rainfall and agriculture in northern Mesopotamia*. Leiden: Nederlands Instituut voor het Nabije Oosten, 3-35.

Wilkinson, T.J., Philip, G., Bradbury, J., Dunford, R., Donoghue, D., Galiatsatos, N., Lawrence, D., Ricci, A. and Smith, S.L. 2014. Contextualizing early urbanization: settlement cores, early states and agropastoral strategies in the Fertile Crescent during the fourth and third millennia BC. *Journal of World Prehistory* 27, 43-109.

Wossink, A. 2009. *Challenging climate change. Competition and cooperation among pastoralists and agriculturalists in northern Mesopotamia (c. 3000-1600 BC)*. Leiden: Sidestone Press.

Yeomans, L., Richter, T. and Martin, L. 2017. Environment, seasonality and hunting strategies as influences on Natufian food procurement: the faunal remains from Shubayqa 1. *Levant* 49, 85-104.

East of Azraq: settlement, burial and chronology from the Chalcolithic to the Bronze Age and Iron Age in the Jebel Qurma region, Black Desert, north-east Jordan

Peter M.M.G. Akkermans and Merel L. Brüning

Abstract

Recent survey and excavation in the Jebel Qurma region in the basalt desert (*harrah*) of north-eastern Jordan have revealed substantial evidence for settlement and burial from the Chalcolithic, Bronze Age and Iron Age. The new data demonstrate considerable diversity in site layout as well as clear shifts in habitation patterns and locational preferences over time. Particularly, the sites from the mid-fifth to fourth millennium BC regularly were of an impressive size, with many dozens or even hundreds of stone-built structures for dwelling at a single site. In later periods, the emphasis seems to be increasingly on small, temporary camp sites. In addition, the fieldwork provided detailed insight into the many cairns for burial in the Jebel Qurma area and the Black Desert at large. These tombs were of different types and sizes, and predominantly date to the first millennium BC.

Keywords: Jebel Qurma, Jordan, Black Desert, nomad, Chalcolithic, Bronze Age, Iron Age, Hellenistic, Roman

Jebel Qurma east of Azraq

The Black Desert begins some 130 km east of Amman, and is characterised by rough and rugged, dark lava fields (*harrah* in Arabic; plural *harrat*). The lava expanse in Jordan has an area of around 11,000 km^2, and is locally known as the Harrat al-Shaba, which itself is part of a much larger basalt plateau (the Harrat al-Sham), stretching from southern Syria through Jordan and into northern Saudi Arabia (cf. Edgell 2006). On the fringes of the volcanic belt is the Jebel Qurma range, situated east of the small oasis town of Azraq, and close to the Jordanian-Saudi border (Fig. 1). This highly arid area (with less than 50 mm of average annual precipitation) comprises basaltic high grounds and table mounds, alternating with stretches of limestone hillocks, gravel plains, and mud flats. A myriad of narrow, shallow wadis carve through the varied desert landscape, and they lead into broad mud flats or into two much larger wadi systems on either side of the Jebel Qurma range: Wadi Rajil in the west and Wadi al-Qattafi in the east (Figs. 2-3). These wadis serve as long, natural corridors through the basalt barrier, and are connected to the flat, shallow depression of the Wadi Sirhan further to the south-west, a major caravan track and communication route between

In: Peter M. M. G. Akkermans (ed.) 2020: *Landscapes of Survival - The Archaeology and Epigraphy of Jordan's North-Eastern Desert and Beyond*, Sidestone Press (Leiden), pp. 185-216.

Figure 1. Map of Jordan showing the location of the Jebel Qurma region (red rectangle) and other principal sites mentioned in the text (source: Terra-MODIS image, adapted from Jacques Descloitres, MODIS Rapid Response Team, NASA/GSFC).

the Levant and Syria on the one hand and the Arabian Peninsula on the other hand.

Annual programs of survey and excavation in the Jebel Qurma region since 2012 (within the framework of the Jebel Qurma Archaeological Landscape Project; cf. Akkermans and Huigens 2018) have sought to examine local settlement and burial from a multi-period, *longue-durée* perspective, and investigate how these relate to the diverse landscape and environment. The fieldwork detected many hundreds of archaeological find spots, ranging from inconspicuous lithic scatters to tombs of different shapes and sizes and concentrations of basalt-built dwellings spread over

Figure 2. Wadi Rajil near Jebel Qurma in the rainy season (April 2017). Most of the water is not collected locally but is directed to Jebel Qurma from the wadi's upper reaches in southern Syria (photograph: Jebel Qurma Project Archive).

Figure 3. The harsh landscape of the Jebel Qurma range. Basalt-covered table mounds alternate with sand dunes and extensive gravel plains, which are cut by erosion gullies and wadis (photograph: Jebel Qurma Project Archive).

several hectares. In addition, the fieldwork has identified several thousands of petroglyphs and inscriptions in Safaitic and Arabic (Akkermans *et al.* 2014; Akkermans 2019; Huigens 2015; 2019; Brusgaard 2019).

This article summarises the evidence for settlement and burial in the Jebel Qurma region from the Chalcolithic to the Bronze Age and Iron Age, roughly between 4500 BC to the beginning of the common era. There were substantial socio-economic transformations and transitions in the southern Levant during this long period, from the development of complex societies in the fifth millennium BC and the onset of urbanism in the fourth millennium, to the rise of highly stratified empires in the Iron Age and afterwards. The effects of these long-term developments on the eastern desert cultures are still poorly understood. There is good evidence for significant population aggregations in the basalt expanse in the Chalcolithic and Early Bronze Age, coupled with the creation of fortifications, extensive pastoralist efforts, and the large-scale, specialist production of cortical flint blanks for tool manufacture. These finds have fuelled hypotheses about the *badia* as an economic centre of considerable significance in these early periods (*e.g.* Philip 2008; Müller-Neuhof 2014). Later periods indicated reduced settlement and a more restrained, localised economy, which served the needs of small and highly mobile groups of herders and hunters. Burial evidence equally suggests that, despite many external ties, the desert communities in the Iron Age displayed a deeply entrenched local character.

Late Chalcolithic and Early Bronze Age I settlement

Jawa is the archetypical Early Bronze Age I (EBA I) site in Jordan's north-eastern *harrah*, dated to the mid/late fourth millennium BC on the basis of pottery finds and recent ^{14}C evidence (Helms 1991; Müller-Neuhof *et al.* 2015). Located on the western fringes of the basalt plateau, close to the Jordanian-Syrian border, the 10-ha site has massive retaining walls for defence, extensive areas with roughly circular domestic architecture, as well as complex hydrological installations (Helms 1981; Betts 1991). Jawa has long been considered as a conundrum, because of its remote and isolated location, situated far from other, concomitant sites. In the southern foothills of Jebel al-Druze, Rukais was the only other broadly contemporary settlement with round EBA I architecture (Betts *et al.* 1996). However, Bernd Müller-Neuhof's recent research in the Jawa hinterland revealed exciting proof for other fortified sites of this period, such as at Khirbet Abu al-Husayn, Tulul al-Ghusayn and Khirbet al-Ja'bariya (Müller-Neuhof 2017; Müller-Neuhof and Abu-Azizeh 2018a; 2018b; see also Müller-Neuhof, this volume). In addition to these hillfort sites with their probably more permanent habitation, there were many temporary camp sites, often with groups of enclosures (Müller-Neuhof *et al.* 2013, 127-128). There is also evidence for garden terraces, irrigation agriculture, herding, and the large-scale exploitation of flint mines (*e.g.* Müller-Neuhof 2013a; 2013b; 2014; Meister *et al.* 2017; 2018). In short, these finds have shown the basalt expanse to be anything but the remote and deserted backwater it is often interpreted to be.

Survey and excavation in the Jebel Qurma region also have identified many sites which we believe to date to the Late Chalcolithic to Early Bronze Age (LC/EBA) periods, primarily on the basis of their lithic assemblages. LC/EBA chronologies in the basalt wasteland still are very preliminary in nature. Both lithic and pottery typologies serve as dating tools, although these have their inherent flaws, particularly in regards to sites known only from their surface materials. A relatively common formal tool type in local lithic assemblages are the tabular scrapers: roughly oval or fan-shaped, unifacial tools, about 5 to 15 cm in size, made on cortical flakes from large flint nodules (cf. Rosen 1997, 71ff.) (Figs. 4-5). These products probably were imported into the Jebel Qurma region, brought from the mining areas in the Jafr basin in the south-west or from the Ruwayshid area in the north-east (*e.g.* Quintero *et al.* 2002; Müller-Neuhof 2013a). However, the scale of the trade must have been limited, as the number of cortical artefacts was restricted to a few pieces from all of the sites in the Jebel Qurma area; the remainder of the lithic assemblages mainly consisted of *ad hoc* tools on flakes.

Although excavations at, for example, Dhuweila and Wisad Pools have suggested that tabular scrapers and other cortical elements first appeared in the Late Neolithic (Betts 1998, 105, 119; Rollefson *et al.* 2013, 16; see also Henry 1995, 372), they are usually considered to be characteristic of the LC to EBA I-III periods, *c.* 4500-2500 BC (see *e.g.* Rosen 1997; Braun 2011; Barket and Bell 2011; Müller-Neuhof 2013a; 2014). Although their chronology needs further detailing through excavation, the occurrence of cortical tools may serve as a *Leitfossil* for dating the find spots in the Jebel Qurma range between the late fifth and early third millennium BC. Pottery from this period is still absent in the Jebel Qurma area, and the earliest ceramics found until now belong to the very end of the third millennium. A caveat is required here, in terms of the very long time frame involved (some 2000 years): we cannot simply assume continual use, let alone contemporaneity, of all recorded sites. Some find spots may fit early in the sequence and others may date later in the period. While the Jebel Qurma area overall has a significant number of recorded LC/EBA sites, their distribution (and hence the size of the population) at any given time was probably much more limited.

Until now, the fieldwork in the Jebel Qurma area has yielded 26 LC/EBA dwelling sites with tabular scrapers and related cortical tools (Fig. 6). The sites predominantly occur

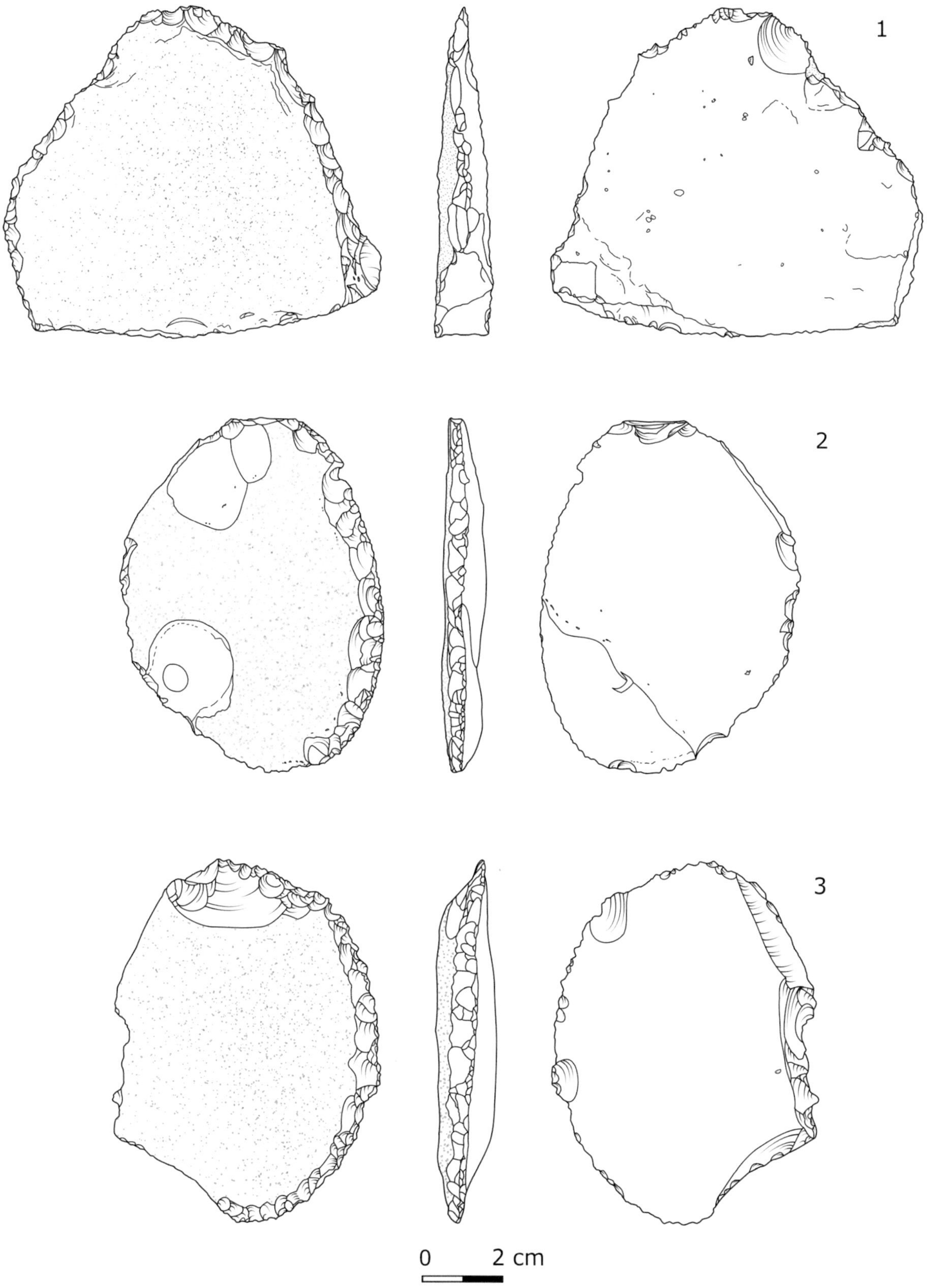

Figure 4. LC/EBA cortical scrapers from sites in the Jebel Qurma region. No. 1: from a double cell building at QUR-619. No. 2: from a wheel at QUR-144. No. 3: from a wheel at QUR-124 (drawing by Keshia Akkermans, Jebel Qurma Project Archive).

Figure 5. LC/EBA cortical tools from sites in the Jebel Qurma region. Nos. 1-3: from the large site of QUR-6. Nos. 1 and 2 were found together in an open area at the site. No. 4: from a wheel at QUR-147 (photographs: Jebel Qurma Project Archive).

on the lower slopes and spurs of the basalt prominences on the edges of the *harrah*, sheltered from the prevailing cold winds in winter and spring by topographic relief. Usually, they were located within close proximity (a few hundred metres at most) to wadis and mud flats that provided water in the wet seasons. In some cases, the sites are found deeper inside the basalt expanse or high on the plateau, where they may appear in association with burials. Find spots also exist in the gravel plains beyond the basalt terrain, often in sheltered locales at the foot of low limestone hillocks.

Some of the find spots were very small and short-lived open-air places with no visible traces except lithic scatters, while others were renewals of pre-existing installations of a much earlier, Neolithic, date. For example, the typical cortical tools repeatedly were found in extensive grouped enclosures that also had substantial amounts of burins of Late Neolithic type, *c.* 6400-6000 BC. A presently unresolved matter with regard to chronology is the regular occurrence of cortical tools in so-called 'wheels': roughly circular arrangements of enclosures, often surrounded by a string of round hut structures (cf. Akkermans *et al.* 2014, 197-200; see also Smith, this volume). The exact dating of these wheels is still uncertain. OSL analysis from two wheels in Wisad Pools suggested that one of them dates to the Late Neolithic period, *c.* 7500-4900 BC, and the other to the LC-EBA I transition, *c.* 4700-3300 BC (Rollefson *et al.* 2016). More recently, charcoal from two fireplaces uncovered in a wheel (QUR-147) in the Jebel Qurma region produced two ^{14}C samples; they dated to 5310-5080 cal BC and 4720-4560 cal BC, respectively. The dates appear to demonstrate that the wheel at QUR-147 was used for a considerable time period, although probably on an intermittent basis. We consider a date for the wheels in the seventh, and certainly the eighth, millennium highly unlikely, given the concurrent site structures and established lithic sequences. We believe it is much more likely that the wheels date from the late sixth to the early fourth millennium BC. In this respect, the cortical tools found in the wheels may date relatively early in the local sequence.[1] An alternative, which cannot be ruled out, is to consider the tabular scrapers as evidence for the later re-use of the wheels.

An arrangement perhaps typical of the fourth millennium BC were the groups of small, free-standing, single or double cell *ghura* huts,[2] usually found in association with enclosures of different sizes (Rowan *et al.* 2015). While some of these places had a handful of buildings distributed over a few hundred square metres, others consisted of many dozens or even hundreds of structures over several hectares. The sites themselves tend to cluster in several places in the basalt expanse, although there were single, isolated occurrences as well (Fig. 6). Highly relevant with regard to their dating is the excavated site of Tulul al-Ghusayn in the Jawa hinterland, where great numbers of small *ghura* dwellings, including 126 double cell structures, were situated in and around the crater of a volcano, about nine km north of the Amman-Baghdad road (Müller-Neuhof and Abu-Azizeh 2016; 2018a; see also Müller-Neuhof, this volume). The double cell installations were about 3 to 4.5 m long and 1 to 1.4 m wide, with their walls preserved to a height of about 0.5 m. Charcoal samples from a fireplace in an excavated double cell building (TAG 209) dated to 3640-3525 cal BC and 3760-3640 cal BC, respectively (Müller-Neuhof and

1 Importantly, a cortical scraper lay next to one of the aforementioned fireplaces at QUR-147, dated to 4720-4560 cal BC.

2 The name is derived from the basalt-capped table-mounds, locally referred to as *ghura*; see Rowan *et al.* 2015, 177.

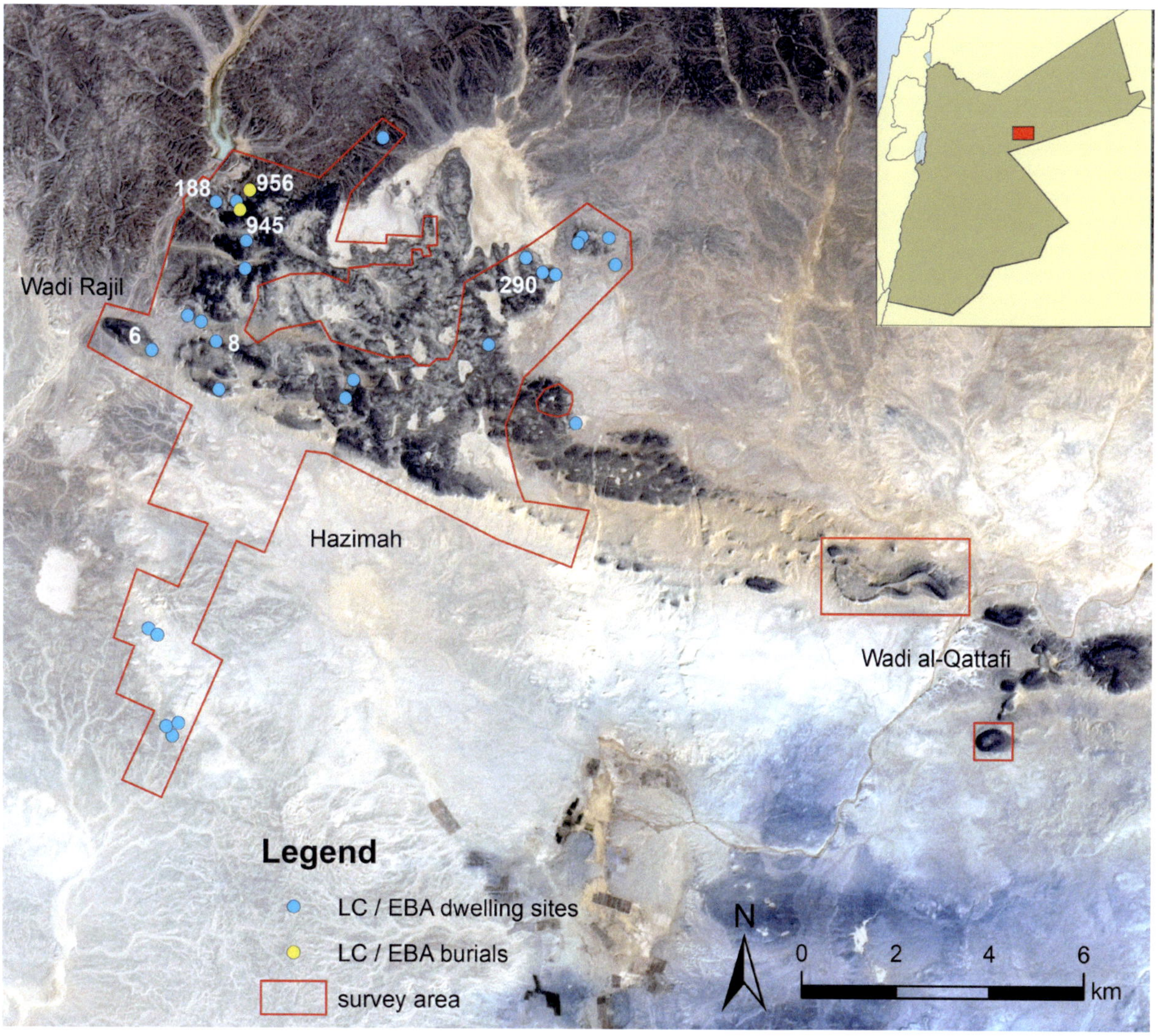

Figure 6. Distribution of LC/EBA sites in the Jebel Qurma region (base map: Landsat 7 – United States Geological Survey; Jebel Qurma Project Archive).

Abu-Azizeh 2016, 5ff.). Another excavated structure at the site (TAG 181) consisted of a single large room (about 4.5 by 1.7 m) with a central fireplace as well as a sizeable, non-local, pottery jar *in situ* nearby. Two AMS dates from the hearth dated to 3640-3350 cal BC (*ibid*., 4).

Two round buildings at the hillfort of Khirbet al-Ja'bariya, west of Tulul al-Ghusayn, have produced additional ^{14}C evidence with dates ranging between 4450-3715 cal BC (Müller-Neuhof and Abu-Azizeh 2016, 10). Importantly, the double cell *ghura* huts appear to be absent at the site. Excavation also took place at Maitland's Mesa in Wadi al-Qattafi, north-east of Jebel Qurma. The site had approximately 200 structures across the entire top of the table mound. Soundings focused on two small circular buildings, one a single cell installation and the other a double cell structure with haphazardly piled walls and small compartments about 2-3 m^2 in extent (Rowan *et al*. 2015, 179-180). Unfortunately, their shallow deposits produced no dateable artefacts, although a LC/EBA date is assumed on the basis of some cortical scrapers scattered across the wider site (Wasse *et al*. 2012, 17, 23).

In the case of Jebel Qurma, the largest sites with both single and double cell *ghura* huts occur in the surroundings of its eponymous hill, where Wadi Rajil debouches out of the basalt into stretches of gravel plains. Low on the northern slope of Jebel Qurma itself is the 8-ha site of QUR-6 with some 225 structures of different shapes and sizes (Fig. 7).[3] The site may even cover a much larger area

3 An earlier paper tentatively dated the site of QUR-6 to the sixth millennium BC (Akkermans *et al*. 2014, 195) but this view now appears to be incorrect.

Figure 7. The numerous *ghura* dwellings and enclosures at the site of QUR-6, at the foot of the hill of Jebel Qurma. View to the north. The area of settlement visible here is about 450 m wide (photograph: APAAME_20111027_DLK-0091, David Kennedy).

(up to 12 ha) if we include an outlying area with dozens of cell buildings on the far eastern portion of the hill. Similar spatially separate arrays of dwellings *within* sites are found elsewhere in the Jebel Qurma area and may be due to internal chronological variation or social constraints (Akkermans *et al.* 2014, 192-197).

High on the slope above the settlement at QUR-6 was a string of round or oval, stone-walled enclosures up to 21 m in diameter. Several enclosures also stood in the settlement area lower down, with sometimes one or more double cell structures or other installations attached to them. The enclosures may have been used for keeping livestock or, perhaps, the intermittent storage of run-off water in the rainy season (Akkermans *et al.* 2014, 196; Philip and Bradbury 2010, 141, 145).

The lower-situated area for habitation had roughly 70 circular or horseshoe-shaped, single cell structures as well as a similar amount of double cell buildings; in addition, there were 17 structures with three or four compartments. The single cell dwellings were generally small, measuring *c.* 2-3 m across. The double cell installations were between 2.8 and 9.5 m long and between 2.4 and 6.5 m wide. The compartments in each building varied from 1 to 4 m in diameter and were divided from each other by narrow partitioning walls. One or more passages *c.* 0.7 m wide gave access to the buildings. The walls of the features were made of piled basalt boulders, with a maximum preserved height of about 1 m. The roofing of the structures remains unknown; it was probably low and made of perishable materials, such as hides and branches. Narrow pathways (40-50 cm wide) ran through the settlement and connected the various structures in the rocky basalt setting, which is otherwise difficult to move across.

In 2017 we excavated two examples of the double cell buildings at QUR-6, both of which were filled with up to 0.5 m of thick wind-blown deposits and loose basalt rocks (probably wall collapse) (Figs. 8-11). They had an exterior measuring about 7 m in length and 4.3-4.6 m in width. Their walls were remarkably wide (1-1.6 m) and low (0.5-0.7 m), and were made of irregularly piled basalt blocks that gently sloped on the outside; in contrast, the cell interiors had straight and carefully stacked facades.[4] A single doorway *c.* 0.4-0.65 m wide gave access to each of the buildings, located either at its long side (structure 126) or at its end (structure 144). A passage in the middle connected the two cells. In structure 126, a large flat upright stone served as wall facade in the opening that connected the compartments. Since it protrudes 20-30 cm above the top of this connecting wall, it may have been placed there to reinforce the passage or, perhaps, to support the roof. When compared to the wall sizes, the interior cells in the buildings were astonishingly small: about 1.1 x 1.6 m and 1.3 x 1.5 (in structure 126) and about 1.6 x 2 m and 2 x 2.8 m (in structure 144). The floor in structure 144 simply consisted of bedrock (cleared of its natural basalt carpet), while structure 126 had an irregular pavement made of small, flat basalt stones. One or more small, shallow hollows (lacking any finds) were sunk into the floors, next to walls. The buildings were practically empty, with the exception of a cortical scraper identified in the main entrance of structure 144 and a worked flint nodule in each of its cells; hence, the exact date of the buildings remains uncertain. Nothing is known about the superstructure of these cell

4 None of the walls showed the "double-faced masonry" said to be characteristic of the sites in the Jawa hinterland; cf. Müller-Neuhof 2017, 125.

Figure 8 (right). The double cell *ghura* structure 126 at QUR-6, prior to excavation (photograph: Jebel Qurma Project Archive).

Figure 9 (below). The double cell *ghura* structure 126 at QUR-6, after excavation (photograph: Jebel Qurma Project Archive).

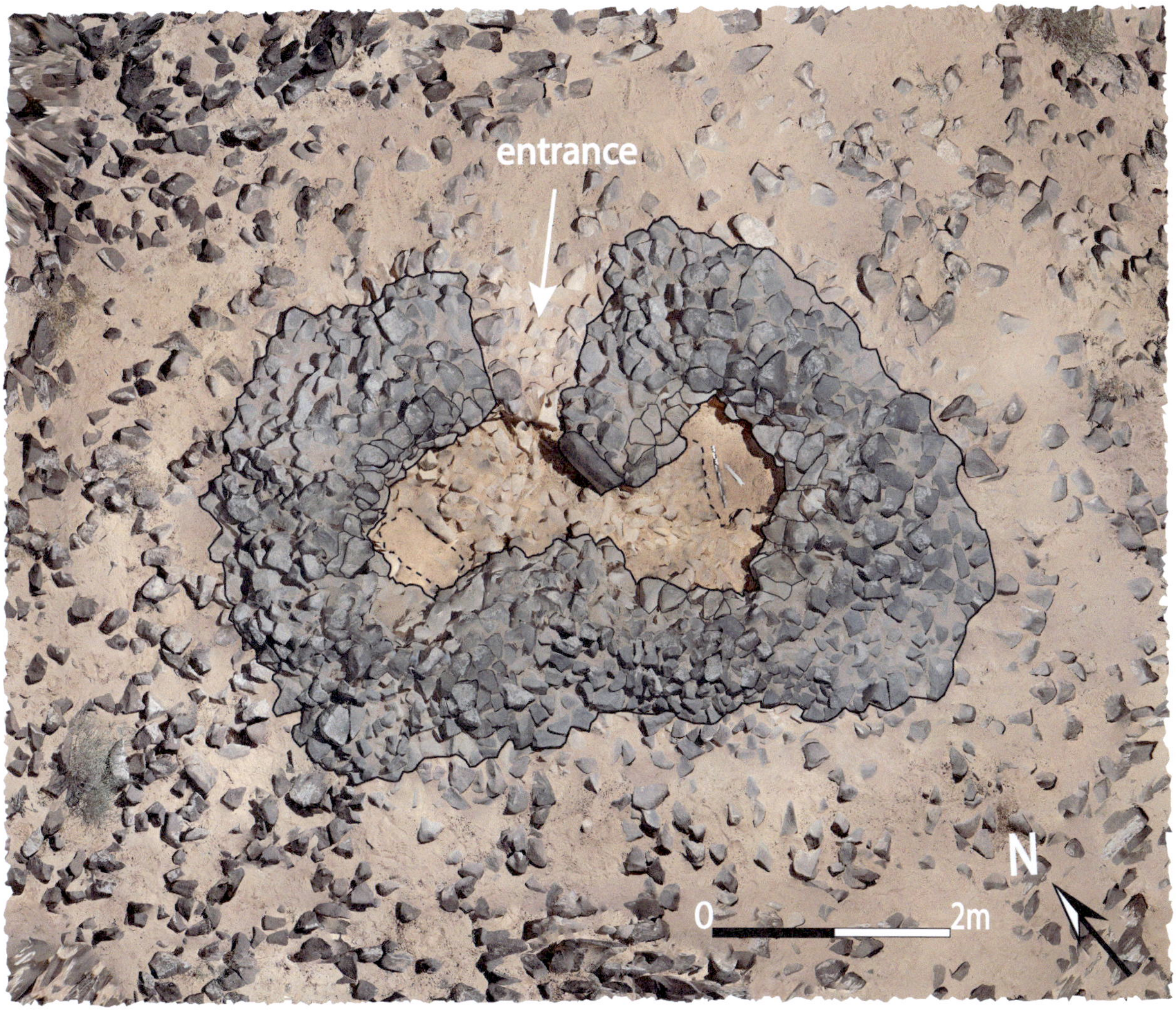

Figure 10. The double cell *ghura* structure 144 at QUR-6, after excavation (photograph: Jebel Qurma Project Archive).

buildings, although the use of perishable materials is likely. For example, Rowan *et al.* (2015, 180) suggest that the low walls were used to hold down the edges of hide roofs. Such – highly portable – roofing materials would suit the interpretation that there was a relatively short-lived, temporary occupation of the buildings, with people perhaps seasonally returning to them. Actually, one of the excavated cell buildings (structure 144) contained evidence for possible re-use, although in only one of its cells: part of the wall of the other cell had collapsed, after which the passage originally connecting both cells seems to have been blocked.

Another major LC/EBA I site (QUR-371) is found at the foot of a basalt-strewn hill at a distance of only 700 m from QUR-6. This site is about 2 ha in areal extent. In its core area, measuring about 160 by 80 m, stood some 50 single and double cell structures, clustered in three distinct groupings around several large enclosures. An outlying area to the north-west had a much more dispersed alignment of seven round dwellings. About 900 m east of QUR-371 is yet another concentration of 37 *ghura* structures, spread over an area of about 100 by 50 m (QUR-8). Most of the buildings were round, although there is also a double cell structure. Significantly, the site is located next to a so-called kite (an extensive installation primarily related to hunting activities; on kites in the Jebel Qurma area, see Akkermans *et al.* 2014, 188-190); it is not excluded that this kite is the main reason for the presence of the *ghura* structures here. If so, the hunting of large game must have been highly important to the local LC/EBA community. Two other (very) small occupations with only two or three circular installations were close by this kite.

Although unambiguous contemporaneity cannot be established yet, we suggest that the agglomeration of sites at the mouth of Wadi Rajil, in close proximity and in clear sight of each other, maintained intimate reciprocal relations. Their specific location – at the interface of the basalt uplands and the vast gravel plains – probably had strategic advantages, such as the availability of

Figure 11. The double cell *ghura* structure 144 at QUR-6, after excavation. The entrance is on the right (photograph: Jebel Qurma Project Archive).

seasonally abundant sources of water and extensive grounds for herding, hunting, as well as (small-scale) agriculture. The attractive setting also facilitated access to, and control of, the major north-south route that ran through Wadi Rajil that helped to connect Arabia and Syria. Given the size of the sites and the large number of structures in them,[5] many hundreds of people may have lived and worked together in the area.

The importance of Wadi Rajil for local settlement is emphasised by a second sizeable cluster of LC/EBA sites only 2.5 km upstream. These are spread in places along Wadi Rajil and several smaller tributaries in the hinterland. The main site was QUR-188, with dozens of round, single cell structures as well as a few double cell buildings, which were all set in rows along the foot of a basalt promontory. Such a position offered a sweeping view over Wadi Rajil below and Qa' Mejalla in the distance. The site stretches over an area of at least 350 by 60 m (*c.* 2 ha), although its original extent and layout of settlement remains unknown, due to severe bulldozing associated with the construction of a dam in Wadi Rajil in the late 1980s.[6] There were three other occupations situated on high ground in the sheltered wadi valleys deeper inside the basalt terrain, less than 1.5 km to the east of QUR-188: all of these were made in pre-existing enclosures and wheels. Maximum site sizes were between 0.1 and 0.5 ha. One site also had a few LC/EBA I tombs (see below).

A third major group of LC/EBA sites is situated in an entirely different setting of the Jebel Qurma region, on two low, basalt-covered rises that lie opposite each other on the edge of an extensive mud flat (cf. Fig. 6). The southern rise had a 200 by 50 m concentration of several dozen single cell 'huts' on the slope (QUR-290), in addition to a wheel about 60 m across. A similar-sized wheel stood 2 km to the south and contained cortical scrapers. Still further to the south-east were several enclosures measuring about 100 by 60 m, partly surrounded by single cell buildings. The northern rise opposite QUR-290 had three typical wheels *c.* 60-70 m in diameter with cortical

5 These structures, we suggest, were in use more or less simultaneously, as most of them seem not to overlie each other or to have been modified in later phases.

6 Corona satellite imagery from the 1970s shows several large enclosures and what may be many more *ghura* structures that do not exist anymore today. See: https://corona.cast.uark.edu/

Figure 12. LC/EBA enclosure at the site of HAZ-47 in the Hazimah plains. The single-row structure is about 18 m in diameter (photograph: Jebel Qurma Project Archive).

tool assemblages, as well as some enclosures about 45 m across at the foot of the mound.

There are also a few small LC/EBA enclosures found on terraces along an inland wadi that sliced through the steep-sloped southern rim of the basalt plateau. Still further south, survey in selected parts of the Hazimah plains identified several enclosure sites situated in sheltered areas near limestone hillocks with relatively dense LC/EBA lithic assemblages (Fig. 12). LC/EBA material was not confined to the enclosures but also extended to the limestone hilltops and their slopes, characterised by small but distinctive lithic scatters without architecture. At least one of these places served as a knapping site, with many cores and debitage pieces loosely spread over an area of roughly one hectare (cf. Akkermans *et al.* 2014, 200-202; Huigens 2015). The sites clearly show that LC/EBA settlement was not restricted to the basalt expanse proper but also included the persistent exploitation of the adjacent *hamad* plains. This conclusion is in agreement with finds from, for example, the Ruwayshid region in north-eastern Jordan (Betts 2013; Müller-Neuhof 2013a). It also aligns with contemporary developments in steppe and desert settings south of Jebel Qurma, such as in the Jafr basin (Fujii 2013) and, still further to the south, the Thulaythuwat area (Abu-Azizeh 2013).

Fourth millennium burials

While there is an abundance of LC/EBA domestic sites in the Jebel Qurma area, the number of associated burials is still astonishingly low: only two cairns can be securely dated to this period at present (cf. Fig. 6). Although we may expect many more tombs in the region, issues such as frequent re-use and looting have led to often substantial adjustments, which limit their identification. To a very large extent, typological quantification depends on excavation (cf. Akkermans *et al.* 2020). Additionally, it cannot be excluded that the practice of burial in cairns was selective in the period under consideration. Perhaps the majority of the dead were disposed of in ways that are still elusive to us. In a recent paper, Bradbury and Philip (2017, 89) state: "...the EB I, at least in the Southern Levant, is characterised by a distinct peak in burial activity; the dead, as well as the living, in this period would appear to be highly visible, rather than invisible, to us." They hasten to add, however, that there also was a substantial degree of spatial and temporal variation, with certain areas intensively used for burial, whilst others were restricted to specific individuals or groups (*ibid.*). Perhaps the latter option is valid for the Jebel Qurma area in the LC/EBA I period.

Figure 13. The fourth millennium BC tower tomb at QUR-956 (photograph: Jebel Qurma Project Archive).

One of the two tombs from the Jebel Qurma area is a round tower tomb, about 3.4 m in diameter and 1 m in height, which is free-standing at a prominent high point along Wadi Rajil (QUR-956; Fig. 13). The tomb was built of large, flat basalt slabs, resulting in a relatively straight and even façade. Its original contents were not preserved, due to the re-use of the tower for burial in the second century AD.[7] Material from underneath the outer wall of the tomb gave an OSL date of 5580 ± 420 BP, or roughly 3980-3140 BC (see Table 2).

The other tomb, about 4.8 m across and 1.1 m high, was partly set on the walls of an earlier built wheel (QUR-945; see Fig. 6). It had a central burial chamber that was oval to rectangular in shape, and had a pavement of flat, unworked stones. Large, flat capstones covered the chamber, with another layer of basalt blocks of different sizes deposited on top. Inside the chamber were the poorly preserved remains of an adult individual *in situ*. Significantly, a flint tabular scraper was laid next to the head of the deceased, apparently as a burial gift. In addition, there were two stone beads, one made of carnelian and the other of a green translucent stone.

7 By the time of re-use, the tomb was partially ruined. However, its interior chamber was renovated and the entire tomb was given a new covering of basalt blocks, which entirely hid the original structure from view.

Late third millennium developments

Although the occurrence of cortical scrapers and related tools formally allows for a date in the middle of the third millennium BC (Rosen 1997, 75), evidence for an extended, early third millennium presence at the *harrah* sites in Jordan is conspicuously absent until now (cf. Smith, this volume). The nearest EBA II-III occupations are found at the sites of Khirbet al-Umbashi and Khirbet ed-Dab'a some 100 km further north on the fringes of the basalt zone in Syria (cf. Braemer 1993; Braemer *et al.* 2004). While this may be a reflection of the current, limited state of research in the basalt expanse, it may, alternatively, indicate local abandonments and reorganisations.

Significantly, in addition to its extensive fourth millennium settlement, Jawa had a minor phase of re-use at about 2000 BC, primarily characterised by a single 'citadel' building about 30 by 26 m in extent (Helms 1981; 1989). It is assumed that the planned, symmetrical residence with its many cubicles and upper storey was used for about a century or so, perhaps as an isolated caravanserai (Bourke 2013, 471). It also may have acted as a multi-functional intermediary between desert pastoralists and more sedentary communities in Transjordan and southern Syria (Helms 1989). Jawa's brief resettlement has long been

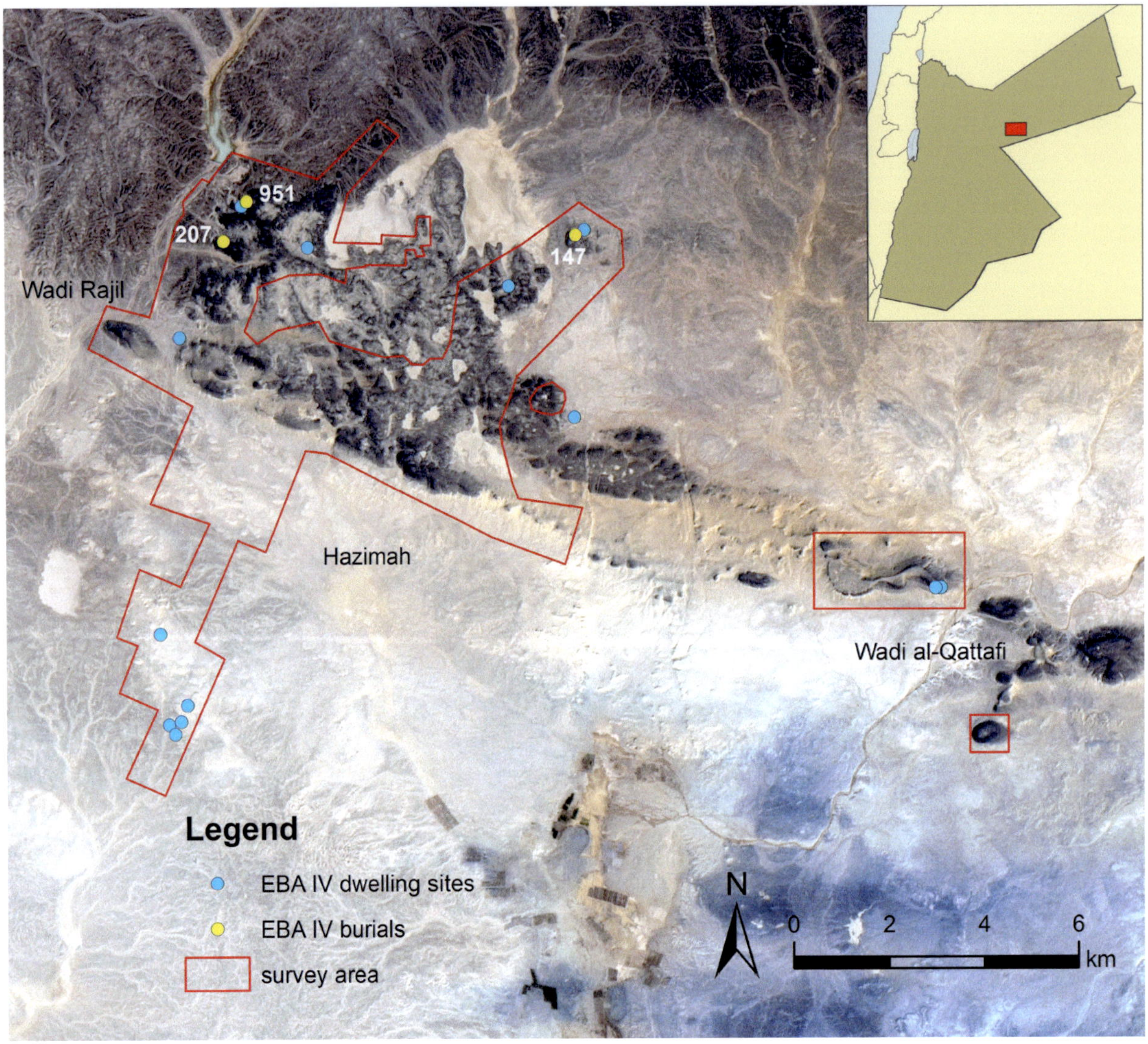

Figure 14. Distribution of EBA IV sites in the Jebel Qurma region (base map: Landsat 7 – United States Geological Survey; Jebel Qurma Project Archive).

deemed to represent an isolated, intermittent phenomenon. Helms (1989, 141) notes that from the EBA IV 'citadel' it is possible to see the southward path of Wadi Rajil, which served as a major north-south route through the otherwise difficult-to-cross basalt uplands. While Jawa is at the upper reaches of Wadi Rajil, the Jebel Qurma range is some 70 km downstream where the wadi emerges from the basalt onto the open plains of Hazimah. In this respect, it perhaps comes as no surprise that recent survey and excavation in the Jebel Qurma area have identified a number of EBA IV domestic sites and cemeteries. Jawa, it appears, stood not on its own but was one of probably many sites in the *harrah* at the end of the third millennium BC, connected through relatively easy-to-travel wadis and mud pans.

The sites of this period, dated to *c.* 2300-2000 BC (EBA IV or Intermediate Bronze Age; cf. Richard 2013; Prag 2013) based on pottery finds and/or ^{14}C and OSL dates, comprise both dwelling sites and burials. Until now, only 13 residential sites have been identified in the Jebel Qurma region (Fig. 14), although we suspect more of the identified sites to date within this period; the sites cannot be dated properly, due to their undiagnostic ceramics. The dwelling sites varied from groups of enclosures at the foot of the basalt promontories to wheels in relatively secluded high grounds, and simple open clearings in the basalt for camping. Several small enclosures sites were also made in sheltered locales in the Hazimah plain, to the south of the basalt range (Fig. 14). None of the sites were newly founded in the EBA IV; without exception, they were palimpsests, with evidence for repeated, periodic use over many centuries or even millennia. The size of the area of habitation in a given period is often difficult to establish, due to the limited number of artefacts. However,

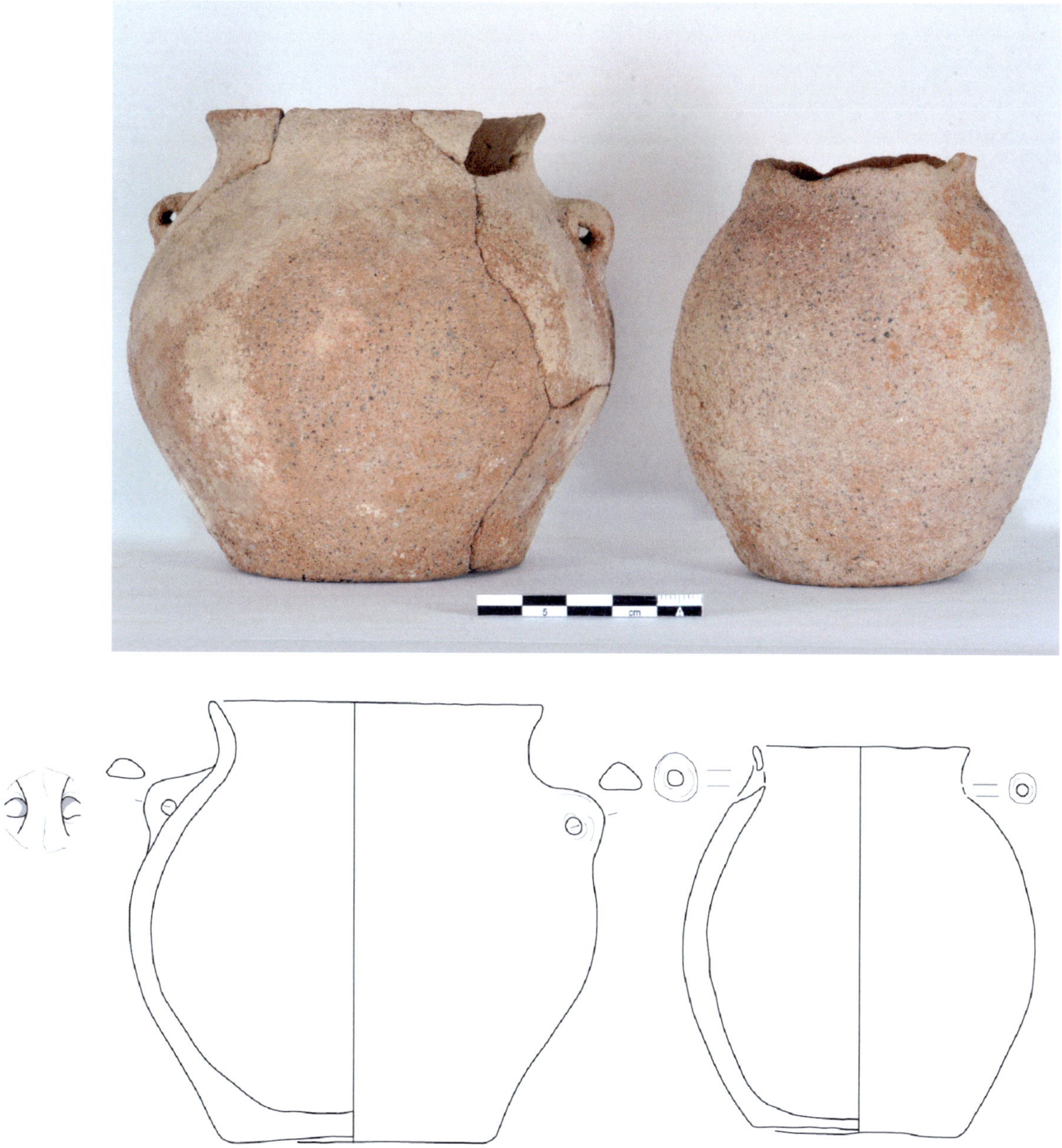

Figure 15. EBA IV *amphoriskoi* from tombs at QUR-951 in the Jebel Qurma region (photograph: Jebel Qurma Project Archive).

most of the identified sites measured only between 40 and 100 m across.

The domestic sites each yielded pottery fragments, although in (very) small quantities. The shapes range from hole-mouth pots and bowls with (sometimes incised) ledge handles to medium-sized jars with everted rims, flat bases and ledge handles low on the body (Figs. 15-16). These types of pottery have parallels at late third to early second millennium sites elsewhere in the southern Levant, such as Jawa (Helms 1989, 152ff.; 1991, 99-100 and Fig. 154), Tiwal esh-Sharqi (Helms 1983; Tubb *et al.* 1990), Bab edh-Dhra' (Schaub and Rast 1989), Jericho (Kenyon and Holland 1983, 168, fig. 66.5; Nigro *et al.* 2005, 174, figs. 182.16-182.17), Khirbet Iskander (*e.g.* Richard *et al.* 1984, 81ff.) and Tell Rukais (Betts *et al.* 1996).

Pottery also occurred in four tombs high up on the slope of a basalt-covered hillock along Wadi Rajil. They each contained a single small, short-necked jar with a flat

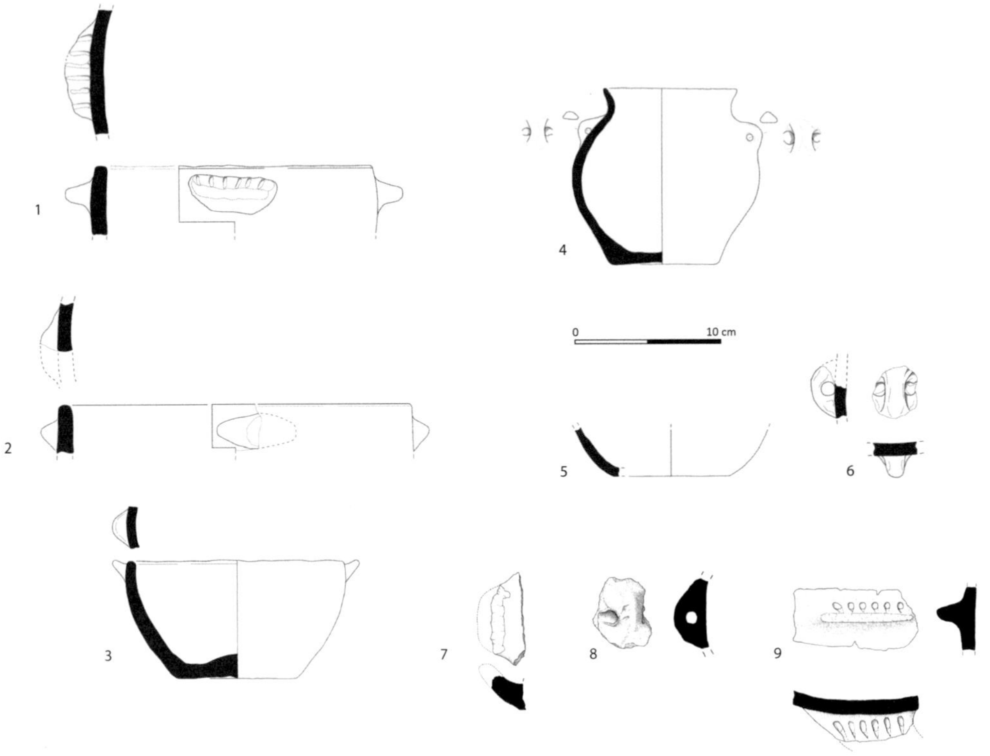

Figure 16. Hand-made, mineral-tempered EBA IV pottery from sites in the Jebel Qurma area. No. 1: pot with incised ledge handle from a wheel at QUR-146. No. 2: pot with ledge handle from an enclosure at QUR-637. No. 3: complete bowl with upward ledge handle from an enclosure at QUR-300. No. 4: *amphoriskos* from a tomb at QUR-951. No. 5: flat base from a tomb at QUR-951. No. 6: loop handle from a wheel at QUR-172. No. 7: ledge handle from an enclosure at HAZ-47 in the Hazimah plain. No. 8: loop handle from a wheel at QUR-147. No. 9: impressed ledge handle from a wheel at QUR-146 (drawings: Jebel Qurma Project Archive).

base and, occasionally, loop handles (Fig. 15). The pots closely resemble the *amphoriskoi* found in the late third millennium cemeteries at Bab edh-Dhra' near the Dead Sea and Tiwal esh-Sharqi in the central Jordan Valley (Helms 1983, 74 and Fig. 18, nos. 4, 12-14; Schaub and Rast 1989, 473ff.; Schaub 1973, Fig. 6; Tubb *et al.* 1990). They also have parallels at EBA IV settlements like Khirbet al-Batrawi near Zarqa (Sala 2006a, 103 and Fig. 3.9) and Tell Umm Hamad (*e.g.* Kennedy 2015, 14 and Fig. 3, nos. 18-25).

The four tombs with ceramics were part of a larger cemetery (QUR-951), consisting of some thirty cairns in total (Akkermans and Brüning 2017; Akkermans *et al.* 2020). The cairn field was high on the slope of a basalt-covered hillock, with a panoramic view over the meandering flood plain of Wadi Rajil below. The area with the cairns was used previously for groupings of stone-walled enclosures from the Late Neolithic period (*c.* 6400-6100 BC). Excavation of 13 cairns showed that they mainly consisted of small and low tower tombs, up to about 3 m across and up to 0.6 m in (preserved) height. Some of these tombs were round and others were more square-shaped with rounded corners (Fig. 17). They were all rather quickly built structures with relatively rough and uneven stacked façades. The interior burial chambers were notably small: round chambers were about 1 m in diameter, while oval chambers were about 0.7-1.4 m in length and 0.5-0.8 m in width. Their internal height varied between 0.3-0.6 m. The chamber walls were either corbelled or straight and covered with capstones. Given their size, the tombs cannot have been used for interment in a supine position but must have facilitated contracted burial, with the deceased resting on the side. Unfortunately, the preservation of the skeletal remains in the tombs was extremely meagre, with at most a few small fragments of bones or teeth remaining. Moreover, most of the tombs were looted, further contributing to the poor preservation of the bones (and other finds). In addition to the tower tombs, there were also a few other small cairns in the cemetery, consisting of conical piles of stones, with a small corbelling burial chamber inside. They resemble the so-called 'ring cairns' (see below) but lack the typical

Figure 17. A typical tower tomb in the EBA IV cemetery at QUR-951, with on the left the tower in 3D and on the right its plan. A: the outer wall of the tomb. B: the interior burial chamber. The area between the outer wall and the central chamber was filled with basalt stones (photograph: Jebel Qurma Project Archive).

Figure 18. The large EBA IV ring cairn at QUR-207. The cairn is about 10 m in diameter and about 1.8 m high (photograph: Jebel Qurma Project Archive).

outer ring of large basalt blocks. No finds are associated with these tombs.

Although none of these other tombs contained pottery,[8] we believe them to be roughly contemporaneous with each other on the basis of the strong similarities in their type and construction. A cemetery of this size suggests a use by a fairly large population, although the sites for the living remain unknown; perhaps the large enclosures on a terrace below the graveyard served this purpose.

Two other tombs, one at the site of QUR-147 and the other at QUR-207, can be ascribed to the EBA IV period,

8 Some tombs had a few beads made of stone and shell, but most of the (excavated) EBA IV cairns yielded no artefacts at all.

Lab. No.	Sample No.	Material	Site	Tomb Type	Date BP	2δ calibrated date (95.4% reliability)
GrM-12051	SN17-051	Human bone bioapatite	QUR-147	Ring Cairn	3701 ± 14	2139-2035 BC
GrM-12076	SN17-051	Human bone bioapatite	QUR-147	Ring Cairn	3740 ± 14	2202-2050 BC
GrM-13351	SN17-051	Human bone bioapatite	QUR-147	Ring Cairn	3593 ± 18	2018-1891 BC
GrM-13204	SN17-093	Human bone bioapatite	QUR-207	Ring Cairn	3586 ± 14	2009-1891 BC
GrM-11920	SN17-092	Human bone bioapatite	QUR-207	Ring Cairn	1585 ± 16	421-537 AD
GrM-12056	SN17-092	Human bone bioapatite	QUR-207	Ring Cairn	1779 ± 12	218-328 AD
GrM-17740	SN18-002	Charcoal	QUR-80	Ring cairn	2240 ± 25	388-206 BC
GrA-67063	SN15-202	Human bone collagen	QUR-215	Ring Cairn	2215 ± 35	380-198 BC
GrA-67032	SN15-096	Human bone collagen	QUR-186	Tower Tomb	1795 ± 35	132-328 AD
GrA-68304	SN16-217	Human bone collagen	QUR-9	Tower Tomb (re-used)	1545 ± 30	425-579 AD
GrA-67035	SN15-201	Human bone collagen	QUR-956	Tower Tomb (re-used)	1890 ± 30	56-217 AD
GrM-12053	SN17-088	Human bone bioapatite	QUR-118	Tower Tomb	1941 ± 12	23-116 AD
GrM-13207	SN17-088	Human bone bioapatite	QUR-118	Tower Tomb	1670 ± 13	341-411 AD
GrM-12054	SN17-090	Human bone bioapatite	QUR-118	Tower Tomb	2222 ± 12	365-207 BC
GrM-13206	SN17-090	Human bone bioapatite	QUR-118	Tower Tomb	2041 ± 13	95 BC- 4 AD
GrA-68436	SN16-208	Human teeth collagen	QUR-2	Tower Tomb	1970 ± 40	50 BC – 125 AD
GrA-68302	SN16-204	Human bone collagen	QUR-2	Cist grave	1905 ± 30	25-211 AD
GrM-13139	SN17-087	Human bone collagen	QUR-148	Cist Grave	1815 ± 25	128-254 AD
GrM-13134	SN17-025	Human bone collagen	QUR-148	Cist Grave	2050 ± 25	164 BC – 16 AD
GrM-11918	SN17-087	Human bone bioapatite	QUR-148	Cist Grave	1957 ± 14	7-77 AD
GrM-11919	SN17-087	Human bone bioapatite	QUR-148	Cist Grave	1996 ± 14	41 BC – 52 AD
GrM-12052	SN17-087	Human bone bioapatite	QUR-148	Cist Grave	2020 ± 12	49 BC – 19 AD
GrM-13139	SN17-087	Human bone collagen	QUR-148	Cist Grave	1815 ± 25	128-254 AD
GrM-13134	SN-17-025	Human bone collagen	QUR-148	Cist grave	2050 ± 25	164 BC -16 AD
GrA-67037	SN14-152	Human bone collagen	QUR-829	Inhumation	1740 ± 30	236-386 AD

Table 1. Radiocarbon dates from tombs in the Jebel Qurma area. Calibration based on OxCal 4.3 (dating carried out by the Centre for Isotope Research, Groningen University, The Netherlands).

Lab. No.	Sample No.	Tomb	Tomb Type	Date BP	Date BC / AD
NLC-8216145	SN16-154	QUR-956	Tower Tomb	5580 ± 420	3985-3095 BC
NLC-8217187	SN17-097	QUR-147	Ring Cairn	3100 ± 1200	2300 BC – 100 AD
NCL-8216141	SN16-040	QUR-215	Ring cairn	2150 ± 450	585 BC – 320 AD
NLC-8216147	SN16-234	QUR-2	Cist Grave	2190 ± 150	325-25 BC
NLC-8216146	SN16-155	QUR-9	Tail	2770 ± 470	1225-285 BC
NLC-8216144	SN16-153	QUR-970	Tail	2690 ± 460	1135-215 BC
NLC-8216142	SN16-041	QUR-215	Tail	2500 ± 460	945-25 BC
NLC-8216143	SN16-075	QUR-32	Tail	2390 ± 380	755 BC -10 AD
NLC-8218140	OSL18-2	QUR-75	Tail	2200 ± 600	780 BC – 420 AD
NLC-8218142	OSL18-7	QUR-80	Tail	2300 ± 600	880 BC – 320 AD
NLC-8218143	OSL18-8	QUR-98	Tail	1400 ± 300	320 AD – 920 AD

Table 2. Optically Stimulated Luminescence (OSL) dates from burial cairns in the Jebel Qurma area (dating carried out by the Netherlands Centre for Luminescence Dating, Wageningen University).

on the basis of ^{14}C data from the skeletal remains in them. Both were so-called 'ring cairns', 8.5-10 m in diameter and up to 1.8 m high. They were conical in shape and characterised by a central burial chamber encircled by an outer ring of large stones (Fig. 18). Each of them contained the poorly preserved remains of a single adult individual in a probably contracted position; importantly, the individual at QUR-207 was associated with a stone wrist-guard (see for a detailed description the contribution by Keshia Akkermans, this volume). Bone material from the cairns gave ^{14}C dates between 2200-1890 cal BC (QUR-147) and 2010-1890 cal BC (QUR-207; see Table 1). An OSL date from underneath the outer ring of the cairn at QUR-147 has very large margins (2300 BC – 100 AD) but is still within the ranges of the ^{14}C dates (Table 2).

The single ring cairn at QUR-147 clearly was a later addition to a (Chalcolithic) wheel which had no other evidence for late third millennium use. However, proof (in the form of a few pottery fragments) for contemporaneous occupation was found at the neighbouring wheel of QUR-146, some 200 m to the north. Perhaps people at the latter site made use of the former wheel for the burial of (some of) their dead.

While the cairn at QUR-147 stood on its own, the ring cairn atop the hill of QUR-207 on Wadi Rajil had three, or possibly four, small tower tombs in its close vicinity, assumedly contemporaneous in date. At the time of construction of these cairns, a series of extensive enclosures dating between the late seventh to fourth millennium stood at the foot of the mound; their additional use for dwelling in the late third millennium BC is likely.

Second millennium BC: absence of pottery

Tell Rukais and other find spots on Wadi al-'Ajib, on the boundary of the dry steppe some 50 km west of Jawa, produced evidence of substantial, lengthy use as well as fortifications in the Middle Bronze Age, after 2000/1900 BC (Betts *et al.* 1996; Sala 2006b). However, proof for settlement in the second millennium BC is extremely sparse in the desert region further to the east. Some late second millennium tombs (after *c.* 1150 BC) have been found in the Jebel Qurma heights and at Wisad Pools (see below). However, the associated areas for dwelling have not yet been identified, despite comprehensive survey of both highly visible places with dense artefact scatters and ephemeral sites with low visibility and few finds. It has been suggested that detrimental climatic conditions contributed to a wholesale evacuation of the basalt wasteland (cf. Akkermans *et al.* 2014, 204; Akkermans and Huigens 2018, 507; Müller-Neuhof 2014, 235), but solid environmental data are absent for the region in the period under consideration.

An alternative, and probably more likely, explanation is that the current absence of sites is predominantly a matter of visibility. Both survey and excavation have demonstrated that the local communities did not use pottery (or any other durable mass artefact, for that matter) from the EBA IV until the late Roman period, making them extremely difficult to detect in the field (Akkermans 2019; see also, *e.g.*, Banning 1996 on site visibility). The absence of ceramics for some 2000 years should not be confused with the lack of people in the basalt desert. Recent excavation of tombs in the Jebel Qurma area have begun to fill the hitherto assumed 'gap' more and more, with the discovery of apsidal tower tombs probably dating to about 1150-800 BC (cf. Akkermans *et al.* 2020). While habitation sites have not yet been identified for the second and early first millennium BC, they are known from the late first millennium, after *c.* 400 BC (see below). Their existence in earlier contexts is, we believe, a matter of increasing investigation in the field.

The absence of pottery in the basalt desert for about two millennia is an intriguing and as of yet unexplained phenomenon. Ceramics were used in great abundance by contemporary (Middle Bronze Age and Iron Age) settled communities in Transjordan as far east as the foothills of Jebel al-Druze, only a few dozen kilometres from the fringes of the basalt expanse. It can hardly be doubted that people in the *harrah* were aware of pottery and its uses, thanks to the good evidence from burials for the import of a variety of exotic products (jewellery, metal; see Akkermans 2019; Akkermans *et al.* 2020). This interregional exchange existed for many centuries, although its scale and intensity remains elusive for the moment. The exclusion of pottery from this trade must have been intentional and meaningful – it was a deliberate, enduring choice to *not* participate in either the production, exchange or use of ceramics. Because of its fragility, pottery is often considered to be inconsistent with nomadic lifeways but this perspective appears to be too simplistic: there were (and still are) many mobile groups that either produced or traded ceramics for their own use (see *e.g.* Cribb 1991; Beck 2009; Gibbs 2012; Grillo 2014; Heitz and Stapfer 2017). That is why a purely functional explanation for the absence of ceramics in the Black Desert over so many generations and centuries is unlikely. In all probability, self-imposed social constraints were in place, which kept the local groups to a very large extent apart from the settled communities of Transjordan and elsewhere. Although the current evidence remains admittedly fragmentary, there is reason to believe that the desert groups of the second and first millennium BC were highly autonomous in their lifeways and deliberately refrained from an overly close affiliation with the urban polities of their time. The often-dismissed dichotomy between the 'desert-and-the-sown' may have been a reality to a very large extent in this period (Akkermans 2019).

Figure 19. 3D image of the apsidal tower tomb at the site of QUR-1075, with its explicitly straight façade oriented to the east. A: outer wall; B: burial chamber. The cairn measures about 3.8 by 3.5 m, with a preserved height of about 1.2 m (photograph: Jebel Qurma Project Archive).

Iron Age cairns

While first millennium settlements are still few and far between, there is an abundance of contemporary cairns for burial in the Jebel Qurma area, particularly for the period after *c.* 700/600 BC (Akkermans and Brüning 2017; Akkermans *et al.* 2020). The cairns were on relatively difficult-to-reach high plateaus and the summits of the basalt hills, above and away from the areas of settlement. They mainly were isolated, single installations; concentrations of graves in Iron Age cemeteries are absent until now, although the latter seem to be present near Qaf and Ithra' at the onset of the Wadi Sirhan, close to Saudi-Jordanian border (Adams *et al.* 1977, 36). Significantly, tombs of this period are not restricted to the basalt terrain proper but also are found on top of low hills in the adjacent *hamad*. People, it appears, exploited the entirety of the north-eastern *badia* and buried their dead at favourable places in a range of different environments. So far, some 45 cairns have been excavated in the Jebel Qurma region, offering new and exciting insights. Very few tombs have been investigated systematically in other parts of the Black Desert (see Harding 1953, 1978; Clark 1981; Richter 2014; Rowan *et al.* 2015; also Kennedy 2012).

Several types of Iron Age cairns can be distinguished, which remained essentially unchanged for hundreds of years. Large, conical ring cairns and round tower tombs occur most frequently, while rectangular cist graves are found only occasionally and only between 300 BC and 200 AD.[9] A fourth type of cairns, apsidal tower tombs, seem to be an exclusively early feature. These tentatively date to *c.* 1150-800 BC, on the basis of artefacts in them, including a typical early Iron Age

9 Dated on the basis of both ^{14}C and OSL samples; see Tables 1 and 2.

Figure 20. The ring cairn at the site of QUR-1078 in the Jebel Qurma area. The typical outer ring of large basalt bocks at the base of the cairn is clearly visible. The cairn is about 4.5 m in diameter and 1.5 m high (photograph: Jebel Qurma Project Archive).

scarab in one tomb and a carnelian axe-shaped pendant in another tomb. These tombs, about 4 m across and up to 1.2 m in height, are roughly hemispherical or squarish in plan with one straight façade which is usually oriented towards the east (Fig. 19). Until now, they occur in groups of two to seven cairns at only two, neighbouring sites in the easternmost part of the Jebel Qurma range. In one instance, the interior burial chamber contained the mixed skeletal remains of three individuals laid next to each other in a single event (a mature adult, flanked on either side by an adolescent and a child, 10-16 years of age). All were placed in a strongly contracted position on their side, roughly oriented east-west, and with the head to the west.

A common type of Iron Age burial was the conical 'ring cairns', with their characteristic outer ring of large basalt blocks that encircled an oval, corbelled burial chamber in the centre (Fig. 20). Basalt blocks filled in the area between the outer ring and the central chamber. These tombs are typically 5-8 m across at their base and about 1-1.5 m high. Occasionally, larger ring cairns do occur, measuring up to 10-12 m in diameter, but these appear to consist of two superimposed cairns (Akkermans *et al.* 2020). Although skeletal preservation is generally poor, it is clear that the deceased consistently were laid to rest in a flexed position on the side – a characteristic which the ring cairns shared with the other types of tombs.

The ring cairns seem to represent an enduring form of burial that was used locally already in the late third millennium BC, if not before (see above). A momentous innovation in the first millennium BC was the attachment of a chain of small cairns off of the head of the main cairn, which could be up to 135 m in length (Figs. 21-22). These chains consisted of five to fifty round or roughly rectangular cairns: some of these were low, inconspicuous heaps of rocks, while others were prominent features (up to 2 m long, 0.8 m wide, and 1 m high). Selected excavation of cairns of twelve chains at different sites yielded no evidence whatsoever of human remains or artefacts within or underneath them. Hence, their function as actual tombs can be excluded. Several researchers have suggested a commemorative role for these chains of cairns (Kennedy 2011, 3190; Rowan *et al.* 2015, 180), but their precise meaning remains elusive.

The tailed tombs are often assumed to be prehistoric in date (*e.g.* Parr *et al.* 1978, 40; Rollefson *et al.* 2016, 941; see also Kennedy 2011, 3195) but this view is incorrect.

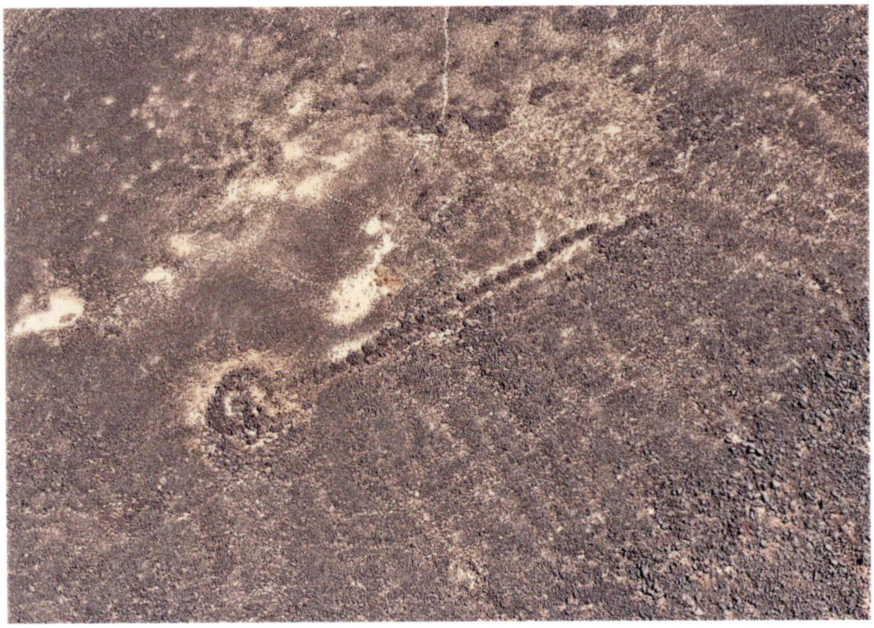

Figure 21. Aerial photograph of the tail of 26 small cairns and its associated tomb at the site of QUR-28 in the Jebel Qurma region. The chain of small cairns is about 67 m long (photograph: APAAME_20081102_DLK-0141, David Kennedy).

Figure 22. Detail of the tail and its individual, rectangular cairns at QUR-28. They measure each about 2 by 1.5 m and are about 0.6 m high (photograph: Jebel Qurma Project Archive).

OSL dates from seven cairn chains, supported by ^{14}C dates and/or artefact assemblages from the associated main tombs, show that these features were made in the first millennium BC and afterwards, probably up to *c.* 300 AD (see Tables 1-2). Additional evidence for a date of the tailed tombs in the first millennium BC comes from excavated burials in Yemen (De Maigret 2009, 331).

Yet another common type of Iron Age burial cairn was the large, round tower tombs, up to 5 m in diameter and 1.5 m high (Fig. 23). In contrast to the early first millennium apsidal tower tombs, the round tower tombs always occur as *single* installations, situated on prominent, eye-catching high grounds with great visibility from the plains below. These tombs, we believe, were "powerful and permanent vehicles for commemorating the dead and linked the past and present in a highly visual and public way. Far from being "secretive" or understood by insiders only, these tombs were easily recognised by locals and foreign visitors to the region alike and may have inspired awe and reverence. These burial grounds must have been liminal places full of social memory; the continual re-use and the repeated burial events at these sites over many centuries confirmed their long-lived role as focal points for social and ritual gatherings of the communities in the area." (Akkermans and Brüning 2017, 139).

Figure 23. Round tower tomb at the site of QUR-1075. The cairn is 3.7 m in diameter and preserved to a height of 1.2 m (photograph: Jebel Qurma Project Archive).

Several round tower tombs had rectangular, east-west oriented cist graves attached to them later, which were up to 2.7 m long, 1.5 m wide, and 1 m high. Originally, these graves were covered with capstones. During later re-use, a layer of basalt rocks replaced the capstones, being laid directly on the remains of one or two individuals in a crouched position. Remarkably, one of the cist graves was accompanied with a small, irregular hollow covered with basalt blocks, containing selective skeletal parts (skulls and long bones) of two individuals. It clearly was a secondary deposit, probably consisting of remains removed from the cist grave in order to facilitate a new burial.

Each type of tomb contained grave goods in various numbers, in the form of either personal jewellery or, occasionally, iron weaponry (arrowheads, javelins; see Keshia Akkermans, this volume). Pottery was entirely absent, but several tombs had fragments of a single, small bronze bowl. An extraordinary find was the discovery of griffon vulture legs in two late first millennium BC tower tombs, which may have served as amulets. Finds from one cist grave included four (weathered) Seleucid bronze coins, one of which could be securely dated to the reign of Antiochus IX Cyzicenus (114-95 BC). Another cairn provided a silver Seleucid *tetradrachme* dated to Antiochus VII Euergetes (138-129 BC; cf. Houghton *et al.* 2008) (Fig. 24).

Interestingly, the Iron Age cairns in the Jebel Qurma area are often found in places that also have substantial numbers of petroglyphs, roughly dated between the third/second century BC and the third/fourth century AD. Although it is tempting the assume a direct, funerary relationship between the cairns and the rock art (*e.g.* Oxtoby 1968; Winnett 1978; see Macdonald 2015 for a recent evaluation), the proof for such an intimate bond is very meagre. Fieldwork in the Jebel Qurma region has identified almost 10,000 petroglyphs and inscriptions at present (cf. Brusgaard 2019; Della Puppa,

Figure 24. Seleucid coins from tombs in the Jebel Qurma region. No. 1: silver Seleucid *tetradrachme* from the surface of a looted cairn at QUR-238. Obverse shows the diademed head of Antiochus VII Euergetes (138-129 BC). Reverse shows an eagle with closed wings standing left on a ship's bow, with a palm branch under the right wing. ΒΑΣΙΛΕΩΣ ΑΝΤΙΟΧΟΥ. Minted in Tyre and dated 130/129 BC. No. 2: bronze coin from the floor of a cist grave at QUR-2, showing the head of Athena with helmet. Reverse shows a ship's bow and ΒΑΣΙΛΕΩΣ ΑΝΤΙΟΧΟΥ ΦΙΛΟΠΑΤΟΡΟΣ. Dated to Antiochus IX Cyzicenus (114-95 BC). No. 3: bronze coin, from the same cist grave as coin no. 2. Obverse wholly weathered. Reverse shows a palm tree. Date unknown (possibly Demetrius II, 129-125 BC) (photographs: Jebel Qurma Project Archive).

Figure 25. Safaitic inscription on the base of the ring cairn at the site of QUR-1078 (cf. Fig. 20). The text reads: *l hs^{1}yb bn ḏkr bn qmhz bn {k}n bn {ḥ}---- bn fḏg w ḥwr*, "By Hs^{1}yb son of Ḏkr son of Qmhz son of {Kn} son of {Ḥ----} son of Fḏg and he wept with grief". It appears to give to name of one of the mourners rather than giving the name of the person buried there (photograph: Jebel Qurma Project Archive. Reading: Michael Macdonald, Oxford).

forthcoming), but only two Safaitic inscriptions explicitly refer to a burial (Fig. 25). Excavation indisputably revealed that many tombs were built with basalt blocks that were previously inscribed with Safaitic rock art, and so they must post-date the rock art.

In her study about the petroglyphs from Jebel Qurma, Nathalie Brusgaard (2019, 166ff.) emphasises the strong connection between the carvings and prominent vantage points in the landscape. The latter, she argues, were vital to the desert nomads and their lifeways, from a strategic, economic, social, and ideological viewpoint. Because of the great importance attached to specific locales, they "perpetuated the production of carvings, continuously creating new visual histories while reinforcing old ones. Through the creation and accumulation of narratives, these places probably took on special socio-economic meanings for the desert nomads." (*ibid.*, 179). It is not a coincidence that these places imbued with great social significance also attracted other powerful visual expressions of reverence and memory, namely the burial cairns. The tombs further enhanced the significance of the rock-art sites in the landscape and *vice versa*.

Settlement in the Hellenistic to Roman and Roman/Byzantine periods

Solid evidence for domestic settlement is available from the fourth/third century BC onwards, with several new strands of data increasing overall site visibility. Some sites had pre-Islamic (Safaitic) inscriptions on stone, dating from perhaps as early as the third century BC up to the fourth century AD, which referred to the construction or the ownership of enclosures and other installations at these places. Others had surface finds of Roman or Roman/Byzantine pottery (no Hellenistic ceramics were found). Excavation in an enclosure at QUR-595 uncovered two fireplaces that were radiocarbon-dated to 400-210 cal BC and 95 cal BC – 60 cal AD, respectively (Table 1); this suggests that the enclosure was used more than once.

In a recent doctoral thesis, Harmen Huigens has extensively discussed the evidence for settlement in the Jebel Qurma region from the Hellenistic to the Early Islamic periods, and we refer to his work for details on this topic (Huigens 2019). Huigens has identified around 30 dwelling sites for the period between *c.* 400/300 BC and 300 AD, ranging from sites with stone-walled enclosures to simple

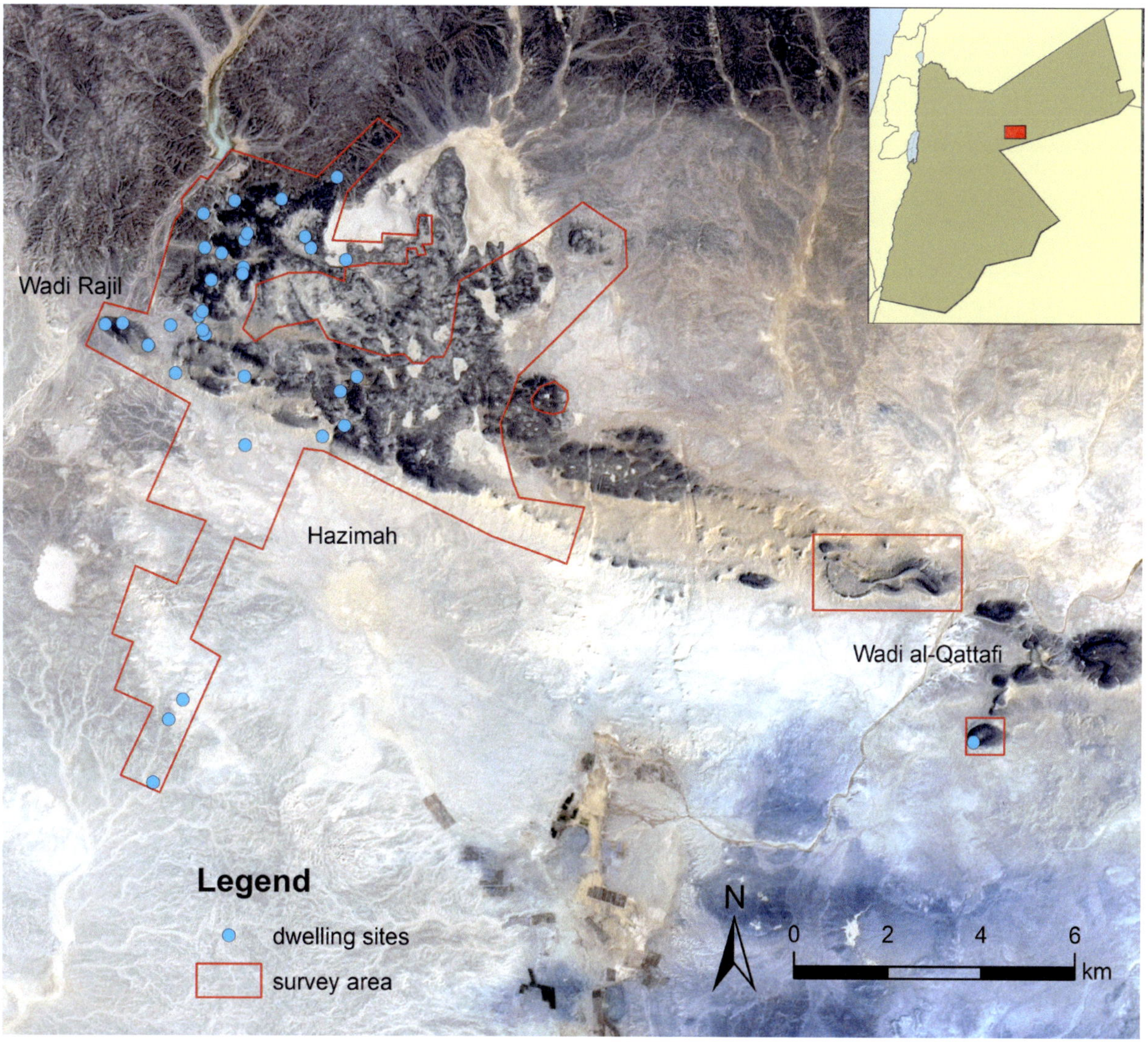

Figure 26. Distribution of Hellenistic to Roman and Roman/Byzantine dwelling sites in the Jebel Qurma region (base map: Landsat 7 – United States Geological Survey; Jebel Qurma Project Archive).

open clearings for camping, or combinations thereof (*ibid.*, 138) (Fig. 26). More recent discoveries also include sites with a single round building about 2-3 m in diameter, provided with a narrow doorway and one or more pre-Islamic inscriptions claiming its ownership (Figs. 27-28). The enclosures also regularly had inscriptions, and consisted either of single structures or of small, irregular groupings several dozen metres across (Fig. 29). The many shallow fireplaces in them suggest domestic and residential activity, in addition to the oft-assumed function of the enclosures for corralling animals. The enclosures may have easily shifted roles according to need and preference, in agreement with their discontinuous, episodic use. Huigens (2019, 140) notes that the enclosures tend to be situated predominantly in rather secluded, inland areas with low visual prominence, perhaps indicating security concerns. Another option is that they were the preferential winter camps of local nomadic groups, with provisions for shelter and protection from the cold for both humans and animals.

The clearings consisted of open areas between 20-50 m in diameter, emptied of their natural basalt cover, probably for camping. Architectural features are absent. The clearings tend to occur in relatively easily accessible, low-lying areas in valleys and near mud pans, close to seasonal water sources and potential grazing areas. There is good evidence for intermittent, *ad hoc* usage of the clearings, in the form of artefact scatters from many different periods.

The different types of sites were small in size, usually little more than several dozen metres across (a few larger sites were around 1 ha in extent), and dispersed over the basalt terrain and adjacent gravel plains. It appears that people were living in single, small groups, in the order of

Figure 27. Small, round installation (Structure 11) in the middle of heavy basalt blocks, high on the slope of QUR-98. The site offers an amazing outlook over the gravel plains below. The structure was surrounded by three Safaitic inscriptions, one of which claimed ownership (photograph: Jebel Qurma Project Archive).

Figure 28. One of three Safaitic inscriptions next to the entrance to Structure 11 at QUR-98, and claiming its ownership. The text reads: *l ʾḥs^{1}n bn gdy h-zrb*, "By ʾḥs^{1}n son of Gdy is the enclosure" (photograph: Jebel Qurma Project Archive. Reading: Michael Macdonald, Oxford).

a few dozen people at most. Given the shallow sediments and their limited artefact assemblages, many sites were used briefly yet repeatedly (perhaps on a seasonal basis). They may have served the recurrent residential needs of highly mobile and nomadic groups, which exploited the basalt landscape and the adjacent plains in a variety of ways (cf. Huigens 2019).

Significantly, the majority of the domestic sites tend to occur along Wadi Rajil and its nearby hinterland, with only a few sites found until now in the basalt uplands further to the east (cf. Fig. 26). Although Wadi Rajil was a main focus of settlement through the ages, we should probably refrain from any overly rigid interpretation. When compared with Wadi Rajil, the paucity of sites in the eastern half of the Jebel Qurma range is at least partially explained by the lower intensity and coverage of surveys here, particularly in the low-lying areas. The occurrence of contemporary tombs in the region also implies the presence of (as yet unrecorded) areas for the living.

Conclusions

Earlier, we strongly interpreted settlement development in the Jebel Qurma range in terms of cyclical shifts and rearrangements, in association with factors such as climate change. For example, in a 2014 report we concluded: "…it is difficult to *not* assume substantial hiatuses in the regional archaeological record; the Jebel Qurma area, it seems, was punctuated by episodes of distinct settlement and regional

Figure 29. Group of enclosures, covering an area of about 80 by 40 m at the foot of Jebel Qurma (QUR-595). They were repeatedly used for domestic activities between 400 BC and 60 AD, according to ^{14}C dates (photograph: Jebel Qurma Project Archive).

abandonment." (Akkermans *et al.* 2014, 203; see also Müller-Neuhof 2013a, 227 for a similar conclusion regarding the Jawa hinterland). To some extent, this perspective needs revision. With increasing research, some hiatuses in the local sequence of occupation have narrowed substantially or even disappeared in their entirety. To give an example: it was long assumed that people were absent in the area in the Iron Age, until the arrival of 'Safaitic' groups in second or first century BC, but this viewpoint has been dismissed by recent excavations of burials. They show that most cairns actually date to the first millennium BC, emphasising an intensive use of the basalt region in this period (Akkermans 2019; Akkermans *et al.* 2020). However, from the data in the present paper, it should be clear that we still cannot claim settlement continuity in the Jebel Qurma region (and perhaps the Black Desert at large) for all periods and ages. There is extensive evidence for habitation in the late fifth and fourth millennium but it is virtually absent for the larger part of the third and the entirety of the second millennium BC. These hiatuses may be realities, although they undoubtedly also reflect research intensities and strategies to a considerable extent. To mention but a few restrictions: the scale of fieldwork is still limited; the problems associated with it are manifold (*e.g.* often restricted site visibility); and the chronological framework needs substantial refinement which can be achieved only through excavation at relevant sites.

Significantly, the data from Jebel Qurma suggest that the main landscapes of Jordan's north-eastern 'panhandle' – the *harrah* and *hamad* – were inhabited and exploited by people in each and every period under consideration (cf. Huigens 2015, 192). From the fifth to fourth millennium onwards (if not before), the sites in the basalt range had their contemporary counterparts in the gravel plains, in the form of both settlement sites and tombs. From the late first millennium BC onwards, we find Safaitic inscriptions also in the *hamad* of Jebel Qurma. However, these are still very rare, perhaps due to matters of preservation, as the few known examples were made on limestone, which is prone to weathering. With regard to the LC/EBA period, the largest (and probably more permanent) sites with their many cell structures were on the fringes of the basalt expanse, whereas the *hamad* hinterland had evidence of small and dispersed locales of enclosures, lithic scatters, and knapping sites. The basalt terrain, it appears, was the preferential area for settlement in this period, because of natural benefits, such as good grounds for shelter and the availability of water, but perhaps also because of social considerations, such as ancestral belongings and attachment to the area. It is likely that the LC/EBA groups were engaged in forms of logistical mobility, pursuing specific tasks in the *hamad* while moving back and forth from their basecamps in the *harrah*.

When compared with the fifth/fourth millennium sites, settlement in the later periods seems to change rather dramatically. Settlement in the late third millennium BC (EBA IV) is restricted to the re-use of older installations in selected places, without the cell buildings of the previous period. First millennium sites seem to predominantly consist of the temporary camp sites of nomadic populations. None of these later sites were ever as large as the late prehistoric settlements. The associated investments in architecture were likewise much more restricted in the later periods, with the exception of cairns for burial.

It should be recalled that the number of LC/EBA tombs is still extremely low. With the onset of the EBA IV, cemeteries with several dozen small and low tower tombs made their appearance, emphasising the importance of the community as a whole; local groups brought their dead to selected, shared grounds instilled with ritual and social memory. In the first millennium BC, the focus shifted from the communal to the individual, with the emphasis on relatively isolated, elaborate tombs of different types on prominent high grounds. These single tombs were in keeping with the small-scale and dispersed nature of settlement in this period. Because of their size and location on panoramic vantage points, they also commemorated the dead in a most visual and spectacular way, and conveyed notions of remembrance and permanence to nomadic people that frequently moved around. The ancestors, it appears, were vital to the first millennium BC communities.

Acknowledgements

We wish to express our gratitude to the Department of Antiquities in Amman, Jordan, and its branch in Azraq, for its continued assistance and encouragement concerning the research in the Jebel Qurma region. Particular thanks go to Dr. Yazid Elayyan (Director-General), Aktham Oweidi (Director of Excavations and Surveys) and Wesam Esaid (Head of the Department's Azraq branch) for their much-valued help. The fieldwork in the Jebel Qurma area was made possible by the support of the Faculty of Archaeology of Leiden University (The Netherlands), the Netherlands Organisation for Scientific Research (NWO), the Leiden University Fund, Migchel Migchelsen and some other private sponsors: we are grateful for their invaluable sponsorship. We are also grateful to the Stichting Nederlands Museum voor Anthropologie en Praehistorie (SNMAP) and the Netherlands Institute for the Near East (NINO) for their generous financial support with regard to radiocarbon and OSL dating. Our great thanks are due to the Centre for Istotope Research of Groningen University (in particular Hans van der Plicht, Michael Dee and Sanne Palstra) for invaluable help with the ^{14}C dating. Paul Beliën (Curator National Numismatic Collection, De Nederlandsche Bank, Amsterdam) is sincerely thanked for his skilful determinations of the coins found in the Jebel Qurma region. We thank the Aerial Photographic Archive for Archaeology in the Middle East (APAAME) for the extremely useful aerial photographs of the Jebel Qurma region. The yearly surveys and excavations in the Jebel Qurma region would not have been possible without the help of a most dedicated team in the field and at home. Many thanks go to all participants in the Jebel Qurma Archaeological Landscape Project in the past years.

References

Abu-Azizeh, W. 2013. The south-eastern Jordan's Chalcolithic-Early Bronze Age pastoral complex: patterns of mobility and interaction. *Paléorient* 39, 149-176.

Adams, R.M., Parr, P.J., Ibrahim, M. and Al-Mughannum, A.S. 1977. Saudi Arabian Archaeological Reconnaissance 1976. *Atlal* 1, 21-40.

Akkermans, P.M.M.G. 2019. Living on the edge or forced into the margins? Hunter-herders in Jordan's north-eastern badlands in the Hellenistic and Roman periods. *Journal of Eastern Mediterranean Archaeology and Heritage Studies* 7, 412-431.

Akkermans, P.M.M.G. and Brüning, M.L. 2017. Nothing but cold ashes? The burial cairns of Jebel Qurma, north-eastern Jordan. *Near Eastern Archaeology* 80, 132-139.

Akkermans, P.M.M.G., Brüning, M.L., Arntz, M., Inskip, S.A. and Akkermans, K.A.N. 2020. Desert tombs: recent research into the Bronze Age and Iron Age cairn burials of Jebel Qurma, north-east Jordan. *Proceedings of the Seminar for Arabian Studies* 50, 1-17.

Akkermans, P.M.M.G. and Huigens, H.O. 2018. Long-term settlement trends in Jordan's north-eastern badia: the Jebel Qurma Archaeological Landscape Project. *Annual of the Department of Antiquities of Jordan* 59, 503-515.

Akkermans, P.M.M.G., Huigens, H.O. and Brüning, M.L. 2014. A landscape of preservation: late prehistoric settlement and sequence in the Jebel Qurma region, north-eastern Jordan. *Levant* 46, 186-205.

Banning, E.B. 1996. Highlands and lowlands: problems and survey frameworks for rural archaeology in the Near East. *Bulletin of the American Schools of Oriental Research* 301, 25-45.

Barket, T.M. and Bell, C.A. 2011. Tabular scrapers: function revisited. *Near Eastern Archaeology* 74, 56-59.

Beck, M.E. 2009. Residential mobility and ceramic exchange: ethnography and archaeological implications. *Journal of Archaeological Method and Theory* 16, 320-356.

Betts, A.V.G. (ed.) 1991. *Excavations at Jawa 1972-1986. Stratigraphy, pottery and other finds*. Edinburgh: Edinburgh University Press.

Betts, A.V.G. (ed.) 1998. *The harra and the hamad – Excavations and surveys in eastern* Jordan (Vol. 1). Sheffield: Sheffield Academic Press.

Betts, A.V.G. (ed.) 2013. *The later prehistory of the badia – Excavations and surveys in eastern Jordan*. Oxford: Oxbow Books.

Betts, A., Eames, S., Hulka, S., Schroder, M., Rust, J. and McLaren, B. 1996. Studies of Bronze Age occupation in the Wadi al-'Ajib, southern Hauran. *Levant* 28, 27-39.

Bourke, S.J. 2013. The southern Levant (Transjordan) during the Middle Bronze Age. In: Killebrew, A.E. and Steiner, M. (eds.). *The Oxford handbook of the archaeology of the Levant, c. 8000-332 BCE*. Oxford: Oxford University Press, 465-481.

Bradbury, J. and Philip, G. 2017. Shifting identities: the human corpse and treatment of the dead in the Levantine Bronze Age. In: Bradbury, J. and Scarre, C. (eds.). *Engaging with the dead: exploring changing human beliefs about death, mortality and the human body*. Oxford: Oxbow Books, 87-102.

Braemer, F. 1993. Khirbet el Umbashi (Syrie). Rapport préliminaire sur les campagnes 1991 et 1992. *Syria* 70, 415-430.

Braemer, F., Échallier, J.-C. and Taraqji, A. 2004. *Khirbet Al Umbashi – Villages et campements de pasteurs dans le 'desert noir' (Syrie) à l'âge du Bronze*. Beyrouth: IFPO.

Braun, E. 2011. The transition from Chalcolithic to Early Bronze I in the southern Levant: a "lost horizon" slowly revealed. In: Lovell, J. and Rowan, Y. (eds.). *Culture, chronology and the Chalcolithic – Theory and transition*. Oxford: Oxbow Books, 160-177.

Brusgaard, N. 2019. *Carving interactions: rock art in the nomadic landscape of the Black Desert, north-eastern Jordan*. Oxford: Archaeopress.

Clark, A. 1981. Archaeological excavations at two burial cairns in the harra region of Jordan. *Annual of the Department of Antiquities of Jordan* 25, 235-265.

Cribb, R. 1991. *Nomads in archaeology*. Cambridge: Cambridge University Press.

De Maigret, A. 2009. *Arabia Felix – An exploration of the archaeological history of Yemen*. London: Stacey International.

Della Puppa, C. Forthcoming. *The Safaitic scripts – An ethno-palaeographic investigation*. Leiden: Leiden University (PhD thesis).

Edgell, H.S. 2006. *Arabian deserts: nature, origin, and evolution*. Dordrecht: Springer.

Fujii, S. 2013. Chronology of the Jafr prehistory and protohistory: a key to the process of pastoral nomadization in the southern Levant. *Syria* 90, 49-125.

Grillo, K.M. 2014. Pastoralism and pottery use: an ethnoarchaeological study in Samburu, Kenya. *African Archaeological Review* 31, 105-130.

Harding, G.L. 1953. The cairn of Hani. *Annual of the Department of Antiquities of Jordan* 2, 8-56.

Harding, G.L. 1978. The cairn of Sa'd. In: Moorey, R. and Parr, P. (eds.). *Archaeology in the Levant*. Warminster: Aris and Phillips, 243-249.

Heitz, K. and Stapfer, R. (eds.) 2017. *Mobility and pottery production: archaeological and anthropological perspectives*. Leiden: Sidestone Press.

Helms, S. 1981. *Jawa: lost city of the Black Desert*. Ithaca, NY: Cornell University Press.

Helms, S. 1983. The EB IV (EB-MB) cemetery at Tiwal esh-Sharqi in the Jordan Valley. *Annual of the Department of Antiquities of Jordan* 28, 55-85.

Helms, S. 1989. Jawa at the beginning of the Middle Bronze Age. *Levant* 21, 141-168.

Helms, S. 1991. The pottery. In: Betts, A.V.G. (ed.). *Excavations at Jawa 1972-1986. Stratigraphy, pottery and other finds*. Edinburgh: Edinburgh University Press, 55-109.

Henry, D.O. 1995. *Prehistoric cultural ecology and evolution: insights from southern Jordan*. New York and London: Plenum Press.

Houghton, A., Lorber, C. and Hoover, O. 2008. *Seleucid coins. A comprehensive catalogue*. Vol. 2. New York: American Numismatic Society.

Huigens, H.O. 2015. Preliminary report on a survey in the Hazimah plains: a hamad landscape in north-eastern Jordan. *Palestine Exploration Quarterly* 147, 180-194.

Huigens, H.O. 2019. *Mobile peoples – Permanent places. Nomadic landscapes and stone architecture from the Hellenistic to Early Islamic periods in north-eastern Jordan*. Oxford: Archaeopress.

Kennedy, D.L. 2011. The "works of the old men" in Arabia: remote sensing in interior Arabia. *Journal of Archaeological Science* 38, 3185-3203.

Kennedy, D.L. 2012. The Cairn of Hani: significance, present condition and context. *Annual of the Department of Antiquities of Jordan* 56, 483-505.

Kennedy, M.A. 2015. Life and death at Tell Umm Hamad, Jordan: a village landscape of the southern Levantine Early Bronze Age IV/Intermediate Bronze Age. *Zeitschrift des Deutschen Palästina-Vereins* 131, 1-28.

Kenyon, K.M.D. and Holland, T.A. 1983. *Excavations at Jericho. Vol. 5: the pottery phases of the tell and other finds*. London: British School of Archaeology in Jerusalem.

Macdonald, M.C.A. 2015. On the uses of writing in ancient Arabia and the role of palaeography in studying them. *Arabian Epigraphic Notes* 1, 1-50.

Meister, J., Krause, J., Müller-Neuhof, B., Portillo, M., Reimann, T. and Schütt, B. 2017. Desert agricultural systems at EBA Jawa (Jordan): integrating archaeological and paleoenvironmental records. *Quarternary International* 434, 33-50.

Meister, J., Rettig, R. and Schütt, B. 2018. Ancient runoff agriculture at Early Bronze Age Jawa (Jordan): water availability, efficiency and food supply capacity. *Journal of Archaeological Science: Reports* 22, 359-371.

Müller-Neuhof, B. 2013a. Southwest Asian Late Chalcolithic/Early Bronze Age demand for "big tools": specialized flint exploitation beyond the fringes of settled regions. *Lithic Technology* 38, 220-236.

Müller-Neuhof, B. 2013b. Nomadische Ressourcennutzungen in den ariden Regionen Jordaniens und der südlichen Levante im 5. bis frühen 3. Jahrtausend v.Chr. *Zeitschrift für Orient-Archäologie* 6, 64-80.

Müller-Neuhof, B. 2014. A 'marginal' region with many options: the diversity of Chalcolithic/Early Bronze Age socio-economic activities in the hinterland of Jawa. *Levant* 46, 230-248.

Müller-Neuhof, B. 2017. The Chalcolithic/Early Bronze Age hillfort phenomenon in the northern badia. *Near Eastern Archaeology 80*, 124-131.

Müller-Neuhof, B. and Abu-Azizeh, W. 2016. Milestones for a tentative chronological framework for the late prehistoric colonization of the basalt desert (north-eastern Jordan). *Levant* 48, 220-235.

Müller-Neuhof, B. and Abu-Azizeh, W. 2018a. Insights into the Chalcolithic/Early Bronze Age colonization of the harra (NE-Jordan) – Part I: Tulul al-Ghusayn. *Annual of the Department of Antiquities of Jordan* 59, 411-424.

Müller-Neuhof, B. and Abu-Azizeh, W. 2018b. Insights into the Chalcolithic/Early Bronze Age colonization of the harra (NE-Jordan) – Part II: Khirbat al-Ja'bariya. *Annual of the Department of Antiquities of Jordan* 59, 425-434.

Müller-Neuhof, B., Abu-Azizeh, L., Abu-Azizeh, W. and Meister, J. 2013. East of Jawa: Chalcolithic/Early Bronze Age settlement activity in al-harra (north-east-Jordan). *Annual of the Department of Antiquities of Jordan* 57, 125-139.

Müller-Neuhof, B., Betts, A. and Wilcox, G. 2015. Jawa, northeastern Jordan: the first ^{14}C dates for the early occupation phase. *Zeitschrift für Orient-Archäologie* 8, 124-131.

Nigro, L., Polcaro, A. and Sala, M. 2005. *Tell es-Sultan/ Gerico alle soglie delle prima urbanizzazione : il villaggio e la necropoli del Bronzo Antico I (3300-3000 a.C.).* Rome: Universit`a di Roma 'La Sapienza'.

Oxtoby, W.G. 1968. *Some inscriptions of the Safaitic bedouin.* New Haven, CT: American Oriental Society.

Parr, P.J., Zarins, J., Ibrahim, M., Waechter, J., Garrard, A., Clarke, C., Bidmead, M. and Al-Badr, H., 1978. Preliminary report on the second phase of the Northern Province Survey 1397/1977. *Atlal* 2, 29-50.

Philip, G. 2008. The Early Bronze Age I-III. In: Adams, R.B. (ed.). *Jordan: an archaeological reader.* Sheffield: Equinox Publishing, 161-226.

Philip, G. and Bradbury, J. 2010. Pre-Classical activity in the basalt landscape of the Homs region, Syria: implications for the development of 'sub-optimal' zones in the Levant during the Chalcolithic-Early Bronze Age. *Levant* 42, 136-169.

Prag, K. 2013. The southern Levant during the Intermediate Bronze Age. In: Killebrew, A.E. and Steiner, M. (eds.). *The Oxford handbook of the archaeology of the Levant, c. 8000-332 BCE.* Oxford: Oxford University Press, 388-400.

Quintero, L.A., Wilke, P.J. and Rollefson, G.O. 2002. From flint mine to fan scraper: the late prehistoric Jafr industrial complex. *Bulletin of the American Schools of Oriental Research* 327, 17-48.

Richard, S. 2013. The southern Levant (Transjordan) during the Early Bronze Age. In: Killebrew, A.E. and Steiner, M. (eds.). *The Oxford handbook of the archaeology of the Levant, c. 8000-332 BCE.* Oxford: Oxford University Press, 330-352.

Richard, S., Boraas, R.S. and Wimmer, D. 1984. Preliminary report of the 1981-82 seasons of the expedition to Khirbet Iskander and its vicinity. *Bulletin of the American Schools of Oriental Research* 254, 63-87.

Richter, T. 2014. Rescue excavations at a Late Neolithic burial cairn in the east Jordanian badya. *Neo-Lithics* 1/14, 18-24.

Rollefson, G.O., Rowan, Y. and Wasse, A. 2013. Neolithic settlement at Wisad Pools, Black Desert. *Neo-Lithics* 1/13, 11-23.

Rollefson, G.O, Athanassas, C.D., Rowan, Y.M. and Wasse, A.M.R. 2016. First chronometric results for 'works of the old men': late prehistoric 'wheels' near Wisad Pools, Black Desert, Jordan. *Antiquity* 90, 939-952.

Rosen, S.A. 1997. *Lithics after the Stone Age.* Walnut Creek, CA: Altamira Press.

Rowan, Y.M., Rollefson, G.O., Wasse, A., Abu-Azizeh, W., Hill, A.C. and Kersel, M.M. 2015. The "land of conjecture": new late prehistoric discoveries at Maitland's Mesa and Wisad Pools, Jordan. *Journal of Field Archaeology* 40, 176-189.

Sala, M. 2006a. Pottery from the Batrawy IV village. In: Nigro, L. (ed.). *Khirbet al-Batrawy: an Early Bronze Age fortified town in north-central Jordan.* Rome: Università di Roma 'La Sapienza', 103-108.

Sala, M. 2006b. Across the desert and the steppe. Ancient tracks from the eastern edges of the az-Zarqa and al-Mafraq districts to the western fringes of the Black Desert of Jordan. In: Nigro, L. (ed.). *Khirbet al-Batrawy: an Early Bronze Age fortified town in north-central Jordan.* Rome: Università di Roma 'La Sapienza', 233-250.

Schaub, R.T. 1973. An Early Bronze IV tomb from Bab edh-Dhra'. *Bulletin of the American Schools of Oriental Research* 210, 2-19.

Schaub, R.T. and Rast, W. 1989. *Bab edh-Dhra': excavations in the cemetery directed by Paul W. Lapp (1965-1967).* Winona Lake, IN: Eisenbrauns.

Tubb, J.N., Henderson, J.D. and Wright, M.M.. 1990. *Excavations at the Early Bronze Age cemetery of Tiwal esh-Sharqi.* London: British Museum Publications.

Wasse A., Rowan, Y. and Rollefson, G.O. 2012. A 7th millennium BC Late Neolithic village at Mesa 4 in Wadi al-Qattafi, eastern Jordan. *Neo-Lithics* 1/12, 15-25.

Winnett, F.V. 1978. *Inscriptions from fifty Safaitic cairns.* Toronto: University of Toronto Press.

Identifying nomadic camp sites from the Classical and Late Antique periods in the Jebel Qurma region, north-eastern Jordan

Harmen O. Huigens

Abstract

This paper discusses the identification of nomadic camp sites in the Black Desert of Jordan between the Hellenistic and Early Islamic periods. It focuses particularly on two features that were studied through surface surveys and excavations in the Jebel Qurma region: enclosures and clearings. The archaeological remains suggest that they were used for residential purposes by short-term visitors to the region. Important in the identification and interpretation of such features are pottery sherds from the Classical and Late Antique periods. The camp sites identified in the Jebel Qurma region vary in morphology and location, and it is suggested that these differences may relate to the use of such features at different times of the year.

Keywords: nomadism, camp sites, Jebel Qurma, landscape archaeology, Classical Antiquity, Late Antiquity, Early Islamic, ceramics

Introduction

One of the most important opportunities offered by newly emerging archaeological data from the *badia* of north-eastern Jordan is the integration of those elements of the ancient societies of the southern Levant that have often been regarded in opposition to one another: the desert versus the sown, the mobile versus the sedentary, and the tribe versus the state. Such dichotomies have been criticised recently on ethnographic and historical grounds, and many researchers (*e.g.* Makarewicz 2013; Porter 2012; Szuchman 2009) have rightly argued that mobile and sedentary communities were much more integrated than previously assumed. This view takes seriously the potential contributions of communities living on the geographic fringe to wider culture-historical trajectories.

This perspective is especially welcome in the case of nomadic communities, who inhabited the *badia* during the Classical and Late Antique periods. These groups have often been marginalised in modern scholarship, for example, by describing them simply as a 'menace' and 'threat' to state systems (*e.g.* Millar 1993, 428-436; Parker 1986, 132), or by regarding them as a more or less static entity largely comparable to 'the Bedouin' as described in nineteenth and early twentieth century travelogues (*e.g.* Donner 1989; Peters 1978). In view of recent criticism (*e.g.* Hoyland 2001, 96-102; Sartre 2005), what needs to be acknowledged instead is the possibility that these nomads formed an integral part of wider socio-economic and political systems on the desert fringe. Equally important in this respect is the exploration of regional and chronological variation in nomadic systems, as these are potentially highly variable and fluid (Barfield 1993; Rosen 2017).

In: Peter M. M. G. Akkermans (ed.) 2020: *Landscapes of Survival - The Archaeology and Epigraphy of Jordan's North-Eastern Desert and Beyond*, Sidestone Press (Leiden), pp. 217-234.

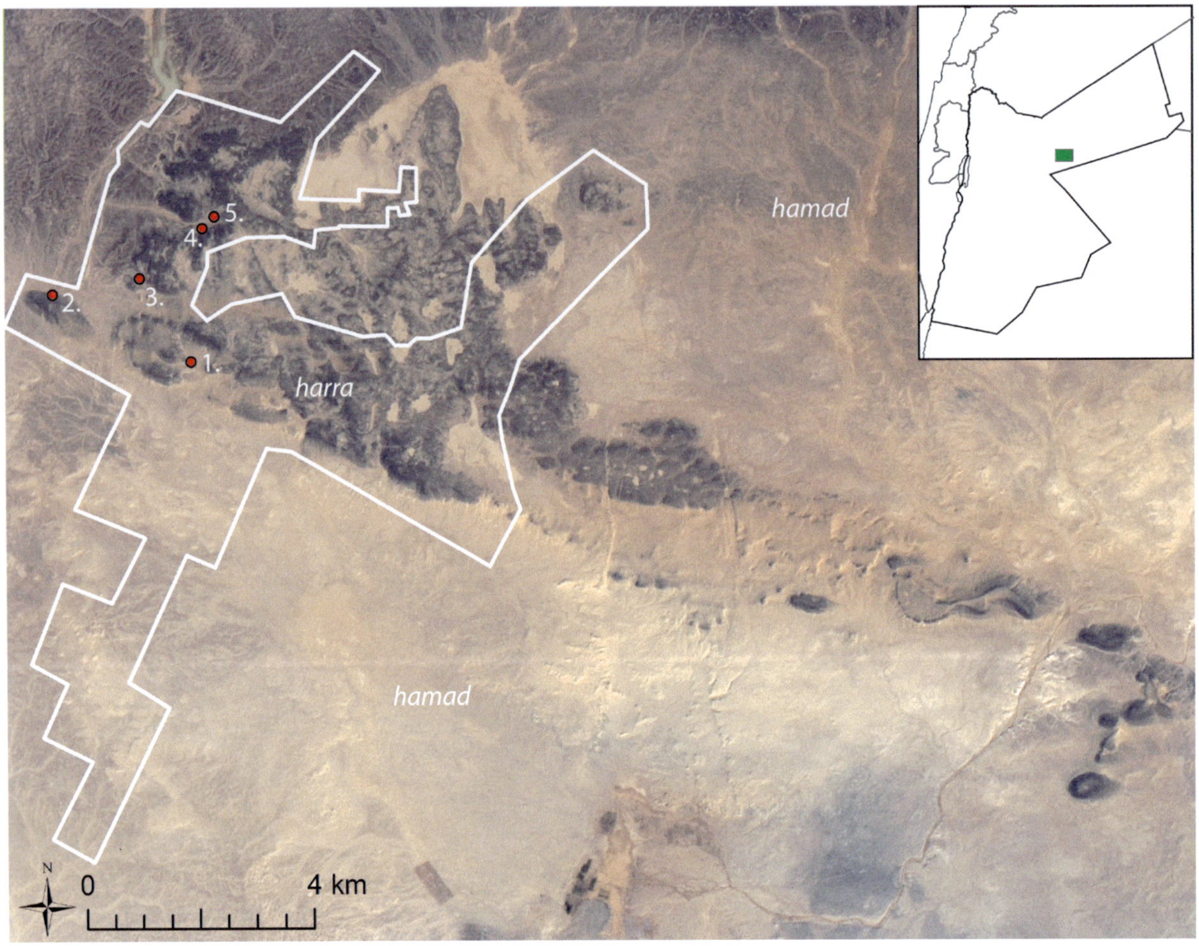

Figure 1. Map of the Jebel Qurma region indicating the surveyed area in white and the sites mentioned in the text in red: (1) QUR-11; (2) QUR-595; (3) QUR-373; (4) QUR-210; (5) QUR-735 (base image: Landsat 7, courtesy of the United States Geological Survey. Inset: location of the study area in Jordan).

Most of what is known about nomadic communities who inhabited the *badia* of north-eastern Jordan comes from Safaitic inscriptions, which are conventionally dated between the first century BC and the fourth century AD, and from ancient literary sources (Hoyland 2001). The archaeological remains of these communities, however, have been very poorly explored. Until recently, the archaeological study of historical-period nomadism in the *badia* was restricted to a handful of excavations of burial cairns (Clark 1981; Harding 1953; 1978) and the documentation of remains that were only incidentally encountered in archaeological projects that focused on the region's prehistoric past (Betts *et al.* 2013). This is unfortunate as relying solely on textual sources is problematic for a number of reasons. Firstly, the Safaitic inscriptions carved out by nomads are difficult to date accurately, are restricted in terms of content, and only provide information on those communities that were actually able to write. Secondly, Roman/Byzantine and Early Islamic literary sources may be heavily biased by their 'outsiders perspective' on nomadic communities (Hoyland 2001). There is therefore a need for a more directly informed perspective on such groups. This can be achieved by more systematic, problem-orientated archaeological research in the *badia* that focuses on nomadic communities of the historical periods.

This has been attempted by the Jebel Qurma Archaeological Landscape Project through annual, two month-long, field campaigns conducted since 2012 (cf. Akkermans and Huigens 2018). The resulting archaeological data sets are studied to explore developments in nomadic lifeways in the Jebel Qurma region during Classical and Late Antiquity (see Huigens 2019). In this paper, I will focus on, and discuss, a particular aspect of this research, namely the way in which camp sites of ancient nomads can be recognised in archaeological terms.

The Jebel Qurma region is situated *c.* 30 km east of the modern town and oasis of Azraq. The desert environment consists of both *harrah* and *hamad* landscapes (Fig. 1). It has witnessed a long yet discontinuous history of occupation characterised by periods of intensive use alternating with phases of apparent abandonment. After a long period of apparent desertion during the second, and possibly much of the first, millennia BC (but see Akkermans and Brüning, this volume), the region was frequented again by nomadic communities between the Hellenistic to Early Islamic periods. Although probably restricted to the Hellenistic and Roman periods, the many thousands of Safaitic inscriptions and petroglyphs from the region provide the first line of evidence for this (see Brusgaard 2019; Della Puppa, forthcoming). Additionally, there are numerous archaeological remains, including camp sites, burial cairns, and artefacts left behind on the ground surface that can be associated with nomads, dateable to between the Hellenistic and Early Islamic periods.

The study of camp sites of mobile peoples of the ancient Near East is not unproblematic. Although numerous studies have shown that camp sites can persist in the archaeological record, archaeological visibility remains a problem which is not easily overcome. Only when such camp sites consist of relatively durable installations, are they fairly easily detectable. This has been the case in, *e.g.*, the Negev, where round, permanent structures (although not necessarily permanently occupied) were part of Late Antique camp sites (Avni 1996; Rosen 2017). Ceramics may also be helpful in the identification of nomadic camp sites (Cribb 1991; Grillo 2014). However, archaeologists sometimes unwarrantedly assume that the use of pottery is common throughout society (Sanders 2016). There are indeed ethnographic examples of mobile pastoralists who used wooden or metal containers rather than pots (Cribb 1991, 72-73).

In the Black Desert, the study of historical-period camp sites is restricted to ethno-archaeological studies of recent Bedouin camps (Betts *et al.* 2013). A number of remote-sensing studies have suggested that stone-built corrals are related to camping activity (Kennedy 2011; 2012; Meister *et al.* 2019) but there is still no comprehensive understanding about the date of construction and the use of these structures due to a lack of rigorous ground-truthing. The morphology of residential units like tents or huts is also completely unknown, as is the nature of other forms of material culture (notably pottery) that can potentially be encountered at camp sites. This makes it difficult to assess the nature of nomadic occupation in the *badia*, even in rather basic terms. Issues such as chronology, mobility, group size and economic practices are difficult to assess when the areas where nomads were actually living remain unexplored. This problem is addressed in this paper by providing an archaeological insight into nomadic camp sites in the Jebel Qurma region from the Classical and Late Antique periods, *c.* 300 BC to 800 AD.

Methods

A number of archaeological correlates are explored here to identify camp sites. These include, firstly, the remains of residential spaces and associated domestic activities. Although tents or huts were probably made largely of perishable materials, their footings can be observed in archaeological contexts (*e.g.* Rosen 1993; Rosen and Avni 1997). Fireplaces, used for cooking, warmth and social activities, and domestic waste, such as broken pottery vessels or remains of other utensils, may also be retrieved. Importantly, the composition of these waste materials may also be used as an indicator for camping activities. For example, limited diversity in the ceramic corpus may be used as an indication for the presence of nomadic communities (Cribb 1991, 75-79). Secondly, the occupational duration of these residential areas is explored, using the degree of architectural investment as an indicator. Although permanent architecture is not necessarily indicative of permanent occupation (cf. Hammer 2014; Kent 1991; Seymour 2009) it seems warranted to regard limited architectural investment as an indicator for short-lived occupation (Binford 1990; Diehl 1992).

A number of field methods were applied for this research. High-resolution satellite imagery was studied in order to pinpoint potential camp sites on a large geographic scale, and intensive pedestrian surveys and excavations were carried out during the field campaigns. The satellite imagery used was 1.8 m resolution Ikonos imagery, covering an area of 172 km^2 of both *harrah* and *hamad* landscapes. The total pedestrian survey area measures *c.* 52 km^2 (Fig. 1). Most of this area was covered by using intensive survey methods. In the open *hamad* landscape a formal transect survey was carried out, complemented by an extensive survey (Huigens 2015), while an intensive yet less rigid survey method was applied in the undulating and rough *harrah* landscape. Excavations focusing on camp sites were carried out at several locations.

Post-fieldwork documentation and analyses included radiocarbon dating, ceramic studies, and spatial analyses carried out in a Geographic Information System (GIS). Radiocarbon dating was carried out by the Groningen Institute for Isotope studies. Ceramic analysis included the formal documentation of pottery sherds and subsequent comparative research based on published ceramic corpora from well-stratified contexts in the southern Levant (Huigens 2019; Vijgen 2019).

Results

Survey activities in the Jebel Qurma region documented a large number of different site types dating between the Hellenistic and Early Islamic periods. These site types were defined on the basis of different archaeological features, artefacts, and epigraphic remains. Rock-art sites, comprising Safaitic inscriptions and petroglyphs, occur widely. In total, about 10,000 inscriptions and petroglyphs have been documented (Brusgaard 2019; Della Puppa, forthcoming); they probably date to the Hellenistic and Roman periods. These carvings are confined almost entirely to the *harrah*, where they mostly occur on high places such as ridges and hill tops. Also from this period, and often spatially associated with the rock art, are a large number of monumental burial cairns (Akkermans and Brüning 2017).

In contrast, the low-lying areas of both the *harrah* and *hamad* landscapes host remains of an entirely different nature, including enclosures and clearings. Enclosures represent a well-known feature type of the Black Desert and are usually interpreted as the remains of ancient camp sites (*e.g.* Kennedy 2011; Meister *et al.* 2019). This has largely remained assumed rather than established. Clearings have remained an even more enigmatic feature type, and many remote sensing and survey projects have actually largely overlooked them (an exception is Kempe and Al-Malabeh 2010). The archaeological research on such features in the Jebel Qurma region sheds new light on the nature and chronology of these features, as presented below.

Enclosures

Numerous stone-built enclosures of different sizes and configurations have been documented through remote sensing and pedestrian surveys in the Jebel Qurma region. Many of these were already constructed in prehistoric times, given the dense lithic scatters with which they are often associated (cf. Akkermans *et al.* 2014; Huigens 2015), but they have been re-used in more recent times. The site of QUR-210 may serve as an example here. This site is situated on the lower slopes of a large valley in the *harrah* (Fig. 2). It had been occupied initially in prehistoric times, when at least some of the enclosures were constructed, given the dense lithic scatters in the enclosed spaces. The site was frequented again during the Hellenistic-Roman period, evidenced by a cluster of Safaitic inscriptions and petroglyphs situated at the centre of the site. One of these inscriptions mentions an enclosure.[1] Ceramics dating to the Late Byzantine and/or Early Islamic period were also collected at the site.

This pattern of re-use of prehistoric enclosures during the Classical and Late Antique periods is paralleled at numerous other sites in the Jebel Qurma region. The presence of domestic waste in the enclosures may indicate that they were not used exclusively for the corralling of herd animals. Instead, the enclosures may have been used, at least in part, as residential areas. In order to test this hypothesis, excavations were carried out at a number of enclosures.

One such enclosure was excavated at the site of QUR-595 – an extensive site comprising multiple enclosures and clearings. The excavated enclosure measured about 13 by 6 m and its walls were preserved up to a height of about 0.5 m (Fig. 3). Sediments had accumulated between the walls. The enclosure had been constructed within a prehistoric structure, although when exactly is uncertain. Numerous small fire pits were found within the enclosure, and radiocarbon dates from these indicate several use phases. The oldest dates were prehistoric, while a younger use phase is provided by two dates, one from the third-fourth century BC and another from the first century BC to the first century AD (Fig. 4). Additionally, a total of 65 ceramics were retrieved from the enclosure's fill. Those fragments that could be dated were from the Late Byzantine or Early Islamic periods.

Another enclosure was excavated at the site of QUR-373. This enclosure consisted of an inner compartment measuring about 7.5 m across and made of low stone walling, and an outer compartment with a diameter of about 18.5 m (Fig. 5). Prior to excavation, the enclosure was filled almost entirely by wind-blown sand deposits up to 0.65 m thick. The excavations revealed a stratigraphic sequence containing a total of 19 fire pits and one larger, ash-filled pit. The deepest and original surface within the enclosure was associated with large, elongated fire pits, which unfortunately could not be dated. However, large amounts of prehistoric chipped-stone artefacts (including many burins) associated with this surface suggest a Neolithic construction date. The sequence of fire pits covering the original surface gave a number of radiocarbon dates between the third and eighth centuries AD (Fig. 4). Also associated with this later phase of re-use were ceramics, including two nearly complete cooking pots, which can be dated on typological grounds to the seventh or early eighth century AD (Fig. 6, no. 9).

Lastly, an enclosure at the site of QUR-11 was excavated. This enclosure measures 22.6 m across and is subdivided into three compartments, enclosed by walls standing to a height of about 0.9 m (Fig. 7). The original construction date of the enclosure is uncertain, given the lack of materials associated with its foundation level, but the youngest occupational deposit in the largest compartment, situated just below the present-day surface, contained a number of small fire pits. Radiocarbon dates from these pits ranged between the seventh and ninth centuries AD (Fig. 4). The ceramics that were collected from the surface during survey activities probably date to broadly the same period (Fig. 6, nos. 4-5). Excavations in the second compartment did not yield such fire pits, while the third compartment remains unexcavated.

1 C. Della Puppa, pers. comm.

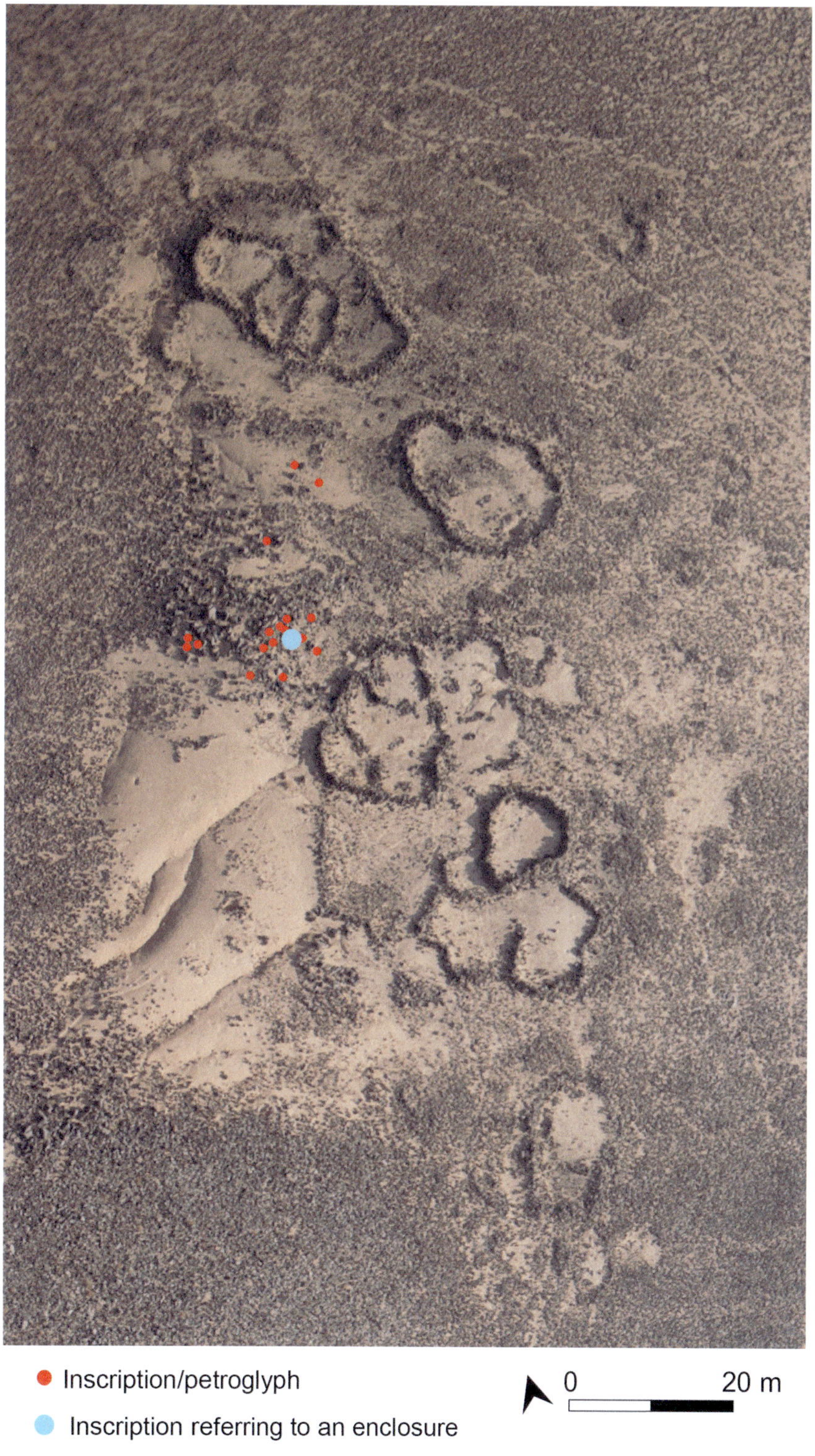

Figure 2. The site of QUR-210 in the Jebel Qurma region: a large multi-period site featuring several stone-built enclosures. One of the Safaitic inscriptions refers to an enclosure (orthorectified aerial photograph by D. Boyer, courtesy of APAAME).

Figure 3. Excavation plan of the enclosure at QUR-595, featuring a series of small fire pits within the enclosed space (drawing by M. Kriek, Jebel Qurma Project Archive).

Clearings

Survey activities in the *harrah* and *hamad* of Jebel Qurma identified a number of potential domestic areas that were defined not on the basis of architectural features but through the presence of ceramic scatters, generally in areas free of the natural rocky surface cover. Such clearings (Fig. 8) were mostly encountered on the bottom of large valleys on the edge of the *harrah* landscape. They vary considerably in size, and they may cover an area as large as one hectare. Clearance heaps were created either within the clearing itself or around its edges. The use of these clearings is not confined to

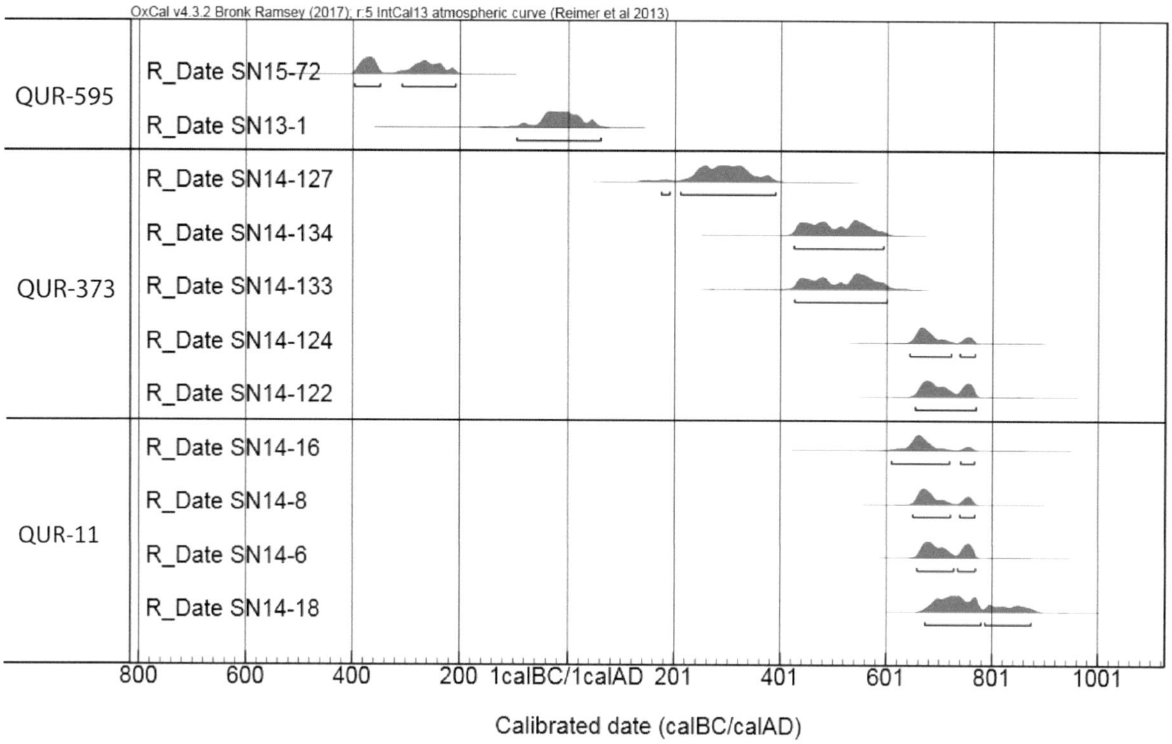

Figure 4. Radiocarbon dates from the excavated enclosures at QUR-595, QUR-373 and QUR-11 (image and calibration produced with OxCal v4.3.2; Bronk Ramsey 2009; Reimer *et al*. 2013).

a single period. Often, the remains of recent Bedouin camps, including tent pitches and associated waste, were encountered in combination with (much) older remains. This palimpsest hampers chronological differentiation between features, as recent activities may have changed the configuration of such sites.

A relevant example is the site of QUR-735, where the main feature is a large clearing situated on a valley floor. This clearing measures *c.* 120 by 90 m and is characterised by a seemingly irregular jumble of clearance heaps (Fig. 8). Prehistoric use of this area is attested by a modest lithic scatter that included a flint arrowhead. Evidence for re-occupation of the site comes from a relatively high density of pottery sherds, which were dated to the Late Byzantine period (Fig. 6, nos. 12-13). Associated outlines of domestic structures, however, were not observed. The latter is not surprising, not only because such structures may leave few architectural traces but also because the site may have been significantly modified by more recent re-use. This phase of occupation is indicated by the presence of modern Bedouin tent remains and associated waste. Similar situations were observed at many of the clearings documented through survey activities.

Ceramics

Many ceramics dating broadly to between the Hellenistic and Early Islamic (Umayyad/Abbasid) periods were collected in the enclosures and clearings (Fig. 6). Material from the preceding Iron Age and subsequent Fatimid periods is rare or absent. Most of the ceramics were highly fragmented. As a result, the number of sherds that could be dated on the basis of comparisons with published corpora is fairly low. A total of 98 diagnostic pottery fragments have been ascribed a Classical or Late Antique date so far (Huigens 2019; see also Vijgen 2019). In addition, the extensive fragmentation hindered attempts to provide a detailed assessment of the types of vessels represented at these sites. In general terms, closed forms, such as cooking pots and jars, are better represented that open forms, such as bowls (Fig. 9a). Only a single oil lamp has been identified. Particularly significant is the absence of high-quality ceramics typical of the period, such as Nabataean Painted Fine wares and *terra sigillata* wares. All of the dated ceramics are the remnants of mass-produced vessels that were brought into the Jebel Qurma region from elsewhere. Although handmade, basalt-tempered, coarse-ware sherds are also present, these are as yet often undateable (cf. Vijgen 2019). At this point, there is no reason to assume that the nomadic communities produced pots locally during this period.

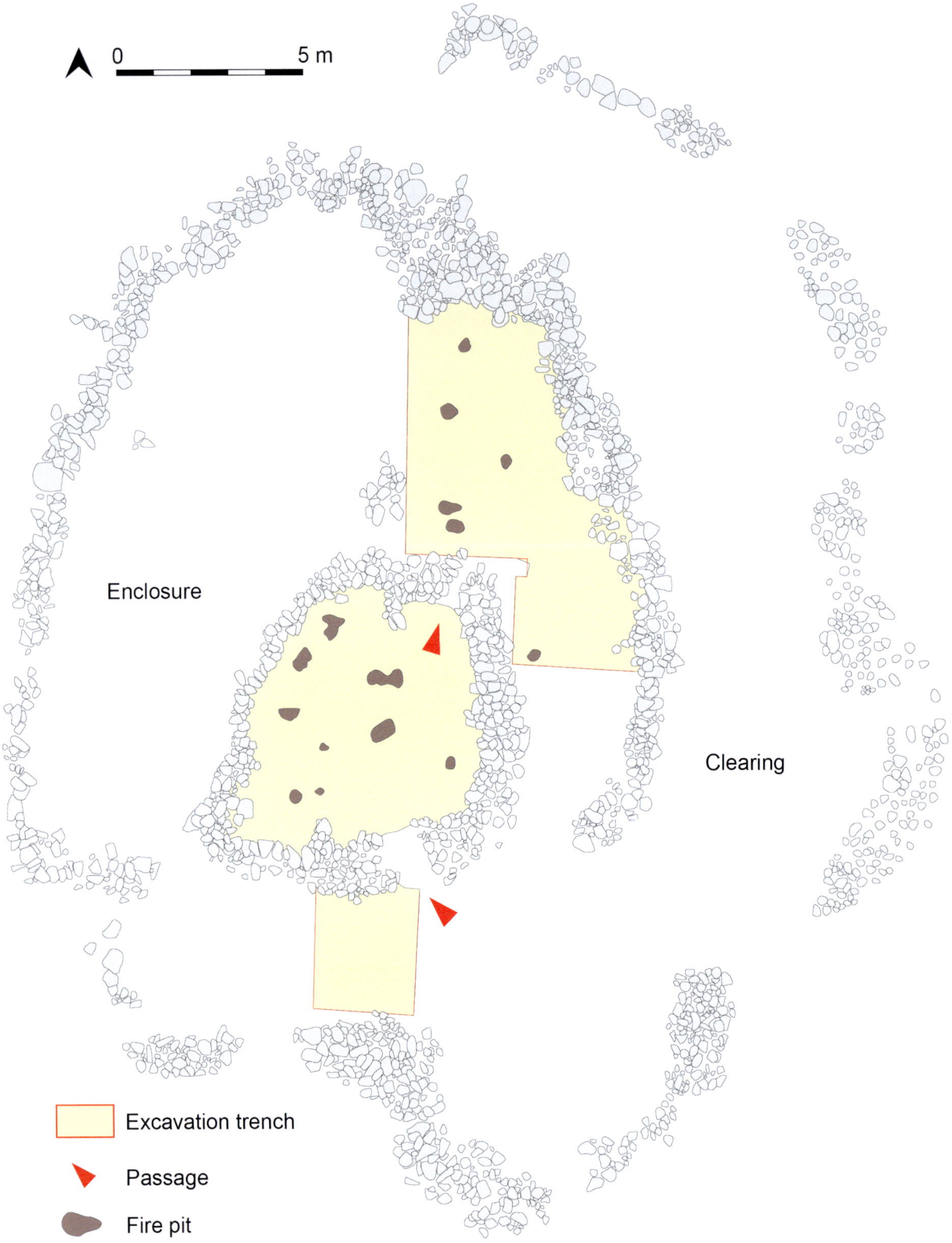

Figure 5. Excavation plan of the enclosure at QUR-373, with a selection of fire pits in the inner and outer compartments of the enclosure (drawing by A. Kaneda, Jebel Qurma Project Archive).

Figure 6 (right). Selection of Hellenistic to Early Islamic-period ceramics from camp sites in the Jebel Qurma region (drawings by A. Kaneda, Jebel Qurma Project Archive).

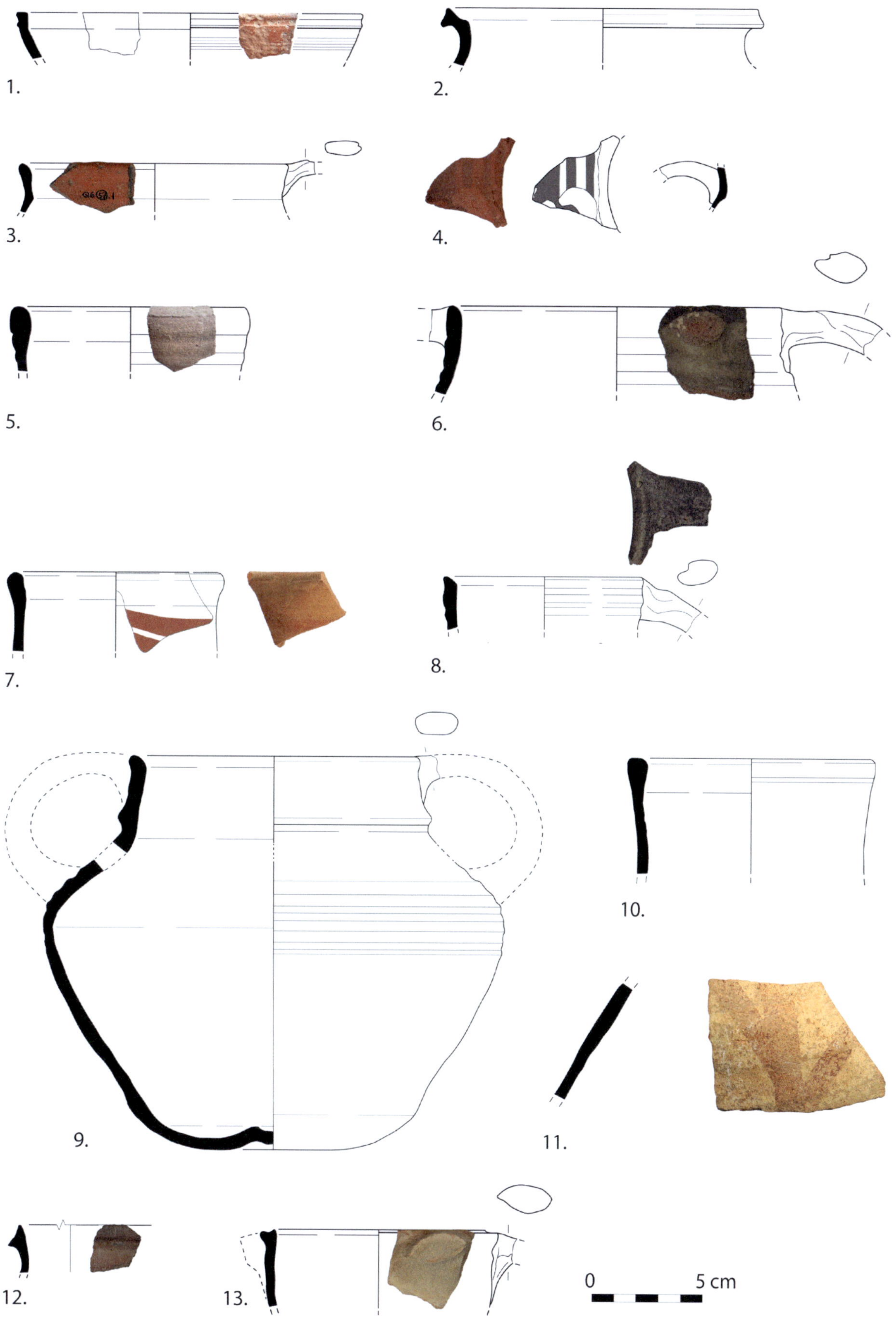
1.
2.
3.
4.
5.
6.
7.
8.
9.
10.
11.
12.
13.
0
5 cm

Figure 7. Aerial view of the excavation trenches in the enclosure at the site of QUR-11 (photograph: Jebel Qurma Project Archive).

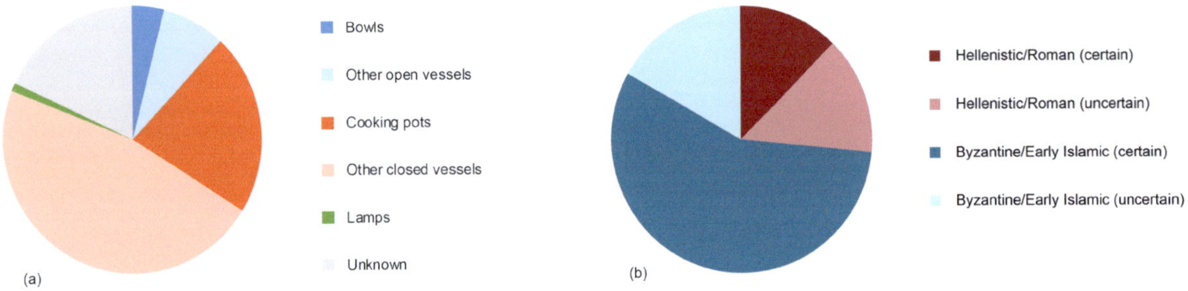

Figure 9. Proportion of (a) vessel types at Classical/Late Antique camp sites and (b) sherds per period found at camp sites.

The majority of dated ceramics belongs to the Byzantine and/or Early Islamic period (Fig. 9b). This is remarkable, since there is ample evidence for significant occupation during other periods, such as the Hellenistic-Roman period, in the form of Safaitic rock carvings and funerary monuments. Therefore, the ceramic trends do not necessarily reflect variation in occupational intensity but rather differences in the use of ceramics. This seems also to be reflected in the occurrence of pottery vessels in funerary contexts in the Jebel Qurma area. While there are many cairn burials from the period between the fourth/third century BC and the third century AD, none of these were accompanied by ceramic vessels (Akkermans and Brüning 2017). Pottery starts to appear in burials from the third century AD onwards (Huigens 2019). Containers made of metal, wood, or leather may have been the more common utensils among the inhabitants of the region during the Hellenistic-Roman

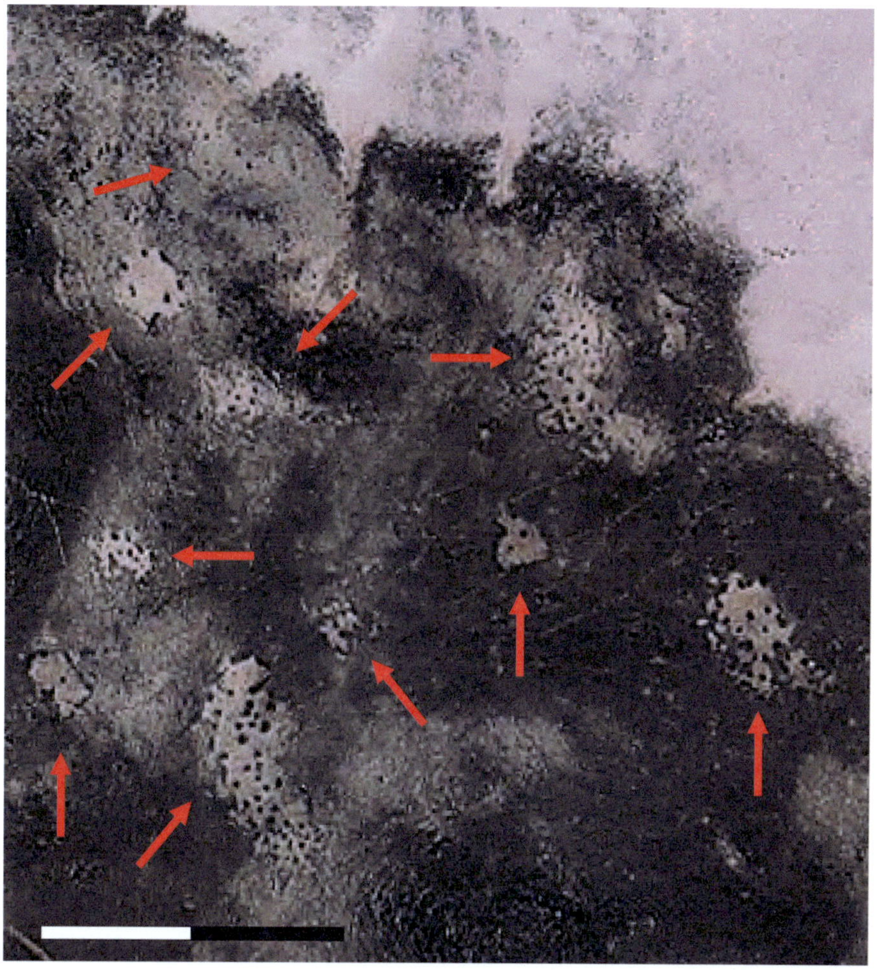

Figure 8. Clearings as visible on Ikonos satellite imagery (above), indicated by the arrows (scale: 100 m), and on the ground at the site of QUR-735 (below).

period, while the use of pottery vessels became more common during the Byzantine-Early Islamic period.

This observation has important implications for the identification and dating of potential camp sites based on surface finds, especially from the periods during which ceramics seem to be uncommon. The use of epigraphic information may to some degree be used as an alternative indicator (see above) but it should be acknowledged that camp sites, especially from the Hellenistic-Roman period, remain difficult to identify on the basis of surface finds alone.

Spatial distribution

Enclosures and clearings usually occur together within the valley systems that run down from the basalt-covered plateau in the centre of the Jebel Qurma region (Fig. 10). These are fairly secluded locations on the boundary between the *harrah* and *hamad* landscapes. Wadis typically run through such valleys, which may provide water in wet seasons. Numerous small paths usually run through these valleys as well. They were created as a natural result of people and animals travelling through the rocky terrain (Huigens 2018). These paths increase the accessibility of the basalt-covered upland, where more sustainable sources of water may be found in the form of mud flats on which surface water may be retained for weeks or even months after the occurrence of rainfall (Huigens 2019).

Although the valleys generally host clearings and enclosures alike, there are some differences between them in terms of their spatial distribution on a smaller scale. Fig. 11 shows a typical situation, in which all of the clearings are situated at the very bottom of a valley, sometimes even in the course of a wadi. However, the enclosures are situated further upslope, and the larger ones are shielded from prevailing westerly winds as they are positioned against natural flanks facing the east. Although situated within the same valley, the distance between the enclosures and clearings is often considerable, *i.e.* over 200 m.

Enclosures and clearings are not confined to large valleys. Sometimes, they are also concentrated around the base of isolated hills, again on the boundary between the *harrah* and *hamad*. In these cases, the enclosures are usually located against the eastern and north-eastern slopes of these hills, protected from westerly winds. Clearings may be situated in more exposed locations, as are the ceramic scatters identified in the Hazimah plains.

Discussion

Based on the above observations, it can be argued that both clearings and enclosures served domestic activities during the period under study. Both types of installations are associated with domestic waste in the form of pottery fragments. Also, the presence of fireplaces in enclosures suggests that these features were used, at least partially, as residential areas. Excavations have not yet been carried out within the clearings, which makes it impossible to assess whether such fireplaces were also present. Nevertheless, discard patterns documented in ethnoarchaeological studies of mobile, pastoralist camp sites show that domestic waste is usually discarded directly around residential units (*e.g.* Cribb 1991; Palmer *et al.* 2007; Simms 1988). It would therefore seem unwarranted to ascribe the ceramics found at clearings situated sometimes over 200 m away from enclosures as waste deriving from domestic activities carried out in said enclosures. Instead, it seems more reasonable that these areas were used, at least in part, for camping as well.

None of these domestic spaces show evidence for permanent occupation. There is hardly any durable architecture present at these sites that can be interpreted as dwellings, houses, and the like. Instead, the occupants of these areas must have been living in tents or huts made of perishable materials, such as wood, cloth, and hides. The characterisation of these area as 'camp sites' therefore seems warranted. The excavations in the enclosures show that at least some of the tents or huts must have been smaller than most Bedouin tents today, which can be as long as 15 m (cf. Simms 1988).

Direct archaeological evidence for pastoral activities is so far lacking. Hardly any faunal remains were retrieved during the excavations, probably due to poor preservation. Macro-remains of animal dung were also not encountered. Microscopic studies for dung remains have not yet been carried out. Therefore, evidence for pastoral production is thus far only circumstantial. If the camp sites can be related to the people who produced the Safaitic inscriptions and petroglyphs (which is possible on the basis of epigraphic evidence in a few cases), we may assume that their herd animals (camel, sheep, goat, perhaps cattle; Macdonald 1993) were also kept at these locations. The enclosures are well-suited for the keeping of animals, and some of the inscriptions indeed refer to these structures. The excavations indicate that not all of the enclosed spaces were necessarily used as residential areas, but that there may have been ample space left to pen animals. The enclosure walls could have provided protection for herd animals (especially the younger ones) against the elements, which is important during the wet and cold months of the year. The use of stone-built enclosures to provide shelter for humans and animals is widely attested ethnographically (Cribb 1991, 95-96). Furthermore, there is no evidence that points towards occupations of a different kind, such as trade caravans or military units. In fact, the ceramic finds from the Jebel Qurma region are largely in line with pottery assemblages described in (ethno-)archaeological studies of nomadic camp sites elsewhere, which are characterised by rather limited variation in vessel types and a scarcity of high-quality fine wares (in comparison with sedentary

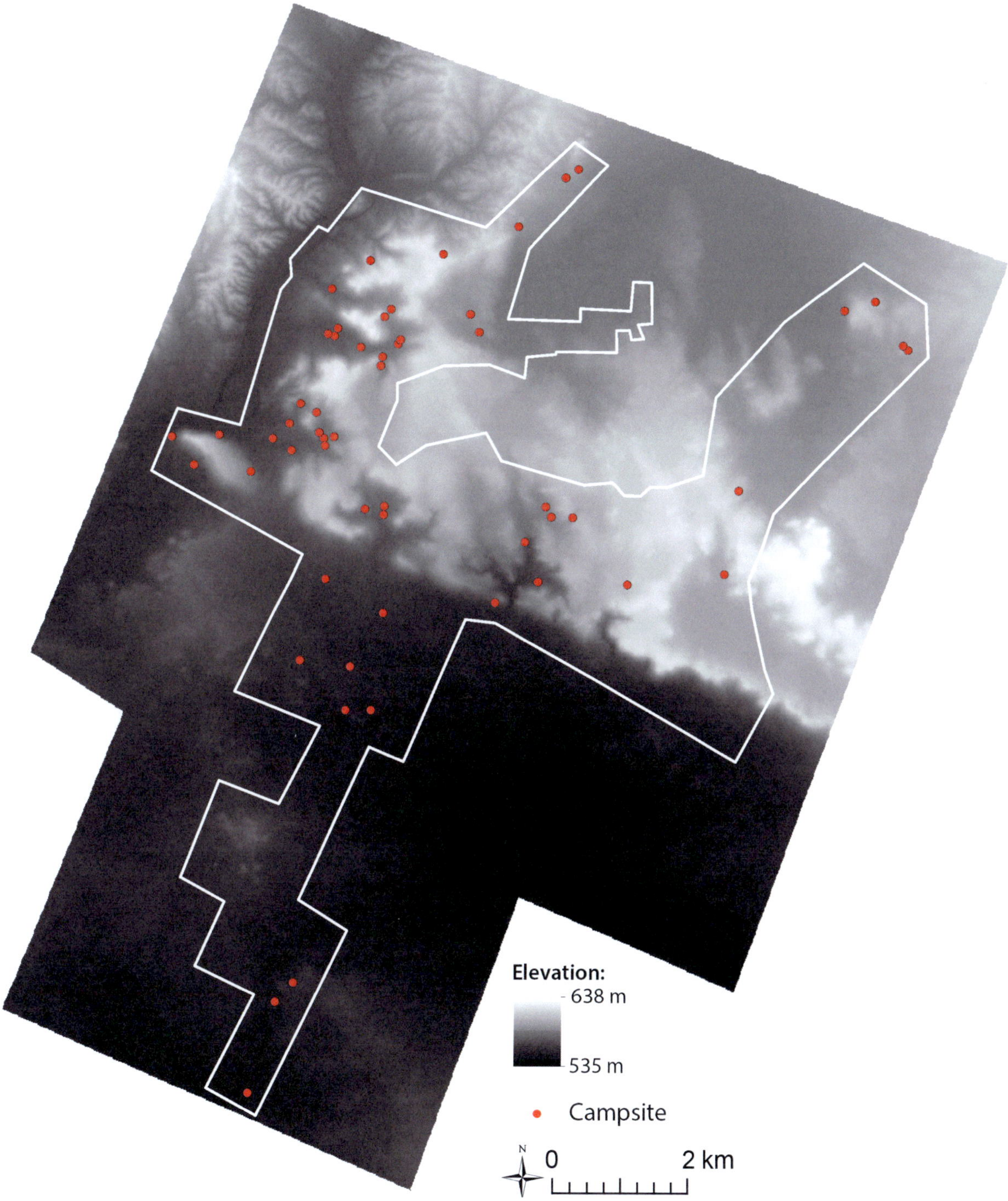

Figure 10. Distribution of Classical/Late Antique camp sites identified in the surveyed area (white) of the Jebel Qurma region (base image: WorldDEM digital elevation model).

contexts; *e.g.* Cribb 1991, 75-79; Rosen 1987; 1993; Rosen and Avni 1997). Given the considerations above, it seems reasonable to suggest that these camp sites were mainly occupied by nomads.

It is proposed here then, that between the Hellenistic and Early Islamic periods nomadic communities used both enclosures and clearings for camping and the keeping of herd animals. However, I also suggest that enclosures and

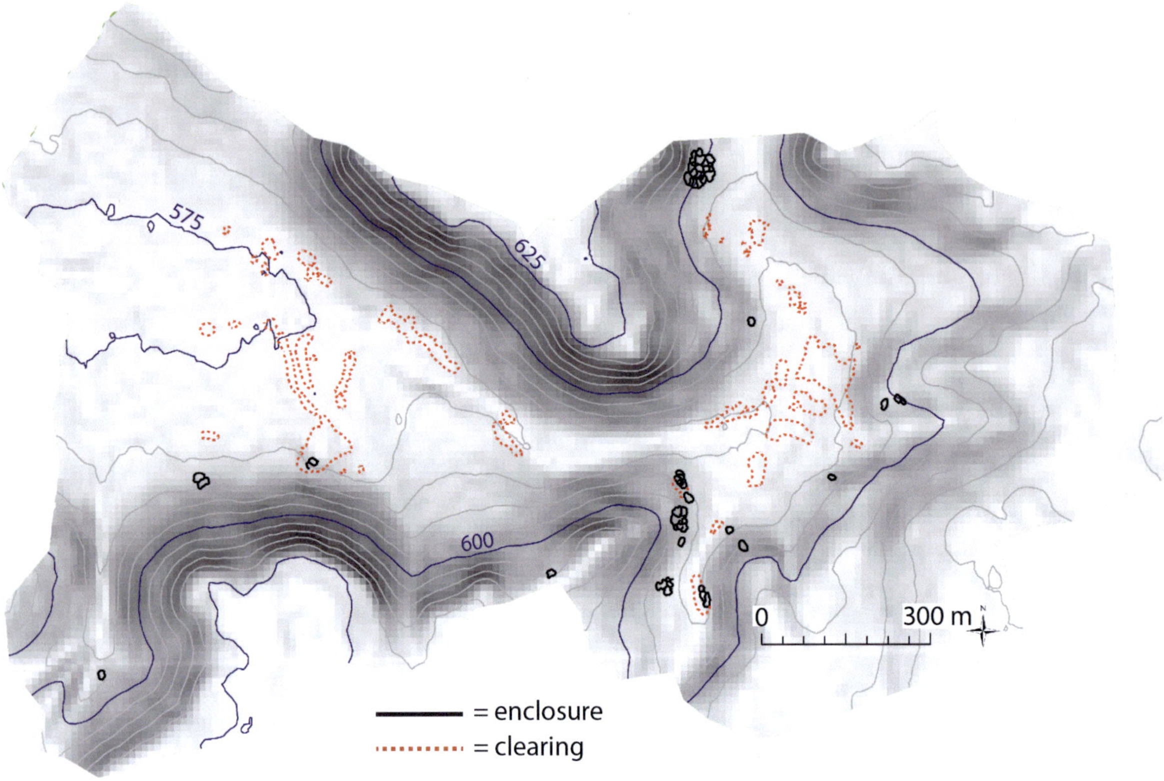

Figure 11. Distribution of enclosures and clearings in a valley in the Jebel Qurma region (base image: WorldDEM slope map with contour lines).

clearings were not necessarily occupied simultaneously. They sometimes differ considerably in terms of morphology and spatial distribution. Enclosures usually occur in rather secluded and slightly elevated locations, whereas clearings predominantly occur in more exposed areas and on the bottom of valleys, sometimes right on the edge of wadis. Enclosures and clearings that were occupied broadly during the same period are nonetheless often separated by several hundred metres, suggesting that they were not part of the same camping unit.

An explanation for this diversity in camp-site location and morphology may be found in the occupation of the Jebel Qurma region during different times of year. The hypothesis put forward here is that camp sites with enclosures were mainly used during the wet and cold winter months, while camp sites lacking enclosures (and mainly consisting of clearings) were occupied mainly during drier, warmer seasons. Diversification in terms of camp-site location and features related to different seasonal requirements is widely documented in ethnographic studies (Cribb 1991, 133-161). In relatively dry and warm periods, the valley floors may have provided easily accessible camping areas, where clearings, once created, could be re-occupied episodically. In wetter and colder conditions, however, people may have preferred using the enclosures, which may have provided protection against the elements for both people and animals, through the enclosure walls and their sheltered location. Also, the location of enclosures somewhat further upslope would keep the camps away from wadis that were prone to flooding. This compares well to nineteenth-century descriptions of Bedouin winter camps in the *harrah* (see *e.g.* von Oppenheim 1899, 219-220).

If this reconstruction is correct, it would imply that nomads frequented the Jebel Qurma region during different times of year, rather than during a single season only. On the basis of the Safaitic inscriptions, Macdonald (1992) has suggested that mobile pastoralists would normally only be present on the edges of the *harrah* at the beginning and at the end of the dry season, but the enclosures in the Jebel Qurma region may suggest that the region was frequented during the wet and cold winter season as well. This pattern is in line with the more recent use of the *badia* by mobile pastoralists, as documented by modern ethnographers. In recent times, some pastoralists preferred to reside in the *harrah* during winter and remained there after the rainy season, until the local natural resources were depleted. Others would spend the winter in the *hamad* and would only frequent the *harrah* at the beginning of the dry season, when surface water was still present, or at the end of the summer, awaiting the first winter rains (Lancaster and Lancaster 1999, 100-102; Musil 1928, 584; Rowe 1999).

A final relevant issue is the stark increase in pottery during the Byzantine and Early Islamic periods, as opposed the Hellenistic-Roman period. I have argued above that this should not necessarily be seen as an indicator of increased activity but as an indicator of increased pottery use among mobile communities. There is at this point no definite explanation for this trend but one possibility is that it reflects closer connections between the nomadic groups in the Jebel Qurma area and sedentary communities beyond the desert during the Byzantine and Early Islamic periods. These periods are characterised by an increasing encroachment of towns and villages onto the nomadic landscapes of the eastern *badia* (see Bartl, this volume), through which contacts between mobile and sedentary communities perhaps intensified.

Conclusion and outlook

The results of the fieldwork carried out in the Jebel Qurma region allow for the archaeological identification of camp sites from Classical and Late Antiquity. It has been argued in this paper that ceramic scatters found in clearings and within enclosures represent domestic waste of occupation. The fireplaces evident within enclosures are also indicative of domestic activities. The limited degree of architectural investment at these places further suggests short-lived occupation. Direct archaeological evidence for pastoral production at these sites is thus far not available but the use of these camp sites by mobile pastoralists is inferred on the basis of epigraphic evidence, the nature of the ceramic assemblages, and the consistent use of enclosures that may have been used partially to pen herd animals. I also hypothesised (based on significant differentiation in morphology and spatial distribution) that camp sites with enclosures were used as winter camps, while sites consisting only of clearings were used in the drier and warmer times of year.

Future research may be geared towards finding more direct archaeological evidence for pastoral production during the period of study. Excavations in the Jebel Qurma region yielded hardly any macroscopic remains of herd animals, perhaps due to poor preservation. However, a number of methods may be applied in future research to find remains of herd animals on the microscopic or chemical level (see Vos, this volume). In addition, the proposed differentiation between winter and summer camps requires further scrutiny, which may be achieved by the study of plant remains from the fire pits in the enclosures. To this end, study of charred plant remains from the enclosures in the Jebel Qurma region is currently underway.

The investigation of archaeological landscapes of the *badia* is necessary to obtain a more comprehensive understanding of nomadic communities who inhabited this region during Classical and Late Antiquity. Camp sites form an important part of these landscapes, as they were the areas of residence and various kinds of economic and social activities. This study has attempted to address the nature of these areas, and may be expanded upon in future research. However, these landscapes not only include camp sites but also rock art, funerary monuments, and other installations. Future research may therefore focus on integrating these different elements in order to come to a more comprehensive understanding of the nature and development of these landscapes and their nomadic inhabitants.

Acknowledgements

This research was carried out within the framework of the Landscapes of Survival project, which was funded by the Netherlands Organisation for Scientific Research (NWO; project number 360-63-100) and directed by Prof. Peter Akkermans. Fieldwork in the Jebel Qurma region (as part of the Jebel Qurma Archaeological Landscape project) was carried out under the auspices of the Department of Antiquities of Jordan. Processed Ikonos satellite imagery was kindly provided by the Jordan Oil Shale Company. Aerial photographs were kindly provided by the Aerial Photographic Archive for Archaeology in the Middle East (APAAME). The editor and Kieran Westley are thanked for proofreading this paper.

References

Akkermans, P.M.M.G. and Brüning, M.L. 2017. Nothing but cold ashes? The cairn burials of Jebel Qurma, north-eastern Jordan. *Near Eastern Archaeology* 80, 132-139.

Akkermans, P.M.M.G. and Huigens, H.O. 2018. Long-term settlement trends in Jordan's north-eastern badia: the Jabal Qurma Archaeological Landscape Project. *Annual of the Department of Antiquities of Jordan* 59, 503-515.

Akkermans, P.M.M.G., Huigens, H.O. and Brüning, M.L. 2014. A landscape of preservation: late prehistoric settlement and sequence in the Jebel Qurma region, north-eastern Jordan. *Levant* 46, 186-205.

Avni, G. 1996. *Nomads, farmers, and town-dwellers: pastoralist-sedentist interaction in the Negev highlands, sixth-eight centuries CE*. Jerusalem: Israel Antiquities Authority.

Barfield, T.J. 1993. *The nomadic alternative*. Englewood Cliffs: Prentice Hall.

Betts, A.V.G., Cropper, D., Martin, L. and McCartney, C. 2013. *The later prehistory of the badia: excavations and surveys in eastern Jordan*. Oxford: Oxbow Books.

Binford, L.R. 1990. Mobility, housing, and environment: a comparative study. *Journal of Anthropological Research* 46, 119-152.

Bronk Ramsey, C. 2009. Bayesian analysis of radiocarbon dates. *Radiocarbon* 51, 337-360.

Brusgaard, N.Ø. 2019. *Carving Interactions. Rock art in the nomadic landscape of the Black Desert, north-eastern Jordan.* Oxford: Archaeopress.

Clark, V. 1981. Archaeological investigations at two burial cairns in the harra region of Jordan. *Annual of the Department of Antiquities of Jordan* 25, 235-265.

Cribb, R. 1991. *Nomads in archaeology.* Cambridge: Cambridge University Press.

Della Puppa, C. Forthcoming. *The Safaitic scripts: a study based on the Jebel Qurma corpus and other editions.* Leiden: Leiden University.

Diehl, M.W. 1992. Architecture as a material correlate of mobility strategies: some implications for archaeological interpretation. *Behavior Science Research* 26, 1-35.

Donner, F.M. 1989. The role of nomads in the Near East in Late Antiquity (400-800 C.E.). In: Clover, F.M. and Humphreys, R.S. (eds.). *Tradition and innovation in Late Antiquity.* Madison, WI: The University of Wisconsin Press, 73-85.

Grillo, K.M. 2014. Pastoralism and pottery use: an ethnoarchaeological study in Samburu, Kenya. *African Archaeological Review* 31, 105-130.

Hammer, E. 2014. Local landscape organization of mobile pastoralists in southeastern Turkey. *Journal of Anthropological Archaeology* 35, 269-288.

Harding, G.L. 1953. The Cairn of Hani'. *Annual of the Department of Antiquities of Jordan* 2, 8-56.

Harding, G.L. 1978. The Cairn of Sa'd. In: Moorey, R. and Parr, P. (eds.). *Archaeology in the Levant: essays for Kathleen Kenyon.* Warminster: Aris and Phillips, 243-249.

Hoyland, R.G. 2001. *Arabia and the Arabs: from the Bronze Age to the coming of Islam.* London and New York: Routledge.

Huigens, H.O. 2015. Preliminary report on a survey in the Hazimah plains: a hamad landscape in north-eastern Jordan. *Palestine Exploration Quarterly* 147, 180-194.

Huigens, H.O. 2018. The identification of pathways on *harra* surfaces in north-eastern Jordan and their relation to ancient human mobility. *Journal of Arid Environments* 155, 73-78.

Huigens, H.O. 2019. *Mobile peoples – permanent places. Nomadic landscapes and stone architecture from the Hellenistic to Early Islamic periods in north-eastern Jordan.* Oxford: Archaeopress.

Kempe, S. and Al-Malabeh, A. 2010. Hunting kites ('desert kites') and associated structures along the eastern rim of the Jordanian harrat. A geo-archaeological Google Earth images survey. *Zeitschrift für Orient-Archäologie* 3, 46-86.

Kennedy, D.L. 2011. The "Works of the Old Men" in Arabia: remote sensing in interior Arabia. *Journal of Archaeological Science* 38, 3185-3203.

Kennedy, D.L. 2012. Wheels in the Harret al-Shaam. *Palestine Exploration Quarterly* 144, 77-81.

Kent, S. 1991. The relationship between mobility strategies and site structure. In: Kroll, E.M. and Price, T.D. (eds.). *The interpretation of archaeological spatial patterning.* New York and London: Plenum Press, 33-59.

Lancaster, W. and Lancaster, F. 1999. *People, land and water in the Arab Middle East. Environments and landscapes in the Bilâd ash-Shâm.* Amsterdam: Harwood Academic Publishers.

Macdonald, M.C.A. 1992. The seasons and transhumance in the Safaitic inscriptions. *Journal of the Royal Asiatic Society* 2, 1-11.

Macdonald, M.C.A. 1993. Nomads and the Hawran in the Late Hellenistic and Roman periods: a reassessment of the epigraphic evidence. *Syria* 70, 303-403.

Makarewicz, C.A. 2013. A pastoralist manifesto: breaking stereotypes and re-conceptualizing pastoralism in the Near Eastern Neolithic. *Levant* 45, 159-174.

Meister, J., Knitter, D., Krause, J., Müller-Neuhof, B. and Schütt, B. 2019. A pastoral landscape for millennia: investigating pastoral mobility in northeastern Jordan using quantitative spatial analyses. *Quaternary International* 501, 364-378.

Millar, F. 1993. *The Roman Near East: 31 BC – AD 337.* Cambridge, MA: Harvard University Press.

Musil, A. 1928. *The manners and customs of the Rwala Bedouins.* New York: American Geographic Society.

Oppenheim, M. von, 1899. *Vom Mittelmeer zum Persischen Golf. Durch den Hauran, die Syrische Wüste und Mesopotamien.* Band 1. Berlin: Dietrich Reimer.

Palmer, C., Smith, H. and Daly, P. 2007. Ethnoarchaeology. In: Baker, G., Gilbertson, D. and Mattingly, D. (eds.). *Archaeology and desertification. The Wadi Faynan landscape survey, southern Jordan.* Oxford: Oxbow Books, 369-395.

Parker, S.T. 1986. *Romans and Saracens: a history of the Arabian frontier.* Winona Lake, IN: Eisenbrauns.

Peters, F.E. 1978. Romans and Bedouin in southern Syria. *Journal of Near Eastern Studies* 37, 315-326.

Porter, A. 2012. *Mobile pastoralism and the formation of Near Eastern civilizations: weaving together society.* Cambridge: Cambridge University Press.

Reimer, P.J., Bard., E., Bayliss, A., Beck, J.W., Blackwell, P.G., Bronk Ramsey, C., Buck, C.E., Cheng, H., Edwards, R.L., Friedrich, M., Grootes, P.M., Guilderson, T.P., Haflidason, H., Hajdas, I., Hatté, C., Heaton, T.J., Hoffmann, D.L., Hogg, A.G., Hughen, K.A., Kaiser, K.F., Kromer, B., Manning, S.W., Niu, M., Reimer, R.W., Richards, D.A., Scott, E.M., Southon, J.R., Staff, R.A., Turney, C.S.M. and Van der Plicht, J. 2013. IntCal13 and Marine13 radiocarbon age calibration curves 0-50,000 years cal BP. *Radiocarbon* 55, 1869-1887.

Rosen, S.A., 1987. Byzantine nomadism in the Negev: results from the emergency survey. *Journal of Field Archaeology* 14: 29-42.

Rosen, S.A. 1993. A Roman-period pastoral tent camp in the Negev, Israel. *Journal of Field Archaeology* 20, 441-451.

Rosen, S.A. 2017. *Revolutions in the desert: the rise of mobile pastoralism in the southern Levant.* New York and London: Routledge.

Rosen, S.A. and Avni, G. 1997. *The 'Oded sites. Investigations of two Early Islamic pastoral camps south of the Ramon Crater*. Beer-Sheva: Ben-Gurion University of the Negev Press.

Rowe, A.G. 1999. The exploitation of an arid landscape by a pastoral society: the contemporary eastern Badia of Jordan. *Applied Geography* 19, 345-361.

Sanders, G. 2016. *Recent finds from ancient Corinth: how little things make big differences*. Leiden: The BABESH Foundation.

Sartre, M. 2005. *The Middle East under Rome*. Cambridge, MA: Harvard University Press.

Seymour, D. 2009. Distinctive places, suitable spaces: conceptualizing mobile group occupational duration and landscape use. *International Journal of Historical Archaeology* 13, 255-281.

Simms, S.R. 1988. The archaeological structure of a Bedouin camp. *Journal of Archaeological Science* 15, 197-211.

Szuchman, J. 2009. Integrating approaches to nomads, tribes, and the state in the ancient Near East. In: Szuchman, J. (ed.). *Nomads, tribes, and the state in the ancient Near East: cross-disciplinary perspectives*. Chicago: The Oriental Institute of the University of Chicago, 1-13.

Vijgen, T. 2019. *Desert pots: studying the technology, morphology, date and distribution of the pottery of the Jebel Qurma region, north-eastern Jordan, from the Bronze Age up until the present*. Leiden: Leiden University (unpublished Research-MA thesis).

The Nabataeans as travellers between the desert and the sown

Will M. Kennedy

Abstract

While it may be misconceived that the Nabataeans followed a linear development from a 'primitive', non-sedentary nomadic origin to a more culturally enriched and sedentary lifestyle, recent historical and material evidence challenges this rather simplistic view of cultural evolution. Archaeological research clearly suggests that Nabataean culture was particularly complex and distinctly characterised by an amalgamation of both Hellenised and Oriental, mobile and sedentary, material culture. The evidence, however, is mostly limited within *urban* boundaries, particularly those of Petra. The question therefore arises whether the archaeological evidence in Petra's rural hinterland allows similar assumptions. Numerous surveys have been carried out in the immediate environment of the Nabataean capital, documenting various archaeological sites ranging from the Iron Age to the Byzantine and Early Islamic periods. Previous works have also discussed rural settlements and land use strategies in the Petra region, laying the focus outside Petra's city limits. Although these studies are immensely important for the understanding of rural Petra, they strongly focus on *sedentary* rural settlements, thus falling short of an overall and in-depth contextualisation of the various other archaeological sites and features documented in the Petra area. This paper deals specifically with more ephemeral and elusive archaeological evidence presumably pertaining to pastoral subsistence strategies in Petra's rural environs and to a more mobile lifestyle of its inhabitants.

Keywords: Petra hinterland, Nabataeans, pastoralism, mobile lifestyle

Introduction

In his comprehensive paper, 'The Nabataeans: travellers between lifestyles', S.G. Schmid (2001) discussed the cultural complexity of the Nabataeans – a once nomadic tribe that settled in and around their future capital Petra as early as the fourth century BC. By presenting various categories of archaeological evidence, Schmid characterises Nabataean material culture and art by the constant 'back-and-forths' between Hellenised and Oriental, mobile and sedentary, traditions and influences. In addition to monumental architecture built in Petra's city centre from the first century BC onward, one particularly illuminating example for the unique nature of Nabataean material culture are the numerous tomb complexes distributed throughout the urban limits of Petra (cf. Schmid 2009; 2012; Wadeson 2011; Petrovszky 2013a; 2013b). Not only do such tomb complexes feature the monumental façade tombs in both Graeco-Roman as well as Oriental fashion, but they also incorporate (sometimes monumental) ritual banqueting installations such as *triclinia* and *biclinia* as well *stibadia.* As such installations are all

In: Peter M. M. G. Akkermans (ed.) 2020: *Landscapes of Survival - The Archaeology and Epigraphy of Jordan's North-Eastern Desert and Beyond,* Sidestone Press (Leiden), pp. 235-254.

linked with the actual tomb by a central courtyard, the conceptual design of Nabataean tomb complexes strongly resembles contemporary Graeco-Roman profane luxury architecture. The complexes are thus considered as the 'houses of the dead' frequently visited by close family, clan or perhaps even tribe members. Clearly, these complexes were reserved for and visited by a very specific and selective group of people. In a later contribution, this has led Schmid to introduce the Foucauldian term *heterotopia* for describing the nature of Nabataean tomb complexes, as these are clearly "closed spaces, where only restricted and well-defined people or groups of people are granted access." (Schmid 2013, 251).

Among other examples of Nabataean *heterotopiai*, Schmid further lists the tribal sanctuary known as the Obodas Chapel, situated along the southern outskirts of Petra. This family or clan-based sanctuary, which dates as early as the second century BC (*e.g.* Tholbecq and Durand 2005; 2013; Tholbecq *et al.* 2008), is also characterised by both rock-cut and freely built *triclinia*, which further highlight the *social* significance of ritual banqueting within Nabataean culture.

Moreover, L. Nehmé's comprehensive analysis of the epigraphic evidence from the urban limits of Petra has resulted in the identification of different social groups that collectively commemorated a specific deity and which were organised within spatially distinct 'districts' within the city (Nehmé 2013). These social groups are also mostly associated with *triclinia*, once more emphasising the social importance of ritual banqueting, particularly for Nabataean fraternal cultic associations known as *marzeah* (Healey 2001, 166-167; Kühn 2005, 75; Wenning 2007, 257; Nehmé 2013; Charloux *et al.* 2016, 14).

Clearly, such already well-explored Nabataean *heterotopiai* from urban Petra bear witness to the distinctly tribal-based social structure of the Nabataeans, that seemingly stems from their nomadic past and continued well after their 'sedentarisation' in and around Petra from the first century BC onward.

While the nomadic 'origin story' of the Nabataeans continues to be subject of scholarly debate (*e.g.* Graf 1990; Macdonald 1991; Wenning 2013), it is generally agreed that the first century BC accounts of Diodorus Siculus (*Diod. Sic.* 19, 94, 1 and 95, 1- 97, 6, in: Fisher 1906) are correct when he describes the early (*i.e.* fourth century BC) Nabataeans as a (semi-)nomadic, pastorally organised people without permanent dwellings (*e.g.* Schmid 2001, 367-371; 2008, 360-361; Hackl *et al.* 2003, 59-61). Indeed, due to the ephemeral nature of early Nabataean material culture, this phase remains extremely elusive in the archaeological record (*e.g.* Graf 2013 with further references) and it is thus only natural that archaeological research on the Nabataeans deals predominantly with traceable material culture (architecture, art, ceramics, coinage, *etc.*) that appears from the first century BC onward. While the continuing nomadic social structure and tribal traditions of later Nabataean culture have been widely acknowledged (as demonstrated by the presented *heterotopiai* above), archaeological research has so much focussed on the sedentary and monumental remains of the Nabataeans, that more 'mobile' aspects of Nabataean culture are largely overshadowed in the scholarly discourse – mainly due to the lack of convincing archaeological evidence.

A similar pattern can also be observed in a recent surge of archaeological studies in Petra's rural hinterland (Kouki 2012; Ladurner 2015; Wenner 2015; Wadeson and Abudanh 2016; Knodell *et al.* 2017). The majority of these studies feature a strong research focus on the development of sedentary rural settlements (*i.e.* villages, hamlets and farms, *etc.*) and the spread of agriculture that occurred in Petra's hinterland simultaneously with the monumentalisation of the city and the assumed sedentarisation of the Nabataeans in the course of the first century BC (Kouki 2012, 84-94; W.M. Kennedy 2018). Particularly considering the archaeological evidence from the first century BC to the second century AD, the dramatic increase of rural settlements and agricultural installations in the Petra hinterland suggests a clear sedentary and agricultural turn, which undoubtedly formed *a* – but arguably not *the* – fundamental socio-economic core of the Petraean hinterland in Nabataean times.

However, a recent re-evaluation of a large archaeological data set derived from various surveys[1] conducted in the Petra region (W.M. Kennedy 2018) has identified a significant number of sites that may be considered as archaeological evidence for the practice of

1 In total, nearly 1800 sites recorded by fourteen survey projects were re-assessed, including agricultural installations, water structures and rural settlements, the regional communication network, industrial sites, funerary and religious structures as well military sites ranging from the Iron Age to the Late Byzantine periods. The re-evaluated archaeological survey data derives from the Edom Survey (Hart 1987), the Beidha Ethnoarchaeological Survey (Banning and Köhler-Rollefson 1983), the Southeast Araba Archaeological Survey (A.M. Smith 2010), the Jabal Shara Survey (Tholbecq 2001; 2013), the Archaeological Survey of the Wadi Musa Water Supply and Wastewater Project ('Amr *et al.* 1998; 'Amr and al-Momani 2001), the Bir Madkhur Project (A.M. Smith 2010), Abudanh's survey of the Udruh region (Abudanh 2006), the Finnish Jabal Harun Project (Kouki and Lavento 2013), the Ayl to Ras an-Naqab Archaeological Survey (MacDonald *et al.* 2012), the Showbak-Dana L2HE Survey (N.G. Smith 2009), the Shammakh to Ayl Archaeological Survey (MacDonald *et al.* 2016), the Petra Area and Wadi Slaysil Survey as well as the Petra Routes Project (Berenfeld *et al.* 2016; Knodell *et al.* 2017), the Petra Hinterland Tombs Project (Wadeson and Abudanh 2016), as well as the Petra Hinterland Survey Project conducted for the author's doctoral research (W.M. Kennedy 2018).

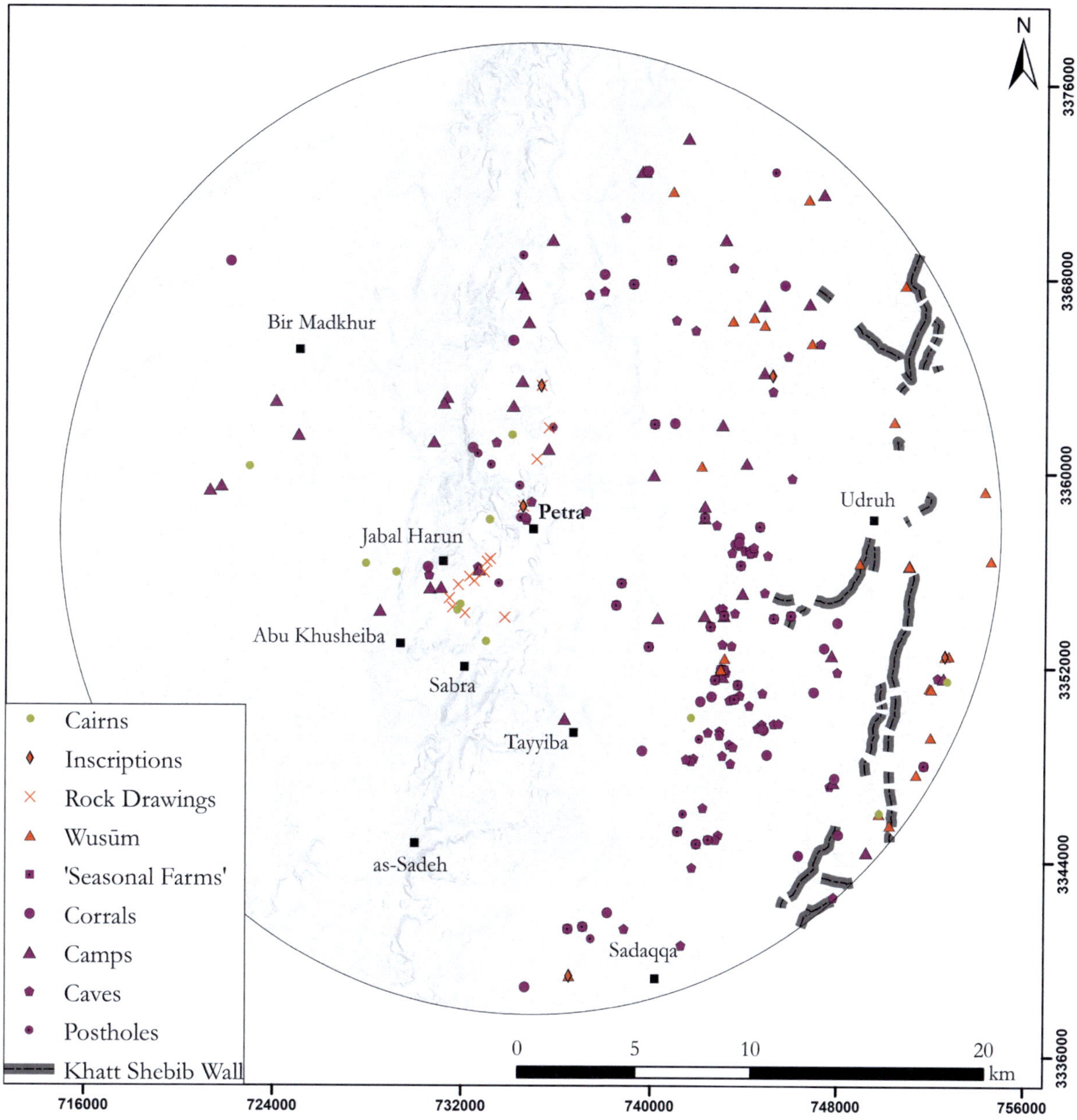

Figure 1. Distribution map of recorded sites pertaining to either pastoral subsistence strategies or to a more mobile way of life in the Petra hinterland (map: W.M. Kennedy; course of the Khatt Shebib after D.L. Kennedy and Banks 2015).

pastoral subsistence strategies in the Petra hinterland[2] as well as evidence pertaining to more general aspects of mobility within Nabataean culture. While it is not possible to discuss the available evidence in full detail, the aim of this paper is nevertheless to emphasise that pastoralism remained a widely practiced subsistence strategy in the Petra hinterland and that a distinctly mobile lifestyle continued despite the aforementioned 'sedentary and agricultural turn' from the first century BC onward.

While this may seem obvious, this particular aspect of Nabataean culture has found little scholarly attention when discussing Petra and its rural environs (*e.g.* Kouki 2012, 98-100). S.A. Rosen's assessment of camp sites in the Negev desert remains the most important work for considering

2 P. Kouki's definition of the Petra hinterland was adopted, which is understood as a 20 km radius around the city (Kouki 2012, 17). This is based on similar assertions concerning the extent of a 'Greater Petra' expressed by M. Lindner (1992, 266) and on the fact that the sixth century AD Petra Papyri list Udruh (Augustopolis) and Saddaqa (Zadacathon) – both situated *c.* 20 km away from Petra – as still being under the city's jurisdiction in the Byzantine period.

the Nabataeans also as mobile pastoralists (most notably Rosen 2007, but also see Rosen 2009; 2017; Rosen and Saidel 2010). The 'non-sedentary' aspects of Nabataean culture have so far not yet been discussed for the immediate Petra area. Indeed, the clear emphasis on 'sedentary' Nabataean material culture in archaeological research may lead to a misleading differentiation between 'the desert and the sown' as famously phrased by Gertrude Bell in the early twentieth century (Bell 1907). As Lenzen (2003, 5-6) points out, the phrase has essentially developed into a concept of distinction between two entities: the 'desert' quickly considered as the vast, uncultivated landscape of the nomad, and the 'sown' being understood as the inhabited space of sedentary peoples. This phrasing implies hidden prejudgments against desert cultures, equating the 'sown' with more developed forms of 'civilisation', and the 'desert' with the 'uncivilised' or primitive world of the roaming nomad. Such distinctions were further emphasised throughout the historiography of nineteenth and early twentieth century Western explorers in the Near East who considered large cities such as Jerusalem or Damascus as central places to gather information, supplies and other necessities, before taking leave into the vast desert landscape and its inhabitants – the Bedouin nomads (*ibid.*, 6). While the contemporary perceptions of the sedentary populations of Arabia were generally negatively connoted, in favour of the 'noble savage' Bedouins of the desert, the supposed cultural distinction between the sedentary and non-sedentary populations was nevertheless accentuated. This understanding impacted archaeological research in the Near East well into the 1980s (*ibid.*, 7).

In order to emphasise that such misconceptions do not apply to Nabataean culture in Petra and its rural environs, this paper aims to demonstrate the existence of a beneficial relationship between the 'desert' and the 'sown' and argues for a constant interplay between the two. The following therefore presents the relevant archaeological evidence derived from the original surveys conducted in the Petra hinterland that point to a more mobile and pastoral way of life in the study area.

The archaeological evidence

The archaeological evidence discussed in this paper includes: conspicuous find clusters (*i.e.* particularly large concentrations of surface pottery), natural and/or rock-cut structures, rock drawings and tribal markings (*wusūm*), short commemorative inscriptions, camp sites, corrals, burial cairns as well as the Khatt Shebib wall (Fig. 1). This evidence is undoubtedly highly debateable, by nature problematic, and suggestive rather than conclusive, and must therefore be considered with caution. It will nevertheless become clear that rural Petra was, at least in part, characterised by a mobile and pastoral way of life in the Nabataean period and beyond.

Find clusters

The various surveys have documented a total amount of 71 'find clusters' that date to the Iron Age as well as from the first century BC until the seventh century AD (Fig. 2). These were further subdivided into 'architectural fragments', 'lithics', 'pottery concentrations' and 'other'. As shown in Fig. 1, the documented find clusters clearly concentrate in the extended Jebel Harun area, around Wadi Musa and to some extent also east of Udruh.

The largest category of all find clusters are concentrations of lithics (n= 34), followed by pottery concentrations (n= 20) and finally by isolated clusters of architectural fragments (n= 10). As most of the lithic scatters date to the Palaeolithic as well as Chalcolithic-Bronze Age periods (*e.g.* Silvonen *et al.* 2013, 350, 352, 370, 378; MacDonald *et al.* 2016, 173, 328), these are of no interest for this contribution. However, it is worth noting that two lithic scatters documented in the Jebel Harun area and along the eastern high plateau are associated with later Nabataean (first and second century AD) concentrations of surface pottery, potentially underlining the longevity of these ephemeral find spots (Silvonen *et al.* 2013, 399; MacDonald *et al.* 2016, 328).

Significant pottery concentrations were recorded mainly along the eastern high plateau, around Beidha and in the Jebel Harun area (Banning and Köhler-Rollefson 1983, 381-382; Killick 1983, 231-236; Hart and Faulkner 1985; Abudanh 2006, 563; N.G. Smith, 2009, 283; MacDonald *et al.* 2012, 173-175; Kouki *et al.* 2013, 2, 4, 11, 16, 33; Silvonen *et al.* 2013, 353, 356-357; MacDonald *et al.* 2016, 117, 272). The recorded pottery concentrations date predominantly to the second century AD (11 out of 20 concentrations) followed by nine sites that belong to the first century AD. While most of the survey reports only rarely comment on the quantity of the documented pottery concentrations, particularly noteworthy are 5300 sherds documented in the Jebel Harun area which homogenously date to the last quarter of the first century AD (Silvonen *et al.* 2013, 353). The local pottery density of this find spot was more than 50 times the average sherd scatter of the entire survey (260 sherds per 100 m^2), indeed rendering this a significant find. Additionally, the site had a large Palaeolithic lithic scatter as well as the probable remains of a recent Bedouin camp. As there is no convincing structural evidence, the exact function of the find must remain undetermined, but the site clearly demonstrates a long history of use. It may be tentatively interpreted as a temporary gathering place for people frequently traversing through, and possibly living in, the area.

Further pottery concentrations were recorded in the Jebel Harun area which may reflect similar activities of groups of people traversing through and/or living in the Petra hinterland. For example, one particular site (FJHP Site Ext040) has a scatter of first century AD Nabataean as well as Late Roman to Early Byzantine (second-fourth

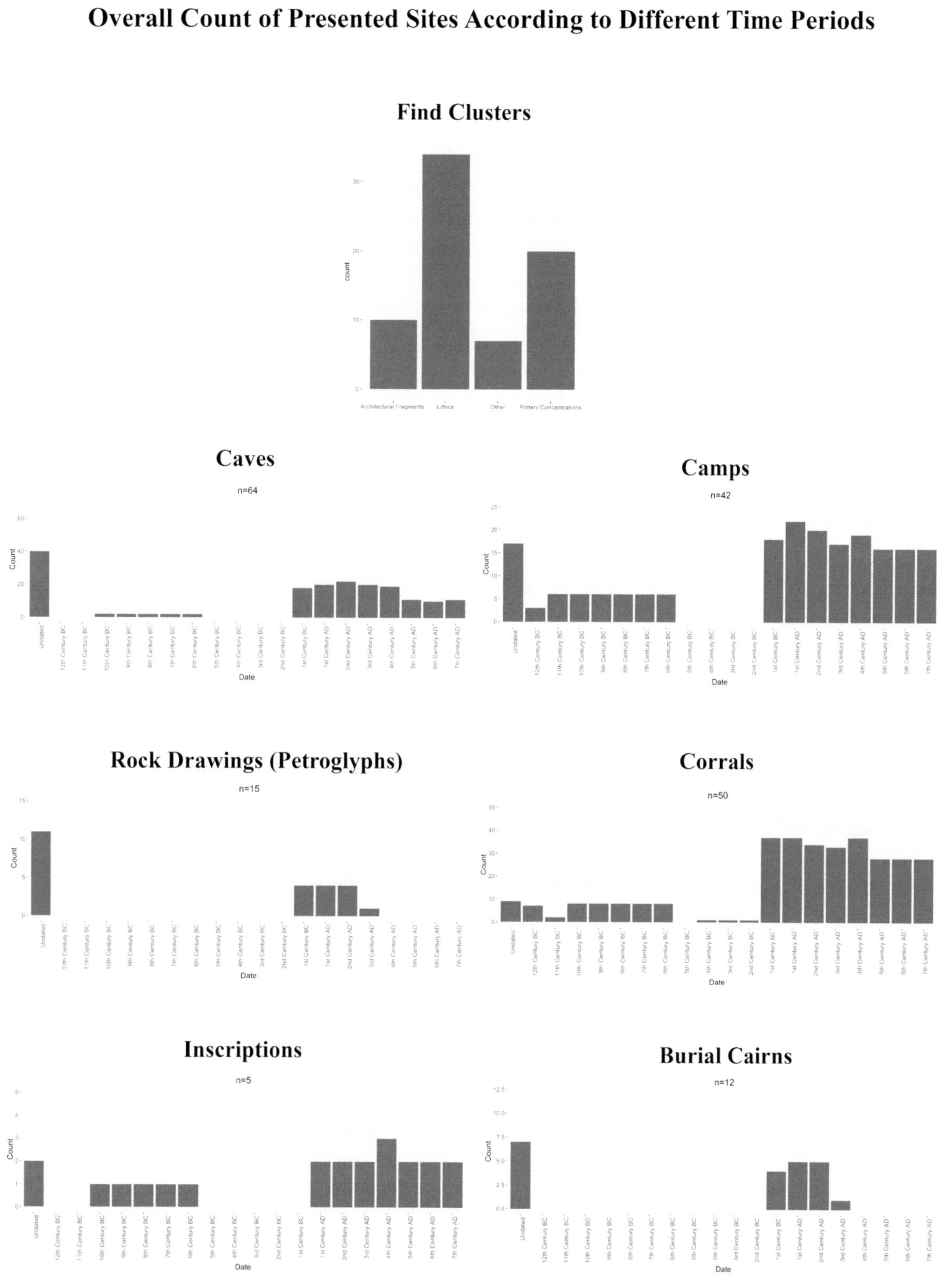

Figure 2. Overall count of the presented sites evidenced in the Petra hinterland pertaining to either pastoral subsistence strategies or to a more mobile way of life (W.M. Kennedy).

Figure 3. Example of possible post-holes for tent-like installations on top of Jabal Umm Zaythuna overlooking Petra's city centre (photograph by W.M. Kennedy).

century AD) pottery, situated on a sheltered plateau of a sandstone outcrop (Kouki *et al.* 2013, 11-12). A possible bedrock mortar was also observed, and a nearby rock-cut cistern presumably supplied the site with water. Because of the availability of water and the presence of the shelter of the plateau, the site was probably an attractive temporary resting place for people traversing through the area.

The Finnish Jabal Harun Project also recorded a pottery concentration of Nabataean coarse ware dating to the second century AD along the banks of Wadi Umm Rattam leading to the Wadi Arabah (Kouki *et al.* 2013, 16). The surveyors also noticed a 'foot print' petroglyph incised into the exposed natural bedrock. Similar 'foot prints' can be frequently observed along pathways and hilltops in the Petra area and are interpreted to have been related to possible ritual pilgrimages (*e.g.* Lindner 1997, 305; 2003, 147, 157, 184-186; Miettunen 2008, 39; Eklund 2013, 284-285; Kouki and Silvonen 2013, 310-311; Fiema 2016, 542). While the observed 'foot print' cannot be associated with such pilgrimages with certainty, it may signify that the site was frequently visited by people passing through and/or living in the area.

Finally, on a prominent hilltop behind the modern settlement of At-Tayyiba south of Petra (which was used by local Bedouins for strategic communication purposes in the early modern era), a significant concentration of surface pottery was recorded (W.M. Kennedy 2016, 149-150; 2018). No built structures were documented.

Admittedly, it is extremely difficult to identify ancient activities on the basis of surface pottery alone and without further archaeological indicators. The pottery concentrations found in the various surveys may nevertheless emphasise the importance of recognising the long habitual tradition of ancient peoples appropriating the dominant natural landscape of the Petra hinterland as gathering places (for presumably various purposes) – particularly when considering the semi-nomadic, mobile background of the Nabataeans.

Natural and/or rock-cut structures

Further support for the claim that ancient peoples in the Petra hinterland exploited the natural landscape can be found in the numerous caves and rock-cut structures documented by the various surveys. Presumably, natural caves and rock shelters were used for (temporary) habitation, for storage of agricultural goods and/or equipment, and for keeping animals. Based on surface pottery, the surveys identified a total of 95 of such sites, dating roughly to the Iron Age as well as between the first century BC and the seventh century AD.

The largest group of such sites consists of natural caves. In total, 64 caves were recorded, predominantly along the eastern high plateau (Figs. 1-2). While some caves are enclosed by low exterior walls, others have simple interior walls suggesting that they were used as seasonal shelters and/or animal pens. However, many of the caves show little trace of ancient human activity – if any at all. Admittedly, it is highly problematic to date the use of natural caves. The majority (40/64) of the caves in the Petra area contain no dateable material. However, 20 caves have surface material or other indirect associations to nearby (dateable) sites, which tentatively indicate a use from the first century BC to the fourth century AD. It is thus plausible that pastoralists made frequent use of such natural features in the past, either as temporary shelters or as animal pens, as they do today. It is likely that far more (and to date unrecorded) natural caves and rock shelters were utilised for such purposes, thus forming an important feature in rural life in the Petra hinterland.

In addition to the caves and rock shelters, there are numerous worked bedrock surfaces throughout the study area that are often situated on hilltops or other places with good visibility over the surrounding landscape. While these cannot be functionally defined with certainty, there are some examples where circular holes were carved into the bedrock surface and which form a quasi-rectangular outline. These can be tentatively interpreted as post holes for tent-like installations (Fig. 3).

The most prominent examples of such installations can be found along a route (Wadi al-Mu'aysirah West) leading from the Al-Begh'ah plain south of Beidha towards Petra, as well as on the hill top of Jebel Umm Zaythuna situated at the immediate outskirts of Petra's city centre (for more on these particular examples, see W.M. Kennedy 2016, 147-149. Several other indications for similar 'post holes' are reported by Abudanh 2006, 493; MacDonald *et al.* 2012, 158, 188; 2016, 365; Kouki *et al.* 2013, 11). The date of these features is still problematic; only four sites yielded surface materials, giving a very rough date between the first century BC and the seventh century AD (MacDonald *et al.* 2012, 158, 188; 2016, 365; W.M. Kennedy 2016, 147-149).

Petroglyphs, *wusūm* and inscriptions

Further evidence for substantial mobility in the Petra hinterland can be found at 54 sites, dating to the Iron Age and between the first century BC and the seventh century AD. These sites feature petroglyphs, *wusūm* and inscriptions, and are distributed mainly along the eastern high plateau and the Jebel Shara escarpment, but also in the eastern Jebel Harun area and some areas immediately north of Petra towards Beidha (Figs. 1-2).

Rock art is commonly observed particularly north of Petra towards Beidha, as well as in the Jebel Harun area where rock surfaces of good quality prevail. The various surveys, however, documented only 15 sites with petroglyphs. Therefore, this brief, selective overview makes no claim to be complete and serves only as an impression of the numerous forms of rock art dispersed throughout rural Petra's landscape. The date of the local rock art is often unclear (cf. Fig. 2) (for recent attempts to date Near Eastern rock art more precisely, see *e.g.* Eklund 2013, 291 with further references; see also Brusgaard, this volume). Surface material and other archaeological indicators at a few sites nevertheless suggest at least a very tentative timeframe for the petroglyphs. One example can be found in the extended Jebel Harun area, where various petroglyphs depicting animals, 'foot prints' and a bow were carved into a *c.* 25 m long, horizontal ledge near a quarry (Kouki 2013, 250; Kouki *et al.* 2013, 6). Based on chisel marks and surface material, it is assumed that the quarry is most likely Nabataean. A similar timeframe (as a *terminus ante quem*) may thus be presumed for the recorded petroglyphs as well. Another large panel of petroglyphs, most likely also associated with a nearby quarry, was recorded further east towards Petra (Kouki *et al.* 2013, 3; Eklund 2013, 283-284). Carved into a horizontal sandstone surface, the petroglyphs include similar 'foot prints' and animals as well as armed human figures and a possible *wasm*. The recorded petroglyphs are most likely of a multi-period date but some of them cover an illegible Nabataean inscription, thus indicating that they were carved in Nabataean times or afterwards.

Many petroglyphs can be observed in the Beidha area as well, including, for example, an artificially cut, large, vertical bedrock surface near Siq al-Amti, depicting several animals, possible *wusūm*, and Greek names (see *e.g.* Banning and Köhler-Rollefson 1983, 381) (Fig. 4). Based on chisel marks, the original surveyors date the

Figure 4. Example of petroglyphs from a vertical bedrock surface near Beidha (Siq al-Amti) (photograph by W.M. Kennedy).

petroglyphs roughly to the Nabataean period (again, as a *terminus ante quem*).

A final example of petroglyphs in the Petra hinterland was observed at a possible Nabataean burial cairn along Naqb Mistalgile – a route leading from the Petra area westwards to the Wadi Arabah (*e.g.* Ben David 2013; W.M. Kennedy 2018). The petroglyphs show anthropomorphic and animal figures carved into a flat sandstone surface immediately next to the site, which suggests that they were associated with the cairn and therefore possibly contemporary with it. Although speculative, they may have been carved by travellers along Naqb Mistalgile, commemorating the deceased in the burial cairn.

While technically to be considered as a specific form of petroglyphs, the various surveys in the Petra hinterland have recorded 24 *wusūm* (singular *wasm*), *i.e.* territorial tribal markings carved into rock surfaces (*e.g.* Macdonald 2012; Eklund 2013, 287; Hayajneh 2016, 516-518; see also Berghuijs, this volume, for a more general discussion of *wusūm*). As for other petroglyphs, the dating of such tribal brands is particularly difficult, explaining why nearly half of all recorded *wusūm* in the study area are undated (Fig. 2). Based on limited surface material, the dates of associated sites, and, rarely, accompanying names and genealogical references in Hismaic or Safaitic script, some *wusūm* in the Petra area may be tentatively dated between the first century BC and the second century AD (compare *e.g.* Abudanh 2006, 465; MacDonald *et al.* 2012, 51; 2016, 116, 174, 243-244, 301-302, 306, 309-310, 334, 343, 376, 391, 400). These feature various signs, including a series of straight and curved lines, inverted L-shaped, key-shaped and hoof-shaped marks as well as circular and other, more abstract *wusūm* (cf. Fig. 5).

The majority of the *wusūm* are situated in the eastern periphery of the Petra hinterland (cf. Fig. 1), although they were also recorded in the Jebel Harun area (Macdonald 2012; Eklund 2013, 287-289; Hayajneh 2016). No pattern or cluster of specific *wasm* types were noticed, but as each *wasm* presumably represents a specific tribal social group, it may be argued that the presence of the various *wusūm* underlines the tribal-based social structure of Petra and its hinterland in antiquity.

Figure 5. Example of *wusūm* in the Petra hinterland. Abudanh Survey No. 325 (Tell Abara, Abu Ar'a Wall) (after Abudanh 2006, 565).

In addition to the petroglyphs and *wusūm*, the various surveys recorded a few Nabataean, Greek and Arabic inscriptions (mostly simple lists of names or commemorative lines). However, there are presumably countless other inscriptions distributed throughout the Petraean hinterland, which still need to be systematically documented (as achieved for urban Petra by L. Nehmé; Nehmé 2012). Several Nabataean inscriptions were noticed near the above-mentioned quarries in the Jabal Harun area, one of which reads: 'May Qayyāmat be safe' (see L. Nehmé's contribution in Eklund 2013, 292-293). Another three Nabataean inscriptions were found along the ancient pilgrimage route from Petra to Jebel Harun (known as the Darb an-Nabi Harun) (Nasarat and Nehmé 2013), consisting of name listings or commemorative lines, such as: 'May Lawdān son of Taymū be safe.' They probably relate to pilgrims traveling to the Nabataean sanctuary on Jebel Harun (Eklund 2013, 290). Similarly, along the route of Wadi al-Mu'aysirah West, a concentration of petroglyphs was documented together with a Nabataean inscription carved in the bedrock surface of the wadi bed and reading: 'Hail Sa'adullahi, son of Salman' (Dalman 1912, 87, no. 45; Berenfeld *et al.* 2016, 83. English translation of Dalman's original translation by the author). The site most likely served as a temporary gathering place for by-passers.

Camp sites and corrals

Although the archaeological evidence is too ambivalent to determine their functions with certainty, the various surveys have recorded a large number of sites (in total 42) which may be interpreted as ancient camp sites (and thus as direct archaeological evidence for pastoral subsistence strategies in the Petra hinterland). Dispersed throughout the entire study area (Fig. 1), these camp sites consist of installations made of low, (semi-)circular or curvilinear stone-built walls with possible openings at one end (Fig. 6).

The geographically closest parallels to the structures in the Petra hinterland can be found in the Negev, where extensive surveys recorded numerous camps of comparable design (*e.g.* Rosen 2007; 2009, 65-68; Saidel 2009; Rosen and Saidel 2010, 68-70). Similar 'Classical'-period camp sites were found in north-eastern Jordan's

Figure 6. Selective overview of possible pastoral nomadic camp sites in the Petra hinterland. A: view of Petra Hinterland Survey Project Site No. 124 . B: Site No. 27-ST038. C: Site No. 123, with associated surface pottery finds (second-third century AD?) (photographs by W.M. Kennedy).

Jebel Qurma region (Huigens 2015, 189-190; see also Huigens, this volume). While the dating of most of the camps in the Negev is based on surface material, many date roughly to the Nabataean and Roman periods (Rosen 2009, 61-68; Rosen and Saidel 2010, 69). The camp site of Giv'ot Reved was excavated, however, yielding exclusively Roman-period pottery (Rosen 1993; 2007, 362-369; Rosen and Saidel 2010, 70).

The Negev sites present convincing archaeological evidence that pastoralism was an important aspect of desert life in the historical periods. In this respect, the region may serve for comparison with similar camp sites found throughout the Petra hinterland. Similar to the Negev, many of these camps (about one-third) remain undated (cf. Fig. 2). However, surface material at the other surveyed camp sites suggests a date to the first and second centuries AD. Moreover, a significant increase of camp sites can be observed in the first century BC, at a time when the number of sedentary rural settlements rises dramatically in the Petra hinterland as well (see Kouki 2012, 84-94; W.M. Kennedy 2018). Pastoralism, it appears, continued to be a vital subsistence strategy in the rural environs of Petra, even at times when the practice of farming accelerated and the Nabataeans increasingly turned towards a sedentary way of life.

The many corrals (n= 50) in the Petra hinterland serve as further direct archaeological indicators for pastoralism. They are predominantly found along the eastern high plateau, where climatic conditions allow for better growth of fields and pastures (Figs. 1-2). Admittedly, without excavation it is difficult to differentiate a corral from a camp site but it is assumed that corrals are far more variable in form than camp sites and their exterior walls appear to be generally higher, as their function is to hold livestock. Surface material from the corrals suggests a predominant dating range from the first century BC to the fourth century AD.

Burial cairns

In the context of this paper, a number of possible burial cairns are relevant (Fig. 7). Some are found along the eastern high plateau and date as early as the first century BC, on the basis of surface pottery material (MacDonald *et al.* 2016, 171, 216) (Fig. 1). However, the majority of all burial cairns (n= 8) were identified in the extended Jebel Harun area (Hertell 2013). One cairn is dated to the Chalcolithic-Bronze Age (*ibid.*, 323) and two between the first and second centuries AD (*ibid.*; Kouki *et al.* 2013, 35). The remaining cairns are undated (Fig. 2).

Figure 7. A possible burial cairn along Wadi Sabra (photograph by W.M. Kennedy).

All cairns in the Jebel Harun area are located on ridges and hilltops, *i.e.* at prominent locations with good visibility to and from their immediate environs (Hertell 2013, 324 and table 1). Similar observations were made for burial cairns recorded by the author along important regional trade routes, such as Naqb ar-Ruba'i, the Umm Qamar pass, Wadi Sabra (Fig. 7) and Naqb Saqqara or Naqb Mistalgile (W.M. Kennedy 2018). Some of these cairns date to the first century BC. Clearly, the cairns not only reflect a continuing 'nomadic' funerary culture in the Petra hinterland during the Nabataean period, but their prominent positions in the landscape also suggest that they served for demarcating tribal territories in Petra's rural environment (more below).

The Khatt Shebib wall

Finally, a highly interesting find in the far eastern periphery of the Petra hinterland is the over 150 km long wall known as the 'Khatt Shebib', which has attracted considerable scholarly interest (see most recently D.L. Kennedy and Banks 2015, with further references). Stretching from Wadi al-Hasa in the north to Ras an-Naqb in the south, only a very short part of the Khatt Shebib runs through the study area. Located immediately east of the Nabataean settlement and Late Roman *castrum* of Udruh, it demarcates a natural border line between the vast desert steppe to the east and the fertile high plateau to the west (Figs. 1 and 8).

The Khatt Shebib is made of unworked, unmortared field stones. Originally, the wall was probably only slightly higher than the preserved height of *c.* 0.5-1 m. It was assumedly *c.* 1-2 m wide. The wall was first observed by Sir A. Kirkbride in the first half of the twentieth century during aerial reconnaissance flights (Kirkbride 1948, 151). More recently, D. Kennedy and R. Banks conducted further aerial studies on the entire stretch of the Khatt Shebib (D.L. Kennedy and Banks 2015). The archaeological evidence clearly indicates that the Khatt Shebib was not a continuous wall. Instead it had large openings and irregular stretches branched off the wall's main direction (*ibid.*, 136, 141, 144). Most notably for the Petra area, there is a 6 km long opening in the vicinity of Udruh (cf. Figs. 1, 8 and 9), where the wall is interrupted near the site of Tell Abara, before continuing again further north at Khirbet Jarba. The dating of the Khatt Shebib remains unresolved (*e.g.* Killick 1986, 436; Findlater 2003, 200-201; MacDonald *et al.* 2012, 466-467). However, Kennedy and Banks are undoubtedly correct in their cautious statement that the wall "is pre-Roman but probably later than the Iron Age" (D.L. Kennedy and Banks 2015, 151).

The function of the Khatt Shebib is also debated. Initially, it was argued that it mainly served to fend off

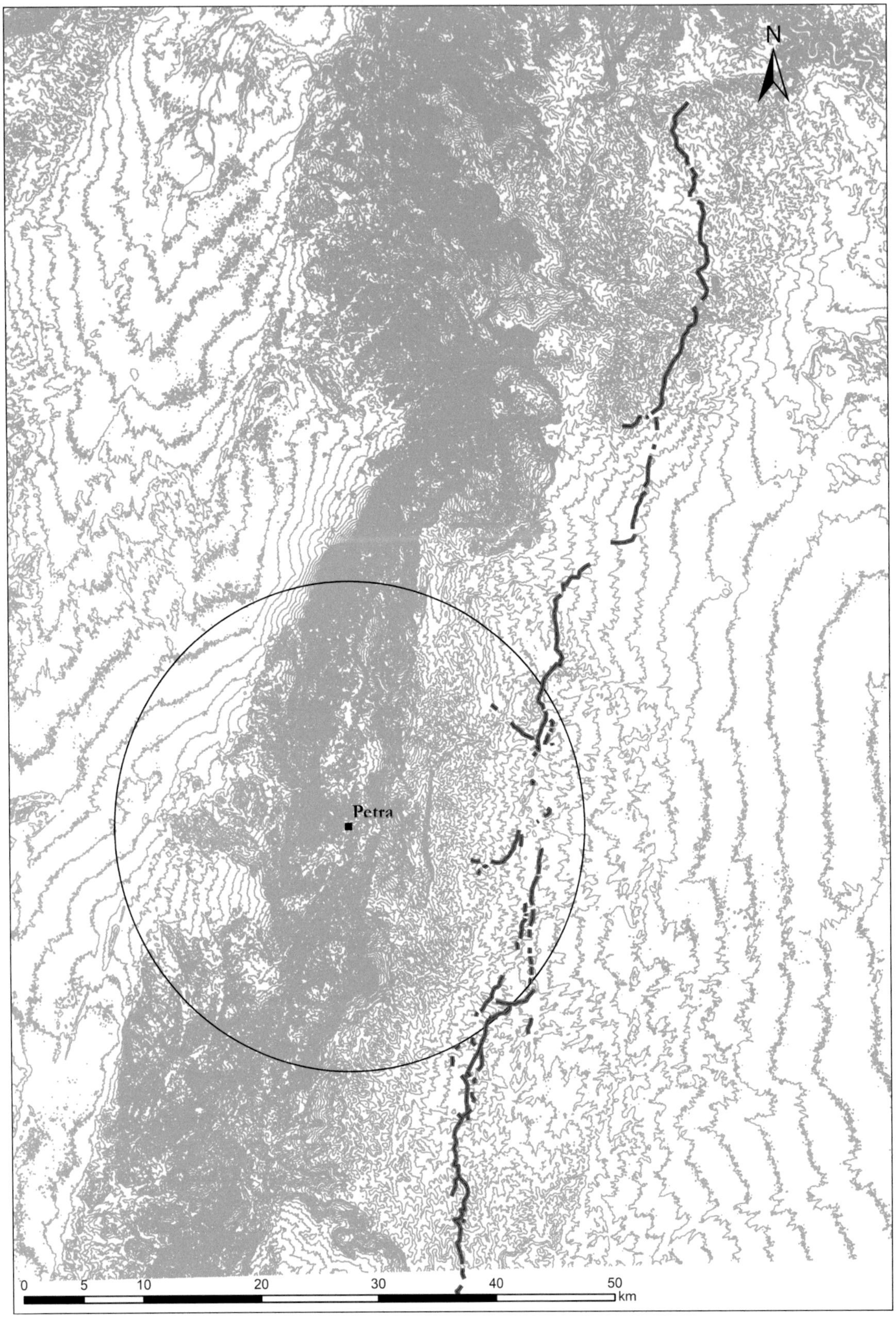

Figure 8. The entire stretch of the Khatt Shebib in relation to the Petra hinterland (map by W.M. Kennedy, with the course of the Khatt Shebib after R. Banks in D.L. Kennedy and Banks 2015).

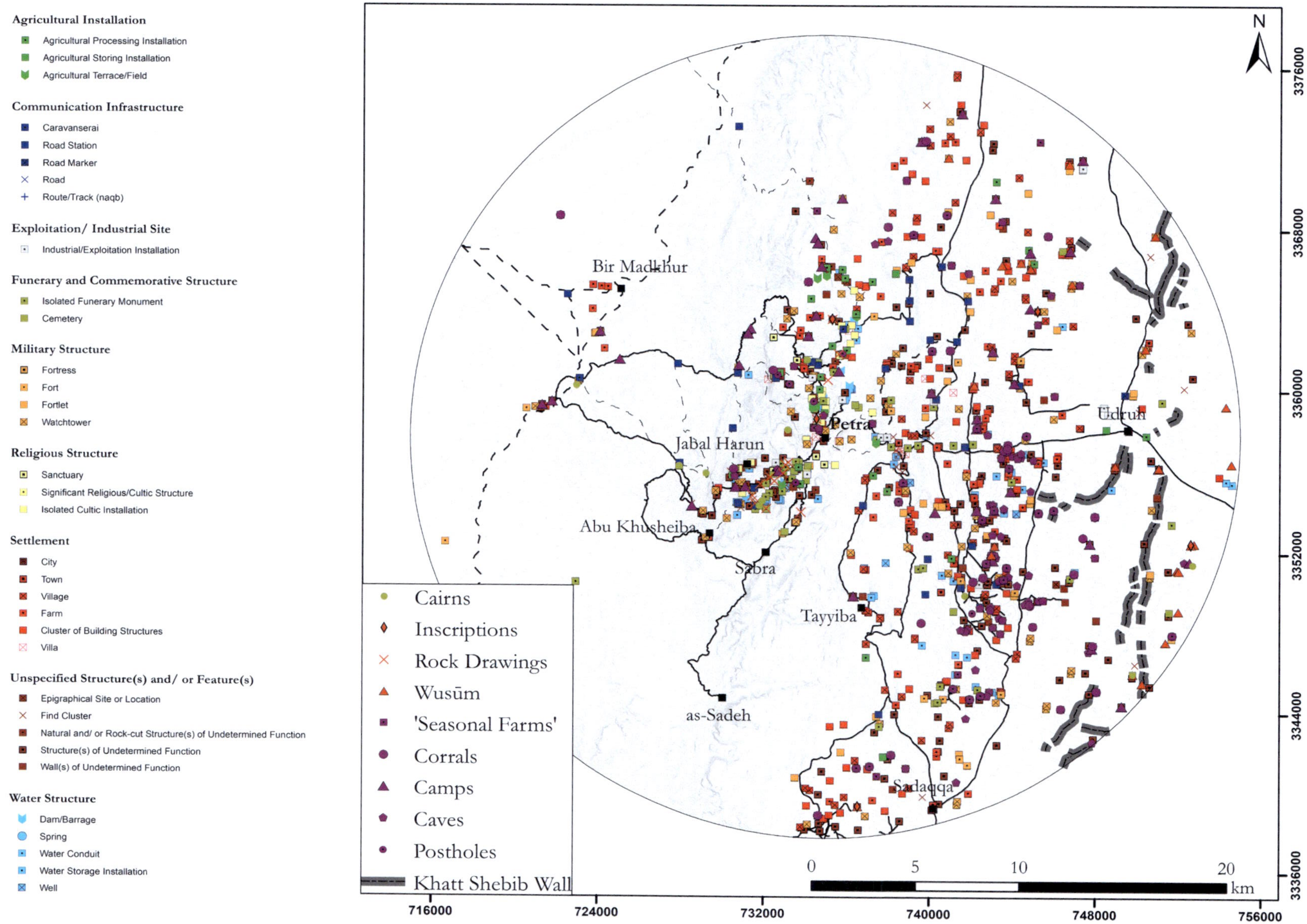

Figure 9. First century AD sites in the Petra hinterland in relation to the Khatt Shebib, with the underlying regional road network (map by W.M. Kennedy, with the course of the Khatt Shebib after R. Banks in D.L. Kennedy and Banks 2015).

nomadic raids from the eastern desert, as it supposedly ran along the line of the eastern Roman frontier (Harding 1967, 154; Killick 1986, 436; Parker 1986, 86). This suggestion was later dismissed by Findlater (2003, 200). Kennedy and Banks recently confirmed Kirkbride's original observation that the wall stretches along the 100 mm rainfall isohyet, and proposed that the Khatt Shebib served as a demarcation wall between the vast desert areas to the east and the cultivable lands to the west (D.L. Kennedy and Banks 2015, 149; also Findlater 2003, 200 who argued that the Khatt Shebib marked a 'boundary area' as well). Indeed, both the distribution maps and the overall counts of all archaeological sites west of the Khatt Shebib confirm earlier claims that agricultural installations and settlements are predominantly situated west of the wall (Fig. 9). The number of sites located east of the wall is significantly smaller than the number of sites recorded west of it. While further research is required to clarify the nature of these sites in more detail, a preliminary assessment of those situated east of the wall includes a conspicuously large number of possible camps, corrals or other structures related to pastoral subsistence strategies (W.M. Kennedy 2018). If this proves to be correct, this would further support the claim that the Khatt Shebib indeed served as a boundary wall between a predominantly settled population to the west and a predominantly pastoral nomadic population to the east.

Similar to the Syrian *Très Long Mur* (Abu Jaber 1995, 740) or the *fossatum africae* in Tripolitania, the Khatt Shebib seemingly demarcated an area "where there was a rapid transition from a predominantly agricultural to a predominantly pastoral way of life" (Mattingly 1995, 171). This would explain the open parts of the wall in the Udruh area. The Khatt Shebib did not serve any major defensive purposes but was as a demarcation line that could be monitored (more below). The wall directed and regulated movement of pastoral nomadic people coming from the eastern desert areas to selected access points into the largely settled area of the Petra hinterland.

Discussion

Although the archaeological evidence is challenging in many ways, this paper has presented data from the Petra hinterland that are rarely discussed within Nabataean archaeology. It is important to realise that these often relatively ephemeral remains are just as important as the 'monumental' material remains for understanding Nabataean culture, as they reflect the widespread use of pastoral subsistence strategies in the Petra hinterland, in addition to farming. They highlight an under-researched, mobile, aspect of rural life in the Petra region, and underline the distinctly tribal-based social structure of the Nabataeans in Petra and its rural surroundings.

The concentrations of pottery, rock art, *etc.*, throughout the diverse Petraean landscape indicate aspects of mobility and the presence of seasonal and/or nomadic pastoralists. These find spots were frequently re-visited and probably used as temporary resting and gathering places. The inscriptions also imply a great deal of mobility, as they were most likely carved by pilgrims on their way to significant religious sites (*e.g.* Jebel Harun). The *wusūm* were made by people who roamed extensively through the Petra area and who were apparently motivated in marking their specific tribal affiliations in the landscape. The 24 *wusūm* show no similarities, suggesting that they represent distinctly different tribal groupings within the study area (although it is, of course, impossible to know whether the *wusūm* were made by locals or foreign visitors to the region).

The burial cairns provide additional evidence for a distinctly tribal social structure of rural Petra.[3] These cairns may have served as territorial markers, since they are located along important regional routes and placed on panoramic high grounds (W.M. Kennedy 2018). For example, the cairns show parallels to the over 400 tombs surveyed in the hinterland of Palmyra (Hesse 2016, 3-4). These are also located along wadis and routes as well as on ridges and hilltops. Surface finds indicate that some of the recorded cairns, particularly in the Jebel Bishri area, were used in the Roman period. Hesse also considers the Palmyrene cairns as indicators for pastoral nomadic activities or prevailing nomadic traditions – as proposed for the cairns in the Petra area. In addition, Hesse (2016, 3) suggests that differences in cairn size and construction effort can be interpreted as a reflection of social stratification. In this respect, the large quantities of Nabataean fine and coarse ware discovered at some of the Petraean cairns (Fig. 7) perhaps indicate that the individual(s) buried there were not of poor social standing.

The fact that large cairns were used for burial of local tribal leaders is attested in a passage from the archive of the Old Babylonian city-state of Mari (Durand 2005, 30; Charpin 2010, 245; Hesse 2016, 3-4). Although of much older date, the text presents an interesting parallel for the significance that cairns (referred to as *hamusûm* or *râmum*) had for demarcating tribal territories. The passage records a complaint addressed to the king of Mari, Zimri-Lim, by a certain Dâdî-hadun, a leader of the Rabbean tribe. Dâdî-hadun had previously allowed the Uprapean tribe to erect a burial monument for their leader Lahun Dagan within the territory of the Rabbeans. However, the burial cairn of Dâdî-hadun's ancestor was then apparently destroyed by members of the Uprapean tribe, for which he has sought revenge and informed Zimri-Lim. Clearly, the passage not only highlights the tribal character of such burial cairns but also emphasises

3 In addition to the monumental Nabataean *hypogea* and façade tombs, which are not discussed in this paper.

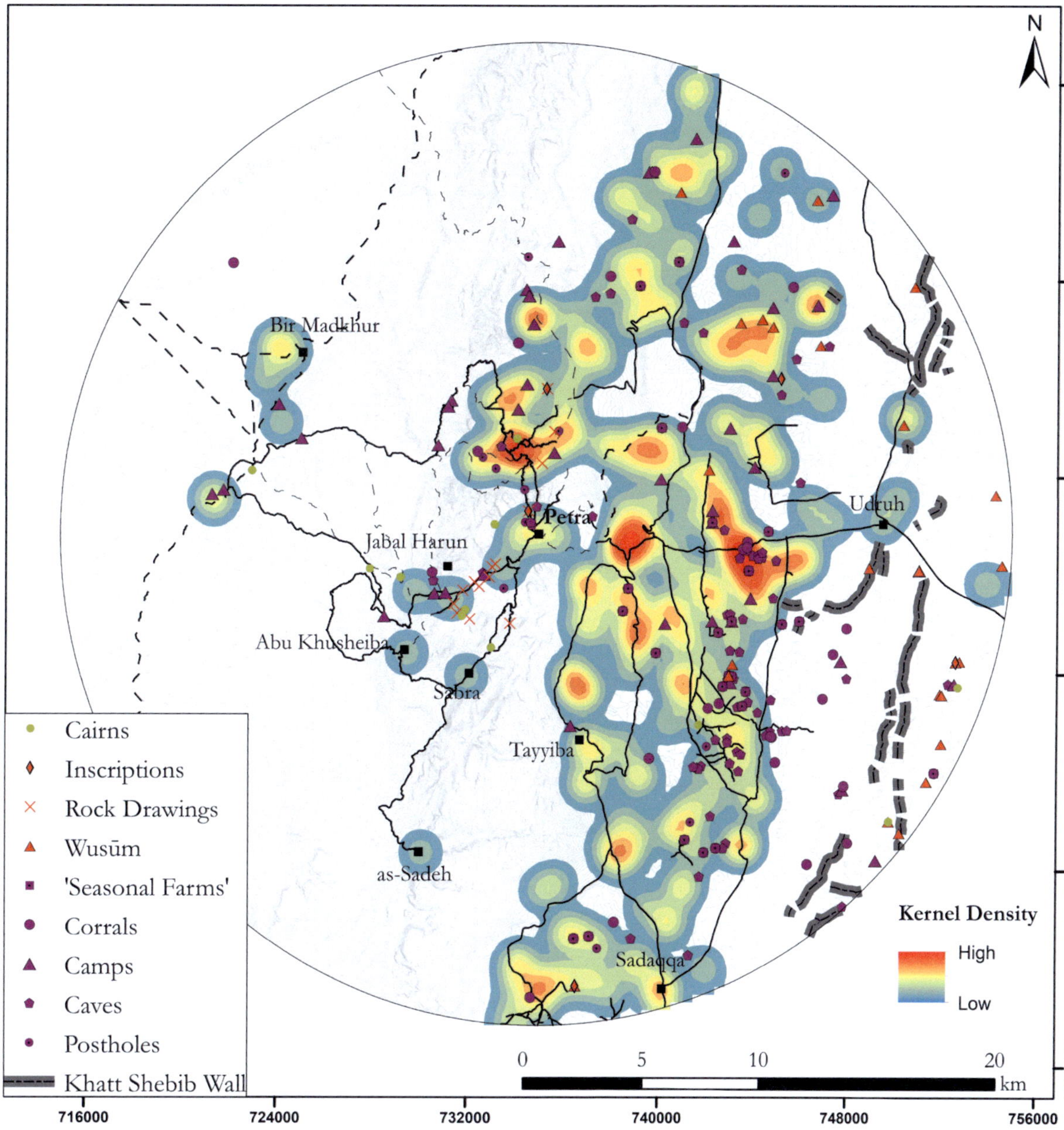

Figure 10. 'Non-sedentary sites' laid over the contemporary kernel density map of all sedentary rural settlements in the Petra hinterland dating to the first century AD, with the underlying regional road network (map by W.M. Kennedy).

the monuments' importance for marking tribal territories in the landscape. It seems reasonable to consider a similar role for the burial cairns in the Petra hinterland.

Furthermore, the camp sites and corrals in the Petra region clearly suggest that pastoral activity was of considerable importance in the Nabataean period (and beyond). Surface material from the numerous camp sites (n= 42) suggests a date mainly in the first and second centuries AD. While it cannot be determined whether the camps were pitched by non-sedentary nomadic pastoralists passing through the Petra hinterland or whether they represent the temporary tent dwellings of local sedentary peoples, the camp sites are – to date unrecognised – direct archaeological evidence from the Petra area that pastoralism was a vital component of the local subsistence strategy that peaked from the first century BC onwards. The 50 corrals distributed throughout the entire Petra hinterland and dating (if dateable at all) predominantly between the first century BC and fourth century AD offer similar conclusions. Potentially, the caves in the region may have also been used by ancient pastoralists, as is still the case today. Twenty-four sites, referred to as 'seasonal farmsteads' or 'seasonal, pastoralist camps' in the original survey reports (*e.g.* MacDonald *et al.* 2012; 2016), were not presented above, as they seem to constitute camps or corrals associated with larger, more substantially built structures, located within cultivable lands and thus best identified as agricultural farms. Surface material dates these sites between the first century BC and fourth century AD. Such sites offer further archaeological evidence that pastoralism was a viable subsistence strategy in the Petra hinterland through time.

Scholars recognise that the sedentary, farming population in the Petra area constituted only one part of the overall rural population and that sedentary groups may have practiced seasonal pastoralism as well. With regard to the Nabataean period, P. Kouki (2012, 99) states that "the sedentary settlement does not represent the whole of the Nabataean society, but a mobile element was retained among the population of the Petra region throughout the existence of the Nabataean kingdom, perhaps specializing in herding in the areas outside the permanent settlement and agricultural land, and/or practicing a form of tethered mobility reminiscent of the historical Petra Bedouin."

Although it is problematic to construe undifferentiated parallels between pre-modern Bedouin societies and ancient mobile peoples (*e.g.* Macdonald 1991), the claim that a 'mobile element' continued to characterise the population of the Petra hinterland is fully supported here. While such an assertion previously lacked direct evidence, there are now many archaeological indicators from Petra's immediate environs suggesting that its rural population indeed remained, at least in part, mobile and followed extensive pastoral subsistence strategies. It is, however, important to note that it is *not* proposed that such a mobile, pastoral rural life took place separately from a sedentary, agriculture-based lifestyle. This becomes most evident when considering the spatial correlation between the presented sites pertaining to a more mobile, pastoral lifestyle and the evidenced sedentary rural settlements (*i.e.* villages, hamlets and farms). When laying all mobile or 'non-sedentary' structures over a GIS-based density map of all sedentary rural settlements dating (as a representative example) to the first century AD (Fig. 10), the situation appears particularly blurry in the Petra area, as no spatial separation can be observed.

Although the dating of the 'non-sedentary' sites is particularly problematic (as emphasised above), it is nevertheless clear that they occur in both strong and weak clusters of rural sedentary settlements during the Nabataean period (and beyond). There is no clear division between a strictly sedentary and non-sedentary population in the Petra hinterland. This supports the argument that pastoralism was practiced along with farming and that a mobile lifestyle persisted, despite the 'sedentary and agricultural turn' of the Nabataeans from the first century BC onwards. It is indeed most likely that a segment of Petra's rural population led a combination of a sedentary and non-sedentary life. This is corroborated when considering the Khatt Shebib wall as a demarcation line between a predominantly settled community to the west and predominantly pastoral nomadic peoples to the east. The wall did not function as a fixed border to fend off externals but served to regulate and monitor activities of pastoral nomadic peoples coming from the vast eastern desert. Importantly for the Petra hinterland, the 6 km long opening of the wall in the Udruh area suggests that it was meant to direct these peoples to selected meeting areas, perhaps for the exchange of livestock from the desert with agricultural and other goods from the settled groups. It is certainly no coincidence that two large stone circles, *c.* 400 m in diameter (circles J5 and J6 in D.L. Kennedy 2013, 53), are situated at both ends of the Khatt Shebib's opening, as they may have served as 'open market areas'. Whether or not this was the case, the Khatt Shebib wall certainly indicates the co-existence and mutually beneficial relationship between sedentary and non-sedentary groups in the Petra hinterland (as also postulated in other regions of the Near East; see *e.g.*, Banning 1986; Kouki 2012, 99-100 with further references).

Conclusion and outlook

This paper has presented a wealth of direct archaeological evidence indicating that life in Petra's rural environs bordered on both sedentary and non-sedentary lifestyles and reflected the tribal-based, nomadic social background of the area. For example, the numerous Nabataean tomb complexes, tribal sanctuaries and other isolated cultic installations within urban Petra, which were presumably used by different social groups (*e.g.* Nabataean *marzeah*), mirror this duality of Nabataean culture. As these installations were frequented by a very selective group of people only, they can be referred to by the Foucauldian term *heterotopiai*. All such Nabataean *heterotopiai* clearly indicate the perpetuating family, clan and/or tribal roots within Nabataean society and culture. This important aspect of Nabataean social structure is supported by the archaeological evidence presented in this paper for a pastoral and more mobile lifestyle in the Petra hinterland.

Due to the limits of this paper, several other heterotopical structures (most notably rural sanctuaries and cultic installations) in the Petra hinterland have not been discussed (see W.M. Kennedy 2018 for details). However, these structures may have served as territorial markers, demarcating specific social landscapes within the wider Petraean hinterland – just as the discussed burial cairns. Similar assumptions may also be made for the larger and more significant settlements in the study area, such as Sabra, Abu Khusheiba, Wadi Musa, Beidha and Udruh (*ibid.*). Comparable to urban Petra, the Petraean hinterland was arguably an intricate patchwork of various social groups that were strongly bound by local 'tribal' affiliations. If so, the Nabataean kings in Petra certainly must have maintained good relations to these different communities, as they played a significant administrative, economic and socio-political role for the survival of the Nabataean capital. However, although these assumptions seem likely, they remain hypothetical at this point and should be elaborated by future research endeavours.

Keeping this in mind, and considering the presented archaeological evidence for the 'non-sedentary' aspect of rural life in the Petra hinterland, it may be very well argued that the Nabataeans were indeed 'travellers between lifestyles' – travellers between the desert and the sown.

Acknowledgements

I would first like to express my sincere gratitude to the organisers of the 'Landscapes of Survival' conference for giving me the opportunity to present and discuss this particular subject matter. Also, many thanks go to Dr Z.T. Fiema for his invaluable comments on the original draft of this paper. E. Conway's linguistic corrections are most appreciated as well. Finally, this work would not have been possible without the kind financial support of the Excellence Cluster Topoi.

References

Abudanh, F. 2006. *Settlement patterns and military organization in the region of Udruh (southern Jordan) in the Roman and Byzantine periods I-II*. Newcastle upon Tyne: University of Newcastle upon Tyne (unpublished PhD thesis).

Abu Jaber, R. 1995. Water collection in a dry farming society. *Studies in the History and Archaeology of Jordan* 5, 737-744.

'Amr, K., Farajat, S., Al-Momani, A., and Falahat, H. 1998. Archaeological survey of the Wadi Musa water supply and wastewater project area. *Annual of the Department of Antiquities of Jordan* 42, 503-548.

'Amr, K. and Al-Momani, A. 2001. Preliminary report on the archaeological component of the Wadi Musa water supply and wastewater project (1998-2000). *Annual of the Department of Antiquities of Jordan* 45, 253-285.

Banning, E.B. 1986. Peasants, pastoralists and 'Pax Romana': mutualism in the southern highlands of Jordan. *Bulletin of the American Schools of Oriental Research* 261, 25-50.

Banning, E.B. and Köhler-Rollefson, I. 1983. Ethnoarchaeological survey in the Beidha area, southern Jordan. *Annual of the Department of Antiquities of Jordan* 27, 375-383.

Bell, G.L. 1907. *The desert and the sown*. London: Heinemann.

Ben David, C. 2013. The ancient routes from Petra to the Wadi 'Araba. In: Kouki, P. and Lavento, M. (eds.). *Petra – The Mountain of Aaron: the Finnish Archaeological Project in Jordan*. Helsinki: Societas Scientiarum Fennica, 273-279.

Berenfeld, M.L., Dufton, J.A. and Rojas, F. 2016. Green Petra: archaeological explorations in the city's northern wadis. *Levant* 48, 79-107.

Charloux, G., Bouchaud, C., Durand, C., Monchot, H. and Thomas, A. 2016. Banqueting in a northern Arabian oasis: a Nabataean triclinium at Dumat al-Jandal. *Bulletin of the American Schools of Oriental Research* 375, 13-34.

Charpin, D. 2010. The desert routes around Djebel Bishri and the Sutean nomads according to the Mari archives. *Al-Rafidan* 2010 (Special Issue), 239-245.

Dalman, G. 1912. *Neue Petra-Forschungen und der Heilige Felsen von Jerusalem*. Leipzig: J.C. Hinrichs'sche Buchhandlung.

Durand, J.-M. 2005. *Le culte de pierres et les monuments commémoratifs en Syrie amorrite*. Paris: Société pour l'Étude du Proche-Orient Ancien.

Eklund, A. 2013. Rock art. In: Kouki, P. and Lavento, M. (eds.). *Petra – The Mountain of Aaron. The Finnish Archaeological Project in Jordan. Vol. 3*. Helsinki: Societas Scientiarum Fennica, 281-296.

Fiema, Z.T. 2016. The Jabal Hārūn site: 1000 years of continuity and change. In: Fiema, Z.T. and Frösén, J. (eds.). *Petra – The Mountain of Aaron: the Finnish Archaeological Project in Jordan. Vol. 2*. Helsinki: Societas Scientiarum Fennica, 539-582.

Findlater, G.M. 2003. *Imperial control in Roman and Byzantine Arabia. A landscape interpretation of archaeological evidence in southern Jordan*. Edinburgh: University of Edinburgh (unpublished PhD thesis).

Fisher, C.T. (ed.) 1906. *Diodori Bibliotheca Historica*. Vol. 3. Leipzig: Teubner.

Graf, D.F. 1990. The origins of the Nabataeans. *ARAM* 2, 45-75.

Graf, D.F. 2013. Petra and the Nabataeans in the Early Hellenistic period: the literary and archaeological evidence. In: Mouton, M. and Schmid, S.G. (eds.). *Men on the rocks. The formation of Nabataean Petra*. Berlin: Logos, 35-56.

Hackl, U., Jenni, H., Schneider, C. and Keller, D. 2003. *Quellen zur Geschichte der Nabatäer. Textsammlung mit Übersetzung und Kommentar*. Freiburg/Göttingen: Universitätsverlag Freiburg / Vandenhoeck & Ruprecht Göttingen.

Harding, G.L. 1967. *The antiquities of Jordan*. London: Praeger.

Hart, S. 1987. The Edom survey project 1984-85. The Iron Age. *Studies in the History and Archaeology of Jordan* 3, 287-290.

Hart, S. and Faulkner, R.K. 1985. Preliminary report on a survey in Edom, 1984. *Annual of the Department of Antiquities of Jordan* 29, 255-277.

Hayajneh, H. 2016. Ancient North Arabian inscriptions, rock drawings and tribal brands (wasms) from the Sammah / 'Ayl ('El) region. In: MacDonald, B., Clark, G.A., Herr, L.G., Quaintance, D.S., Hayajneh, H. and Eggler, J. (eds.). *The Shammakh to Ayl archaeological survey, southern Jordan (2010-2012)*. Boston, MA: American Schools of Oriental Research, 505-541.

Healey, J.F. 2001. *The religion of the Nabataeans. A conspectus*. Leiden: Brill.

Hertell, E. 2013. Cairns. In: Kouki, P. and Lavento, M. (eds.). *Petra – The Mountain of Aaron. The Finnish Archaeological Project in Jordan. Vol. 3*. Helsinki: Societas Scientiarum Fennica, 323-327.

Hesse, K.J. 2016. Palmyra, pastoral nomads, and city-state kings in the Old Babylonian period. Interaction in the semi-arid Syrian landscape. In: Meyer, J.C. (ed.). *Palmyrena: city, hinterland and caravan trade between Orient and Occident*. Oxford: Archaeopress, 1-9.

Huigens, H.O. 2015. Prelminary report on a survey in the Hazimah plains: a hamad landscape in north-eastern Jordan. *Palestine Exploration Quarterly* 147, 180-194.

Kennedy, D.L. 2013. Remote sensing and 'Big Circles'. A new type of prehistoric site in Jordan and Syria. *Zeitschrift für Orient-Archäologie* 6, 44-63.

Kennedy, D.L. and Banks, R. 2015. The Khatt Shebib in Jordan: from the air and space. *Zeitschrift für Orient-Archäologie* 8, 132-154.

Kennedy, W.M. 2016. Reassessing the impact of natural landscape factors on spatial strategies in the Petra hinterland in Nabataean-Roman times. *Proceedings of the Seminar for Arabian Studies* 46, 137-164.

Kennedy, W.M. 2018. *Terra Petraea. An archaeological landscape characterization of the Petra hinterland in Nabataean-Roman times*. Berlin: Humboldt-Universität zu Berlin (PhD Thesis).

Killick, A.C. 1983. Udruh: 1980, 1981, 1982 seasons. A preliminary report. *Annual of the Department of Antiquities of Jordan* 27, 231-243.

Killick, A.C. 1986. Udruh and the southern frontier. In: Freeman, P. and Kennedy, D.L. (eds.). *The defence of the Roman and Byzantine East*. Oxford: Archaeopress, 431-446.

Kirkbride, A. 1948. Shebib's Wall in Transjordan. *Antiquity* 22, 151-154.

Knodell, A.R., Alcock, S.E., Tuttle, C.A., Cloke, C.F., Erickson-Gini, T., Feldman, C., Rollefson, G.O., Sinibaldi, M., Urban, T.M. and Vella, C. 2017. The Brown University Petra Archaeological Project: landscape archaeology in the northern hinterland of Petra, Jordan. *American Journal of Archaeology* 121, 621-683.

Kouki, P. 2012. *The hinterland of a city. Rural settlement and landuse in the Petra region from the Nabataean-Roman to the Early Islamic period*. Helsinki: Helsinki University Press.

Kouki, P. 2013. Production sites. In: Kouki, P. and Lavento, M. (eds.). *Petra – The Mountain of Aaron. The Finnish Archaeological Project in Jordan. Vol. 3*. Helsinki: Societas Scientiarum Fennica, 247-252.

Kouki, P., Eklund, A., Hertell, E., Silvonen, S. and Ynnilä, H. 2013. The FJHP extended survey site catalog. In: Kouki, P. and Lavento, M. (eds.). *Petra – The Mountain of Aaron. The Finnish Archaeological Project in Jordan. Volume III. The Archaeologcial Survey*. Helsinki: Societas Scientiarum Fennica, 1-51. Available on accompanying CD [Accessed 23 March 2020].

Kouki, P. and Lavento, M. (eds.) 2013. *Petra – The Mountain of Aaron: the Finnish archaeological project in Jordan. Vol. 3. The archaeological survey*. Helsinki: Societas Scientiarum Fennica.

Kouki, P. and Silvonen, S. 2013. Ritual sites. In: Kouki, P. and Lavento, M. (eds.). *Petra – The Mountain of Aaron. The Finnish archaeological project in Jordan. Vol. 3. The archaeologcial survey*. Helsinki: Societas Scientiarum Fennica, 301-316.

Kühn, D. 2005. *Totengedanken bei den Nabatäern und im Alten Testament. Eine religionsgeschichtliche und exegetische Studie*. Münster: Ugarit-Verlag.

Ladurner, M. 2015. Jabal ash-Sharah, Jordanien. Nabatäische Wohn- und Wirtschaftsstrukturen im Hinterland von Petra. *e-Forschungsberichte des DAI 2015 Faszikel* 2, 42-45.

Lenzen, C.J. 2003. The desert and the sown: an introduction to the archaeological and historiographic challenge. *Mediterranean Archaeology and Archaeometry* 16, 5-12.

Lindner, M. 1992. Abu Khusheiba – a newly discovered Nabataean settlement and caravan station between Wadi 'Arabah and Petra. *Studies in the History and Archaeology of Jordan* 4, 263-267.

Lindner, M. 1997. Ein christliches Pilgerzeichen auf Umm el-Biyara. In: Lindner, M. (ed.). *Petra und das Königreich der Nabatäer*. Bad Windheim: Delp, 304-306.

Lindner, M. 2003. Über Petra hinaus. Archäologische Erkundungen im südlichen Jordanien. Rahden: Verlag Marie Leidorf.

MacDonald, B., Herr, L.G., Quaintance, D.S., Clark, G.A. and Macdonald, M.C.A. 2012. *The Ayl to Ras an-Naqab archaeological survey, southern Jordan (2005-2007)*. Boston, MA: American Schools of Oriental Research.

MacDonald, B., Clark, G.A., Herr, L.G., Quaintance, D.S., Hayajneh, H. and Eggler, J. 2016. *The Shammakh to Ayl archaeological survey, southern Jordan (2010-2012)*. Boston, MA: American Schools of Oriental Research.

Macdonald, M.C.A. 1991. Was the Nabataean kingdom a 'bedouin state'? *Zeitschrift des Deutschen Palästina-Vereins* 107, 102-119.

Macdonald, M.C.A. 2012. Inscriptions, rock drawings and wusūm from the Ayl to Ras an-Naqab archaeological survey. In: MacDonald, B., Herr, L.G., Quaintance, D.S., Clark, G.A. and Macdonald, M.C.A. *The Ayl to Ras an-Naqab archaeological survey, southern Jordan (2005-2007).* Boston, MA: American Schools of Oriental Research, 433-463.

Mattingly, D.J. 1995. *Tripolitania.* London: B.T. Batsford Ltd.

Miettunen, P. 2008. Jabal Harun: history, past explorations, monuments, and pilgramages. In: Fiema, Z. T. and Frösén, J. (eds.). *Petra – The Mountain of Aaron. The Finnish archaeological project in Jordan. Vol. 1. The church and the chapel.* Helsinki: Societas Scientiarum Fennica, 27-49.

Nasarat, M. and Nehmé, L. 2013. Three Nabataean inscriptions from Jabal Hārūn. In: Kouki, P. and Lavento, M. (eds.). *Petra – The Mountain of Aaron. The Finnish archaeological project in Jordan. Vol. 3. The archaeological survey.* Helsinki: Societas Scientiarum Fennica, 297-298.

Nehmé, L. 2012. *Atlas archéologique et épigraphique de Pétra.* Paris: Académie des Inscriptions et Belles-Lettres.

Nehmé, L. 2013. The installation of social groups in Petra. In: Mouton, M. and Schmid, S.G. (eds.). *Men on the rocks. The formation of Nabataean Petra.* Berlin: Logos, 113-127.

Parker, S.T. 1986. *Romans and Saracens: a history of the Arabian frontier.* Philadelphia, PA: American Schools of Oriental Research.

Petrovszky, K. 2013a. Nabataean tomb complexes in the context of Mediterranean funerary architecture of the Roman era. *Studies in the History and Archaeology of Jordan* 11, 459-465.

Petrovszky, K. 2013b. The infrastructure of the tomb precincts of Petra: preliminary results of the tacheometrical survey in selected areas. In: Mouton, M. and Schmid, S.G. (eds.). *Men on the rocks. The formation of Nabataean Petra.* Berlin: Logos, 189-204.

Rosen, S.A. 1993. A Roman-period pastoral tent camp in the Negev, Israel. *Journal of Field Archaeology* 20, 441-451.

Rosen, S.A. 2007. The Nabateans as pastoral nomads: an archaeological perspective. In: Politis, K.D. (ed.). *The world of the Nabataeans.* Vol. 2. Stuttgart: Franz Steiner Verlag, 345-375.

Rosen, S.A. 2009. History does not repeat itself: cyclicity and particularism in nomad-sedentary relations in the Negev in the long term. In: Szuchman, J. (ed.). *Nomads, tribes, and the state in the ancient Near East.* Chicago, IL: The Oriental Institute of the University of Chicago, 57-86.

Rosen, S.A. 2017. *Revolutions in the desert: the rise of mobile pastoralism in the southern Levant.* New York: Routledge.

Rosen, S.A. and Saidel, B.A. 2010. The camel and the tent: an exploration of technological change among early pastoralists. *Journal of Near Eastern Studies* 69, 63-77.

Saidel, B.A. 2009. Pitching camp: ethnoarchaeologial investigations of inhabited tent camps in the Wadi Hisma, Jordan. In: Szuchman, J. (ed.). *Nomads, tribes, and the state in the ancient Near East.* Chicago, IL: The Oriental Institute of the University of Chicago, 87-104.

Schmid, S.G. 2001. The Nabataeans: travellers between lifestyles. In: MacDonald, B., Adams, R. and Bienkowski, P. (eds.). *The archaeology of Jordan.* Sheffield: Sheffield Academic Press, 367-427.

Schmid, S.G. 2008. The Hellenistic period and the Nabataeans. In: Adams, R.B. (ed.). *Jordan. An archaeological reader.* London: Equinox, 353-411.

Schmid, S.G. 2009. Überlegungen zum Grundriss und zum Funktionieren nabatäischer Grabkomplexe in Petra. *Zeitschrift des Deutschen Palästina-Vereins* 125, 139-170.

Schmid, S.G. 2012. Die dritte Dimension der Felsfassaden: nabatäische Grabkomplexe in Petra. In: Schmid, S.G. and van der Meijden, E. (eds.). *Begleitbuch zur Ausstellung Petra. Wunder in der Wüste. Auf den Spuren von J. L. Burckhardt alias Scheich Ibrahim im Antikenmuseum Basel.* Basel: Verlag Schwabe AG, 182-194.

Schmid, S.G. 2013. Foucault and the Nabataeans or what space has to do with it. In: Mouton, M. and Schmid, S.G. (eds.). *Men on the rocks. The formation of Nabataean Petra.* Berlin: Logos, 251-270.

Silvonen, S., Kouki, P., Eklund, A., Hertell, E., Lavento, M. and Ynillä, H. 2013. The FJHP Survey Site Catalog. In: Kouki, P. and Lavento, M. (eds.). *Petra – The Mountain of Aaron. The Finnish Archaeological Project in Jordan. Vol. 3. The archaeologcial survey.* Helsinki: Societas Scientiarum Fennica, 347-409.

Smith, A.M. 2010. *Wadi Araba in Classical and Late Antiquity: an historical geography.* Oxford: Archaeopress.

Smith, N.G. 2009. *Social boundaries and state formation in ancient Edom: a comparative ceramic approach.* San Diego, CA: University of California, San Diego (unpublished PhD thesis).

Tholbecq, L. 2001. The hinterland of Petra from the Edomite to the Islamic periods: the Jabal ash-Sharaj Survey (1996-1997). *Studies in the History and Archaeology of Jordan* 7, 399-405.

Tholbecq, L. 2013. The hinterland of Petra Jordan and the Jabal Shara during the Nabataean, Roman and Byzantine periods. In: Mouton, M. and Schmid, S.G. (eds.). *Men on the rocks. The formation of Nabataean Petra.* Berlin: Logos, 295-312.

Tholbecq, L. and Durand, C. 2005. A Nabataean rock-cut sanctuary in Petra: second preliminary report on the 'Obodas Chapel" excavation project, Jabal Numayr (2002-2004). *Annual of the Department of Antiquities of Jordan* 49, 299-311.

Tholbecq, L. and Durand, C. 2013. A late second b.c. Nabataean occupation at Jabal Numayr: The earliest phase of the "Obodas Chapel" sanctuary. In: Mouton, M. and Schmid, S.G. (eds.). *Men on the rocks. The formation of Nabataean Petra.* Berlin: Logos, 205-222.

Tholbecq, L., Durand, C. and Bouchaud, C. 2008. A Nabataean rock-cut sanctuary in Petra: second preliminary report on the 'Obodas Chapel' excavation project, Jabal Numayr (2005-2007). *Annual of the Department of Antiquities of Jordan* 52, 235-254.

Wadeson, L. 2011. Nabataean tomb complexes at Petra: new insights in the light of recent fieldwork. *Proceedings of the Australasian Society for Classical Studies* 32, 1-24.

Wadeson, L. and Abudanh, F. 2016. Newly discovered tombs in the hinterland of Petra. *Studies in the History and Archaeology of Jordan* 12, 83-99.

Wenner, S.E. 2015. *Petra's hinterland from the Nabataean through Early Byzantine periods (ca. 63 BC-AD 500).* Raleigh, NC: North Carolina State University (unpublished MA thesis).

Wenning, R. 2007. Nabatäische Votivnischen, Clan-Heiligtümer, Tempel und Votive. In: Frevel, C. and von Hesberg, H. (eds.). *Kult und Kommunikation: Medien in Heiligtümern der Antike.* Wiesbaden: Reichert, 247-277.

Wenning, R. 2013. Towards 'Early Petra': an overview of the early history of the Nabataeans in its context. In: Mouton, M. and Schmid, S.G. (eds.). *Men on the rocks. The formation of Nabataean Petra.* Berlin: Logos, 7-22.

The desert and the sown: Safaitic outsiders in Palmyrene territory

Jørgen Christian Meyer

Abstract

The relationship between nomads and the sedentary population in the Middle East has changed throughout history from conflicts to co-operation. In ancient Palmyra, numerous villages and estates occupied the mountains north of the oasis, which were also an important summer pasture for nomads from the southern dry steppe, the *badia*. There was no contradiction, however, between the desert and the sown. This was not due to the fact that the inhabitants of Palmyra descended from the population of the Syrian dry steppe, bringing with them ties of kinship and patronage. Some of the nomads entering the Palmyrene territory were Safaitic speaking groups, which were not integrated into the social network of the city and were liable to a special tax for their grazing rights. Instead, the Palmyrenes complemented several forts and stations in the immediate hinterland to settle potential conflicts. The forts controlled the main lines of communication, the territory and important water resources. This system continued up into the late Roman and Umayyad periods.

Keywords: relations between nomads and sedentary population in Palmyrena, Safaitic-speaking nomads in Palmyrene territory, Palmyrene control of the territory

Introduction

> "All of the fifteen huts of the hamlet of Arak were deserted. The inhabitants had suffered much from the Bedouins camping in the neighborhood and still more from numerous raiding bands; therefore, as they always do under such circumstances, they had moved in a body with their supplies to Tudmor. The Turkish Government some time in the seventies had ordered a strong barrack to be built halfway between the hamlet and the springs, with five gendarmes for a regular garrison; but this was now deserted, because the Bedouins only made fun of it." (Alois Musil 1928, 85-86. March 1912).

When the Czech theologian, orientalist and explorer Alois Musil carried out his expeditions in Syria, Jordan and the Arabian Peninsula from 1908 to 1915, he navigated in a very complicated political landscape. The Ottoman authorities had only limited control of the Syrian interior, and competing tribes harassed agricultural settlements and made travel a great challenge. This was undoubtedly partly due to general weakening of the Ottoman Empire in the nineteenth century (Lewis 2000; Reilly 2015), but also in the eighteenth century the governor of Aleppo had to negotiate with hostile Bedouin tribes

In: Peter M. M. G. Akkermans (ed.) 2020: *Landscapes of Survival - The Archaeology and Epigraphy of Jordan's North-Eastern Desert and Beyond*, Sidestone Press (Leiden), pp. 255-264.

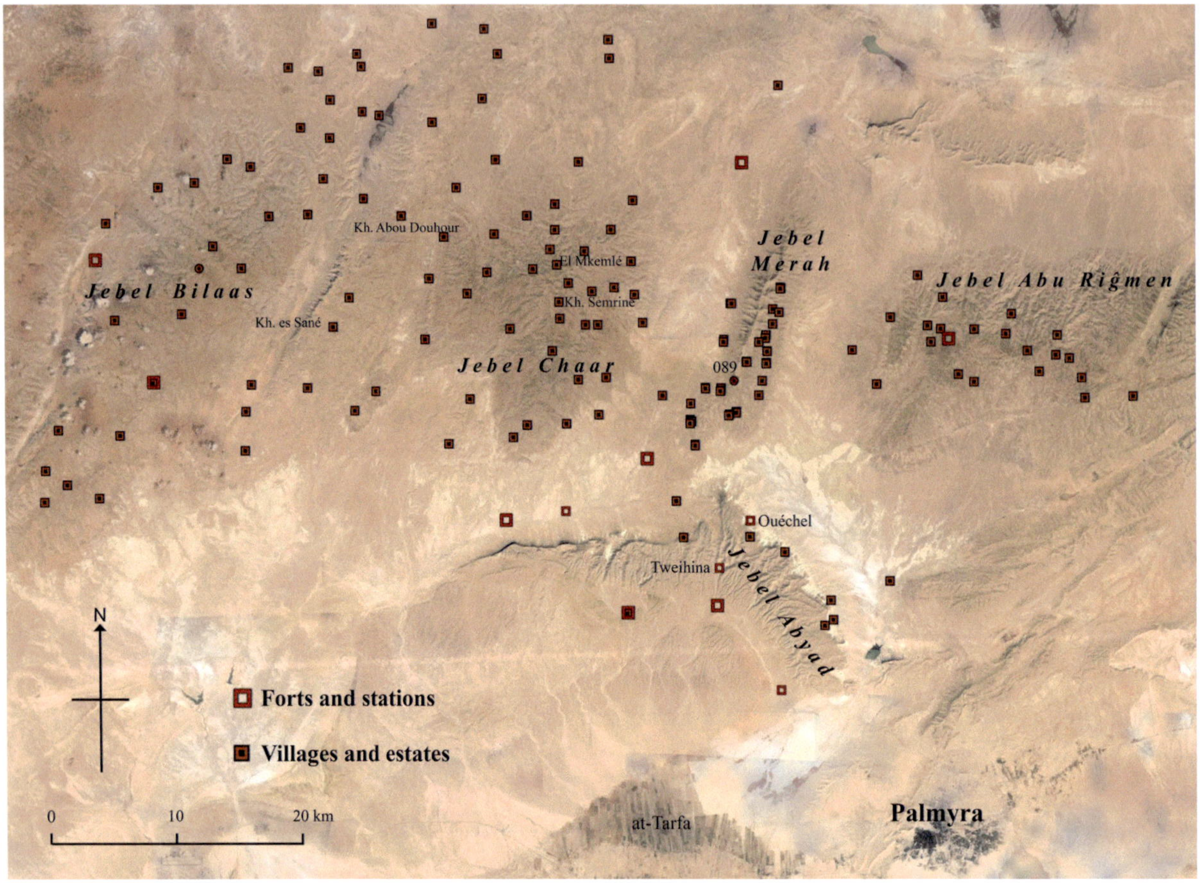

Figure 1. Distribution of settlements, forts and stations north of Palmyra (J.C. Meyer, based on Google Earth).

who occasionally raided villages, attacked caravans, and skirmished with one another (Marcus 1989, 140-141). This was not a continuous conflict. There were long periods with mutual understanding and sharing of the tribute from the caravans from Aleppo to Baghdad and Basra (Grant 1937, 137-139), but the relationship between the desert and the sown was a problematic issue. The conflicts were rooted far back in time. Already in the twelfth century, the dwindling population of the oasis in Palmyra had moved behind the protecting walls of the precinct of the ancient temple of Bel, which was converted into a fortress. Important agricultural districts, like the At-Tarfa depression west of Palmyra (Fig. 1), were abandoned (Meyer 2017, 41-46), and the population of Tadmur was compelled to import cereals from the western part of Syria in exchange for salt from the *salina* south-east of the oasis (Musil 1928, 145-146).

Palmyra and its hinterland

The relationship between Palmyra and the hinterland was completely different in the Roman period. The Palmyrenes opened up a lucrative caravan route to the south-east across the dry steppe (*badia*) to Hit on a regular basis (Fig. 2). The former smaller settlement in the oasis became a caravan city between the Mediterranean and the Indian Ocean, and it grew into a great city – a metropolis in the proper sense of the word – that needed provisions from the surrounding territory to feed the population (Meyer 2017, 1). Numerous villages and estates with gardens and fields now occupied the northern hinterland (*ibid.*, 28-54) (Fig. 1).

The Palmyrenes obviously had much more control over the nomadic groups on the dry steppe than the Ottomans had. The Palmyrenes descended from the Aramaic-speaking population of the *badia*, and they brought with them ties of kinship, friendship and patronage (Sommer 2005, 170-183). Michael Sommer even argues that the tribes of the city of Palmyra reflected the actual nomadic tribes of the *badia*, and that tribal solidarity between the population of the city and the nomads on the dry steppe remained a strong link, also after the habitation in the oasis grew into a large city, which in many ways could be compared with the largest cities in the eastern part of the Roman Empire (*ibid.*, 175-178; Sommer 2017, 98-99). Whatever the exact nature of the Palmyrene tribes, we know from modern nomadic groups, that some families may have houses and

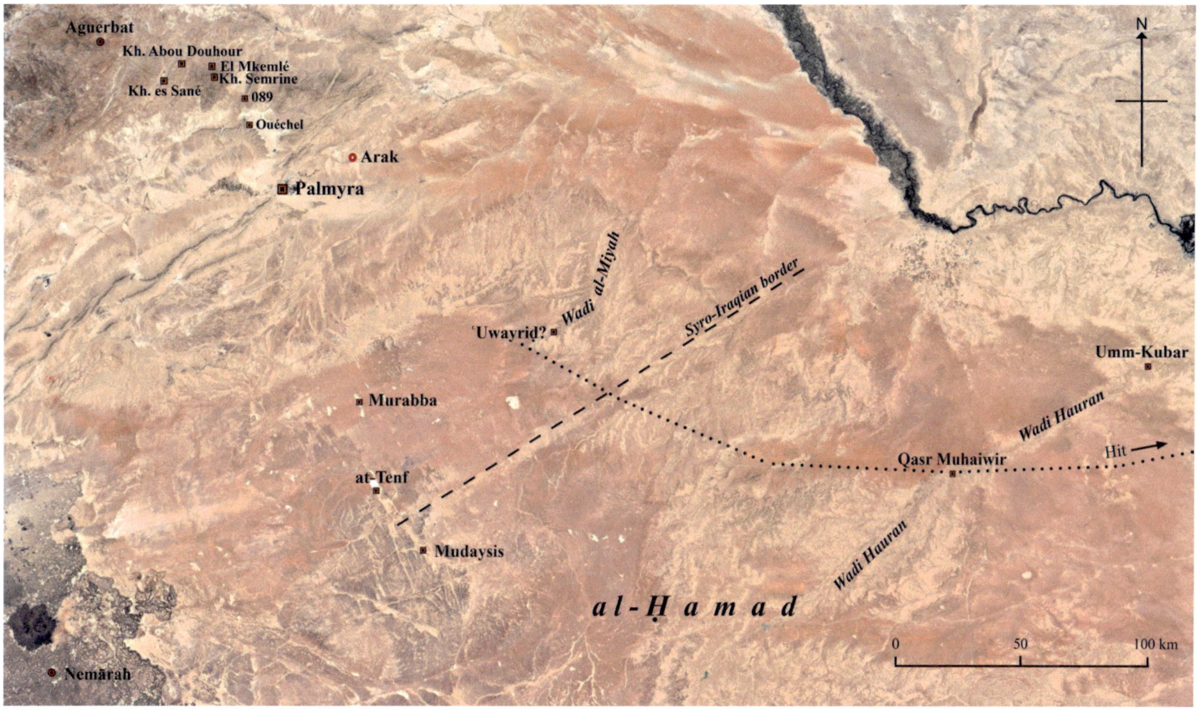

Figure 2. Distribution of Safaitic inscriptions north and south of Palmyra (J.C. Meyer, based on Google Earth).

gardens in the cities (O'Brien 2000, 121-122; Bell, Diaries 22/4 1914; MacDonald 2014, 162). Close social relations in themselves do not guarantee peaceful conditions for the caravan merchants and the sedentary population, but they make it much easier to settle disputes.

Inscriptions from Palmyra tell us that problems may occur (Gawlikowski 1994). In AD 199, the four tribes of Palmyra honoured Ogeilu son of Maqqai for continually raising commands against the Bedouins and providing safety to merchants and caravans (Yon 2012, no. 222; PAT 1378). Soades, son of Boliades, was honoured for saving a caravan in great danger in 132 (Yon 2012, no. 150; PAT 0197), for defending a caravan against robbers led by an ʿAbdallat from the otherwise unknown place Eheîteî in 144 (Yon 2012, no. 127), and for his generosity toward merchants and caravans in 145 (Fox *et al.* 2005, 88-89; PAT 1062). Other inscriptions honour people who had assisted the caravans or brought them back at their own expense, which probably means that they paid tribute or protection money out of their own pockets to the nomadic groups along the route in return for immunity from raiding (Yon 2012, nos. 67, 74, 87). We do not know how often and why these extraordinary measures were necessary, and perhaps the inscriptions represent an exception to normal conditions (Young 2001, 147; Seland 2016, 68), but it raises the question of how far into the *badia* the Palmyrene social and political network extended.

Palmyra seems to have had firm territorial control as far as Wadi al-Miyah about 120 km south-east of the oasis (Meyer 2017, 63-65; Meyer and Seland 2016) (Fig. 2). Beyond that, the control was probably sporadic and the political and tribal landscape much more complicated. The Palmyra tariff has an important section about nomadic groups outside Palmyrene territory (Matthews 1984, 180; Fox *et al.* 2005, 46, 54; Healey 2009, 173, 184-185):

> Ἐννόμιον συνεφωνήθη μὴ deῖν πράσσε[ιν ἐκτὸς τῶν] τελῶν· [τ]ῶν δὲ ἐπὶ νομὴν μεταγομένων [εἰς Παλ]μυρηνὴν θρεμμάτων ὀφείλεσθαι· χαρα[κτη]ρίσασθαι τὰ θρέμματα ἐὰν θέλῃ ὁ δημο[σιώνης,] ἐξέστω.

> "It has been agreed that payment for grazing rights is not to be exacted [as distinct from the normal?] taxes; but for animals brought into Palmyrene territory for the purpose of grazing, the payment is due. The animals may be branded, if the tax-collector so desires." (Greek: 233-237, Aramaic: 145-149. Translated by Fox *et al.* 2005)

The section differentiates between animals belonging to the territory and animals brought in from outside. As the seasonal migrations cut across political borders, the differentiation is not strictly geographical but based on a social and political affiliation. Some nomadic groups coming from outside with their herds were obviously not integrated into the Palmyrene social and political network, and they were liable to pay a special tax. Who were these groups and where did they come from?

Safaitic inscriptions

Our knowledge of the ancient nomadic population around Palmyra is scanty, and apart from drawing ethnographic parallels to attested tribes and their migration patterns in the Ottoman Empire, the only source is a few inscriptions. Of particular interest are several inscriptions with Safaitic text (Fig. 2), the language spoken by nomadic groups in southern Syria, north-eastern Jordan and north-western Saudi Arabia.

The Safaitic inscriptions have been found in the mountains north-west of Palmyra, where Daniel Schlumberger carried out excavations in the 1930s, in Palmyra itself, and south and south-east of the oasis. Schlumberger and Ryckmans published ten Safaitic inscriptions:[1] two from Kheurbet Semrine (PNO no. 2 quarter; PNO no. 21 bis), one from el-Mkemlé (PNO no. 34 ter), one from Kheurbet al-Sane (PNO no. 54), one from Kheurbet Abou Douhour (PNO no. 60), two from the small fort close to the springs at Ouéchel in Jebel Abyad (PNO no. 63 bis, PAT 1730, OCIANA Damascus Museum 4205; PNO no. 63 quarter, OCIANA Damascus Museum 4207), and three from the area north-west of Palmyra with unknown exact provenance, gathered by the inhabitants of Aguerbat, a small village north-west of Jebel Bilaas (PNO no. 80, OCIANA Damascus Museum 2746; PNO no. 81, OCIANA Damascus Museum 1669; PNO no. 82). One large rock-cut inscription has been registered at site 089 in Jebel Merah by the Syrian-Norwegian survey (Meyer 2017, 171-172). Three inscriptions have been found in Palmyra in the Allat sanctuary by the Polish mission (Gawlikowski 1995, 107).[2] A relief from the Louvre (AO 19801), with the triad of Bel-Shamin, Alglibol and Malakbel, has both an Aramaic and Safaitic text (Seyrig 1949, 29-33, 35-40; Dentzer-Feydy and Teixidor 1993, 144-145). Another inscription has been found in the quarries north-east of Palmyra.[3]

Many of the inscriptions are very short or fragmentary, but seven of them provide more information:

1. Ouéchel. PNO no. 63 quarter; OCIANA Damascus Museum 4207.
   ---- fty brd ḏ- ʾl ʿ---
   "---- young slave of Brd of the lineage of ʿ----"

2. Ouéchel. PNO no. 63 bis; OCIANA Damascus Museum 4205.
   l whb bn kn ḏ- ʾl ṣwkt w wgm ʿl- ʾbgr bn s^{2}kr
   "By Whb son of Kn of the lineage of Ṣwkt and he grieved for ʾbgr son of S^{2}kr"

3. Kheurbet Abou Douhour. PNO no. 60.
   ----ʿwḏy ḏ ʾ[l] ----ḏ----
   "----ʿwḏy of the {lineage of}----"

4. Unknown provenance. PNO no. 80; OCIANA Damascus Museum 2746.[4]
   ----rmyn ḏ- ʾl mʿs^{1} w rʿy bt ---- s f h lt w ds^{2}r ḥwr l- ʾhl w t----
   "----rmyn of the lineage of Mʿs^{1} and he pastured at T---- and so O Lt and Ds^{2}r [grant] a return to the family of Wt ----"

5. Unknown provenance. PNO no. 81; OCIANA Damascus Museum 1669.
   l ms^{1}k bn ʿbṭ ḏ- ʾl nmr w wgm ʿl- ʿbṭ w ʿl- ms^{1}k bn kd ---- trḥ
   "By Ms^{1}k son of ʿbṭ of the lineage of Nmr and he grieved for ʿbṭ and for Ms^{1}k son of Kd ---- perished"

6. Unknown provenance. PNO no. 82.
   l ʿ{w}{ḏ} bn zbd bn ʾkm {ḏ} ʾl {s^{2}}bʿr
   "By {ʿwḏ} son of Zbd son of Km of the lineage of {S^{2}}bʿr"

7. Jebel Merah, site 89. Meyer 2017, 171–172. (Fig. 3)
   1) l kmd bn ʾs^{2}g ḏ ʾl gḏl w mṭy f h lt s^{1}lm w h-ḫṭṭ
   2) l {k}md w {r}{ʿ}{y} {b} …
   3) w ds^{2}r s^{1}lm {m} {h/ʾ}ḏ ʿʿwr w wgm ʿl m{.}k{.} bn mk…
   1) "By Kmd son of ʾs^{2}g of the lineage of {Gḏl} and he was on a journey so, O Lāt, may he and this writing be secure"
   2) "By {Kmd} and {he pastured} {at/during}…"
   3) "so Dusharā may it (the writing) be secure {against} {him} {who} would efface (it) and he grieved for {M.k.} son of Mk…"

In seven of the eleven inscriptions from north of Palmyra, the authors specify their affiliation, not by long descent chains, son *(bn)* of, son of, *etc.*, but by stating a lineage group (ʾl): *ṣwkt, gḏl, nmr, Mʿs^{1}, {S^{2}}bʿr*. One of the unpublished inscriptions from the Allat sanctuary also has a lineage group, but the name is missing. The lineage group of *gḏl* is attested in two inscriptions in the OCIANA database (OCIANA C 321 and C 2268), *nmr* in two other inscriptions (OCIANA HCH 82 and HCH 126). *Nmr* may refer not to a lineage group as such, but to a geographical affiliation, *i.e.* Nemarah. Nemarah was one of the most important locations in Hauran with easy access to water and pasture, and the area is mentioned in 46 inscriptions (OCIANA C

1 Some of the inscriptions in the National Museum in Damascus have later been published in the OCIANA database with corrected translations. Michael MacDonald has revised the translation of the remaining inscription.

2 I am grateful to Michał Gawlikowsky for having given me access to photos of the graffiti.

3 According to personal information from Khaled al-As'ad and Jean-Baptiste Yon.

4 The rendering of the text in *PNO*, mentioning Hauran, is not correct; see Macdonald 1993, 364.

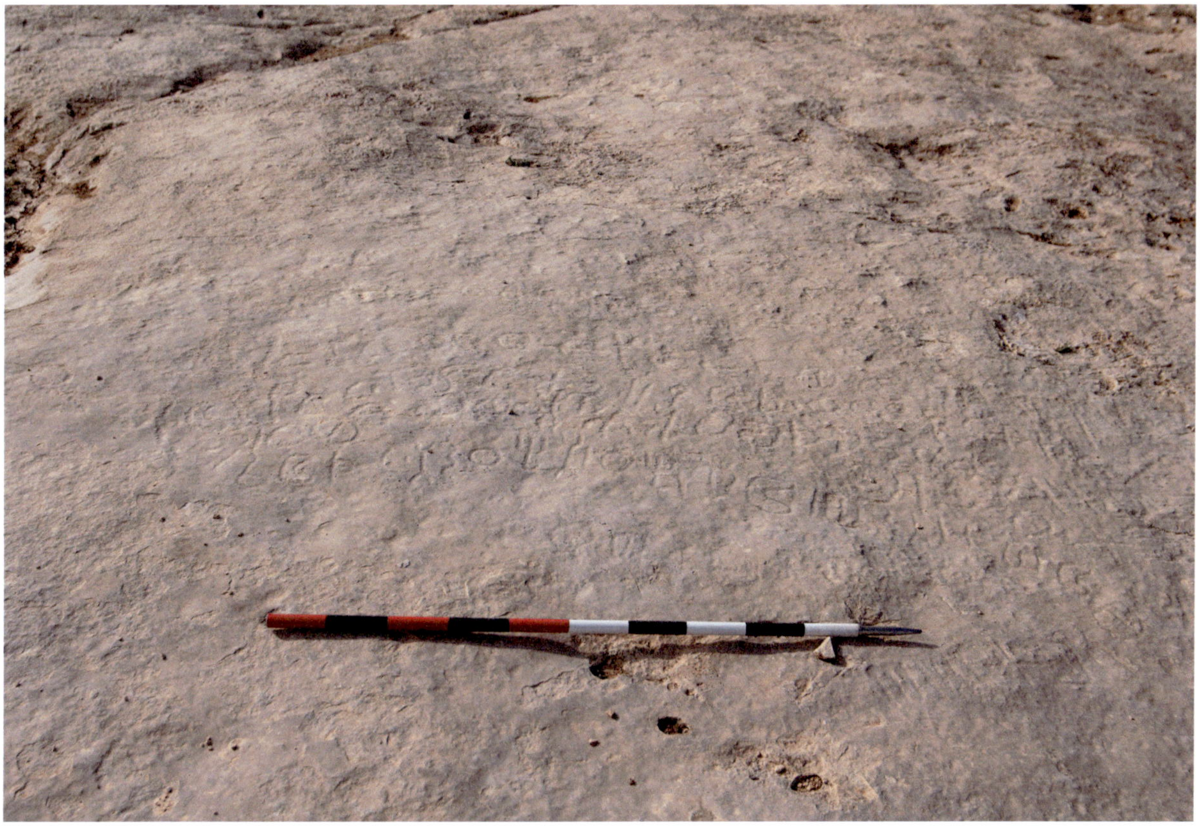

Figure 3. Safaitic rock-cut inscription at site 089, Jebel Merah (J.C. Meyer).

523, C 1656, C 1894, C 1895, C 2732, C 2803, C 3143, C 3878, C 4355, LP 330, LP 426, LP 532, LP 540, LP675, WH 2060, HaNSB 133, KRS 896, KRS900, KRS 901, MKWS 33, Is.M 148, Is.Mu 168, Is.Mu 321, Is.H 162, Is.K 276, Is.H 583, Is.M 867, Is.M 358, AWS 2, AWS 48, AWS 52, AWS 53, AWS 54, AWS 76, AWS 109, AWS 131, AWS 221, AWS 275, RSIS 199, RSIS 339, BRenv.K 1, NLB.C 1, SESP.M 2, SESP.D 22, NRW.F 1, Al-Namārah.H 108).

The Safaitic inscriptions south and south-east of Palmyra are relatively few,[5] but they show the same tendency. Three of the seven inscriptions from Mudaysis 30 km south-east of At-Tanf have an affiliation to a lineage group, *i.e.* OCIANA Damascus Museum 177748 (*rks¹*), 17751 (*ntg*) and 17752 (*rks¹*), and one from the At-Tanf area, *i.e.* OCIANA Palmyra Museum no.1 (*gfft*). *rks¹* is a relatively common affiliation in the OCIANA database (OCIANA HCH 104, NTSB 2.1, WH 2837, CSI.S 9, SHS 10, RSIS 110, THSB 4). All the three inscriptions at Umm Kubar at the eastern end of Wadi Hauran, two of them partially bilingual Aramaic/Safaitic (Safar 1964), have an affiliation: OCIANA SIWH 3 (*nʿmn*), SIWH 6 (*nʿmn*) and SIWH 7 (*ṣḥ*).

The total number of Safaitic inscriptions in the Palmyrene area is low, but the tendency is clear. About two-third of the registered inscriptions have an affiliation. This is remarkable compared to the Safaitic core area. Of 33,164 registered inscriptions in the OCIANA database, only 2.7% have an affiliation to a lineage group. Nothing seems to indicate that the Safaitic speaking people in the Palmyrene territory belonged to an isolated group based around Palmyra, using the villages and estates as a more permanent seasonal base, as suggested by Schlumberger (PNO 129-132), with a different custom of stating their affiliation. None of the reliefs or altars found in the villages of Jebel Chaar have Safaitic dedications. Apart from the relief in the Louvre, mentioned above, and a small dedication together with some Aramaic dedications on the rim of a krater at the sanctuary of Abgal at Kheurbet Semrine (PNO no. 21 bis), the Safaitic inscriptions are graffiti in the proper sense of the word. The rock-cut inscription from site 089 in Jebel Merah and PNO no. 82 mention travel, and PNO no. 80/OCIANA 2476 "return to family". Some of the affiliations point to the Safaitic area and several graffiti from northern Jordan and southern Syria mention travel to and from Palmyra (*tdmr*) (Meyer

5 Apart from the graffiti with an affiliation: two at Qasr Muhaiwir (Gregory and Kennedy 1985, 201, 409, note 201. The texts are illegible), several at At-Tanf (Poidebard 1934, 126, pl. XCVIII a-b), and one at ʿUwayriḍ in Wadi al-Miyah (OCIANA Palmyra Museum 1357). The exact location of the latter is uncertain.

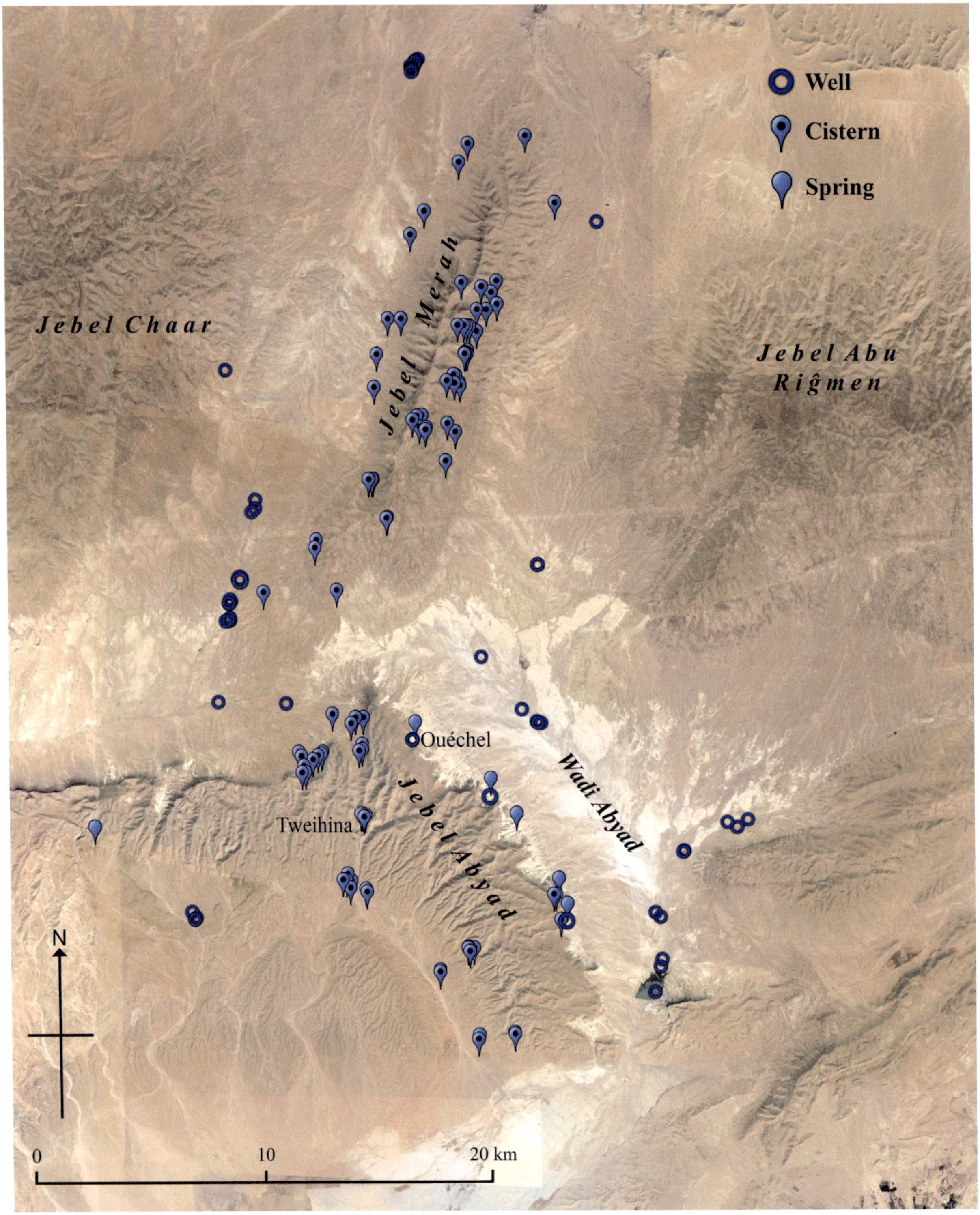

Figure 4. Distribution of springs, wells and cisterns in Jebel Abyad and Jebel Merah (J.C. Meyer, based on Google Earth).

2017, 210-211), as this one from Wadi Salman in northern Jordan (OCIANA LP 717):

l ʿm bn šmt bn ġnm bn *ʾnʿm w ʾty m tdmr f h lt s*^{1}*lm.*
"By ʿm son of S^{2}mt son of Ġnm son of ʾnʿm and he came from Tadmur and so O Lt [grant] security"

The extremely high occurrence of affiliations in the Palmyrene Safaitic graffiti is most probably due to the fact that the authors were far from home and felt a need to state their relationship to a larger group. They were not an integrated part of the population of the Palmyrene territory but occasional visitors with their herds. The territory north of Palmyra with its mountain chains has a much higher precipitation than the huge dry steppe (*badia*) to the south. The mountains lie within the 200-250 mm isohyet, which in years with adequate rain even allows the cultivation of barley (Meyer 2017, 19-20), and they offered an attractive summer grazing ground for wildlife and pasture for sheep and goats when the vegetation on the *badia* south of Palmyra had wilted, as mentioned by Musil (1928, 149): "Our guide asserted that in summer the districts north of Tudmor swarm with gazelles, but that, on the other hand, these animals are rarely seen in winter or spring, when they remain in the neighborhood of al-Ḥamâd. Not until all the grass there has wilted and dried up do they return to Palmyrena, which, with its more adequate moisture, keeps the grass in good condition until late in the autumn."

Conclusion: nomads and sedentaries

In antiquity, villages and estates occupied the northern territory, but there does not seem to have been any competition or conflict between horticulture and agriculture on one hand and pastoralism on the other. The harvest took place from the beginning of May, and the stubble fields were an excellent grazing ground at the approach of the hot season. In return, the herds fertilised the fields with their manure before the ploughing. There are only a few spring areas and wells in the hinterland but the villages and estates relied on cisterns with elaborate systems of long supply channels harvesting the surface run-off that would otherwise have entered the wadis during heavy showers, if not detained for gardens and fields (Meyer 2017, 24-27) (Fig. 4). The construction of cisterns thus increased the amount of water for human and animal consumption dramatically. In addition, the city was a good market for wool, animals for slaughter, hides, animal fat, all mentioned in the Palmyra tariff as liable to tax, and the market offered the nomads cereals, salt, oil, clothes, products from the copper- and blacksmith, and items of prestige.

How large were the Safaitic speaking groups that entered the northern Palmyrene area at the approach of the hot season with their herds? The amount of Safaitic graffiti is negligible, compared to the Safaitic core area, but the registered finds must represent a much larger number. In the Safaitic area, the graffiti are engraved on hard, black volcanic basalt stones, and they stand up well to erosion over the centuries. Outside this area the rock is softer, often limestone. Sharp grinding grains will soon sand-blast engravings exposed to the wind. Most of the inscriptions from the Palmyrene area come from Schlumberger's excavations of villages in Jebel Chaar. The abandoned buildings were partly covered by wind-blown material protecting the graffiti, and there can be no doubt that future excavations of the new sites will increase the number of inscriptions. The scarcity of finds applies not only to the Safaitic, but also to the Aramaic graffiti. It is remarkable that the Safaitic graffiti north of Palmyra amount to about 40% of all graffiti found in an otherwise Aramaic speaking area.

There can be no doubt that Safaitic speaking groups made up a large part of the pastoralists that brought their animals into the Palmyrene territory from outside and that they were liable to a tax for the grazing rights mentioned in the tariff. They did not have a permanent seasonal base in the mountains north of Palmyra. They were far from home, as mentioned above. Occasionally they entered the Palmyrene area at the approach of the hot season from the south-eastern limestone desert (*al-hamad*) rather than returning to the well-watered region in the west and south-west around Hauran (MacDonald 2014, 149). Some groups may even have migrated to the Euphrates valley along Wadi Hauran, where they left their visiting cards at Umm Kubar (cf. Fig. 2).

Even if the nomadic and the sedentary ways of life can be regarded as two complementary economic systems, years with insufficient rain in the *hamad* would have given rise to conflicts, when the nomads entered agricultural districts before the harvest, and sheep and goat can inflict great damage on gardens. Further, the tribal composition of the *badia* is not static. New nomadic groups entering from the south, competition for the best grazing grounds and water resources, or skirmishes between nomadic groups may disturb the seasonal pattern of migrations. Whatever the reason, the Safaitic groups could not be controlled by old ties of patronage or kinship, and so the Palmyrenes needed to exercise more direct control of the territory and also to administer a complicated tax system regarding different grazing rights.

The Palmyrenes had at their disposal mobile dromedary units depicted on several reliefs in the Palmyra museum (Fig. 5). It is a matter of dispute whether they functioned as military escorts of the caravans across the *badia* (Seland 2016, 69, with references), but it certainly gave the authorities the ability to mobilise military force quickly in areas of conflict within its own territory. Palmyra also had a more permanent military presence in the northern hinterland within a short distance of the

Figure 5. Member of the Palmyrene dromedary corps. From the precinct of the Bel temple (J.C. Meyer).

oasis (Meyer 2017, 58-59) (Fig. 1). At Tweihina, a small fort overlooked an important Y-junction in Jebel Abyad only one day's travel from the oasis (*ibid.*, 103-111; PNO 48-50, 86-88) (Fig. 6). It controlled the only communication line through Jebel Abyad from the southern *badia* to the attractive summer pastures north-west of the city. From Greek and Latin inscriptions we know that Roman military personnel complemented the fort (PNO 86–87). A small fort at Ouéchel, also with Roman military personnel, overlooked one of the most important spring areas at the edge of Jebel Abyad (Meyer 2017, 86-89; PNO 46-48, 85-86). Other forts and stations controlled the territory and important water resources, like spring and well areas (Meyer 2017, 62). Some of them may also have functioned as centres of tax farming administering the complicated grazing rights (*ibid.*, 69). The forts and stations were probably not complemented all the year round, only in the season when large flocks of sheep and goats entered the territory. They prevented the grazing season from developing into a chaotic situation with conflicts between nomads and villages, and between different nomadic groups, especially in years with insufficient precipitation in the winter and spring pasture grounds of the *hamad*.

This system of control over the nomadic groups, which ensured peaceful relations between the desert and the sown, did not depend on specific Palmyrene social or cultural relations with the population of the *badia*. It survived the fall of Zenobia in the end of the third century and the subsequent transformation of Palmyra into an important Roman military and administrative stronghold along *Strata Diocletiana*. The villages and estates continued to provide provisions to the city in the oasis up to the Umayyad period (Meyer 2017, 13-16). In the late Roman period, new forts and stations were built along *Strata Diocletiana* expanding the territorial control and safeguarding the route from Damascus to Resafa. In addition, the Byzantines and the Umayyads had the advantage of being able to negotiate peaceful relations with the powerful Banu Kalb tribe, who controlled the population of the *badia* in the northern part of the Arabian Peninsula. In the following centuries, the habitation in the oasis declined and the villages in the mountains disappeared. In the Ottoman period, a Bedouin strategy of survival took over in the hinterland of Palmyra. They exploited and maintained the cisterns with supply channels in the former villages, but the relationship between the desert and the sown had changed.

Figure 6. The fort at Tweihina on a small hilltop overlooking the route through Jebel Abyad (J.C. Meyer).

Sigla

PAT — Hillers, D.R. and Cussini, E. 1996. *Palmyrene Aramaic Texts.* Baltimore: Johns Hopkins University Press.

PNO — Schlumberger, D. 1951. *La Palmyrène du nord-ouest. Villages et lieux de culte de l'époque impériale. Recherches archéologiques sur la mise en valeur d'une région du désert par les Palmyréniens*. Paris: Librairie orientaliste Paul Geuthner.

References

Bell, G.M.L. 'Diaries'. New Castle: Newcastle University Gertrude Bell Archive. Online archival material: http://gertrudebell.ncl.ac.uk/index.php.

Dentzer-Feydy, J. and Teixidor, H. 1993. *Les antiquités de Palmyre au musée du Louvre*. Paris: Réunion des Musées Nationaux.

Fox, G., Lieu, S. and Ricklefs, N. 2005. Select Palmyrene inscriptions. In: Gardner, I., Lieu, S. and Parry, K. (eds.). *From Palmyra to Zayton: epigraphy and iconography*. Turnhout: Brepols, 27-188.

Gawlikowski, M. 1994. Palmyra as a trading centre. *Iraq* 56, 27-33.

Gawlikowski, M. 1995. Les arabes en Palmyrène. In: Lozachmeur, H. (ed.). *Présence arabe dans le croissant fertile avant l'Hégire*. Paris: Éditions Recherche sur les Civilisations, 103-108.

Grant, C.P. 1937. *The Syrian desert. Caravans, travel and exploration*. London: A. & C. Black.

Gregory, S. and Kennedy, D. 1985. *Sir Aurel Stein's limes report*. Oxford: British Archaeological Reports (International Series 272).

Healey, J.F. 2009. *Aramaic inscriptions and documents of the Roman period*. Oxford: Oxford University Press.

Lewis, N. 2000. The Syrian steppe during the last century of Ottoman rule: Hawran and the Palmyrena. In: Mundy, M. and Musallam, B. (eds.). *The transformation of nomadic society in the Arab East*. Cambridge: Cambridge University Press, 33-43.

Macdonald, M.C.A. 1993. Nomads and the Hawran in the late Hellenistic and Roman periods: a reassessment of the epigraphic evidence. *Syria* 70, 303-403.

Macdonald, M.C.A. 2014. 'Romans go home'. Rome and other 'outsiders' as viewed from the Syro-Arabian desert. In: Dijkstra, J.H.F. and Fisher, G. (eds.). *Inside and out. Interactions between Rome and the peoples on*

the Arabian and Egyptian frontiers in Late Antiquity. Leuven: Peeters, 145-164.
Marcus, A. 1989. *The Middle East on the eve of modernity. Aleppo in the eighteenth century*. New York: Columbia University Press.
Matthews, J.F. 1984. The tax law of Palmyra: evidence for economic history in a city of the Roman East. *The Journal of Roman Studies* 74, 157-180.
Meyer, J.C. and Seland, E.H. 2016. Palmyra and the trade route to Euphrates (Hatra, Palmyra and Edessa). *ARAM* 28, 497-523.
Meyer, J.C. 2017. *Palmyrena. Palmyra and the surrounding territory from the Roman to the Early Islamic period*. Oxford: Archaeopress.
Musil, A. 1928. *Palmyrena. A topographical itinerary*. New York: American Geographical Society.
O'Brien, R. (ed.) 2000. *Gertrude Bell. The Arabian diaries, 1913-1914*. New York: Syracuse University Press.
Poidebard, A. 1934. *La trace de Rome dans le désert de Syrie. Le limes de Trajan à la conquète arabe. Recherches aériennes (1925-1932)*. Paris: Librairie Orientaliste Paul Geuthner.
Reilly, J.A. 2015. Town and steppe in Ottoman Syria. *Der Islam* 92, 148-160.
Safar, F. 1964. Inscriptions from Wadi Hauran. *Sumer* 20, 9-27.
Seland, E.H. 2016. *Ships of the desert and ships of the sea. Palmyra in the world trade of the first-third centuries CE*. München: Harrassowitz Verlag.
Seyrig, H. 1949. Antiquités syriennes. *Syria* 26, 17-41.
Sommer, M. 2005. *Roms orientalische Steppengrenze. Palmyra – Edessa – Dura-Europos – Hatra. Eine Kulturgeschichte von Pompeius bis Diocletian*. Stuttgart: Franz Steiner Verlag.
Sommer, M. 2017. *Palmyra. A history*. London: Routledge.
Yon, J.-P. 2012. *Inscriptions grecques et latines de la Syrie. Tome 17, fasc. 1. Palmyre*. Beyrouth: Presses de l'Ifpo.
Young, G.K. 2001. *Rome's eastern trade. International commerce and imperial policy, 31 BC – AD 305*. London: Routledge.

The north-eastern *badia* in Early Islamic times

Karin Bartl

Abstract

The settlement of the north-eastern *badia* in Early Islamic times is mainly characterised by the emergence of various impressive monuments, known as 'desert castles' or *quṣūr*,[1] which have long been the subject of intense discussions about their original function. They form a group of monuments that covers the entire region of Bilad al-Sham and includes a variety of different types of buildings and settlements. In addition to representative buildings, various hydraulic installations form defining structural units in many sites and suggest an at least temporary agricultural use of the semi-arid steppe. Another group consists of larger villages with scattered buildings, which apparently arise at the same time. Overall, the number of settlement sites from the seventh to ninth/tenth centuries is rather small. However, it can be assumed that in addition to the partially long-known structures, further isolated farmsteads and small settlements existed, which can only be recognised by intensive prospecting and will hopefully complement the current settlement image in the future.

Keywords: desert castles, quṣūr, Early Islamic period, Umayyads, Abbasids

The region

Jordan includes a number of very different landscapes, among which the desert and steppes (Arabic: *badia*) make up by far the largest area. The *badia* comprises about 80% of the territory of Jordan and is divided into the three areas of north, central and south, each of a different size and structure. The north-eastern *badia* includes in addition to the so-called 'panhandle', *i.e.* the area between the town of Safawi and the Jordanian-Iraqi border, also the entire region between Azraq and the north-south line between Mafraq and Sahab, about 20 km east of Amman. To the west of this border is the cultivated land with an annual isohyet of approximatively 200 mm (Ababsa 2013, Fig. I.12).

One of the essential characteristics of all areas of the *badia* is the complete lack of perennial water courses. The water supply is based therefore mainly on the winter rainfall through which the ground-water reservoirs are replenished and for whose long-term use a sophisticated water management had already been developed in prehistoric periods. Dams, canals and reservoirs for water regulation and distribution have been detected particularly in many areas of the basalt desert (Rollefson *et al.* 2011; Rollefson 2016; Müller-Neuhof 2016; see also Müller-Neuhof, this volume).

---

1 Where transliterated Arabic texts or words are cited, the system of transliteration specified in the *International Journal of Middle Eastern Studies* has been followed.

In: Peter M. M. G. Akkermans (ed.) 2020: *Landscapes of Survival - The Archaeology and Epigraphy of Jordan's North-Eastern Desert and Beyond*, Sidestone Press (Leiden), pp. 265-286.

An exception in terms of water supply in the north-eastern *badia* is the oasis of Azraq, one of the largest oases in south-west Asia. Here, aquifers and springs such as ʿAin Sawda and ʿAin al-Assad provide a permanent water supply, which forms the basis for a correspondingly extensive flora and fauna (Amr *et al.* 2011, 188ff.). In addition, numerous *sabkhas* or *qa*ʾ areas (Arabic *qa*ʾ: depression, mud flat), often remnants of palaeolakes, are reservoirs for temporary precipitation (Barth and Böer 2002).[2]

However, the current unfavourable settlement conditions caused by climatic and anthropogenic impacts are not to be equated with the conditions in the prehistoric and historical periods (after the end of the Umayyad period). The results of recent research, especially in the basalt region of the 'panhandle', show an almost continuous settlement of the *badia* from the Epipalaeolithic period (Richter *et al.* 2014; see also Richter, this volume) until the Early Islamic period.

For the period between 100 and 700 AD, data of Soreq Cave near Jerusalem show increasing aridity (Orland *et al.* 2009). Moreover, based on the evaluation of various palaeoclimatic data as well as historical and archaeological sources from south Anatolia and the Levant, a sophisticated model was recently presented for the Late Roman-Early Byzantine to Early Islamic periods. Thereafter, in the Levant a short-term drought between 350 and 470 AD is followed by a period (470 to 670) with more humid climate, which is followed by increasing aridity (Izdebski *et al.* 2016).[3]

The high density of settlement between the fifth and seventh centuries in the region to the west of the *badia* provides clear evidence of the favourable settlement conditions (see *e.g.* Kerestes *et al.* 1977-78, Tab.1; Al-Khoury *et al.* 2006; Bartl *et al.* 2001). Even after the Islamic conquest, the positive settlement development continued here. On the basis of historical and archaeological sources, the time span between the eighth to tenth centuries has been divided into three sections: continuation of urban settlement in the eighth century, contraction in the second to third quarter of the ninth century, and a strong rural economy in the tenth century (Kennedy 1985; Walmsley 1992; 2007, 352). The earthquake at the beginning of the year 749, which led to destruction in many places (Sbeinati *et al.* 2005, 362ff.), and the relocation of the caliphal centre under the Abbasids, formed an incision in settlement development. For the region of Bilad al-Sham, the time around 750 was therefore also occasionally defined as the actual end of Late Antiquity (Kennedy 1999, 235).

Early Islamic settlement

The term 'Early Islamic' describes the first two periods after the Islamic conquest: the Umayyad caliphate (661-750) and the first centuries of the Abbasid dynasty (750-1257). The centre during the Umayyad rule was Damascus; after the seizure of power by the Abbasids in 750, the centre of the empire moved first to Al-Anbar on the Euphrates and later to Baghdad, where under the second caliph Al-Mansūr the Abbasid capital was established from 762 onwards. From the end of the ninth century, the south-western part of the empire was under the control of the Tulunids (878-905) and Ikhshidids (935-969), who were replaced by the Fatimids in 969.

With the Umayyad conquest a new division into military provinces, the *ajnād* (singular *jund*) took place. The region to the north, east and south-east of Damascus as far as Ayla/ʿAqaba now formed the jund Dimashq, which encompassed the entire area of central Syria and eastern Jordan including large parts of the *badia*. The western border areas of the coastal zone between Sidon/Saidā and Gaza to the Jordan valley as well as the Golan formed the ajnād al-ʿUrdunn in the north and Filastīn in the south. The provinces were divided into different districts, among which the units of jund Dimashq, Balqāʾ, Bathaniyya and Hauran, with the capitals Damascus/Dimashq, ʿAmmān, Darʿā and Buṣrā, bordered immediately west and north of the north-eastern *badia* (Le Strange 1890, 24ff.).

The Umayyad reign (661-750) is characterised in all parts of the empire by dynamic construction activity (Bacharach 1996). In the north-eastern *badia*, this period represents the last settlement phase characterised by significant building activities with numerous new foundations before the beginning of the re-settlement of the region, which has been observed for some twenty years. Contrary to this development, up to now very few building activities are documented for the early Abbasid period.

Generally, based on the previously known data for all post-Iron Age periods since the second half of the first century BC, the north-eastern *badia* is characterised by a very low settlement density. An exception is the south-western area of the foothills of the Hawran with comparatively dense rural settlement (De Vries 1998; 2000). In the areas to the east and to the south-east, in Roman and late Roman times, settlements are manifested above all in watch posts, which were built within the framework of border control (Parker 1986; Kennedy 2004).

The Early Islamic period in the *badia* and its peripheral regions is characterised by the emergence of a number of individual buildings and small settlements, whose significance and function are still controversial and which occasionally are based on the extension and remodelling

2 In general, however, the overuse of water resources in recent decades has led to a large reduction in the oasis landscape. Nevertheless, this area still has the densest vegetation in the *badia*. For some years, the restoration of parts of the dried-up oasis area has been promoted (Alraggad and Jasem 2010).

3 For general climate trends in the southern Levant, see *e.g.* Rambeau 2010.

of Roman-late Roman sites. They are subsumed under the term 'desert castles' (Arabic: *quṣūr*), which encompasses a multitude of formally different structures. This is therefore to be understood as a *term of concept* rather than an architectural definition (Bloch 2014). It is avoided in recent analyses and replaced by more neutral definitions such as *établissement* (establishment, settlement) (Genequand 2012).

On the basis of older interpretations and new field research, the Early Islamic sites have recently been linked to three primary functions, which, however, are not all necessarily relevant at the same time in the various sites: as a meeting place for caliphal elites and local tribal leaders, as agricultural estates to increase state revenues, and as a permanent or temporary residence for elites (Genequand 2012, 395ff.). In addition, other uses such as hunting lodges or halting stations on roads and caravan routes are conceivable.

Altogether, less than twenty sites of the Early Islamic period are located in the northern and central *badia* of Jordan. Most of them consist of substantial architecture, either made of limestone or of basalt. All sites are characterised by specific features and differ significantly in terms of size and structure.

It is unclear whether and to what extent further small settlements, camp sites or farmsteads existed in the *badia* next to these exceptional settlement sites. However, the scattered references to smaller places exclusively identified by ceramic surface material indicate a more complex use of the *badia* since the Roman period than previously assumed (see Kennedy 2014, 108; Huigens 2015, 189; 2019; see also Huigens, this volume).

In view of the numerous gaps between the surveyed areas of the north-eastern *badia*, no definite statement on settlement development over large areas is currently possible. However, recent surveys, with a focus explicitly on the documentation of Late Antique and Early Islamic sites, have shown that the density of sites is rather low for these periods (Bartl *et al.* 2014).

Brief research history

The archaeological exploration of the Early Islamic periods in the north-eastern *badia* began at the end of the nineteenth century by the records of J. Gray Hill (1896); R.E. Brünnow and A. von Domaszewski (1905); A. Musil (1907); M. Moritz (1908); G. Bell (1907); and A. Jaussen and R. Savignac (1922). Despite some adverse circumstances such as tribal disputes, a large amount of fundamental data on settlement, especially in the region west of Azraq, was obtained in the period between about 1895 and 1914. At this time, the aim of the research was exclusively the documentation of existing stone buildings, which were called *qaṣr* (castle).

The early descriptions were supplemented from the end of the 1930s by various documentations of researchers such as A. Stein (Gregory and Kennedy 1985), N. Glueck (1944) and R.W. Hamilton (1946), and occasionally were expanded by new discoveries. However, systematic site documentation (Kennedy 1982; King 1982; King *et al.* 1983; Helms 1990), excavations (Urice 1981; 1987; Bisheh 1989; Al-Najjar *et al.* 1989; Waheeb 1993), building research, and restoration work (Almagro *et al.* 1975; Vibert-Guigue and Bisheh 2007) did not begin until the 1970s.

Settlements

The well-known Early Islamic settlement of the steppe mainly focused on the western part of the northern and central *badia* up to the area of Azraq (Fig. 1). Only two sites are known from the eastern area on the edge of the basalt region: Qaṣr Burquʿ (Gaube 1974; Helms 1991) and Ar-Risha (Helms 1990).

The sites to the west of Azraq include Muwaqqar (Brünnow and von Domaszewski 1905; Musil 1907), Qaṣr Mushash (Bisheh 1989; Bartl *et al.* 2013; 2014; Bartl 2015; 2016), Qaṣr Kharana (Urice 1981; 1987), Quṣayr ʿAmra (Musil 1907; Almagro *et al.* 1975; Vibert-Guigue and Bisheh 2007; De Palma *et al.* 2013), Azraq (Kennedy 1982), Khirbat al-ʿUmari (Glueck 1944; Bartl and Akkermans 2016), Hibabiya (Kennedy 2014), Qaṣr al-Hallabat, Hammam al-Sarah (Bisheh 1982, 1985; Arce 2008a) and Fedein/ Mafraq (Humbert 1986, 1989; Bisheh 2018). Also in the *badia*, but further south, are Qaṣr al-Tuba (Musil 1907; Jaussen and Savignac 1922; Vibert-Guigue 2008) and the dam of Wadi al-Jilat (Politis 1993).

Other Early Islamic places are located in the cultivated region. These include Amman (Northedge 1992; Arce 2008b), Qaṣr al-Mshatta (Cramer *et al.*, 2016), Umm el-Walīd (Bujard and Trillen 1997; Bujard *et al.* 2001; Bujard and Schweizer 1992; Genequand 2008), Qastal (Carlier and Morin 1984; 1986; 1989; Bisheh 2000, 2007), and Khan al-Zabīb (Bujard and Trillen 1997; Kennedy 2004, 135ff.; Genequand 2008).

Early Islamic settlement traces are also known from Late Roman to Early Byzantine sites of the Hawran region, for example, Dayr al-Kahf (Parker 1986, 178; Kennedy 2004, 76ff.), Umm al-Quttein (Kennedy 2004, 84) and Umm al-Jimal (De Vries 1998; 2000, 43ff.). Here usually only a few changes to Late Antique buildings were observed and the continuous settlement into the Umayyad and early Abbasid periods can often be demonstrated only on the basis of ceramic finds.

All sites of the *badia* mentioned before may be divided into two groups (Table 1). The first consist of sites whose occupations go back to Roman and late Roman times, whereas the settlements of the second group were founded in the Umayyad period. Besides the above-mentioned places, the former group includes the square structures at Fedein, Qaṣr al-Hallabat, Qaṣr Burquʿ, Qaṣr al-Azraq, and Qaṣr Mushash.

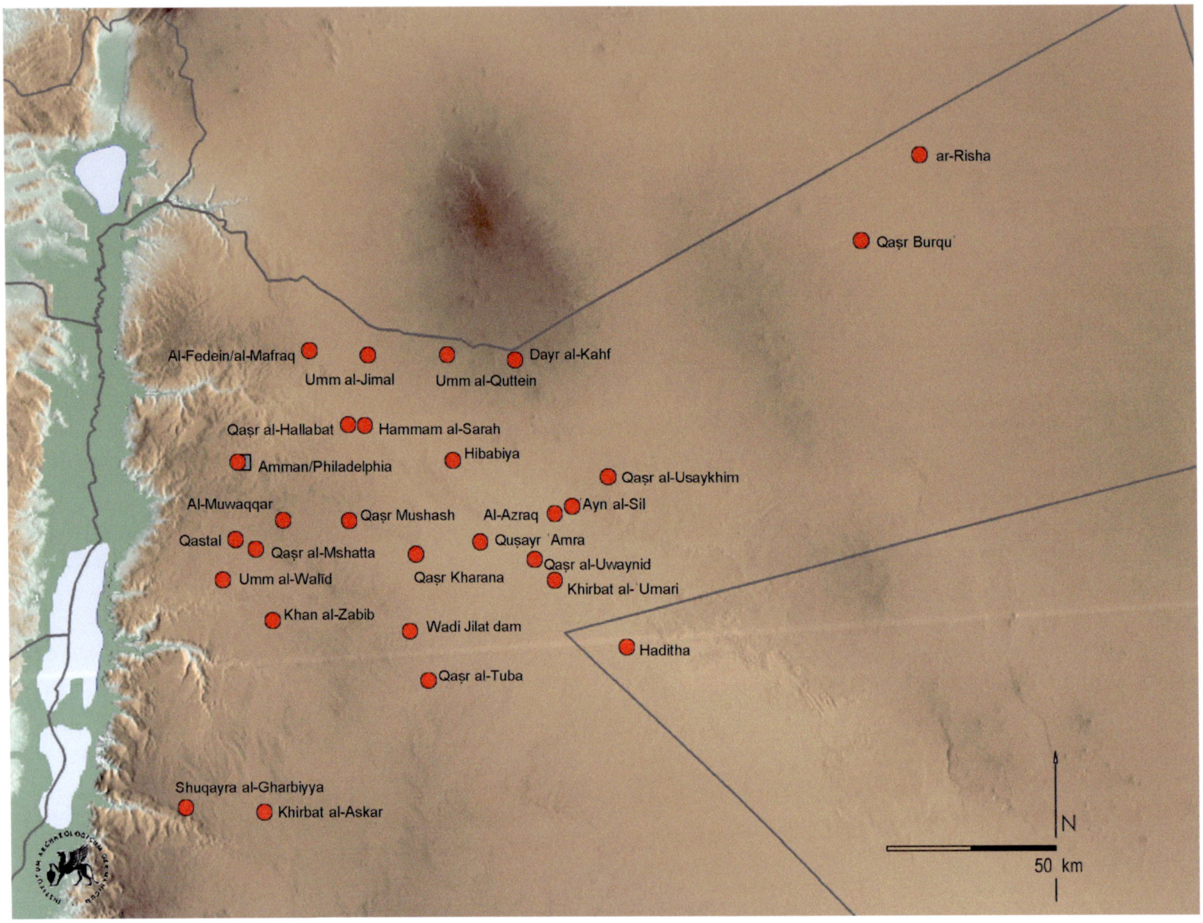

Figure 1. Early Islamic sites mentioned in the text (map by Th. Urban, German Archaeological Institute, Orient Department, using USGS/NASA 3-arc second SRTM data).

Two other prominent Roman to late Roman sites, however, show no signs of Early Islamic settlement: Qaṣr al-Uwaynid (Kennedy 2004, 62-66) and Qaṣr Usaykhim (Kennedy 2004, 66ff.; Al-Khoury and Infranca 2005). These are – like the ones mentioned before – connected with the protection of the Roman border from the second/third century onwards.

The second group of sites whose occupation most probably began in the Early Islamic period is composed of the following site complexes: Muwaqqar, Qaṣr Kharana, Quṣayr ʿAmra, ʿAin al-Sil, Qaṣr al-Tuba, Ar-Risha, Hibabiya and Khirbat al-ʿUmari.

In general, two different forms of settlement can be distinguished: sites with isolated individual buildings and sites with a larger number of different building structures, including domestic units, utility buildings, hydraulic installations, and special buildings of the *qaṣr* type.[4] The latter type of building is usually of square shape and characterised by a simple scheme in which a central courtyard is surrounded on all four sides by rows of rooms that often open individually to the courtyard but can also be interconnected (Genequand 2006). This shape corresponds largely to the building type of the *castrum* already known from Roman times (Hanel 2007). However, this building form is not only connected to the western Mediterranean, but is also known from the Arabian Peninsula (Northedge 2008), so that the Early Islamic *quṣūr* may represent a "fusion or interaction of Arabian and Antique form" (Kennedy 1999, 233).

A novelty of the Early Islamic *qaṣr* in Bilad al-Sham is a specific form of internal structuring. It often follows the so-called *bayt* scheme, *i.e.* the 'house in the house' and consists of several multi-room apartments around the central courtyard.[5]

4 For a systematic listing of all building types and installations of almost all Umayyad sites, see Genequand 2012, Tabs.1-3.

5 *Bayt* sequences of rooms form self-contained units of five- to six-room groups, most of which consist of a central hall with adjoining rooms on both sides (Creswell 1969, 515-518).

Site	Individual buildings	Settlement	Precursor	Early Islamic foundation	Hydraulic installation	Dating
'Ain al-Sil	x		?	?	x	
Muwaqqar		x		x	x	inscription
Azraq north (qaṣr)	x		x			
Azraq south				x?	x	
Hibabiya		x		x		
Fedein		x	x		x	
Khirbat al-'Umari		x		x		
Nomad village, Azraq north		x		x		
Qaṣr al-Hallabat and Hammam al-Sarah		x	x		x	
Qaṣr Kharana	x			x		inscription
Qaṣr Mushash		x	x		x	
Quṣayr 'Amra		x		x	x	inscription
Ar-Risha		x		x		
Qaṣr Burqu'	x		x		x	inscription
Qaṣr al-Tuba	x			x	x	

Table 1. Early Islamic sites in the north-eastern *badia.*

With regard to the research of the *quṣūr*, it should be emphasised that the investigations mostly concentrated on the visible, preserved architectural structures, and only in a few cases were the closer and wider surroundings of the prominent buildings taken into account. In addition, only in very few places were geophysical prospections carried out that could provide information on originally existing but no further visible structures. It can therefore be assumed that in most cases the actual structure of the known sites is more complex than previously known.

Among the above mentioned places, sites characterised by a multitude of different buildings and facilities form the majority, while only three complexes, Qaṣr Kharana, Qaṣr Burquʿ and Qaṣr al-Tuba, are isolated individual buildings with no visible settlement context. Among the more complex sites are Qaṣr al-Hallabat, Quṣayr ʿAmra, the buildings in the oasis of Azraq, Ar-Risha, Hibabiya, Muwaqqar, Qaṣr Mushash and Khirbat al-ʿUmari. The following overview is particularly focused on the three last, lesser known sites.

Qaṣr Kharana

Qaṣr Kharana is located about 50 km east of Amman and consists solely of the fortress in a compact square layout comprising a small courtyard and a *bayt*-like internal structure. The *buyūt* (singular: *bayt*) of Kharana consist of a central *iwān*, *i.e.* a hall with an open front, and various rooms on both side of it. The building has two storeys but there is so far no known indication as to any additional buildings or other hydraulic installations in the immediate vicinity. In the centre of the courtyard, the foundations of a cistern were discovered during exploratory work, and in the entrance area of the *qaṣr* also a drainage pipe (Urice 1981, 9ff., Fig.4). A graffito in one of the interior rooms dates its functional period to around 710 AD, *i.e.* to the rule of Walīd I (705-715) (Abbott 1946; Gaube 1977, 85; Imbert 1995, 404ff.). The form of the building suggests a representative purpose, perhaps as a meeting point of elites with tribal leaders (Gaube 1979, 205; Imbert 1995).

Qaṣr Burquʿ

Qaṣr Burquʿ lies in the northern part of the Jordanian *badia*, about 35 km south of today's Jordanian-Syrian border and 135 km north-east of Azraq. The structure dates back to Roman times, probably to the third/fourth century. It consists of a tower, probably originally used for military purposes, a square building complex with a large courtyard, and a barrage located north-west of it, through which winter precipitation is dammed. At the enclosure wall, a building inscription of Walīd I dating to the year 700 points to the renewal or expansion of the site. Various functions have been proposed for the place: on the one hand as *badiya, i.e.* as a meeting place (of the prince) with local tribes and as a place for hunting (Gaube 1974, 100); on the other hand the use as a caravanserai or as a communication place for tribal leaders has been considered possible (Helms 1991, 209). An inscription from the year 1380 AD points to a re-use in medieval times, possibly as a *khan* (Gaube 1974, 100).

Qaṣr al-Tuba

Qaṣr al-Tuba is the most remote 'desert castle' and is located 65 km south-west of Azraq. It is a large complex of 140 x 75 m, of which, however, only the western part has been completed. It is assumed that it was built by Walīd II (743-744) at the end of his reign. The site was visited and described first by A. Musil (1907), later by Jaussen and Savignac (1922). The water sources of the *qaṣr* and the gardens in the vicinity were possibly wells nearby the wadi (Musil 1907, 111 and Figs. 99-100; Vibert-Guigue 2008).

Qaṣr al-Hallabat

Qaṣr al-Hallabat, located about 40 km north-east of Amman, was founded as a Roman fort on the Via Nova Traiana in the second/third centuries. It was later converted into a *quadriburgium* and destroyed by an earthquake in the middle of the sixth century. The first *quadriburgium* was replaced by another one, which was used as a monastery with a chapel in the seventh century. In Umayyad time this building was converted into a representative *qaṣr*. No specific ruler could be linked to the Early Islamic transformation (Bisheh 1982; 1985; Arce 2008). There are numerous other buildings and water storage facilities in the surroundings (Bisheh 1985, Figs. 2, 5; Kennedy 1982, Figs. 10-12).

Fedein

Fedein is located on a hill in the centre of today's city of Mafraq. The core of the site is an Iron Age structure with cyclopean masonry. In the sixth century AD a monastery was established. For the Umayyad period, historical sources document a representative building that may have been destroyed at the beginning of the ninth century. Excavations have exposed the monastery to the west of the citadel. This is a long rectangular building with a central courtyard and surrounding rows of rooms and a chapel in the north-east corner. Adjacent to the east is a square building of the *qaṣr* type from the Umayyad period, of which the stucco panels are reminiscent of Abbasid types. In a walkway to the then abandoned monastery, numerous objects were found in a cache, including outstanding bronze objects dating back to the Umayyad period (Humbert 1989; Bisheh 2018; von Gladiss 2004, 244). East of it is a bath of square shape. North-east of the building was originally a large reservoir.[6]

Quṣayr ʿAmra

Quṣayr ʿAmra is located 60 km east of Amman and is well known for its reception hall and bath with impressive mural paintings. The site has a large areal extent and is characterised by prestigious as well as smaller buildings that are dispersed over a large expanse in which also hydraulic installations are noticeable. It includes the following structures: a *qaṣr*, a small mosque, water reservoirs and two wells of the *saqiyyah* type (Vibert-Guigue and Bisheh 2007, Pl. 2b). According to an inscription in the reception hall, it is thought that "(...) the lodge was constructed while Walīd II was still a prince, during the early half of the reign of Caliph Hishām bin ʿAbd al-Malik (725 and 743 AD)." (Palumbo and Atzori 2014, 633).

During the ongoing restoration work on the murals, soundings were also conducted around the reception hall/bath, exposing several rooms directly beneath the recent surface. They prove an originally much more complex structure of the site than is recognisable today (De Palma *et al.* 2013, 435ff.).

Azraq

The oasis at Azraq is located 90 km to the east of Amman. The *qaṣr* is situated in the northern part of the modern settlement (= Azraq al-Duruz) and represents the largest Roman complex to the east of Amman (ancient Philadelphia). The surface pottery predominantly dates to the third and fourth centuries, and the building inscriptions also belong to this period (Kennedy 2004, 56ff.). Although evidence of an Early Islamic occupation is lacking, it has to be assumed that the complex continued to exist in this period and that it was even used in medieval times, as suggested by a later inscription at the western tower mentioning construction work in the year 634 h, *i.e.* 1237 AD.

Early Islamic building activity is known from both the southern and northern quarters of the settlement in the Azraq oasis. In the south (= Azraq Shishan/Shamali), near the modern Wetland Reserve, there is a large water reservoir whose façade consists of carved basalt slabs (Vibert-Guigue 2004; Abu-Azizeh and Vibert-Guigue 2014). Several decades ago, a *qaṣr*-like structure nearby was documented but disappeared meanwhile (Kennedy 1982, 103ff.).

ʿAin al-Sil

In the northern part of the Azraq oasis, ʿAin al-Sil represents an Early Islamic residential building in the centre of other structures, of which one has an impressive enclosure wall. This former courtyard house with a connected bath was built with basalt slabs and was two storeys high. Recent excavations in the north and east of the *qaṣr* have yielded more building remains (Elter and Al-Jbour 2013). The complex was defined as the core of a small estate (Bisheh 1989).

Ar-Risha

Ar-Risha, about 40 km north-east of Qaṣr Burquʿ, is an open settlement consisting of eighteen buildings of different sizes and shapes, among others a *qaṣr*, houses or room groups of the *bayt* type, and a mosque. It is assumed that the site was founded in the middle of the seventh century.

---

6 See map collection Jordan 1:50.000, Washington D.C., edition 3-DMA, series K737, sheet 3254 IV/al-Mafraq (1977).

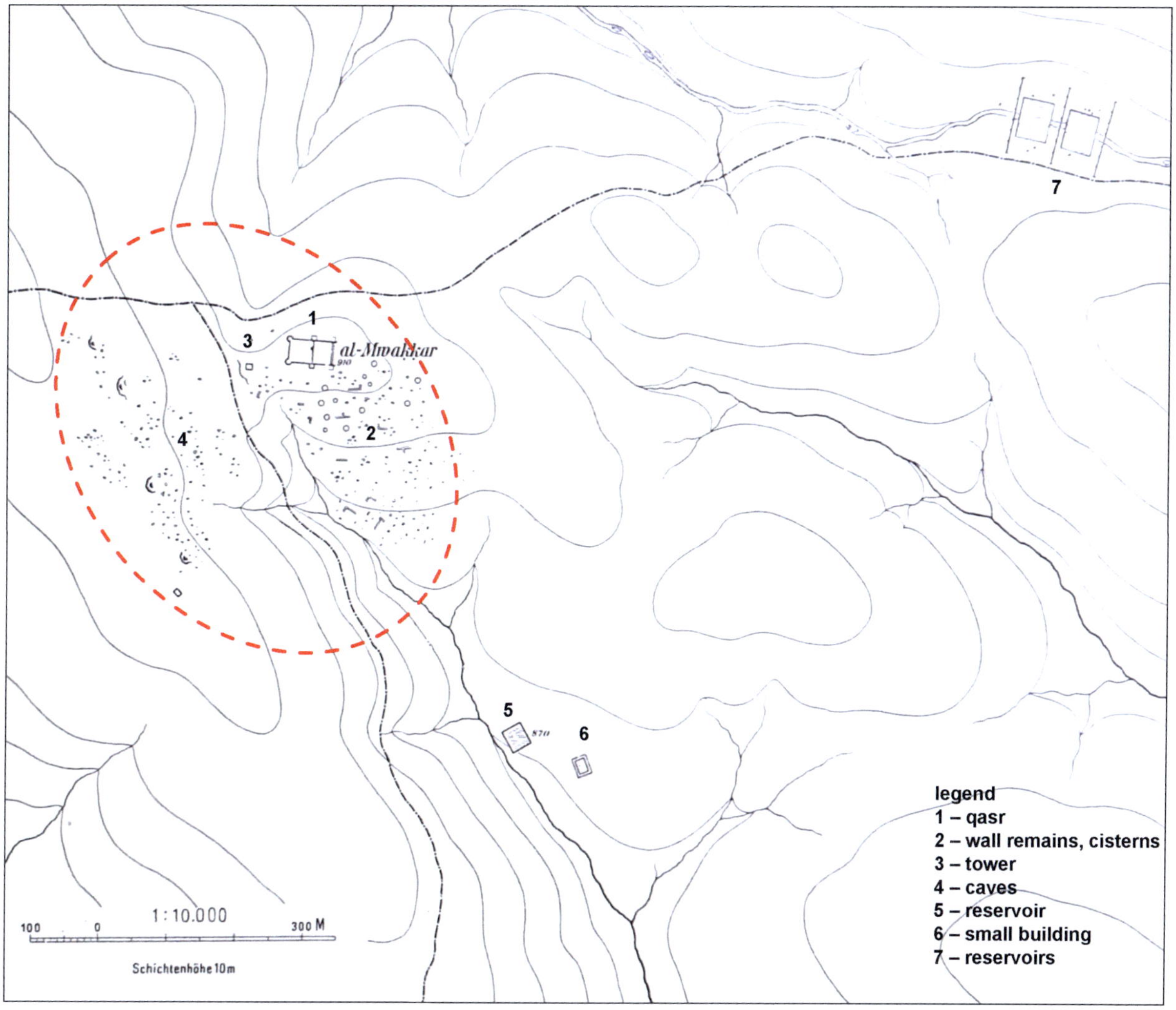

Figure 2. The Early Islamic site of Muwaqqar (A. Musil 1907, Fig. 20; numbering by K. Bartl).

The site was defined as a permanent *parembole nomadon* (*i.e.* tent camp), in other words, a permanent Bedouin settlement. This can hypothetically be considered as a kind of desert variant of the *amsar* (camp cities), that is, urban centres of the Early Islamic period that grew out of military camps. These nomad settlements could possibly have served as meeting places between Umayyad and tribal elites (Helms 1990).

Hibabiya

The now almost completely destroyed site of Hibabiya is located about 30 km north-west of Azraq on the edge of a mud flat and consisted originally of about thirty buildings of different shape and size, including some of the *qaṣr* type. The elongated settlement was dated by surface finds to the Late Roman-Early Byzantine and Early Islamic period. This site also corresponds to the type of nomad settlement (Kennedy 2014, 100ff.).[7]

Muwaqqar

One of the most interesting Early Islamic sites of the north-eastern *badia* is Muwaqqar. The place is located approximately 20 km east of Amman and has been almost completely overbuilt by a modern village. The village lies on a ridge that allows a wide view of the eastern limestone desert and at the same time represents a kind of landscape

7 Further settlements of this type have been defined at Qaṣr al-Hallabat and Khirbat al-Askar south-east of Karak as well as at Jabal Says in Syria and Qaṣr Swab in Iraq (Kennedy 2014). In contrast to this type of settlement, the Early Islamic site of Shuqayra al-Gharbiyya, 14 km south-east of Karak and 18 km west of Khirbat al-Askar, appears to be of a more prestigious character (Shdaifat *et al.* 2006; Shdaifat and Ben Badhann 2008).

border between the cultivated land in the west and the eastern arid steppe.[8]

Muwaqqar was first described by Sir J. Gray Hill, who reached the site on one of his Jerusalem-based trips in March 1895, but apparently had already seen it as early as 1891. Muwaqqar bears here the name Umm Moghr (Gray Hill 1896, 30ff.). R. Brünnow and A. von Domaszewski (1905, 182ff.) visited the place in spring and summer 1898 as well as A. Musil. G. Bell (1907, 52ff.), B. Moritz (1908, 418), A. Jaussen and R. Savignac (1922, 8) followed in 1900, 1905 and 1909.

The site map created by A. Musil (1907, Fig. 20) is the most important basis for understanding the original situation of the place. It depicts the following structures, of which nos. 1-5 are also mentioned or described in the text (*ibid.*, 27ff.) (Fig. 2):

1. The '*qaṣr*' (dimensions approx. 65 m E-W x 39 m N-S)
2. Individual remains of walls as well as numerous onion-shaped cisterns to the south and south-east of the *qaṣr* (extent of the area approx. 270 m N-S x 190 m E-W).
3. A tower and caves west of the *qaṣr* (size of the tower about 5 x 5 m, distance from the *qaṣr* about 65 m, distance of the caves *c.* 130 m, extent of the area about 250 m E-W x 300 m N-S).
4. A rectangular reservoir located approximately 580 m south-east of the *qaṣr* in a south-east to north-west orientation (size: 34 m E-W x 31.5 m N-S).
5. A smaller rectangular building south-east of the reservoir (13.5-17 m N-S x 13.26 m E-W), surrounded by a wall on all sides, resulting in a total size of 20.5 x 27 m.
6. Two large, adjacent reservoirs, north/north-east of the *qaṣr*.

1. The *qaṣr* is the largest building complex, of which the eastern part has survived to this day. This is the basement of a long rectangular building, whose barrel-vaulted rooms are currently used as stables. Originally these rooms formed a kind of terrace to create a plain ground for the construction of the rooms built on it. At the time of its discovery, a massive enclosure wall with two round corner towers and a square tower in the longitudinal wall were visible above the substructure. A north-south wall divided this wall into two areas. Additions were found in the north and south-east. In the north-east corner, remains of an upright standing wall were preserved. On the north wall was a staircase to the upper floor. Above the substructure, fragments of pillars and capitals of limestone were found at the eastern edge in the rubble. Musil therefore assumed a colonnade was here, *i.e.* a portico open to the east, which originally should have extended even further to the east.

Excavations in Muwaqqar took place for the first time in the late 1980s and early 1990s (Al-Najjar *et al.* 1989; Waheeb 1993), and the eastern area of the western courtyard could be clarified. Here, the north-south dividing wall between the western and eastern parts of the palace was uncovered, followed by a series of eight rooms. The northern end of the dividing wall forms a round corner bastion with a diameter of three metres. Fragments of mosaic floors, decorated stucco elements, and glass panes were found in the filling debris of the rooms (Waheeb 1993, Fig. P. 8, 10). In addition, the central part of the western outer wall of the western courtyard was uncovered and several rooms were found in the eastern courtyard.

An important feature represents a stone slab with the inscription of a *surah* of the Quran in Kufic capitals, which is now integrated into the façade of a recent building east of the *qaṣr*, but probably originated from it. It bears the date 137 h (= 754 AD), so it belongs to the beginning of the Abbasid period (Waheeb 1993, Fig. 2b).[9] An early Abbasid period of use was also assumed on the basis of ceramics analysed from several soundings at the *qaṣr* and to the east of it (Al-Najjar 1989).[10] Whether and to what extent the Umayyad palace of Muwaqqar was destroyed by order of the Abbasid caliph Abul-Abbas is unclear (see Bosworth 1993, 807; Musil 1928, 283, with references to the sources). The ceramic evidence points to the continuous use or re-use until the ninth or tenth century.

Today, the eastern half of the *qaṣr* is the only visible part of the Early Islamic settlement. All other parts mentioned above have now been overbuilt or disappeared due to the recent settlement activities. As mentioned above about this building only the substructure of the eastern court with the vaulted rooms is preserved. The eastern façades of the rooms are constructed with stones, probably originating from the *qaṣr*. The rooms are accessible through wooden doors of recent date and are today used as stables (Figs. 3a-b).

Above the ceiling, an original floor of red limestone slabs has been preserved to this day, on which column bases and drums as well as building parts with reliefs are found (Fig. 4). Nowadays, the western part of the *qaṣr* area is sealed by a road and an adjoining residential building and was probably largely destroyed. However, during a

8 This situation can also be seen in the recorded number of ruins on older maps, where only very few ancient sites are listed east of Muwaqqar (see map collection Jordan 1:50.000, Washington D.C., edition 3-DMA, series K737, sheet 3253 IV/Sahab (1970) and Jordan 1:50.000, edition 3-DMA, series K737, sheet 3253 I/Qaṣr Mushash (1974).

9 Unfortunately, the promising studies of 1989 and 1992 did not continue. A brief description of the condition of the site in the year 2001 can be found in Genequand (2001, 10).

10 Concerning the dating, some of the ceramic findings are equivocal. However, at least one piece is definitely of early Abbasid date. It is a bowl fragment with splash-glaze, a type that is well known from the Abbasid sites of Samarra in Iraq and Raqqa in Syria (Al-Najjar 1989, Fig. 9, no. 37).

Figure 3a-b. Al-Muwaqqar, eastern part of the *qaṣr*, vaulted rooms today used as stables (photographs by K. Bartl, German Archaeological Institute, Orient Department).

Figure 4. Muwaqqar, column fragment and decorated stone slab in the debris of the *qaṣr* (photograph by K. Bartl, German Archaeological Institute, Orient Department).

visit in 2015, parts of walls and lime plaster floors as well as some *tesserae* were visible *c.* 50 m to the west of the road. Adjacent to the palace to the north and south there are also modern residential buildings, and in one of them some sculptured capitals are kept by the owner.

2. There is no further information about the tower and the caves.

3. The building remnants south of the *qaṣr* are documented by two aerial photographs taken in 1939 by A. Stein.[11] They show a series of flat, rectangular elevations with room-like interior structure.

4. The reservoir to the south-east was accessible through staircases at the north-west and south-west corner. Two aerial photographs of A. Stein of 1939 show the reservoir, which was apparently completely intact and filled up with soil to the edge.[12]

5. At a distance of about 560 m was another building, of which around 1900 still quite significant remains were recognisable. It was a three-aisled structure in which the two side aisles were separated from the main room by two rows of arches. Its function is unclear. G. Bell considered a subsequent re-use as a stable probably due to mounting holes on the arcades (Bell 1907, 52ff.), while D. Genequand (2001, 10) suggested an original function as a reception hall. The three-aisled building is no longer preserved today. However, about 700 m south-east of the *qaṣr* a wall and floor parts of mosaic are still visible. They were interpreted by D. Kennedy as possible evidence for the south-east settlement area with the reservoir and three-aisled building.[13]

6. There is no information in the text of Musil for the two reservoirs to the north-east of the settlement centre, which he marked on his plan. As shown on the map, two basins were located about 1.1 km from the *qaṣr* and traversed a large wadi running in east-west direction, *i.e.* the basins were built directly into the wadi. Both reservoirs of about 53 x 46 m in size are oriented south-west to north-east and are located at a distance of 25 m to each other.

As the plan suggests, they are separated by a north-south wall of *c.* 120 m length, which also crosses the wadi. Corresponding walls can also be found near the eastern and western long sides of the ensemble. With regard to the water supply, either drains from the east, *i.e.* the flow direction of the wadi, are conceivable or a water overflow from the wadi into the basin. The longitudinal walls may have been used to hold back the sediments of the wadi (see

11 APAAME_19390327_Stein-BA-ASA-3-0487;APAAME_19390327_Stein-BA-ASA-3-0489.

12 APAAME_19390509_Stein-BA-ASA-3-0650;APAAME_19390509_Stein-BA-ASA-3-0651. The reservoir, however, looks rather different here than in a photograph of 1898 in the archive of Brünnow and von Domaszewski, in which the backfilling does not seem to have reached the edge of the pool (http://vrc.princeton.edu/archives/files/original/3/18450/565.jpg).

13 APAAME_20130414_DLK-13. See http://www.apaame.org/2013/04/flight-20130414-field-trip-20130416-al.html; http://www.apaame.org/2013/10/seminar-david-kennedy-al-muwaqqar.html.

Figure 5. Al-Muwaqqar, extension of the Early Islamic site (satellite image: Google Earth © 2017 DigitalGlobe; numbering by K. Bartl).

Sauvaget 1967, 37). It is unclear whether it was actually two adjacent reservoirs or a single one. A 1948 RAF aerial photograph[14] shows one single complex. Today, there is at this point a modern reservoir with concrete walls and about 60 x 60 m in size, in which apparently the existing ancient structure was used and expanded.

Particularly striking features of Muwaqqar are the numerous, elaborately decorated stone capitals, scattered in the central settlement area but especially at and around the *qaṣr* (Musil 1907, Fig.89; Brünnow and von Domaszewski 1905, Fig.760; Bell 1907, Fig. P. 53; Jaussen and Savignac 1922, pl. I-II).[15]

The most important object is a water gauge, which originally came from the water reservoir south-east of the *qaṣr* (Hamilton 1946b). This bears an inscription of Yazīd II (720-724), with which this ruler could be verified as founder or user of the palace (Mayer 1946).[16] The name of this ruler is also found in a description of the site by Yaqūt, who mentions Muwaqqar as Yazīd´s country estate, which he frequently visited from Damascus (Musil 1928, 283).

The settlement of Muwaqqar thus comprises two areas: the centre with the *qaṣr* and the surrounding buildings and installations (cisterns) that describes a size of about 1.5 km E-W x 1.1 km N-S, and an outer area with larger water storage facilities, which are located at a distance of about 600 to 1100 m to the centre. The total size of the place would therefore be about 3.8 km SE-NE x 2 km SW-NE (Fig. 5).

The central building forms the *qaṣr*, which was situated on an elevation. A compilation of various drawings (Genequand 2012, Fig. 199) suggest that the building is a modified *qaṣr* type. It consists of two courtyards, of which the eastern rests on a basement with barrel-vaulted rooms. This was possibly completed in the east by an arcade. The

14 APAAME_19480914_RAF-5073 Crop.

15 The capitals were not systematically recorded on any of the visits. On the basis of the few published photos, however, E. Herzfeld (1910, 129ff.) pointed early to two examples (Brünnow and von Domaszewski 1905, 760, 763) of Iranian or Sasanian influence (see also Kröger 1982, Pl. 167; Talgam 2004, 34).

16 However, based on the analysis of historical sources, G. Fowden (2004, 151, FN 54) concludes that the place already existed under ʿAbd al-Malik (685-705), as it is known as his retreat from the plague.

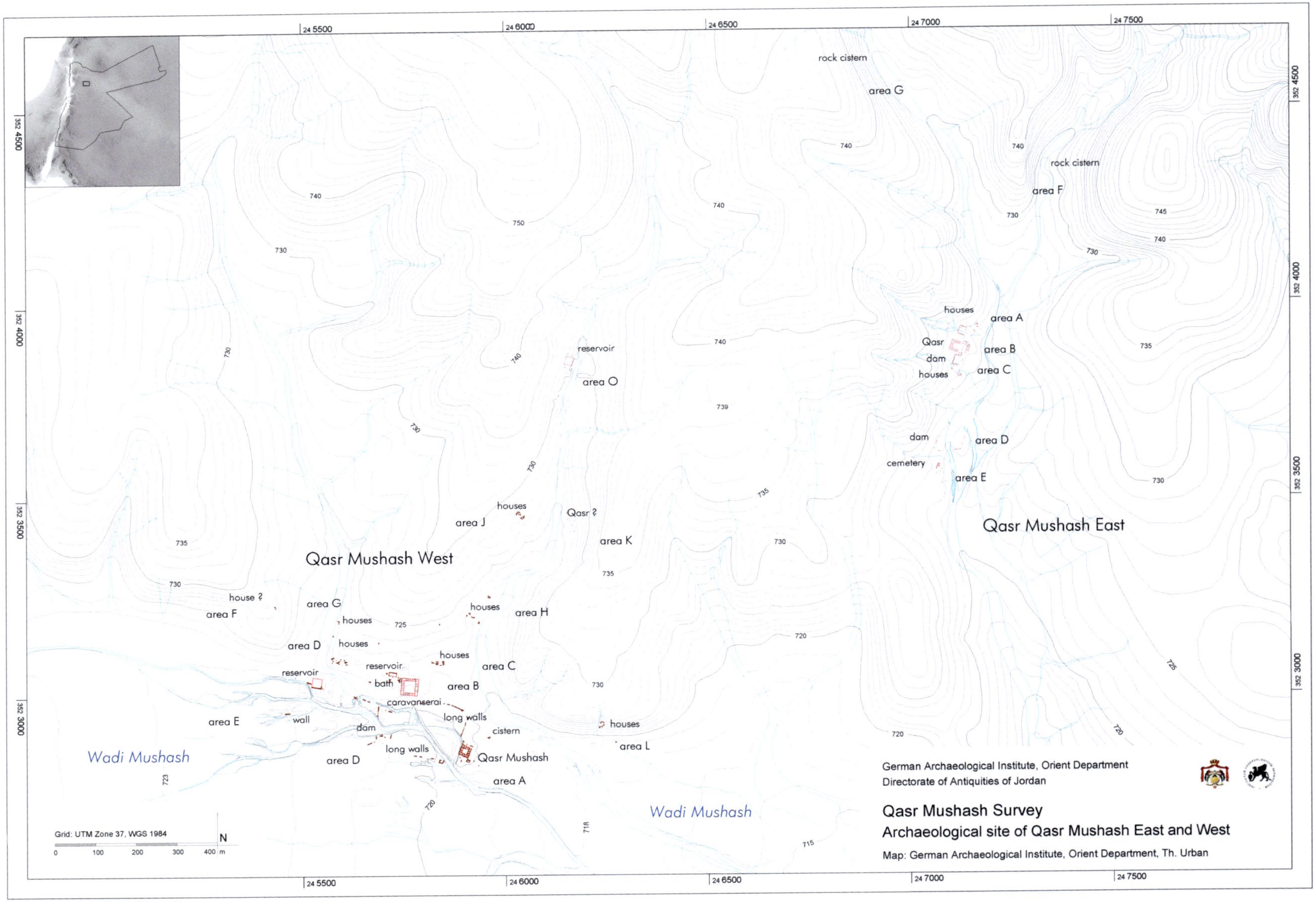

Figure 6. Topographical map of the Early Islamic site of Qaṣr Mushash (map by Th. Urban, German Archaeological Institute, Orient Department).

excavations at the west wall of this courtyard point to a row of small rooms.

However, the date of the construction phase of these units is unclear, as is the dating of the various walls on the floor in the south-eastern part of the courtyard. The western courtyard has a row of rooms to the east. In analogy to comparable buildings such as the eastern palace in Umm el-Walīd (Haldimann 1992, fig. 2), corresponding room units would also be conceivable on the other courtyard-surrounding sides.

The pottery from sounding I and possibly also from the remote sounding IV might indicate that the entire central settlement area was used until the ninth-tenth centuries. However, it must remain open to discussion whether the place was destroyed at the end of the Umayyad period by the devastating earthquake in 749 or during the Abbasid conquest and later resettled.

Among the well-known Early Islamic sites of the *badia*, Muwaqqar forms a prominent place, mainly due to the building decoration of its central building (*qaṣr*). Like the mosaic elements, stucco, and glass window pane fragments, they prove its original prestigious character and suggest a probable use as a residence. The area to the south of the *qaṣr* could have included service buildings for the palace or residential buildings.

Moreover, in addition to official functions, the documented complexes suggest possible agricultural activities, of which the large reservoirs away from the settlement provide evidence (Sauvaget 1967, 37) and a use as a halt on the postal route to inner Arabia (Fowden 2004, 276), which probably at least partly followed today's route of the Amman-Azraq highway.[17]

Qaṣr Mushash

Qaṣr Mushash lies about 20 km north-east of Muwaqqar. This site is one of the few places that could preserve the original impression of a completely isolated settlement in the arid steppe due to its remote locality to this day. The site is of impressive spatial dimensions and, in terms of its existing building structures, more complex but less prestigious than Muwaqqar. Its total area is about 2 x 2 km, but only about 15% was used as a settlement. It is divided into western and eastern parts, of which the former forms the settlement centre (Fig. 6).

In contrast to Muwaqqar, no historically known personality can be associated with Qaṣr Mushash, as dated inscriptions or coins were not found, nor is the place mentioned in written sources. However, pottery sherds and ^{14}C data show a pre-Islamic foundation, probably a watch post.[18]

Like Muwaqqar, Mushash was also visited in 1898 by Musil. The *qaṣr* and its surroundings were briefly described and sketched (Musil 1907, Figs. 104-105). Later visitors included B. Moritz in 1905 and A. Stein in 1938. In the early 1980s, G. King visited the site as part of his regional survey on late Antique and Early Islamic settlement in northern Jordan and created a first site map (King 1982; King *et al.* 1983). G. Bisheh carried out soundings and excavations in the entrance area of the *qaṣr* and in the bath in 1982 and 1983 (Bisheh 1989). On-site inspections were repeatedly carried out by D. Kennedy as part of the APAAME project (Kennedy and Bewley 2004, 220ff.). A short visit in 2001 was documented by D. Genequand (2001, 14ff.).

Between 2011 and 2017, the site was the subject of extensive research as part of the cooperation project 'Qaṣr Mushash Survey' of the Orient Department of the German Archaeological Institute and the Department of Antiquities of Jordan. This included the archaeological prospection of the entire site and its surroundings, the detailed recording of all visible architectural structures including the *qaṣr*, soundings at the bath and at the *qaṣr*, and the creation of a topographical map of the site. In addition, there was continuous monitoring of the place that is strongly threatened by illegal looting and vandalism (Bartl *et al.* 2013, 2014; Bartl 2014; 2015; 2016).

Qaṣr Mushash West consists of a variety of architectural structures. At the heart of this area and the settlement as a whole is the *qaṣr*, which is located directly on the eastern bank of Wadi Mushash and consists of a simple square structure of 26 x 26 m with a central courtyard and 14 surrounding rooms (Fig. 7). This complex dates back to Roman times and was apparently renewed in the Early Islamic period. Evidence of this is a graffito inscription at the entrance to the *qaṣr*, which mentions building activities but not the name of a ruler to be associated with them (Bartl 2015, Fig. 6).[19]

About 200 m to the west of the *qaṣr* is the central area of the settlement with a bath and a reservoir to the west of it. A large, now completely destroyed, square building of 40 x 40 m dimensions with a central courtyard corresponds formally to the caravanserai type, located to the east of the bath. Some small structures to be interpreted as residential buildings occur south of the reservoir (Fig. 8).

17 For the traffic routes, see King 1987.

18 Based on the evaluation of surface pottery, D. Kennedy suggested a dating to the fourth to the seventh centuries (Kennedy and Bewley 2004, 221). However, an ESA Hayes 111 sherd from the vicinity of the *qaṣr* indicates a possibly earlier date (M. Gschwind, München, pers.comm.).

19 Special thanks go to F. Imbert, Aix-en-Provence, for the translation and assessment of the inscription.

Figure 7. Landscape and *qaṣr* at the site of Qaṣr Mushash (photograph by Th. Urban, German Archaeological Institute, Orient Department).

Figure 8. Qaṣr Mushash West, view of the central area with reservoir, bath and caravanserai (photograph by APAAME 1997_DLK-4088034495_78b5cd679f_o, D.L. Kennedy, Aerial Photographic Archive for Archaeology in the Middle East).

In addition, more areas to the north of the central part show residential buildings of different shape and size, a round building, a cistern, and other reservoirs as well as numerous walls, which are to be interpreted as remnants of dams or barrages. Moreover, A. Stein reported about the (later) Bedouin use of the ruins as a burial ground (Gregory and Kennedy 1985, 286); individual burials were also found during recent investigations.

The western part of the settlement thus has several building agglomerations that probably must be associated with different functions. An important aspect of the site is water management. Obviously, winter precipitation was stored in open and closed tanks, and barrages were used for damming the wadi floods. As an additional simple technique, the digging of holes to reach groundwater is conceivable – a simple water extraction method, which is still used today by Bedouins for watering their flocks and which was documented at the site by Musil at the end of the nineteenth century (Musil 1907, Fig. 104).

Hypothetically, the use as a caravan stop can be defined as the most important function of the site. The interconnected building ensemble of reservoir, bath, and caravanserai forms, as well as a large reservoir to the west of the settlement, confirm this interpretation. The former complex then served as accommodation, and the latter could have been a watering place for animals.

The use of the *qaṣr* as a residence, which is the only building completely built of stones, seems rather unlikely in view of the simple construction, especially if compared with the neighbouring desert castles at Muwaqqar, Qaṣr Kharana and Quṣayr ʿAmra and their prestigious shapes and facilities. The *qaṣr* in Mushash is a purely functional building with only small rooms and simple amenities, of which only remains of the lime plaster of the walls have been preserved. The only building with more elaborate features is the bath, where fragments of marble wall and floor tiles, window panes of glass and stucco frames, as well as glass decorations were found.

Moreover, the different groups of buildings, as well as the remnants of dams preserved in several places, point to agricultural activities. It is conceivable that the damming of the wadis created flooded areas in the bordering areas, in which grain could be sowed.[20]

A special feature of Qaṣr Mushash is a second settlement area located about 1.2 km to the north-east of the western part of the site (= Qaṣr Mushash East). This complex, with a size of about 1100 m N-S x 500 m E-W, also consists of several units. These include, in addition to a small *qaṣr* building of 16 m side length and an eastern courtyard, a small quarter with residential buildings to the north of it, several dams, a cemetery, and two large rock cisterns. The dating based on ceramic finds indicates a use in the Umayyad period, but it remains to be seen whether this settlement is exactly of the same date as Qaṣr Mushash West.

Both settlements seem to have been abandoned at the end of the Umayyad period. One possible reason could have been the 749 earthquake, which also might have led to the destruction of Muwaqqar (Waheeb 1993). At Qaṣr Mushash West, evidence of this could be a dislocated wall on the west side of the *qaṣr*[21] as well as the accumulated stone rubble in the interior. A comprehensive re-settling did not take place at any later time. However, isolated finds such as a glazed tile fragment made of frit ware or clay pipe fragments prove the occasional presence of people in medieval and Ottoman times.

Qaṣr Mushash, in comparison to the surrounding Early Islamic sites, is a 'simple' settlement site whose structures are most likely to be used as a halt on the route between the cultivated region of the central Jordanian plateau and the eastern steppe regions,[22] but less as a temporary accommodation of the Umayyad elites. The documented buildings and installations, however, indicate at least the temporary presence of larger groups of people as well as planning components with regard to the construction and maintenance of hydraulic facilities. These aspects can be considered as indirect evidence of 'state planning' as material procurement, necessary manpower, and planning may have gone beyond the possibilities of private persons. A simultaneous use as an agricultural estate would therefore also be conceivable.

The period of expansion of the Roman settlement may date back to the first half of the eighth century, during which the neighbouring Quṣayr ʿAmra, Qaṣr Kharana and Muwaqqar were also in use. Unlike in the last place, however, Qaṣr Mushash does not show comprehensive post-Umayyad use.

Khirbat al-ʿUmari

Khirbat al-ʿUmari is an Early Islamic site located about 26 km south-east of Azraq South and 17 km north of today's Jordanian-Saudi border. It was first documented in 1944 by N. Glueck as part of his surveys on Nabataean colonisation and then largely forgotten (Glueck 1944, 14ff.).[23] Through the evaluation of satellite photographs, the rediscovery of this large settlement site was possible in 2015 (Bartl and Akkermans 2016).

The place is located in a slightly hilly steppe landscape and extends over a length of about 1100 m SW-NE and a width of about 200 m NW-SE on both banks of a small wadi, which is about 3 km east of the great Wadi Shuʾeib al-Jashsha (Fig. 9). Both wadis open into the Qaʾ al-ʿUmari, a *sabkha* area of about 6 km in length and 2 km in width, which is the main seasonal water resource of the area. Between the settlement and the south bank of the *qaʾ* is a distance of about 3.3 km. Both the *qaʾ* and its margins and the wadi banks have comparatively dense vegetation even in the dry periods.

The settlement consists of several groups of buildings. The recent episodes of destruction of the site, however, do not allow a more precise definition of the individual buildings. The most densely populated area is in the south-east. Here, over an area of 450 m N-S x 230 m E-W, about fifteen buildings can be detected, many of them of elongated, rectangular shape. Another settlement area with an extension of 110 m N-S x 60 m E-W, consisting of several houses, is located south-west of it on the western bank of the wadi. Some isolated smaller buildings are located north of these two complexes.

Apart from some overturned stone slabs on the edge of a side wadi, which may be parts of a small dam system, no hydraulic installations could be determined. The water supply could, however, have been through simply tapping groundwater in the wadi bottom, so-called *mshash*,

20 The current increase in agricultural production in the wadi areas around Qaṣr Mushash proves that even in fairly dry years, cereal crops in the region are still worthwhile.

21 T. Niemi, pers. comm.

22 For the possible route, see MacAdam 1994, Fig. 11.

23 However, the place can be found in the digital monuments register of Jordan (MEGA-Jordan) under the name Amari (MEGA -J, 2521, JADIS 3311001). The place mark is located about 3 km north-west of the actual locality.

30 6600
30 6700
30 6800
30 6900
30 7000
30 7100
350 1900
350 1800
350 1700
350 1600
350 1500
350 1400
350 1300
350 1200
350 1100
UTM data WGS 84, Zone 37R
N
0
150 m
Area J
Area H
Area G
Area F
Area E
Area D
Area C
Area B
Area A
© DAI 2016

by which the water level can be easily reached at high groundwater level (see above) or through constructed wells that are filled in today.

In the surface material the variety of typical red painted ceramics of the Umayyad period is striking. The comparatively large number of glass fragments may be regarded as a special feature, since these are generally very rare at Early Islamic sites.

The function of the settlement can only be determined hypothetically due to the poor state of preservation. However, it is likely that the most prominent and/or best-preserved buildings were originally located in the heavily damaged south-eastern settlement district. This area is also the highest part of the settlement, allowing a wide view in all directions. However, it is unclear whether the central settlement area originally contained one or more larger buildings of the *qaṣr* type, which may be an indication of a particular use.

The location of the settlement near the Qaʾ al-ʿUmari, which until today is an important source of water during the winter months and a popular pasture for camel herds in summertime, is likely to have been a caravan stop. The proximity to the two nearest towns in the north-west and south-east, the oasis of Azraq in Jordan and Haditha in Saudi Arabia, which can also be considered as caravan stations due to their water resources (Abu-Azizeh and Vibert-Guigue 2014; Rees 1929), forms another indication of this assumption. Both places are 28 and 26 km away from Khirbat al-ʿUmari, which is about a day's journey with camels.

Both the founding date of the settlement and the time of use of the place are unclear. Even if a pre-Islamic foundation cannot be completely ruled out, the surface material rather points to the establishment in the Umayyad period. The absence of later, *i.e.* glazed fritwares or geometrically painted ceramics from the Middle and Late Islamic periods, also indicate a relatively short period of use, which perhaps only included the Umayyad period and the beginning of the Abbasid period.

In terms of settlement type, Khirbat al-ʿUmari corresponds to other isolated places of the *badia*, such as Ar-Risha or Hibabiya, which represent scattered agglomerations of buildings (Helms 1990; Kennedy 2014). The apparent absence of prestigious buildings and installations for communal use, *e.g.* large reservoirs or cisterns, as they are to be expected in 'state' foundations, seems to support this classification of the place.

Discussion

After a kind of settlement hiatus in the second and, perhaps, first millennium BC, settlement activities in the north-eastern *badia* started again in Roman times with the protection of the border (see Akkermans and Brüning, this volume, for an alternative view). During the Early Islamic period some of the Roman structures were re-used and extended. Additionally, several new foundations are attested.

Figure 9 (left). Schematic map of Khirbat al-ʿUmari (map by Th. Urban, German Archaeological Institute using Image © 2016 CNES/Astrium, © 2016 ORION-ME).

site	residence	agricultural estate	caravan halt	settlement
ʿAyn al-Sil		x		
Al-Muwaqqar	x	x?	x	
Al-Azraq north, qaṣr	x?		x	
Al-Azraq south		x		
Hibabiya				x
Al-Fedein	x	x?		
Khirbat al-ʿUmari			x	x
nomad village, Azraq north				x
Qaṣr al-Hallabat und Hammam al-Sarah	x	x	x	
Qaṣr Kharana	x			
Qaṣr Mushash		x	x	
Quṣayr ʿAmra	x			
Ar-Risha				x
Qaṣr Burquʿ			x	
Qaṣr al-Tuba	x			

Table 2. Early Islamic sites in the north-eastern *badia*: possible functions.

The Umayyad period represents the last heyday of the settlement in the region. It is mainly characterised by the emergence of facilities of a more official nature, which are probably abandoned mostly in the second half of the eighth century. The function of the individual sites can usually only be determined hypothetically (Table 2).

However, due to their sizes and features, some of the sites may definitely be associated with official purposes, and therefore with the use by the Umayyad elites. These include Qaṣr al-Hallabat, Fedein, Qaṣr Kharana, Quṣayr ʿAmra and Muwaqqar. For Muwaqqar, the use as a temporary residence of Yazīd II is also documented by historical sources but not in a contemporary context. Whether his frequent visits were similar to the 'systematised' sequence of seasonal use of various residences, as evidenced by ʿAbd al-Malik (685-705) (Whitcomb 2016, 99ff.), remains an open question.

The use as agricultural domain could be attributed to water installations, dams, and enclosure walls for Qaṣr al-Hallabat, Fedein, ʿAin al-Sil, Azraq South, Muwaqqar, and Qaṣr Mushash (Gilbertson and Kennedy 1984). However, it can be assumed that the dimensions of the estates were quite small and served the exclusive subsistence economy for the individual settlements rather than the production of significant surpluses for markets as a source of income for the elites.[24]

A primary or secondary function as a way station or caravan stop is conceivable for most places. With the exception of Ar-Risha and Qaṣr al-Tuba, all sites are at or very close to major traffic routes. Also, the distances to the nearest places would mostly correspond to the distances of day marches.

Several sites belong to the type of 'nomad settlements': Ar-Risha, Hibabiya, Khirbat al-ʿUmari, and a settlement recently discovered north of Azraq as part of the APAAME project (Kennedy 2014). These sites may have been meeting places between caliphal and tribal elites, but they may also have been simple nomadic settlements or way stations. The intensified archaeological research in the *badia*, especially in the basalt region, suggests the discovery of further settlements of this type.

The temporal allocation of the known buildings due to inscriptions is only possible in a few cases. Three rulers are associated with construction activities in north-eastern *badia*: Walīd I (705-715), Yazīd II (720-724) and Walīd II (743-744). Thereafter, the construction activities at Qaṣr Burquʾ and the construction of Quṣayr ʿAmra by Walīd I (705-715) in his function as *amir*, *i.e.* before his reign as a caliph, is attested. Qaṣr Kharana was in use under Walīd I. The time of use of Muwaqqar is associated with Yazīd II (720-724) and the two unfinished buildings Qaṣr Mshatta and Qaṣr al-Tuba are assigned to the caliph Walīd II (743-744) (Bacharach 1996). However, based on written sources, it is sometimes assumed that occasionally, as in Muwaqqar and Qaṣr Kharana, construction activities had already taken place under ʿAbd al-Malik (685-705) (Fowden 2004; Urice 1987; Imbert 1995). For Qaṣr Burquʾ a reactivation of the site from the middle of the seventh century onwards is considered possible due to ceramic finds.

Essentially, however, the Umayyad construction activity in the north-eastern *badia* is likely to fall in the first half of the eighth century. The end of the use of these sites can usually not be clearly determined. The sudden abandonment of most of the settlements at the end of the Umayyad period could be connected with the often assumed shift of significance towards Iraq as a result of the takeover of power by the Abbasids. But possibly negative climatic changes or natural catastrophes such as earthquakes could also have had an impact on settlement activities.

The post-Umayyad settlement history of the *badia* is so far largely unclear. A continuation of use after 750 is known only for Muwaqqar, which was probably inhabited until the early Abbasid period, and for Qaṣr Uwaynid, which was not documented in Early Islamic times, but which is mentioned in the tenth century under the name Al-Awnid and was possibly also used as a caravan stop (Kennedy 2004). Qaṣr al-Azraq and Qaṣr Burquʿ were settled, as the inscriptions of the thirteenth and fourteenth centuries show, even in the Middle Ages. A function as caravan stations is most likely. As palaeographic analyses of some graffiti suggest, Qaṣr Kharana was visited occasionally in the fourteenth century, at least (Imbert 1995). Whether a more extensive post-Umayyad use of the *badia* can be assumed remains an open question for the time being.

24 Small-scale agricultural activities were previously suspected for Qaṣr al-Hallabat (Bisheh 1985, 265).

References

Abbott, N. 1946. The Kasr Kharāna inscription of 92 H. (710 A.D.). A new reading. *Ars Islamica* 11-12, 190-195.

Ababsa, M. (ed.) 2013. *Atlas of Jordan. History, territories and society*. Beyrouth: Presses de l'IFPO.

Abu-Azizeh, L. and Vibert-Guigue, C. 2014. Azraq: 'Ayn Sawda Reservoir. *American Journal of Archaeology* 118, 632.

Al-Khoury, M. and Infranca, G.C. 2005. The archaeological site of Qaṣr al-Usaykhim. *Annual of the Department of Antiquities of Jordan* 49, 351-364.

Al-Khoury, M., Abu-Azizeh, W., Steimer-Herbet, T. and Tarboush, M. 2006. West Irbid survey 2005 (WIS), preliminary report. *Annual of the Department of Antiquities of Jordan* 50, 121-138.

Almagro, M., Caballero, L., Zozaya, J. and Almagro, A. 1975. *Qusayr 'Amra. Residenca y baños omeyas en el desierto de Jordania*. Madrid: Instituto Hispano-Árabe de Cultura & Ministerio de Asuntos Exteriores.

Al-Najjar, M. 1989. Abbasid pottery from el-Muwaqqar. *Annual of the Department of Antiquities of Jordan* 33, 305-321.

Al-Najjar, M., Azar, H. and Qusous, R. 1989. Preliminary report on the results of the excavations at al-Muwaqqar. *Annual of the Department of Antiquities of Jordan* 33, 5-12.

Alraggad, M. and Jasem, H. 2010. Managed aquifer recharge (MAR) through surface infiltration in the Azraq Basin/Jordan. *Journal of Water Resource and Protection* 2, 1057-1070.

Amr, Z.S., Mordy, D. and Al-Shudiefat, M.F. 2011. *Badia – The living desert*. Amman: Al Rai Printing Press.

Arce, I. 2008a. Hallabat: castellum, coenobium, praetorium, Qaṣr. The construction of palatine architecture under the Umayyads (I). In: Bartl, K. and Moaz, A. (eds.). *Residences, castles, settlements. Transformation processes from Late Antiquity to Early Islam in Bilad ash-Sham. Proceedings of the International Conference held at Damascus, 5-9 November, 2006*. Rahden: Verlag Marie Leidorf, 153-182.

Arce, I. 2008b. The palatine city at Amman citadel. The construction of a palatine architecture under the Umayyads (II). In: Bartl, K. and Moaz, A. (eds.). *Residences, castles, settlements. Transformation processes from Late Antiquity to Early Islam in Bilad ash-Sham.*

Proceedings of the International Conference held at Damascus, 5-9 November, 2006. Rahden: Verlag Marie Leidorf, 183-203.

Bacharach, J. 1996. Marwanid Umayyad building activities: speculations on patronage. *Muqarnas* 13, 27-44.

Barth, H.-J. and Böer, B. (eds.) 2002. *Sabkha ccosystems. Vol. I: the Arabian Peninsula and adjacent countries.* Dordrecht: Kluwer Academic Publishers.

Bartl, K. 2014. Qaṣr Mushash Survey, Jordanien. Die Arbeiten der Jahre 2012 und 2013. *e-Forschungsberichte des Deutschen Archäologischen Instituts* 3, 57-61. https://publications.dainst.org/journals/efb/1724/4616.

Bartl, K. 2015. Qaṣr Mushash Survey, Jordanien. Die Arbeiten des Jahres 2014. *e-Forschungsberichte des Deutschen Archäologischen Instituts* 2, 50-56. https://publications.dainst.org/journals/efb/1662/4578.

Bartl, K. 2016. Qaṣr Mushash Survey, Jordanien. Die Arbeiten der Jahre 2015 und 2016. *e-Forschungsberichte des Deutschen Archäologischen Instituts* 3, 124-128. https://publications.dainst.org/journals/efb/1553/4463.

Bartl, K. and Akkermans, P.M.M.G. 2016. Khirbat al-'Umari. The rediscovery of an Early Islamic site south of Azraq. *Zeitschrift für Orient-Archäologie* 9, 200-221.

Bartl, K., Bisheh, G., Bloch, F., Bührig, C., Saleh, H. and Urban, Th. 2014. Qaṣr Mushash: 'Wüstenschloss' oder Karawanenhalt? *Zeitschrift für Orient-Archäologie* 7, 222-245.

Bartl, K., Bisheh, G., Bloch, F. and Richter, T. 2013. Qaṣr Mushash survey: first results of archaeological fieldwork in 2011 and 2012. *Annual of the Department of Antiquities of Jordan* 57, 179-193.

Bartl, K., Khraysheh, F., Eichmann, R., Ghunima, K.A., Rollefson, G.O. and Müller-Neuhof, B. 2001. Palaeoenvironmental and archaeological studies in the Khanasiri region, northern Jordan. Preliminary results of the archaeological survey 1999. *Annual of the Department of Antiquities of Jordan* 45, 119-134.

Bell, G. 1907. *Syria. The desert and the sown.* London: William Heinemann.

Bisheh, G. 1982. The second season of excavations at Hallabat. *Annual of the Department of Antiquities of Jordan* 26, 133-144.

Bisheh, G. 1985. Qaṣr al-Hallabat: an Umayyad desert retreat or farm-land? *Studies in the History and Archaeology of Jordan* 2, 263-265.

Bisheh, G. 1989. Qaṣr Mshash and Qaṣr ´Ayn al-Sil: two Umayyad sites in Jordan. In: Bakhit, A. and Schick, R. (eds.). *The history of Bilad al-Sham during the Umayyad period.* Amman: Bilad al-Sham History Committee, 81-103.

Bisheh, G. 2000. Two mosaic floors from Qastal. *Liber Annuus* 50, 431-437.

Bisheh, G. 2007. The Umayyad minaret at al-Qaṣtal and its significance. *Studies in the History and Archaeology of Jordan* 9, 263-267.

Bisheh, G. 2018. Al-Fudayn, Mafraq, Jordanie. *Museum With No Frontiers*, http://islamicart.museumwnf.org/database_item.php?id=monument;isl;jo;mon01;6;en

Bloch, F. 2014. Einleitung. In: Bartl, K., Bisheh, G., Bloch, F., Bührig, C., Saleh, H. and Urban, T. Qaṣr Mushash: 'Wüstenschloss' oder Karawanenhalt? *Zeitschrift für Orient-Archäologie* 7, 223-226.

Bosworth, C.E. 1993. Al-Muwakkar. In: *Encyclopaedia of Islam* 7. Leiden: Brill, 807.

Brünnow, R.-E. and von Domaszewski, A. 1905. *Die Provincia Arabia. Bd. 2. Der Äussere Limes und die Römerstrassen von El-Ma´an bis Bosra.* Strassburg: Verlag Karl J. Trübner.

Bujard, J. and Schweizer, F. (eds.) 1992. *Entre Byzance et l´Islam. Umm er-Rasas et Umm el-Walid. Fouilles genevoises en Jordanie.* Genève: Musée d´Art et d´Histoire.

Bujard, J. and Trillen, W. 1997. Umm el-Walīd et Khān az-Zabīb, cinq qusūr omeyyades et leurs mosquées revisités. *Annual of the Department of Antiquities of Jordan* 41, 351-374.

Bujard, J., Genequand, D. and Trillen, W. 2001. Umm el-Walid et Khan az-Zabib, deux établissements omeyyades en limite du desert jordanien. In: B. Geyer (ed.). *Conquête de la steppe et appropriation des terres sur les marges arides du Croissant Fertile.* Lyon: Maison de l´Orient Méditerranéen, 189-218.

Carlier, P. and Morin, F. 1984. Récherches archéologiques au chateau de Qastal. *Annual of the Department of Antiquities of Jordan* 28, 343-383, 491-493.

Carlier, P. and Morin, F. 1986. *Qastal al-Balqa´(Jordanie). Un site umayyade complet.* Taulignan: S.C.P. Plein-Cintre.

Carlier, P. and Morin, F. 1989. Qastal al-Balqā: an Umayyad site in Jordan. In: Bakhit, A. and Schick, R. (eds.). *The history of Bilad al-Sham during the Umayyad period.* Vol. 2. Amman: History of Bilād al-Shām Committee, 104-139.

Cramer, J., Perlich, B. and Schauerte, G. (eds.) 2016. *Qaṣr al-Mschatta. Ein frühislamischer Palast in Jordanien und Berlin.* Bd. 2. Petersberg: Michael Imhof Verlag.

Creswell, K.A.C. 1969. *Early Muslim architecture, Volume I, Part I, Umayyads A.D. 622 – 750.* Oxford: Clarendon Press.

De Palma, G., Palumbo, G., Shhaltoug, A., Arce, I., Arrighi, C., Atzori, A., Birrozi, C., De Vivo, G.S., Di Marcello, S., Esaid, W., Gaetani, M.C., Ghraib, R., Haron, J., Hjazeen, H., Khirfan, H., Lash, A., Mano, M.-J., Mariani, F., Meschini, A., Sarra, A., Tomassetti, C. 2013. Qusayr 'Amra World Heritage Site: preliminary report on documentation, conservation and site management activities in 2012-2013. *Annual of the Department of Antiquities of Jordan* 57, 425-440.

De Vries, B. 1998. *Umm el-Jimal. A frontier town and its landscape in northern Jordan*. Portsmouth, RI: Journal of Roman Archaeology.

De Vries, B. 2000. Continuity and change in the urban character of the southern Hauran from the fifth to the ninth century: the archaeological evidence at Umm el-Jimal. *Mediterranean Archaeology* 13, 39-45.

Elter, R. and Al-Jbour, K. 2013. Un ensemble architectural omeyyade au nord de Azraq: Qaṣr ʿAyn as-Sil. *Studies in the History and Archaeology of Jordan* 11, 639-649.

Fowden, G. 2004. *Qusayr ʹAmra: art and the Umayyad elite in Late Antique Syria*. Berkeley, CA: University of California Press.

Gaube, H. 1974. An examination of the ruins of Qaṣr Burqu. *Annual of the Department of Antiquities of Jordan* 19, 93-100.

Gaube, H. 1977. Amman, Harane und Qastal. Vier frühislamische Bauwerke in Mitteljordanien. *Zeitschrift des Deutschen Palästina-Vereins* 93, 52-86.

Gaube, H. 1979. Die syrischen Wüstenschlösser. Einige wirtschaftliche und politische Gesichtspunkte zu ihrer Entstehung. *Zeitschrift des Deutschen Palästina-Vereins* 95, 182-209.

Genequand, D. 2001. Projet "Implantations umayyades de Syrie et de Jordanie". Rapport sur une campagne de prospection et reconnaissance. *Schweizerisch-Liechtensteinische Stiftung für archäologische Forschungen im Ausland, Jahresbericht* 2001, 121-161.

Genequand, D. 2003. Maʹan, an Early Islamic settlement in southern Jordan: preliminary report on a survey in 2002. *Annual of the Department of Antiquities of Jordan* 47, 25-35.

Genequand, D. 2006. Umayyad castles: the shift from Late Antique military architecture to Early Islamic palatial building. In: Kennedy, H. (ed.). *Muslim architecture in greater Syria. From the coming of Islam to the Ottoman period*. Leiden: Brill, 3-25.

Genequand, D. 2008. Trois sites omeyyades de Jordanie centrale: Umm el-Walid, Khan ez-Zebib et Qaṣr al-Mshatta. In: Bartl, K. and Moaz, A. (eds.). *Residences, castles, settlements. Transformation processes from Late Antiquity to Early Islam in Bilad ash-Sham. Proceedings of the International Conference held at Damascus, 5-9 November, 2006*. Rahden: Verlag Marie Leidorf, 125-151.

Genequand, D. 2012. *Les établissements des élites Omeyyades en Palmyrène et au Proche-Orient*. Beyrouth: IFPO.

Gilbertson, D.D. and Kennedy, D.L. 1984. An archaeological reconnaissance of water harvesting structures and wadi walls in the Jordanian desert, north of Azraq oasis. *Annual of the Department of Antiquities of Jordan* 28, 151-162.

Gladiss, A. von. 2004. Die antike Welt im Kulturwandel – Der Beginn der Islamisierung. In: *Gesichter des Orients. 10.000 Jahre Kunst und Kultur aus Jordanien* (exh. cat., Berlin, Stiftung Preussischer Kulturbesitz). Mainz: Verlag Philipp von Zabern, 236-247.

Glueck, N. 1944. Wadi Sirhan in north Arabia. *Bulletin of the American Schools of Oriental Research* 96, 7-17.

Gray Hill, J., 1896. A journey east of the Jordan and the Dead Sea, 1895. *Palestine Explorarion Fund, Quarterly Statement* 27, 24-46.

Gregory, S. and Kennedy, D.L. (eds.) 1985. *Sir Aurel Steinʹs limes report*. Oxford: British Archaeological Reports (International Series 272).

Haldimann, M.-A. 1992. Les implantation omeyyades dans la Balqa: lʹapport dʹUmm el-Walid. *Annual of the Department of Antiquities of Jordan* 36, 307-323.

Hamilton, R.W. 1946a. Some eighth-century capitals from al-Muwaqqar. *The Quarterly of the Department of Antiquities in Palestine* 12, 63-69.

Hamilton, R.W. 1946b. An eighth-century water-gauge at al-Muwaqqar. *The Quarterly of the Department of Antiquities in Palestine* 12, 70-72.

Hanel, N. 2007. Military camps, canabae and vici. The archaeological evidence. In: Erdkamp, P. (ed.). *The companion to the Roman army*. Oxford: Blackwell, 395-416.

Helms, S. 1990. *Early Islamic architecture of the desert: a Bedouin station in eastern Jordan*. Edinburgh: Edinburgh University Press.

Helms, S. 1991. A new architectural survey of Qaṣr Burquʿ, eastern Jordan. *The Antiquaries Journal* 71, 191-215.

Herzfeld, E. 1910. Die Genesis der islamischen Kunst und das Mshattā-Problem. *Der Islam* 1, 27-63, 105-144.

Huigens, H.O. 2015. Preliminary report on a survey in the Hazimah plains: a hamad landscape in north-eastern Jordan. *Palestine Exploration Quarterly* 147, 180-194.

Huigens, H.O. 2019. *Mobile peoples – Permanent places. Nomadic landscapes and stone architecture from the Hellenistic to Early Islamic periods in north-eastern Jordan*. Oxford: Archaeopress.

Humbert, J.-B. 1986. El-Fedein-Mafraq. *Liber Annuus*, 36, 1986, 354-358.

Humbert, J.-B. 1989. El-Fedein-Mafraq. In: Ministère des Affaires Etrangères (France) (ed.). *Contribution française à l'archéologie jordanienne*. Beyrout, Amman, Damas: IFPO, 125-131.

Imbert, F. 1995. Inscriptions et espaces dʹécriture du palais dʹal-Karranā en Jordanie. *Studies in the History and Archaeology of Jordan* 5, 403-416.

Izdebski, A., Pickett, J., Roberts, N. and Waliszewski, T. 2016. The environmental, archaeological and historical evidence for regional climate changes and their societal impacts in the eastern Mediterranean in Late Antiquity. *Quaternary Science Reviews* 136, 189-208.

Jaussen, A. and Savignac, R. 1922. *Mission en Arabie*. Vol. III. Paris: Paul Geuthner.

Kennedy, D.L. 1982. *Archaeological explorations on the Roman frontier in north-east Jordan. The Roman and Byzantine military installations and road network on the ground and from the air.* Oxford: British Archaeological Reports (International Series 134).

Kennedy, D.L. 2004. *The Roman army in Jordan*. 2nd rev. edition. London: The Council for British Research in the Levant.

Kennedy, D.L. 2014. 'Nomad villages' in north-eastern Jordan: from Roman Arabia to Umayyad Urdunn. *Arabian Archaeology and Epigraphy* 25, 96-109.

Kennedy, D.L. and Bewley, R. 2004. *Ancient Jordan from the air*. London: The Council for British Research in the Levant.

Kennedy, H. 1985. From polis to madina: urban change in Late Antique and Early Islamic Syria. *Past & Present* 106, 3-27.

Kennedy, H. 1999. Islam. In: Bowersock, G.W., Brown, P. and Grabar, O. (eds.). *Late Antiquity. A guide to the postclassical world*. Cambridge, MA: Harvard University Press, 219-237.

Kerestes, T.M., Ludquist, J.M., Wood, B.G. and Yassine, K. 1977-78. An archaeological survey of three reservoir areas in northern Jordan. *Annual of the Department of Antiquities of Jordan* 22, 108-135, 251-269.

King, G.R.D. 1982. Preliminary report on a survey of Byzantine and Islamic sites in Jordan. *Annual of the Department of Antiquities of Jordan* 26, 85-95.

King, G.R.D. 1987. The distribution of sites and routes in the Jordanian and Syrian deserts in the Early Islamic period. *Proceedings of the Seminar for Arabian Studies* 17, 91-105.

King, G.R.D. 1992. Settlements patterns in Islamic Jordan: the Umayyads and their use of the land. *Studies in the History and Archaeology of Jordan* 4, 369-375.

King, G.R.D., Lenzen, C. and Rollefson, G.O. 1983. Survey of Byzantine and Islamic sites in Jordan, second season report, 1981. *Annual of the Department of Antiquities of Jordan* 27, 387-436.

Kröger, J. 1982. *Sasanidischer und frühislamischer Stuckdekor*. Mainz: Verlag Philipp von Zabern.

Le Strange, G.1890. *Palestine under the Moslems*. London: Alexander P. Watt.

MacAdam, H.I. 1994. Settlements and settlement patterns in northern and central Transjordania, ca. 550 – ca. 750. In: King, G.R.D. and Cameron, A. (eds.). *The Byzantine and Early Islamic Near East*. Vol. 2. Princeton, NJ: The Darwin Press Inc., 49-93.

Mayer, L.A. 1946. Note on the inscription from al-Muwaqqar. *The Quarterly of the Department of Antiquities in Palestine* 12, 73-74.

Moritz, M. 1908. Ausflüge in der Arabia Petraea. *Mélanges de la Faculté Orientale, Université Saint-Joseph III, Fasc.* I, 416-433.

Müller-Neuhof, B. 2016. Prehistoric settlements in the northern badia (Jordan). In: Reindel, M., Bartl, K., Lüth, F. and Benecke, N. (eds.). *Palaeoenvironment and the development of early settlements*. Rahden: Verlag Marie Leidorf, 149-160.

Musil, A. 1907. *Kusejr ʹAmra*. Bd. I. Wien: Kaiserliche Akademie der Wissenschaften.

Musil, A. 1928. *Palmyrena. A topographical itinerary*. New York: American Geographical Society.

Northedge, A. 1992. *Studies on Roman and Islamic ʹAmmān*. Vol. 1. Oxford: Oxford University Press.

Northedge, A. 2008. The Umayyad desert castles and pre-Islamic Arabia. In: Bartl, K. and Moaz, A. (eds.). *Residences, castles, settlements. Transformation processes from Late Antiquity to Early Islam in Bilad ash-Sham. Proceedings of the International Conference held at Damascus, 5-9 November, 2006*. Rahden: Verlag Marie Leidorf, 243-259.

Orland, I.J., Bar-Matthews, M., Kita, N.T. and Ayalon, A. 2009. Climate deterioration in the eastern Mediterranean as revealed by ion microprobe analysis of a speleothem that grew from 2.2 to 0.9 ka in Soreq Cave, Israel. *Quaternary Research* 71, 27-35.

Palumbo, G. and Atzori, A. (eds.) 2014. *Qusayr ʿAmra. Site management plan*. Amman: Department of Antiquities of Jordan.

Parker, S.T. 1986. *Romans and Saracens: a history of the Arabian frontier*. Winona Lake, IN: Eisenbrauns.

Politis, K.D. 1993. The stepped dam at Wadi el-Jilat. *Palestine Exploration Fund* 125, 43-49.

Rambeau, C.M. 2010. Palaeoenvironmental reconstruction in the southern Levant: synthesis, challenges, recent developments and perspectives. *Philosophical Transactions of the Royal Society A* 368, 5225-5248.

Rees, L.W.B. 1929. The Transjordan desert. *Antiquity* 3, 389-406.

Richter, T., Otaegui, A.A., House, M., Rafaiah, A.M. and Yeomans, L.M. 2014. Preliminary report on the second season of excavations at Shubayqa 1. *Neo-Lithics* 12/1, 3-10.

Rollefson, G.O. 2016. Greener pastures: 7th and 6th millennia pastoral potentials in Jordanʹs eastern badia. In: Reindel, M., Bartl, K., Lüth, F. and Benecke, N. (eds.). *Palaeoenvironment and the development of early settlements*. Rahden: Verlag Marie Leidorf, 161-170.

Rollefson, G.O., Rowan, Y. and Wasse, A. 2011. The deep-time necropolis at Wisad Pools, eastern Badia, Jordan. *Annual of the Department of Antiquities of Jordan* 55, 267-285.

Sauvaget, J. 1967. Chateaux umayyades de Syrie. Contribution a lʹétude de la colonization arabe aux Ier et IIème siècle de lʹhégire. *Revue des Études Islamiques* 35, 1-49.

Sbeinati, R., Darawcheh, R. and Mouty, M. 2005. The historical earthquakes of Syria: an analysis of large and moderate earthquakes from 1365 B.C. to 1900 A.D. *Annals of Geophysics 48*, 347-435.

Shdaifat, Y.M. and Ben Badhann, Z.N. 2008. Shuqayra al-Gharbiyya: a new Early Islamic compound in central Jordan. *Near Eastern Archaeology* 71, 185-188.

Shdaifat, Y.M., Tarawneh, K.F. and Ben Badhann, Z.N. 2006. Mu´tah University excavations at Shuqayra al-Gharbiyya: preliminary report on the 2005 season. *Annual of the Department of Antiquities of Jordan* 50, 205-216.

Talgam, R. 2004. *The stylistic origins of Umayyad sculpture and architectural decoration.* Wiesbaden: Harrassowitz Verlag.

Urice, S.K. 1981. The Qaṣr Kharana Project, 1979. *Annual of the Department of Antiquities of Jordan* 25, 5-19.

Urice, S.K. 1987. *Qaṣr Kharana in the Transjordan.* Durham, NC: American Schools of Oriental Research.

Vibert-Guigue, C. 2004. Le reservoir monumental de l´oasis d´Azraq ash-Shīshān et la découverte de blocs sculptés: un défi ècologiques technique et iconographique. *Studies in the History and Archaeology of Jordan* 11, 165-185.

Vibert-Guigue, C. 2008. Les sâqiya de Qaṣr at-Tûba: culture de l'eau et reflet iconographique. *Syria* 89, 129-174.

Vibert-Guigue, C. and Bisheh, G. 2007. *Les peintures de Qusayr ´Amra. Un bain omeyyade dans le bâdiya jordanienne.* Beyrouth: IFPO.

Waheeb, M. 1993. The second season of excavations at al-Muwaqqar. *Annual of the Department of Antiquities of Jordan* 37, 5-22.

Walmsley, A. 1992. Fihl (Pella) and the cities of north Jordan in the Umayyad and Abbasid periods. *Studies in the History and Archaeology of Jordan* 4, 377-384.

Walmsley, A. 2007. Economic developments and the nature of settlement in the towns and countryside of Syria-Palestine, ca. 565-800. *Dumbarton Oaks Papers* 61, 319-352.

Whitcomb, D. 2016. Periodic palaces: an economic approach toward understanding the "desert Castles". In: Sack, D., Spiegel, D. and Gussone, M. (eds.). *Wohnen – Reisen – Residieren. Herrschaftliche Repräsentation zwischen temporärer Hofhaltung und dauerhafter Residenz in Orient und Okzident.* Petersberg: Michael Imhoff Verlag, 95-101.

Depicting the camel: representations of the dromedary in the Black Desert rock art of Jordan

Nathalie Østerled Brusgaard

Abstract

The dromedary camel is Arabia's most iconic animal and it features prominently in the rock art of north Arabia. However, few studies have been conducted on these images and their significance, with the main interpretations focussing on either highly functional or highly symbolic meanings. This article presents the results of the first in-depth study of the dromedary camel figures in Safaitic rock art from the Black Desert in northern Jordan. It aims to form a new understanding of this prevalent motif through studying the figures' form and production process and to contribute to our understanding of the role of the dromedary in the ancient Near East. This study reveals that the dromedary images from the Jebel Qurma area have a standard form with proportions and features that are not naturalistic, but do have accurate anatomical details. The production process used to carve them followed a series of steps and shows evidence for planning. Both the functional and ritual interpretations of these images have their limitations. However, the cultural-historical context of the role of the dromedary indicates that it is highly likely that this animal played an important economic and ritual part in the desert societies, but that these are only two elements of a complex socio-ideological relationship between nomads and their camels. The prominent dromedary carvings should be interpreted in light of this framework.

Keywords: dromedary camel, rock art, chaîne opératoire, pastoral ideology, Black Desert, Jebel Qurma, Safaitic

Introduction

The dromedary camel is often seen as synonymous with the history of ancient Arabia. As the backbone of the long-distance caravan trade and by facilitating the opening up of new, marginal places to human exploitation, the dromedary was essential for the development of the region (Almathen *et al.* 2016; Bulliet 1975; Köhler-Rollefson 1993). Although its role in the development of the ancient caravan cities such as Palmyra is unmistakable (Seland 2015), it is especially the dromedary's use in past and present nomadic societies for which it is iconic. Historical, archaeological, and epigraphic sources attest to its significance in the societies of the desert from the first millennium BC onwards. It is therefore perhaps unsurprising that the dromedary camel is ubiquitous in the desert rock art of Arabia. It dominates rock art corpora such as the petroglyphs from Shuwaymis in Saudi Arabia (Guagnin *et al.* 2016; see also Guagnin, this volume), Hismaic rock art from southern Jordan (Corbett 2010), and Safaitic rock art from the Syro-Jordanian desert (Brusgaard 2019;

In: Peter M. M. G. Akkermans (ed.) 2020: *Landscapes of Survival - The Archaeology and Epigraphy of Jordan's North-Eastern Desert and Beyond*, Sidestone Press (Leiden), pp. 287-304.

Macdonald 1993). The dominant presence of the dromedary in Arabian rock art was noted as early as 1932 (Rostovtzeff 1932) and its frequent mention in Safaitic inscriptions has been remarked upon as well (Macdonald 1993).

However, little further investigation has been made into the nature of these dromedary representations and what they can tell us about the animal's role in ancient Arabian societies. The studies that have considered it have tended to place the carvings in either a simple, utilitarian framework or a highly symbolic one. Additionally, previous data sets have been small and the focus has been on the images themselves and not on other important aspects of the carvings, such as their form or production. Now the 'Landscapes of Survival' project has made it possible to assemble a thorough data set of Safaitic rock art, documented in the Jebel Qurma region of north-eastern Jordan, allowing for a systematic analysis of the desert petroglyphs (Brusgaard 2019).[1] As such, it is now possible to examine in detail the nature of the dromedary depictions.

This article will discuss the images of dromedary camels in the Safaitic rock art of the Jebel Qurma region, including their quantity and form. Subsequently, it asks what these depictions can tell us about the significance of the dromedary camel in the ancient Near East. Relatively little is known about the ideology of the nomads of the ancient Near Eastern deserts, compared to the older and contemporary empires of this region. In particular, little is known about the so-called 'pastoral ideology', the belief systems so distinct to peoples whose lives are (semi-) dependent on herd ownership (cf. Parkes 1987; Rosen 2008). Understanding the role and position of dromedaries is essential to furthering this discussion. This chapter does not endeavour to resolve so complex an issue, but it endeavours to contribute to the debate through the investigation of the representational evidence from the Black Desert in north Arabia. By placing the engraved images of dromedaries in their cultural-historical context, this chapter intends to shed light on one of the most iconic animals of the Near East. Additionally, it aims to contribute to our knowledge on the Black Desert rock art as a whole.

The dromedary camel in the ancient Near East

Many articles have been written about the dromedary camel's biological qualities that make it uniquely suited for desert life and for the activities for which humans have used it since its domestication (*e.g.* Gauthier-Pilters and Dagg 1981; Köhler-Rollefson 1993; Rosen and Saidel 2010; Seland 2015). I will therefore not summarise these here and instead focus on the current knowledge on the development of the dromedary's role in the ancient Near East. The domesticated dromedary camel, or one-humped camel (*Camelus dromedarius*),[2] has had a dominant presence in Arabia for over two millennia, but many questions still surround the nature of its domestication process. Many aspects of this issue have remained largely unexplored until recently, for several reasons. Ilse Köhler, writing in 1984, states that the archaeological record on the domestication process is poor due partly to a sample bias; research in the region has focused on the settled communities in the Fertile Crescent rather than on desert areas where it is more likely that the dromedary was domesticated (Köhler 1984, 201). Although this imbalance has changed somewhat in the last thirty years, there are still some complications. Among other things, there is a lack of zooarchaeological evidence from well-dated contexts and little is known about the distribution of the wild one-humped camel (Almathen *et al.* 2016). However, new evidence is shedding light on the matter.

Representational evidence in the form of figurines and reliefs, the nature of dromedary bone assemblages, including their demographic profile and a decrease in bone size, and the context of these faunal finds indicate that the domesticated dromedary was not widely present in the ancient Near East before 1000 BC (Almathen *et al.* 2016; Magee 2015; Sapir-Hen and Ben-Yosef 2013). It is likely that domestication occurred in the late second millennium BC (Almathen *et al.* 2016; Magee 2015). Additionally, recent genetic evidence "support[s] a scenario with an initial domestication followed by consecutive introgression from wild populations." (Almathen *et al.* 2016, 6711). There is also good reason to suggest that the wild ancestral population was already limited to the south-east coast of Arabia, providing a tentative place for domestication (*ibid.*, 6710).

The question of *why* domestication took place remains more elusive. Magee (2014; 2015) has argued, based on zooarchaeological evidence, that the driving force was probably the need for a reliable food source, which would have included meat but probably milk as well. However, it seems that using dromedaries as mounts and pack animals followed shortly afterwards (Magee 2015, 273). Whatever the initial motivations, the dromedary would have quickly become an essential provider, providing secondary products such as wool, dung for fuel, and milk, which has more health benefits and is available for more months per year than sheep or goat's milk (Magee 2014; Rosen and Saidel 2010). And, perhaps more iconically, its conversion into a pack and riding animal transformed the region. Often aptly called the 'ship of the desert', the dromedary's physiological characteristics allowed people to traverse

1 See Akkermans, this volume, for a brief introduction of both the Jebel Qurma Archaeological Landscape Project and the Landscapes of Survival Project.

2 In this chapter, the terms dromedary and camel will be used interchangeably to denote the dromedary camel (*Camelus dromedarius*).

regions and distances hitherto impossible, opening up trade routes and new areas for exploitation (Köhler-Rollefson 1993; Magee 2014; Seland 2015). It is thus unsurprising that the dromedary features in the iconography of ancient Arabia, from Iron Age camel figurines from Saudi Arabia, Oman, U.A.E., and Yemen (Magee 2015) to Palmyrene reliefs (Seland 2015) and Nabataean reliefs and figurines (Corbett 2010) to the desert petroglyphs mentioned above and discussed further below.

Additionally, there is interesting archaeological evidence on the possible symbolic role of the dromedary from the southern part of the peninsula. A number of camel burials have been excavated here that show evidence for the sacrifice of animals in funerary rituals, possibly representing *balîya* graves. The *balîya* ritual is described in later textual sources and is regarded as "the sacrifice of an animal for a deceased individual to use in the afterlife as it was conceived in the pre-Islamic period in Arabia." (King 2009, 81). The textual sources write that usually a female animal, commonly a dromedary, is chosen for the ritual. King (2009, 87) notes that "the choice of the female camel for a *balîya* to provide the dead with a riding animal in the hereafter corresponds with practice in life. The female is preferred to the male camel for riding because of its more benign temperament." There is a myriad of evidence for *balîya*-like camel immolations from U.A.E., Oman, Yemen, and Bahrain, most of which date from between the fourth century BC and the third century AD (Curci and Maini 2017; King 2009). Additionally, a possible *balîya* burial was also found in Wadi Rum, southern Jordan, where the burnt remains of a camel was found in a pit, accompanied by a Nabataean inscription referring to *blw'*, which probably relates to the classical Arabic term *balîya* (Hayajneh 2006).

As King (2009, 91) rightly points out, we cannot assume that the archaeological evidence for these practices in south and south-eastern Arabia equates with similar practices in the rest of the region. However, there may be some epigraphic evidence indicating similar rituals, to which I will return later. From the eastern *badia*, the textual and pictorial engravings are one of few sources that can provide insights into the role of dromedaries in the societies of this region. Textual references and iconography from the sedentary centres in the early first millennium BC refer to the use of dromedaries by (semi-)nomadic groups in northern Arabia (Magee 2014, 210). However, the indigenous perspective, how these nomadic groups used and related to their dromedaries, is missing from these accounts (*ibid.*). Figurines like those from southern Arabia have not been found and bone preservation is poor, meaning that as of yet (zoo) archaeological assemblages cannot help in reconstructing the symbolic or economic role of the dromedary. The desert inscriptions and rock art can therefore provide important insights. As mentioned in the introduction, the Safaitic inscriptions sometimes refer to dromedaries. The majority of the references are authors that sign an image of a camel (see below), but inscriptions with a narrative component mention pasturing and watering camels, keeping watch for them, and migrating with camel herds (Al-Manaser and Macdonald 2017; Macdonald 1993). These form the impression that these societies were camel pastoralists, although the form and scale cannot be deduced from the inscriptions. Their mention in the 'day-to-day' activities described in the inscriptions suggests that dromedaries were significant enough to warrant mention, but what kind of significance they had is unclear.

For this reason, rock art is another potential valuable source of information. Various brief considerations on the dromedary in ancient north Arabian rock art have focused on mostly utilitarian interpretations of these motifs, including them being a representation of the daily life of the desert societies and expressions of ownership of dromedaries (cf. Clark 1980; Oxtoby 1968; Winnett and Harding 1978). In contrast, Corbett (2010), working on Hismaic rock art, argues against these interpretations and suggests that the images may have been sacrificial offerings. Drawing on pre-Islamic poetry and the evidence for *balîya* graves, Corbett (2010, 150) proposes that the camel was "an important symbolic mediator within pre-Islamic Arabian society." Similarly, Eksell (2002, 140), studying Thamudic and Safaitic carvings, argues that the camel was "endowed with sacrality" and that the carvings of dromedaries were symbolic gifts to either a person or deity. This conclusion is also inspired by pre-Islamic poetry and based on various, somewhat random, sources.

Although admirable challenges to the earlier, highly functional interpretations of the camel motifs, both studies are problematic due their loose use of different sources and their heavy reliance on pre-Islamic poetry as an interpretative framework. This poetry is believed to have been composed orally in the sixth and seventh centuries AD and later written down in the eighth century. It is interesting for a comparative perspective of early Arabian desert life due to the many similarities in the fauna mentioned in the poetry and depicted in the rock art. However, there are many difficulties in using it as a reflection of pre-Islamic society due to it being written down much later in an Islamic society (McDonald 1978), let alone using it as a reflection of the different societies carving inscriptions and images across northern Arabia at least a few centuries earlier. Therefore, this poetry cannot be our main basis for understanding the Safaitic (or other ancient north Arabian) rock art. It is thus time to revisit the evidence and form a new framework for interpreting the camel images, especially in light of new theoretical developments in our understanding of past human-animal interactions (*e.g.* Russell 2012; Sykes 2014).

Figure 1. This image of a female camel is signed by the author. He states his name and refers to 'the young she-camel'. The camel has been incised and then filled in by hammering. A mistake appears to have been made when carving the foreleg, or the camel is unfinished (photograph: Jebel Qurma Archaeological Landscape Project).

Safaitic rock art

It is not known how much rock art can be found across the Black Desert (*harrah*) of northern Arabia, but tens of thousands is not an exaggerated estimate. Each survey in even small areas of the basalt desert recovers several thousand Safaitic inscriptions and over 30,000 have been recorded in the ongoing project 'Online Corpus of the Inscriptions of Ancient North Arabia' (OCIANA) (Al-Manaser and Macdonald 2017; OCIANA 2017). Many inscriptions are associated with a pictorial engraving and images occur on their own as well, so it is likely that there are nearly as many Safaitic pictorial as textual engravings. The rock art and the inscriptions are fundamentally linked to one another, with inscriptions often signing an image and referring to it. A common find is, for example, an image of a dromedary camel accompanied by an inscription stating 'By [name] son of [name] is the camel' (Fig. 1).

Various inscriptions from the *harrah* make reference to known dates that make it possible to roughly place the engravings in the Late Hellenistic to Early Roman period, but this conventional dating can only be seen as a guideline (Al-Jallad 2015, 18). Rock art from other periods can be found in

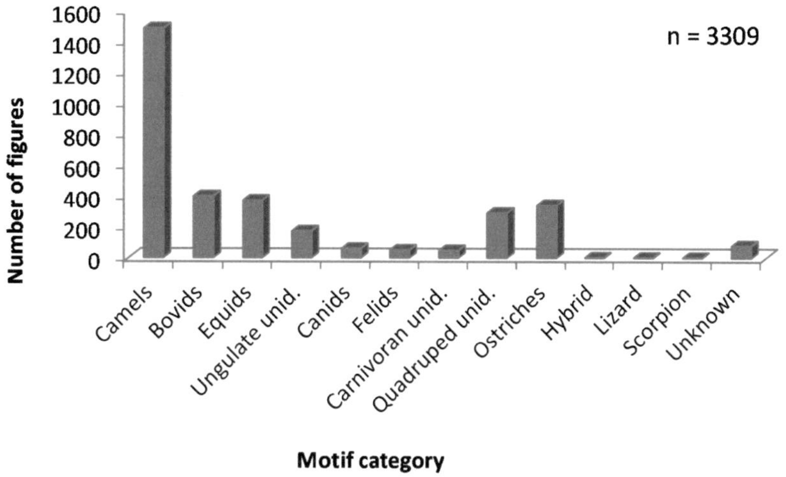

Figure 2. The frequency of the different types of zoomorphic motifs, categorised by family, that occur in the Jebel Qurma corpus. The camels are by far the most frequent.

this region as well, including a few early Neolithic engravings (Betts 1987) and much more recent carvings from the last few centuries, such as those associated with Arabic inscriptions and the many *wusūm* (cf. Berghuijs, this volume). Since 2012, the Jebel Qurma Archaeological Landscape Project has been documenting all of the pictorial and textual engravings and the many archaeological structures in the Jebel Qurma area of the Black Desert, situated in north-eastern Jordan, approximately 30 km east of the oasis town of Azraq. The petroglyphs dating to the Late Hellenistic/Early Roman period, the 'Safaitic rock art', have been researched by the author, which resulted in the first-ever systematic study of Safaitic rock art (Brusgaard 2015; 2019). It is to this rock art that the vast majority of the dromedary camel motifs belong.

Dromedary depictions in the Jebel Qurma region

The fieldwork carried out in the Jebel Qurma area has yielded several thousand Safaitic inscriptions and petroglyphs. In total, 4511 rock art figures were recorded that can be ascribed to the same period of production as the Safaitic texts. The vast majority (73%) are zoomorphic figures, representing domestic animals, such as camels, horses, and dogs, and wild animals, such as lions, ibexes, and wild asses (for all results, see Brusgaard 2019). A small percentage (7%) are anthropomorphic figures, of which the majority are 'archers', *i.e.* human figures holding a bow-and-arrow. The rest are geometric figures, such as the frequent set of seven dots or lines, sun motifs, and unidentifiable figures. The dromedary camel motif is a dominant part of the corpus. There are 1486 representations of dromedary camels, making up a third of the entire corpus and almost half of the 3309 zoomorphic figures (Fig. 2).

Camels are also mentioned a total of 312 times in the Safaitic inscriptions of Jebel Qurma, of which all but two are inscriptions signing a depiction of a camel, *i.e.* 'By [name] son of [name] is the camel'. The other two inscriptions are inscriptions with content (*i.e.*, having a narrative component). One mentions pasturing the she-camels and the other keeping watch for the camel (Della Puppa, forthcoming). Both inscriptions are associated with an image of a dromedary camel.

In Safaitic, several different words are used for dromedary, depending on age, sex, and number. For example, *nqt(n)* is used for 'the (two) she-camel(s)' and *gml(n)* for 'the (two) he-camel(s)', while the 'young (she) (he)camel' is denoted by *bkr(t)*. The inscriptions signing an image of a dromedary can therefore help to identify the age and sex of the depicted camel. Additionally, they can also be indicated by various anatomical features and the context in which the dromedary is portrayed. Based on these elements, a distinction can be made in the rock art between 'young camel' and 'camel' and 'male' and 'female'.

The 'young' camels are usually indicated by the accompanying inscription. However, in a number of instances, the context can aid identification as well. There are 16 scenic compositions where a small dromedary is depicted nursing from a larger female dromedary, whose udders are usually depicted (Fig. 3) (Brusgaard 2019; Brusgaard and Akkermans, in press). These small dromedaries are also identified as young camels. In total there are 162 young dromedaries in the Jebel Qurma rock art.

While the Safaitic text indicates the sex of a number of the dromedary figures, the main method of identification is the anatomical features. The presence of a phallus or udders indicates a male or female camel, respectively. Additionally, the position of the camel's tail is also generally taken to be an indication of the camel's gender in Arabian rock art. Macdonald and Searight (1983, 575) first recognised the relationship between the sex of the camel as given in the inscription and the position of the tail: male

Figure 3. On the right of the panel is a female camel nursing a young camel. An anthropomorphic figure is holding the mother camel. On the left of the panel are various wild animals: a felid, four ibexes, and an ostrich. All of the figures appear to have been carved as part of the same composition (photograph: Jebel Qurma Archaeological Landscape Project).

camels' tails hang down while females' tails curl up. This seems to hold true for different kinds of Arabian rock art, such as Safaitic and Hismaic. Indeed, the Jebel Qurma material also supports this interpretation; inscriptions referring to 'the (young) she-camel' are always associated with an image of a camel with its tail curled up (Fig. 4). The use of the up-curled tail to indicate a female camel may stem from the fact that when a female camel is in heat, she will raise her tail when near a male (Khanvilkar *et al.* 2009, 72). Additionally, the position of the tail is also used to judge whether the conception was successful; the she-camel cocks her tail when she is pregnant (*ibid.*). It is unclear whether the camels with up-curled tails are meant to represent pregnant females, females in heat, or whether it is just a convention used to depict a female camel. However, the latter interpretation seems most likely considering the large number of camels depicted with their tails curling up and the frequent association of this type of depiction with an inscription referring to 'the she-camel' or 'the young she-camel'. Moreover, some of the female camels are depicted with udders, which may instead be an extra emphasis to indicate that the camel is pregnant or nursing.

The position of the tail, the presence of udders or a phallus, and the word used in the inscription were all used to determine whether a camel is female or male. In contrast to the other zoomorphic figures, the majority of dromedaries have a clear sex indicated – 1090 in total – and the majority of these are female. There are 760 females, making up 51% of all dromedary camel figures and 70% of the sexed camels (Table 1). A much smaller percentage is male: 330 in total. Of these figures, 148 are young female dromedaries, but only eight are young males. The (young) female dromedary is thus the most common depiction in the Jebel Qurma rock art.

In terms of composition, the dromedary is often a lonesome figure. In contrast to the other zoomorphic figures, they are rarely depicted in scenes. Besides the above-mentioned nursing scenes, dromedary camels generally feature in two types of interactions with other figures: being led or held by an anthropomorphic figure (11 scenes) (Fig. 3) and conflict scenes between humans (17 scenes). Interestingly, there are also two depictions of what might be bull camels fighting. In both images two male camels are facing each other with their necks crossed, perhaps indicating the manner in which bull camels fight (cf. Gauthier-Pilters and Dagg 1981, 89).

Figure 4. The dromedary camel has a curled up tail indicating that it is a female camel. The inscription refers to 'the young she-camel'. The camel is completely incised and has hairs on her hump and tail. Her udders are also depicted and her muzzle defined (photograph: Jebel Qurma Archaeological Landscape Project).

Additionally, there is one scene in which two lions are attacking a dromedary and one scene where a rider on a dromedary is attacking a large felid. The conflict scenes featuring camels are interesting because they might depict conflict in the context of a raid. The occurrence of raids and the 'activity' of raiding are mentioned in Safaitic inscriptions both from Jebel Qurma and elsewhere in the *harrah* (Al-Jallad 2015; Al-Manaser and Macdonald 2017; Della Puppa, forthcoming). The representation of camel raids in the rock art has also been convincingly argued by Macdonald (1990) for scenes whereby riders on horseback are touching a dromedary camel with their

Age and sex	Number of figures
Female	611
Young female	149
Male	322
Young Male	8
Unknown	391
Young unknown	5
Total	**1486**

Table 1. Number of dromedary figures according to their sex and whether they are young or of unspecified age.

Figure 5. Commonly the dromedary figures are carved together with only a geometric motif and the accompanying Safaitic inscription. The composition tends to take up the entire panel. This male dromedary is carved with incised hairs on its tail and hump, it has an exaggerated hump, and it has the typical 'Jebel Qurma style' curved body and long, straight neck (photograph: Jebel Qurma Archaeological Landscape Project).

spear, thus 'claiming' it. One such scene occurs in Jebel Qurma, although here the rider is also on a dromedary camel (Brusgaard and Akkermans, in press, fig. 5b). Additionally, I have argued elsewhere that there is another type of conflict scene that appears to depict camel raiding (Brusgaard and Akkermans, in press). In these scenes, the dromedary camel is the centre focus and humans fight each other around it, sometimes on horseback. Often there is also one person holding the camel, as if holding it back from the attackers *(ibid.*, figs. 5c-d).

In total, 84 of the dromedary camels are depicted in scenes. In addition, there are 47 camels that have a rider. The rest are 'individual' dromedaries, with or without an associated inscription. It is common to find them on a large panel where there is an accumulation of individual images or a larger composition (Fig. 3). However, even more frequent, and perhaps most interesting, are the hundreds of dromedary figures depicted as the sole animal on a relatively small boulder. These panels are commonly composed of the dromedary as the central figure, an associated inscription next to or surrounding it, and a geometric motif (Fig. 5).

Producing the camel

Besides looking at numbers of camel depictions and composition, the production of them and their form can also reveal interesting insights. The form or style of rock art motifs has interested researchers for decades and its

Figure 6. This large male dromedary figure shows one of the recurring variations on the body and abdomen shape: an elongated body and a straight abdomen. It also has a 'bell curve' shaped hump. The neck has intentionally been left unpecked. The outline of the figure was carved first; this groove stands out from the in-fill of the body (photograph: Jebel Qurma Archaeological Landscape Project).

study has often yielded new perceptions on typologies and chronologies (*e.g.* Domingo Sanz 2012). Additionally, it can be seen as an indicator for the societies that created the rock art and their identity (Domingo Sanz 2009). In contrast, the study of production techniques and *chaîne opératoire* has only started gaining momentum in rock art research in the last ten years. However, various studies have shown the many insights that this line of research can reveal (*e.g.* Fahlander 2012; Lødøen 2010). This study of the dromedary depictions looked at the form of the motifs and examined how they were produced. As such, the motifs can be understood as more than just their end product, but as a rock art production from beginning to end. Additionally, the large dataset of dromedary motifs can provide insights into the general production and form of the Jebel Qurma rock art.

The form of the dromedary motifs

Rollefson *et al.* (2008) remarked in their study of rock art from Wisad Pools and surrounding areas in northern Jordan that the dromedary motifs show a large variation in style. To some extent, this is the case for the camel corpus from Jebel Qurma as well, although they all share the appearance of being depicted in profile and all but two are depicted as if standing still rather than in motion. Of the 1486 camel figures, 216 are simple with few distinguishing features. However, the rest is more elaborate and well executed, allowing for a few remarks to be made about their form. Interestingly, the detailed and elaborate (the 'well-executed') dromedary depictions are also all images signed by the carver (*i.e.*, those associated with an inscription stating 'By [name] is the camel'). There is indeed variation within these camel figures but a recurring variation whereby a number of typical features could be identified for the majority of the camel figures and others which appear to occur in particular 'types' of figures. In all of the figures the most striking at first glance is the exaggerated hump, which in the majority of camels makes up a third or more than the camel's total height. Another typical feature of all of these dromedary figures is the position of the head and neck. They are almost always depicted with a long, straight neck, holding their head up high. They are usually looking forward, but in 95 cases the camels' heads are facing up, as if looking upwards and in 11 instances they are looking backwards.

These features are typical for almost all of the dromedary figures and all of those signed by an inscription. Additionally,

Figure 7. The hump of this dromedary has been enlarged. The figure has been carved with a thick incised outline and incisions to fill in the body (photograph: Jebel Qurma Archaeological Landscape Project).

there are a few attributes that appear to have a number of variations. The hump is generally exaggerated in size but its shape varies between three recurring types: round, rounded triangular, and a shape that can best be described as a bell curve. The first shape is the most common. Additionally, the body tends to be curved and quite short, paired with a concave abdomen. However, there is a recurring variation on this whereby the body is curved but elongated, paired with a straight abdomen (Fig. 6). Lastly, the number of legs varies. The vast majority of the figures (1066 figures) are depicted with two legs, but a small number have all four legs depicted (234 figures). The rest have either three legs, whereby one is hobbled, or the legs are not visible due to weathering, effacing, or superimposition. There are thus indeed some variations on the style of the motif, but recurring ones. A thorough qualitative and quantitative analysis of these features and their occurrences should make it possible in further research to identify stylistic camel types in the Safaitic rock art.

These features and general appearance of the dromedary depictions result in a motif that in many ways is not naturalistic and has features and proportions contrasting with that of a real dromedary. A real dromedary camel's most characteristic feature is probably also its hump. Its single hump is also one of the features distinguishing it from its cold climate counterpart, the Bactrian camel. However, the hump is generally quite small and is only pronounced when vegetation is abundant, allowing for a large storage of fat (Gauthier-Pilters and Dagg 1981). Even when this is the case, the hump never grows very large either in width or height in relation to the camel's body (*ibid.*, Pl. 5). The hump can disappear almost all together when food is scarce (*ibid.*, 71). In comparison, the petroglyph camels' humps are sometimes hugely exaggerated. In one case, the camel's hump was even modified to be even taller (Fig. 7).

Real dromedaries also differ in the position and shape of their neck, the shape of their body, and their overall proportions. A dromedary generally always holds its head in the extension of its body when standing or on the move. Its long neck has a characteristic curve in it. The tall, straight neck portrayed in the camels in the rock art is usually only seen in real camels when they are stretching their necks to reach vegetation above them (cf. Gauthier-Pilters and Dagg 1981, 34, Pl. 17). The dromedary's body is long with a convex ('hanging') abdomen and a convex back. The latter feature allows the camel to carry a heavier load than a horse *(ibid.*, 109). As a result of its long body and its low, extended neck, the camel's body is longer than it is tall. As mentioned above, the carved camels of Jebel Qurma have a relatively short body and although their backs are convex, their abdomens are almost always either straight or concave. In combination with the long, straight neck, this way of depicting camels produces a dromedary that appears almost as tall as it is long.

A final qualitative remark about the form of the camels provides an interesting contrast to the seemingly 'unnaturalistic' depictions. Although their general form appears to deviate from that of a real camel in some aspects, there are a number of specific anatomical details often added to the carvings of the dromedaries. For example, the muzzle is often depicted as curving down in the typical hanging lip of a camel. Another detail added to 216 of the dromedary images is the definition of the hind leg(s), depicting the hock clearly. Additionally, a number of the camel are depicted with hair on their tail (40 camels), hump (45), neck (7), head (2), throat (9), or a combination thereof (40). The hairs on the tail, hump, neck and head are represented by small incised lines while the throat hairs are represented by a pounded 'lump' on the underside of their neck. All dromedaries have hair on their tails, but they can also grow a substantial amount of hair on the rest of their body in the winter to both protect them from the cold and to prevent dissipation of body heat (Gauthier-Pilters and Dagg 1981, 72). These 'hairy camels' may therefore be depictions of camels in the winter. However, it must be noted, not all camels grow hair in the winter, so the absence of them does not mean it represents a camel in summer. It is interesting to note that there is almost no variation in how the hairs are depicted. The throat hairs are always depicted by a pounded lump while the rest are almost always depicted by thin incised lines sticking straight out; in a few cases they are pounded. It is also a fairly accurate representation of how the hair grows on dromedaries and, like the other features, adds interesting anatomical detail to the depictions.

There are thus naturalistic and unnaturalistic qualities to the dromedary depictions. Overall, there appears to be a particular standard appearance, which includes the long, straight neck, the exaggerated hump, and a slightly curved body. A few different variations on some features are present, such as the hump and abdomen shape, and other features are present in only a part of the camel corpus, such as the depiction of hairs and the defined hock. However, these variations are not infinite and it is clear that there were one or more specific styles according to which these images were made.

Technique	Number of figures
Incised	89
Pecked	53
Pounded	1042
Percussion undefined	89
Pecked and incised	24
Pounded and incised	162
Pounded and pecked	2
Percussion undefined and incised	22
Pounded and pecked and incised	3
Total	**1486**

Table 2. The types of techniques that are used to carve the dromedary figures of the Jebel Qurma corpus and the number of occurrences. The techniques are also used in different combinations.

Production process

If the final image has a fairly standard appearance, then it follows that the production process by which it is made should show signs of some standardisation as well. Therefore, analysing the techniques used to carve the dromedaries and the *chaîne opératoire* by which they were carved can provide insight into their form. Additionally, it has been proposed that it may not always be the final rock art image that was of most significance; the actual process of creating the carving may have been equally important (Fahlander 2012, 100). This is an interesting perspective to consider for the dromedary carvings, especially in light of the theories that these were carved as votive offerings. Therefore, understanding the process by which they were created could shed light on their significance.

Three carvings techniques can be recognised in the production of the rock art of Jebel Qurma: pounding, pecking, and incising. Pounding and pecking are both percussion techniques, with carving using a hammer stone. Pounding is the process of direct percussion, carving directly with a hammer stone and pecking is the process of indirect percussion, using a hammer stone and a chisel to carve. The incision technique uses a sharp, pointed tool directly on the rock and results in narrow, often deep grooves. Percussion technique can be recognised by its broader, often shallower marks. On basalt, pounding tends to result in irregular lines and a more uneven appearance as it is not possible to align each blow precisely with the previous one (Keyser and Rabiega 1999). Pecking tends to produce more regular, neater lines because the carver is able to control and align the lines

Figure 8. Left: the carver made a mistake when sketching this female dromedary figure, starting too far to the right on the panel, indicated by the thin incision marks. Right: the incised sketch marks have been traced for clarity (photograph: Jebel Qurma Archaeological Landscape Project; tracing by the author).

more carefully (*ibid.*). In some cases it was not possible to differentiate between the two types of percussion technique.

All three techniques were observable in the camel depictions, but pounding is by far the most frequent (Table 2). Pecking is the least common. 213 of the depictions were made using a combination of techniques, of which pounding and incising is the most common. A combination of techniques is used in a few different ways. In 57 dromedary images, the ears are incised while the rest of the body is pounded or pecked. Additionally, as stated above, the hair depicted on the head, neck, hump, and tails of the dromedaries is almost always incised. However, perhaps the most interesting use of more than one technique is when one technique is used to create an outline of the image and another is used to fill it in.

The use of outlines gives insight into the *chaîne opératoire* of carving. How they are used becomes clear when we study the figures that are unfinished or where mistakes have been made. These motifs "may hint of the sequence in which the different elements were cut – especially if we can assume that the most important aspects also set the frames for the whole composition of a motif." (Fahlander 2012, 102). The latter can be considered for the figure itself, but also for the whole composition of the figure and accompanying inscription. In the Jebel Qurma corpus, there are 23 unfinished camel figures and a couple where clear mistakes have been made during the production process. Two of these demonstrate the carving process very well. One figure is a male dromedary where to the right of the figure is an incised line that is clearly a first attempt at carving the foreleg and neck (Fig. 8). The incision stops at the top of the neck. It appears that the carver started on an outline of the leg and neck before realising that the head would not fit on the panel. The dromedary was then instead carved more to the left. This figure indicates that the motifs may have been made in outline first, sketched with thin incisions, before being pounded or pecked over. The other figure shows evidence of a similar mistake and carving process (Fig. 9). The outline of this dromedary has also been incised and in some places even made with several thin incisions, very much resembling a sketch. The majority of its body has been filled in with incisions but it is clearly unfinished as the hind leg has not been carved properly. In several areas, there is also pounding on the body. It is not clear whether this was done by the original carver, but the pounding has not been finished either. Interestingly, upon closer inspection, more than one attempt was also made on this image. Above the neck and head is a very faintly incised neck and head, as if the carver made a thin outline of the body before deciding to carve the neck lower down.

These two interesting figures where mistakes have been made and/or have been left unfinished, provide insights into the production process. They indicate that an outline,

Figure 9. The dromedary figure is unfinished and a mistake was made in the initial sketch. The original incised outline of its head and neck is barely visible (left), unless viewed from very close or traced (right). The inscription refers to a 'young she-camel' (photograph: Jebel Qurma Archaeological Landscape Project; tracing by the author).

or 'sketch', was made of the figure first using thin incisions before the figure was pounded or pecked over. And many of the other figures corroborate this. 116 of the dromedary figures show clear signs of having been made in outline first. The majority of these have been made using a combination of techniques whereby there is an incised outline, which is usually hardly visible. In most cases, there are only faint traces of incisions in a few places, not a clear outline around the whole image. This suggests that the incised outline was made first and was intended as a sketch to be hammered over. It is therefore likely that many of the other figures were made in outline first too, but that there are no traces of the sketch because it has been hammered over completely. In 35 figures, only one technique was used to carve them and the outline is more visible. Approximately a third are only incised, with a single outline and an in-fill consisting of incisions (cf. Fig. 4). The others have a clearly pounded or pecked outline and are then filled in using the same technique (Fig. 6).

The majority of the dromedary figures are filled in completely. However, there are 176 dromedary figures that consist of only an outline. Some are only a simple, incised figure, but many consist of a well-executed outline. 80 dromedary figures are partially filled in, whereby often the whole body is filled in except the hump or neck and head (Fig. 6). This is clearly a deliberate, perhaps aesthetic, choice and not a result of an unfinished carving. Concerning the entire composition, there are indications that the image was usually carved first and then the inscription. In most compositions, the inscription curves around small elements of the image, suggesting it was carved last (*e.g.* Figs. 4-5). There are some exceptions, such as shown in Fig. 9, where the inscription might have been carved first.

These figures suggest that the carving process consisted of first making either an outline, which was subsequently usually filled in, or a sketch, which was pounded or pecked over. In the latter cases, it appears that the original sketch was not meant to be visible in the final image. It is therefore likely that this process was used in many of the figures. This sequence of carving suggests that the images and their final appearance were planned. This is unsurprising considering the relatively standard appearance of the dromedary motif, outlined above. When the form is meant to appear a certain way, with a specific shape and particular proportions, it makes sense that an element of planning should be used to achieve this. Interestingly, it appears that this technique was more commonly employed for the creation of the dromedary motif. Only 56 of the 1823 other zoomorphs show clear signs of an outline or sketch. The majority of these are equids, either domestic or wild. Indeed, many of the wild ass, horse, and mule figures show a similar level of detail and skilled execution as the dromedary figures, although they are much fewer in number. Further investigation should show whether there is a similar standard appearance.

On a final note, the typical appearance of the Jebel Qurma camels is more apparent when they are compared to Safaitic camel carvings from other areas in the *harrah*. For example, some dromedary figures found further north in the basalt desert are more naturalistic, exhibiting a less exaggerated hump, a more elongated body, and often a curved neck (*e.g.* KRS 1153, KRS 2502, and IBS 425 in OCIANA). Additionally, there is a contrast in technique as well; the dromedaries are generally incised and consist of only an outline. Interestingly, these figures are associated with inscriptions in the so-called 'fine script', a type of inscriptions rarely found in Jebel Qurma (cf. Della Puppa, forthcoming). The form and technique of the dromedary figures in the Jebel Qurma area are therefore not necessarily representative for an overall 'Safaitic style'. They provide insights into the camel carvings from this area, which may have differed temporally, geographically, and/or culturally from other rock art corpora in the Black Desert.

Discussion

Reviewing the results of this study, a number of points can be made in summary. The dromedary motif is the most common of all motifs in the Jebel Qurma corpus and dominates the range of zoomorphic motifs. Of the sexed dromedary figures, the female dromedary is most frequently depicted. Most dromedary figures occur on their own, either on a large panel where other figures have been carved as well or as the single image on a panel, often accompanied by an inscription and sometimes a geometric motif. When the dromedary is depicted interacting with other figures, it is usually being ridden, being led, nursing a young camel, or depicted in a conflict scene. However, these are the minority of camel images. There is some variety in how the figures are depicted and some are very simple in form. However, in those that are not, there appear to be some typical characteristics and a fairly standard form. It is not a naturalistic one, with an exaggerated hump, a long, straight neck, and a curved body and concave abdomen. Yet it often includes anatomical details, such as hair, a curved muzzle, and a defined hind leg. Some features vary a bit in shape and it may be possible to recognise stylistic dromedary types. However, overall there appears to be a recurring appearance in the Jebel Qurma camels, which contrasts in some aspects with, for example, the camel figure associated with the Safaitic fine script. They contrast in technique as well, with the Jebel Qurma depictions revealing an interesting process of production. The majority of the dromedary figures are pounded, but pecked and incised figures occur as well. Additionally, some of them are made using a combination of techniques. The most striking of these are the ones whereby incisions are used to create a preliminary outline or sketch after which the figure is pounded or pecked over. These carvings indicate that the images were planned, which is unsurprising considering the specific features and proportions of the dromedary depictions. Following on these results, the question is then how to interpret these many dromedary images.

The every-day camel

As outlined above, previous interpretations have generally been functional; the carvings express claims of ownership of a dromedary or merely represent the carver's own camel. In this way they show the daily life of the desert nomads. This is not a surprising interpretation considering the content of the Safaitic inscriptions and considering the cultural-historical context. It is clear that by the mid-Iron Age, the domesticated dromedary was well-established in the Near East and played an important role in this region. More specifically for the desert societies, the inscriptions give the impression that these societies were camel pastoralists, possibly with goats and sheep too. However, sheep and goats are not mentioned in the Jebel Qurma corpus of inscriptions (Della Puppa, forthcoming) and it is of course possible that there were differences in subsistence between groups whereby some were camel breeders, others were mixed pastoralists, and others owned mostly sheep and goats (Macdonald 1993, 319).

Additionally, it must be considered that the references to dromedaries and pastoral activities are not an exact reflection of the daily life of these societies. As Al-Jallad (2015, 3) has shown, the subject matters in the texts are selective and limited. It is therefore more interesting to consider why activities involving dromedaries are written about, just as, among others, camping, grieving, and raiding are, while other aspects of daily life are not. And while the dromedary is a common topic in the inscriptions, it is even more pervasive in the rock art.

A functional interpretation in which these peoples are carving the dromedaries that were an important part of their subsistence or are expressing ownership of their dromedaries is thus logical. However, it is a limited approach on account of two things. Firstly, it does not explain the sheer number of camels and the selective nature of motifs depicted. If these images are mere expressions of daily life, one would expect to find a larger variety of subject matters, reflecting the whole world around them. Yet, like the inscriptions, the types of motifs and the scenes portrayed in the rock art are limited and selective (Brusgaard 2019; Brusgaard and Akkermans, in press). Additionally, different things appear to have been important as content in the rock art versus the inscriptions. The references to pastoral activities with dromedaries in the texts and the depiction of dromedaries and pastoral activities such as leading and nursing match well. However, although signed inscriptions of wild animals are common (*e.g.* 'By [name] is the wild ass'), there are only a few rare mentions of hunting. In contrast, wild animals are abundant in the imagery and hunting is the most dominant theme in the scenes depicted in the Jebel Qurma rock art, exceeding pastoral scenes by far (Brusgaard 2019; Brusgaard and Akkermans, in press). Combined, the two types of engravings provide insight into the world of these peoples, but each reflects a different selection of the world.

Similar disparities between daily life and what is portrayed in rock art have been shown to exist for rock art across the globe (for an overview, see Russell 2012, 14).

Most famously, the sympathetic magic or hunting magic theory, which was developed for European Palaeolithic cave art but has been applied to many different rock art corpora, has been discredited on various accounts but most convincingly by the failure to find faunal assemblages in the archaeological record that matched what was depicted in the cave art (Keyser and Whitley 2006; Russell 2012). Here, too, the depictions are selective and it is clear that these selections in what to depict and how often and what not to depict are significant ones. Equally, the style of depiction is an important factor to consider. Style "combines personal interpretation and choices within the rules regulating the artistic expressions of a specific period and context." (Domingo Sanz 2009, 54). It can express social information about the authors and their social groups, including identification and perceptions of the world around them. This can be encoded in the form of the motif or object, but also mode of production (*ibid.*). The dromedary motif in the Jebel Qurma region does not portray a naturalistic camel and has a number of common, interesting features, such as the exaggerated hump and tall neck. The recurrence of this particular portrayal of the camel and the specific production process used to create it suggest certain cultural conventions about dromedaries and these images. They likely reflect a shared ideal image of the camel, rather than portraying real-life individual camels.

Secondly and following on from the first point, the choice to represent some animals more than others cannot be explained merely in terms of their everyday role in these societies. This reduces the significance of the iconography to a merely functional one and the significance of the animal to a purely economic one. Neither finds any merit in the many forms of animal representation and animal use in the archaeological and ethnographic record from across the world. As Russell (2012, 14) has succinctly stated: "People depict animals because they are food for thought rather than just food."

Although it is not clear on what scale the Black Desert nomads owned and herded dromedary camels and what exact role these animals played in their subsistence economies, it is clear that they had dromedaries. And the role they played warranted their depiction more than any other animal. Parkes (1987) has argued that imagery in pastoral societies focuses on the main animal being herded, which plays a central part in the symbolic and ritual sphere of the society. The economic importance of the herd animal is only part of a wider cultural significance. This can be seen in, for example, the pastoral 'cattle societies' of eastern Africa (Herskovits 1926; Insoll *et al.* 2015; Lincoln 1981) and societies herding goats in central Asia (Parkes 1987).

The ritual camel

Corbett (2010, 127) recognises that the functional interpretation is limited and argues instead that the dromedary may have gained significance beyond its exploitation. Specifically, he argues that the she-camel had an important symbolic role and that "the image, and its offering or sacrifice, were employed to convey a range of ideas about death, the sacred, and the solidarity of the tribal community." (*ibid.*, 150).

Eksell (2002, 140) investigates the topic in less detail but arrives at a similar conclusion, stating that the images are forms of symbolic giving or sacrifice. Eksell interprets the carvings in general from this perspective, basing her conclusion partly on her interpretation of the formulaic content and syntax of Safaitic inscriptions. The inscriptions customarily start with '*l*', the *lām auctoris*. This is generally seen as a 'mark of authorship', usually translated as 'by' (*i.e.* 'By [name] is the camel') (Macdonald 2006, 294). This can be interpreted as the image is by this person (Macdonald 2006, 295). However, Eksell suggests it should be interpreted as 'for', which can denote a sacral meaning for the texts and associated image (cf. Al-Jallad 2015, 4; Eksell 2002, 115-116). However, as Macdonald (2006) and Al-Jallad (2015) have shown, the *l* is simply an introductory particle to the phrase and its translation depends on the context. Therefore, it does not reveal anything about how to interpret the image associated with the text.

Although Corbett and Eksell arrive at their conclusions on shaky grounds, there may be merit to the 'sacrificial image' perspective. Already as early as 1932, Rostovtzeff (1932, 111) suggested that the camel carvings in the Arabian and Sinai deserts were "dedications or recommendations to preserve the camel from harm." Rostovtzeff likens them to the camel figurines that have been found in Arabia. Many more figurines have been found since, as mentioned above. However, there is no consensus on the meaning of these figurines. Magee (2014, 212) has suggested that the representations of dromedaries being used for trade and transport could be seen "as a reflection of their relative novelty within the region."

The dromedary petroglyphs from northern Arabia date to a much later period, when the domesticated dromedary is already firmly established in Arabia. Interestingly, they are contemporary with the *balîya* camel burials from the southern part of the peninsula. Only one possible *balîya* burial has been found in north Arabia, *i.e.* the grave and inscription from Wadi Rum. However, there is epigraphic evidence that it might have been a wider spread practice, possibly in a different form. There are Safaitic inscriptions referring to the *bly*, which, like the Nabataean *blw'*, is interpreted as *balîya*. For example, the inscription WH16 reads 'By ʾtm son of ʿn son of Ẓʿn and he set up this Baliyya for his brother' (WH16; Al-Manaser and Macdonald 2017). The use of this term in Safaitic inscriptions suggests that the *balîya* ritual, perhaps involving camels, occurred in north Arabia too, either in practice or symbolically. That the petroglyphs functioned as symbolic sacrificial camels is one such possibility, for example as a votive offering for a deity, as suggested by previous scholars, or as a sacrifice for a deceased in the afterlife, like the *balîya* ritual.

In this case, it may not have been merely that the image of the camel was intended as a symbolic sacrifice or offering, but that carrying out the ritual, carving the image, was just as significant. It is a plausible explanation for the camel carvings studied here, which show a series of steps and considerable care in the execution, including planning and preparation through sketching and outlining. On the one hand, it can be argued that this process was taken to ensure that the final image achieved the correct proportions and appearance. On the other hand, the practice of planning, sketching the outline, carving the camel, and finally adding small details like the hairs might have been important elements in a long symbolic act that ended with the image of the camel.

However, this raises two questions. First, does this mean that all of the carvings should be interpreted within this framework? Other zoomorphic figures show signs of planning and careful execution, albeit on a smaller scale. Was it irrelevant whether one carved a camel or an oryx as the process of carving was most important? This seems unlikely, especially considering the selective nature of the rock art. The type of motif and the appearance were clearly of importance too. The 'votive offering' theory therefore has its limitations too; it does not explain the significance of the other motifs. Here more research is needed to develop a framework that encompasses the rock art material as a whole (cf. Brusgaard 2019). Moreover, this theory does not explain the prevalence of the dromedary camel motif.

This brings me to the second question: what was the significance of the dromedary camel? If the dromedary camels were carved symbolically as a part of a ritual, why were these animals specifically used? The use of animals in rituals does not stem from their ritual importance. That certain animals are specifically selected for, and deemed important for, certain rituals usually has its grounds in a tightly interwoven combination of economic (everyday) importance, prestige value, and social significance of the animal. This is apparent from a myriad of widespread ethnographic accounts on the ritual use of domestic animals, such as cattle in the abovementioned east African examples (Herskovits 1926; Insoll *et al.* 2015; Lincoln 1981), goats in the central Asian pastoralist communities described by Parkes (1987), water buffalo in Thailand (Tambiah 1969), and pigs in east Asia (Russell 2012). The global archaeological record also illustrates that mostly domestic animals are sacrificed, because they are "sufficiently identified with the sacrificer to serve as a substitute in communications with the divine." (*ibid.*, 125).

If dromedary camels were used (symbolically) in ritual practices in Arabia, it says as much about their social relationship with their herders as it does about their economic value. To understand the nature of this relationship, and subsequently the camel's social and economic importance, it is clear that further research is needed on the dromedary's position in the ancient Near Eastern societies and in the future this will hopefully include new investigations from archaeological, zooarchaeological, anthropological, epigraphic, and representational perspectives.

Returning to the Safaitic camel carvings, it is clear that these images need to be studied and interpreted in light of the role of the dromedary in the ancient Near East. Their prominence in the rock art of the Jebel Qurma region, and probably the Black Desert rock art in general, fits into a wider pattern of camel carvings in the rock art of the Arabian Peninsula, as well as other representational and symbolic evidence. While the data sets from other areas do not yet permit quantitative and detailed qualitative comparisons, the Jebel Qurma corpus allows for a number of inferences to be made about the 'local' material. The dromedary camel motif dominates the material and its presence and its inclusion in pastoral scenes matches the pastoral subsistence that can be deduced from the inscriptions. It has a fairly standard form and careful planning and execution has gone into many of the carvings to create this form. Although not naturalistic, the camels also have small, anatomical details that suggest an intimacy of the carvers with these animals. The female dromedary is most prevalent, which suggests that the she-camel held a greater importance. This is in line with the position of the female dromedary in modern camel-herding societies. While the camels sometimes feature in scenes or with a rider, the majority are depicted on their own, indicating that the importance of this motif lies in its own value, not in its interaction or relationship with other figures. Until now, a few interpretations have been made about the camel carvings, all of which fall broadly in two categories: a functional interpretation in terms of the dromedary's economic, everyday exploitation and a ritual interpretation in terms of the dromedary's ritual exploitation in ancient Arabian practices. Both have their merits but also limitations. Moreover, trying to interpret the carvings within either of these frameworks assumes a dichotomy between the economic and ritual spheres that likely did not exist in these past societies (cf. Brusgaard 2016). To move forward, a new framework is needed in which the complete and complex picture of past human-camel relationships is considered, which includes the dromedary's economic value, ritual importance, and social significance. This will shed new light on the significance of the dromedary in real life as well as the many thousands of dromedary carvings. Further research on the Safaitic rock art as a whole and the role it played in these desert societies can also aid in understanding the significance of individual motifs. Additionally, analyses of the form and production of the carvings can provide new insights into the style of the Safaitic carvings, including possible geographical, temporal, or cultural differences.

Conclusion

This study has endeavoured to investigate the dromedary camel carvings from the Jebel Qurma area in the Black Desert of Jordan. It has done so in an effort to shed new light on the significance of this prevalent motif and further our understanding of the role of dromedaries in the ancient Near East in general. Although this study has perhaps raised more questions than answers, it is my hope that it has shown the value of detailed study of a large rock art dataset and placing animal representations in their cultural historical context. This investigation is part of ongoing research on the rock art of the Jebel Qurma area and new insights will be gained for both the dromedary motif and the rock art as a whole. However, it is clear that the dromedary was an essential concept in the ancient desert societies. These peoples were evidently depicting what was significant in their societies, not what was just present in their societies. Furthermore, the role that the dromedary and the pictorial and textual engravings played in these societies is complex and multi-layered. Researching these issues further can shed light on these societies and on the so far understudied pastoral ideology.

Acknowledgements

This research was carried out as part of the Landscapes of Survival project, directed by Peter Akkermans at Leiden University and funded by the Netherlands Organisation for Scientific Research (NWO).

References

Al-Jallad, A. (2015). *An outline of the grammar of the Safaitic inscriptions*. Leiden: Brill.

Al-Manaser, A.Y.K. and Macdonald, M.C.A. 2017. *The OCIANA corpus of Safaitic inscriptions, preliminary edition*. Oxford: The Khalili Research Centre.

Almathen, F., Charruau, P., Mohandesan, E., Mwacharo, J. M., Orozco-terWengel, P., Pitt, D., Abdussamad, A. M., Uerpmann, M., Uerpmann, H.-P. and De Cupere, B. 2016. Ancient and modern DNA reveal dynamics of domestication and cross-continental dispersal of the dromedary. *Proceedings of the National Academy of Sciences* 113, 6707-6712.

Betts, A.V.G. 1987. The hunter's perspective: 7th millennium BC rock carvings from eastern Jordan. *World Archaeology* 19, 214-225.

Brusgaard, N.Ø. 2015. Pastoralist rock art in the Black Desert of Jordan. In: Giraldo, H. and García Arranz, J.J. (eds.). *Symbols in the landscape: rock art and its context*. Proceedings of the XIX International Rock Art Conference IFRAO 2015. Tomar: Instituto Terra e Memória, 761-767.

Brusgaard, N.Ø. 2016. 'Wives for cattle'? Bridewealth in the Bronze Age. In: Müller, A. and Jansen, R. (eds.). *Metaaltijden. Bijdragen in de studie van de metaaltijden*. Leiden: Sidestone Press, 9-19.

Brusgaard, N.Ø. 2019. *Carving interactions: rock art in the nomadic landscape of the Black Desert, north-eastern Jordan*. Oxford: Archaeopress.

Brusgaard, N.Ø. and Akkermans, K.A.N. In press. Hunting and havoc: scenes depicted in Black Desert rock art of Jebel Qurma, Jordan. In: Davidson, I. and Nowell, A. (eds.). *Making scenes: global perspectives on scenes in rock art*. New York: Berghahn.

Bulliet, R.W. 1975. *The camel and the wheel*. Cambridge, MA: Harvard University Press.

Clark, V.A. 1980. *A study of new Safaitic inscriptions from Jordan*. Melbourne: University of Melbourne.

Corbett, G.J. 2010. *Mapping the mute immortals: a location and contextual analysis of Thamudic E/Hismaic inscriptions and rock drawings from the Wādī Hafīr of southern Jordan*. Chicago: University of Chicago (unpublished PhD thesis).

Curci, A. and Maini, E. 2017. Zooarchaeological analysis of two dromedaries (*Camelus dromedarius L.*) from late Iron Age graves in Wādī Uyūn at Sināw (al-Sharqiyyah. Sultanate of Oman). *Proceedings of the Seminar for Arabian Studies* 47, 67-74.

Della Puppa, C. Forthcoming. *The Safaitic scripts – An ethno-palaeographic investigation*. Leiden: Leiden University (PhD Thesis).

Domingo Sanz, I. 2009. From the form to the artists: changing identities in Levantine rock art (Spain). In: Domingo Sanz, I., Fiore, D. and May, S.K. (eds.). *Archaeologies of art: time, place, and identity*. Walnut Creek, CA: Left Coast Press, 99-129.

Domingo Sanz, I. 2012. A theoretical approach to style in Levantine rock art. In: McDonald, J. and Veth, P. (eds.). *A companion to rock art*. Hoboken, NY: Wiley-Blackwell, 306-321.

Eksell, K. 2002. *Meaning in ancient north Arabian carvings*. Stockholm: Almqvist and Wiksell.

Fahlander, F. 2012. Articulating stone. The material practice of petroglyphing. In: Back-Danielsson, I.-M., Fahlander, F. and Sjöstrand, Y. (eds.). *Encountering imagery: materialities, perceptions, relations*. Stockholm: Stockholm University, 97-115.

Gauthier-Pilters, H. and Dagg, A.I. 1981. *The camel. Its evolution, ecology, behavior, and relationship to man*. Chicago, IL: The University of Chicago Press.

Guagnin, M., Jennings, R., Eager, H., Parton, A., Stimpson, C., Stepanek, C., Pfeiffer, M., Groucutt, H.S., Drake, N.A., Alsharekh, A. and Petraglia, M.D. 2016. Rock art imagery as a proxy for Holocene environmental change: a view from Shuwaymis, NW Saudi Arabia. *The Holocene* 26, 1822-1834.

Hayajneh, H. (2006). The Nabataean camel burial inscription from Wādī Ram/Jordan (based on a drawing from the archive of Professor John Strugnell). *Die Welt des Orients* 36, 104-115.

Herskovits, M.J. 1926. The cattle complex in East Africa. *American Anthropologist* 28, 230-664.

Insoll, T., Clack, T. and Rege, O. 2015. Mursi ox modification in the Lower Omo Valley and the interpretation of cattle rock art in Ethiopia. *Antiquity* 89, 91-105.

Keyser, J.D. and Rabiega, G. 1999. Petroglyph manufacture by indirect percussion: the potential occurrence of tools and debitage in datable ontext. *Journal of California and Great Basin Anthropology* 21, 124-136.

Keyser, J.D. and Whitley, D.S. 2006. Sympathetic magic in western North American rock art. *American Antiquity* 71, 3-26.

Khanvilkar, A.V., Samant, S.R., Ambore, B.N. and Shirval, D.S. 2009. Reproduction in camel. *Veterinary World* 2, 72-73.

King, G. 2009. Camels and Arabian baliya and other forms of sacrifice: a review of archaeological and literary evidence. *Arabian Archaeology and Epigraphy* 20, 81-93.

Köhler, I. 1984. The dromedary in modern pastoral societies and implications for its progress of domestication. In: Clutton-Brock, J. and Grigson, C. (eds.). *Animals and archaeology: early herders and their flocks.* Oxford: British Archaeological Reports (International Series 202), 201-206.

Köhler-Rollefson, I. 1993. Camels and camel pastoralism in Arabia. *The Biblical Archaeologist* 56, 180-188.

Lincoln, B. 1981. *Priests, warriors, and cattle: a study in the ecology of religions.* Berkeley, CA: University of California Press.

Lødøen, T. 2010. Concepts of rock in late Mesolithic western Norway. In: Goldhahn, J., Fuglestvedt, I. and Jones, A. (eds.). *Changing pictures: rock art traditions and visions in northern Europe.* Oxford: Oxbow Books, 35-47.

Macdonald, M.C.A. 1990. Camel hunting or camel raiding? *Arabian Archaeology and Epigraphy* 1, 24-28.

Macdonald, M.C.A. 1993. Nomads and the Ḥawrān in the late Hellenistic and Roman periods: a reassessment of the epigraphic evidence. *Syria* 70, 303-413.

Macdonald, M.C.A. 2006. Burial between the desert and the sown: cave-tombs and inscriptions near Dayr al-Kahf in Jordan. *Damaszener Mitteilungen* 15, 273-301.

Macdonald, M.C.A. and Searight, A. 1983. Inscriptions and rock-art of the Jawa area, 1982. A preliminary report. *Annual of the Department of Antiquities of Jordan* 27, 571-576.

Magee, P. 2014. *The archaeology of prehistoric Arabia: adaptation and social formation from the Neolithic to the Iron Age*. Cambridge: Cambridge University Press.

Magee, P. 2015. When was the dromedary domesticated in the ancient Near East? *Zeitschrift für Orient-Archaeologie* 8, 253-278.

McDonald, M.V. 1978. Orally transmitted poetry in pre-Islamic Arabia and other pre-literate societies. *Journal of Arabic Literature* 9, 14-31.

OCIANA. 2017. The Online Corpus of the Inscriptions of Ancient North Arabia. Oxford. Online: http://krc.orient.ox.ac.uk/ociana/index.php/database [Accessed 6/12/2017].

Oxtoby, W.G. 1968. *Some inscriptions of the Safaitic bedouin*. New Haven, CT: American Oriental Society.

Parkes, P. 1987. Livestock symbolism and pastoral ideology among the Kafirs of the Hindu Kush. *Man* 22, 637-660.

Rollefson, G.O., Wasse, A. and Rowan, Y. 2008. Images of the environment: rock art and the exploitation of the Jordanian badiah. *Journal of Epigraphy and Rock Drawings* 2, 17-51.

Rosen, S.A. 2008. Desert pastoral nomadism in the longue durée. A case study from the Negev and the southern Levantine deserts. In: Barnard, H. and Wendrich, W. (eds.). *The archaeology of mobility: Old World and New world nomadism*. Los Angeles, CA: Cotsen Institute of Archaeology, UCLA, 115 140.

Rosen, S.A. and Saidel, B.A. 2010. The camel and the tent: an exploration of technological change among early pastoralists. *Journal of Near Eastern Studies* 69, 63-77.

Rostovtzeff, M.I. 1932. The caravan-gods of Palmyra. *Journal of Roman Studies* 22, 107-116.

Russell, N. 2012. *Social zooarchaeology: humans and animals in prehistory*. Cambridge: Cambridge University Press.

Sapir-Hen, L. and Ben-Yosef, E. 2013. The introduction of domestic camels to the southern Levant: evidence from the Aravah valley. *Tel Aviv* 40, 277-285.

Seland, E.H. 2015. Camels, camel nomadism and the practicalities of Palmyrene caravan trade. *ARAM* 27, 45-54.

Sykes, N. 2014. *Beastly questions: animal answers to archaeological issues.* New York: Bloomsbury Publishing.

Tambiah, S.J. 1969. Animals are good to think and good to prohibit. *Ethnology* 8, 423-459.

Winnett, F.V. and Harding, G.L. 1978. *Inscriptions from fifty Safaitic cairns*. Toronto: University of Toronto Press.

Bows on basalt boulders: weaponry in Safaitic rock art from Jebel Qurma, Black Desert, Jordan

Keshia A.N. Akkermans

Abstract

The Safaitic rock art of Jordan's Black Desert is a fascinating yet under-examined subject. In this contribution, I discuss the representations of weapons in the rock art of the Jebel Qurma region in north-east Jordan. Additionally, I will give an overview of the material evidence of weaponry produced by recent excavations in the region's burial cairns. Detailed visual analysis distinguished four categories related to weaponry in the rock art: bows, pole weapons, swords/daggers, and shields. Patterns in the use of these objects vary for each category. Most notable are the firm association of lances with riders on animal-back, and the archers that are predominantly depicted on foot.

Keywords: weaponry, rock art, Jordan, Jebel Qurma, desert communities, bow-and-arrow, Near East

Introduction

The north-east of Jordan is home to the 'Black Desert', which is an extensive, arid region dominated by dark basalt uplands (*harrah*), alternating with vast gravel plains (*hamad*). Despite the region's forlorn ambience and harsh climate, it holds a wealth of archaeological remnants of many different periods. Inscriptions in Safaitic and petroglyphs were carved into many of the dark basalt boulders that litter the landscape, and these show a variety of figurative and geometric motifs (Brusgaard 2015; 2019; Brusgaard and Akkermans, in press). Whereas the Safaitic texts have received considerable attention (see *e.g.* Al-Jallad 2015 and references therein; Littmann 1943; Macdonald 1993), research on the contemporaneous petroglyphs has remained limited. A notable advancement is the recent doctoral dissertation focussing on the rock art of the Jebel Qurma area by N. Brusgaard (2019), which comprises the first systematic, contextual study of the contents and motifs of the rock art of the Black Desert (see also Brusgaard, this volume). Other important studies have focused mainly on specific motifs, such as women and chariots (*e.g.* Macdonald 1993; 2009).

The majority of figurative depictions in the rock art of the Jebel Qurma region consists of zoomorphic figures, while anthropomorphic representations make up a relatively small share (Brusgaard 2015; 2019). The main category of material culture depicted in the petroglyphs is that of weaponry. Studies on the motifs and objects in rock art can provide valuable insights into the material culture used by its makers (May *et al.* 2017), especially in regions where the archaeological record is complicated by numerous issues. For example, artefacts found in the burial cairns in the Jebel Qurma region are often

In: Peter M. M. G. Akkermans (ed.) 2020: *Landscapes of Survival - The Archaeology and Epigraphy of Jordan's North-Eastern Desert and Beyond*, Sidestone Press (Leiden), pp. 305-316.

Figure 1. Weaponry found during the 2017-2018 fieldwork campaigns in cairn burials in the Jebel Qurma region. First row: severely fragmented blade, possibly a lance head. Second row: front and side view of a javelin blade with a long tang. Third row: front and side views of three arrowheads. The central rib is clearly visible on the second point from the left. Two possible rounded tangs are lacking adjoining blades (photograph: Jebel Qurma Project Archive).

Figure 2. Armour found in cairn burials of the Jebel Qurma region during the 2017-2018 fieldwork campaigns. First row: front and side views of four armour scales of various sizes. Note the curvature at the bottom of the plates. The two scales on the left are still joined with their rivets. Second row: front and side view of a simple rectangular wrist-guard made of sandstone (photograph: Jebel Qurma Project Archive).

poorly preserved, due to the porous construction of the cairns, their continuous re-use, looting, and the region's extreme climatic conditions (Akkermans and Brüning 2017, 135). Such conditions also impacted the preservation of weaponry in these tombs, and excavations therein retrieved only a handful of examples. In this article, I will discuss both the material and rock-art manifestations of weaponry in the Jebel Qurma region in the Classical and Late Antique periods.

Weaponry in the material record

In the Jebel Qurma area, burial tombs were the only places that contained preserved weaponry. Two tombs yielded some weapons made of iron, in the form of at least five arrowheads, a javelin, and a potential spearhead (Fig. 1) (Akkermans *et al.* 2020). The javelin and four of the arrowheads were found in the original (but collapsed) grave chamber of a so-called 'ring cairn' at the site of QUR-80. The cairn had undergone several phases of re-use and alteration since its initial construction. Looting activity in antiquity left the original grave heavily disturbed. Moreover, a subsequent phase of re-use levelled the top half of the ring cairn and used it as a foundation to build a straight-walled 'tower tomb' on top of it. A fifth arrowhead and the possible spearhead were found in a burial in a round tower tomb at the site of QUR-98. Iron weaponry similar to the pieces presented here was found earlier at sites like Lachish, Al Khadr, and Beer-Sheba, and suggest the use of the two Jebel Qurma cairns between the late ninth and seventh century BC (see Cross and Milik 1956; Gottlieb 2004; 2016).

The arrowheads have simple, leaf-shaped blades and lozenged tangs. Unfortunately, they are all severely corroded, which makes precise typological identification difficult. Most of them have a mid-rib that is strongly pronounced. Nearly all of the arrowheads are fragmented: only one has its tang still attached. The arrowheads range between 4.5 cm and 6 cm in length, 1.0 cm and 1.5 cm width, and, depending on the presence or lack of a mid-rib, between 0.3 cm and 0.8 cm in thickness. The javelin (found in the tomb at QUR-80 together with four arrowheads) has a short and relatively thick leaf-shaped blade with a long, rounded tang. It measures 10 cm by 1.5 cm and is 1.0 cm thick, but the object's corroded exterior belies its dimensions. A central rib on the blade is either fully lacking or masked by corrosion as well. Eight iron fragments were also found together, all of which have a central rib on one side of the blade. They could be re-joined to form the long, narrow blade of a single spearhead about 17 cm in length, 2 cm in width, and 0.6 cm in thickness. Alternatively, the pieces may come from two somewhat shorter blades. The corrosion of the iron does not rule out either of these options.

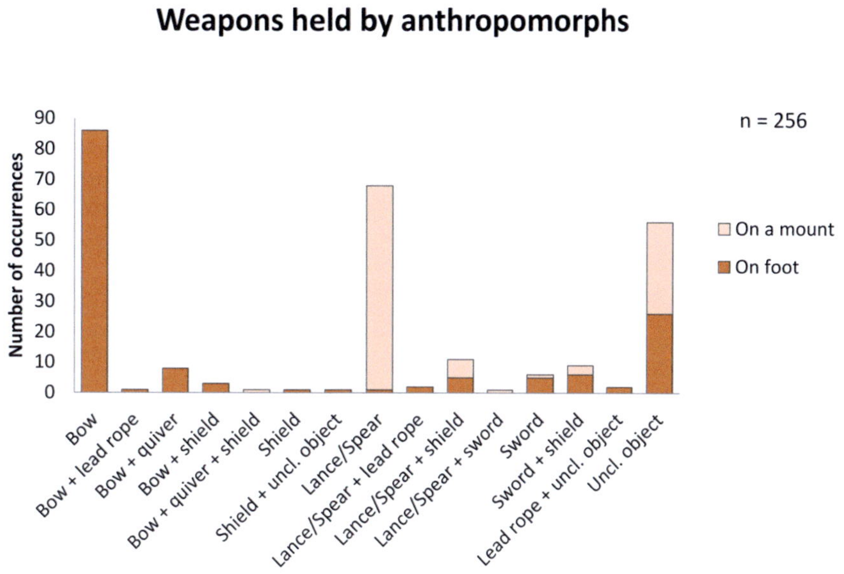

Figure 3. Stacked bar chart showing the types of weapons and weapon combinations, as well as the number of anthropomorphic figures holding them. The unidentifiable objects ('uncl. object') are also included. Figures mounted on an animal are displayed in light ochre and figures on foot are indicated by dark ochre.

A more common find in the burial cairns were the bronze armour scales, which were about 2.5 cm long, 1.5 cm wide and 0.1 cm thick (Fig. 2). They came from graves dated to the mid and late first millennium BC (Akkermans and Brüning 2017; Akkermans *et al.* 2020), but never were present together with iron weapons or with any other weaponry. Generally, their shape is either polygonal or rounded, with a wider top that tapers at the base. On multiple occasions the scales still possessed their rivets, used for attaching the plate to the fabric corselet underneath. Most of the scales are curved at the wider end, presumably to further aid adherence to the under-cloth. The scales were found in several tombs but, curiously, their quantity per grave is very low. Most scales have been found as single pieces or, occasionally, in small groups of only two or three per cairn. Perhaps they were segments that fell off during looting activities in antiquity which removed the armour corselets from the tombs. Alternately (and probably more likely), the armour scales may have served a ritual purpose in the mortuary practices of the region. Maran (2004, 23) ascribes an apotropaic element to individual appearances of armour scales; a single scale would have contained protective qualities equal to a complete, plated armour coat.

Excavations in another burial chamber identified a remarkable example of non-metallic weaponry, a sandstone archery wrist-guard. The object measures about 7.3 cm long, 2.5 cm wide, and 1.0 cm thick, and has a straight-sided outline with rounded corners. A single perforation is present on either end of it. The object appears to be made out of a layered type of sandstone. The layers have disintegrated in places, with the surface of one side of the object starting to crumble and flake off. From what is left of the brace, its cross-section appears to have been straight on both sides.

The wrist-guard stems from a much earlier context than the other objects discussed so far. It was found in the burial chamber of a ring cairn at the site of QUR-207, which radiocarbon evidence has dated to the late Early Bronze Age IV period, *c.* 2010-1890 cal BC (cf. Akkermans and Brüning, this volume, Table 1). Remarkably, the grave has yielded no other objects related to archery, such as metal or stone arrowheads. In fact, this context fits well into the wider discussion regarding the functionality of these stone wrist-guards; some scholars ascribe a more decorative or symbolic significance to them rather than a practical use, based on their decoration and their intended position on the arms (Woodward *et al.* 2006; Fokkens *et al.* 2008). The burial at QUR-207 offers little help in interpreting the object: the relatively poor preservation of the skeletal remains did not allow for a reconstruction of how and where the object was worn on the body.

Weaponry in rock art

Extensive fieldwork in the Jebel Qurma area since 2012 has yielded many thousands of Safaitic inscriptions and petroglyphs (Brusgaard 2019; see also Brusgaard, this volume). The present analysis is based on rock art documented in the region during surveys between 2012 and 2016. In this time span, more than 300 sites were meticulously searched for petroglyphs. The survey documented all encountered rock art, forming an extensive corpus of both ancient and modern petroglyphs that portrayed a large variety of subjects. Most of Jebel Qurma's rock art can be found on large, isolated basalt boulders in the landscape. The petroglyphs tend to cluster in specific locations high up on the hills, which offer a good overview of the landscape (Brusgaard 2019). Carving, pounding, or pecking removed parts of the dark upper surface of the rock, revealing a lighter stone base underneath.

As with most rock art, dating the petroglyphs in the Jebel Qurma region has proven to be difficult. Conventionally, they are dated between the first century BC and the fourth century AD, although this dating is based on meagre evidence and informed speculation; the chronological boundaries are, in fact, unknown (Al-Jallad 2015, 17). The surveys and excavations in the Jebel Qurma area have shown that rocks with petroglyphs were regularly re-used as building material for tombs that probably date as early as the third and second centuries BC (Akkermans and Brüning 2017; Akkermans *et al.* 2020). This observation strengthens the presumption that the Safaitic rock art of Jebel Qurma is several centuries older than previously believed.

The 2012-2016 fieldwork yielded a total of 4637 rock-art motifs. Anthropomorphic figures take up 562 (about 12%) of this amount, including those mounted on animals. Of these, 268 figures are shown holding objects, most of which are weapons. A number of these figures hold 'lead ropes' and 'whips'; since these are not classed under weaponry, they are excluded from this study. 56 objects could not be identified. In general, the petroglyphs of the Jebel Qurma area display little detail, making it difficult to identify weapons. Altogether there were 198 figures with a clearly depicted weapon or a combination of weapons, divided over 172 individual rock-art panels (Fig. 3).

The depicted weaponry can be divided into four categories: (1) bow-and-arrow; (2) pole weapons; (3) swords and/or daggers; and (4) shields. These types of weaponry could be used individually and in any combination. The shields, although technically not a weapon, are considered part of the weaponry equipment as defensive gear.

The bow-and-arrow

The largest share of the petroglyph weapon assemblage from Jebel Qurma is made up by the bow-and-arrow, with 99 anthropomorphic figures holding this weapon. In rare occasions the quiver is indicated as well, being worn on the back with the fletchings of the arrows pointing upwards (Fig. 4). The arrows are generally depicted as simple lines that are occasionally crowned by undefined arrowheads. The shape of the bows, on the other hand, is clear and remarkably uniform throughout all its depictions. Its appearance corresponds with that of the composite bow, which is recognisable by its relatively small size and its distinct arch, reminding one of a cupid's lip (Bowden 2012, 48). Several bows show a slight curvature at their outer ends, classifying these as double-convex composite bows.

The shape of the convex bow is the result of its structural composition: the body is built upon a thin wooden core that functions as its 'skeleton', onto which other components are fixed (Zutterman 2003, 126; Miller *et al.* 1986, 183). The materials for these additional components are selected for their capability to withstand and adapt to the pressure and tension on the bow's body when it is drawn (Zutterman 2003, 121). In antiquity, materials for bows generally encompassed wood, bone, horn, sinew, and glue made from the swimming bladders of fish (Paterson 1966, 70-77; Bowden 2012, 44). Sinew was applied to the outer sides of the bow and would be stretched when the bow was spun (*i.e.* flexed and held tight by a bowstring). The 'belly' of the bow, which faced the archer, was made of horn and would contract when the bow was used.

The materials used and the bow's design made the composite bow highly efficient. Performance equivalent to that of the 'self-bow' or 'laminated bow' could be achieved with less effort. A self-bow required much more strength to launch an arrow of similar dimensions and weight at the same speed as if one used a composite bow (Loades 2016, 5). In practical terms, this means that the composite bow can be kept spun for a longer time, culminating in a more precise aim (Miller *et al.* 1986, 185). The upward curling notches of recurve composite bows would help to reduce the recoil shock generated upon release of the arrow. As a result, it causes the arrow to deviate less from its intended flight path (Bowden 2012, 44).

The technology of the composite bow was by no means restricted to the Black Desert, nor was it a 'new' technology. The earliest appearances of composite bows in Mesopotamia date to the third millennium BC. Later, the Hittites, Egyptians and Assyrians (among others) all used composite bows of a rather triangular shape (Miller *et al.* 1986, 180; Zutterman 2003, 120-123). With the foundation of the Persian empire by Cyrus II around 550 BC, the triangular composite bow slowly faded out of usage. The recurve composite bow replaced the triangular composite bow across a wide geographic area, which can also be seen in the rock art found in the Jebel Qurma area, and continued to be used for another 1800 years (Rimer 2004, 85).

It is important to realise that the manufacture of a good composite bow would have taken considerable time and craftsmanship, and it could take a couple of months up to two years before one was completely finished (Paterson 1966; Loades 2016, 5). The addition of an element to the bow's wooden base required a thorough period of drying before adding the next element. Due to its lengthy manufacturing time, previous research has proposed that professional bowyers in sedentary communities produced composite bows in large batches of a couple of hundred at a time (McEwen 1978; Miller *et al.* 1986, 184, who point out: "A completed composite bow was a tour de force of precision engineering and bonding ...").

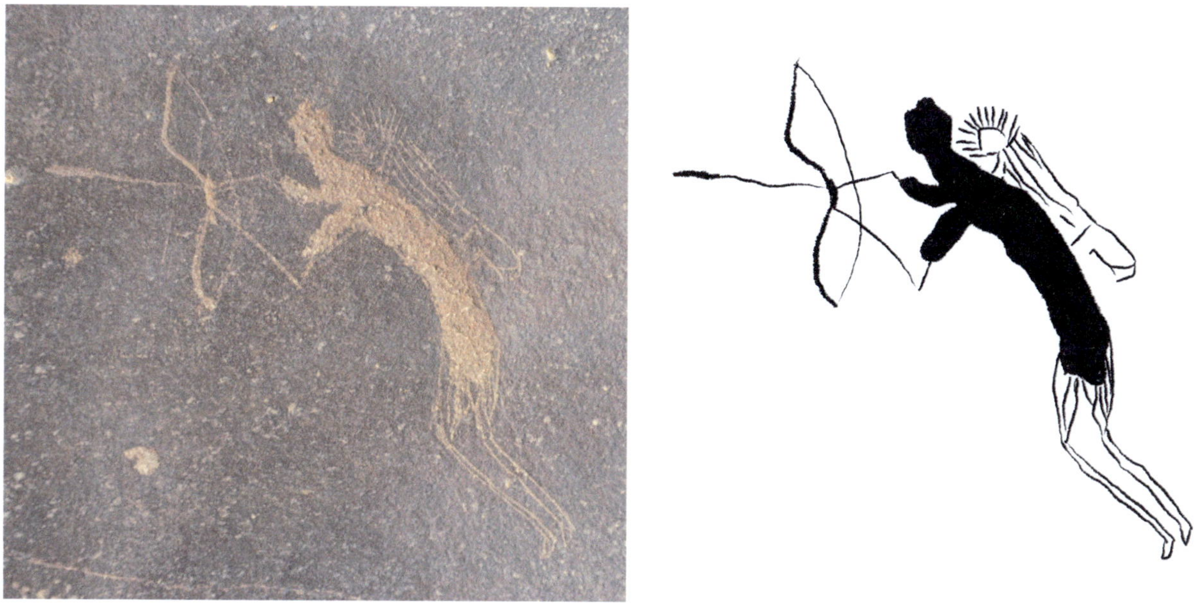

Figure 4. Depiction of an archer with a composite bow. Note the quiver carried on the figure's back, with the arrows sticking out from the top (photograph: Jebel Qurma Project Archive; tracing by the author).

The current evidence suggests that the desert populations of the Jebel Qurma area were nomadic or semi-nomadic hunters and pastoralists (see *e.g.* Akkermans and Huigens 2018; Akkermans 2019; Huigens 2019). Therefore, it is unlikely that they made the bows themselves. If the rock art scenes are accurate depictions of nomadic life in the desert and of its associated material culture, we may wonder where the desert groups obtained these bows from and whether they indicate specific types of interaction between the desert and the settled regions (which undoubtedly existed; cf. Macdonald 2014).

While the nature of such trade requires further definition, a more direct interpretation is that the remaining weapon categories depicted in the corpus of rock art from the Jebel Qurma area were locally fabricated; its required craftsmanship notwithstanding, the production process of weapons such as lances, arrows, and spears is less lengthy and requires a lesser variety of raw materials. A small, temporary metal workshop identified at the site QUR-595 demonstrates that at least some part of the pastoralist population in the region could work iron in a rudimentary way (Huigens 2019; Akkermans and Brüning 2020). However, radiocarbon samples of charcoal inside the installation at QUR-595 produced a date between 100 to 385 AD, which is considerably later than the suggested date of the physical weaponry in the area. Additionally, there is a clear disparity in the level of specialism needed to manufacture or repair simple tools for everyday use and to forge high-quality blades. With only one metal-working installation identified at present, it would be rather premature to make assumptions on the local people's proficiency in metal working.

Pole weapons

At Jebel Qurma, 81 out of the 256 (*c.* 32%) weapon depictions represent pole weapons. Pole weapons are part of the class of 'melee weapons' or close-combat weapons, and fit within the sub-category of the pointed arms (Woosnam-Savage 2004, 417). Divergences in the length and mode of use in the rock art suggest that they depict more than one kind of pole weapon. Presumably spears, lances, and javelins were all part of the repertoire of pole weapons in the *harrah*. Differentiating between these three weapons often results in a semantic discussion, because it is unclear what the conventional terms 'lance' and 'spear' actually define (Potts 1998, 183). The modern Arabic word *rumḥ'* can refer both to the lance and the spear. The Safaitic word *rmy*, in turn, is connected to the Arabic verb *ramā*, meaning 'to throw'. Gordon (1953, 68) describes a spear as "just a dagger at the end of a shaft", which functions as "a thrusting weapon with a longer reach". Potts (1998, 183) uses the term 'spear' to refer to a "light projectile which could be thrown over a considerable distance at an enemy and for which the term 'javelin' is sometimes employed". The term 'lance' refers to "a much heavier and longer weapon which, although it could be thrown a short distance, was more commonly hand-held and used for thrusting in close combat." (*ibid.*). Most of the pole weapons in the rock art of Jebel Qurma range from fairly long to very long in relation to the figure that holds them, and are therefore most likely to be lances (Fig. 5). In addition, they have a strong association with depictions of anthropomorphic figures on equids and camel-back (see below). The shorter pole weapons are often held by figures on foot, and are regularly accompanied by a shield. These are interpreted as spears.

Figure 5. Scene with swordsmen/spearmen, archers, and a rider with a lance, all attacking a carnivorous mammal. At least four figures carry a sword at the waist. The panel shows a clear divide in weapon choice between those on foot and those on animal-back. The Safaitic inscription and later additions are not traced (photograph: Jebel Qurma Project Archive; tracing by the author).

Swords and/or daggers

Depictions of swords and/or daggers are quite rare in the Jebel Qurma region. Altogether only 15 swords/daggers have been identified, although the actual number may be higher. Apart from cross-guards and pommels, the swords/daggers offer little visual clues for recognition in the rock art. For example, the distinct lunate pommel daggers, so characteristic for rock carvings in the Arabian Peninsula (Newton and Zarins 2003; Aksoy 2017, 6), do not occur in the Jebel Qurma region and, perhaps, in the Harrat al-Sham at large.

The basis for identifying a sword or dagger in the Jebel Qurma petroglyphs is if the object is held at an outer end and seems relatively short. The most clear-cut example in the corpus is that of a horseman carrying a stick-shaped object at the waist (Fig. 6). A shorter bar that diagonally crosses the front side of the rider's body could portray the cross-guard of the sword. While the carving does not clearly depict the belt itself, the position of the objects suggests that it is hanging from the hips, just as a sword would when carried in a sheath. This coincides with Michael Macdonald's observation that Safaitic rock art very rarely depicts horseman figures wielding a sword, but they are shown occasionally with a sword at their belts (Macdonald 2007, 282).

A second notable depiction of a sword comes from a carving of a figure holding a long object in front of a camel (Fig. 7). This object slightly widens before turning back into a point at the outer end. What is most remarkable about this depiction, however, is the clear curvature in the blade. Its sickle-shaped blade is reminiscent to that of the Assyrian *sappara* and the Egyptian *khopesh*. As early as 1917, Sir William Flinders-Petrie noted with regard to the *khopesh*: "the peculiarity of the type is the deep hollowing of the back, and the projecting of the edge far in advance of the handle. By its great curvature it was intended for a wiping cut." (Flinders-Petrie 1917, 27). The *khopesh* is not identical to a sickle sword, although both types of weapons have considerable similarities in shape. The sickle sword was sharpened on the concave inner side of the curving blade, like that of an agricultural sickle. The *khopesh*, on the other hand, was sharpened on the convex outer side or, on rare occasions, sharpened on both sides. There is evidence for the use of *khopesh*-like arms or sickle swords in the third to first millennium BC in Assyria, Babylonia, Phoenicia, Anatolia, Palestine, and Egypt (Gordon 1958, 23-24).

If the figure at Jebel Qurma is indeed holding a *khopesh* or a sickle sword, it would be the only curved sword represented in the local repertoire. The image is not accompanied by a Safaitic inscription and therefore cannot

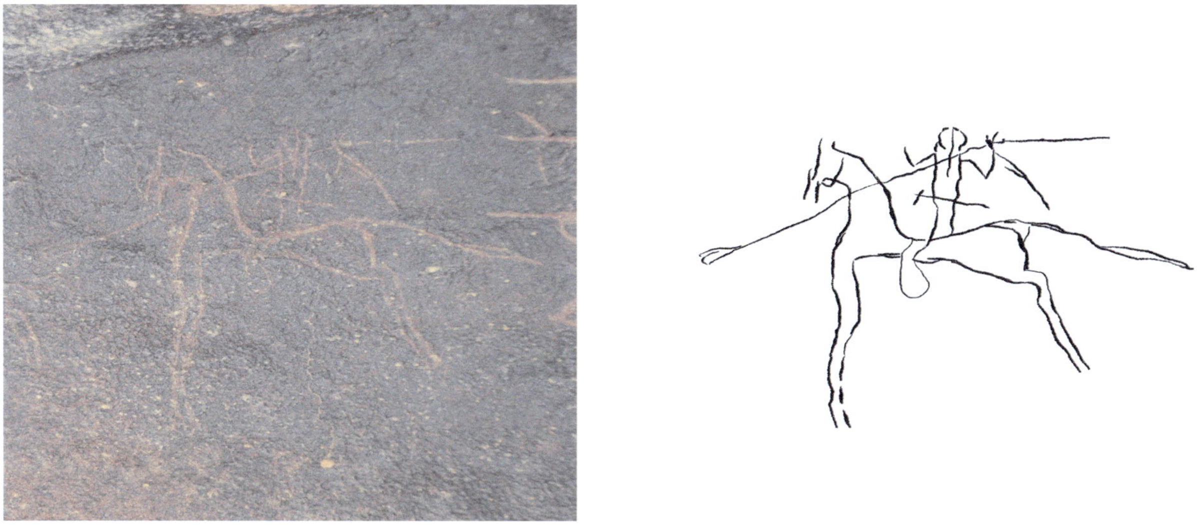

Figure 6. Figure holding a lance and carrying a sword at the waist, while seated on an equid. Note the diagonal bar crossing the shaft of the sword and the way in which it is carried at the waist (photograph: Jebel Qurma Project Archive; tracing by the author).

Figure 7. A figure standing next to a camel carries a shield and curved sword. The sword curves alongside the edge of the boulder face (indicated in grey), while the head of the camel is not affected by this curvature (photograph: Jebel Qurma Project Archive; tracing by the author).

be unambiguously dated to the Safaitic period. The style of the image also deviates slightly from the general Safaitic rock-art corpus at Jebel Qurma, strengthening the suspicion that the image belongs to a different time period. It is also worth mentioning that the curvature of the sword perfectly follows the curvature of the rock onto which the image was made. Since it follows its curvature so well, it is possible that the depiction did not intend to represent a curved sword at all but formed into a curve because of the uneven surface of the rock face. However, the neck of the camel depicted next to the figure holding the discussed object also stretches over the curvature of the rock but remains unaffected by it.

Defensive gear

At least 28 figures are depicted holding shields. One panel shows three figures holding shields while drawing bows, but most shields are accompanied by spears and swords (see Fig. 8). All the shields are relatively small and round, but

Figure 8. Two figures on foot holding short spears and small round shields. A stripe pattern decorates the torso of each figure (photograph: Jebel Qurma Project Archive; tracing by the author).

their patterning varies: nine have a circle in the middle, four bear a central cross, two show feather-like decorations, one has cross-hatching, and one has several lines radiating from the centre towards the shield's outer edges. The remainder of shields either have decoration that is unclear or no ornamentation at all. It is unclear what the shield adornment represented. Circles on shields could be depictions of the shield boss (*umbo*), which attaches the grip of the shield to the shield itself using a convex, round piece of material in the centre of the shield. Linear decoration such as cross-hatching might imply leather strips on the front.

Lines and cross-hatching are not exclusive to shields. Lines cover the bodies of a (very) small number of anthropomorphic figures (Fig. 9). Among the nine known patterned figures, the number of figures also handling weapons is relatively high: three carry a sword and a shield, and two hold a spear and a shield. Three scenes depict figures both with and without patterned torsos. The patina of the patterns in these scenes does not stand out from the rest of the image, ruling out that they were later additions. As with the case of the shields, it is unclear whether the patterns are of functional or decorative nature. Stripes and other lines also adorn some carved dromedaries, equids, bovines and other animal species. Such patterns do not fit the animal's coat colours or armour, suggesting they serve as decoration (Brusgaard 2019). As for lines added to the human figures, they could represent (plate) armour, straps carrying swords, bows or quivers, or even clothing details (although Safaitic engravings rarely indicate clothing; Macdonald 2007, 274).

Modes of weapon use

Analysis of the employment of weaponry in the rock art of Jebel Qurma shows distinct relational patterns between weapon use and figures mounted on animals. The majority of the 79 riders that carry weaponry ride equids, probably either horses or mules/hinnies. Some ride dromedary camels. The category of pole weapons has strong affiliations with mounted figures, as 62% of weapon-wielding riders carry a spear or lance. On the other hand, the share of pole-weapon use for figures on foot is about 5%.

Bow use shows quite the opposite trend: although bows make up almost 66% of all weapons handled on foot, they comprise less than 1% of the weapons used on animal-back. The latter observation is surprising, as the composite bow is highly suited for use on horse-back for several reasons. Firstly, the composite bow has an elongated drawing time, resulting in an improved aim. This is especially beneficial when the bow is operated while riding a moving animal. Secondly, the size of the bow can be kept to a small size. A small bow allows a mounted archer to move fluidly and turn his upper body all the way to the rear of the horse if needed, known as the 'Parthian shot' manoeuvre (Overtoom 2017, 103; Zutterman 2003, 134). It is hard to explain the lack of

Figure 9. A possible raiding scene revolving around two camels. Two figures on foot use composite bows, while two riders seated on a single horse wield a long lance. Both riders have horizontally striped bodies. The Safaitic inscription is not traced (photograph: Jebel Qurma Project Archive; tracing by the author).

riders carrying bows from a functional perspective. Instead, the recurring depiction of archers on foot, apparently involved in short-range combat, might point to specific cultural conventions (K. Akkermans 2017). The preference for pole weapons seems to confirm a statement by Potts (1998, 185) on the lance being "the principal weapon of most ancient and indeed much modern cavalry."

Conclusion

Several conclusions can be drawn from the study of weapons in the material record and the rock art of the Jebel Qurma region. Recent excavations of burial cairns have yielded a few iron arrowheads, an iron javelin, a possible iron spear, a stone wrist guard, and some bronze armour scales (cf. Akkermans *et al.* 2020). The distribution of the bronze armour scales is rather curious, with only one to three scales per grave. This pattern can be explained as residual evidence of looting activities or, more likely, as a testament to the apotropaic use of armour scales in funerary rituals. The majority of these objects are poorly preserved, which makes more specific typological identification difficult. In terms of chronology, the majority of the iron weapons come from burial contexts that are likely to pre-date the Safaitic-period petroglyphs. However, the temporal relationship between the material and iconographic evidence of weapons requires further research.

It is possible to discern two main categories of weapons in the rock art: the bow-and-arrow, of which the composite bow is especially prevalent, and pole weaponry. Of the latter, the majority probably represent lances, which were used primarily by riders. Spears only make up a small share, and are used by pedestrian figures. Apart from the unclear objects, 198 anthropomorphic figures can be recognised as holding weaponry. Of these, 99 hold a bow (or a combination of a bow and another weapon) and 82 hold a lance/spear (or a combination of a lance/spear and another weapon).

There is a clear distinction between the use of the bow-and-arrow and the use of pole weapons. The bow is almost exclusively depicted with figures on foot. Pole weapons are shown primarily with anthropomorphic figures riding animals, in which case these weapons likely portray lances. Despite being difficult to identify, there are at least 15 recognised instances of swords.

Shields are commonly depicted together with other weapons. All shields are small and round or ovoid in shape. The various patterns engraved on the shields are particularly intriguing, yet their meaning remains uncertain at present. However, there are rare instances of similar patterns depicted on the torsos of anthropomorphic figures as well, perhaps signifying clothing, armour, or decoration.

It is still difficult to interpret these findings on the theme of weapons in the rock art of Jebel Qurma. The discovery of new weapons by future fieldwork campaigns will aid in comparing between the weapons shown in petroglyphs and those in the material record. As of now, chronological discrepancies between the rock-art weaponry and its physical counterparts impede any direct conclusions.

Moreover, it is important to establish the degree of realism in the images. For instance, the lack of archers riding animals may be the result of societal customs of weapon use, if these depictions can be considered representative or expressive of the reality of these societies. Like all symbolic systems, rock art has a social function (Layton 2001). The rock-art tradition of the Jebel Qurma region is highly repetitive. It effectively forms a visual language based on the standardised shapes of animals and anthropomorphic figures. The panels display high levels of naturalism: scenes related to weapons show no apparent signs of otherworldly activity, mythical narration, or ritual or religious practices. Instead, the rock-art scenes seem to remain fairly mundane, depicting events that were likely more exciting than the general activities of everyday life, such as a raid or a successful hunting party. At the same time, the scenes do not portray domestic activities, apart from two highly unclear depictions of what may be human intercourse. There is a complete lack of carvings of domestic structures, vegetation, trading caravans, *etc*.. The rock art of the Jebel Qurma area is an expression of local cultural and social norms, rather than an assemblage of random depictions of interest to individual carvers (cf. Brusgaard 2019). These observations about petroglyphs compliment previous statements by Al-Jallad (2015, 3) regarding the Safaitic inscriptions, in which he notes that the texts are decidedly formulaic and uniform. The range of subjects presented in these carvings is limited and highly selective, and thus by no means an emanation of 'unstructured self-expression'.

Indeed, it would be incorrect to simply "read [rock art] as a mirror of society" (Walderhaug 1998, 298). Nevertheless, the insights that rock-art analysis can provide about past worldviews and interactions should not be understated. For example, the persistent depiction of double-convex composite bows suggests that these were physically present in the Jebel Qurma region. However, as mentioned earlier, producing composite bows is a delicate, time-consuming task that is best executed in a sedentary context, which suggests that the mobile or semi-mobile pastoralist societies of the *harrah* maintained intimate relationships with settled communities elsewhere.

A final word of caution is warranted: the outcomes of this analysis should not necessarily be taken as representative of the Black Desert as a whole (cf. Brusgaard, this volume). Weapon assemblages and patterns of weapon use may vary considerably throughout the wider *harrah*. The populace that used the Safaitic script and produced the petroglyphs likely comprised several individual cultural or ethnic communities dispersed over a broad geographic area, each with their own practices, preferences, traditions, and material culture (cf. Al-Jallad 2015; Macdonald 2009).

Acknowledgements

This research was carried out as part of the Jebel Qurma Archaeological Landscape Project, directed by Prof. Peter Akkermans (Leiden University). I would like to thank Peter Akkermans for his kind invitation to contribute to this book, and for all the motivating, understanding, and insightful conversations we shared about the material. I also express my gratitude to Timothy Stikkelorum for providing me with many fascinating notes on archery. My special thanks go to Nathalie Brusgaard, who played a quintessential role in the execution of this study with her kind and useful feedback.

References

Akkermans, K.A.N. 2017. *Battles in basalt: a study of the weaponry depicted in the Safaitic rock art of Jebel Qurma, north-eastern Jordan*. Leiden: Leiden University.

Akkermans, P.M.M.G. 2019. Living on the edge of forced into the margins? Hunter-herders in Jordan's north-eastern badlands in the Hellenistic and Roman periods. *Journal of Eastern Mediterranean Archaeology and Heritage Studies* 7, 412-431.

Akkermans, P.M.M.G. and Brüning, M.L. 2017. Nothing but cold ashes? The cairn burials of Jebel Qurma, north-eastern Jordan. *Near Eastern Archaeology* 80, 132-139.

Akkermans, P.M.M.G. and Brüning, M.L. 2020. A Late Roman pastoralist ironworking site in the north-eastern Black Desert, Jordan. In: Ahrens, A., Rokitta-Krumnow, D., Bloch, F. and C. Bührig, C. (eds.). *Drawing the threads together: studies on archaeology in honour of Karin Bartl*. Münster: Zaphon, 103-119.

Akkermans, P.M.M.G. and Huigens, H.O. 2018. Long-term settlement trends in Jordan's northeastern badia: the Jabal Qurma Archaeological Landscape Project. *Annual of the Department of Antiquities of Jordan* 59, 503-515.

Akkermans, P.M.M.G., Brüning, M.L., Arntz, M., Inskip, S.A. and Akkermans, K.A.N. 2020. Desert tombs: recent research into the Bronze Age and Iron Age cairn burials of Jebel Qurma, north-east Jordan. *Proceedings of the Seminar for Arabian Studies* 50, 1-17.

Aksoy, Ö.C. 2017. A combat archaeology viewpoint on weapon representations in northwestern Arabian rock art. *Mediterranean Archaeology and Archaeometry* 17, 1-17.

Al-Jallad, A. 2015. *An outline of the grammar of the Safaitic inscriptions*. Leiden: Brill.

Bowden, J. 2012. The origin and the role of the composite bow in the ancient Near East. *Ancient Warfare Magazine* 5, 42-54.

Brusgaard, N.Ø. 2015. Pastoralist rock art in the Black Desert of Jordan. In: Giraldo, H. and García Arranz, J.J. (eds.). *Symbols in the landscape: rock art and its context*. Proceedings of the XIX International Rock Art Conference IFRAO 2015. Tomar: Instituto Terra e Memória, 761-767.

Brusgaard, N.Ø. 2019. *Carving interactions: rock art in the nomadic landscape of the Black Desert, north-eastern Jordan*. Oxford: Archaeopress.

Brusgaard, N.Ø. and Akkermans, K.A.N. In press. Hunting and havoc: narrative scenes in the Black Desert rock art of Jebel Qurma, Jordan. In: Davidson, I. and Nowell, A. (eds.). *Making scenes: global perspectives on scenes in rock art*. New York: Berghahn Books.

Cross, F.M. and Milik, J.T. 1956. A typological study of the El Khadr javelin and arrow-heads. *Annual of the Department of Antiquities of Jordan* 3, 15-23.

Flinders-Petrie, W. 1917. *Tools and weapons, illustrated by the Egyptian collection in University College, London, and 2,000 outlines from other sources*. London: British School of Archaeology in Egypt.

Fokkens, H., Achterkamp, Y. and Kuijpers, M. 2008. Bracelets or bracers? About the functionality and meaning of Bell Beaker wrist-guards. *Proceedings of the Prehistoric Society* 74, 109-40.

Gordon, D.H. 1953. Swords, rapiers and horse-riders. *Antiquity* 27, 67-78.

Gordon, D.H. 1958. Scimitars, sabres and falchions. *Man* 58, 22-27.

Gottlieb, Y. 2004. The weaponry of the Assyrian attack. Section A: the arrowheads and some aspects of the course of the Lachish siege battle. In: Ussishkin, D. (ed.). *The renewed excavations at Lachish, 1974-1994*. Tel Aviv: Institute of Archaeology, Tel Aviv University, 1907-1969.

Gottlieb, Y. 2016. Beer-Sheba under attack: a study of arrowheads and the story of the destruction of the Iron Age settlement. In: Herzog, Z. and Singer-Avitz, L. (eds.). *Beer-Sheba III: the early Iron Age IIA enclosed settlement and the late Iron IIA-Iron IIB cities*. Winona Lake, IN: Eisenbrauns, 1192-1228.

Huigens, H. 2019. *Mobile peoples – permanent places: nomadic landscapes and stone architecture from the Hellenistic to Early Islamic periods in north-eastern Jordan*. Oxford: Archaeopress.

Layton, R. 2001. Ethnographic study and symbolic analysis. In: Whitley, D.S. (ed.). *Handbook of rock art research*. Walnut Creek, CA: AltaMira Press, 311-332.

Littmann, E. 1943. *Safaitic inscriptions*. Leiden: Brill.

Loades, M. 2016. *The composite bow*. Oxford: Osprey Publishing.

Macdonald, M.C.A. 1993. Nomads and the Ḥawrān in the late Hellenistic and Roman periods: a reassessment of the epigraphic evidence. *Syria* 70, 303-403.

Macdonald, M.C.A. 2007. Goddesses, dancing girls or cheerleaders? Perceptions of the divine and the female form in the rock art of pre-Islamic North Arabia. In: Sachet, I. and Robin, C.J. (eds.). *Dieux et déesses d'Arabie. Images et representations*. Paris: De Bocard, 261-297.

Macdonald, M.C.A. 2009. Wheels in a land of camels: another look at the chariot in Arabia. *Arabian Archaeology and Epigraphy* 20, 156-184.

Macdonald, M.C.A. 2014. Romans go home? Rome and other 'outsiders' as viewed from the Syro-Arabian desert. In: Dijkstra, J.H.F. and Fisher, G. (eds.). *Inside and out. Interactions between Rome and the peoples on the Arabian and Egyptian frontiers in Late Antiquity*. Leuven: Peeters, 145-16.

Maran, J. 2004. The spreading of objects and ideas in the Late Bronze Age eastern Mediterranean: two case examples from the Argolid of the 13th and 12th centuries B.C. *Bulletin of the American Schools of Oriental Research* 336, 11-30.

May, S. K., Wesley, D., Goldhahn, J., Litster, M. and Manera, B. 2017. Symbols of power: the firearm paintings of Madjedbebe (Malakunanja II). *International Journal of Historical Archaeology* 21, 690-707.

McEwen, E. 1978. Nomadic archery: some observations on composite bow design and construction. In: Denwoord, P. (ed.). *Arts of the Eurasian steppelands*. London: SOAS, 188-202.

Miller, R., McEwen, E. and Bergman, C. 1986. Experimental approaches to ancient Near Eastern archery. *World Archaeology* 18, 178-195.

Newton, L. and Zarins, J. 2003. Aspects of Bronze Age art of southern Arabia: the pictorial landscape and its relation to economic and socio-political status. *Arabian Archaeology and Epigraphy* 11, 154-179.

Overtoom, N.L. 2017. The Parthians' unique mode of warfare: a tradition of Parthian militarism and the battle of Carrhae. *Anabasis* 8, 95-122.

Paterson, W.F. 1966. The archers of Islam. *Journal of the Economic and Social History of the Orient* 9, 69-87.

Potts, D.T. 1998. Some issues in the study of the pre-Islamic weaponry of southeastern Arabia. *Arabian Archaeology and Epigraphy* 9, 182-208.

Rimer, G. 2004. Archers and archery. In: Bradford, J.C. (ed.). *International encyclopedia of military history*. Oxford: Routledge.

Walderhaug, E.M. 1998. Changing art in a changing society: the hunters' rock-art of western Norway. In: Chippindale, C. and Taçon, P.S.C. (eds.). *The archaeology of rock-art*. Cambridge: Cambridge University Press, 285-301.

Woodward, A., Hunter, J., Ixer, R., Roe, F., Potts, P., Webb, P.C., Watson, J.S. and Jones, M.C. 2006. Beaker Age bracers in England. *Antiquity* 80, 530-543.

Woosnam-Savage, R.C. 2004. Edged weapons. In: Bradford, J.C. (ed.). *International encyclopedia of military history*. Oxford: Routledge.

Zutterman, C. 2003. The bow in the ancient Near East. *Iranica Antiqua* 38, 119-165.

'Your own mark for all time': on *wusūm* marking practices in the Near East (*c*. 1800-1960 AD)

Koen Berghuijs

Abstract

Wusūm form a historical system of markings used by largely mobile pastoralist groups throughout the Near East, and are commonly encountered during archaeological surveys. Despite their ubiquity, our current understanding of the underlying marking practices is extremely limited. Therefore, this paper investigates the economic and socio-political dimensions of *wusūm* brands and petroglyphs from an ethnohistorical perspective, and delineates the potential role of archaeology in the pursuit of interpreting and explaining *wusūm*. Bringing together a multitude of relevant primary sources and archaeological data from the Jebel Qurma region in north-eastern Jordan, this paper aims to provide a comprehensive perspective on the phenomenon of *wusūm* marking systems in the Near East and to move towards bridging the gap between ethnographic sources on the one hand, and archaeological data on the other.

Keywords: wusūm, Bedouin, petroglyphs, Black Desert, Jebel Qurma, Jordan

Introduction

Wusūm (singular: *wasm*) form a system of markings used by Bedouin groups throughout the Middle East. Frequently encountered in the form of animal brands (Fig. 1) or petroglyphs (Fig. 2), *wusūm* often consist of lines or dots, or a combination of the two, and may take on geometric shapes such as circles and squares. Comparable to the concept of a signature, *wusūm* operate on, and refer to, various levels of different but closely intertwined modes of political organisation, including families and tribes.

Wusūm are frequently encountered in archaeological surveys throughout the Near East but have so far not received much scholarly attention. Recent publications mention *wusūm* found on Ahl al-Jebel tombs in north-eastern Jordan (Lancaster and Lancaster 1993, 160-161), on a solitary rock in Wadi Hafir in southern Jordan (Corbett 2010, 355), on rock facades near the Jubbah palaeolake in the Nefud in Saudi Arabia (Jennings *et al.* 2013, 671), and at many (unspecified) locations throughout the Arabian Peninsula (Khan 2000; Nayeem 2000). Additional *wusūm* petroglyphs have been reported from the Negev (Eisenberg-Degen *et al.* 2016) and even from the island of Socotra, Yemen (Jansen van Rensburg 2016, 150).

To most archaeologists, *wusūm* form a type of serendipitous petroglyphs that appear to be idle derivatives of modern-day camel brands and, because of that, are deemed the result of Bedouin pastime practices not worthy of further investigation. William and Fidelity Lancaster, for example, have argued that since "the shape of the mark was no

In: Peter M. M. G. Akkermans (ed.) 2020: *Landscapes of Survival - The Archaeology and Epigraphy of Jordan's North-Eastern Desert and Beyond*, Sidestone Press (Leiden), pp. 317-332.

Figure 1. Branded *wasm* on the right cheek of a camel, Jordan (photography by Koen Berghuijs).

Figure 2. *Wusūm* petroglyphs on basalt stone (QUR-575, RA 14) (photograph: Jebel Qurma Project Archive).

more or less important than the position on the animal" (1993, 161), *wusūm* "lose half their meaning immediately (...) when such marks are placed on stones." (*ibid.*). Their somewhat pessimistic conclusion is characteristic of most publications on *wusūm*: "In general, all that can be said about *wusūm* is that they are there; their meaning is open to too many interpretations, in the majority of cases, for their significance to be more than indicative." (*ibid.*). Although true to some extent (see below), their verdict on *wusūm* petroglyphs does not do justice to the perceived importance of the markings among Bedouins themselves. During his pioneering field research into oral poetry and narratives from central and southern Nejd, Dutch scholar and diplomat Marcel Kurpershoek documented a Bedouin sheikh's perspective on the nature of contemporary *wusūm*: "The brand ... as we, the Arabs, say 'The brand is fore-ordained,' that is, it is like one's name, one's name is like the brand. It is like writing your name, it is what you are known by, it distinguishes you from all others. Each tribe has its own brand and the brands are all different. But each tribe knows the brand of the others." (Kurpershoek 2002, 607).

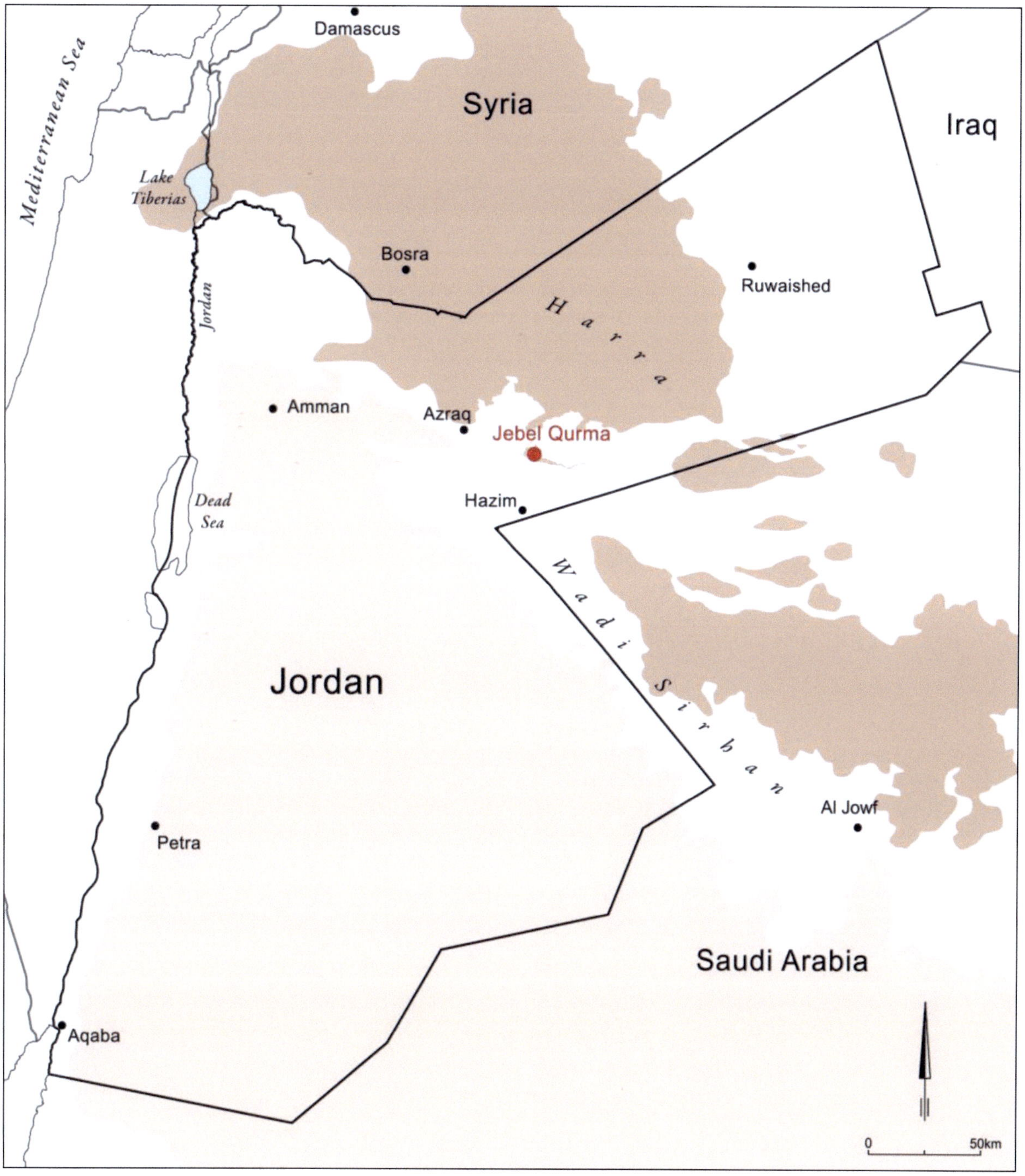

Figure 3. The Jebel Qurma research area in north-eastern Jordan (image: Jebel Qurma Project Archive).

The noun *al-wasm* belongs to a semantic field pertaining to 'marking' and 'branding', and was translated by Kurpershoek as animal 'brand', since the ensuing part of his recorded narrative deals with such permanent skin markings on camels. Nowadays, *wusūm* indeed occur mainly as animal brands to indicate ownership; examples of such brands can still be observed among camel herds in the region (see Fig. 1).

In the nineteenth century, however, *wusūm* were utilised in a much wider range of contexts and appear to have played important roles in Bedouin societies. Ethnographic sources, oral traditions, and travellers' accounts not only feature a rich variety of marking applications but also demonstrate the economic and socio-political dimensions of *wusūm*. Now largely lost, the marking practices recorded by these early explorers and scholars demonstrate that *wusūm* were indelibly connected with various aspects of Bedouin life, including territoriality, access to natural resources, and matters of identity.

This article sets out to make a contribution to the study of *wusūm* from three separate, but interconnected, perspectives. First, it will introduce several important ethnographic sources that shed light on the internal workings of *wusūm* marking systems as well as their economic and socio-political relevance within historical Bedouin societies. Second, since most information about *wusūm* marking practices necessarily derives from the ethnographic record, it is important to delineate the potential role of archaeology in the pursuit of interpreting and explaining *wusūm*. Vice versa, acknowledging that ethnographic and archaeological examples of *wusūm* are both different exponents of the same system goes a long way to bridge the interpretive gap between these respective data sets. The third and last section of this paper provides some suggestions as to what role *wusūm* petroglyphs can play in the archaeological enterprise.

These last two sections will draw on relevant insights from the extensive *wusūm* data set from the Jebel Qurma region in north-eastern Jordan. Since 2012, Leiden University's Jebel Qurma Archaeological Landscape Project (directed by Professor Peter Akkermans) has conducted intensive surveys and excavations in a research area situated on the western edge of the Black Desert, approximately 30 km east of the oasis town of Al-Azraq (Fig. 3). The region under study measures 21 x 16 km (336 km^2) and is situated between Wadi Rajil and Wadi al-Qattafi, two large wadi systems mainly fed by seasonal precipitation from the Jebel al-'Arab region in southern Syria. Significantly, the Jebel Qurma region is situated near the northern end of the vast natural depression of Wadi Sirhan, which was an important route between the Levant and inner Arabia. So far, many hundreds of *wusūm* petroglyphs have been documented in the Jebel Qurma region; these offer a timely opportunity to study afresh the phenomenon of *wusūm* marking practices in the Near East.

The politics of *wusūm*: animal brands

The first European to note the existence and use of *wusūm* as animal brands was the Swiss explorer Johann Ludwig Burckhardt. During his travels in the deserts of Arabia and Syria between 1810 and 1812, Burckhardt observed that: "all the Bedouin camels are marked with a hot iron, that they may be recognised if they straggle away, or should be stolen. Every tribe and every taifé, or family of a tribe, has its own particular mark. This is generally placed on the camel's left shoulder, or its neck." (Burckhardt 1831, 198-199). Not only the shape of the *wasm* was indicative of the associated social group but also its position on the animal in question. Common locations included the camel's nose, forehead, cheeks, neck, shoulders, hips, legs, and buttocks (Burckhardt 1831, 198-199; Musil 1908, 28-123). The belly and the hump were never marked by *wusūm* as these areas tend to be covered by thick fur during the winter, thus obscuring any underlying brands (Wetzstein 1877, 14).

The branded *wasm* consisted either of a single shape applied to a particular position on the body, or of multiple shapes applied to different parts of the body. Alois Musil (1908, 32), for example, recorded the multi-shaped *wasm* of Al-Terabin, a tribe residing in south-central Jordan, which consisted of a circle on the cheek, an L-shape on the neck, and a straight line on the front right leg; more examples are included in his detailed overview of the tribal groups in 'Arabia Petraea' (*ibid.*, 22-123).

Combining shapes was necessary to prevent repetitive use of the same brands by different groups or individuals. Theoretically, the many thousands of tribes, clans, and families each required a brand unique to their group, but since central coordination for such an immense synchronising enterprise was obviously absent, repetition was inevitable. Nonetheless, sufficient varieties of *wusūm* shapes were continuously created, and so a valid and practical marking system was maintained. Adding new *wusūm* to the existing corpus (or discarding or modifying old markings) was necessary for several reasons. The French explorer Charles Huber informs us that: "[l]orsqu'un proche parent fait pâturer ses chameaux avec un autre, pour qu'il ne puisse pas y avoir confusion, il ajoute souvent à l'ousm familial un matraq, un halaqah, ou tout autre signe pour les distinguer. Ce signe ainsi ajouté s'appelle šâhid." (Huber 1891, 178).

Thus, Bedouins distinguished between a hereditary shape (the general brand of a tribe or family) and a *shahid* ("witness"; one or more shapes added to this general brand), the latter being used only by a son to distinguish his own camels from his father's. The observations of a traveller in Egypt suggest that the son enjoyed a certain freedom in his choice of a *shahid*: "The camels are all branded, both with the tribal mark, of which those of the great tribes are well known to all, and with the family mark, which are, of course, much more difficult to remember. (...). If the tribal mark happened to be three parallel lines, the family mark may be a dot at the end of one of these lines, or some variation equally trifling." (Jennings-Bramley 1907, 33). After his father's death, the son would normally return to using the hereditary *wasm* and discard the temporary *shahid*, even when he had formed his own family. In the new situation it was no longer necessary to make a distinction between his camels and his father's, since these were undoubtedly part of the inheritance (Huber 1891, 178).

The individual shapes that make up such complex *wusūm* were locally referred to on the basis of their visual appearance. Charles Reignier Conder (1883, 179) informs us that one of the Beni Sakhr *wusūm* was known as "the coffee-spoon" and another as "the necklace". Contemporary explorers sought the origins of these shapes in ancient scripts, drawing attention to the similarities between *wusūm* and various "Himyaritic letters" (Palmer 1871, 354-355). Richard Burton suggested that *wusūm* "probably preserved the primitive form of the local alphabets" and that the marking practice "doubtless dates from remotest antiquity" (Burton and Tyrwhitt Drake 1872, 341-342). This assumption not only seems unlikely but also differs markedly from local folklore, which explained the invention and subsequent adoption of some *wusūm* as symbolic, politically charged, acts taking place in more recent times. Orientalist and diplomat Johann Gottfried Wetzstein, for example, recorded the following etiological passage about the origins of the Fuheilia *wasm* in 1862: "(…) die Keule (debbûsa) der Fuheilîa, eines jetzt ebenfalls sehr geschwächten Stammes, dessen Fürst früher (noch Anfangs dieses Jahrhunderts) bei seiner Investitur vertragsmässig eine stählerne, mit eingelegten goldenen Arabesken gezierte Schlachtkeule von der osmanischen Regierung erhielt. Er führte den Titel "Fürst der syrischen Nomaden" (emîr Arab es-Schâm), und die Keule, das Symbol der Herrschaft, wurde zum Wesm der Völkerschaft." (Wetzstein 1877, 15).

In these situations, in which leadership was challenged or newly acquired, the appropriation of a new *wasm* was a break with the past and a consolidation of the new political status. The act's transformative nature is particularly evident in an oral narrative about Slewih ibn Ma'iz al-'Atawi, a legendary desert warrior who lived in central Nejd in the second half of the nineteenth century (Kurpershoek 1995, 3). The storyteller, Khalid ibn Slewih, was Slewih's great-grandson and a sheikh of the Duwi Atiyyah of the Rugah, the northern division of the Tebah: "Slewih raided black and white camels from al-Midla. One day, Slewih was visited by the camels' former owner in his tent: "Slewih received and entertained him hospitably and let him sleep in his tent. He did not put any question to him nor did he ask for the reason of his visit. The next morning he said, 'Slewih, I have been your guest since yesterday evening. You have received me honourably, you spared no effort to make me feel welcome, nor did you ask any questions.' He replied, 'You are most welcome, no matter whether you want to stay longer or to depart early.' 'I am Gasi al-Midla of Ghatan, the owner of those camels resting near the watering-place, that herd of white camels.' He said, 'May God preserve your life, al-Midla, from the moment you set out until your safe return home. Don't worry, you'll get what you came to see me for. Walk over to your camels right away and drive them from their resting-place at the water. Take them all! He said, No, I have not come to retrieve my camels, son of Ma'iz. You have captured them fair and square, as is the custom between our tribes. Sometimes we are robbed, the next time it is your turn. No, I have not come for my camels, I have come to make a very humble request.' 'Speak and it is yours,' was the reply. 'It's about the staff, the brand on their neck. Please do not remove it, but take it as your own mark for all time.' 'Your wish is granted,' Slewih said." Ever since that day, from the time when he captured the camels of al-Midla till this very day, this has remained the brand of Slewih: the staff with the curved grip between the two slashes of Duwi 'Atiyyah on the small of the neck, next to the cheek on the left side. That is the brand I now use on my camels, yes, indeed. He never used his old brand, the brand he had inherited from his father, Ma'iz. Between the two slashes of Duwi 'Atiyyah, one on the cheek and the other on the broad lower part of the neck, he now branded the staff on the small of the neck on the left side." (Kurpershoek 1995, 158-159).

Regardless of whether the recited narrative approximates the true course of events it claims to describe (see Shryock 1997), the text is particularly insightful regarding the relevance of hereditary *wusūm* and *shahid* systems. Splitting away from his family to establish an independent sheikhly lineage, Slewih adopted the new *wasm* to stress the break with his father. Moreover, the transgenerational oral narrative ties the legendary origins of the family's *wasm* to the current position of the family, and to Khalid ibn Slewih's status in particular: "whatever economic value these herds represent is irrelevant besides their paramount importance as the bearers of the marks testifying to their owners' blue-blooded antecedents." (Kurpershoek 1995, 69).

Khalid's recital also introduces several additional complexities of branded *wusūm* systems. In the context of raiding (Arabic: *ghazzu*), camel brands played a somewhat obscure role. Well-documented events like the one involving Slewih are unfortunately not found in other ethnographic sources. Raiding expeditions were a salient element of Bedouin societies prior to the establishment of the British Mandate (Fletcher 2015). Long regarded as the expression of the supposed warlike nature of Bedouins (Sweet 1965, 1132), it is now generally understood to have been an intricate exchange and (re) distribution mechanism of camels and other livestock, and thus of wealth and natural resources (*ibid.*; Van der Steen 2013, 111-112). Raids were formally organised and often undertaken by men from different families and tribes (Sweet 1965, 1141). Among the Sinai Bedouin, "word is sent round that on a certain day those willing to join must assemble at such a well. It is only there, and on seeing how many have come together, that plans are laid and the direction to be taken decided on. Joining, or not joining, is voluntary and it is therefore difficult, until it is ascertained how many fighting men it can number, to decide what foe the company is fit to face." (Jennings-Bramley 1906, 199).

Valuable information concerning the functionality of *wusūm* systems in raiding contexts can be found in the letters and diaries of the British traveller and political officer Gertrude Bell. In February 1914, while preparing for a journey from the basalt region of northern Arabia to Nejd, Bell's Huweitat guides received information about a group of Rwala Bedouins residing in Wadi Sirhan in northern Saudi Arabia, an important part of the extensive route networks throughout the region. As a result, Bell (1914b) writes, "we had changed all our plans, given up Jof [Al-Jauf] and the [Wadi] Sirhan, fearing that the Ruwalla might stop us". The Huweitat and Rwala were traditional adversaries, and Bell's Huweitat host Awwad had the unfortunate privilege of being at risk of "any man of the Ruwalla whom he might chance to meet [as he] would cut his throat at sight." "Moreover," she continued, "we have Ruwalla camels bought from the Miri with the Wasm upon them". Though not Huweitat raiding booty, to confront the Rwala with camels bearing their own *wasm* (besides the Miri and Huweitat brands) would not have benefitted the already hostile diplomatic ties between both tribes (see Jennings-Bramley 1908, 35). In Bell's anecdote, therefore, the Rwala would undoubtedly have reclaimed the camels that had obviously once belonged to them – not necessarily without violence.

Once camels had been collected from a raid, they were usually driven back to the new owners' camp and marked with their *wasm*. This custom is recorded in another passage from Gertrude Bell's same diary entry. Bell repeatedly detailed the legendary achievements of Muhammad Abu Tayi, the brother of her Huweitat host Awwad (and a cousin of T.E. Lawrence's later ally in the Arab Revolt, Auda Abu Tayi) and wrote that "[he] has brought back immense numbers of camels from his raids. He puts his wasm on them and confides them to different people that if a ghazzu comes they may not all be found round his tents." (Bell 1914b).

A similar narrative, though one with a less joyful ending, was documented by Kurpershoek in central Arabia: "He [the chief of Al Arja] was told, 'Your camels are here! Right now the sons of al-Hazimi are busy dividing them and branding them, as you can hear from their roars.' It is said that when he saw his camels he fell unconscious, Abd ar-Rahman. While he lay there, unconscious, the people sprinkled him with water, but to no avail. His intestines burst open and he bled to death." (Kurpershoek 2002, 453).

In these contexts, the application of the new brand can perhaps be understood as the finalisation of the raiding expedition and as staking a temporary claim to natural resources rather than one of full and permanent ownership. The temporary character of camel ownership is illustrated by the fact that many camels bore multiple *wusūm* of their respective owners as a permanent 'skin biography': the mere presence of the various *wusūm* of the Rwala, Miri, and Huweitat (and perhaps even more) on the camels in Bell's excerpt shows that any expression of a claim had, in fact, merely been a temporary one. Moreover, Muhammad Abu Tayi himself prepared for an expected reciprocal *ghazzu* by distributing the camels around his camp; the animals would undoubtedly be driven off as raiding booty in the near future by one tribe or another. Economic wealth in the form of camels or any other livestock was not regarded as a significant trait of character. Rather, it was the manner in which that wealth – however temporary – had been acquired and the reputation built on fruitful raiding expeditions that signalled an individual's achieved status. Indeed, "(...) the reputation of having been a great raider in his day will confer some consideration on even the oldest." (Jennings-Bramley 1907, 26). The temporary nature of ownership was well-understood among the Sinai Bedouins: "A probable result of these raids is the small social importance given to a man by his riches. The rich man of to-day may be the beggar of to-morrow, and vice versâ [sic], so that no special importance is given by wealth, so easily lost or regained." (*ibid.*).

Wusūm, however, were indeed effective as ownership claims in cases of mistaken identity. Allied tribes or individuals were generally not raided but if such an unfortunate event were to occur (at night, for example), then the *wusūm* on the raided animals would ensure their safe return to their previous owners, as Gertrude Bell observed: "The Howaitat raid came to very little – we have met stragglers of Audeh's band on their way home and heard the tale. They fell by night upon the Swaid, who are Shammar, and drove off their camels. But when dawn came and they say [sic] the brands on the camels, the wasms, they found that they belonged to Ibn al Rummal, another Shammar shaikh, who had been camping with the Swaid. Now Ibn al Rummal is a friend of 'Audeh's, and his father-in-law foreby, and he therefore returned all the camels. But a band of the Sukhur, who had raided with Audeh, were untrammelled by any ties of friendship, and they kept the camels they had taken." (Bell 1914c).

It becomes clear from this anecdote that the political decision-making of the Huweitat differed from that of the Sukhur: the first group returned the raided camels to the Shammar, whereas the second did not. The validity of the *wasm* as a temporary claim was thus only accepted by allies as a ways of avoiding repercussions that could damage the political alliances between groups. If the raiding party was "untrammelled by any ties of friendship" with the raided party, the validity of the *wasm* would not be acknowledged. *Wusūm* on camels, therefore, exerted substantial influence on raiding processes: the markings expressed a temporary claim to the animal in question, and simultaneously indicated which animals could and could not be taken by foes and allies alike during reciprocal raiding expeditions. Moreover, an individual's reputation fused with his *wasm*, linking the mark to his mythical raiding exploits and connecting subsequent generations with their legendary ancestors.

Economy and identity: *wusūm* petroglyphs

As mentioned above, *wusūm* were not only used as animal brands but can also be found as petroglyphs throughout the Near East. A large part of these *wusūm* petroglyphs appears to have been utilised to express claims on natural resources, such as wells and territory for camping and pasturing herds. During a meeting of the *Berliner Gesellschaft für Anthropologie, Ethnologie und Urgeschichte* in 1877, Wetzstein reminisced about the ubiquity of *wusūm* petroglyphs and stressed their economic importance: "Man findet dieselben [Eigenthumszeichens der syrischen Nomaden] sehr häufig an den Thoren und Mauern der alten verlassenen Städte, auf den Säulen und steinernen Wassertrögen der Ruinenorte, an glatten Felswänden, bei den Brunnen und alten Cisternen mit grosser Sorgfalt tief in den Stein eingegraben, um anzuzeigen, dass das Recht, bei diesen Oertlichkeiten zu weiden und die Herden zu tränken, oder Ansiedlern daselbst den Feldbau zu gestatten, ausschliesslich denjenigen Stämmen oder Stammzweigen zustehe, welche die dort eingegrabenen Eigenthumszeichen führen." (Wetzstein 1877, 14).

A more cautious observation on the marking of well heads was made by English author and traveller James Silk Buckingham. On March 2, 1816, Buckingham passed through Wadi al-Thamad, a valley discharging into the Dead Sea and close to the present-day city of Madaba, Jordan, and recorded the following: "Along its banks are many wells of a moderate depth, with hewn cisterns and drinking troughs for cattle, which in the present day, as they did in the patriarchal ages, form the principal strength as well as the wealth of a tribe, the possession of these securing the necessary supplies, without which no Arab camp, with their numerous flocks and herds, could long exist. On many of the wells and cisterns I observed the following characters [see original publication for image], which are said to be the work of Arabs, but whether for mere pastime, or with a view to mark the property of particular tribes, or of individuals belonging to such tribes, I could not learn." (Buckingham 1825, 98).

Wells formed a crucial and valuable asset in a tribal territory or *dirah* and were often claimed by a specific tribe, family, or even individual (Van der Steen 2013, 87). Although the claimant of a well was undoubtedly known among rivalling groups, the application of his *wasm* onto the well-head may have reinforced his claim with a sense of permanence – the extent to which his claim to a well was respected, is another matter.

Wusūm on solitary stones have sometimes been interpreted by travellers (see also Van der Steen 2013, 20) as demarcations of a tribal territory. William Lancaster has convincingly argued that no tribe legally 'owned' or governed a territory (Lancaster 1981; also Van der Steen 2013, 84) but rather dominated a particular area by their mere presence. If a sheikh's claim to any tract of land remained uncontested by rival groups (Van der Steen 2013, 85), its validity was possibly consolidated by applying *wusūm* to landmarks, demarcating the geographical extent of such a claim: "It is, however, a fact that the Bedawín do mark their borders with stones, and often inscribe rude symbols of their tribe upon them. We found several other stones with similar marks in various parts of the [Sinai] Peninsula (...)." (Palmer 1871, 147-148).

It is likely that Palmer here refers to the boundaries of a *dirah*. However, the practical implications of any such demarcation are unknown and seem quite limited, if effective at all. It is certainly the case that the crossing of another's *dirah* or the sharing of natural resources present within were often regulated through the payment of usage rights to the dominant tribe or through negotiated conventions (Van der Steen 2013, 84). The concept of *dirah* and associated *wusūm* conventions are touched upon in several poems and narratives from the Dawasir in Nejd:

> "If a traveler carries a stick carved with our tribal mark, neither he nor his mount need anyone to accompany him for his safety." (Kurpershoek 1999, 155).
>
> "A land where a protected neighbour's herds pasture unfettered, and our marking on his stick allows a stranger to travel safely in the wilderness." (*ibid.*, 207).
>
> "If a stranger's stick has been carved with our brand, even alone, he will not be disturbed within our tribal borders." (Kurpershoek 2002, 479).

The *wasm* of the Dawasir on the camel stick of a member of another tribe provided the latter "with the necessary 'passport' to travel freely and pasture his camels wherever he liked in their *dirah* without being harassed by anyone" (Kurpershoek 1999, 201). It may well be possible that a marked camel stick also allowed its bearer to travel through the *dirah* of a tribe allied with the owners of the *wasm*, although there is no supporting evidence for this.

Sporadic snippets of ethnographic data furthermore indicate that tomb stones of Bedouin burials were commonly engraved with the *wasm* of the (family of the) deceased. Cemeteries scattered with such tomb stones were reported by Eduard Sachau at Khunasara, Syria, in the 1880s. Situated just south of the city wall, "man sieht auf einigen grabsteinen (...) ein Wusm, d.h. das Zeichen eines Kamels, das der Besitzer desselben hier eingegraben hat" (Sachau 1883, 119). A dozen years later Reverend William Ewing encountered several *wusūm* engraved on head stones of burial tombs in the Hawran. In April 1895, he published his diaries of his journey, along with drawings and sketches of architectural elements and inscriptions. On the top of a hill called Tell el Talaya, an extensive cemetery was located: "On many of the stones," Ewing observed, "were the (...) brandmarks of the Arabs." (Ewing 1895, 163). In north-eastern Jordan, Gertrude Bell also observed a *wasm* on a burial tomb in a somewhat macabre setting: "Inside the qasr there is a grave with a Sherari wasm on it. Sayyah looked at it and pointed out that the man had been killed. The red cotton keffiyyeh and a bit of white cotton clothing thrown upon it were steeped in blood." (Bell 1914a). In fact, the occurrence of *wusūm* on Bedouin burial tombs was apparently so common, that Bell even registered their absence: "Not far from the wells is an Awaji burying place with a central tomb, with no wasm on it however. It is the tomb of the father of their chief shaikh, Mish'an. He was killed in a ghazzu." (Bell 1914d).

While in south-eastern Syria, Wetzstein documented the story behind a rather remarkable burial with a double *wasm*: "Auf dem Berge Muntâr bei dem Dorfe El-hîgâna, 6 Stunden östlich von Damask, steht auf einem Grabhügel ein Stein mit dem Doppelzeichen [see original publication for image]. Dasjenige rechter Hand ist das Wesm der Gemâilïa, das andere der No'eim. Beide Stämme gehören zu den Trachoniten. Zwei befreundete Jünglinge, welche, der eine dieser, der andere jener Völkerschaft angehörend, in einer Stammfehde dort gegen einander kämpfen mussten und auf den Tod verwundet wurden, verlangten, in einem Grabe beerdigt zu werden." (Wetzstein 1877, 15).

In these contexts *wusūm* clearly functioned as an expression of identity of the deceased, and as a ways of stressing the connection between the dead and the living members of a family – in the largely illiterate Bedouin societies of the nineteenth century, a perfect substitute for more elaborate epitaphs.

Finally, in two rare examples, the political dimensions of *wusūm* petroglyphs extended into the realm of settled areas as well. Johann Gottfried Wetzstein (1877, 15) wrote that *wusūm* found on a column just outside of a Hawrani settlement indicated that the village inhabitants were tributaries of the four Bedouin tribes represented by the markings, possibly in exchange for protection. Part of a similar configuration of petroglyphs at the city of Bosra was found to have been intentionally effaced – according to Wetzstein's somewhat enigmatic words, this was done in order to show that "ihre Inhaber keine Anrechte mehr auf die Stadt haben" (*ibid.*). These *wusūm* thus acted as physical testimonies to the interdependent, but often uneasy, relationship between villages and Bedouin groups.

Carving a role for archaeology

The examples of intricate *wasm* practices outlined above derived from living, ethnographic contexts. Unfortunately, archaeological data cannot be expected to provide the same degree of detailed insights: the practices themselves have largely died out, as have many potential informants. Instead, archaeologists in the field are left with marked rocks in a landscape in which *wusūm* no longer seem to signify anything, but, according to the ethnographic sources, *could* signify almost anything. Such a different data set requires a different set of questions altogether. This section discusses two such questions: firstly, to what extent can archaeological data contribute to our understanding of the workings of *wusūm* systems? And secondly, how can *wusūm* be utilised in answering broader archaeological questions? Building on just some of the insights gleaned from the rich Jebel Qurma data set, both questions will be addressed below.

The role of archaeology in wusūm studies

The archaeological analysis of any large corpus of *wusūm* petroglyphs will involve sincere but ultimately unsuccessful attempts at classifying the documented markings. The first reflexes of typology-oriented researchers (the author included) would be to convert field photographs into neat digital vector images, to group these according to their main hereditary *wasm*, and to produce an unambiguous catalogue of *wusūm* from the surveyed area with cross-references to the associated tribes or families – in short, to create order. This, however, is impossible. This rather bold statement is supported by the users of *wusūm* themselves and the findings of several scholars before me, as we shall see below.

The specificity of *wusūm* petroglyphs is fairly opaque compared to that of animal brands, where the same motif may be used on specific locations on the body to signify different objects, *i.e.* socio-political groups. This lack of specificity in petroglyphs obstructs a mathematical approach to the documentation of *wusūm* systems, as the required 'fixedness' is simply absent. This notion instantly explains two situations commonly encountered in *wusūm* studies. First, there is the issue of misinterpretation of a *wasm*'s object. During his survey of the north Arabian desert in the first half of the twentieth century, Henry Field collected hundreds of *wusūm* tracings. He subsequently invited a diverse group of "Arabs from the Baghdad, Mosul and Hilla Liwas, three Beduins from western Iraq, and two Wahabis from Saudi Arabia" (Field 1952, 2) to identify the objects of the recorded *wusūm*: "As each *wasm* was drawn on the ground in the center of this assembled group, discussion was invited. The consensus of expert opinion regarding the ownership of this tribal mark was recorded (...)." (*ibid.*).

The word 'consensus' is crucial here, as it suggests that the interpretations of the various *wusūm* were not always unanimously agreed upon. Moreover, the *wusūm* tracings were sent to Colonel F.G. Peake in Amman who, with the help of several Ageyli Bedouin, prepared a second list of identifications. Perhaps not surprisingly, discrepancies

Figure 4. Petroglyph of a branded camel with associated Safaitic inscription (QUR-186, RA 21) (photograph: Jebel Qurma Project Archive).

occurred between the two lists: the objects of the *wusūm* were interpreted differently by different users of the same marking system. Field provides several possible explanations for the differences in interpretations: "The same wasm may be used by widely separated sub-tribes or even by different tribes. Particularly confusing is the custom among the tribes of using the same wasm but placing it on different parts of the camel's body in order to make the distinction as to ownership. As for the marks copied from well-heads, buildings or stones, weathering had often obliterated essential lines or curves in the wasm. Moreover, we could not always be sure of details when the mark had been hammered originally on a rough surface with any stone lying conveniently at hand. Furthermore, many items in this list are at least several hundred years old, there being no way to estimate the age of a weathered mark on an ancient, but still used, well-head such as those at Bayir and Al Jidd." (Field 1952, 2).

Thus, producing classifications of *wusūm* is inherently fraught with difficulties, precisely because of the issues outlined by Field. The users of *wusūm* systems have different opinions about the object to which a *wasm* may refer, and there is virtually no possibility of establishing the 'real' object without first-hand experience. In the context of *wusūm*, such experience consists of a lifelong learning-process of familiarising oneself with as wide a variety of *wusūm* as possible, for example by handling branded camels (in pasturing, breeding or raiding contexts), via oral transmissions, and perhaps even by "drawing wasms [sic] in the sand" (Bell 1914d; Field 1952, 2). But then, the ambiguity of *wusūm* is and was an inherent element of the marking system, as noted by Richard Burton in the 1870s: "(...) the custom is dying out: the modern Midianites have forgotten the art and mystery of tribal signs (Wusúm). In many places the people cannot distinguish between inscriptions and "Bill Snooks his mark," and they can interpret very few of the latter." (Burton 1879, 321).

So how exactly, then, can archaeology contribute to our understanding of *wusūm* systems at large? The answer lies in the chronology of *wasm* marking practices. Written sources provide ample evidence for the practice of *wusūm* marking in the nineteenth and twentieth centuries. It is possible that the use of *wusūm* goes back several centuries but evidence for earlier *wusūm* markings has yet to be produced. The Jebel Qurma corpus does not provide radically new information with regard to the chronology of *wusūm* systems but does demonstrate that there certainly is research potential. Through association

Figure 5. *Wasm* petroglyph associated with a figure holding a spear (QUR-779, RA 29) (photograph: Jebel Qurma Project Archive).

of carved *wusūm* with adjacent, similarly patinated and executed petroglyphs, it is possible to deduce (relative) chronological information about *wusūm*. A petroglyph possibly depicting a branded camel (Fig. 4) is associated with, and referred to, the adjacent Safaitic inscription (Chiara Della Puppa, pers. comm.), providing a tentative pre-Islamic date for the practice of branding camels (rather than for the application of *wusūm* onto rocks).

Another rock-art scene (Fig. 5) links a *wasm* to an era in which spears were in use; spears are frequently mentioned in nineteenth-century travel reports and appear to have been in use until the First World War, well after the widespread introduction of European guns. On the other end of the temporal spectrum are a group of *wusūm* petroglyphs that occur in conjunction with a rather recent appearance in the deserts: trucks. These petroglyphs (Fig. 6) bear a striking resemblance to the 1960s Mercedes-Benz trucks that are still used today in the region around Al-Azraq for transporting water, vegetables, fruits (particularly water melons), animals, and fodder. These petroglyphs represent a class of machinery that was probably introduced in the region in the third quarter of the twentieth century, providing a relative recent date for the practice of *wasm* petroglyph marking (see also Müller-Neuhof 2019).

On the basis of the ethnographic information presented above and the relative dating of associated carvings, the *wusūm* petroglyphs in the Jebel Qurma region can thus be tentatively consigned to the period between the early nineteenth and late twentieth century. It is certainly possible that the marking practice existed before but until new data become available, the lower temporal extreme is only open to speculation.

The role of wusūm in archaeology

The data set from the Jebel Qurma region is the first systematically documented *wusūm* corpus in the Near East and provides an opportunity to assess the research potential of such petroglyphs in archaeological contexts. As we have seen, under favourable circumstances, *wusūm* petroglyphs may provide opportunities to establish a chronological framework for the marking practice itself. In turn, the distribution and immediate context of *wusūm* petroglyphs may add to our understanding of the use of landscape, modes of mobility, as well as the meaning of place in the (post-)Ottoman era.

At a macro-level, the distribution of *wusūm* (n= 737) in the Jebel Qurma region (Fig. 7) demonstrates that the majority of *wusūm* petroglyphs is found along the outer edges of the basalt upland. The Jebel Qurma region is strategically located at the northern end of the Wadi Sirhan, a vast natural gateway to central and eastern Arabia. A plausible explanation for the presence of *wusūm* in the region is that it reflects the routes via which nomadic groups, caravans, and other travellers moved through the landscape, typically between the Hawran and northern Arabia. Indeed, the itineraries of several European travellers indicate that the region was part of a large nineteenth-century route network that may well have been in use for centuries (Blunt 1881; Huber 1891).

Wusūm are typically found on slopes (n= 174) and hill tops (n= 521) in the region, rather than at hill bases or in low-lying wadi valleys (n= 42). Most of the wadi systems in the Jebel Qurma region provide access to the upper basalt plateau and surrounding hills, as do the vast mud flats in the north and east. Furthermore, networks of interwoven pathways and animal tracks enable movement through the

Figure 6. Petroglyph of a modern truck with a *wasm* superimposed onto an earlier rock-art scene with camels (QUR-639, RA 7) (photograph: Jebel Qurma Project Archive).

dense basalt as well, and generally lead from the lower wadis and valleys to the basalt plateau higher up. The distribution of *wusūm*, like other petroglyphs, thus functions as a rough proxy for modes of mobility at large. Bedouin tribes exploited particular areas according to unwritten rules of territoriality (see above), but the Jebel Qurma region is too small an area to identify any differentiation in tribal movements. Moreover, the as yet unknown chronological differences between individual *wusūm* petroglyphs hinder our understanding of such movements significantly: we cannot be sure that similarities in *wasm* composition denote a chronological overlap, or even refer to one and the same socio-political group.

On a smaller scale, the presence of *wusūm* in the basalt upland, a part of the *harrah* that is difficult to cross, can be explained in similar terms. The basalt-covered hill tops and slopes provide an abundance of green areas during late winter and early spring ideally suited for pasturing animals, more so than the surrounding plains. Access to the basalt is provided by networks of small pathways,

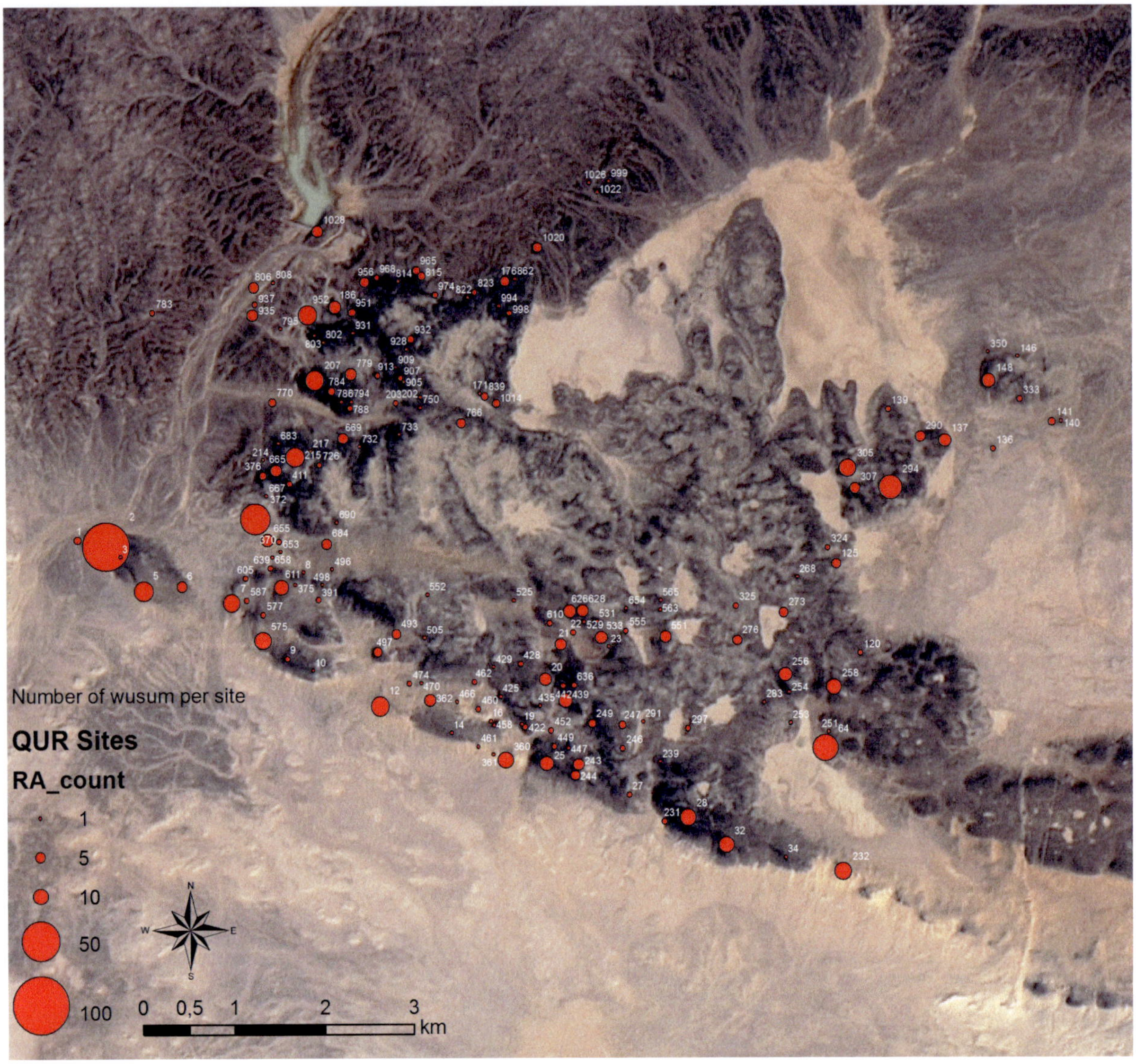

Figure 7. Distribution map of *wusūm* petroglyphs per site in the Jebel Qurma region. Background image: LANDSAT 7 (map: Jebel Qurma Project Archive).

running from the lower valleys into the plateau, and connecting structures and sites with one another.

The archaeological contexts of *wusūm* support the earlier ethnographic observations, in that *wasm* marking appears to be linked to particular social practices, such as marking burials. Although the sample size of the *wusūm* corpus from the Jebel Qurma region is significant, too few unique associations can be made between *wusūm* and archaeological structures as to identify specific social practices not already found in ethnographic reports.

The Jebel Qurma corpus does show, however, that *wusūm* petroglyphs do not 'behave' differently from other inscriptions or rock-art panels in the region. Virtually all *wusūm* are part of larger rock-art concentrations at high locations, and sometimes even superimposed onto earlier petroglyphs. Factors in the selection of favourable locations for making *wusūm* petroglyphs thus appear to be: (1) relatively high locations; (2) the presence of a structure; and (3) the presence of other petroglyphs.

The majority of *wusūm* occur in large rock-art concentrations of up to 600 petroglyphs (*e.g.* the site of QUR-2). Here, *wusūm* are situated among a variety of other petroglyphs with a chronological depth of at least two millennia. In several instances, the markings are placed on the same rock panel as Safaitic inscriptions or even superimposed over older engravings. It would thus seem reasonable to argue that petroglyph concentrations have the tendency to attract the creation of more petroglyphs, by means of which the spatial extent of a rock-art site expands over time. The potential of petroglyphs to prompt additional petroglyphs has only recently been acknowledged in archaeology. In a study of northern European rock-art panels, for example, this process was termed 'the act of accumulation,' referring to the long-term addition of images to a rock panel through which a petroglyphic palimpsest is created (Sapwell and Janik 2015). The recognised 'act of accumulation' provides an additional explanation for the occurrence of rock-art clusters in the Jebel Qurma region – in particular for the presence of *wusūm* within these clusters.

In archaeological landscapes, where particular 'places' are more difficult to delineate (especially in the absence of archaeological structures), this practice can perhaps best be understood as a ways of signing the landscape: the act of creating a petroglyph necessarily transforms an otherwise empty space into a socialised, meaningful place – both to the maker and future interpreters of the petroglyph.

Discussion

The internal ambiguity (the fluid, dynamic, and non-fixed nature) of *wusūm* and the multiplicity of contexts in which the markings operated, appear to derail any systematic investigations into the phenomenon of *wusūm* marking. It was, however, precisely because of their ambiguity that *wusūm* were able to function within a largely oral and tribal society. In daily life, the ambiguity of the markings was dealt with through negotiation on the basis of memory, rather than encyclopedic, external referents – much like the socio-political organisation itself was explained.

If the proposed dating for *wusūm* petroglyphs in the Jebel Qurma area bears any merit, it is in the fact that it allows us to examine the theory and practice of *wusūm* marking systems in relation to several general aspects of the Bedouin societies that produced these petroglyphs, so as to establish how *wusūm* were able to operate in spite of their ambiguity; of particular interest here are the themes of tribalism and kinship. It is argued that both the socio-political make-up and the illiterate nature of historical Bedouin societies were prerequisites for *wusūm* systems to operate effectively, at least on a practical level.

The self-identification of Bedouin groups was "not based on a specific trait, such as shared religion, ancestry or territory, but on a combination of shared memories, values, symbols and myths." (Van der Steen 2013, 198). The excerpt from the oral narrative about Slewih discussed above indicated the inextricable links between *wusūm* and mythical origins, legendary ancestors, and legitimisation of current socio-political statuses. A Rwala informant in the Jebel Qurma region, now living in Al-Azraq, informed me that the *wasm* of his family has been used for at least six generations, and then continued to recall oral narratives of the exploits of his 'grandfathers', a generic term for his real or imagined ancestors associated with the *wasm*. In this instance, the *wasm* functioned as an *aide-mémoire*, simultaneously providing ascribed status to the informant in question.

It becomes clear, then, that *wusūm* fulfilled extremely important roles in daily lives of users of the marking system – not only as a ways of indicating material wealth and status derived from raided camels, but also as a materialised connection to a tribe's mythical past and heroic ancestors. These connections between a mythical past and modern-day members of a tribe lie at the core of Bedouin society, in which "all social and political relationships are conceived, expressed, and explained in genealogical terms" (Macdonald 2005, 47). Among the Rwala, this extremely complex socio-political system has been termed "generative genealogy" (Lancaster 1981, 24-35). It basically consists of a particular 'bottom' group of individuals (called *ibn amm*) who share mutual responsibility for each other's safety and needs encountered in daily life. The upper part of the genealogy consists of a non-rigid, theoretical, or imagined map of relations between different sections of a single tribe, and even between tribes (*ibid.*; Macdonald 2005, 48). The lower part of the genealogy consists of living or recently deceased individuals, and is therefore fixed. The upper part is also well-known among members of a tribe but, in contrast to the lower part, is not fixed – to some extent, manipulation of ancestry and socio-political relations may therefore take place. The conceptual break between the lower and upper parts of the genealogy signals the obscure joins between the two, and establishing the exact relationship between an individual in the lower group and an 'ancestor' from the upper part is therefore impossible. These relationships, however, are claimed, consolidated, explained and even manipulated through continuous negotiation between different members of the lower part of the genealogy; for in societies with a vibrant oral culture, "a 'historical fact' is only what a sufficient number of people agree they remember" (Macdonald 2005, 48).

The ambiguous, living nature of the generative genealogy (at least, among the Rwala) matches that of intergenerational *wusūm* systems which, as we have seen above, are equally fluid and subjective. The upper part of the genealogy expresses claim to an imagined past and descent, or a 'sense of belonging' known as *asabiyyeh* (Van der Steen 2013, 105-106). The socialising of the landscape through applying *wusūm* petroglyphs assigns that same sense of belonging to a particular place, ranging from a single rock to an archaeological structure, and even to a concentration of other petroglyphs. *Wusūm* thus form materialised representations of diachronic kinship, whether real or imagined.

The generative genealogy is only possible in the absence of an external written referent: people can disagree about what they care to remember, but not about fixed, written records. This explains why discussions arose about the signification of *wusūm* during Henry Field's investigations into the markings; in 2014, I observed a similar scenario during interviews with three members of the Rwala (a father and his two sons). In cases of dispute, the elderly father with status and authority was deemed correct in his interpretation of the *wasm*, and his sons were submissive to his judgement.

Concluding remarks

Wusūm marking practices formed an important part of nineteenth-century Bedouin life. The markings, whether applied to animals or burials, were simultaneously physical representations of abstract socio-political relations and identities; a direct link with mythical ancestors from which hereditary status was derived; and expressions of (temporary) claims to resources, all bound together in a single *wasm*. The tentative dating of *wusūm* petroglyphs from the Jebel Qurma region offers a new way of identifying (post-) Ottoman-era human presence and activity in the area, and perhaps the Near East at large. It is hoped that this paper not only draws attention to the historical importance of *wusūm*, but also encourages Near Eastern scholars to contribute to our understanding of *wusūm* marking practices.

References

Bell, G.M.L. 1914a. Gertrude Bell Archive, diary entry 21/01/1914. Online: http://gertrudebell.ncl.ac.uk/diary_details.php?diary_id=1074 (Accessed: 12 October 2016).

Bell, G.M.L. 1914b. Gertrude Bell Archive, diary entry 01/02/1914. Online: http://gertrudebell.ncl.ac.uk/diary_details.php?diary_id=1093 (Accessed: 12 October 2016).

Bell, G.M.L. 1914c. Gertrude Bell Archive, diary entry 13/02/1914. Online: http://gertrudebell.ncl.ac.uk/diary_details.php?diary_id=1115 (Accessed: 12 October 2016).

Bell, G.M.L. 1914d. Gertrude Bell Archive, diary entry 14/02/1914. Online: http://gertrudebell.ncl.ac.uk/diary_details.php?diary_id=1116 (Accessed: 12 October 2016).

Blunt, A. 1881. *A pilgrimage to Nejd*. London: John Murray.

Buckingham, J.S. 1825. *Travels among the Arab tribes inhabiting the countries east of Syria and Palestine*. London: Longman, Hurst, Rees, Orme, Brown, and Green.

Burckhardt, J.L. 1831. *Notes on the Bedouins and Wahabys*. London: Henry Colburn and Richard Bentley.

Burton, R.F. and Tyrwhitt Drake, *C.F.* 1872. *Unexplored Syria*. London: Tinsley Brothers.

Burton, R.F. 1879. *The Land of Midian (revisited)*. London: C. Kegan Paul & Co.

Conder, C.R. 1883. Arab tribe marks (*ausam*). *Palestine Exploration Fund Quarterly* 15, 178-180.

Corbett, G.J. 2010. *Mapping the mute immortals: a locational and contextual analysis of Thamudic E/Hismaic inscriptions and rock drawings from the Wādī Ḥafīr of southern Jordan*. Chicago, IL: University of Chicago (Unpublished PhD thesis).

Eisenberg-Degen, D., Nash, G.H. and Schmidt, J. 2016. Inscribing history: the complex geographies of Bedouin tribal markings in the Negev desert, southern Israel. In: Brady, L.M. and Taçon, P.S.C. (eds.). *Relating to rock-art in the contemporary world. Navigating symbolism, meaning, and significance*. Boulder, CO: University Press of Colorado, 157-187.

Ewing, W. 1895. A journey in the Hauran. *Palestine Exploration Fund Quarterly* 27, 161-184.

Field, H. 1952. *Camel brands and graffiti from Iraq, Syria, Jordan, Iran, and Arabia*. Baltimore, MD: American Oriental Society.

Fletcher, R.S.G. 2015. *British imperialism and 'the tribal question': desert administration and nomadic societies in the Middle East, 1919-1936*. Oxford: Oxford University Press.

Huber, C. 1891. *Journal d'un voyage en Arabie (1883-1884)*. Paris: Imprimerie Nationale.

Jansen van Rensburg, J. 2016. Rock art on Socotra, Yemen: the discovery of a petroglyph site on the island's south coast. *Arabian Archaeology and Epigraphy* 27, 143-152.

Jennings, R.P., Shipton, C., Al-Omari, A., Alsharekh, A.M., Crassard, R., Groucutt, H. and Petraglia, M.D. 2013. Rock art landscapes besides the Jubbah palaeolake, Saudi Arabia. *Antiquity* 87, 666-683.

Jennings-Bramley, W.E., 1906. The Bedouin of the Sinaitic Peninsula. *Palestine Exploration Quarterly* 38, 197-205.

Jennings-Bramley, W.E., 1907. The Bedouin of the Sinaitic Peninsula. *Palestine Exploration Quarterly* 39, 22-33.

Jennings-Bramley, W.E., 1908. The Bedouin of the Sinaitic Peninsula. *Palestine Exploration Quarterly* 40, 22-35.

Khan, M. 2000. *Wusum. The tribal symbols of Saudi Arabia*. Riyadh: Ministry of Education.

Kurpershoek, P.M. 1995. *The story of a desert knight: the legend of Šlēwīḥ al'Aṭāwi and other 'Utaybah heroes*. Leiden: Brill.

Kurpershoek, P.M. 1999. *Bedouin poets of the Dawasir tribe: between nomadism and settlement in southern Najd*. Leiden: Brill.

Kurpershoek, P.M. 2002. *A Saudi tribal history. Honour and faith in the traditions of the Dawasir*. Leiden: Brill.

Lancaster, W. 1981. *The Rwala bedouin today*. Long Grove: Waveland Press.

Lancaster, W. and Lancaster, F. 1993. Graves and funerary monuments of the Ahl al-Ǧabal, Jordan. *Arabian Archaeology and Epigraphy* 4, 151-169.

Macdonald, M.C.A. 2005. Literacy in an oral environment. In: Bienkowski, P., Mee, C. and Slater, E. (eds). *Writing and ancient Near Eastern society. Papers in honour of Alan R. Millard*. New York and London: T&T Clark, 45-113.

Müller-Neuhof, B. 2019. Was Kilroy a truck driver? Modern petroglyphs in the basalt desert of NE-Jordan. In: Nakamura, S., Adachi, T. and Abe, M. (eds.). *Decades in deserts: eassays on Near Eastern archaeology in honour of Sumio Fujii*. Tokyo: Rokuichi Syobou, 169-177.

Musil, A. 1908. *Arabia Petraea: ethnologischer Reisebericht*. Vienna: A. Holder.
Nayeem, M.A. 2000. *The rock art of Arabia: Saudi Arabia, Oman, Qatar, The Emirates and Yemen*. Hyderabad: Hyderabad Publishers.
Palmer, E.H. 1871. *The desert of the exodus: journeys on foot in the wilderness of the forty years' wanderings*. London: Bell and Daldy.
Sachau, E. 1883. *Reise in Syrien und Mesopotamien*. Leipzig: F.A. Brockhaus.
Sapwell, M. and Janik, L. 2015. Making community: rock art and the creative acts of accumulation. In: Stebergløkken, H., Berge, R., Lindgaard, E. and Stuedal, H.V. (eds.). *Ritual landscapes and borders within rock art research*. Oxford: Archaeopress, 47-58.
Shryock, A. 1997. *Nationalism and the genealogical imagination*. Berkeley, CA: University of California Press.
Sweet, L.E. 1965. Camel raiding of north Arabian bedouin: a mechanism of ecological adaptation. *American Anthropologist* 67, 1132-1150.
Van der Steen, E. 2013. *Near Eastern tribal societies during the nineteenth century: economy, society and politics between tent and town*. Sheffield: Equinox.
Wetzstein, J.G. 1877. Über die Arten des arabischen Wesm. *Zeitschrift für Ethnologie* 9, 14-16.

Rock art in Saudi Arabia: a window into the past? First insights of a comparative study of rock art sites in the Riyadh and Najrān regions

Charly Poliakoff

Abstract

Rock art has been recorded in the south of Saudi Arabia since the beginning of the twentieth century. One should bear in mind that the first expeditions in this country focused only on epigraphic material. Nevertheless, there has been a recent surge of interest in rock art in terms of what it can say about local cultural identity. In fact, on the fringes of the Saudi deserts, there are many engravings on rocky hills and isolated boulders, which challenge the limited archaeological finds. The time span of these engraving practices ranges from the Neolithic until today. Most of the time they depict animal life, hunting activities, and warriors brandishing their weapons. The date of these scenes can be inferred from multiple data. First, one can compare the patina colours and the superimpositions which occur between figures to give elements of relative chronology. Second, the surrounding ancient inscriptions (in Thamudic, South Semitic, Nabataean, and Old Arabian) as well as some details in the scenes (*e.g.* weaponry, animal species, and practices) provide a chronological indicator. Furthermore, recording the location of the petroglyphs helps to understand the regional distribution of the motifs. After a brief presentation of the background in this region, I will try to answer a critical question: 'Does Saudi Arabian rock art help us to understand how these people once lived, conceived, and perceived their own reality?' I will take in account the possibilities and the limits of such an exercise in two vast regions: Riyadh and Najrān.

Keywords: archaeological survey, archaeology, cognition, petroglyphs, rock art, Saudi Arabia

Introduction

In the south of Saudi Arabia, one can drive off-road for hours through rocky deserts, small oases, and sand-filled wadis. When crossing this vast landscape, field archaeologists record all the data which do not seem natural but anthropogenic. Indeed, three obvious objects of study are available for archaeological surveys in these regions: dry-stone structures, inscriptions, and rock art. This rock art consists of figurative and abstract motifs, engraved in a dark-black to reddish-brown sandstone patina. For the figurative part of the assemblage, the 'phytomorphic' figures are exceptional, unlike the common 'zoomorphic', 'anthropomorphic', and 'technomorphic' figures. Taken together, they emphasise that these landscapes were once the theatre of animal

In: Peter M. M. G. Akkermans (ed.) 2020: *Landscapes of Survival - The Archaeology and Epigraphy of Jordan's North-Eastern Desert and Beyond*, Sidestone Press (Leiden), pp. 333-342.

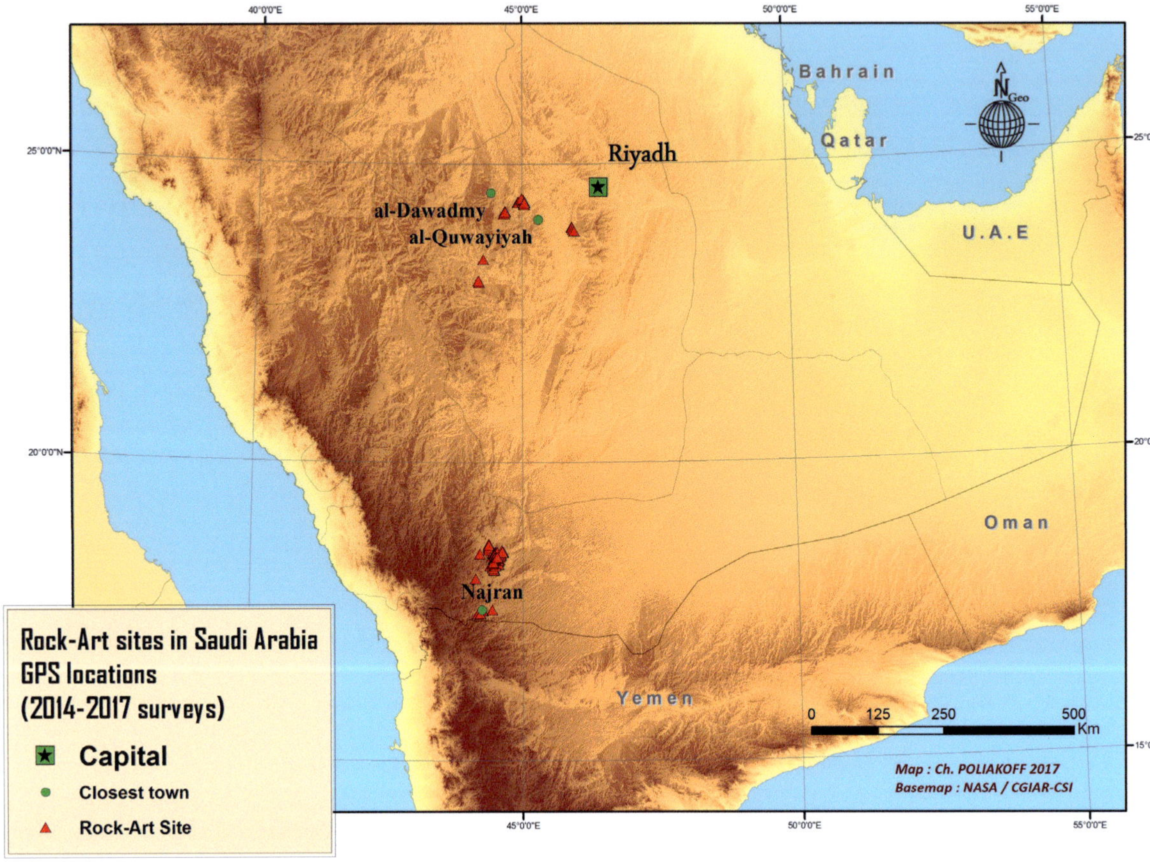

Figure 1. Map of rock-art sites in Saudi Arabia (2014-2017 surveys).

life and human activities. Indeed, the figurative motifs gathered into scenes provide information about the people who lived in this area in the remote past. The following paragraphs propose some insights on my main research question: does the rock art of Saudi Arabia help us to understand how people once lived, conceived, and perceived their own reality? In other words, what is the size and the nature of this window into the past? The data introduced here come from three years of investigations on pre-Islamic engravings undertaken since 2014, in collaboration with the Saudi-French team in the Najrān and Riyadh areas.[1] Fig. 1 shows the rock-art sites which were recently found in the south-west of Saudi Arabia. The most interesting petroglyphs have been discovered in the northern part of the Najrān region, in the area of Bi'r Ḥimā, Jabal al-Qārah, Jabal al-Kawkab, and Al-Kurmah. A few petroglyphs have also been found between Al-Dawādimi and Al-Quwayʿiyah in the Najd.

1 This contribution is part of a doctoral thesis at the Paris 1 Pantheon Sorbonne University, under the supervision of Prof. Pascal Butterlin. See Poliakoff 2017.

Rock art in south-west and central Saudi Arabia: a brief overview

The first collection of photographs of petroglyphs from the south-west and centre of Saudi Arabia comes from the Ryckmans-Philby-Lippens expedition in 1951. Study of this material was only completed after almost twenty years (see Anati 1968a; 1968b; 1972; Anati and Tchenov 1974). In 1988, after ten years of survey by various teams in Saudi Arabia, Majeed Khan (1993) criticised Anati's interpretations and drew attention to the petroglyphs in Arabia. In 2003, a third phase of investigation began with a Japanese epigraphic campaign in Jabal al-Kawkab (Kawatoko *et al.* 2005). In 2007, a Saudi-French mission began to make an inventory of rock art at the site of ʿĀn al-Jamal in Najrān, although its emphasis was on the inscriptions (Arbach *et al.* 2007; 2008). In 2017, three members[2] of the Najrān team joined a project of the Saudi Commission for Tourism and National Heritage (SCTH), focusing on the rock art and other remains of Najrān for inclusion in the UNESCO cultural heritage list.

2 Mounir Arbach, Anaïs Chevalier, Charly Poliakoff.

Figure 2. A petroglyph from the Najd area, west of Riyadh. This scene depicts the hunt of a bovid and a dog. Note the details, such as the isolated quiver and the bow aimed at the animal.

This brief overview of the research suggests that the evidence for petroglyphs is based on the time and attention that each survey dedicated to rock art. To document most of the small- to medium-sized panels encountered in Saudi Arabia, there is no need for experts of heavy and expensive documentation techniques, such as giga-pan imaging, robotised cameras, and laser scanners. Nevertheless, if an archaeological mission aims to study rock art, a few critical rules should be followed in order to properly document the art, which can be done with a simple GPS device and a good camera. For a rock-art site, the GPS accuracy should be within 20 m or (preferably) less, and for a rock-art panel on a boulder better than 5 m. Indeed, the recording requires a dGPS and/or a set of low-altitude aerial photos (using a drone or kite). Photography in the field should comprise the locational context, the rock-art panel itself, and details of the engravings. To capture a panel or a scene, one should carefully position the optical axis at a 90° angle to the rock surface and use a 50-80 mm lens to avoid major distortions (Moore 1991, 138). Adding a well-oriented scale makes the resulting photograph suitable for measurements. Last but not least, the archaeologist should be able to understand how natural light in the field can undermine or sharpen the final photograph for analysis.

Following these minimum requirements, every rock-art panel should receive a single GPS point, linked to the accurate visual records. This gives a bird's-eye view on the data and allows researchers to conduct various spatial GIS analyses. Then the establishment of a chronology of the petroglyphs can be attempted, which can be inferred from various data. On the one hand, comparing the patina colours and the superimpositions from one figure to another can offer elements of relative chronology. This comparison should be done using figures on the same rock surface. Moreover, this comparison is not allowed when two petroglyphs are too far away from each other or show local alterations and patina inversion processes. On the other hand, absolute dating can give chronological markers. Though direct dating is neither available nor reliable for petroglyphs, it is possible to use indirect dating techniques. In this article, I used inscriptions to set roughly a *terminus post quem* and *ante quem* from the first millennium BC to the first millennium AD (but see Macdonald 2015). Nevertheless, at least in the Najrān region, it seems possible to sort 'monumental'[3] and South

3 One should note that this word is problematic in itself. Not all the inscriptions found in south-western Arabia are monumental, but at least they are very far away from the *ductus* of cursive writing.

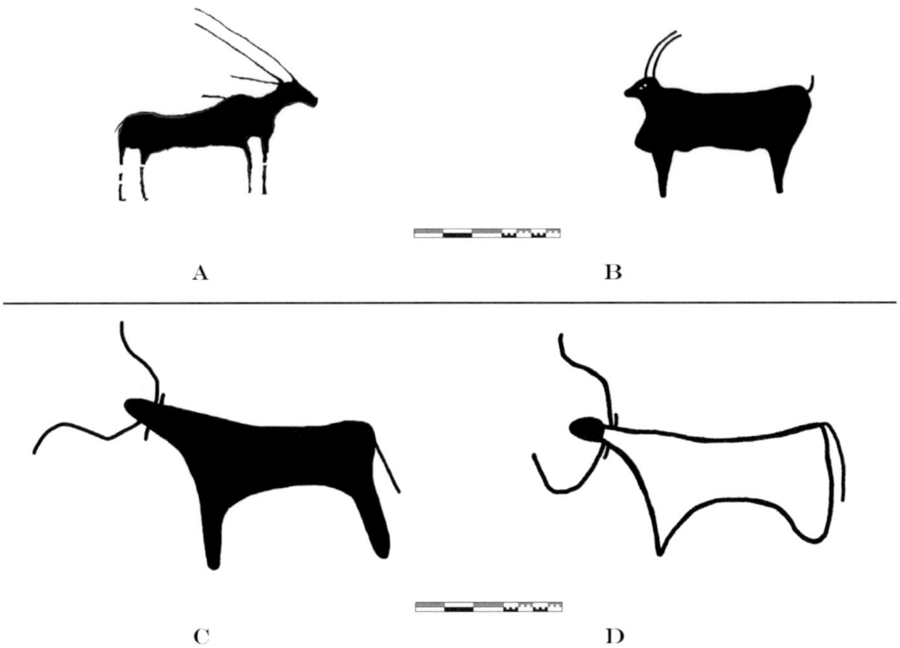

Figure 3. Comparison of the size of Bovidae depiction and the issues of identifying sub-species (scale: 50 cm). A: naturalistic oryx from the north of Najrān. B: naturalistic ibex (?) from the north of Najrān. C: schematic bovine (?) from the Najd. D: schematic bovine (?) from the north of Najrān.

Pre-Islamic zoomorphic engravings						
First level of interpretation	Aves	Bovidae	Camelidae	Carnivora	Equidae	Squamata
Second level of interpretation	Ostrich Unidentified	Antelope Bovinae Caprinae Goat/Sheep Unidentified	Camel Dromedary Unidentified	Canidae Felidae Unidentified	Ass/Donkey Horse Unidentified	Lizard Snake Unidentified
Frequency	Frequent	Very Frequent	Very Frequent	Frequent	Rare	Very Rare

Table 1. Overview of the zoomorphic categories and sub-categories and their frequency in the petroglyphs of south-western Saudi Arabia.

Semitic[4] petroglyphs into very rough time periods based on their ornamentations (Stein 2013, 193).

Some cognitive insights from zoomorphic figures

The setup of a scene has very few variations. Most animals are depicted in full profile or a slight bi-angular perspective along the horizontal to sub-horizontal axis. The animal representation can be singular or belong to a herd, or it can be facing or fleeing another human or animal figure (Fig. 2). The depiction of hunting dogs (or a wild canine pack) has a specific variation as they jump and surround their prey (cf. Fig. 8A). In some cases, caprines or antelopes are engraved along a vertical or subvertical axis, with their heads toward the sky. If one analyses this choice on a cognitive basis, the format could be a way for the engraver to express the animal's escaping behaviour towards the top of a hill, or the depiction of motion in the pictorial space from the foreground to the background. However, in some cases this vertical arrangement of an animal or a bunch of figures can be explained by the crumbling of a boulder or a part of the cliff through processes of erosion. In this case, the vertical axis is not intentional but related to a natural event.

Although the general species are recognisable, interpreting different sub-species is tentative and risky. In order to avoid a hasty interpretation, I defined large categories. The vast majority of pre-Islamic animal depictions are Bovidae, Camelidae, and (large) Aves. This choice to depict large animals and the lack of micro-fauna, other mammals and small birds, emphasises that medium- to large-sized animals truly mattered for pre-Islamic engravers. Nowadays, the only real large-sized animal which can be seen in the landscape is the (domestic) dromedary. Carnivora are less common in the depictions. This is a large convenient group because

4 I prefer using this label for South Arabian inscriptions, following Macdonald (2015, 1).

Figure 4. Half life-sized bovine engraved on a large boulder in the north of Najrān (scale: 50 cm).

Figure 5. Three half life-sized and one large life-sized depictions of dromedaries on a high cliff in the northern Najrān region.

distinguishing between the Canidae or Felidae is impossible in half of the cases. Various species of predators can be found in these regions during pre-Islamic times, such as the caracal, lion, cheetah, panther, wolf, and, of course, the domestic or feral dog. Equidae and Squamata are very rare. For the Squamata category, one can find lizards and snakes but these representations do not offer enough details to allow for further identification. An unexpected last category depicting Elephantidae has been found at a site near Ḥimā, but consists of only three depictions (Monchot and Poliakoff 2016, 87).

The size of the depiction of one particular animal is also of importance. Petroglyphs can vary in size from a dozen centimetres to more than two metres. In the overall corpus of representations, Bovidae, even from different subspecies, seem to have a roughly medium life-size (Figs. 3A-3B). Two schematic bovines from the Najd and the north of the Najrān province have the same overall size and features and have been depicted from a similar bi-angular perspective (Figs. 3C-3D). By contrast, half- to life-sized representation of Bovinae seem to be a specificity of the Najrān region (Fig. 4). Given the known faunal remains, it is obvious that large wild bovines such as *Bos primigenus* and *Bos syncerus* once lived in southern Arabia (Hadjoui 2007, 53), but there also should be later domesticated variants. The direct link between the size of one type of depiction and its state of domestication has been accurately avoided. Moreover, until now in the south, the details of the scenes are difficult to interpret, because these Bovinae are not clearly hunted or

Figure 6. The author showing how large camel petroglyphs were pecked by the engravers, by standing on two climbing holds.

used for other apparent purposes. Most of the time, except for some representations in the Najd (see Fig. 2), marks or details of hunting come from a later pecking phase, which indicate that a later engraver altered the meaning of the scene. The identification of petroglyphs as Bovidae and subsequently as Bovinae is still an unsolved or ignored issue in southern Arabia.

Speaking of huge representations, many half to fully life-sized depictions of dromedaries have been found in the north of the Najrān region (Fig. 5). They are located high on the slope of a mountain, far away from the ground level but massive enough to be seen from a few hundred metres. These three dromedaries are depicted from a bi-angular perspective with their mouths open, and the largest one has a detail added to its head. These two small dotted circles could be a way of representing eyes from a frontal viewpoint or they could be a so-called 'tribal marker' (*wasm*). If the latter interpretation is used to understand the carving, then this does not depict a wild animal. I do not intend to discuss here the domestication of the camel in the Arabian Peninsula. Nevertheless, in the Najrān region, scenes of speared or stabbed camels are very common and emphasise local hunting practices (Monchot and Poliakoff 2016, 78-79). However, there are also many carvings of small camels with curvy tails, sometimes mounted by a warrior. Another panel on the same mountain as the large dromedaries emphasises the engravers' use of natural or handmade staircases and climbing holds to reach the highest surfaces of the cliff (Fig. 6). The energy cost and the risks of such an activity should now be investigated at other locations to understand who these people were and why they performed these actions. One should also note that these engravings, at difficult to reach spots, have been better at surviving intentional damage until the advent of firearms. This could partially explain the desire to spend this energy and take these risks in their creation. Despite the impossibility of giving an exact date to this kind of engraving, it should belong to a time period ranging from prehistory to the Late Bronze Age. Moreover, large camel depictions seem to be a widespread phenomenon in Saudi Arabia, given the recent discovery of animal reliefs in the Jawf region (Charloux *et al.* 2018).

The pre-Islamic Equidae figures are very rare and mostly depict asses or donkeys. Indeed, the introduction of the domesticated horse to south-western Arabia came very late (Ryckmans 1963). Nevertheless, very rare examples of mounted horses do occur in close association with South Semitic inscriptions (see Alexander 1996, 6-7). A recent discovery on the northern fringes of the Najrān region sheds new light on the way equids could have been

Figure 7. A recently discovered petroglyph of a life-sized equid in the north of Najrān. Note the sensorial details and the considerable attention given to the outline (scale: 50 cm).

perceived by pre-Islamic engravers (Fig. 7). This life-sized depiction of probably an ass or donkey shows naturalistic details of the sensory organs: the long ears, the tip of the nose, the lips, and the eye shown *en profile*. The body of the animal has been cut into the rock surface by a large line of pecking, which gives the illusion of relief. This is the first example of a petroglyph of this size in the Najrān region with such a level of accuracy. A comparison can be made with the recent discovery of a large relief of an equid in the Jawf area (Charloux *et al.* 2018).

Combinations of anthropomorphic and technomorphic figures

Most of the time, the anthropomorphic petroglyphs belong to a vertical scheme. This category depicts a group of human figures involved in hunting, skirmishing activities, or individuals that raise their arms or brandish weapons (Newton and Zarins 2000; Poliakoff 2014; 2017). Most of the human depictions come from the south-west of Saudi Arabia. The limited repertory of the scenes depicting human practices is surprising: their stereotyped stances, the few sensorial details, and the lack of facial expression make them anonymous people. Indeed, only the engravers and their community would know who each character was in these scenes; it is also possible that they did not intend to represent real people and scenes. Thus, it is unclear whether these engravings of anthropomorphic figures represent real events, or mere fantasies. Although a clear answer is not possible given the current state of knowledge, some insights are given by the technomorphic figures.

'Technomorphic' refers to non-living, human-made and technological objects such as weapons, adornment, and architecture. Although some can be depicted in isolation, the vast majority is linked to human representations. The main technomorphic depictions consist of weapons. There are four main variations within the outfit of the pre-Islamic warriors/hunters. The first variation shows very small to small archers on foot, which chase antelopes and dromedaries (Fig. 8A; see also Fig. 2). They are outfitted with a double-convex bow, a horizontal quiver, and a crescent-pommel dagger. The blade of this dagger is believed to be made of a copper alloy (Newton and Zarins 2000, 155). Moreover, real metal artefacts of these weapons have been unearthed in Mesopotamia and in the eastern part of the Arabian Peninsula (Potts 1998; 2007; see also the contribution by Keshia Akkermans, this volume). Similarly-shaped daggers were also found in Yemen, although in looted burials, which are not very

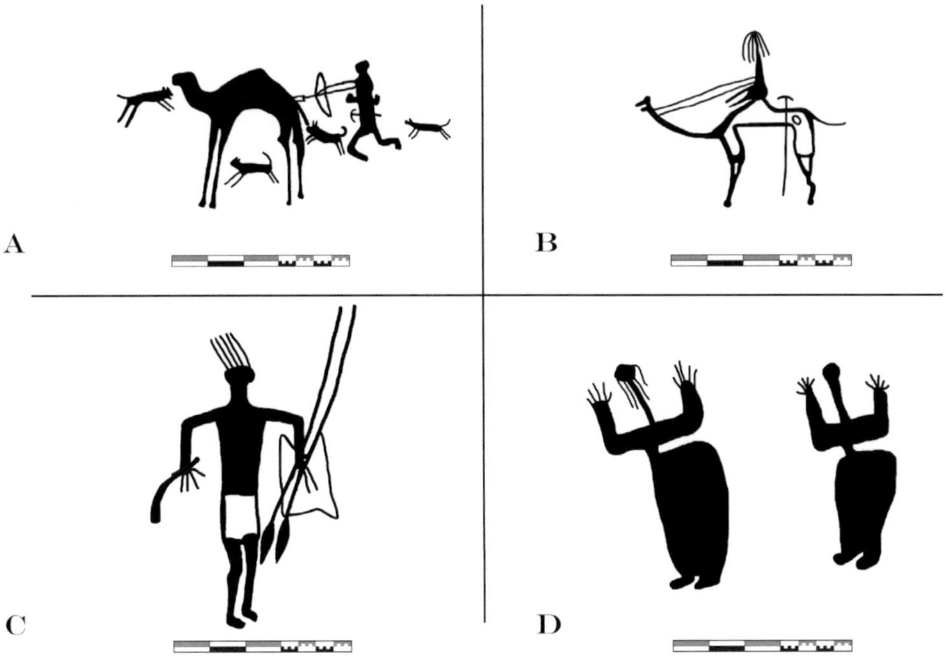

Figure 8. Comparison of anthropomorphic figures in the north of the Najrān region (scale: 50 cm). A: hunting scene of a dromedary surrounded by dogs. The hunter-warrior is outfitted with a bow, a quiver, and a crescent-pommel dagger at his waist. B: a human figure riding a camel. Note the long crescent-pommel sword hung at the camel's side. C: a hunter/warrior holding a sickle sword, spears, and rectangular shield. D: figures with large hips and long hair, raising their arms toward the sky in a stereotyped stance.

helpful for understanding their precise period of use. No examples of such daggers have been reported from the interior of the Arabian Peninsula so far.

The second variation depicts the same kind of small hunters or warriors who no longer hunt but ride camels. In this case, the crescent-pommel dagger becomes a crescent-pommel elongated sword. Sometimes, these long weapons hang at the camel's side (Fig. 8B).

The third variation emphasises medium- to large-sized warriors who grasp a sickle-shaped sword or a spear in the right hand, and a shield and two spears in the left hand (Fig. 8C). The sickle-blade sword is well-known and was used for a very long period in the Near East. This is confusing when trying to sort rock art representations into a chronological framework. Nevertheless, one unique copper-alloy exemplar has been found recently in the centre of Saudi Arabia (Schiettecatte 2013, 59). Even if its date requires greater precision, this artefact was unearthed in a grave dated from the second to the first millennium BC. In relation to rock art, it seems interesting to find very few representations that combine crescent-pommel daggers and sickle swords. There is great hope that inferential statistics will give a conclusive indication on this issue.

The fourth variation depicts small- to medium-sized human figures with long hair, large hips, and bare hands raised toward the sky (Fig. 8D). The interpretation and the possible meanings of these figures are problematic. Macdonald (2012) has addressed several issues on the nature of these representations. In the south-west, there are not enough details to give a clear interpretation of the gender of all these figures. Moreover, if one takes into account the possibility of transvestism, an interpretation becomes highly uncertain.

However, a recent discovery in the north of Najrān sheds new light on this sensitive issue of gender (Fig. 9). Indeed, this anthropomorphic figure looks like a female, with curvy hair and large hips. But the figure carries two swords and a round shield. Next to this representation, is an inscription that mentions a woman named 'Ḥayat' by the engraver. It could be the name of the figure in the representation and/or the name of the engraver herself.[5] This kind of female warrior has been found in a few cases in Najrān, but it seems that it relies on a misunderstanding or a will to change the original gender of the figure by another engraver. Another clue comes from the later engravers who are very confused with these representations. They sometimes try to add a phallus or peck a deep hole into the pelvis, suggesting that these figures are no longer understood.

5 I thank Mounir Arbach and Jerôme Norris for suggesting this interpretation.

Figure 9. An anthropomorphic figure with a sophisticated headdress and the hands raised toward the sky (scale: 10 cm). The figure is outfitted with two swords and a round shield. The inscription to the left reads: "Ḥayat taqar".

Conclusion

For a long time, the investigation of rock art in the south of Saudi Arabia was mainly the by-product of epigraphic and archaeological projects. However, this has begun to change with the enthusiastic and increasing appreciation of rock art as a highly valuable category of cultural heritage in the country. This paper offered some insights into recent discoveries in the centre and the south-west of Saudi Arabia. The quality of this material emphasises the fact that the petroglyphs of these regions need more attention. Then, by studying three categories of engravings (zoomorphic, anthropomorphic, and technomorphic figures) this contribution re-assessed existing interpretations of rock art. If animals are depicted in a limited fashion, I insisted on the interpretation that the engravers made the deliberate choice to depict large animals and not any other elements of the existing fauna (see also Brusgaard, this volume). These large creatures were highly visible in the local environment, as well as objects of hunting or breeding, or even both at the same time, such as during the domestication process of the camel. The size of a zoomorphic figure could depend not only on a shared cultural standard of representations, but also on the nature of the relationship between a group of humans and the particular species. Nevertheless, the overall picture of past human societies in these places is blurred through the windows of rock art. In fact, research in this region lacks vast survey evidence and rock art is not able to fill all these archaeological gaps. Although anthropomorphic figures can be of various sizes or positions, most of the time they are equipped with distinctive weapons. Even if it is possible to understand this gear as a valuable cultural marker, it remains difficult to find these objects in well-dated archaeological context. Filling in the missing links between the technomorphic representations and the real artefacts would help to address the need for projects that combine rock art and archaeology in a balanced way.

Acknowledgements

I thank Christian Julien Robin, director of the Najrān mission, who allowed me to participate in the surveys in the Najrān and Riyadh regions. I also thank Mounir Arbach and Anaïs Chevalier, who kindly helped me with the recording of the rock art. Thanks also go to the members of the Saudi Commission for Tourism and National Heritage, who supported me in my research and contributed to the UNESCO inventory, especially Dr. Ali al-Ghabban, Salih al-Murrah and Bassam al-Harithi.

References

Alexander, D. (ed.). 1996. *Furusiyya: the horse in the art of the Near East*. Riyadh: King Abdulaziz Public Library.

Anati, E. 1968a. *Rock art in central Arabia. Vol. 1: The 'oval-headed' people of Arabia*. Louvain: Institut Orientaliste.

Anati, E. 1968b. *Rock art in central Arabia Vol. 2/2: The 'realistic-dynamic' style of rock art in central Arabia*. Louvain: Institut Orientaliste.

Anati, E. 1972. *Rock art in central Arabia. Vol. 3: Corpus of the rock engravings*. Louvain: Institut Orientaliste.

Anati, E. and Tchernov, E. 1974. *Rock art in central Arabia .Vol. 4: Corpus of the rock engravings*. Louvain: Institut Orientaliste.

Arbach, M., Dridi, H., Gajda, I. and Robin, C.J. 2007. Première campagne de la mission archéologique franco-saoudienne dans la région de Najrān (8- 20 avril 2007). Rapport préliminaire. Online: https://halshs.archives-ouvertes.fr/halshs-00581431.

Arbach, M., Charloux, G., Robin, C.J. and Schiettecatte, J. 2008. Deuxième campagne de la mission archéologique franco-saoudienne dans la région de Najrān. Rapport préliminaire. Online: https://halshs.archives-ouvertes.fr/halshs-00581438.

Charloux, G., Al-Khalifa, H., Al-Malki, T., Mensan, R. and Schwerdtner, R. 2018. The art of rock relief in ancient Arabia: new evidence from the Jawf province. *Antiquity* 92, 165-182.

Hadjoui, D. 2007. La faune des grandes mammifères. In: Inizan, M-L. and Madiha, R. (eds.). *Art rupestre et peuplements préhistoriques du Yémen*. Sanaa: CEFAS, 51-60.

Kawatoko, M., Tokunaga, R. and Iizuka, M. 2005. *Ancient and Islamic rock inscriptions of southwest Saudi Arabia I: Wādī Khushayba*. Tokyo: University of Foreign Studies.

Khan, M. 1993. *Prehistoric rock art of northern Saudi Arabia*. Riyadh: Ministry of Education, Department of Antiquities and Museums.

Macdonald, M.C.A. 2012. Goddesses, dancing girls or cheerleaders? Perceptions of the divine and the female form in the rock art of pre-Islamic north Arabia. In: Sachet, I. and Robin, C.J. (eds.). *Dieux et déesses d'Arabie: images et représentations*. Paris: De Boccard, 261-297.

Macdonald, M.C.A. 2015. On the uses of writing in ancient Arabia and the role of palaeography in studying them. *Arabian Epigraphic Notes* 1, 1-50.

Monchot, H. and Poliakoff, C. 2016. La faune dans la roche: de l'iconographie rupestre aux restes osseux entre Dûmat al Jandal et Najrān (Arabie Saoudite). *Routes de l'Orient, Hors-série* 2, 74-93.

Moore, E.A. 1991. A comparative study of two prehistoric artistic recording localities. In: Pearson, C. and Swartz, B.K. (eds.). *Rock art and posterity: conserving, managing and recording rock art*. Melbourne: Australian Rock Art Research Association.

Newton, L.S. and Zarins, J. 2000. Aspects of Bronze Age art of southern Arabia: the pictorial landscape and its relation to economic and socio-political status. *Arabian Archaeology and Epigraphy* 11, 154-179.

Poliakoff, C. 2014. *Étude des gravures rupestres du Jabal al-Kawkab (Arabie saoudite)*. Paris: Université Paris 1 Panthéon-Sorbonne (unpublished MA thesis).

Poliakoff, C. 2017. *L'évolution de l'armement par le prisme de l'art rupestre dans la région de Najrān (Arabie saoudite): tentative de synchronisation de l'iconographie et du matériel archéologique*. Paris: Éditions de la Sorbonne. Online: https://books.openedition.org/psorbonne/6787?lang=en.

Potts, D.T. 1998. Some issues in the study of the pre-Islamic weaponry of southeastern Arabia. *Arabian Archaeology and Epigraphy* 9, 182-208.

Potts, D.T. 2007. Meskalamdug's dagger. In: Al-Zayla'I, A.U. (ed.). *Studies on the history and civilization of Arabia*. Riyadh: Ministry of Culture and Information, 35-40.

Ryckmans, J. 1963. L'apparition du cheval en Arabie ancienne. *Ex Oriente Lux* 17, 211-226.

Schiettecatte, J. (ed.). 2013. *Third season of the Saudi French mission in al-Kharj. Preliminary report*. Online: https://halshs.archives-ouvertes.fr/halshs-01062149.

Stein, P. 2013. *Palaeography of the Ancient South Arabian script. New evidence for an absolute chronology*. Singapore: John Wiley and Sons.

Graffiti and complexity: ways-of-life and languages in the Hellenistic and Roman *ḥarrah*

Michael C.A. Macdonald

Abstract

As is well-known, the *ḥarrah* or basalt desert of southern Syria, north-eastern Jordan and northern Saudi Arabia is full of inscriptions and the content of the vast majority (the so-called 'Safaitic' inscriptions) suggest that they were carved by nomads. But there are also Greek graffiti carved by members of nomadic social groups as well as by people who had travelled from the settled areas of the Ḥawrān, or further. Others carved their graffiti in Safaitic, while claiming to be Nabataeans. There are even references to settlements of nomadic tribes in the desert. What does all this mean? In this paper, I will explore the interaction of nomadism and sedentarism in this part of the *badia* during the late Hellenistic and early Roman periods through the lens of the casual writings of individuals and will attempt to show how the traditional antithesis of the 'Desert and the Sown' hinders, rather than helps, our understanding of the ancient societies there.

Keywords: badia, ḥarrah, nomads, sedentaries, writing, Safaitic, Greek, Latin, Roman army

Introduction

A *ḥarrah* (from an Arabic root meaning 'to burn') is an area of ancient lava flows which, over millions of years, have broken up into billions of basalt stones and boulders. Through the interaction of the chemicals in the basalt and those in the atmosphere the exposed parts of these rocks have developed a black patina, giving the overall impression of a 'Black Desert'. There are many *ḥarrāṭ* in the Middle East, but it is the one stretching from southern Syria, across north-eastern Jordan and into northern Saudi Arabia, which is part of the *badia*, or 'desert region' of these countries, that I shall be discussing here.

Since scholarly explorers in the nineteenth century first recorded inscriptions in this *ḥarrah*, it has come to be recognised as a hotbed of ancient literacy. Tens of thousands of graffiti in Safaitic, Thamudic B,[1] Greek, Latin, and Arabic adorn the rocks with which the ground is so liberally covered. For, when the thin black patina is pierced, the light grey pumice colour of the lava shows through and this looks almost white against the black background. A graffito therefore shows up very well and is worth the considerable effort required to carve it on this extremely hard rock. Over the millennia the carving itself gradually patinates to black, like the surface around it, but the fact that graffiti some

1 For these terms see Macdonald 2000, 33, 35, 43-46.

In: Peter M. M. G. Akkermans (ed.) 2020: *Landscapes of Survival - The Archaeology and Epigraphy of Jordan's North-Eastern Desert and Beyond*, Sidestone Press (Leiden), pp. 343-354.

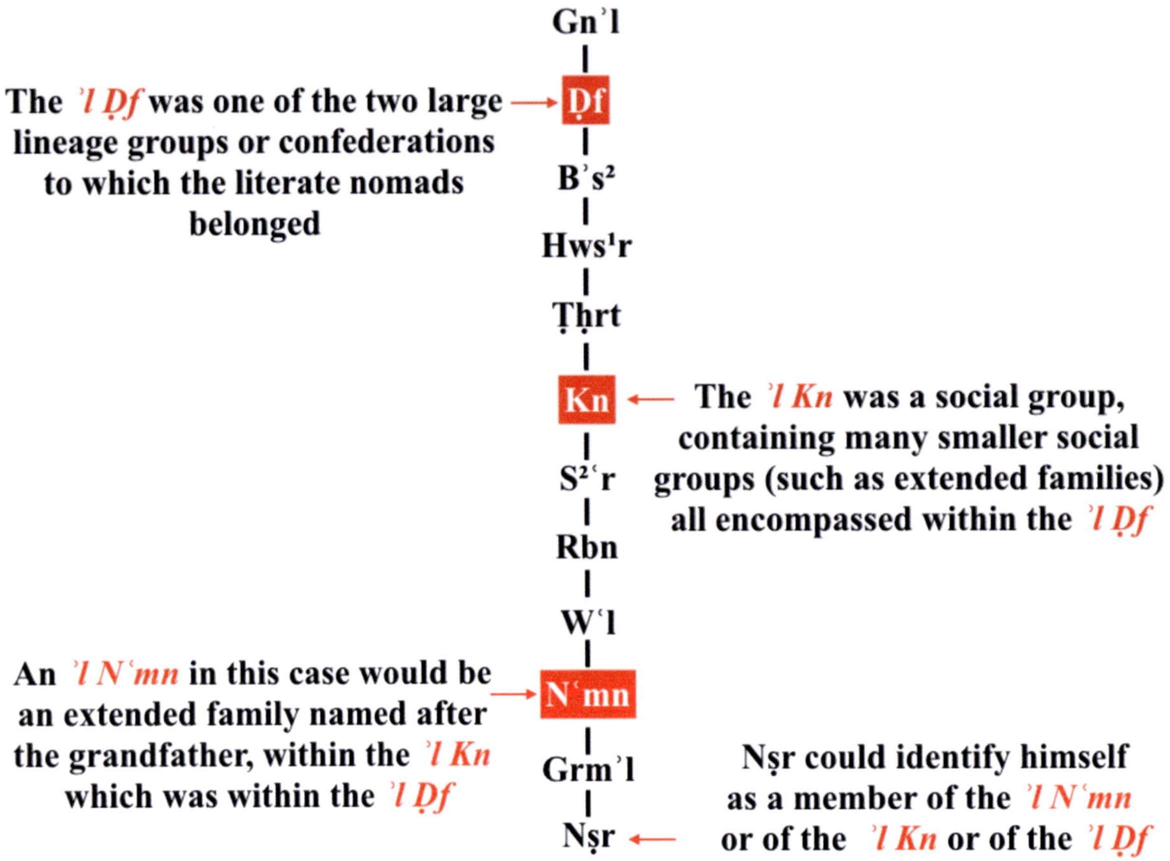
Gnʾl
Ḍf
Bʾs^2
Hws^1r
Ṭḥrt
Kn
S^2ʿr
Rbn
Wʿl
Nʿmn
Grmʾl
Nṣr
The ʾl Ḍf was one of the two large lineage groups or confederations to which the literate nomads belonged
The ʾl Kn was a social group, containing many smaller social groups (such as extended families) all encompassed within the ʾl Ḍf
An ʾl Nʿmn in this case would be an extended family named after the grandfather, within the ʾl Kn which was within the ʾl Ḍf
Nṣr could identify himself as a member of the ʾl Nʿmn or of the ʾl Kn or of the ʾl Ḍf

Figure 1. An example of the use of the word *ʾl* to designate different levels of lineage group.

2000 years old are today mostly orangey-red, gives some idea of how slow the process is.

However, those who carved graffiti only represent the *ḥarrah*'s most recent inhabitants and visitors – during the last 2300 years or so. Earlier, non-literate populations left their mark with rock drawings, and this tradition was maintained by the literate peoples who succeeded them. Moreover, the literacy of the inhabitants, or their desire to use it to carve graffiti here, fluctuated at different periods. Thus, the Ancient North Arabian graffiti – in this case mainly Safaitic and Thamudic B – appear to have ceased some time in or before the fourth century AD, as does most of the Greek and Latin. There is then a burst of Arabic graffiti in the first two centuries of the hijra (seventh to early ninth centuries AD), then another apparent pause before further burst in the Ayyubid-Mamluk period (twelfth to early sixteenth centuries AD), and finally very few texts until the late twentieth and twenty-first centuries.

We have no way of telling how long the nomads who carved their graffiti here, had been living in this 'Black Desert' before they learnt to write. Nor can we be certain that they were the only nomads in the area. There may well have been other groups all, or most, of whose members did not leave inscriptions. Certainly, there are lineage groups which are mentioned in the inscriptions, but whose members have apparently left no graffiti of their own – or rather none which have been found so far. These groups fall into two categories: on the one hand, there are those, like the tribes of Ṭayyiʾ and Liḥyān, which are known from other sources and appear in the graffiti as raiders from outside the *ḥarrah*.[2] On the other hand, there are lineage groups (for which the word in Safaitic is *ʾl*) whose names only occur in graffiti by members of other groups. Thus for instance, the *ʾl* ms^1b (C 2702), *ʾl ḏʾb gnʾl* (C 4039), *ʾl* hs^1k (C 4388), *ʾl ḥrm* (C 4438, SESP.S 9), *ʾl ʾ*s^1hm (SESP.S 1), *ʾl* $s^2rṯ$ (SESP.S 1), *ʾl* ms^2*ʾr* (SESP.S 14).[3] It is possible that members of these apparently 'silent' groups were not literate, but there are also other possible explanations. The nomads who carved the Safaitic inscriptions perceived all social

2 Note however that members of another such group, the *Ḥwlt* (see Macdonald 1993, 308; 2009, 18) have left at least one Safaitic and two Hismaic graffiti; see Lemaire and Macdonald 2018, 298–302.

3 All these inscriptions can be found in the Online Corpus of the Inscriptions of Ancient North Arabia (OCIANA) at http://krcfm.orient.ox.ac.uk/fmi/webd#ociana, where all the other Safaitic inscriptions mentioned in this paper can also be found.

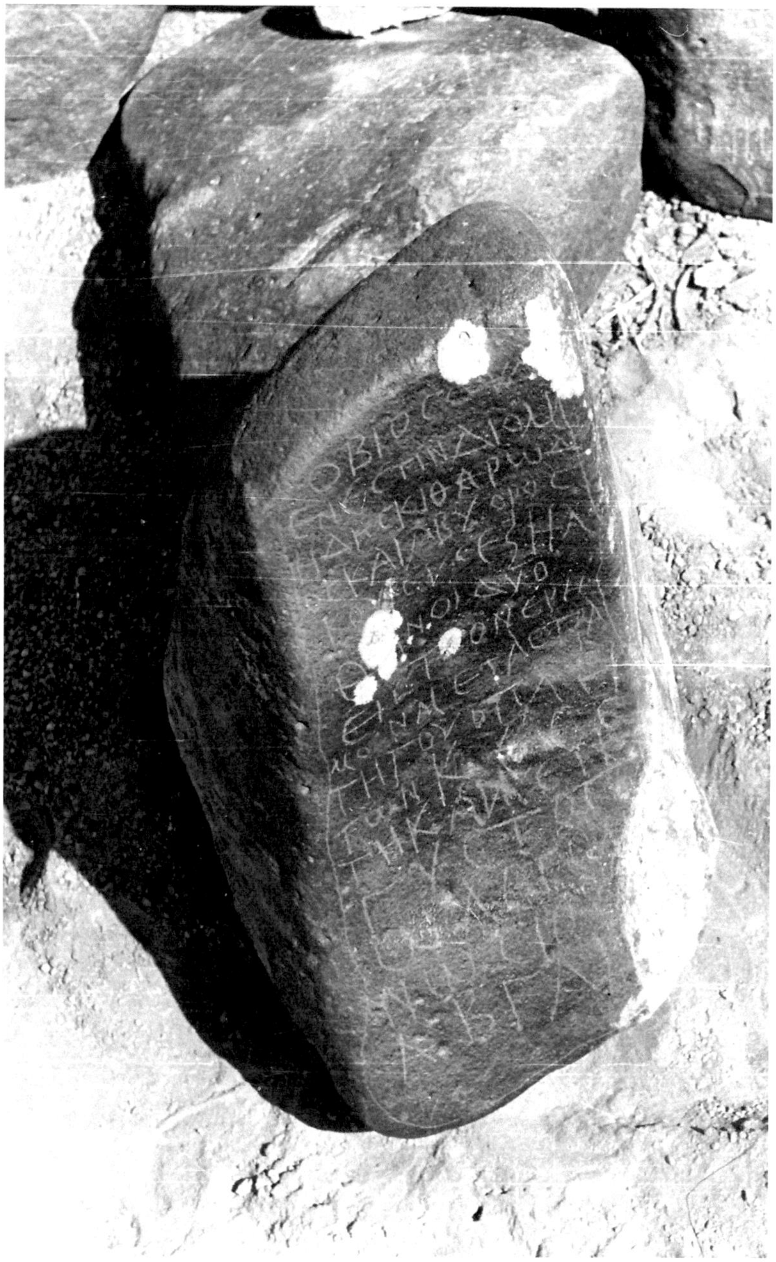

Figure 2. Greek graffito by Diomedes the kithara-player and Abchoros the barber at Jathūm, north-eastern Jordan (photograph by the late F.V. Winnett).

groups as genealogically-based and used the word *ʾl* to refer to all of them, from extended families to nations, such as the Romans or the Jews. Each family was part of a succession of 'higher' genealogically-based groups, and each author could choose which of these he used to identify himself, in any particular circumstances (Fig. 1). So a member of one of these apparently 'silent' *ʾl*s may have carved inscriptions and either simply given his name and patronym, or genealogy, without stating his *ʾl*, or he may have used an identification higher up or lower down the hierarchy of social groups (see Macdonald 1993, 367 for a fuller explanation of this).

But nomads were not, of course, the only people living in the *ḥarrah* at this period. There were Roman soldiers posted at various places and there is a famous pathetic cry from two camp followers which says: "Life is worthless! Diomedes the kithara-player and Abchoros the barber, the two of them, went out into the desert with the commander of the foot soldiers and were stationed near a place called the Cairn of Abgar." (Mowry 1953; Schwabe 1954) (Fig. 2). Anyone who has visited the site of Jathūm (Fig. 3), where this text was found, can sympathise with their distress and their yearning for the bright lights of Boṣrā.

Figure 3. Jathūm, north-eastern Jordan (photograph by the late F.V. Winnett).

Nomads and Romans at Al-Namārah

Naturally, there were Roman soldiers at other places as well. For instance, at Al-Namārah in southern Syria, the Roman army built a fort on an 'island' in a basin in the Wādī al-Shām. Thanks to its natural position and to channels and wells made over the millennia, water from the flash floods in the winter and spring was retained here all the year round (Fig. 4). This made it an important camping place for the nomads during the dry season, and we have many Safaitic graffiti which say that the author was coming to, or was already camped at, the permanent water sources of Al-Namārah.[4] One author dates his inscription to "the year he escaped from Al-Namārah-of-the-government to the lineage group of ʿwḏ" (LP 540).

This policy of building small forts at places of permanent water in an effort to control the nomads was one the Romans employed elsewhere in the *ḥarrah*, for instance at Azraq and probably Jabal Says and Burqūʿ. Interestingly, it was one re-employed by Glubb Pasha, the founder and commander of the Jordanian army, two thousand years later in the 1930s, for whom it worked equally well (Glubb 1983, 104-105).

Al-Namārah is indeed an extraordinary place and when in 1996 I discovered that part of it had been damaged by bull-dozing, I was given permission by the Syrian authorities to mount a rescue survey there, together with my Syrian colleague Hussein Zeinaddin and my French archaeologist colleagues Frank Braemer and Jean-Claude Echallier. What we found were not only the hydrological works I have just mentioned but fields cleared of stones and watered by channels from the dams and diversions in the wadi (Fig. 5). In some cases these ran for a kilometre or more. There are bands of these fields of approximately 3.25 ha, side by side and often with sophisticated arrangements for irrigation. Some fields are much larger. One, for instance, is approximately 500 m long by 150 m wide. While it is not possible to date them, we know that at least some of them were cleared a very long time ago, since the wind has by now blown away most of the soil, revealing the stones lying beneath it. Field and irrigation systems like this are relatively rare in the *ḥarrah*, though there is one 3 km north of Al-Namārah on the left bank of the Wadi al-Shām, and others, I believe, near Jawa in Jordan (Macdonald 2009b).

The Roman fort at Al-Namārah was later demolished to provide stones to build a mediaeval mausoleum, including stone doors and an unfinished Greek inscription used as a lintel. The inscription (Fig. 6) is the beginning of a dedication to either the Roman emperor Caracalla (r. AD 198-217) or his short-lived successor, from Syria, Elagabalus (r. AD 218-222). On the rocks below the fort and on the walls of the wadi

4 See for instance, *w wrd h-nmrt* "and he watered at al-Namārah" (C 523, 1894, 2803, *etc.*); *w qyẓ ʾl-h-nmrt* "and he spent the dry season near al-Namārah" (LP 330, Is.Mu 168), *etc.*

Figure 4. Al-Namārah 'island' in a basin formed by the Wadi al-Shām and a tributary wadi, looking west (photograph by M.C.A. Macdonald).

Figure 5. Al-Namārah: part of a series of fields irrigated by a long channel bringing water from dams in the Wadi al-Shām (photograph by M.C.A. Macdonald).

Figure 6. Al-Namārah: the unfinished Greek inscription re-used as a lintel in the mausoleum: ΕΠΙ Α(ὐτοκράτορος) Μ. Αυρ(ηλίου) ΑΝΤΩΝΕΙΝΟΥ ΚΟΚΩ ... (IGR III, no. 1255) (photograph by M.C.A. Macdonald).

Figure 7. Al-Namārah: an inscription of the Legio II Parthica.

basin, there are graffiti in Greek by soldiers (one of them a *dromedarius*) and Safaitic graffiti by nomads. There are also two Latin graffiti marking the presence of the III Legion Cyrenaica, which was transferred from Egypt to Boṣrā when the Roman Province of Arabia was established in AD 105/106, and remained there for almost two centuries. In addition, in 1860 a Latin graffito was found (Fig. 7), which has since disappeared, naming the II Legion *Parthica*. This legion was created by Septimius Severus and accompanied his son Caracalla to the eastern frontier in AD 216-217 to fight the Parthians. It was involved in the assassination of Caracalla and that of his successor Macrinus, and in the eventual accession of Elagabalus, a priest from Emesa (modern Homs). All this happened around Antioch and Apamea, a long way from Al-Namārah, and, as far as I know, we have no record of the legion coming as far south as this. On the other hand, since the inscription on the lintel mentions either Caracalla or Elagabalus, and it was only under Elagabalus that the legion took the titles included in this graffito, it seems likely that a detachment from the Third Cohort of this legion was stationed here, if only briefly.

Some of the Greek graffiti by Roman soldiers at Al-Namārah give the author's village and his lineage group, a reminder that settled peoples in ancient Syria and Arabia belonged to lineage groups just as much as nomads did. Thus, we have, for instance, Αζωοc Βορδου κώμηc Cοδαλαc φυλῆc Χαχαβηνων "Azōos son Bordos of the village of Sodala and the lineage group of Kakab" (Waddington 1870, no. 2265, re-read by Dussaud and Macler 1901, 96, no. 263). Whether this lineage group is the same as the *ʾl kkb* mentioned in several Safaitic inscriptions (*e.g.* C 65, WH 2828, KRS 4456, *etc.*), as some writers have assumed, is impossible to say (see the discussion of such insecure identifications in Macdonald 1993, 352-367).

The names of the soldiers at Al-Namārah are an interesting mix of Semitic names in Greek form such as *Gadd*, *Ṣubayḥ*, *Taym*, *etc.*, and Latin names such as *Adrianus*, *Lucianus*, *Flavius*, *etc.*, with only occasional etymologically Greek names, such as *Dōsitheos*. This might suggest that the recruits would have been local and the officers Romans. But the situation was almost certainly considerably more complex than this. For we have a number of Safaitic inscriptions which mention that their authors served in the Roman army, or possibly even the Nabataean or Herodian armies. These inscriptions are often dated to 'the year so-and-so was appointed commander' (Macdonald 2014, 155-156) and, given that the names of these commanders are all etymologically Semitic, they are likely to have been not Romans but locals, possibly from the settled areas but also possibly nomads. It rather depends on what these units, known as *ms¹rt*, were required to do. If the unit was to be deployed in the settled lands, an officer from there would be more likely to have the necessary know-how. On the other hand, if it was desert work it would have been sensible to have a nomad as commander (despite the problems of tribal loyalties outweighing loyalty to the command structure of the army). From one graffito (Ms 64), it seems that these units were raised from particular lineage groups: "By ʿqrb son of ʾbgr, a horseman in the unit of the ʿmrt lineage, in the year that Ġwṯ son of Rḍwt was appointed," *i.e.* the author uses the unit raised from his lineage group as his identification – though it should be said that this inscription is so far unique. We also have graffiti by nomads who had mutinied or deserted and were on the run from the Romans.

It may well have been this connection with the Roman army that prompted some nomads to give their children (or even to adopt themselves) Latin and Greek personal names, such *ʾqlds¹* (Claudius),[5] *tts¹* (Titus),[6] perhaps *wrqns¹* (Ὑρκανός),[7] *ʾftnyʾ* (Αφθόνιος),[8] *grgṣ* (Γεώργιος).[9] Of these the name *tts¹*, which is borne by ten different men, is by the far the most popular and there are even two inscriptions in which the authors give their lineage group as the *ʾl tts¹* (CEDS 322, SIAM 42). The latter presumably refers to a family or extended family, the patriarch of which was called *tts¹*, as explained above. Given Titus's military exploits in Syria and Judaea, it is perhaps not surprising that his name was sufficiently famous for some nomads to adopt it. We have indications – though alas none absolutely certain – that some of the nomads joined Agrippa II in helping Titus put down the First Jewish Revolt (Macdonald 2014, 152-53). This may be the reason why some of them adopted his name but, if so, it is interesting that while Philip the Tetrarch, Agrippa I and Agrippa II who ruled the settled lands closest to some of these nomads at this period, are mentioned in descriptions of events and in dating formulae, we have no evidence of their names being adopted by the nomads.[10]

The spoken language of the local recruits from the settled areas may well have been a form of Arabic. We know that the Arabic spoken in the settled regions of Syria and Arabia was an unwritten language and so those who wished to write had to do so in Aramaic or Greek. The soldiers at the military outpost at Al-Namārah probably chose Greek because it was the unofficial language used by the Roman army in the East (the official language, of course, being Latin). Ironically, the dialect of Arabic used by the nomads *was* a written language and had its own script (Safaitic).[11] So, we have two groups

5 In KRS 1507 and BS 1130 (apparently the same man, see no. 9), SSWS 177 and WH 837 (possibly the same man), and BS 1100.

6 BS 10, C 2308, 2309, HaNS 665, HaNSB 293, ISB 176, KRS 3160, 3161, 3162, 3244, MKJS 22, NSR 44, 47, THSaf 40. Of these NSR 44 and 47 refer to the same man, while HaNS 665 and MKHS 22 refer to the same man, and KRS 3160, 3161, and 3162 almost certainly refer to the same man since they are all carved on the same rock.

7 TIJ 208.

8 WH 2833a, see Müller 1980, 73.

9 KRS 1507, BS 1130, both apparently by the same man *Grgs¹* son of *ʾqlds¹*.

10 There are ten Safaitic graffiti mentioning *grfṣ* (Ἀγρίππας, almost certainly Agrippa II, since Agrippa I spent most of his time in Rome): Al-Namārah.H 91; HSNS 1, 2, 4-7; KRS 1023, 1039; SESP.U.8; The name *flfṣ* (Φίλιππος) occurs only once in a context where it probably refers to Philip the Tetrarch, Ms 44.

11 In the past (see most recently Macdonald 2010, 16-17), I have grouped the languages normally expressed in the Ancient North Arabian [ANA] scripts on the one hand and "Old Arabic" on the other as "two mutually comprehensible dialect bundles, most strikingly distinguished by the form of the definite article...". However, Ahmad Al-Jallad has convincingly shown that (a) the dialects expressed in the ANA scripts were very different from each other and do not form a dialect bundle; (b) that the criteria I had used to distinguish ANA dialects from Old Arabic, including the definite article, were not valid in historical linguistic terms; and (c) that the language used in the Safaitic inscriptions was in fact Old Arabic (see, for instance, Al-Jallad 2015, 11-17; also Kootstra 2016).

who almost certainly could have understood each other when speaking, but who *wrote* in different languages and scripts.

A wonderful example of this is the inscription recently published by Ahmad Al-Jallad and Ali Al-Manaser (2015), which is carved in Greek letters but, apart from the representation of the author's name and those of his father and grandfather, is in the Old Arabic language. The writer clearly knew the Greek alphabet and had an inkling of the formulae for expressing his name in Greek – so he knew that his name should end in -ος and his father's name should end in -ου (though he probably did not think of it as being in the genitive). However, he did not realise that his grandfather's name should take the definite article before it (του Βαναου), or that the way to express the *nisbah* in Greek would have been ὁ ἰδαμήνος ("the Idamite"), not αλ-ιδαμι which is of course just a transcription of the Arabic *nisbah*. Clearly the author of this text was an Arabic-speaker (either from the settled lands or a nomad). Clearly, also, he knew the Greek alphabet but not the language. However, what we cannot know is whether he also knew the Safaitic alphabet but was experimenting by trying to write Arabic in the Greek script for fun or to show off, or whether the Greek alphabet was the only script he knew and he just had to struggle to fit his spoken language to its constraints. Understandably, there are inconsistencies in his transliteration, such as the different spellings of the verb *atawa in lines 3-4 as αθαοα, and in lines 4-5 as αθαοευα, or the curious rendering of the verb *raʿiyau "they pastured" as ειραυ, if, as seems likely, this is the right interpretation (see the detailed analysis and discussion of this text in Al-Jallad and Al-Manaser 2015, 52-59).

Figure 8. A graffito in Old Arabic expressed in Greek letters from Wadi Salmā, north-eastern Jordan. From Al-Jallad and Al-Manaser 2015, Fig. 2 (photograph by Sabri Abbadi).

One could contrast the author of this inscription with a man from a lineage well-known in the Safaitic inscriptions, who wrote his name and lineage twice in good Greek, again in the middle of the *ḥarrah*, once in Syria (Fig. 9) and once in Jordan.[12] Here, he gives his name, and that of his father, his *nisbah* and his clan using the correct Greek formulae. In the process he gives us the vocalisation of the famous lineage group *ḍf* as Ḍayf and of the less famous *kn* as Kawn. Even more extraordinary is the fact that, immediately below this graffito, there is another, this time in Safaitic, apparently complimenting the author on his Greek (Macdonald *et al.* 1996, 484-485)! At the same spot in Syria, another man gives his name in both Greek and Safaitic (WR.D 1 and 2). Both cases demonstrate a greater degree of experience in the transliteration of Arabic in the Greek alphabet than the inscription discussed in the previous paragraph.[13]

Figure 9. Graffiti at the confluence of the Wadi Rushaydah and the Wadi al-Shām, southern Syria. The longer Greek text starts in the second line and reads Σααρος Χεσεμανου Σαιφηνος φυλης καυνηνων. Below it within a cartouche is the Safaitic inscription apparently complimenting Saaros on his Greek (photograph by M.C.A. Macdonald).

12 In Syria at the confluence of the Wadi Rushaydah and the Wadi al-Shām (WR.C 4) and in Jordan in the upper part of Ghadīr al-Ghuṣayn (Mg 1). See Macdonald *et al.* 1996, 480-484.

13 On this subject, see Al-Jallad and Al-Manaser 2015; 2016; Al-Jallad, in press.

Other sedentaries, in the southern parts of the Roman province of Syria and the north of the Nabataean kingdom, seem to have spoken Aramaic and some also *wrote* it in the Ḥawrān Aramaic and the Nabataean scripts respectively (Macdonald 2003, 44-46, 54-56, figs. 30-36). However, further south it seems that many Nabataeans spoke Arabic but used Aramaic as their written language. We find surprisingly few Nabataean graffiti in the *ḥarrah* but we do have the curious phenomenon of three people who carved their graffiti in Safaitic calling themselves "the Nabataean" (*h-nbṭy*).[14] Possibly, they found it easier to express themselves in their spoken language using a script which had letters for all the consonants needed to write it, or possibly they did so just for fun. However, whatever their motives, in each case the author had mastered the orthography of Safaitic and employed the normal 'Safaitic' definite article *h-* rather than *ʾl-* which all the evidence available to us suggests was the form used in the Arabic spoken by the Nabataeans.[15]

In the Badia Epigraphic Survey of 2015, we also found a nine-line Palmyrene graffito at a cairn covered with Safaitic graffiti.[16] Presumably, its author was a traveller or merchant passing through the region. Although this is only the second Palmyrene text to be found in the *ḥarrah*, it is not really surprising given that merchants from Palmyra left inscriptions in places as far away as north-eastern England and the island of Soqotra in the Indian Ocean.

Returning to Al-Namārah, for a moment, it is worth looking at one other example of the use of language. The site is most famous for the tomb of Marʾ al-Qays called "king of all the Arabs" or "king of all the region called ʿArab".[17] This tomb is a kilometre due east of the 'island' with the Roman fort and had on its lintel (now in the Louvre) a five-line inscription in the Arabic language transcribed in the Nabataean Aramaic script (Macdonald 2009b, 321-322). It has always been thought a very odd place to build an elaborate mausoleum over a leader whose family would later rule Al-Ḥīrah in southern Iraq, and act as a client state for Iran. Why bury this "king" in an empty piece of basalt desert within the Roman province of Arabia and a kilometre from a Roman fort? In the most recent reading of the inscription, I hope to have shown that the most difficult crux in the text reads "and he gave his sons [rule over] the (settled) peoples, and they were appointed agents for Persia and for Rome" (Macdonald 2015, 408).

Figure 10. Al-Namārah: gravestones in the cemetery (photograph by M.C.A. Macdonald).

That would mean that Marʾ al-Qays was working with both the superpowers of the time, not just for Iran.

It seems that the most reasonable explanation for the siting of his mausoleum, near a Roman fort in the middle of nowhere, would be that he was killed in battle near this place. The inscription is dated to AD 328, and we do not know whether the Roman fort was still in use at that time. If it was, he would have to have been fighting on the Roman side, but if it was not he could have been fighting for anybody or simply for himself. The idea that he died in battle in this place is strengthened, though alas not proved, by the presence of a cemetery on the plateau to the north of the island where, within a low enclosure wall, there are hundreds of single upright stones roughly 1.75 to 2 m apart which look very much as though they are marking the graves of a large number of people who died at the same time, for instance in a battle (Fig. 10). Of course, only excavation will show whether or not these are contemporary with Marʾ al-Qays's mausoleum.

As is well known, Marʾ al-Qays's epitaph was composed in the Arabic language expressed in the Nabataean

---

14 See CSNS 661, RMenv.C 1 and 2. This is a good example of how mistaken it is to define ethnicity by the script an individual uses. See Macdonald 1998, 186; 1993, 306-310, on the error of calling people 'Safaites'.

15 See for instance the examples in the Ain Avdat inscription, JSNab 17, and the Namārah inscription (Macdonald 2015, 399-409).

16 This is in preparation for publication by Maria Gorea.

17 For the latest reading and interpretation of the inscription, see Macdonald 2015, 405-409.

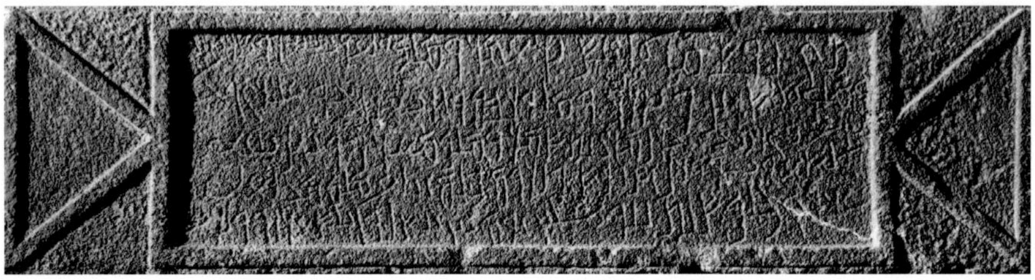

Figure 11. The epitaph of Marʾ al-Qays (the 'Namārah inscription') (by courtesy of the Musée du Louvre. Photograph by M.C.A. Macdonald).

Aramaic script. Here, the Arabic is written with great confidence and consistency and is clearly the work of a professional with long experience of using this script to express Arabic, despite the fact that there were only sixteen different letter shapes to represent Arabic's twenty-eight consonants (Macdonald 2008, 219-220). Indeed, the use of a special sign for *lām-alif*, which is a combination common in Arabic but rare in Aramaic, again suggests that Arabic had been written in ink using the Nabataean script for a considerable time before this inscription was carved.[18]

Conclusion: interaction

There was clearly a great deal of interaction between the nomads of the *ḥarrah* and the populations of the settled areas, not only by nomads serving in the armies of the settled states or by merchants passing through the *ḥarrah*, but from personal relations between members of lineage groups with settled and nomadic sections, just as there are today. The authors of several Safaitic graffiti record that they had spent time in the Ḥawrān either visiting it or pasturing their animals. For centuries, nomads have brought their camels to eat the stubble in the fields of the Ḥawrān after the harvest, with the camels manuring the fields in return. Although we have no explicit reference to it in Safaitic, this sensible arrangement may well have been practised at the time. Indeed, the presence of a nomad with his camels in the Ḥawrān at the end of the dry season is specifically mentioned in several graffiti in which the author says that he migrated with them from the Ḥawrān to the inner desert.[19]

Some authors of Safaitic graffiti went further afield and we have texts saying that they spent time in the lush area of Gilead in northern Jordan and several who said they were on their way to Palmyra, near which Safaitic inscriptions have also been found. It is therefore not surprising that the graffiti of the nomads are often well-informed about events in the wider world, such as the Nabataean and Herodian rulers, the death of Tiberius' adopted son, Germanicus, near Antioch, Persian invasions of the Roman Province of Syria, *etc.*

There are also some graffiti that mention the herding of cattle, which is surprising so far away from the settled areas. One, from north of the Ṣafā, apparently says that the cattle died of cold because there was a sudden cold spell in the early summer (*ṣayf*), though the copy is not entirely clear.[20] We find cattle mentioned at Al-Ḥifneh[21] some 20 km from Jabal al-ʿArab, but also at Al-Namārah[22] some 50 km, and at Al-ʿUdaysiyyah[23] and Al-ʿIsāwī[24] some 60 km away as the crow flies. Indeed, one says that he drove the cattle from the Ḥawrān to Al-Namārah.[25] Another, of unknown location, says that he "spent the season of the later rains with two flocks of sheep and some cattle".[26]

Finally, there is a curious puzzle. There are three references in the Safaitic graffiti to an *iskān*. One says that the author "found refuge for the night in the *is^{1}kān* of the *ʾl Dʿf*",[27] while the other two are by members of the *ʾl Dʿf* and date their texts by the year the Lihyanites (from north-west Arabia) made a sudden attack on the *is^{1}kan* (presumably of their *ʾl*).[28] It is not exactly clear what the term *iskān* means in this context. In Arabic, the word can mean the settlement of all or part of a tribe, with the implication that they settled under pressure from, or with the help of, an outside party (Lane 1863-1893, 1393b). The basic meaning of the root is to become stationary, to stop moving for good, which suggests that it is a permanent settlement rather than a nomadic encampment (*ibid.*, 1392c-1393b). If this is the meaning of *is^{1}kān* in Safaitic – and there is no guarantee that it is – it would suggest that part, at least, of the *ʾl dʿf* had settled, either voluntarily or under pressure. The fact that all three

18 This is because it is only by extensive writing in ink that the letter-forms and ligatures of a script develop, since the writer finds faster and easier ways of writing. By contrast, if a script is used exclusively for carving inscriptions on stone, there is no reason for it to develop and differences between forms of the script used in inscriptions are due to changes in fashion (as in the *musnad*, or monumental Ancient South Arabian script) or greater or lesser ability on the part of individual masons, or in the case of scripts used exclusively for carving graffiti on rocks, like Safaitic and Hismaic, the taste and ability of the individual author, or how well he/she had learnt the letter shapes.

19 HaNSB 197, 218, AbSWS 84.

20 C 860.

21 C 3791 (=LP 90), LP 155, LP 159.

22 C 3531.

23 C 974.

24 LP 968 (= Is.H.6).

25 Al-Namārah.H 75.

26 SIAM 34.

27 C 777, *w bt (ʾ)s^{1}kn ʾl dʿf* at Haǧar al-Helle 30 km west of Ruǧm al-Marʾah.

28 BRenv.B 1 and A 2 near Biʾr al-Ruṣayʿī.

inscriptions which mention the *is¹kān* are a long way out in the desert, does not necessarily mean that the *is¹kān* was there and it could have been – indeed almost certainly must have been – in the settled areas where agriculture could be practised, since pastoralists in the conditions of southern Syria and north-eastern Jordan need to be mobile in order to follow the pasture. One should also remember, of course, that when the authors of these graffiti refer to the *ʾl dḟ*, or any other *ʾl*, they are not referring to the whole group but to members of it (Macdonald 2009a, 333-334). So far, we do not have sufficient evidence to decide whether the *is¹kān* of the *ʾl dḟ* was a settlement of some members of an otherwise nomadic tribe, or of more or less the whole tribe? Was it voluntary or forced? And was it in Jabal al-ʿArab, or another part of the settled lands? Is the fact that it is only mentioned three times in more than 33,000 Safaitic inscriptions a sign that it was short-lived, perhaps utterly destroyed by the Lihyanite attack? Or was it not a settlement at all?

As we have seen, in the late Hellenistic and the Roman periods, the *ḥarrah* of southern Syria, north-eastern Jordan, and northern Saudi Arabia, was the scene of extensive interaction between nomads, sedentaries and state forces, with the use and mixing of several different languages and scripts. We are fortunate that so many of the individuals involved had the desire to leave their mark in informal texts, for these often tell us much more about their ways of life and private feelings than formal, public inscriptions would. Of course, even this gives us a very fragmentary view of their activities and relationships, but even fragments are better than nothing! When compared to our ignorance of such personal and linguistic relationships in the towns and villages of Nabataea, the Herodian kingdoms of the Ḥawrān, and the provinces of Syria and later Arabia, we can be very grateful for the desire and opportunity to carve graffiti in the desert!

Sigla

IGR	Inscriptions in Cagnat *et al.* 1906-1927.
Wadd	Inscriptions in Waddington 1870.

Inscriptions with other sigla can be found in the Online Corpus of the Inscriptions of Ancient North Arabia (OCIANA), http://krcfm.orient.ox.ac.uk/fmi/webd/ociana

References

Al-Jallad, A.M. 2015. *An outline of the grammar of the Safaitic inscriptions*. Leiden: Brill.

Al-Jallad, A.M. 2020. *The Damascus psalm fragment. Middle Arabic and the legacy of Old Ḥigāzī*. Chicago, IL: Oriental Institute, University of Chicago.

Al-Jallad, A.M. and Al-Manaser, A. 2015. New epigraphica from Jordan I: a pre-Islamic Arabic inscription in Greek letters and a Greek inscription from north-eastern Jordan. *Arabian Epigraphic Notes* 1, 51-70.

Al-Jallad, A.M. and Al-Manaser, A. 2016. New epigraphica from Jordan II: three Safaitic-Greek partial bilingual inscriptions. *Arabian Epigraphic Notes* 2, 55-66.

Cagnat, R., Toutain, J., Jouguet, P. and Lafaye, G. 1906-1927. *Inscriptiones Graecae ad res Romanas pertinentes* (4 vols). Paris: Leroux.

Dussaud, R. and Macler, F. 1901. *Voyage archéologique au Ṣafâ et dans le Djebel ed-Drûz*. Paris: Leroux.

Glubb, J.B. 1983. *The changing scenes of life. An autobiography*. London: Quartet.

Kootstra, F. 2016. The language of the Taymanitic inscriptions and its classification. *Arabian Epigraphic Notes* 2, 67-140.

Lane, E.W. 1863-1893. *An Arabic-English lexicon, derived from the best and most copious eastern sources*. London: Williams and Norgate.

Lemaire, A. and Macdonald, M.C.A. 2018. Some Ancient North Arabian notes. *Semitica* 60, 295-308.

Macdonald, M.C.A. 1993. Nomads and the Ḥawrān in the late Hellenistic and Roman periods: a reassessment of the epigraphic evidence. *Syria* 70, 303-403.

Macdonald, M.C.A. 1998. Some reflections on epigraphy and ethnicity in the Roman Near East. *Mediterranean Archaeology* 11, 177-190.

Macdonald, M.C.A. 2000. Reflections on the linguistic map of pre-Islamic Arabia. *Arabian Archaeology and Epigraphy* 11, 28-79.

Macdonald, M.C.A. 2003. Languages, scripts, and the uses of writing among the Nabataeans. In: Markoe, G. (ed.). *Petra rediscovered: lost city of the Nabataeans*. New York: Abrams, 36-56, 264-266, 274-282.

Macdonald, M.C.A. 2008. The phoenix of Phoinikēia: alphabetic reincarnation in Arabia. In: Baines, J., Bennet, J., and Houston, S. (eds.). *The disappearance of writing systems: perspectives on literacy and communication*. London: Equinox, 207-229.

Macdonald, M.C.A. 2009a. Arabians, Arabias, and the Greeks: contact and perceptions. In: Macdonald, M.C.A. *Literacy and identity in Pre-Islamic Arabia*. Variorum Collected Studies 906. Farnham: Ashgate, 1-33.

Macdonald, M.C.A. 2009b. Transformation and continuity at al-Namarā: camps, settlements, forts, and tombs. In: Bartl, K. and Moaz, A. (eds.). *Residences, castles, settlements. Transformation processes from Late Antiquity to Early Islam in Bilad al-Sham*. Rahden: Verlag Marie Leidorf, 317-332.

Macdonald, M.C.A. 2010. Ancient Arabia and the written word. In: Macdonald, M.C.A. (ed.). *The development of Arabic as a written language*. Oxford: Archaeopress, 5-28.

Macdonald, M.C.A. 2014. 'Romans go home'? Rome and other 'outsiders' as viewed from the Syro-Arabian desert. In: Dijkstra, H.F. and Fisher, G. (eds.). *Inside and out: interactions between Rome and the peoples on the Arabian and Egyptian frontiers in Late Antiquity.* Leuven: Peeters, 145-163.

Macdonald, M.C.A. 2015. The emergence of Arabic as a written language. In: Fisher, G. (ed.). *Arabs and empires before Islam.* Oxford: Oxford University Press, 395-417.

Macdonald, M.C.A., Al-Mu'azzin, L. and Nehmé, L. 1996. Les inscriptions safaïtiques de Syrie, cent quarante ans après leur découverte. *Comptes rendus des séances de l'Académie des Inscriptions et Belles-Lettres* 140, 435-494.

Mowry, L. 1953. A Greek inscription at Jathum in Transjordan. *Bulletin of the American Schools of Oriental Research* 132, 34-41.

Müller, W.W. 1980. Some remarks on the Safaitic inscriptions. *Proceedings of the Seminar for Arabian Studies* 10, 67-74.

Schwabe, M. 1954. Note on the Jathum inscription. *Bulletin of the American Schools of Oriental Research* 135, 38.

Waddington, W.H. 1870. *Inscriptions grecques et latines de la Syrie recueillies et expliquées.* Paris: Firmin Didot.

Gaius the Roman and the Kawnites: inscriptional evidence for Roman auxiliary units raised from the nomads of the *ḥarrah*

Ahmad Al-Jallad, Zeyad Al-Salameen, Yunus Shdeifat and Rafe Harahsheh

Abstract

The paper provides epigraphic evidence for mixed military troops in the *ḥarrah*, consisting of both Romans and local nomads. The Romans may have deployed these units against incursions by nomadic groups from north Arabia, or against the Nabataeans.

Keywords: Jordan, ḥarrah, Romans, nomads, Safaitic epigraphy

Introduction

The relationship between the Roman empire and the nomads to the east of the Ḥawrān was recently the subject of a fascinating paper by Michael Macdonald (2014). In this very closely argued essay, Macdonald concludes, based on the epigraphic evidence from settled areas of the Ḥawrān and ingenious solutions to the enigmatic terms *ngy*, *hdy*, and *s¹rt* in the Safaitic inscriptions, that the Romans raised auxiliary military units from the nomadic tribes of the *ḥarrah*. While all of the pieces of the puzzle fit together, proof that the nomads who produced the Safaitic inscriptions belonged to such units was still lacking. No inscriptions discovered so far stated in unambiguous terms: so-and-so *s¹rt* 'served in the military' for *rm* 'Rome'.

In 2017, the Wādī El-Khḍerī project, led by Z. Al-Salameen, Y. Shdeifat and R. Ḥaraḥsheh, discovered a remarkable set of inscriptions in north-eastern Jordan. These texts – four Safaitic inscriptions and a Greek text enclosed in cartouches – were carved on a protruding rock face located at site K38. While Safaitic-Greek texts are not unknown in the *ḥarrah* (Al-Jallad and Al-Manaser 2016), what makes this collection unique is that it contains the first unambiguous Safaitic text composed by or for a Roman soldier. As such, the collection constitutes our first direct documentation of mixed military units in the *ḥarrah*, consisting of both Romans and local nomads.[1] This paper will edit these new texts and discuss their historical context in light of Macdonald's hypothesis.

1 There are examples of Greek graffiti composed by Roman soldiers in the desert (Mowry 1953), but none of these indicate any sort of cooperation with local nomads.

In: Peter M. M. G. Akkermans (ed.) 2020: *Landscapes of Survival - The Archaeology and Epigraphy of Jordan's North-Eastern Desert and Beyond*, Sidestone Press (Leiden), pp. 355-362.

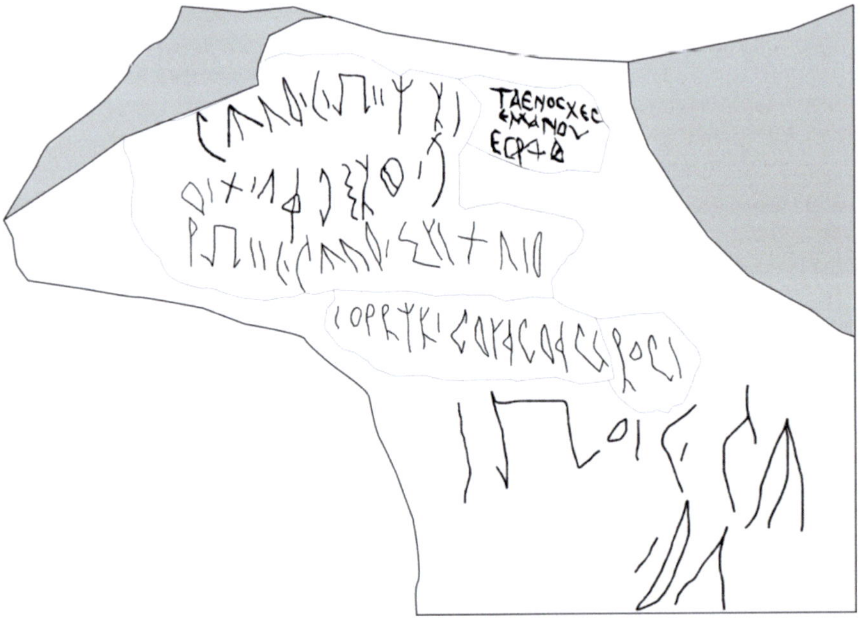

Figure 1a-b. Basalt rock with set of inscriptions from site K38 in Wādī El-Khḍerī, north-east Jordan (photograph by Zeyad al-Salameen. Tracing: Ahmad Al-Jallad).

The texts

Safaitic 1

... bn kḥs^{1}mn bn ẓnn ḏ ʾl kn w ʿs^{2}rq s^{1}nt ngy ẓnn bn kḥsmn f h lt s^{1}lm

'.. son of Khs^{1}mn son of Ẓnn of the lineage of Kawn and he set off to the inner desert the year Ẓnn son of Khs^{1}mn was announced commander so O Lt may he be secure'

The text contains no new vocabulary or personal names. The name before *kḥs^{1}mn* is weathered beyond recognition but the sequence of *kḥs^{1}mn* son of *ẓnn* is attested in twelve inscriptions. An individual named *s^{2}ʿr* son of *kḥs^{1}mn* produced a Greek inscription in which he claimed affiliation with both the *ḍf* (Ḍayf) and *kn* (Kawn) lineages.[2] The same individual may have produced the inscription RMSK 1, but the name *ḳḥs^{1}mn* is written twice in that text, perhaps a result of dittography.[3] Kḥs^{1}mn son of Ẓnn also appears as the father and grandfather of an individual named *ẓʿn*, who may have been the brother of *s^{2}ʿr*.[4]

Macdonald (2014) recently discussed the meaning of the verb *ngy* and convincingly interpreted it as 'to announce, declare', especially when used with the noun *hdy*[5] – the entire phrase, he explains, refers to the appointment of an individual as commander of a raiding party or military troop.[6] The present inscription is therefore dated to the year Ẓnn son of Kḥs^{1}mn was announced (commander).

The translation of *ʾs^{2}rq* requires two remarks. First, its translation as 'to set off to' rather than a simple perfective 'he has migrated' is supported by the fact that the present text was not discovered in the inner desert itself. While it is possible that the author wrote this text upon his return, the final prayer makes more sense as a request for security on the journey to the *madbar*. Second, while the verb is unquestionably used in the context of the seasonal migrations of the nomads from the *ḥarrah* to the *ḥamād* (*madbar*) (Macdonald 1992), it also occurs frequently in the context of military movement was well. Following Al-Jallad and Jaworska (2019), it is possible to render this verb in the context of non-migratory activities as a simple verb of movement to or towards the inner desert.

Safaitic 2

l gyṣ ḏ ʾl rm h-dr m-dr{b}

'By Gaius of the people of Rome, at this place, from [the] {road}'

The name following the *lām auctōris* is attested for the first time in Safaitic. It does not seem to have a Semitic etymology; rather, it appears to spell the Latin name 'Gaius'. The representation of Latin [g] with Safaitic *g* is expected, and Greek and Latin [s] are represented with *s^1* and *ṣ* freely (Al-Jallad 2015, 41-42). This identification is further supported by his affiliation with the *ʾl rm*, that is, the Romans.[7] The terse expression that follows is so far unique. The translation *h-dr* as 'this place, region', rather than 'camp site', is discussed in Al-Jallad (2015, 311), although the present context permits both translations. The phrase *m-drb* is attested for the first time. The final three letters probably comprise the word *darb* 'path, trail, road' – a hapax legomenon in Safaitic.[8] There is damage on the rock near the final *b*, causing it to appear as if there is an arm protruding from one end, resembling an *ʾ*. However, such an *ʾ* would appear rather different from the previous ones carved.

While admittedly awkward in its phrasing, the text is best interpreted as describing Gaius' momentary halting in this place, having come from a road nearby, perhaps on patrol. The absence of the definite article is

---

2 This inscription is MISS.I 1, discussed in detail in Macdonald *et al.* 1996.

3 RMSK 1: *l s^{2}ʿr bn kḥs^{1}mn bn ḳḥs^{1}mn bn ẓnn bn s^{2}ʿr bn gnʾl ḏ- ʾl kn w s^{1}rt s^{1}nt ngy ʿmd bn ʾs^1 hdy w s^{1}nt drg-h ṣmkrn h- mḏ f h gdḍf s^{1}lm w ġnmt l-ḏ dʿy h-s^{1}fr w nqʿt l-ḏ mḥy h- s^{1}fr* 'By S^{2}ʿr son of Kḥs^{1}mn son of Kḥs^{1}mn son of Ẓnn son of S^{2}ʿr son of Gnʾl of the lineage of Kawn and he served in the military the year ʿmd son of ʾs^1 was announced commander and the year Ṣmkrn the Persian made him (?) surrender so, O Gdḍf, may he be secure and may he who reads this writing have spoils but may he who erases the writing be thrown out (of the grave)'.

4 SIJ 88: *l ẓʿn bn kḥs^{1}mn bn ẓnn ḏ-ʾl ḍf w hdy s^{1}nt ngy qṣr h-mḏ f h lt s^{1}lm* 'By Ẓʿn son of Kḥs^{1}mn son of Ẓnn of the lineage of Ḍf and he served as commander the year Caesar put the Persians to flight so, O Lt, may he be secure'.

5 Macdonald (2014, 156) compares this word to Palmyrene *hdyʾ* 'commander'. The absence of the article is not explained; perhaps it was loaned as definite from Aramaic, pronounced *haddāyā* or was indefinite, 'a commander'.

6 In other contexts, *ngy* can be translated as 'to escape', as it has been traditionally (Al-Jallad and Jaworska 2019). There are cases, however, in which both translations do not suffice, for example, SIJ 88 which is dated to the year *ngy qṣr h-mḏ* 'the year Caesar *ngy* the Persians'. We follow Clark's suggestion (1979, 100, no. 86) that *ngy* here should be interpreted as a D-stem /naggaya/, meaning 'to eject', that is, 'to make retreat'; 'to put to flight'.

7 There is only one other Safaitic inscription where the author expresses affiliation with the Romans, C 319, which was read by the edition as *l s^{1}wr bn ʿly bn s^{1}ʿd ḏ-ʾl rm w {.}{.}ẓr f h lt s^{1}lm*. Unlike the present inscription, the author has three Arabic names, but it should be said that these names are common in the onomasticon of settled peoples as well and each is found in Greek transcription. This text is only known from a hand copy, and given its poor quality, we must also admit the possibility that the author was claiming affiliation with the social group *hrm* (LP 435).The verb after *w* should probably be reconstructed as *nẓr* 'to keep watch'. If *ʾl rm* 'of the people of Rome' is correct, then, in light of the present inscription, we can tentatively suggest that the author of C 319 was written by a for a settled person, who was perhaps keeping watch as part of a Roman military unit.

8 Compare to Classical Arabic *darbun* and Aramaic *darbā*, with the same meaning.

difficult to explain. It is possible that the author used the *al*-article, with assimilation of the coda to the [d] of *darb* and the loss off the onset, producing something similar to modern Arabic *mid-darb* 'from the road'.

Safaitic 3

l rʿṣ
'By Rʿṣ'

Now, the following letters are enclosed in their own cartouche, suggesting that they comprise a separate text. But if this is the case, then it should be read from right to left in contrast with the others written left to right. It contains only a single name *rʿṣ*, which is attested for the first time here. The root is known in Arabic, *rʿṣ* meaning 'to shake, writhe' (Lisān 1671b).

Given these difficulties, it is possible that the cartouche is secondary and these four letters in fact complete Safaitic 2. If so, the word would read as *ṣʿrl*, and should probably be understood as a toponym. The translation of the entire phrase would therefore be: 'By Gaius the Roman, in this place, from the road to Ṣrʿl'. No toponym is known to us by this name, but that is not unexpected as many toponyms, unknown from other sources, are attested in the Safaitic corpus.

It is unlikely that Gaius wrote his own Safaitic inscription. The hand is virtually identical to the other Safaitic texts on the rock, suggesting that all were written by the same man, perhaps Ẓʿn son of Kḥs¹mn (see inscription 4).

Safaitic 4

l ẓʿn bn kḥs¹mn
'By Ẓʿn son of Kḥs¹mn'

Since the first name of the author of inscription 1 is missing, this text could contain the name of the same man. This name is attested in the following inscriptions: AbMNS 1; SIJ 93; SIJ 88; SIJ 46; SIJ 24, although the final one seems to refer to a different man. This text and Greek 1 (below) make up a partial bilingual inscription.

Greek 1

Ταενος Χεσεμανου
Εσραθ
"ṭāʿen (son) of Keḥsemān
serves in the military'

This Greek inscription transcribes the name found in inscription 4, indicating the following pronunciation for Safaitic *ẓʿn* /ṭāʿen/. The name *kḥs¹mn* has appeared previously in transcription, in MISS.I 1. It is spelled identically here, indicating that its vocalisation is [keḥsemān]. The name has no satisfying etymology; it is not attested in the Islamic-period onomasticon, nor is it found in the onomasticon of other Semitic languages. Macdonald suggests that it is a prepositional name comprising *ka-* 'like' and the name Ḥašmōn, the eponymous ancestor of the Hasmonean dynasty. The vocalisation, however, suggests that the name Ḥašmōn was Arabicised, with a redistribution of vowels and a reversal of the Canaanite shift.[9] A connection with Aramaic *kḥsn* 'forcefully' (DNWSI, 393) or *kḥšw* 'emaciation' (Jastrow 1903, 629) may also be considered, but in both cases the *m* of the Safaitic requires an explanation.

The final word of the inscription is not a personal name or a Greek word. The letters ε σ ρ θ are clear, but the third letter from the left appears as a cross that intersects with a large crack in the stone. Perhaps the author decided to utilize the preexisting crack to form an Alpha. We suggest that this reflects an attempt to transcribe the prefix conjugation of the Safaitic *s¹rt* 'to serve in a troop'. The spelling εσραθ likely represents [jesrat], with an /i/ preformative vowel (Al-Jallad and Jaworska 2019; Macdonald 2014, 159). The same phenomenon occurs in the A1 (Al-Jallad and Al-Manaser 2015), where the prefix conjugation of *rʿy* is rendered ειραυ [jirʕaw].[10] The absence of an initial Iota in this word is unexpected but not necessarily problematic. Greek lacked an established way of representing word-initial [j], and the proximity of the glide to the high vowel may have caused a degree of confusion. The same thing seems to be at play in the spelling of [jirʕaw] in A1 – the writer rendered this word in Greek as ειραυ rather than the expected ιεραυ. It is also possible to take this verb as a first person, suggesting that the author switched subjects. In this case, it would render [ʔesrat] 'I serve in the military'.[11]

Since Ẓʿn seems to have written Gaius' text, one may wonder whether Gaius returned the favour, writing the

9 Ka-ḥašmōn > ka-ḥašmān > ka-ḥšamān, and then the shifts of /a/ to /e/ (Macdonald *et al.* 1996, 484). The corpus of Safaitic-Greek texts is too small to make definitive statements about the phonologies of the dialects of the *harrah*, but it should be mentioned that the preposition k is written χα in the Greek epigraphy of the Ḥawrān, *e.g.* χααμος = kʿmh /ka-ʿammoh/ 'like his ancestor' (Al-Jallad 2017, 178).

10 As with A1, this may suggest that Barth-Ginsberg's law – where the vowel of the preformative prefix is /i/ if the verbal theme vowel is /a/ but /a/ otherwise – was active.

11 We should also point out that the mood of this verb is unclear. If we take it as a reflex of the Proto-Arabic indicative, -Arabic indicative, *yafʿalu*, with the expected loss of the final /u/ vowel, then the translation as a present tense is possible. If, however, it reflects the short prefix conjugation, the Classical Arabic jussive *yafʿal*, then it may be possible to translate it as a preterite, 'he served in the military'. We prefer the former interpretation as the presence of Gyṣ suggests that both men were still deployed on their mission.

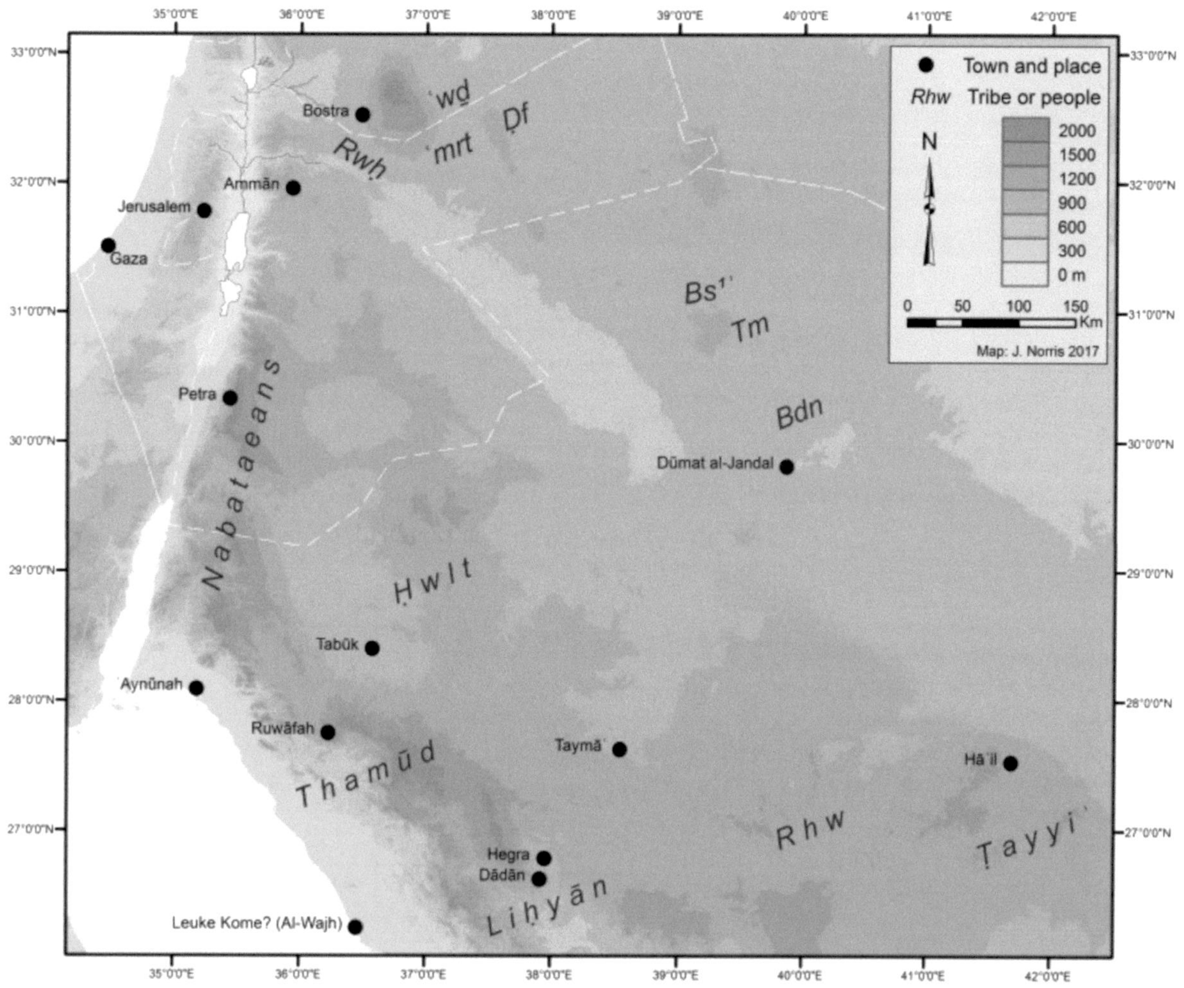

Figure 2. Distribution of lineage groups and peoples in north Arabia (map: Norris and Al-Manaser 2018).

present Greek text for his colleague. We find this to be unlikely as it is hard to understand why a Greek speaker would write an Arabic word in transcription. Rather, Ẓ‘n son of Kḥs^{1}mn may have been the only literate person, and only partially so in Greek, among the men mentioned in these texts; or perhaps he was the only one who cared to carve inscriptions.

Historical context

How do these texts connect to each other, and what circumstances could have led to their production? The Greek inscription suggests that Ẓā‘en was deployed in a military unit; thus, it is reasonable to posit that *‘$s^{2}rq$* in text 1 refers to military movement towards or into the desert. It is further possible that Ẓā‘en was serving in a unit commanded by Ẓnn son of Kḥs^{1}mn, a possible kinsman of Ẓā‘en. The presence of the Roman Gaius suggests that this unit was allied with Rome and, indeed, possibly raised by the Romans. Gaius could have been a Roman soldier stationed with this unit or perhaps someone charged with liaison with the nomads, an interpretation of the title στρατηγὸς νομάδων that Macdonald (1993) has argued for convincingly.[12]

If we are correct and these inscriptions were written by members of an auxiliary Roman military unit in the

12 In remarks on a draft of this paper, Mr. Bloomfield (Oxford) remarked that it would be unlikely that a Roman officer would refer to himself with a *praenomen* rather than his family *nomen*. We thank him for this valuable observation, and while a valid point, we must remember that the structure of Safaitic inscriptions begins with the *praenomen* and not the family name. Moreover, it is not likely that the Roman wrote this text himself. Mr. Bloomfield also suggests the possibility that Gaius was a nomad who took a Roman name and the remark ‘of the people of Rome’ was a humorous remark. It is true that some nomads had Greek and Roman names – and there are examples such as Tts^{1} Titus, Grgs^{1}, Mrṭs^{1} – but it is unclear if these were taken later in life or given to them by their parents for one reason or another. In the present case, it is the combination of both the affiliation with ‘the people of Rome’ and the military context that makes such a coincidence less likely.

desert, the next question is: against whom were they deployed? If Ẓāʿen was indeed the brother of S²aʿār, author of MISS.I 1, then it seems likely that this unit was raised from the large confederacy of Ḍayf. Norris and Al-Manaser (2018) have reconstructed the territory of this lineage group based on the concentration of their inscriptions; see Fig. 2.

The Safaitic inscriptions mention several conflicts between the Ḍayfites and other groups in the region; some of these conflicts may have been connected to their alliance with Rome. The Romans could have raised such units to defend against incursions by nomadic groups from north Arabia. This hypothesis is supported by the Safaitic inscription Khunp 2, which states:

l ʿwḏ bn gmr bn qnʾl w rʿy h-mʿzy s¹nt ḥrb ʾl ḍf ʾl ḥṣd b-bṣry f h lt s¹lm

'By ʿwḏ son of Gmr son of Qnʾl and he pastured the goats the year the lineage of Ḍf and the lineage of Ḥṣd went to war near Bostrā so, O Allāt, may he be secure'

No Safaitic inscriptions are known to have been produced by individuals from the lineage of Ḥṣd but Winnett and Reed (1970) have published a Hismaic inscription from Sakākā in northern Saudi Arabia by a man from this group:

WTI 11

l yʿly bn rs² ḏ ʾl ḥṣd w wgm ʾl-hnʾ w ʿ{l}-gdy

'By Yʿly son of Rs² of the lineage of Ḥṣd and he grieved for Hnʾ and {for} Gdy'

While the Khunp 2 does not explicitly mention the Romans, the battle between the Ḍayfites and the Ḥṣd so close to Bostrā could suggest that the latter were a threat to Roman interests; the Ḍayfites then acted as the first line of defense against incursions from north Arabia into this region. If this reconstruction of events is correct, then these texts would have been produced after AD 106, following the Roman annexation of Nabataea.

Ḍayfite military units may have been deployed against the Nabataeans, either before the annexation of the kingdom or against Nabataean rebels after the fall of Petra. The inscription RWQ 334 seems to record the defeat of the Ḍayfites at the hands of the Nabataeans and concludes with a prayer for the deliverance of the province:

RWQ 334 (portion)

wgd ṯr ʾl ḍf glyn m-ḥrb nbṭ (flṭ)t l-mdnt

'he found the traces of the lineage of Ḍayf, who were exiled on account of the Nabataean war; may the province be delivered'

It is impossible to determine the date of this event. The Ḍayfites could have served the Romans before the annexation of the Nabataea. Units raised from this tribe would have acted as a buffer between the Nabataeans and the Roman empire, and their defeat may have represented a direct threat to the Roman province of Syria, perhaps the province signified by *mdnt* in RWQ 334. On the other hand, there seem to have been rebellions against the Romans following the annexation of the Nabataean kingdom. The inscription WH 2815 is dated as follows: *s¹nt mrdt nbṭ ʿl-ʾl {r}m* 'the year the Nabataeans rebelled against the Romans'. Nomadic auxiliary military units may have been deployed to put down such rebellions. In the case of RWQ 334, however, it would seem that the Ḍayfite unit was defeated, and as a result the author felt that the province, whether Syria or Arabia, was threatened.

Conclusions

A precise understanding of the chronology and circumstances under which these texts were produced remains impossible; there are several interpretive possibilities available and context does not allow us to arbitrate between them. Nevertheless, this fascinating group of inscriptions provides our first concrete evidence for the activities of Roman auxiliary military units raised from the nomadic tribes of the *ḥarrah*, thereby confirming Macdonald's (2014) hypothesis.

Sigla

AbMNS	Safaitic inscriptions in Abbadi 2012
C	Safaitic inscriptions in Ryckmans 1950
DNWSI	Hoftijzer and Jongeling 2015
MISS.I	Safaitic inscriptions in Macdonald *et al.* 1996
RMSK	Safaitic inscriptions in Al-Rousan 2005b
RWQ	Safaitic inscriptions in Al-Rousan 2005a
SIJ	Safaitic inscriptions in Winnett 1957
WH	Safaitic inscriptions in Winnett and Harding 1978
WTI	Safaitic inscriptions in Winnett and Reed 1970

References

Abbadi, S. 2012. Al-māʾ fī ʾl-nuqūš al-ʿarabiyyah al-šamāliyyah al-qadīmah (al-ṣafawiyyah) "dirāsah taḥlīliyyah li-nuqūš ṣafawiyyah ǧadīdah. *Jordan Journal for History and Archaeology* 6, 103-23.

Al-Jallad, A. 2015. *An outline of the grammar of the Safaitic inscriptions*. Leiden: Brill.

Al-Jallad, A. 2017. Graeco-Arabica I: the southern Levant. In: Al-Jallad, A. (ed.). *Arabic in context: celebrating 400 years of Arabic at Leiden University*. Leiden: Brill, 99-186.

Al-Jallad, A. and Al-Manaser, A. 2015. New epigraphica from Jordan I: a pre-Islamic Arabic inscription in Greek letters and a Greek inscription from north-eastern Jordan. *Arabian Epigraphic Notes* 1, 51-70.

Al-Jallad, A. and Al-Manaser, A. 2016. New epigraphica from Jordan II: three Safaitic-Greek partial bilingual inscriptions. *Arabian Epigraphic Notes* 2, 55-66.

Al-Jallad, A. and Jaworska, K. 2019. *A dictionary and grammar of the Safaitic inscriptions*. Leiden: Brill.

Al-Rousan, M.. 2005a. *Nuqūš Ṣafāwiyyah Min Wādī Qaṣṣāb Bi-l-ʾUrdunn*. Riyadh: King Saud University (PhD thesis).

Al-Rousan, M. 2005b. Naqš Šuʿayyit Bin Kaḥsimān. *Adumatu* 11, 45-52.

Clark, V.A. 1979. *A study of new Safaitic inscriptions from Jordan*. Ann Arbor, MI: University Microfilms International.

Hoftijzer, J. and Jongeling, K. 2015. *Dictionary of the north-west Semitic inscriptions*. Leiden: Brill.

Jastrow, M. 1903. *Dictionary of the Targumim, Talmud Babli, Yerushalmi, and Midrashic literature*. London and New York: Luzac and Co.

Macdonald, M.C.A. 1992. The seasons and transhumance in the Safaitic inscriptions. *Journal of the Royal Asiatic Society of Great Britain and Ireland* 2, 1-11.

Macdonald, M.C.A. 1993. Nomads and the Ḥawrān in the late Hellenistic and Roman periods: a reassessment of the epigraphic evidence. *Syria* 70, 303-403.

Macdonald, M.C.A. 2014. Romans go home? Rome and other 'outsiders' as viewed from the Syro-Arabian desert. In: Dijkstra, J.H.F. and Fisher, G. (eds.). *Inside and out. Interactions between Rome and the peoples on the Arabian and Egyptian frontiers in Late Antiquity*. Leuven: Peeters, 145-163.

Macdonald, M.C.A, Al-Muʾazzin, M. and Nehmé, L. 1996. Les inscriptions safaïtic de Syrie, cent quarante ans après leur découverte. *Comptes-rendus des Séances de l'Académie des Inscriptions et Belles-Lettres* 140, 435-494.

Mowry, L. 1953. A Greek inscription at Jathum in Transjordan. *Bulletin of the American Society of Oriental Research* 132, 34-41.

Norris, J. and Al-Manaser, A. 2018. The Nabataeans against the Ḥwlt – once again. An edition of new Safaitic inscriptions from the Jordanian harrah desert. *Arabian Epigraphic Notes* 4, 1-24.

Ryckmans, G. 1950. *Corpus Inscriptionum Semiticarum: Pars Quinta, Inscriptiones Saracenicae Continens: Tomus I, Fasciculus I, Inscriptiones Safaiticae*. Paris: E Reipublicae Typographeo.

Winnett, F.V. 1957. *Safaitic inscriptions from Jordan*. Toronto: University of Toronto Press.

Winnett, F.V. and Lankester Harding, G. 1978. *Inscriptions from fifty Safaitic cairns*. Toronto: University of Toronto Press.

Winnett, F.V. and Reed, W.L. 1970. *Ancient records from north Arabia*. Toronto: University of Toronto Press.

Remarks on some recently published inscriptions from the *ḥarrah* referring to the Nabataeans and the 'revolt of Damasī'

Jérôme Norris

Abstract

A recently published group of new Safaitic and Nabataean inscriptions from north-eastern Jordan contains interesting references to the Nabataeans and the history of their kingdom. Some texts date to the kings 'Aretas' and 'Rabbel' and others refer to a historical figure bearing the rare name of *Dmṣy*. Four texts provide new evidence on the meaning of the Safaitic word *s¹lṭn*, showing that this could equally mean 'governor' besides the traditional translation of 'authorities'. The present contribution re-examines one Nabataean and eight Safaitic inscriptions from this collection on which a number of alternative interpretations or translation improvements seem possible. This gives us the opportunity to produce further epigraphic and historical comments on these important documents. Special attention is given to the mention of a probably Nabataean governor of 'Gilead' and to the so-called 'revolt of Damaṣī', a mysterious event about which Winnett has formulated a number of hypotheses in 1973, which, however, no longer seem tenable in light of a close review of the available evidence.

Keywords: Ancient North Arabian, Safaitic, Nabataean, revolt of Damaṣī (Dmṣy), ḥarrah, Gilead/Galaad, Jordan

Introduction

Zeyad Al-Salameen, Younis Al-Shdaifat and Rafe Harahsheh have recently published a selection of very interesting Safaitic and Nabataean inscriptions (NEH 1-16) recorded during a survey they carried out in 2017 in the *ḥarrah* (basalt desert) of north-eastern Jordan (Al-Salameen *et al.* 2018).[1] The texts come from eleven sites located in the eastern part of the lava field and at the limit of the *hamād* steppe, immediately west of Wādī Miqāṭ and Qaṣr Burquʿ and south of the Tell al-Mismā hill (Fig. 1).

The Nabataean texts (NEH 15A-B, 16) are no more than short signatures accompanied by words of blessing, but they remain significant since very few Nabataean inscriptions have been found so far in this region (see Al-Manaser and Norris 2019). The Safaitic ones, on the other hand, present some much more developed content with important information relating to the history of the Nabataean realm. Firstly, these include new references to the Nabataeans

1 Several of these inscriptions were also recorded by the members of the OCIANA Badia Survey project (2015-2018), directed by Michael C.A. Macdonald.

In: Peter M. M. G. Akkermans (ed.) 2020: *Landscapes of Survival - The Archaeology and Epigraphy of Jordan's North-Eastern Desert and Beyond*, Sidestone Press (Leiden), pp. 363-390.

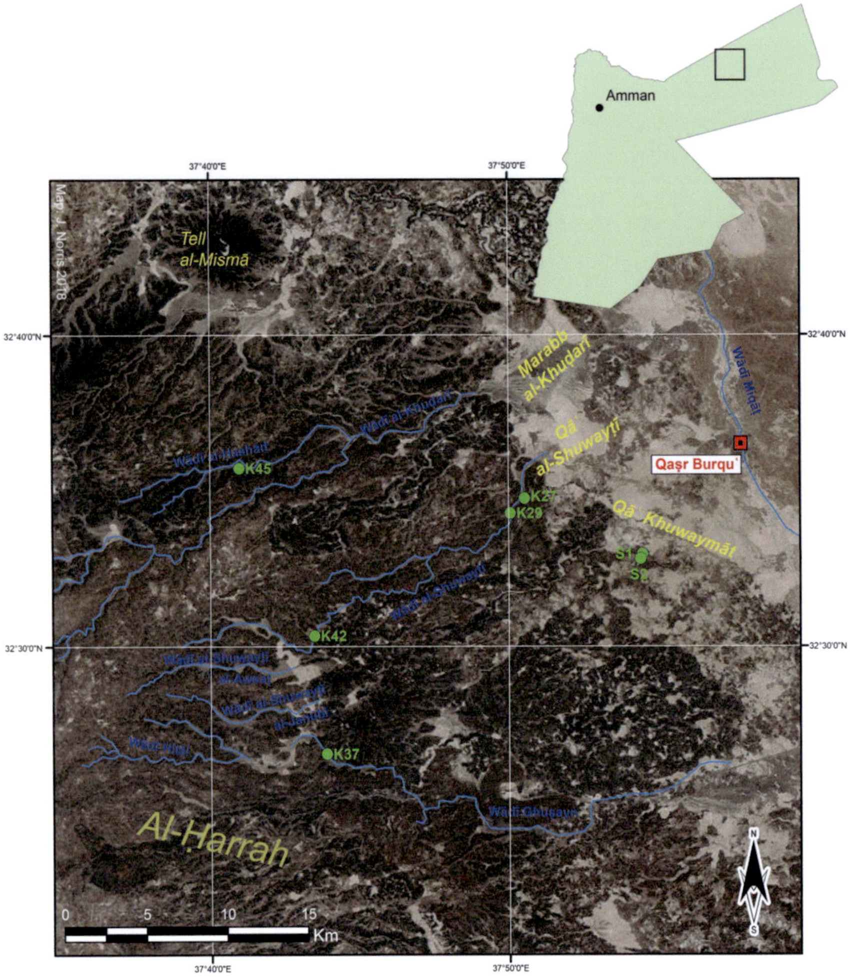

Figure 1. Map of the eastern part of the Jordanian *harrah* showing the sites from which the inscriptions under study come (J. Norris 2018).

themselves and to their rulers. Besides two texts referring to an intriguing and previously unknown war conducted by one 'Aretas' (NEH 1-2), three documents are firmly dated to "the year Aretas was made king" (*s^1nt mlk ḥrṯt*; NEH 3), "the year Aretas died" (*s^1nt mt ḥrṯt*; NEH 4) and "the year Rabbel was made king" (*s^1nt mlk rbʾl*; NEH 9). These correspond respectively to 9/8 BC, AD 40 and AD 70/71 if, as seems likely, the kings in question are Aretas IV and Rabbel II.[2] While three other texts mention the Nabataeans either as enemies (NEH 12) or, on the contrary, as those for whom the authors perform military services (NEH 10-11), a fourth appears of particular interest as its carver proclaims himself as being from "the Nabataean people" (*l ʿdy ḏ-ʾl nbṭ*; NEH 13). Although Safaitic inscriptions written by persons identifying themselves as 'Nabataean' have long been recognised, this appears to be the first time that an author does so in using the common Ancient North Arabian phrase *ḏ-ʾl* , while this is systematically expressed with the *nisbah h-nbṭy* "so and so, the Nabataean" in previously known documents (see Macdonald *et al.* 1996, 444-449).

Secondly, three inscriptions provide new evidence on the meaning of the word *s^1lṭn* in Safaitic. To date, this was commonly taken as an abstract noun meaning "authorities", presumably referring to that of the neighbouring settled powers of the Nabataeans and the Romans.[3] The phrase *ḥrb ḥrṯt s^1lṭn-h* occurring in NEH 1 and 2 and the texts NEH 7 and 8 in which the word also refers to a person, reveal that this can actually and equally be a title applied to men, perfectly matching with Classical Arabic *sulṭān* "ruler, governor" (Lane 1405c). As the editors brilliantly suggest on the basis of the Nabataean epitaph CIS II 196 from Madaba that contains the construction *šlṭwn-hm* "their rule" (Al-Salameen *et al.* 2018, 63), this is likely to be nothing more than the Safaitic equivalent of

2 However, the editors cautiously emphasise that one cannot exclude the possibility that this *Ḥrṯt* could be Aretas III (85-62 BC), who was equally an important figure in the Levant during the first century BC, having ruled Damascus from 84 to 72 BC (Al-Salameen *et al.* 2018, 66). It is also important to note that the person mentioned in NEH 4 is not explicitly described as a 'king'.

3 Ryckmans 1942, 135; Winnett 1973, 54; Graf 1989, 363, 376-377; Scharrer 2010, 275; Al-Jallad 2015, 342.

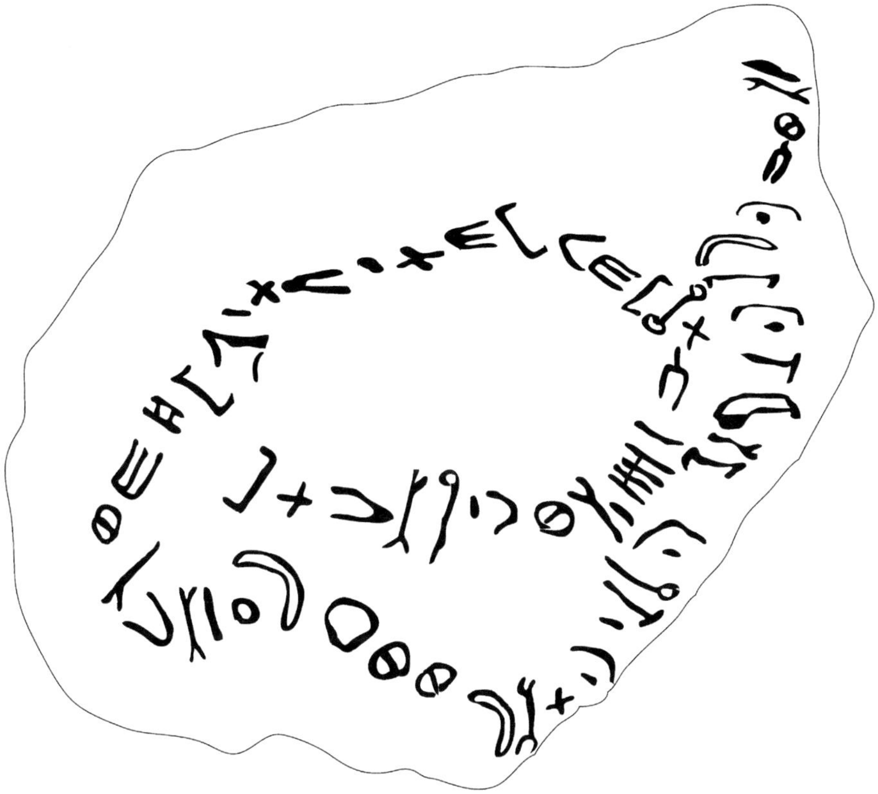

Figure 2. The Safaitic inscription NEH 1 from site K37 (tracing: J. Norris 2018).

the Nabataean *'srtg'* (Greek *strategos*), an official in the Nabataean provincial system in charge of both military and civil affairs (Nehmé 2015).

Finally, two texts contain some new references to a person bearing the rare name of *Dmṣy* (NEH 5 and 6). This might be the same figure as the one described as the leader of a rebellion in two Safaitic inscriptions (SIJ 287 and 823), a mysterious event in the Syro-Arabian desert's history frequently referred to as the 'revolt of Damaṣī' in the modern literature, about which much has been speculated.

While the editors provide a very good analysis of this new and valuable epigraphic material, some of the texts, however, allow for alternative interpretations, slight corrections or translation improvements, a task which the present contribution aims to undertake. Organised in two sections, this paper will focus first on eight Safaitic inscriptions (NEH 1, 2, 5, 6, 7, 8, 9, 14) on which a number of remarks appear possible. The second section will briefly discuss the Nabataean texts NEH 15A and 15B, which, to my mind, must be treated as a single inscription and not as two as the *editio princeps* suggests. The re-examination of these nine texts offers an opportunity to produce further epigraphic and historical comments on them, intending to complete the editors' argumentation or, when judged necessary, to formulate some alternative views. Special attention will be given to the so-called 'revolt of Damaṣī' and the mention of an apparent Nabataean governor of 'Gilead', a region of northern Transjordan which has only been briefly dominated by the kings of Petra.

Epigraphic conventions

{ } enclose letters and words of which the reading is doubtful
{/} indicates alternative interpretations of a letter or a word
[ ] enclose letters or words which are restored or added to facilitate the comprehension
< > enclose additional letters inscribed in error
--- indicates a damaged area in which an unknown number of letters have been lost
* marks reconstructed forms
*/ /enclose proposed vocalisations
DN: divine name
N: name
PN: personal name

Safaitic section

NEH 1

The first text (Fig. 2) of the collection was found at the site numbered K37 by the surveyors, which lies in a tributary of Wādī Ghuṣayn that feeds into the mud flat of Marabb Hilāl, about 28 km south-west from Qaṣr Burquʿ. The editors

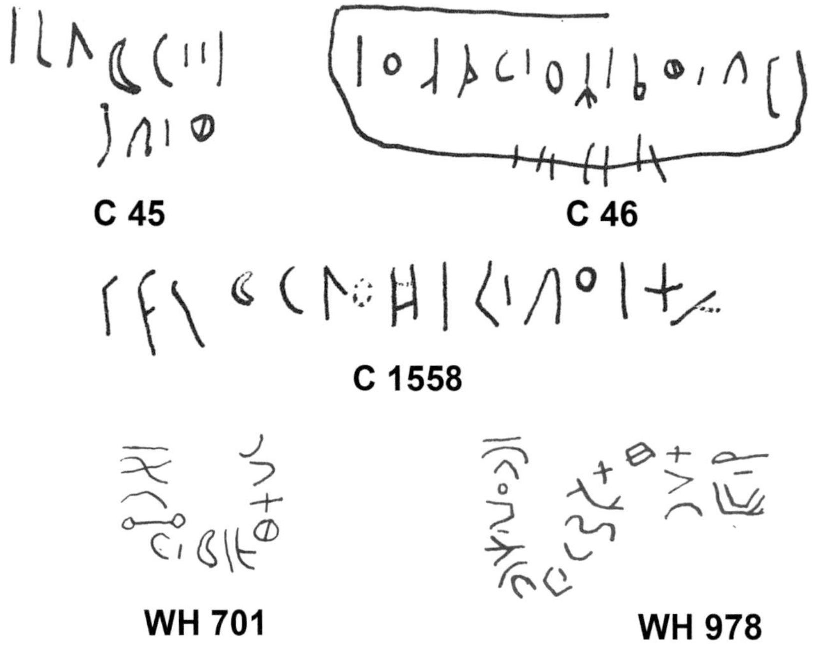

Figure 3. Examples of *ẓ* with a 'triangular' shape. C 45: *l rs^{1}m bn nl w nẓr* "By Rs^{1}m son of Nl and he stood guard". C 46: *l ʿhd bn gḏly w nẓr* "By ʿhd son of Gḏly and he stood guard"; *l krm b{n} {ʿ}ḍl bn ẓʿnt h*-[gml] "By Krm {son of} {ʿ}ḍl son of Ẓʿnt is the [camel]". WH 701: *l ẖbt bn mlk w tẓr* "By H̱bt son of Mlk and he lay in wait". WH 978: *l rbʿ bn ḥrb bn rfʾt w tẓr mny* "By Rbʿ son of Ḥrb son of Rfʾt and he awaited fate."

read *l ʾws bn mr bn zmhr bn yzn bn tʿm wwgm ʿl ʾbh wḥḍr {s}nt snt ḥrb ḥrṯt slṭnh wbny hstr* "By ʾws son of Mr son of Zmhr son of Yzn son of Tʿm and he grieved for his father and he was present [in] Fnt the year [that] Ḥrṯt fought his governor, and he built the shelter" (Al-Salameen *et al.* 2018, 62). Three points need to be made here. The first relates to the verb *ḥḍr*. Although it is true that the basic meaning of *ḥaḍara* in Classical Arabic is "to be present" (Lane 588c), the authors of the Safaitic inscriptions generally do not make use of a specific verb to describe their presence somewhere. The locative predications are simply introduced by the preposition *b*- (LP 653: *s^{1}nt lgy*[n] *b-nqʿt* "the year the legions of Germanicus were at Nqʿt"), while being at a place is usually expressed with the common formulae *l PN* + place name and *l PN h-dr* "By PN, at this place" (C 110, 167, 171, 179, 263, etc.; cf. Al-Jallad 2015, 168-169, 201). With respect to the Safaitic verb *ḥḍr*, it is now accepted that it should most frequently be translated as "to camp by permanent water", bearing a closer connection with the sense of the Classical Arabic active participle *ḥāḍir* "any people staying, or dwelling, by waters" (Lane 590b; Al-Jallad 2015, 321-322).[4] This refers to the periods during which the nomads gather at places with permanent sources of water (*maḥḍar*), in particular at the end of the summer, to await the first rains and in times of droughts (Macdonald 1992, 9). Echoes of this are found in a number of texts whose authors declare that they were *ḥḍr*-ing while *tẓr h-s^{1}my* "awaiting the rains" (*e.g.* C 1926, 1927). Note also WH 3559.1 and 3584 that contain the phrase *tẓr h-s^{1}my b-ḥḍr* "he awaited the rains while camping by permanent water", WSRBZ.B 1 which reads *ḥḍr h-dr w qyẓ* "he camped by permanent water at this place while he spent the dry season" and Is.H 506, *ḥḍr s^{1}nt ḥgz bʿls^{1}mn* "he camped by permanent water the year Bʿls^{1}mn withheld (the rain)".[5] It seems therefore reasonable to deduce that this is what the inscription under discussion deals with, in particular since site K37 from which it comes, is extremely close (4 km) from Biyār al-Ghuṣayn, whose wells may well have been a vital *maḥḍar* of the nomads throughout the ages.

The second point concerns the word appearing immediately after *ḥḍr*. The editors' analysis of it appears somewhat confused. They read it as *{s^1}nt* in the transcription and *Fnt* in the translation (Al-Salameen *et al.* 2018, 62). When commenting on the inscription, they firstly take the reading as *Fnt*, which they identify as a place-name already known from four Safaitic inscriptions of the OCIANA corpus, after which they claim later in the paragraph that the word *s^{1}nt* is mistakenly "repeated twice" (*ibid.*). Whereas the reading *Fnt* appears tempting in view of two texts reading *w ḥḍr ʿl-fnt* "and he camped near permanent water on the edge of Fnt" (Internet 7 and 9) and of NEH 3 that seems to also refers to this place (Al-Salameen *et al.* 2018, 65), this interpretation appears problematic. The first letter of the word consists of an open triangle facing up, whose right line has at its foot a curious mark, possibly extraneous, which is a stroke going left at roughly 230°. This glyph is very unlikely to be *f*, since it would lack its usual wavy shape and its two peripheral undulations, appearing moreover in an uncommon position as the Safaitic *f*

4 One apparent exception is C 4985, where the verb *ḥḍr* describes the sun's presence in Aquarius. See Al-Jallad 2014, 219.

5 For some readings and comments on these inscriptions, *see* Macdonald 1992, 9, no. 55; Al-Jallad 2015, 186, 211.

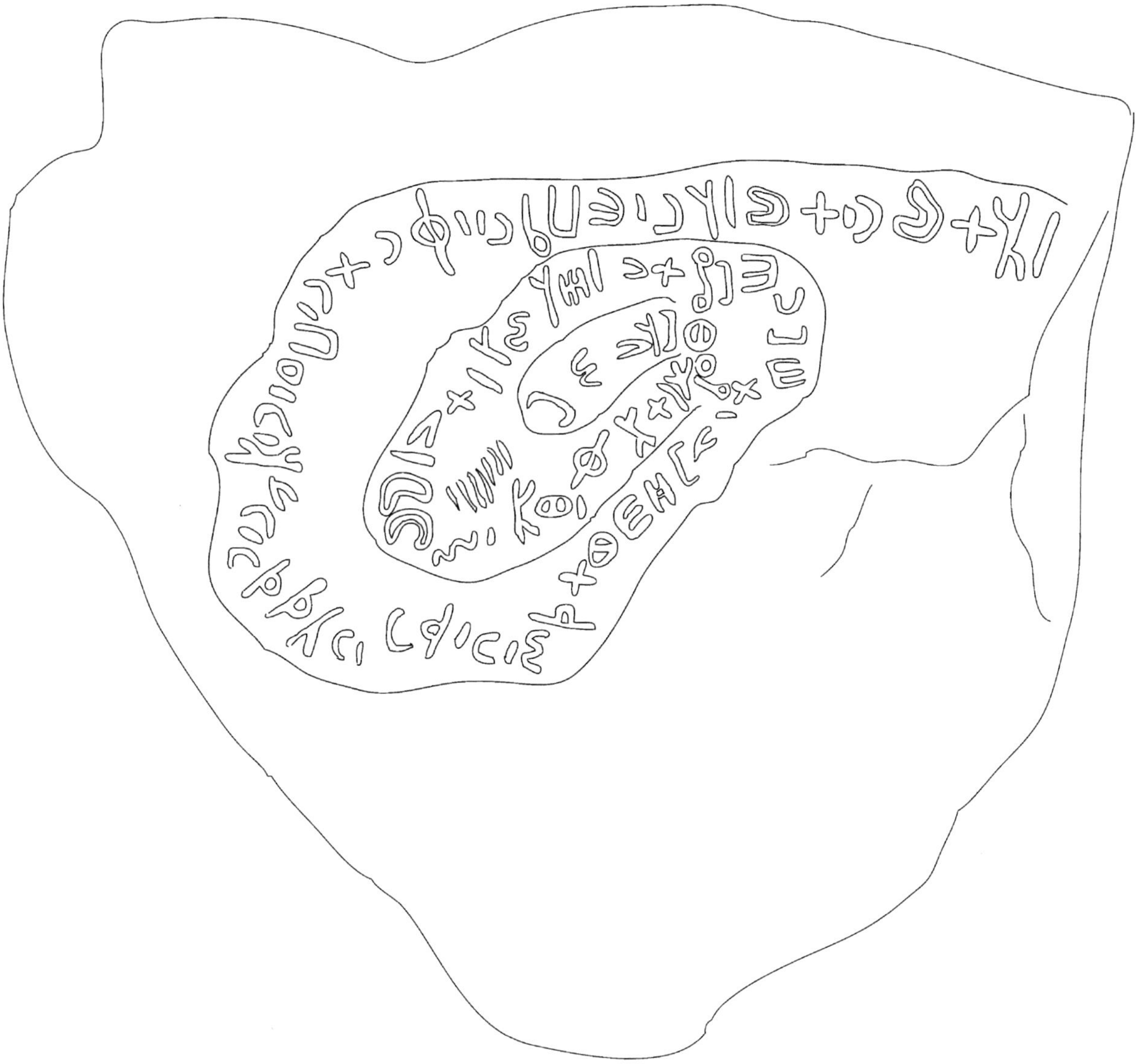

Figure 4. The Safaitic inscription NEH 2 from site K37 (tracing: J. Norris 2018).

usually has a horizontal stance.[6] At first sight, the reading of a s^1 would therefore seem more attractive as this letter frequently takes a V-shape in some varieties of the Safaitic script (C 57, 399, 435, 1415, *etc.*).[7] If so, one can posit that the first s^1*nt* represents a temporal adverb (cf. Classical Arabic *sanatan*) and the second the element of the dating formula, "and he camped near permanent water for a year, the year Ḥrṯt waged war upon his governor". Nevertheless, this is not without problems. The four other s^1-s of the inscription appear very different from the glyph under question, as they are all written horizontally and facing the direction of the text. Moreover, three of them bear a clear tail (in *ʾws*^{1}, s^1*nt*, and s^1*lṭn*), which contrasts once again with this sign. Since it appears unlikely to be *f* or s^1, I would tentatively suggest that it represents *ẓ*. Compare it with a very similar *ẓ* that one can observe in C 45, 46, 1558, 2140, WH 701, 978, *etc.* (Fig. 3).[8] However, such an interpretation appears only possible if one takes the lower line of the glyph as an extraneous mark, which appears difficult to ascertain from the available photograph (Al-Salameen *et al.* 2018, 62,

6 Note the occasional occurrence of some *f*-s with a vertical stance (as in Thamudic B) in the variety of the Safaitic alphabet which V. Clark has labelled the '90° script'. This script appears, however, clearly different from the one from our inscription, which corresponds to Clark's 'common script' (Clark 1979, 67-71).

7 This happens mostly in the so-called 'fine script' according to Clark's 1979 classification, which is once again a variety of the Safaitic script different from the one from the inscription under study.

8 I am most grateful to C. Della Puppa who informed me about the presence of a similar shape of the *ẓ* in Is.K 291 where the verb read as *nfr* by the editor should be corrected to *nẓr*.

Fig. 2).[9] The reading *{ẓ}nt*, which I propose, should therefore remain purely a suggestion. As it stands between the verb and the dating formula without a preposition, the most natural interpretation is that this represents a toponym.[10]

The final point concerns the last two words of the inscription, read *h-s¹tr* by the editors. However, both the photograph and the tracing make it clear that the text definitively reads *ʾ-s¹tr*. This is a good example of a Safaitic inscription whose dialect does not make use of the common *h*-definite article but exhibits instead a form which could either correspond to the *ʾ*- or the *ʾl*-article with an assimilation of the *l* to the following coronal (see Al-Jallad 2015, 74-76). Note that the phrase *ʾ-s¹tr* is already known from at least two other Safaitic inscriptions (ASFF 385; BS 894). If the preceding remarks are correct, I would therefore suggest to read the whole inscription as follows:

l ʾws¹ bn mr bn zmhr bn yzn bn tʿm w wgm ʿl-ʾb-h w ḥḍr {ẓ}nt s¹nt ḥrb ḥrṯt s¹lṭn-h w bny ʾ-s¹tr

"By ʾws¹ son of Mr son of Zmhr son of Yzn son of Tʿm and he grieved for his father while he camped by permanent water {at Ẓnt} the year Ḥrṯt waged war upon his governor, and he built the shelter."

NEH 2

This text comes from the same site (K37) as that of the previous one and presents a very similar content, being dated according to the same political event (Fig. 4).[11] The editors read: *l ʾtm bn tmlh bn ḥẓy bn nqbt bn ẓʿn bn ʾs bn bdd h bn bdn bn šdt wḥḍr snt ḥrb ḥrṯt slṭnh f h lt slm mšnʾ w nqʿt lḍ yʿwr hsfr* "By ʾtm son of Tmlh bn Ḥẓy son of Nqbt son of Ẓʿn son of ʾs son of Bddh son of Bdn son of Šdt and he was present [here in] the year [that] Ḥrṯt fought his governor, O Allāt [grant] security from enemies and [inflict] ejection from the grave on whoever scratches out the inscription" (Al-Salameen *et al.* 2018, 64). Taking into account the remarks formulated above, one can slightly modify the translation to read:

l ʾtm bn tmlh bn ḥẓy bn nqbt bn ẓʿn bn ʾs¹ bn bddh bn bdn bn s²dt w ḥḍr s¹nt ḥrb ḥrṯt s¹lṭn-h f h lt s¹lm m-s²nʾ w nqʿt l-ḏ yʿwr h-s¹fr

"By ʾtm son of Tmlh son of Ḥẓy son of Nqbt son of Ẓʿn son of ʾs¹ son of Bddh son of Bdn son of S²dt and he camped by permanent water the year Ḥrṯt waged war upon his governor, so, O Lt, may he be secure from enemies and may he who would efface this writing be thrown out of the grave."[12]

The fact that the author declares to have camped by permanent water in the same year as the carver of NEH 1 does reinforce the hypothesis that the area of K37 was the location where this took place. Camping at the same place and during the same period, it seems probable that the authors of both texts were kinsmen or that they came from two related groups sharing the same territory and water sources. While the letter-shapes of the two texts appear extremely similar,[13] it is of particular interest to note that NEH 1 makes use of the *ʾ(l)*-definite article, whereas NEH 2 employs for its part the form *h*-. Attention has recently been drawn to the occurrence of dialectal differences within some lineage groups who employed the Safaitic alphabet such as the *ʿmrt* tribe, certain members of which use the *ʾ(l)*-definite article and others the *h*-morpheme (Al-Jallad 2015, 15). Nevertheless, it is not clear whether the two texts examined here offer a new example of this situation. The *h*- obviously functions as a proximal demonstrative in NEH 2, *h-s¹fr* "this writing", while this appears difficult to determine about *ʾ*- in NEH 1. In other words, it remains perfectly possible that the carver of NEH 2 was also a speaker of an *ʾ(l)*-dialect who would have used a separate demonstrative morpheme *h*-.

Of course, NEH 1 and 2 appear of great interest from a historical point of view, as they highlight an event so far unknown of the Nabataean kingdom's history. The event seems to have been a war conducted by one of the four rulers bearing the dynastic name of 'Aretas' against one of his officers or provincial governors. This indication of apparent civil war within Nabataea is somewhat reminiscent of the troubled context in which the accession of Aretas IV took place. For the record, Aretas was crowned king after the death of Obodas III in the winter of 9/8 BC, when the ambitious minister Syllaeus was present in Rome to justify his action following a conflict that broke out with Herod the Great. The new king immediately sent a letter to Augustus in which he accused Syllaeus of different crimes, including having killed Obodas himself with poison (Josephus, *AJ* 16.296). Once back in Petra, Syllaeus is said to have started a

9 The reading of a *ẓ* seems also possible if one takes the problematic mark as one of the two converging arms of the letter which would face backward with an uncommon oblique stance and would have a flat 'roof'. I am most grateful to C. Della Puppa for this suggestion.

10 I am not aware of any place in the Syro-Arabian desert to which the name *{Ẓ}nt* could correspond. However, compare it with the Jabal al-Ẓannah in the United Arab Emirates or the name of the Banī Ẓannah, a Yemeni tribe from Ḥaḍramawt.

11 Al-Salameen *et al.* 2018, 64, Figure 3.

12 On the curse formula *nqʿt (b-wdd-h)* and its translation, see Al-Jallad 2015, 136; Al-Jallad and Macdonald 2015, 155-156.

13 If the reading *{Ẓ}nt* suggested above is correct, then the only notable difference in the letter-forms of both texts would be *ẓ*, having a rectangular shape in NEH 2 and a triangular one in NEH 1. The first two *n*-s in NEH 1 look like dots, though they occur as short vertical lines in the rest in the text, exactly as in NEH 2.

Figure 5. Bronze coin of Aretas IV's first year of reign (9 BC), which presents the laureate head of the king uncommonly turned left on the observe and two cornucopiae crossed on the reverse with the monogram of the name *Šly* (Meshorer 1975, no. 44. Photograph by CNG www.cngcoins.com, electronic auction 351 [2015], lot 376).

campaign of political conspiracies and assassinations, having eliminated a number of Nabataean nobles and some of the king's friends, among whom a certain Soaimos (*Suhaym*) who is described as "one of the most powerful personages in Petra" (Josephus, *AJ* 17.54; *BJ* 1.574-575). Syllaeus' political manoeuvring and power aspirations are also reflected in a number of coins struck during the first year of Aretas IV that bear the monogram of *Šly* or the abbreviation *Š*(*ly*) (Meshorer 1975, nos. 44-45) (Fig. 5).

While it is difficult to think that Syllaeus acted without any retaliation from Aretas IV, the available documents make no mention of an armed conflict between the king and the minister. There is consequently no apparent reason to link the war described in our two Safaitic inscriptions to these events, more particularly since the Nabataean minister was known as the king's *ʾḥ* "brother" (RES 657 = 1100; MP 685; Strabo 16.4.21) and not as something to which the Safaitic *s¹lṭn* may correspond. On the other hand, Josephus mentions two other "Arabs" involved in Syllaeus' machinations, one of whom is described as a φύλαρχον "chieftain" (*AJ* 17.56; *BJ* 1.577). Could this stand for a Nabataean *ʾsrtgʾ*/ Safaitic *s¹lṭn*? There is, once again, nothing in support of this, as Josephus expressly mentions some Nabataean provincial governors as *strategoi* in another passage (*AJ* 18.112), which makes it clear that he would not have used the title of *phylarch* to name these officials. We must therefore admit that this war conducted by Aretas against one of this governors has left no traces in the historical records. This is not surprising at all as it must be remembered that the literary sources are completely silent concerning the historical events that took place within the Nabataean kingdom from approximately AD 1 to 30, primarily because of the death of Nicolaus of Damascus, which divested Josephus of his main informant (Starcky 1966, 914; Bowersock 1983, 60).

NEH 5

This graffito and the one which follows are the two texts referring to *Dmṣy* (Fig. 6).[14] They are carved next to each other on a stone that was found on site S1, which lies to the south-east of Qāʿ Khuwaymāt. The editors read *l ḥnʾl bn ḥgy ḏʾl ṭsm wnẓr ʾl dmṣy w tšwq ʾl ʾsd bn yẓr* "By Ḥnʾl son of Ḥgy of the tribe of Ṭsm and he was on the look-out for Dmṣy and he longed for ʾl ʾsd son of Yẓr" (Al-Salameen *et al.* 2018, 64). Although the text is correctly transcribed, the preposition *ʾl*-, which always stands before the object of the verb *ts²wq*, seems to have been incorrectly taken as an element of the personal name that follows it or mistakenly left in the translation.

Regarding the word *nẓr*, the editors opt to interpret it by conforming to how this verb was traditionally translated in the earlier editions of Safaitic inscriptions, namely "to be on the look-out" on the basis of the primary meaning of the Arabic root √nẓr "to look at" (Lane 2810c). However, this choice appears somewhat questionable since the phrase *nẓr ʾl-dmṣy* appears understood here in the same way that Littmann and Winnett translated the formula *nẓr s²nʾ* "he was on the look-out for enemies" (see LP 1263; SIJ 808 and 858). But, in fact, these two sentences have two opposite meanings. Thanks to the many discoveries that greatly enriched the Safaitic corpus in the past few years, it is now clear that the verb *nẓr* is largely used, though not exclusively, in military contexts to refer to guarding activities (cf. Classical Arabic *nāẓir* "guardian, watcher"; Lane 2813b). Interestingly, this is also the case in other varieties of Ancient North Arabian. In Dadanitic, the verb *nẓr* and its variant *nṭr* occur in soldiers' inscriptions stationed at Madāʾin Ṣāliḥ who record to have "guarded Dadan" (*nẓr*/*nṭr ddn*; AH 312, 315, 328, *etc.*). Note also some

14 In their conclusion, Al-Salameen *et al.* (2018, 76) indicate that these two inscriptions "are the first known texts linking the name of *Dmṣy* with the verb *nfr* 'hastened." This is undoubtedly a mistake as none of their two texts, nor any other Safaitic inscription, exhibit the name *Dmṣy* and the verb *nfr* together.

Figure 6. The Safaitic inscriptions NEH 5 and 6 from site S1 (photograph by Ali Al-Manaser, OCIANA Badia Survey Project).

Taymanitic texts from the vicinity of Taymāʾ that exhibit the cognate forms *nṭr* and *nṣr* as references to different military services of guarding.[15]

To return to the Safaitic case, Ahmad Al-Jallad has recently noted that there are three different types of *nẓr*-texts. These are the texts where the verb occurs independently, *l PN w nẓr* "By PN and he stood guard"; those where it precedes a direct object, *w nẓr N* "he stood guard against N", and those in which it appears with a benefactive introduced by a preposition, *w nẓr ʾl-N* and *nẓr bʿd N* "he stood guard for, on behalf of N" (Al-Jallad 2015, 218-219).[16] This therefore allows us to define more precisely the sense of the present inscription, the author of which actually declares to have served as a guard for or under the command of *Dmṣy*:

l ḥnʾl bn ḥgy ḏ-ʾl ṭs¹m w nẓr ʾl-dmṣy w ts²wq ʾl-ʾs¹d bn yẓr
"By Ḥnʾl son of Ḥgy of the lineage of Ṭs¹m and he stood guard on behalf of Dmṣy and he longed for ʾs¹d son of Yẓr."

15 See Kootstra 2016, 79-80; 2018, 205-208.

16 To the prepositions *ʾl* and *bʿd* listed by Al-Jallad (2015, 218), one may had *l-* which introduces the benefactive of the verb *nẓr* in WH 610 and 1027 which read *w nẓr l-(h)-ms¹rt* "and he stood guard for the troop" as well as in CSNS 628, *nẓr l-rbʾl* "while standing guard for Rbʾl".

NEH 6

Carved right next to the previous inscription (Fig. 6), NEH 6 is read by the editors as *l šmr bn bmr wnẓr ʾl dmṣy* "By Šmr son of Bmr and he was on the look-out for Dmṣy" (Al-Salameen *et al.* 2018, 67). On the basis of the remarks formulated in the previous paragraph, one may read:

l s²mr bn bmr w nẓr ʾl-dmṣy
"By S²mr son of Bmr and he stood guard on behalf of Dmṣy"

It is of particular interest to note that the phrase *w nẓr ʾl-dmṣy* recurs a third time in a recently published inscription from Wādī Usaykhim, *l ns²l bn mʿn bn mṭl ḏ-ʾl tm w nẓr ʾl-dmṣy b-ḫms¹ mʾt frs¹ s¹nt ḥrb ʿmm* "By Ns²l son of Mʿn son of Mṭl of the lineage of Tm and he served as a guard under Dmṣy in (a troop) of five cavalry units, the year of the war of ʿmm" (MM 47 = Al-Housan 2017, no. INS-NO-21; Al-Husan and Al-Rawabdeh 2018). With men under his command standing guard on the western and eastern extremities of the *ḥarrah*, this makes it clear that the person mentioned in these three texts must be an important military chief of some kind. Since his personal name, *Dmṣy*, appears extremely rare, the editors have therefore good reasons for assuming that this could be the same person as the one described as the leader of a rebellion (*mrd*) in SIJ 287 from Jawa and SIJ 823 from Tell al-ʿAbd (Al-Salameen *et al.* 2018, 67). If this is so, these

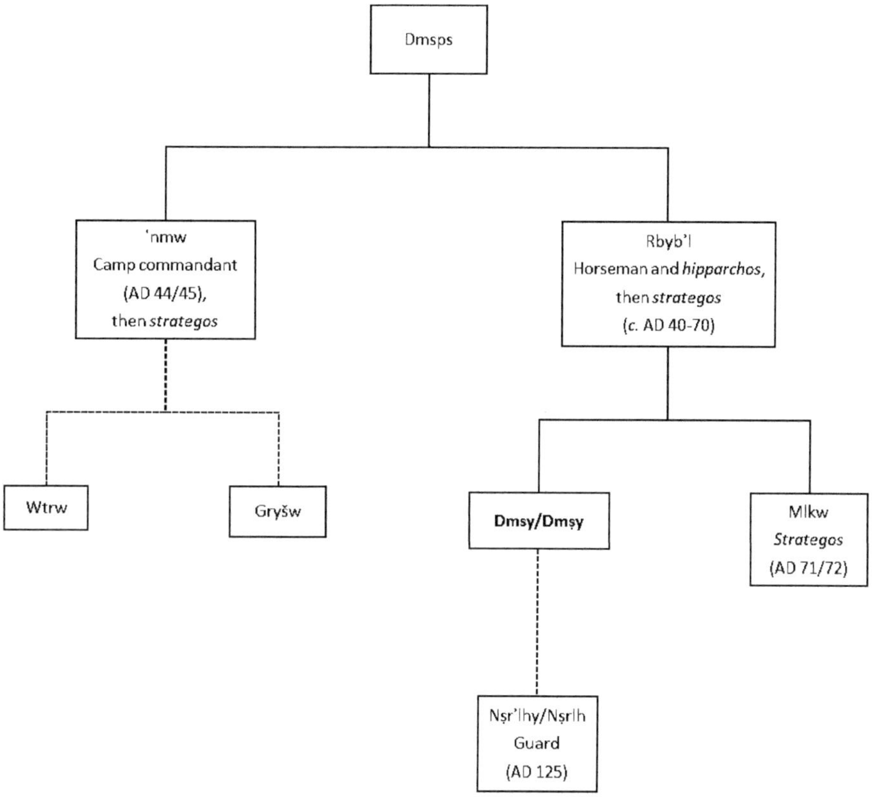

Figure 7. Partial genealogical tree of the Damasippos' family with the military and administrative titles of its members.

three references to guarding activities on the edges of the *ḥarrah* could likely be operations related to his "revolt".

This mysterious event has been the subject of a famous article published in 1973 by Winnett, who analysed it as a great revolt, lead by a member of a well-known family of Nabataean governors, that broke out against Rabbel II when he acceded to the throne in AD 70/71 (Winnett 1973, 54-57). Winnett's main argument was that the name spelt *Dmṣy* in Safaitic is likely to represent a Greek anthroponym and to correspond to the one spelt *Dmsy* in Nabataean Aramaic, which appears to be borne by only one man within the Nabataean documents, the son of the *strategos* of Ḥegrā *Rbyb'l* son of *Dmsps* (Damasippos) (Figs. 7-8).[17] According to Winnett, the cause of the revolt was because Rabbel II would have denied *Dmsy/Dmṣy's* right to inherit the office of governor of Ḥegrā in giving it instead to the youngest son of *Rbyb'l*, *Mlkw*.[18] *Dmsy/Dmṣy* would have then responded in starting a rebellion immediately after his brother's appointment and won the support of three tribes of the *ḥarrah* desert, the *Ḍf*, the *Ms^{1}kt* and the *Mḥrb* (*ibid.*, 55-56). This scenario has widely been accepted among the scientific community and subsequent scholars have even taken Winnett's case further in suggesting that Rabbel II's title *dy 'ḥyy w šyzb 'm-h* "he who brought life and deliverance to his people" might refer to this crisis and his success in crushing *Dmsy's* revolt.[19] More recently, however, an alternative explanation of *Dmṣy/Dmsy's* motivations was suggested by F.M. Al-Otaibi (2011, 89-94) for whom the revolt would have been instead an attempt by southern Nabataea to achieve independence from Petra and the north.

One part of Winnett's theory might receive some credit from a dated inscription recently discovered during a survey of the Al-Jawf area undertaken by the Saudi-Italian-French project at Dūmat al-Jandal, northern

17 I cannot understand why Al-Salameen, Al-Shdaifat and Harahsheh raise the issue of whether *Dmsy/Dmṣy* "might have been of the royal family, or may have been the son of Malichus II" (2018, 67). The only known inscription referring to the Nabataean *Dmsy* makes it clear that he is the son of the *strategos* of Ḥegrā: CIS II 287 = JSNab 84, *dkyr dmsy br rbyb'l 'srtg' b-ṭb* "May be commemorated Dmsy son Rbyb'l, the *strategos*, in well-being". There is absolutely nothing which allows us to link him nor the other members of Damasippos' family to Malichos II and the royal dynasty.

18 Given the very common practice of papponymy (naming the *eldest* male child after his grandfather) in north-west Arabia, Winnett (1973, 55) is certainly right in assuming that *Dmsy* could be the oldest son of *Rbyb'l* since his name is manifestly a hypocoristic, or contracted form, of the grandfather's name, *Dmsps* (Damasippos).

19 Bowersock 1983, 72, no. 48, 156; Graf 1989, 363; Al-Otaibi 2011, 89-94; Al-Housan 2017, 35-42; Al-Salameen *et al.* 2018, 67; Al-Husan and Al-Rawabdeh 2018.

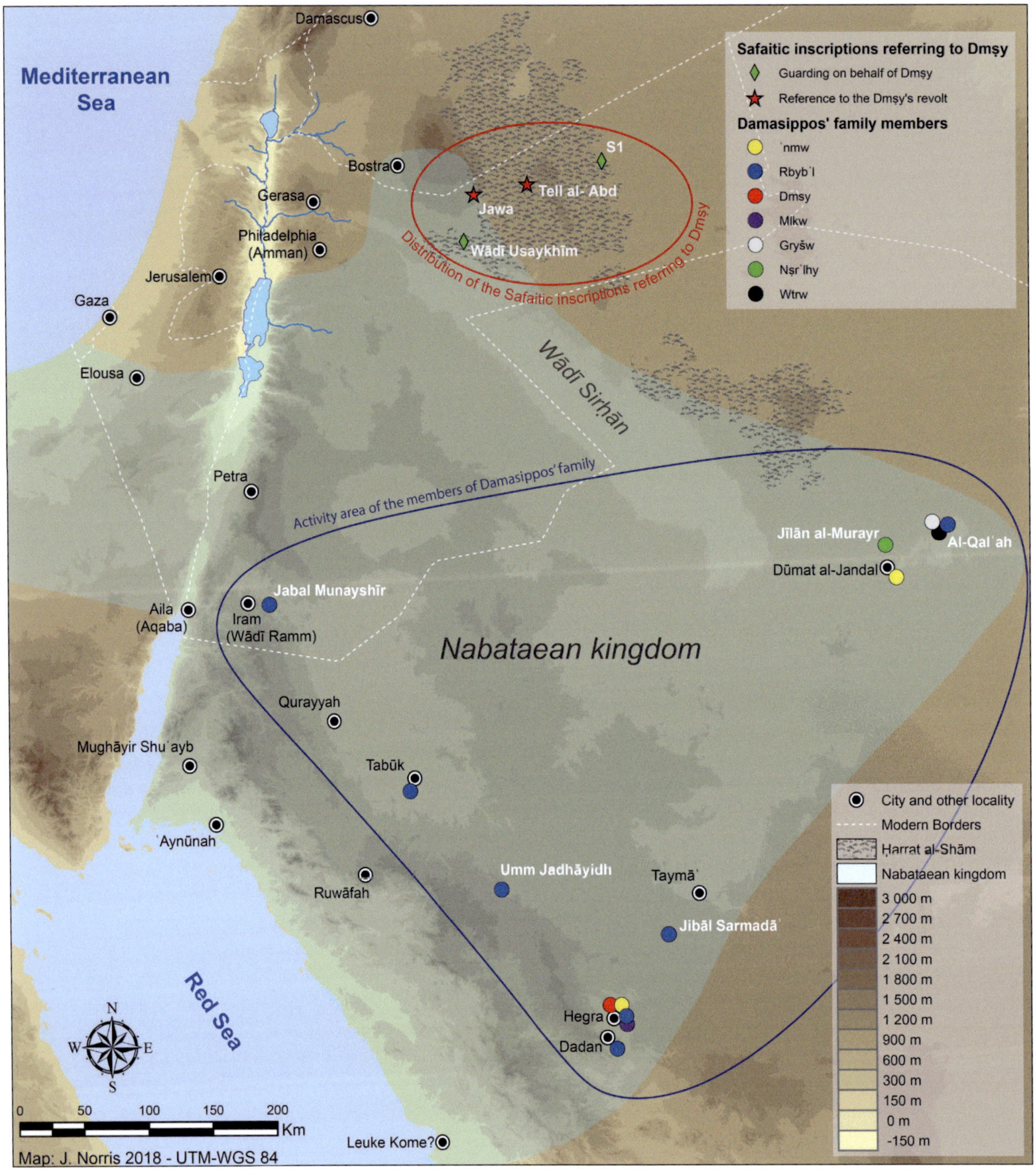

Figure 8. Map of north-west Arabia and the southern Levant showing the distribution of the Safaitic inscriptions referring to Dmṣy and the places where the members of the Nabataean family of Damasippos are attested (J. Norris 2018).

Saudi Arabia.[20] This consists of a bilingual Nabataean/Ancient North Arabian text carved by a guard named *Nṣrʾlhy*/*Nṣrlh* in AD 125, who spells his patronym as *Dmṣy* in Mixed Safaitic/Hismaic and as *Dmsy* in the Nabataean counterpart (DaJ34ANA3 + DaJ34Nab4; Norris 2018,

20 This project is directed by Guillaume Charloux (CNRS, Orient & Méditerranée, France) and Romolo Loreto (University of Naples "L'Orientale", Italy).

86-88).[21] Before that, Winnett's propositions that the Safaitic *Dmṣy* could correspond to the Nabataean Aramaic *Dmsy* was indeed nothing else than a likely hypothesis since, as M.C.A. Macdonald has rightly pointed out, the Safaitic *Dmṣy* could well have also represented a name derived from the Arabic root √dmṣ (Macdonald 1980, 186; 1993, 360). This new inscription therefore allows us to rule out these legitimate doubts, confirming moreover Winnett's assumption that *Dmsy* and *Dmṣy* transcribe a name of Greek origin, presumably a hypocoristic of Δαμάσιππος, since the Ancient North Arabian has *ṣ* where the Nabataean has *s*, disqualifying a Semitic etymology and pointing obviously towards a Greek *sigma.*

While it remains possible that the father of the author of the bilingual text from Dūmah is *Dmsy* son of *Rbyb'l*, considering that several members the Damasippos' family were active in the Al-Jawf area during the first century AD (Norris 2018, 88), the common claim that the Safaitic texts from north-eastern Jordan also deal with this Nabataean figure remains to be proven (Fig. 8 and Table 1). On the one hand, none of these Safaitic texts reveal *Dmṣy's* genealogy, which makes it impossible to confirm whether the leader of the revolt and the son of the Nabataean *strategos* are really the same person. There is, on the other hand, a further Safaitic inscription known for a long time that contains a seventh reference to the name *Dmṣy* in Ancient North Arabian where it appears borne, not by a military chief but by the author's father who is a member of the *'mrt* tribe (SIAM 36). Unless assuming that the family of the Nabataean governors descending from Damasippos were of *'mrt* origin, for which there is no apparent evidence nor geographical connections,[22] this may indicate that different persons of the Syro-Arabian desert could well have borne this foreign and rare name.

Finally, some serious doubts can be raised about Winnett's chronological reconstruction and his attempt to show that the Safaitic texts referring to the revolt are contemporaneous with the activity period of the Nabataean *Dmsy*, namely sometime between the end of Malichos II's reign (40-70) and Rabbel II's crowning. Since SIJ 287 is dated according to a double event, the revolt of *Mḥrb* and the revolt of *Dmṣy* (*s¹nt mrd mḥrb w s¹nt mrd dmṣy*), Winnett thought that he could determine when the first one took place on the basis of an inscription carved by a *Mḥrb* tribesman reading *w qṣṣ s¹nt mlk rb'l* (ISB 57). According to Winnett, the verb *qṣṣ* would mean here "to follow" and be used by the author to indicate that he has joined his tribe in an exceptional event that could have been the *Mḥrb* revolt, deducing that the simultaneous rebellion of *Dmṣy* took place in AD 70/71, when Rabbel II acceded to the throne (Winnett 1973, 56). Clearly this is an entirely circular argument, the basis of which is no longer tenable. First, one should admit that it is extremely difficult to follow Winnett in assuming that the context of the *qṣṣ* action described in ISB 57 might be the revolt of *Mḥrb* simply because its author is from the *'l mḥrb!* On the other hand, the meaning of the verb *qṣṣ* is now far better understood than it was when Winnett was writing, as it has been demonstrated that it bears a strong military connotation and should most frequently be translated as "to patrol" rather than "to follow, track" (Al-Jallad 2015, 334-335). In other words, the author of ISB 57 was simply saying that "he patrolled the year Rabbel was made king", with no apparent relationship with the revolt of *Mḥrb* nor that of *Dmṣy*. Similarly, the text SIJ 823 which reads *w qṣṣ b'd ḍf s¹nt mrd dmṣy* is in no way implying that the author took part in a military expedition with his *Ḍf* tribe "for, or against, Damaṣī" (Winnett 1973, 56), but simply that his author "patrolled on behalf of the *Ḍf* the year of the revolt of Dmṣy". Although it seems possible that this patrolling operation was related to the event by which the texts is dated, what the *Ḍf's* involvement in *Dmṣy's* rebellion was, is however impossible to know, at least for now. All in all, one should admit that there is not a single document suggesting that *Dmṣy's* revolt actually took place in AD 70/71 and that the tribes of *Ms¹kt*, *Mḥrb* and *Ḍf* had joined his cause. The only element that could eventually support Winnett's theory is therefore the onomastic evidence and the correct equivalence of the Ancient North Arabian *Dmṣy* with the Nabataean *Dmsy*, which, to my mind, remains an extremely weak argument. Pending the discovery of new material that could indicate *Dmṣy's* genealogy, I would therefore say that one cannot not exclude the possibility that the so-called 'revolt of Damaṣī' was nothing else than a purely local event within the *ḥarrah* that broke out at an unknown date, instead of being that huge revolt lead by a Nabataean aristocrat from Ḥegrā against Rabbel II, as reconstructed by Winnett.[23]

21 Interestingly, the texts NEH 5 and 6 are composed in a variant of the Safaitic alphabet which seems mid-way between the 'Safaitic square script' and the 'Mixed Safaitic/Hismaic script' used in the Dūmat al-Jandal area and employed by *Nṣr'lhy/Nṣrlh* son of *Dmsy/Dmṣy*. On the Mixed Safaitic/Hismaic texts from Dūmah, see Norris 2018, 79-88.

22 J.T. Milik (1980, 41-48) argued convincingly that the *'mrt* tribal group was involved in the Nabataean sphere, certain members of which had occasionally composed Nabataean graffiti (MNT 2a, 2b, 2c) and Nabataean-Safaitic partial bilinguals in the *ḥarrah* (MNT 1+MSTJ 12; MNT 2d+MSTJ 10), as well as one Nabataean-Greek memorial inscription at Madaba (MNIN 6). That being said, I am aware of only one Safaitic text from the Dūmat al-Jandal area (ThNTJ 7) attesting of the presence of a man from the *'l 'mrt* in north-west Arabia, which can hardly be taken as an evidence to establish a connection between this social group and the Damasippos' family members.

23 If so, this would not be an isolated situation. Compare with the "revolt" of *bn "ẓmy* according to which ASWS 59 is dated.

Text	*Script*	*Reference*	*Provenance*
Dmṣy as a military chief			
l ḫr bn ʾs^1 bn ḫr ḏ-ʾl ms^1kt w wld b-h-dr s^1nt mrd mḥrb w s^1nt mrd dmṣy w ḫrṣ h-s^2nʾ f h lt w ds^2r s^1lm w mwgd "By Ḫr son of ʾs^1 son of Ḫr of the lineage of Ms^1kt and he helped to give birth in this place [Jawa] the year of the rebellion of Mḥrb and the year of the rebellion of Dmṣy, and he kept watch for the enemies, so, O Lt and Ds^2r, let there be security and {abundance/glories}"	Safaitic (Clark's "common script")	SIJ 287	Jawa (Ḥarrat al-Shām)
l mgd bn zd bn qdm bn mrʾ ḏ-ʾl ḍf w qṣṣ bʿd ḍf s^1nt mrd dmṣy {l} ---m {n} ʾ s^1lm f {ʾ} "By Mgd son of Zd son of Qdm son of Mrʾ of the lineage of Ḍf and patrolled on behalf of the Ḍf the year of the revolt of Dmṣy"	Safaitic (Clark's "fine script")	SIJ 823	Tell al-ʿAbd (Ḥarrat al-Shām)
l ns^2l bn mʿn bn mṭl ḏ-ʾl tm w nẓr ʾl-dmṣy b-ḫms^1 mʾt frs^1 s^1nt ḥrb ʿmm "By Ns^2l son of Mʿn son of Mṭl of the lineage of Tm and he served as a guard under Dmṣy in (a troop) of five cavalry units, the year of the war of ʿmm"	Safaitic (square script)	MM 47 (= Al-Housan 2017, 35-43; Al-Husan and Al-Rawabdeh 2018	Wādī Usaykhim (Ḥarrat al-Shām)
l ḥnʾl bn ḥgy ḏ-ʾl ṭs^1m w nẓr ʾl-dmṣy w ts^2wq ʾl-ʾs^1d bn yẓr "By Ḥnʾl son of Ḥgy of the lineage of Ṭs^1m and he stood guard on behalf of Dmṣy and he longed for ʾs^1d son of Yẓr"	Safaitic (square script)	NEH 5	S1, near Qāʿ Khuwaymāt (Ḥarrat al-Shām)
l s^2mr bn bmr w nẓr ʾl-dmṣy "By S^2mr son of Bmr and he stood guard on behalf of Dmṣy"	Safaitic (square script)	NEH 6	S1, near Qāʿ Khuwaymāt (Ḥarrat al-Shām)
The name Dmṣy/Dmsy in genealogies			
l grm bn dmṣy ḏ-ʾl ʿmrt w ndm ʾl-ʾb-h w ʾl-grm bn ʿqrb bn ʿm "By Grm son of Dmṣy of the lineage of ʿmrt and he was devastated by grief for his father and for Grm son of ʿqrb son of ʿm"	Safaitic (square script)	SIAM 36	Unknown place (Ḥarrat al-Shām?)
l nṣr{l}{h} bn dmṣy "By Nṣr{l}{h} son of Dmṣy"	Mixed Safaitic/Hismaic	DaJ34ANA3	Jīlān al-Murayr near Dūmat al-Jandal (north-west Arabia)
dkyr nṣrʾlhy br dmsy w grm[w] *{b}{r} tymw b-ṭb l-ʿlm nṭryʾ šnt 10+5+4 b-yrḥ ʾdr* "May be commemorated Nṣrʾlhy son of Dmsy and Grm[w] {son} of Tymw for good forever, the guards, the year 19 in the month of Adār"	Nabataean	DaJ34Nab4	Jīlān al-Murayr near Dūmat al-Jandal (north-west Arabia)
Dmsy son of *Rbybʾl*			
dkyr dmsy br rbybʾl ʾsrtgʾ b-ṭb "May be commemorated Dmsy son Rbybʾl, the *strategos*, in well-being"	Nabataean	CIS II 287 (= JSNab 84)	Jabal Ithlib, Ḥegrā (north-west Arabia)

Table 1. List of the Ancient North Arabian and Nabataean inscriptions in which the name Dmṣy/Dmsy occurs.

NEH 7

This text was found on site K27, which lies in Wādī al-Shuwayṭī, a north-east to south-west valley stretching out to Qāʿ al-Shuwayṭī, approximately 11 km west from Qaṣr Burquʿ. It is written on the same stone and immediately next to a second Safaitic inscription (NEH 7bis) left uncommented (Fig. 9).[24] The editors read *l rks bn šmrʾl bn ʾdʿg w wwgd ṯr ʿmh wqṣṣ snt ngy sʿd bn rbʿl slṭn fhlt slm wgty lnšb rhq* "By Rks son of Šmrʾl son of ʾdʿg and he found the trace of his uncle and patrolled [in] the year [that] the governor Sʿd son of Rbʿl escaped. So, O Allāt [grant] security *wgty lnšb rhq*" (Al-Salameen *et al.* 2018, 67-68).

The author's name is obscured by a later drawing representing two human figures facing each other. This makes it difficult to confirm whether *rks^1* is its correct reading, since only the first letter appears clear on the photograph. From what is discernible, one may as well suggest reading it as *R{g}{l}*, which represents a much more common name than *Rks^1* in Safaitic (HIn 271 and 285).[25] The patronym which follows is read *S^2mrʾl* by the *editio princeps*, a theophoric compound that would appear here for the first time with no Ancient North or South Arabian parallels. However, in contrast to what the editors' tracing shows, the first letter is not undulated on its entire length but only on its lower part, which suggests that it could represent *ġ* instead of *s^2*. Similarly, the identification of the second sign as *m* is questionable since its inner line cannot be fully distinguished. It would, moreover, be small in size and present quite a straight spine, whereas the *m*-s in *ʿm* and *s^1lm* both take an elongated form resembling a sort of boomerang with a curved spine. I therefore wonder whether this is not a *y* bearing its circle on its bottom, as is the case with the other *y*-s in the text, whose appearance would be somewhat distorted by a number of scratches and damage. Compare it with the *g* in the grandfather's name which has a line across it, which gives it the appearance of a *w* or a *q*. If this is so, we would be dealing with the extremely common name *Ġyrʾl*. This is of interest since the second graffito on the stone reads *l ġyrʾl bn ʾʿdg bn ḫl bn ḍhdt bn kṯbt bn ḥmyn bn ġḍḍt w rʿy f h lt ġnyt w ʿwr l-ḏ yʿwr h-s^1fr* "By Ġyrʾl son of ʾʿdg son of Ḫl son of Ḍhdt son of Kṯbt son of Ḥmyn son of Ġḍḍt and he pastured so, O Lt, let there be abundance and may he who would efface this writing go blind" (NEH 7bis). If *{Ġ}{y}rʾl* is the correct reading

24 For the photograph, see Al-Salameen *et al.* 2018, 68, fig. 7.

25 Research on OCIANA gives three attestations of *Rks^1* against 140 for *Rgl*.

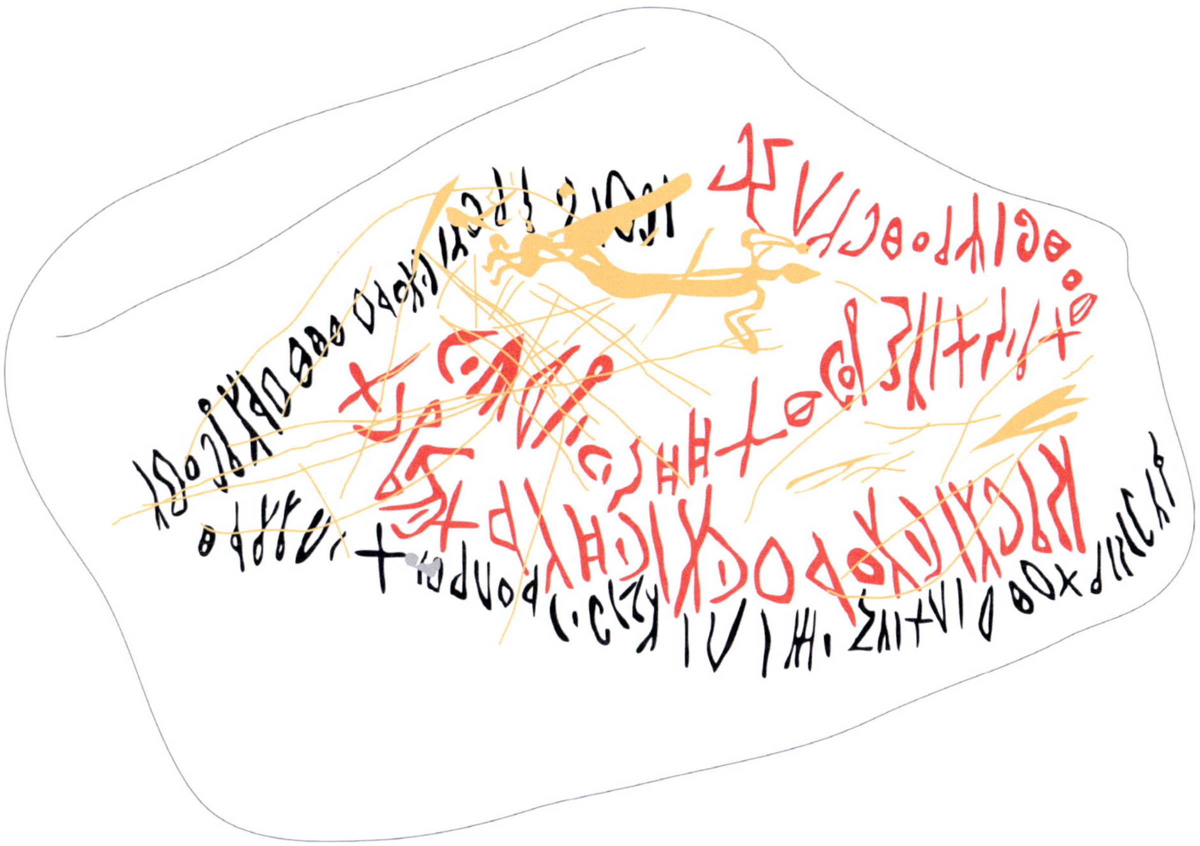

Figure 9. The Safaitic inscriptions NEH 7 (in black) and NEH 7bis (in red) from site K27 (tracing: J. Norris 2018).

in NEH 7, then we would have here the inscriptions of a father and his son. Indeed, the grandfather's name in NEH 7 is *ʾʾdg*, not *ʾdʿg* as read by Al-Salameen *et al.* 2018, 67-68,[26] which is patronym in NEH 7bis. This allows us to note that the authors of these two texts actually belong to a well-known family descending from *Ḍf* son of *Gnʾl*, the eponymous ancestor of the *Ḍf* tribe, whose members have left more than about fifty Safaitic inscriptions in both the Jordanian and the Syrian parts of the *ḥarrah* (Fig. 10).

There is an unexpected *w* between *ʾʾdg* and *w wgd* that is unlikely to represent the Nabataean ending *-w*, corresponding more certainly to a dittography of the conjunction *w*. The editors translate the word *ʿm* as "uncle", assuming a meaning identical with Classical Arabic ʿamm "paternal uncle". Although this opinion has prevailed for some time (Ryckmans 1951, 388-390), Littmann and subsequent scholars have demonstrated that this kinship term actually refers to the "paternal grandfather" in Safaitic, while the normal term for "paternal uncle" is *dd*.[27] This is a good example of certain Safaitic lexical items which find better cognates in North-West Semitic languages such as Hebrew and Aramaic, rather than Classical Arabic, as is also the case with the words *nẖl* "valley" and *mdbr* "desert, steppe" (Macdonald 1993, 381). As a result, the "traces" to which the text refers can hardly correspond to the second Safaitic inscription on the stone which was carved by *Ġyrʾl*, not *ʾʾdg*. One should therefore wait for the final publication of the surveyors' material to verify whether an inscription by a certain *ʾʾdg* can be found in the immediate vicinity of where NEH 7 lies or if these "traces" refer to something else.

Let us turn now to the dating formula s^1*nt ngy* s^1*ʿd bn rbʾl* s^1*lṭn*, which the *editio princeps* translates as "the year [that] the governor S^{1}ʿd son of Rbʾl escaped" (Al-Salameen *et al.* 2018, 68). It should be noted, however, that an interpretation "the governor S^{1}ʿd son of Rbʾl" would require the word s^1*lṭn* to be defined with one of the three prefix articles usually employed in Safaitic (*h-*, *ʾ-* or *ʾl-*), which is not the case. If one follows the editors' translation of the verb *ngy*, this would consequently rather

26 The name *ʾdʿg* is known from only one Safaitic inscription (WH 1326). For its part, *ʾʾdg* occurs 33 times in the OCIANA corpus, although virtually always referring to the same person, *ʾʾdg* son of *Ḫl* son of *Ḍhdt* (C 3850, 4972, KRS 1085, 1090, *etc.*) (Fig. 10).

27 Littmann 1943, 335; Al-Jallad 2015, 56, 206, 303; 2017a, 163.

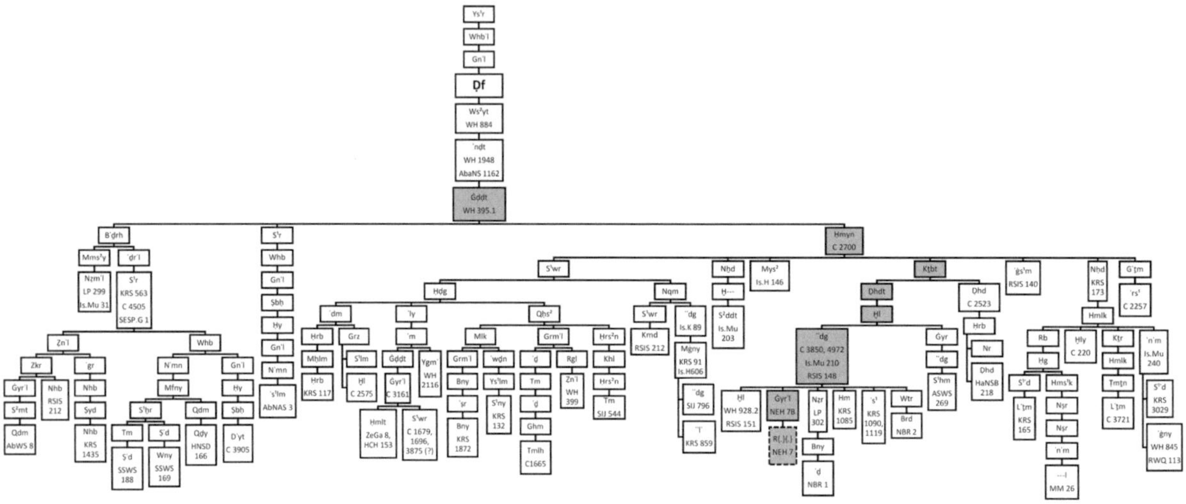

Figure 10. Partial genealogical tree of the author of NEH 7bis and possibly of the author of NEH 7 (for the epigraphic sigla which are not listed in the references below, see OCIANA: http://krcfm.orient.ox.ac.uk/fmi/webd#ociana).

mean something such as "S^{1}ʿd son of Rbʾl, a governor, escaped" or possibly "S^{1}ʿd son of Rbʾl saved a governor" if *ngy* is conjugated in the causative (cf. Classical Arabic forms II and IV, *naǧǧā-hu* and *ʾanǧā-hu* "he saved him"; Lane 3028c). On the other hand, Michael Macdonald has recently and convincingly demonstrated that there are two different verbs falling together as *ngy* in Safaitic. The first derives from the root √ngy and indeed means "to escape", being generally followed by an object introduced by the preposition *m(n)*. See, for instance, *w ngy m-ḥwlt* "and he escaped from the Ḥwlt" (WH 153); *w ngy m-nbṭ* "and he escaped from the Nabataeans (AbaNS 349); *s^1nt ngy h-lgy{n} m-bṣry* "the year the legion fled from Boṣrā" (SESP.U 1). The second actually comes from the root √ngw and occurs in several dating formulae with the sense of "announce, declare, appoint", the most common of these being *s^1nt ngy PN hdy* "the year PN was announced leader" as a reference to the appointment of a commander to a military unit drawn from the nomads (Macdonald 2014, 154-156). Since the corpus also attests of the phrases *s^1nt ngyt ʾ-mlkt* "the year the queen was announced" (SIJ 786; translation: Al-Jallad 2015, 282) and *s^1nt ngy ḥrṯt* "the year Aretas was announced (king)" (HASI 23), we can safely deduce that the formula *s^1nt PN s^1lṭn* occurring in NEH 7 as well as in LP 424 (= Is.M 106) refers in fact to the appointment of governors, certainly some Nabataean ones if one takes into account their onomastic and the content of NEH 1-2 discussed above (Fig. 11).[28]

The editors read the governor's patronym as *Rbʾl*. Nevertheless, one should note that there is a clear bow-shaped line occurring between *b* and *ʾ*, which may lead us to re-read this as *Rbbʾl*. However, this mark presents a slightly different orientation from the first *b* so it remains difficult to be sure whether it belongs or not to the text. It therefore seems best to consider *Rbʾl* and *Rb{b}ʾl* as two probable readings. Whichever the correct reading is, this person has clearly nothing to do with the Nabataean king Rabbel II. I therefore see no reasons to justify the editors' view, who class this inscription among "the Safaitic texts mentioning events dated to the reign of Rbʾl" (Al-Salameen *et al.* 2018, 61, 67-68). On the other hand, if *Rbbʾl* is the correct reading and this person is the same as the one described as a *s^1lṭn* in NEH 8 (see below), one may be tempted to link him with the Nabataean *strategos Rbybʾl* son of *Dmsps* to whom we referred earlier, knowing that his name would precisely appear as *Rbbʾl* in Safaitic. However, the Nabataean documents give no information at all about a son of *Rbybʾl*

28 The reading and translation of LP 424 proposed by Littmann was *l mlk bn ġyrʾl bn nyr w nẓr s^1nt ngy mn-h-s^1lṭn* "By Mlk son of Ġyrʾl son of Nyr and he was on the look-out in the year in which he escaped from the government" (Littmann 1943, 114). The modern photograph taken by the team of the Safaitic Epigraphic Survey Programme (SESP), which is available on OCIANA under the *siglum* Is.M 106, allows us to re-read this as *l mlk bn ġyrʾl bn n{h/y}b w nẓr s^1nt ngy mlk s^1lṭn* "By Mlk son of Ġyrʾl son of N{h/y}b and he stood guard the year Mlk was announced governor" (Fig. 11). There is maybe a second reference to this governor in RQ.A 10, if one interprets the phrase *s^1rt ʾl-mlk h-s^1lṭn* as "he served in a troop under Mlk the governor" instead as "he served in a troop under the authority of the governor", as Al-Jallad (2015, 149) suggests.

Figure 11. Inscription LP 424 (= Is.M 106) from Al-ʿĪsawī in southern Syria (photograph: OCIANA, http://krc.orient.ox.ac.uk/ociana/corpus/pages/OCIANA_0026876.html).

whose name would be spelt as S^1*ʿd* in Safaitic, his only two sons of whom we are so far aware of are *Dmsy* (JSNab 84) and *Mlkw* (JSNab 34). Moreover, NEH 8 describes *Rbbʾl* as the governor of "Gilead" while *Rbybʾl* son of *Dmsps* seems to have exercised his functions exclusively in the southern sector of the Nabataean kingdom (cf. Fig. 8), from Wādī Ramm and Dūmat al-Jandal up to Ḥegrā (Nehmé 2015, 119). These lead us to admit that there is no demonstrable connection between these two homonymous governors.[29]

The final section of the inscription appears extremely difficult to interpret. As rightly noted by the editors, who cautiously decided to leave it untranslated, it does not conform to one of the usual formulae used to end an inscription or an invocation in Safaitic. Very tentatively, I would however propose the following. As emphasised by Al-Salameen *et al.* (2018, 68), the root √gty is unproductive in the Semitic languages. I therefore suggest that the first letter does not represent *g* but *m*. The glyph consists of an elongated ellipse which appears extremely different from the three *g*-s in *ʾdg*, *wgd* and *ngy* that present a much more larger and rounded shape. Although it is true that it is not perfectly identical with the *m* in s^1*lm* standing just before, it presents a similar bent spine and finds a very good parallel with the first *m* occurring in LP 424 mentioned above. If correct, one may have the substantive *mt* "death" followed by a word that I would read *yl{ġ}b*, not *lns²b* as the editors do. The small dash that they take as *n* is clearly a natural mark on the stone as others occur in this area, showing a whitish colour which has nothing to do with the reddish patina of the two Safaitic inscriptions. Similarly as with the patronym *S²mrʾl/{Ġ}{y}rʾl* discussed above, I believe that the following glyph does not represent s^2 but *ġ*, appearing only bended on its top. If I am correct, this could represent a D-stem verb in the third person masculine singular of the prefix conjugation corresponding in meaning with the Classical Arabic Gt *talaġġaba* "he chased, hunted, pursued" (*talaġġaba-nī dahr* "Fortune long pursued me"; Lane 2663c). The prefix conjugation can probably be used here to express an independent volitive sentence, as is the case in KRS 583 which ends in *w hmr ygy h lh* "so let the rain flow, O Lh!" (translation: Al-Jallad 2017b, 84).[30] The final word of the text, *rhq*, is so far only attested as a personal name in Safaitic (HIn 290). It is likely to represent here a G-stem active participle of the Arabic root √rhq, the basis meaning of which is "to come upon, cover, a

29 Similarly, one may be tempted to see a connection between the s^1*lṭn* Mlk to whom LP 424 and RQ.A 10 (?) refer and the Nabataean *strategos* Mlkw son of *Rbybʾl* (JSNab 34). Nevertheless, Ḥegrā where the latter was in service appears separated by about 680 km from Al-ʿĪsawī where LP 424 was discovered, which makes it difficult to admit that the appointment of a new governor in this distant city of the Ḥijāz would have represented an event of importance for a nomad of the *ḥarrah* desert.

30 On the use of the prefix conjugation, either independently or with the asseverative *l*, as a volitive, see Al-Jallad 2015, 109-110.

Figure 12. The Safaitic inscription NEH 8 from site K45 (tracing: J. Norris 2018).

thing" (Lane 1170b; cf. *Qur'ān* 80, 41: *tarhaqu-hā qatarah* "darkness will cover them"). A fair assumption might be that this refers to the potential vandals who would recover or overwrite the inscription, as an alternative to the basic curse formula *ḏ/m(n) yʿwr* "him/whosoever would efface (the inscription)". If my preceding remarks are correct, they would permit the following re-interpretation:

l r{.}{.} bn {ġ}{y}r'l bn ʺdg w <w> wgd 'ṯr ʿm-h w qṣṣ s^{1}nt ngy s^{1}ʿd bn rb{b}'l s^{1}lṭn f h lt s^{1}lm w {m}t yl{ġ}b rhq

"By R{.}{.} son of {Ġ}{y}r'l son of ʺdg and he found the traces of his paternal grandfather while he patrolled the year S^{1}ʿd son of {Rb'l/Rbb'l} was announced governor, so, O Lt, may he be secure and let death pursue any obscurer (of this inscription)."

NEH 8

This text was recorded on site K45, which consists of an ancient camp site composed of several stone enclosures located on the summit of a hill on the southern bank of Wādī al-Ḥashād, 15 km north-west of K27 where the previous inscription comes from. According to the editors, the text reads *l ysm'l bn rgl bn ġyr'l wr'y hḍ'n wwrd hrb snt qttl rb'l slṭn glʿd* "By Ysm'l son of Rgl son of Ġyr'l and he pastured the sheep and watered [in] Hrb [in] the year Rb'l the governor of Glʿd, waged war" (Al-Salameen *et al.* 2018, 68). The only problem in the reading is the governor's name. In contrast to NEH 7, here there is no ambiguity in the presence of two successive *b*-s (Fig. 12), which leads us to correct it as *Rbb'l*. Thus, one should read:

l ys^{1}m'l bn rgl bn ġyr'l w r'y h-ḍ'n w wrd hrb s^{1}nt qttl rbb'l s^{1}lṭn glʿd

"By Ys^{1}m'l son of Rgl son of Ġyr'l and he pastured the sheep and then went to water at Hrb, the year Rbb'l the governor of Gilead waged war."

The author's name and genealogy appear of interest as I noted that *Rgl* could well represent the name of the author of NEH 7 and *Ġyr'l* his patronym. Nevertheless, I would refrain from identifying the author of NEH 8 as the son of the one of NEH 7, because of the uncertainties in the reading and the absence of a much more detailed genealogy for NEH 8. I have followed the editors' suggestion in interpreting the word *Hrb* as a place name, since this is precisely what one may expect after the verb *wrd*. Compare with the phrase *w wrd h-nmrt* "and he went to water at Namārah" which occurs in about thirty Safaitic inscriptions of the OCIANA corpus. Note that the G-stem active participle of root √hrb in Classical Arabic means "one who returns from water" (Lane 2889c) and that the Palmyrene Arabic attests the word *hrubbe* "cittern" (DRS 447), which may offer a suitable etymology to explain the name of this place of water whose location remains unknown.[31]

Of course, the great interest of this text lies in the mention of a "governor of Gilead" who, according to his personal name, which clearly echoes the Nabataean *Rbyb'l*, is likely to be a Nabataean official rather than a

31 An alternative interpretation of the phrase *w wrd hrb* could be to take *hrb* either as an infinitive expressing purpose or as an active participle used adverbially. Given the opposite meanings of *wrd* "go to water" and *hrb* "returning from water", this appears however less likely than the place name hypothesis suggested by Al-Salameen *et al.* (2018, 68). Otherwise, one could suggest taking *wrd* as "going down" (cf. Al-Jallad 2015, 353) and *hrb* as "to flee/ fleeing" or "to return/returning from water", but this does not make satisfactory sense. Finally, one can also posit a reading *w wrd h-rb* "and he went to water at the Rb" in assuming that *rb* could represent a word or a toponym referring to a place with abundance of herbage or water, similarly as the term *mrb* (cf. Classical Arabic *marabb*) which derives from the same root. I thank C. Della Puppa for this brilliant suggestion.

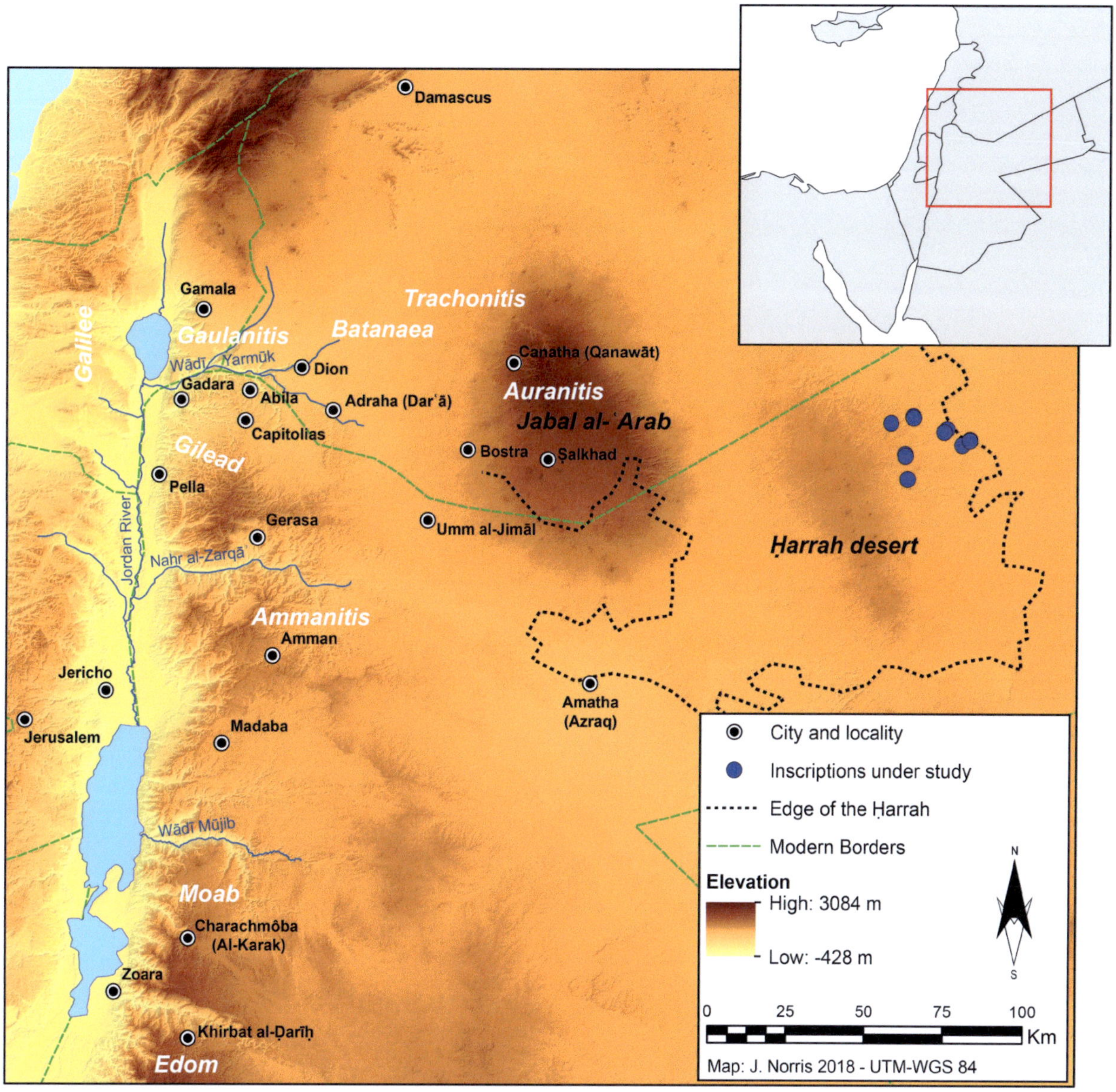

Figure 13. Map of northern Transjordan with the location of the Gilead region and the main regional towns (J. Norris 2018).

Seleucid, Hasmonean/Herodian or Roman one.[32] This is quite intriguing, since the rulers of Petra have only briefly dominated this area of northern Transjordan (Fig. 13), to which three other Safaitic texts refer (C 2473, KRS 15, ASWS 168). The editors recall two well-known episodes where the Nabataeans are explicitly mentioned in relation with Gilead, though they do not discuss the historical implications of the inscription further. The first is the meeting which the Maccabee brothers had with the Nabataeans in AD 164/163 somewhere in the desert east of the Jordan, where they were informed by the latter of the misfortunes of which the Jewish inhabitants of "Galaad" (the Hellenised form of the Semitic *Glʿd*) were the victims (1Macc. 5.25-26). The second is the conquest by Alexander Jannaeus of some Nabataean territories in the lands of Galaad and Moab around 90 BC (Al-Salameen *et al.* 2018, 69). Our mention of a presumably Nabataean governor of Gilead can hardly have something to do with the first episode. Indeed, this took place when the kings of Petra were not yet representing the authority in northern Transjordan and southern Syria, which were parts of the Seleucid kingdom in that period (Sartre 2001, 412-413). The

32 See the discussion on NEH 7 above about the absence of evidence to connect this governor of northern Transjordan to *Rbybʾl* son of *Dmsps*, unless we assume, without any proof, that *Rbybʾl* was in service in northern Nabataea before having been appointed to the south.

context of the second episode, prior to which the Nabataean penetration in the north had greatly progressed, appears however more suitable. We know that around 100 BC, Jannaeus attacked Moab and Galaad and that these two regions are during this event described as belonging to 'Arabia' and to be inhabited by 'Arabs' (*AJ* 13.374; *BJ* 1.89). This might imply that the Nabataeans had extended their influence there sometime before, though this cannot be confirmed, because other 'Arab' groups besides the Nabataeans are occasionally found in Gilead and because this area has actually been considered as an 'Arabia' at least since the third century BC.[33]

If Nabataean control of Gilead during the late second century BC cannot be confirmed, it is known for sure that the area was integrated into the Nabataean kingdom after the victory won over Jannaeus by Obodas I (*c.* 96-85 BC) in 93 BC at Garada in the Golan (*AJ* 13.375; *BJ* 1.90). According to Josephus, this reduced the Hasmonean to surrender "to the king of the Arabs the territory which he had conquered in Moab and Galaaditis" (*AJ* 13.382). The activity of the governor to whom NEH 8 refers is therefore likely to have taken place sometime between 93 BC and the subsequent campaigns of Jannaeus in 83/82 BC and those of the Romans in 64/63 BC, which pushed the Nabataeans back from the western and fertile area of northern Transjordan and resulted in the integration of the Greek cities of the area (Pella, Gadara, Abila, Capitolias, Dion, Gerasa and Philadelphia) into the Roman province of Syria.[34] If so, one of the different battles in which the Nabataeans fought against the Hasmoneans and the Romans for the control of Galaad and the Golan might even have been the war that *Rbyb'l* is said to have conducted (*s¹nt qttl rbb'l*).

Another possibility would be to link this mention of a governor of Gilead to the campaign waged by Aretas IV against Herod Antipas around AD 37/38 near Gamala, north of the Yarmūk (*AJ* 18.113-114). According to Josephus, the pretext of the conflict was the repudiation ten years before of Aretas' daughter to whom Antipas was married (*AJ* 18.110-114). However, G.W. Bowersock (1983, 36-37) has formulated the hypothesis that Aretas' intention could instead have been to regain areas of former Nabataean influences in the north, the opportunity of which was provided by the death of Philipp and the integration of his tetrarchy to the province of Syria as well as by the apparent absence of a Roman governor at Antioch during this period. According to this theory, Aretas' remarkable victory over Antipas in the Golan (*AJ* 18.114-116) would have persuaded him to push farther north and to even annex Damascus for a short time, explaining the famous reference in 2Cor. 11.32 to an *ethnarch* of Aretas in this city (Bowersock 1983, 68). If this hypothesis is correct, the Gilead may also have temporally fallen back under Nabataean control, but there is unfortunately no way to confirm this, since the available sources give no detail about the exact places which were won by the Nabataeans after Antipas' defeat. However, it remains once again tempting to see a connection between the war described in our inscription and this battle taking place immediately north of Gilead during which some *strategoi* are said to have been dispatched at the head of both the Herodian and the Nabataean troops.[35]

It should be remembered, on the other hand, that the geographical limits of the Gilead region are somewhat vague and appear to have been shifting throughout the ages, which could make it impossible to determine when a Nabataean governor was actually in activity there. Mentioned almost a hundred times in the Old Testament, the term 'Gilead' has indeed both a narrow and a broad sense in the Biblical texts. Originally, it was restricted to the mountainous area located between the modern Al-Ṣalṭ in the south and the Jabbok river (Nahr al-Zarqāʾ) in the north, still known today as 'Ǧabal Ǧilʿad' among the Christian communities. Over time, the toponym came to be used for the lands both south and north of the Jabbok, generally from the Yarmūk up to the Wādī Ḥisbān/Wādī Kafrayn, but extending in some instances even beyond the Yarmūk in the north and up to the Arnon river (Wādī Mūjib) in the south.[36] Regarding its eastern limits, the Gilead land is generally considered as being bordered by the steppe region beyond Gerasa (Jerash), though these are occasionally extended as far east as to the Jabal Ḥawrān/Jabal al-ʿArab and the desert towards the Euphrates (Heidet 1903, 48-49). In considering such a wider sense of the term, one should recognise that this "governor of Gilead" could therefore well have been in office in a city such as Boṣrā or Umm al-Jimāl, which remained under Nabataean

33 See Bowersock (1983, 19) and Sartre (2001, 413) about the surprise attack of the Maccabee brothers by some non-Nabataean Arabs somewhere between Gilead and the Ḥawrān, and Macdonald (2003, 314) about the description of southern Gilead as an "Arabia" during the Syrian campaign of Antiochos III in 218 BC. This relationship between Gilead and Arabs, or Arabians, appears quite old if one thinks back to the Biblical account describing the caravan of the Ishmaelites/Madianites who, coming from Galaad, brought Josephus to Egypt (Gen. 37.25-36).

34 The campaign of Alexander Jannaeus resulted, in Gilead, in the reconquest of Pella, Dion and Gadara (*AJ* 13.393-397). See Bowersock 1983, 25; Sartre 2001, 394. See also Gatier (1988, 159-163) who demonstrates that the last regional cities in the hands of the Nabataeans were Gerasa and Philadelphia, which Pompey's officer, M. Aemilius Scaurus, seized from them in 63 BC. According to the same scholar, the epithet of "Philippian" borne by the inhabitants of Pella may go back to L. Marcius Philippus, propraetor of Syria in 59, for having defended their "freedom" against the Nabataean Arabs (Gatier 1988, 163).

35 *AJ* 18.113: καὶ δυνάμεως ἑκατέρῳ συλλεγείσης εἰς πόλεμον καθίσταντο στρατηγοὺς ἀπεσταλκότες ἀνθ᾽ ἑαυτῶν.

36 Heidet 1903, 48-49; Ottosson 1992, 1020; Macdonald 2000, 195-208.

Figure 14. The Safaitic inscription NEH 9 from site K42 (tracing: J. Norris 2018).

control throughout the first century BC to first century AD. This is not at all unlikely in light of the text 1Macc. 5.24-27 mentioned above, which lists Boṣrā among the cities of "Galaaditis" where the Jews were threatened (see Sartre 1985, 46-47). Of course, this proposition appears possible only if one admits that the nomads of the *ḥarrah* had also developed such a wide definition of the term 'Gilead', which is at present impossible to know. Among the three other Safaitic texts which refer to the *Glʿd* land (C 2473, KRS 15, ASWS 168), only C 2473 gives vague geographical information when its author indicates that he *ʾhmd ʾbl f rdf m-glʿd* "stayed at Abila and then returned from Gilead". This implies that the cities located south of the Yarmūk were indeed perceived as belonging to Gilead, though this cannot help in determining what the limits of this district were according to the author and where he situated the border between Gilead and the Ḥawrān *(ḥrn).*

NEH 9

Found at K42, a site located in the south-western part of Wādī al-Shuwayṭī, this inscription represents a beautiful composition written in the so-called 'Safaitic square script' which appears, moreover, precisely dated according to Rabbel II's crowning in AD 70/71 (Fig. 14).[37] The editors read *l ʿzz bn ṣyd bn qdm ḏʾl kkb wyʿmr bṣlḫd wdṯʾ snt mlk rbʾl* "By ʿzz son of Ṣyd son of Qdm of the tribe of Kkb and he dwells in Ṣlḫd and he spent the season of later rains [in] the year of king Rbʾl" (Al-Salameen *et al.* 2018, 69). We have here a rare and interesting case where the narrative section makes use of a verb in the prefix conjugation, which the *editio princeps* suggests to render by the indicative imperfect "he dwells". However, since the action is said to take place in Ṣalkhad,

37 For the photograph, see Al-Salameen *et al.* 2018, 69, fig. 9.

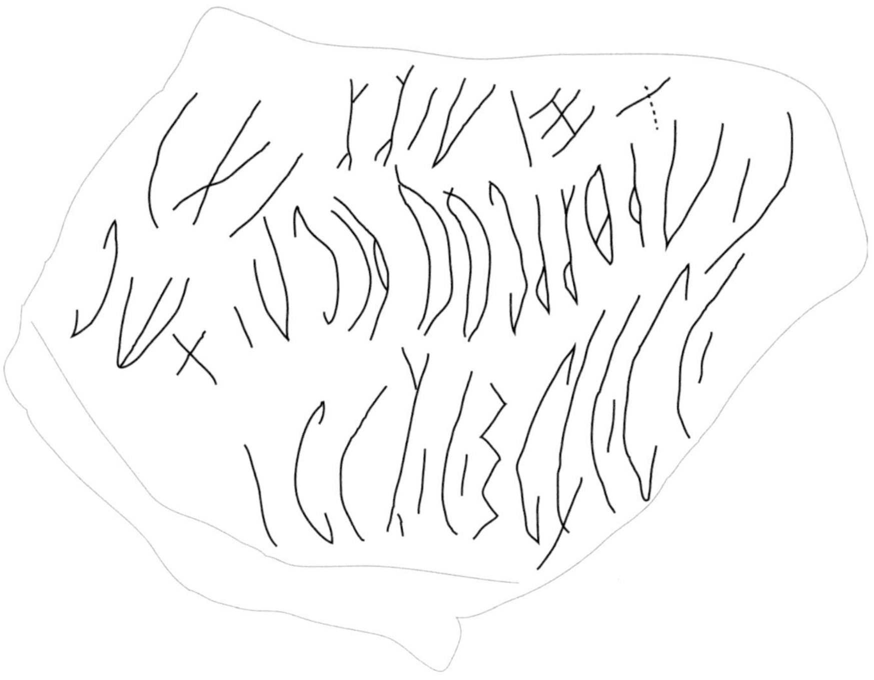

Figure 15. The Safaitic inscription NEH 14 from site K29 (tracing J. Norris 2018).

an important city located in the southern part of the Jabal al-ʿArab (about 95 km west of K42), one has to admit that this can hardly refer to the present circumstances of the writer. Two alternative interpretations seem therefore possible. First, one can analyse this phrase as making a preterit use of the unaugmented prefix conjugation, which would explain why the following verb, *dṯʾ*, occurs for its part in the suffix conjugation. Thus, "and he dwelt in Ṣalkhad and then spent the season of the later rains (here)". If this is correct, NEH 9 would find a good parallel with the Graeco-Arabic inscription A1 that mixes the prefix and suffix conjugated verbs **yrʿw* (ειραυ) and **ʾtw* (αθαοα) within the same sentence, both used to express a past tense, "they pastured" and "he came" respectively (Al-Jallad and Al-Manaser 2015, 52-57; Al-Jallad 2015, 111). A second possibility is to interpret *yʿmr* as expressing a future tense and to take the suffix conjugated *dṯʾ* as an anterior adverbial clause introduced by the conjunction *w*.

If one considers the former suggestion, we note that the carver would have spent the rainy season of the winter at Ṣalkhad, after which he returned into the desert only for the mid-February to mid-April period, just before the summer. This sounds like quite a strange scenario with respect to the normal migration patterns of the *ḥarrah* nomads (Macdonald 1992, 9-10). Many of the Safaitic texts which give us evidence of some exceptional journeys taken by their authors to the Ḥawrān exhibit the formula *ʾs²rq m-ḥrn* "he migrated to the inner desert from the Ḥawrān" (C 2021, 3339; RWQ 331), which implies that these had taken place before, not during, the winter. This is also reflected in HaNSB 197 which places the departure from the Ḥawrān to the inner desert during the rising of Capricorn (mid-December to mid-January)[38] and KRS 1706 whose author says that he returned, or was present, with his sheep in the Ḥawrān during the rising of Taurus (mid-April to mid-May).[39] It is clear that this is the long dry season of *qyẓ* (mid-April to early October), in which these individuals were privileged to go to the Ḥawrān, when the nomads were forced to abandon the inner desert and camp near permanent sources of water. This leads me to believe that the second proposition is the most likely and that the verb *yʿmr* should refer to a future time, Ṣalkhad being the place where the author intended to stay during the summer. As a great deal of uncertainty still surrounds the different functions of the prefix conjugation in Safaitic (see Al-Jallad 2015, 107-112), this should however remain a working hypothesis. If one accepts it, then we can read:

l ʿzz bn ṣyd bn qdm ḏ-l kkb w yʿmr b-ṣlḫd w dṯʾ s¹nt mlk rbʾl
"By ʿzz son of Ṣyd son of Qdm of the lineage of Kkb and he will dwell in Ṣalkhad after having spent the season of the later rains (here) the year Rabbel was made king."

38 HaNSB 197: *w ʾs²rq m-ḥrn b-ʾbl-h s²ʿr b-rʾy yʾm{r}* "and he migrated to the inner desert from the Ḥawrān with his camels to herbage during the rising of {Capricorn}". On the Safaitic zodiac, see Al-Jallad 2014.

39 KRS 1706: *w ʿyd h-ḍʾn b-ḥrn b-rʾy ʾʾly* "and he put the sheep in an enclosure in the Ḥawrān during the rising of Taurus" (translation: Al-Jallad 2015, 261).

NEH 14

This text comes from K29, a site located in the eastern part of Wādī al-Shuwaiṭī, one kilometre south from K27. The *editio princeps* reads *l rb'l bn šrk bn rbn bn sd wṣyr mmdbr snt ḥrb ḫl 'hl slṭt* "By Rb'l son of šrk son of Rbn bn Sd and he returned from the desert [in] the year [that] Ḫl fought the family of the authority" (Al-Salameen *et al.* 2018, 73, fig. 14) (Fig. 15). If one accepts this interpretation, the text would contain a new word derived from the root √s^1lṭ that the editors translate as "authority", representing the exact equivalent of Classical Arabic *sulṭah* (pl. *suluṭāt*) which bears precisely that meaning. Nevertheless, some doubts may be raised about a reading *'hl s^1lṭt* and the suggested translation of "the family of the authority" that does not seem to make very much sense.

A careful examination of the photograph leads us to recognise that the word *'hl* "family, people" is a misreading. The letter interpreted as *h* by the editors clearly bears a fork on both its extremities, indicating that it actually corresponds to '. Although other interpretations seem possible, the simplest way to explain these two successive '-s is to assume that we are dealing with the personal name *Ḫl'* (HIn 225), followed by the *'l*-definite article whose coda would not be assimilated to the following sun-letter, as this occurs in C 5137 (*w ḥll 'l-dr*) and C 2446 (*'l-nbṭy*).[40] An alternative, though more speculative, possibility would be to ascribe one of the two '-s to a dittographic mistake made by the carver. *Ḫl* indeed represents a far more common personal name than *Ḫl'* in Safaitic, appearing about 69 times in the OCIANA corpus whereas *Ḫl'* is attested only four times. Moreover, there is another Safaitic inscription dated according to an event that one might be tempted to connect to the one described here, *ṣyr m-mdbr s^1nt qtl 'l ḫl 'l nqd h-s^1lṭn* "and he returned to a place of water from the inner desert, the year the lineage of Ḫl [and] the lineage of Nqd fought the governor/authorities" (AWS 341).

With regard to the definite word, its final letter is not as clear as it is represented on the editor's tracing. One of its two lines appears rather faint on the photograph and does not clearly extend past the other one, raising the question of whether it really belongs to the text. If not, this should represent a *n* whose orientation, which points right at roughly 30°, would align on that of the previous *ṭ*. In view of these uncertainties, I would therefore consider both *s^1lṭn* "governor" and *s^1lṭt* "authority" as probable readings.

l rb'{n/l} bn s^2rk bn rbn bn s^1d w ṣyr m-mdbr s^1nt ḥrb {ḫl'/ḫl <'>} 'l-s^1lṭ{n/t}

"By Rb'{n/l} son of S^2rk son of Rbn son of S^1d and he returned to a place of water from the inner desert, the year {Ḫl'/Ḫl} waged war upon the {governor/authority}."[41]

Nabataean section

NEH 15 A-B

The last document which deserves discussion is the Nabataean text(s) NEH 15A-B, recorded on site S2, a hill located south-west of Qāʿ al-Khuwaymāt. This consists of two beautiful lines written in the Classical Nabataean script (Fig. 16). The editors interpret them as two separate texts surprisingly introduced with the *lām auctoris*, which leads them to characterise their contents as "mixed Nabataean-Safaitic". They read the first line as *l qymtw dkrwn* "Remembrance of Qymtw" (NEH 15 A) and the second as *l mlkt* "By Mlkt" (NEH 15 B) (Al-Salameen *et al.* 2018, 74-75). However, two reasons support the idea that we are instead dealing with a single inscription.

Firstly, in contrast to what the editors say in claiming that "the names of these two persons are attested previously in Nabataean", the anthroponym occurring in the upper line has actually never been encountered before with the *-w* ending, being so far known exclusively in the form *Qymt*.[42] This is not surprising as the Nabataean names constructed with a *-t* morpheme, which belong to the class of the so-called "diptotic" names, normally do not take *wāwation* (*Bgrt, Hn't, W'lt, Ḥrtt, Klybt, Mnʿt, ʿbdt, ʿmyrt, Šqylt, etc.*),[43] with the exception of extremely rare cases occurring after the loss of the Nabataean case system (Al-Jallad, in press). This *w* can consequently better be explained as representing the coordinating conjunction "and". The fact that it appears ligatured to the preceding

40 For discussions on the unassimilating *'l*-article, see Macdonald 2000, 51; Al-Jallad 2015, 74-75; 2017a, 167-168.

41 I have marked the last letter of the author's name as doubtful and being capable of representing both *l* and *n* since it appears rather small for a *l*, presenting almost the same size as the *n* in *bn*. Nevertheless, *Rb'n* is so far known from only two Safaitic inscriptions from the OCIANA corpus, one of which the reading is doubtful, so the editor's reading *Rb'l* is certainly the most likely. Note that a brother of the author has left C 4121 and that C 4123 appears to have been carved by his father.

42 See CIS II 366 (doubtful), 368, 371, 400 (= MP 170), 404 (= MP 172); RES 1398; JSNab 324 and SSAI 8, references taken from Cantineau (1932, 142) and Negev (1991, 58). Note that in Negev's notice, the figure 7 is misplaced and should be in the EM column. To these, one should now add Jobling (1982, 203) from the Wādī Ramm area; TANI.Nab 73 from near Tabūk; MP 135.1, 156 and 164 from Petra. From the Greek transcriptions, it seems that there are two different names which the Nabataean and the Ancient North Arabian form *Qymt* could recover. These are Καιαμαθου */Qayāmat/ (PPUAES IIIA, 5,751) and Καεμαθος */Qayyimat/ (IGLS XXI, 5.1, 326).

43 Among other examples, see CIS II 238 where the *p'l*-name *Tymw* takes *wāwation* but not the *p'lt*-one *Hn't*.

Figure 16. The Nabataean inscription NEH 15A-B from site S2 (photograph by Ali Al-Manaser, OCIANA Badia Survey Project).

word does not represent a specific obstacle to this interpretation since there is a number of texts in which this situation can precisely be observed.[44] Secondly, the word *dkrwn*, which represents an abstract noun in the absolute state attested in different varieties of Middle and Late Aramaic with the meaning of "remembrance, memory, commemoration",[45] occurs systematically in Nabataean before (not after) a benefactive that is introduced by the preposition *l-*.[46] This is within some very standardised formulae for commemoration that present seven types of variants expressed or not "in the presence" of a deity (Table 2). Whereas the editors are certainly right in taking the first *l* as a particle used to introduce the author's name, which remains to be quite an exceptional occurrence in Nabataean epigraphy,[47] the *l* before *Mlkt* should definitely represent the dative preposition "to, for". Consequently, I would suggest it be read as:

l qymt w dkrwn
l-mlkt
"By Qymt, and may Mlkt have remembrance."

44 See, for instance, Savignac and Starcky 1957, line 3 where the conjunction is attached to the *t* of *[dwm]t*; ThNS 7, line 2 where it is attached to the š of *prš*; ThNUJ 217 (= ThMNN 692), line 1 and ThMNN 828, line 2 where it is attached to the *b* of *ṭb, etc.*

45 Schulthess 1903, 46; DNWSI 322, 330-331.

46 The only document presenting an exception is CIS II 338, which seems to exhibit a juxtaposition of *dkrwn* to a personal name read *ʾryš* by the editor. However, this may either be the result of a writing error or of a miscopying of the text which remains so far known from only a bad drawing made by Huber in 1884, knowing that there is no other attestation of *ʾryš* in the Nabataean onomasticon.

47 Al-Salameen *et al.* (2018, 74) list five Nabataean inscriptions from north-west Arabia that would begin with the *lām auctoris*: ThMNN 6 (= JSNab 45), 7 (= JSNab 46), 564 (= ThNUJ 92), 616 (= ThNUJ 142), 678 (= ThNUJ 204). Unfortunately, the presence of a *l* appears extremely difficult to verify in almost all of these texts. Although it seems clear in ThMNN 6 and 7 (JSNab 45 and 46), one should recognise that it could well function as the dative preposition instead of as an introductive particle in JSNab 45, *l rmʾl br ḥyw šlm* "May Rmʾl son of Ḥyw have security". In ThMNN 678 (= ThNUJ 204), I read the personal name *Yšlm* (cf. the Ancient North Arabian form *Ys¹lm*) and not the phrase *l šlm*. There is a number of texts from the Ḥawrān which are, or seem to be, introduced by *l-* (CIS II 172, 187, LPNab 105), but this is in dative phrases expressing possession of funeral monuments. A comparable situation applies to the Nabataean inscription from Kharazah (MNIN 8b), in the Wādī Ramm area, where the *l* by which the text begins is used to express the property of a dam near to which the text is carved, *l-šbʿ br ʾlh ʿtyd šnt ʾrbʿyn w ḥdh l-ḥrtt mlk nbṭw* "to Šbʿ son of ʾlh [belongs this dam] constructed in the year forty-one of Ḥrtt, king of the Nabataeans".

Type		**Text**	
Transcription	*Translation*	*Reference*	*Provenance*
Simple memorials			
dkrwn l-PN	May PN have remembrance	TANI.Nab 14	Jabal Abū Makhrūq (north-west Arabia)
		NEH 15A-B	Qāʿ al-Khuwaymāt area (Ḥarrat al-Shām)
dkrwn ṭb l-PN	May PN have good remembrance	CIS II 163	Sīʿ (Ḥawrān)
		TANI.Nab 68	Tabūk area (north-west Arabia)
		CIS II 426e	Al-Maʿaytharah (Petra)
		CIS II 478	Al-Bayḍāʾ (Petra)
		MP 154	Jabal al-Madhbaḥ (Petra)
dkrwn ṭb w šlm l-PN	May PN have good remembrance and security	NNIA 5	Oboda/Avdat (Negev)
		CIS II 3072	Wādī Naṣb (Sinai)
		CIS II 476	Al-Bayḍāʾ (Petra)
		MP 24	Al-Madras (Petra)
		MP 114	Jabal al-Madhbaḥ (Petra)
		MP 160	Jabal al-Madhbaḥ (Petra)
		MP 135.1	Wādī Farasah (Petra)
		RSIWU 3	Wādī Abū ʿUllayqah (Petra)
		CIS II 316	Mabrak al-Nāqah (north-west Arabia)
Memorials before a deity			
dkrwn l-PN mn qdm DN	May PN have remembrance in the presence of DN	CIS II 338	Laqaṭ (north-west Arabia)
dkrwn ṭb l-PN mn qdm DN	May PN have good remembrance in the presence of DN	MIRP 7.1	Shuʿab Qays (Petra)
		CIS II 236	Jabal Ithlib (Ḥegrā)
dkrwn ṭb w šlm mn qdm DN l-PN	May PN have good remembrance and security in the presence of DN	MP 156	Jabal al-Madhbaḥ (Petra)
		ThNUJ 217	Umm Jadhāyidh (north-west Arabia)
dkrwn ṭb l-PN b-ʿydn DN	May PN have good remembrance by the protection of DN	RSIWU 1	Wādī Abū ʿUllayqah (Petra)

Table 2. List of the different types of Nabataean texts for remembrance expressed with the noun *dkrwn*.

Conclusive remarks

As Zeyad Al-Salameen, Younis Al-Shdaifat and Rafe Harahsheh (2018, 76) emphasise, this group of texts, which contain some "Nabataean echoes", constitutes a remarkable addition to the corpus of Safaitic and Nabataean texts from north-eastern Jordan. They provide a further confirmation of a striking situation to which Michael Macdonald has recently brought attention to, which is the surprising amount of information that the nomads of the *ḥarrah* had about important political events taking place in the neighbouring settled areas of Syria and Transjordan (Macdonald 2014, 163), some of which had not even left any traces within the historical sources! They also illustrate the complex relationships which their writers had developed with the inhabitants and powers of these regions, prominently the Nabataeans, towards whom the nomads expressed ambivalent attitudes which include the assertion of a 'Nabataean identity' for some individuals and the request for booty or victory against them by others.

Having re-examined and suggested a number of alternative interpretations on eight Safaitic and one Nabataean inscriptions from this collection, I would like to conclude the present study with three remarks. First, the re-interpretation of the phrase *s^1nt ngy PN s^1lṭn* occurring in NEH 7 and LP 424 as "the year PN was announced *s^1lṭn*" provides further evidence supporting the editors' explanation of the term *s^1lṭn* as the title of an official in charge of a territorial district who could be the same as the Nabataean *ʾsrtg* (*strategos*), the best translation of which should be "governor". Whereas the names borne by two of these *s^1lṭn*-s, *Mlk* (LP 424, RQ.A 10?) and *Rbbʾl* (NEH 7? and 8), obviously evoke the figures of two Nabataean *strategoi* of Ḥegrā belonging to the well-known family of Damasippos, *Rbybʾl* and his son *Mlkw*, we have pointed out that there is at present no demonstrable evidence to suggest a connection between them.

Second, if the identification of the "governor of Gilead" mentioned in NEH 8 as a Nabataean is correct, we have noted two suitable episodes during which the appointment of a Nabataean official in this region could have taken place. These are the periods of 93-64/63 BC during which the kings of Petra took control of northern Transjordan and the Golan after the battle of Garada, and the year AD 37/38 when a war broke out between Herod Antipas and Aretas IV after which a number of northern territories could have been briefly re-occupied by the Nabataeans. The broader sense which the term "Gilead/ Galaad" occasionally bears in the sources and the fact that this land lacks any real eastern boundaries have, however, lead to emphasise that one cannot exclude that this official could simply have been in service in regions such as that of Umm al-Jimāl and Boṣrā, especially since we are ignorant of the limits the nomads of the *ḥarrah*, as well as the Nabataeans, were giving to this toponym.

Finally, the re-examination of the available documents referring to the so-called 'revolt of Damaṣī' leads one to express some serious doubts about the scenario reconstructed by Winnett in 1973 which has generally been accepted among the scientific community, although it appears to be based on a number of circular arguments and a series of misunderstandings about the verb *qṣṣ*. There is not a single inscription indicating that the insurrection actually took place in AD 70/71 and that this was directed against Rabbel II. The only argument supporting the hypothesis that the military chief called *Dmṣy* in the Safaitic inscriptions may be the Nabataean *Dmsy* son of *Rbybʾl* is the equivalence of the Ancient North Arabian form *Dmṣy* with the Greek-like name spelt *Dmsy* in Nabataean Aramaic that has recently been confirmed in light of a new inscription from Dūmat al-Jandal. However, this appears far from being sufficient proof and I would say that more material is needed to evaluate the exact extent of this mysterious event and determine when, where and against whom it really broke out.

Acknowledgements

I am very grateful to Peter Akkermans and Ahmad Al-Jallad for the invitation to the Landscapes of Survival international conference held at Leiden on 17-18 March 2017 to which I was honoured to participate; what a great moment of inspiration and discussion. My lecture was about a Thamudic B inscription from southern Jordan, though I decided to change the subject matter of my contribution as I felt that the present discussion on the epigraphy of the *ḥarrah* would fit better with the topic of the proceedings. I therefore owe additional thanks to the organisers for having accepted this modification and for all the fruitful exchanges I had with them about the archaeology and epigraphy of north-eastern Jordan. I would also like express my gratitude to all the persons involved in the organisation of the conference, especially to Merel Brüning who considerably facilitated my travel and stay at Leiden. My sincere thanks also go Hani Hayajneh and Michael C.A. Macdonald for their insightful remarks on the presentation I made at the conference, and to Ali Al-Manaser, Chiara Della Puppa and Guillaume Charloux for their most valuable comments and suggestions. I own additional thanks to Ali Al-Manaser and Michael C.A. Macdonald for the photographs of NEH 5-6 and NEH 15A-B they have generously sent me and for having authorised me to publish them. All errors are, of course, my own.

Sigla

A1	Graeco-Arabic inscription from Wādī Salmā published by Al-Jallad and Al-Manaser 2015
AbaNS	Safaitic inscriptions in Ababneh 2005
AH	Dadanitic inscriptions in Abū al-Ḥasan 2002
AJ	Flavius Josephus, *Antiquitates Judaicae* (Jewish Antiquities)
ASFF	Safaitic inscriptions recorded by S. Al-Abbadi at Qāʿ Fahadah and Tell al-Fāhdawī (published on OCIANA)
ASWS	Safaitic inscriptions published in Banī ʿAwād 1999
AWS	Safaitic inscriptions published in ʿAlūlū 1996
BJ	Flavius Josephus, *Bellum Judaicum* (Jewish War)
BS	Inscriptions recorded by M.C.A. Macdonald and A. Al-Manaser during the 2015 Badia Survey in north-eastern Jordan (published on OCIANA)
C	Safaitic inscriptions in *Corpus Inscriptionum Semiticarum. Pars V. Inscriptiones Saracenicae continens, Tomus 1. Inscriptiones Safaiticae.* Paris: Imprimerie nationale. (2 volumes). 1950-1951
CIS II	Nabataean inscriptions in *Corpus Inscriptionum Semiticarum. Pars II. Inscriptiones aramaicas continens*. Paris: Imprimerie nationale, 1889-1954.
CNG	Classical Numismatic Group
CSNS	Safaitic inscriptions in Clark 1979
DaJ ANA	Ancient North Arabian inscriptions recorded during the 2010-2017 seasons of the Dūmat al-Jandal Archaeological project
DaJ Nab	Nabataean inscriptions recorded during the 2010-2017 seasons of the Dūmat al-Jandal Archaeological project
DNWSI	Hoftijzer and Jongeling 1995
DRS	Cohen *et al.* 1970-1995
HaNSB	Safaitic inscriptions in Ḥarāḥšah 2010
HASI	Safaitic inscriptions recorded by A.Q. Al-Huṣan (published on OCIANA)
HIn	Harding 1971
IGLS XXI, 5.1	Greek and Latin inscriptions published in Bader 2009

Internet	Safaitic inscriptions whose photographs were found on Internet (published on OCIANA)
ISB	Safaitic inscriptions in Oxtoby 1968
Is.H	Safaitic inscriptions from Al-ʿĪsāwī recorded by Hussein Zeinaddin during the Safaitic Epigraphic Survey Programme (SESP) (published on OCIANA)
Is.M	Safaitic inscriptions from Al-ʿĪsāwī recorded by Michael C.A. Macdonald during the Safaitic Epigraphic Survey Programme (SESP) (published on OCIANA)
JSNab	Nabataean inscriptions in Jaussen and Savignac 1909-1922
KRS	Safaitic inscriptions recorded by G.M.H. King during the Basalt Desert Rescue Survey (published on OCIANA)
Lane	Lane 1863-1893
LP	Safaitic inscriptions in Littmann 1943
LPNab	Nabataean inscriptions in Littmann 1914
MIRP	Nabataean inscriptions in Milik and Starcky 1975
MM	Safaitic inscriptions from the Mafraq Museum which are published on OCIANA with the numbering system Al-Mafraq Museum 1-107
MNIN	Nabataean inscriptions in Milik 1958
MNT	Nabataean inscriptions published in Milik 1980
MP	Nabataean inscriptions from Petra published in Nehmé 2012
MSTJ	Safaitic inscriptions in Macdonald and Harding 1976
NEH	Safaitic and Nabataean inscriptions published in Al-Salameen, Al-Shdaifat and Harahsheh 2018
NNIA	Nabataean inscriptions published in Negev 1961
OCIANA	Online Corpus of the Inscriptions of Ancient North Arabia. http://krc.orient.ox.ac.uk/ociana/index.php
PPUAES IIIA	Greek and Latin inscriptions published in Littmann *et al.* 1921
RES	*Répertoire d'Épigraphie Sémitique*. Publié par la Commission du *Corpus Inscriptionum Semiticarum*, Académie des Inscriptions et Belles-Lettres. (8 volumes). Paris: Imprimerie Nationale, 1900-1968.
RQ.A	Safaitic inscriptions from Rijm Qaʿqūl A recorded by the Safaitic Epigraphic Survey Programme (SESP) (published on OCIANA)
RSIWU	Nabataean inscriptions published in Roche 2013
RWQ	Safaitic inscriptions published in Al-Rūsān 2004
SESP.U	Safaitic inscriptions from an unknown site recorded by the Safaitic Epigraphic Survey Programme (SESP) (published on OCIANA)
SIAM	Safaitic inscriptions in Macdonald 1980
SIJ	Safaitic inscriptions in Winnett 1957
SSAI	Nabataean inscriptions in Savignac 1933
TANI.Nab	Nabataean inscriptions in Al-Theeb 1993
ThMNN	Nabataean inscriptions in Al-Theeb 2010
ThNS	Nabataean inscriptions in Al-Theeb 2014
ThNUJ	Ancient North Arabian inscriptions in Al-Theeb 2003
WSRBZ.B	Nabataean inscriptions in Al-Theeb 2002
ThNTJ	Safaitic inscriptions from an unspecified site which are published on OCIANA
WH	Safaitic inscriptions in Winnett and Harding 1978

References

Ababneh, M.I. 2005. *Neue safaitische Inschriften und deren bildliche Darstellungen*. Aachen: Shaker Verlag.

Abū al-Ḥasan, Ḥ. 2002. *Nuquš liḥyaniyyah min minṭaqat al-ʿUlā. Dirāsah taḥlīliyyah muqārinah.* Riyadh: Wizārat al-maʿārif wakālat al-āṯār wa-ʾl-matāḥif.

Al-Housan, A.Q. 2017. A selection of Safaitic inscriptions from Al-Mafraq, Jordan: II. *Arabian Epigraphic Notes* 3, 19-46.

Al-Husan, A. and Al-Rawabdeh, N. 2018. Revolt of the Nabataean Damasī in the light of new epigraphic material evidence. *Acta Orientalia Academiae Scientiarum Hungaricae* 71, 467-478.

Al-Jallad, A. 2014. An ancient Arabian zodiac. The constellations in the Safaitic inscriptions, Part I. *Arabian Archaeology and Epigraphy* 25, 214-230.

Al-Jallad, A. 2015. *An outline of the grammar of the Safaitic inscriptions*. Leiden: Brill.

Al-Jallad, A. 2017a. Graeco-Arabica I: the southern Levant. In: Al-Jallad, A. (ed.). *Arabic in context. Celebrating 400 years of Arabic at Leiden University*. Leiden: Brill, 99-186.

Al-Jallad, A. 2017b. Marginal notes on and additions to *An outline of the grammar of the Safaitic inscriptions* (ssll 80; Leiden: Brill, 2015), with a supplement to the dictionary. *Arabian Epigraphic Notes* 3, 75-96.

Al-Jallad, A. In press. One *wāw* to rule them all: the origins and fate of wawation in Arabic and its orthography. In: Donner, F. and Hasselbach, R. (eds.). *Scripts and scripture*. Chicago, IL: University of Chicago Press.

Al-Jallad, A. and Al-Manaser, A. 2015. New epigraphica from Jordan I: a pre-Islamic Arabic inscription in Greek letters and a Greek inscription from north-eastern Jordan. *Arabian Epigraphic Notes* 1, 51-70.

Al-Jallad, A. and Macdonald, M.C.A. 2015. A few notes on the alleged occurrence of the group name 'Ghassan' in a Safaitic inscription. *Archiv für Orientforschung* 53, 152-157.

Al-Manaser, A. and Norris, J. 2019. Two more Nabataean inscriptions from the Syro-Jordanian ḥarrah desert. *Palestine Exploration Quarterly* 15, 69-86.
Al-Otaibi, F.M. 2011. *From Nabataea to Roman Arabia: acquisition or conquest?* Oxford: Archaeopress.
Al-Rūsān, M.M. 2004. *Nuqūš ṣafawīyah min Wādī Qiṣṣāb bi-'l-'urdun. Dirāsah maydāniyyah taḥlīliyyah muqārinah.* Riyadh: King Sa'ūd University (unpublished doctoral thesis).
Al-Salameen, Z., Al-Shdaifat, Y. and Harahsheh, R. 2018. Nabataean echoes in al-harrah: new evidence in light of recent field work. *Palestine Exploration Quarterly* 150, 60-79.
Al-Theeb, S.A. 1993. *Aramaic and Nabataean inscriptions from north-west Saudi Arabia.* Riyadh: King Fahd National Library.
Al-Theeb [Al-Ḏīyīb], S.A. 2002. *Nuqūš ğabal Umm Ğaḏāyiḏ al-nabaṭiyyah.* Riyadh: Maktabat al-Malik Fahd al-waṭaniyyah.
Al-Theeb [Al-Ḏīyīb], S.A. 2003. *Nuqūš ṯamūdiyyah ğadīdah min al-Ğawf. Al-mamlakat al-'arabiyyat al-sa'ūdiyyah.* Riyadh: Maktabat al-malik Fahd al-waṭaniyyah.
Al-Theeb [Al-Ḏīyīb], S.A. 2010. *Mudawwanat al-nuqūš al-nabaṭiyyah fī-'l-mamlakat al-'arabiyyat al-sa'ūdiyyah.* 2 vols. Riyadh: Dārat al-malik 'Abd al-'Azīz.
Al-Theeb [Al-Ḏīyīb], S.A. 2014. *Nuqūš mawqi' Sarmadā' muḥāfaẓah Taymā'.* Riyadh: Fahrasah maktabat al-malik Fahd al-waṭaniyyah/Ğāmi'ah al-malik Sa'ūd (Dirāsāt Āṯāriyyah 12).
'Alūlū, Ġ.M.Y. 1996. *Dirāsah nuqūš ṣafawīyyah ğadīdah min Wādī al-Sū' ğanūb Sūrīyyah.*, Irbid: Yarmouk University (unpublished MA thesis).
Bader, N. 2009. *Inscriptions grecques et latines de la Syrie. Tome XXI, Inscriptions de la Jordanie. Tome 5, La Jordanie du Nord-Est. Fascicule 1.* Beyrouth: IFPO
Banī 'Awād, 'A. R. 1999. *Dirāsah nuqūš ṣafawiyyah ğadīdah min ğanūb wādī sārah/al-bādiyyah al-'urdunniyyah aš-šamāliyyah.* Irbid: Yarmouk University (unpublished MA thesis).
La Bible de Jérusalem. Traduite en français sous la direction de l'École Biblique de Jérusalem, ed. 2009. Paris: Les Éditions du Cerf.
Bowersock, G.W. 1983. *Roman Arabia.* Cambridge, MA: Harvard University Press.
Cantineau, J. 1930-1932. *Le Nabatéen.* 2 vol. Paris: Librairie Ernest Leroux.
Clark, V.A. 1979. *A study of new Safaitic inscriptions from Jordan.* Ann Arbor, MI: University Microfilms International (unpublished doctoral thesis, University of Melbourne).
Cohen, D., Bron, F. and Lonnet, A. 1970-1995. *Dictionnaire des racines sémitiques ou attestées dans les langues sémitiques.* Fasc. 1-5. Paris: Mouton.
Gatier, P.L. 1988. Philadelphie et Gérasa du royaume nabatéen à la province d'Arabie. In : Gatier, P.L., Helly, B. and Rey-Coquais, J.P. (eds.). *Géographie historique au Proche-Orient: Syrie, Phénicie, Arabie, grecques, romaines, byzantines. Actes de la table ronde de Valbonne, 16-18 septembre 1985.* Paris: Éditions du CNRS, 159-170.
Graf, D. 1989. Rome and the Saracens: reassessing the nomadic menace. In: Fahd, T. (ed.). *L'Arabie préislamique et son environnement historique et culturel. Actes du Colloque de Strasbourg 24-27 juin 1987.* Leiden: Brill, 343-400.
Ḥarāḥšah, R.M.A. 2010. *Nuqūš ṣafā'iyyah min al-bādiyyat al-'Urduniyyah al-šamāliyyah al-šarqiyyah, dirāsah wa taḥlīl.* Amman: Ward Books.
Harding, G.L. 1971. *An index and concordance of pre-Islamic Arabian names and inscriptions.* Toronto: University of Toronto Press.
Heidet, L. 1903. Galaad. *Dictionnaire de la Bible* 3, 47-59.
Hoftijzer, J. and Jongeling, K. 1995. *Dictionary of the north-west Semitic inscriptions,* 2 vols. Leiden: E.J. Brill.
Jaussen, A. and Savignac, R. 1909-1922. *Mission archéologique en Arabie,* 5 vols. Paris: Leroux/Geuthner.
Jobling, W.J. 1982. Aqaba-Ma'an survey, Jan.-Feb. 1981. *Annual of the Department of Antiquities of Jordan* 26, 199-209.
Josephus, *Jewish Antiquities, Books 1-20.* Ed. and transl. Thackeray, H.St.J., Marcus, R. and Feldman, L.H. 1926-1965 (9 volumes). Cambridge, MA: Harvard University Press.
Josephus, *Jewish War, Books 1-7.* Ed. and transl. Thackeray, H.St.J. 1927-1928 (3 volumes). Cambridge, MA: Harvard University Press.
Kootstra, F. 2016. The language of the Taymanitic inscriptions and its classification. *Arabian Epigraphic Notes* 2, 67-140.
Kootstra, F. 2018. The phonemes ẓ and ṭ in the Dadanitic inscriptions. In: Nehmé, L. and Al-Jallad, A. (eds.). *To the madbar and back again. Studies in the languages, archaeology, and cultures of Arabia dedicated to Michael C.A. Macdonald.* Leiden: Brill, 202-217.
Lane, E.W. 1863-1893. *An Arabic-English lexicon, derived from the best and most copious eastern sources.* London: Williams and Norgate.
Littmann, E. 1914. *Nabataean inscriptions from the southern Ḥawrân. Publications of the Princeton University archaeological expeditions to Syria in 1904-1905 and 1909, Division IV, Semitic Inscriptions, Section A.* Leiden: Brill.
Littmann, E. 1943. *Safaïtic inscriptions. Syria. Publications of the Princeton University archaeological expeditions to Syria in 1904-1905 and 1909. Division IV, Section C.* Leiden: Brill.

Littmann, E., Magie Jr, D. and Stuart, D.R. 1921. *Syria. Publications of the Princeton University Archaeological Expeditions to Syria in 1904-5 and 1909. Division III: The Greek and Latin Inscriptions. Section A: Southern Syria*. Leiden: Brill.

MacDonald, B. 2000. *East of the Jordan: territories and sites of the Hebrew scriptures*. Boston, MA: American Schools of Oriental Research.

Macdonald, M.C.A. 1980. Safaitic inscriptions in the Amman Museum and other collections II. *Annual of the Department of Antiquities of Jordan* 24, 185-208.

Macdonald, M.C.A. 1992. The seasons and transhumance in the Safaitic inscriptions. *Journal of the Royal Asiatic Society* 2, 1-11.

Macdonald, M.C.A. 1993. Nomads and the Hawran in the late Hellenistic and Roman periods: a reassessment of the epigraphic evidence. *Syria* 70, 303-403.

Macdonald, M.C.A. 2000. Reflections on the linguistic map of pre-Islamic Arabia. *Arabian Archaeology and Epigraphy* 11, 28-79.

Macdonald, M.C.A. 2003. "Les Arabes en Syrie" or "la pénétration des Arabes en Syrie". A question of perceptions? *Topoi Supplement* 4, 303-318.

Macdonald, M.C.A. 2014. 'Romans go home'? Rome and other 'outsiders' as viewed from the Syro-Arabian desert. In: Dijkstra, J.H.F. and Fisher, G. (eds.). *Inside and out. Interactions between Rome and the peoples on the Arabian and Egyptian frontiers in Late Antiquity*. Leuven: Peeters, 145-163.

Macdonald, M.C.A. and Harding, G.J. 1976. More Safaitic texts from Jordan. *Annual of the Department of Antiquities of Jordan* 21, 119-133.

Macdonald, M.C.A., Al-Mu'azzin, L. and Nehmé, L. 1996. Les inscriptions safaïtiques de Syrie, cent quarante ans après leur découverte. *Comptes-rendus des Séances de l'Académie des Inscriptions et Belles-Lettres* 140, 435-494.

Meshorer, Y. 1975. *Nabataean coins*. Jerusalem: Hebrew University of Jerusalem.

Milik, J.T. 1958. Nouvelles inscriptions nabatéennes. *Syria* 35, 227-251.

Milik, J.T. 1980. La tribu des Bani ʿAmrat en Jordanie à l'époque grecque et romaine. *Annual of the Department of Antiquities of Jordan* 24, 41-54.

Milik, J.T. and Starcky, J. 1975. Inscriptions récemment découvertes à Pétra. *Annual of the Department of Antiquities of Jordan* 20, 111-130.

Negev, A. 1961. Nabatean inscriptions from ʿAvdat (Oboda). *Israel Exploration Journal* 11, 127-138.

Negev, A. 1991. *Personal names in the Nabataean realm*. Jerusalem: Hebrew University of Jerusalem.

Nehmé, L. 2012. *Pétra: atlas archéologique et épigraphique. 1. De Bāb as-Sīq au Wādī al-Farasah*. Paris: Académie des Inscriptions et Belles-Lettres.

Nehmé, L. 2015. *Strategoi* in the Nabataean kingdom: a reflection of central places? *Arabian Epigraphic Notes* 1, 103-122.

Norris, J. 2018. A survey of the Ancient North Arabian inscriptions from the Dūmat al-Jandal area (Saudi Arabia). In: Macdonald, M.C.A. (ed.). *Languages, scripts and their uses in Ancient North Arabia*. Oxford: Archaeopress, 71-93.

Ottosson, M. 1992. Gilead. *The Anchor Bible Dictionary* 2, 1020-1022.

Oxtoby W.G. 1968. *Some inscriptions of the Safaitic bedouin*. New Haven, CT: American Oriental Society.

Roche, M.J. 2013. Le sanctuaire d'Isis du Wādī Abū ʿUllayqa, au sud de Pétra. *Studies in the History and Archaeology of Jordan* 11, 543-555.

Ryckmans, G. 1942. Les inscriptions safaïtiques relevées par M. et Mme Dunand. *Comptes-rendus des Séances de l'Académie des Inscriptions et Belles-Lettres* 86, 127-136.

Ryckmans, G. 1951, Les noms de parenté en safaïtique. *Revue Biblique* 58, 377-392.

Sartre, M. 1985. *Bostra: des origines à l'Islam*. Paris: P. Geuthner.

Sartre, M. 2001. *D'Alexandre à Zénobie. Histoire du Levant antique. IVe siècle av. J.-C. – IIIe siècle ap. J.-C.* Paris, Fayard.

Savignac, R. 1933. Le sanctuaire d'Allat à Iram (1). *Revue Biblique* 42, 405-422.

Savignac, R., and Starcky, J. 1957. Une inscription nabatéenne provenant du Djôf. *Revue Biblique* 64, 196-217.

Scharrer, U. 2010. The problem of nomadic allies in the Near East. In: Facella, M. and Kaizer, T. (eds.). *Kingdoms and principalities in the Roman Near East*. Stuttgart: F. Steiner, 241-335.

Schulthess, F. 1903. *Lexicon syropalaestinum. Adiuvante Academia Literarum Regia Borussica*. Berlin: Reimer.

Starcky, J. 1966. Pétra et la Nabatène. *Supplément au Dictionnaire de la Bible* 7, 886-1017.

Strabo, *Geography*. Ed. and transl. Jones, H.L. 1961, 8 vol. Cambridge, MA: Harvard University Press.

Winnett, F.V. 1957. *Safaitic inscriptions from Jordan*. Toronto: University of Toronto Press.

Winnett, F.V. 1973. The revolt of Damasī: Safaitic and Nabataean evidence. *Bulletin of the American Schools of Oriental Research* 211, 54-57.

Winnett, F.V. and Harding, G.L. 1978. *Inscriptions from fifty Safaitic cairns*. Toronto: University of Toronto Press.

Two new Safaitic inscriptions and the Arabic and Semitic plural demonstrative base

Phillip W. Stokes

Abstract

This article presents two inscriptions discovered in north-east Jordan, published here for the first time. These two inscriptions provide the first unambiguous attestations of a plural demonstrative pronoun *ʾly* in the pre-Islamic epigraphic corpora. Following a discussion of the inscriptions themselves, as well as the morphology and syntax of the forms as attested, I will situate the forms in the context of attested ones in other Arabic corpora. The ultimate goal is to provide a reconstruction of the proto-Arabic plural demonstrative base, which I reconstruct as **ʾulay*.

Keywords: Safaitic, epigraphy, philology Arabic linguistics, historical linguistics, language reconstruction

Introduction

The archaeology and epigraphy of the Jordanian *badia* continues to produce significant discoveries that fundamentally change our understanding of the Late Antique Near East, the peoples who populated it, and the languages they used to express themselves. The largest corpus of inscriptions written in a North Arabian script is that of the Safaitic corpus (Macdonald 2000, 35). Since Ahmad Al-Jallad's (2015) foundational study of the grammar of the varieties written in this script, the potential linguistic importance of these inscriptions has increasingly become a topic of study (*e.g.* Van Putten 2017; Al-Jallad 2018). This article presents two inscriptions discovered at Marabb Aš-Šurafāt in the *badia* of north-east Jordan, published here for the first time. These two inscriptions provide the first unambiguous attestations of a plural demonstrative pronoun in the pre-Islamic epigraphic corpora. Following a discussion of the inscriptions themselves, I will discuss interesting aspects of the attestations of the newly-attested demonstratives. I will then propose a reconstruction for the proto-Arabic plural demonstrative base, contextualising the attested Safaitic form within the history of both Arabic and the other Semitic languages. It is hoped that such a contribution will further illustrate the potential of the Safaitic inscriptions for illuminating the linguistic history of Arabic and Semitic, in addition to the social and political facets of life in the Late Antique Near East.

In: Peter M. M. G. Akkermans (ed.) 2020: *Landscapes of Survival - The Archaeology and Epigraphy of Jordan's North-Eastern Desert and Beyond*, Sidestone Press (Leiden), pp. 391-398.

Figure 1. Inscription 1: SMS 1

Inscriptions

The two inscriptions published here were discovered in Wadi Marabb Aš-Šurafāt in north-eastern Jordan during a survey conducted by members of the OCIANA team and the Leiden Center for the Study of Ancient Arabia in April, 2015.[1]

Inscription 1: SMS1

l grmt(n) bn xr w-gls^1 ʾly ʾ-rgm f-wgd s^{1}fr ʾb-h f-ndm
"By Grmt(n) son of Xr, and he halted/stopped at these cairns and found the inscription of his father, and he grieved"

The script in which the inscription (Fig. 1) is written is unremarkable. The surface of the rock is somewhat faded, and there are a few spots where damage obscures part of a letter, but, overall, the writing is clear and easily read. The main aspect of the inscription around which there is uncertainty is the name of the author of the inscription, *Grmt*. Following the *t* glyph, it is possible to read a *n*, but it is also possible that the vertical mark is damage to the rock face. If it is a *n*, then this name would attest an occurrence of *nunation* (Arabic *tanwīn*), otherwise rare in the Safaitic corpus (Al-Jallad 2015, §4.5.1).

The inscription is composed of Genealogy + Inscription Finding and Grieving, characteristic of so many Safaitic inscriptions (Al-Jallad 2015, §22.1, 5). The phrase *w-gls^1 ʾly ʾ-rgm*, however, is so far unique in the Safaitic corpus and, thus, requires discussion. The verb *gls^1* "he sat, halted," occurs frequently in Safaitic. There are a few possible interpretations of the following word, *ʾly*. It is possible that it represents underlying *ʾilay*, and is cognate with the preposition *ʾilā* "to, toward," well-known from later forms of Arabic. In such a reading, the phrase *gls^1 ʾly ʾ-rgm* could be understand as "he halted in front of the cairns." Further, in Classical Arabic (henceforth ClAr), the verb *jalasa* does occur with *ʾilā* with the meaning "he sat by" (Lane 1863-1893, I, 443, column 3). However, while this is certainly possible, two factors suggest otherwise. First, the verb *gls^1* does not normally occur with a preposition: cf. JaS 159.2 *l hnʾ bn ʾk{l}{l}t {f} gls^1 ḥlt f h lt w ds^2{r} s^{1}lm* "By Hnʾ son of {Kllt} {and so} he camped ḥlt and so O Lt and {Ds^{2}r} [grant] security."[2] Further, in Safaitic, verbs of motion in general often do not take prepositions; rather, it is probable that the accusative case functioned to mark location and goals of travel (Al-Jallad 2015, §4.6.1). Second, and more problematic for a reading of *ʾilā* here, is the fact that the cognate of *ʾilā* in Safaitic is everywhere else *ʾl*, and not ***ʾly* (*ibid.*, 144). This is also the case for, *e.g.*, ClAr *ʿalā*, which in Safaitic is *ʿl* and not ***ʿly* (*ibid.*, 148-149).

1 The survey was AHRC-funded. Participants in the survey were Michael C.E. Macdonald, Ali Al-Manaser and Chiara Della Puppa.

2 Inscription accessed on OCIANA online database, 9/14/2018: http://krc.orient.ox.ac.uk/ociana/corpus/pages/OCIANA_0010230.html

Figure 2. Inscription 2: SMS2

Another possibility, argued for here, is to read *ʾly* as a plural proximal demonstrative. That the final glide [y] strongly suggests that the final sequence is a diphthong, and that *ʾly* is written to represent underlying /ʾulay(v)/. Reading *ʾly* fits the context nicely, and produces "and he halted at these cairns." If this is correct, a further of its occurrence here is noteworthy, namely that a plural demonstrative is used with an inanimate plural noun. Al-Jallad (2015, §6.1) notes the paucity of clear examples of adjectival agreement, but reports that plural non-human nouns perhaps trigger singular agreement. In early ClAr (Fischer 2002, §113-4), as well as many modern dialects, however, the agreement patterns are more complex, with some non-human nouns triggering plural agreement (cf. also Brustad 2000, §2.3, who argues for a cline of individuation that governs the choice of singular or plural agreement). If the reading proposed here is correct, it suggests that in Safaitic, as in later forms of Arabic, both were possible.

Inscription 2: SMS2

l ḥrs^1 bn ʾglḥ w-wgm ʿl bdbl trḥ w-l-h ʾ-ʾfs ʾly w-s^{2}ʾm
"By Ḥrs^1 son of ʾglḥ and he grieved for Bdbl who had perished, and for him are these monuments; and he traveled north"

The present inscription (Fig. 2) is the third of three inscriptions on the same rock face, with two others written above it. The rock face upon which this inscription was carved, along with two others above it, is relatively smooth and undamaged, and the inscription is thus clear and the reading quite certain. While the patina is clear and even throughout the inscription, the patina of the last two words of the inscription, *ʾly w-s^{2}ʾm* are noticeably thinner than the previous words. Nevertheless, there is no significant difference between the letter shapes that suggests a different hand. For that reason, it seems probable that the same hand produced the entire inscription. Further, the similarities between the hand that produced the third inscriptions, and the one that produced the other two are so great that it seems likely that the same person inscribed all three.

Two aspects of this inscription are noteworthy from grammatical and lexicographic perspectives. First, the phrase *ʾ-ʾfs^1 ʾly* is unique so far in the Safaitic inscriptions. The word *ʾ-ʾfs* is a combination of the article ʾ, and the plural of *nfs^1* "funerary monument; stele" (Al-Jallad 2015, 330). I have read the following word, *ʾly*, as another example of the plural proximal demonstrative *ʾulay(v)*. There are several pieces of evidence that support such a reading. First, the phrase *h-nfs^1* "the funerary monument" is regularly followed by a *t*, *i.e.*, *nfs^{1}t*. Al-Jallad (2015, 82) has argued, based on the fact that in no attested Semitic language does the cognate of *nfs^1* occur with the feminine *at*, but that it is treated as feminine elsewhere (in, *e.g.*, the Namārah inscription), that this *t* in Safaitic is to be interpreted as a feminine singular demonstrative. The attestation of the plural *ʾfs^1* + plural demonstrative *ʾly* in this inscription would thus directly parallel these other, attested, singular forms. The fact that the demonstrative is post-nominal, instead of the more common pre-nominal, could either reflect a relatively restricted construction, limited to phrases with *nfs*^{1}, *mdbr* and *ʾrḍ* (*ibid.*, §4.9 for a discussion of these), or perhaps it reflects a development like what took place in, *e.g.*, Cairene Arabic.

Another first from this inscription is the apparent use of *s*^{2}*ʾm* as a verb "to go north." The noun *s*^{2}*ʾm* "north" is attested in Safaitic (AbNS 1128; Al-Jallad 2015, 343), but not, so far, as a verb. The context of the present inscription, following the phrase *ʾ-ʾfs*^{1} *ʾly* "these monuments," and preceded by the conjunction *w* is incompatible with a nominal usage. A verb *šaʾʾama* "He made them to go, or journey, to *aš-Šaʾm*," is attested in ClAr (Lane 1863-1893, II, 1400, column 1). In this inscription, however, there is nothing to suggest a causative meaning; thus I have translated it simply "he traveled north."

Plural demonstratives in Arabic

A number of plural demonstrative forms are attested in early Islamic Arabic corpora, including the Qurʾān and ClAr. These include the following (Wright 2005, I, §339-344):

	Classical Pronunciation	Spelling
Near Pl. Dem.	*ʾulā*	ألى \ أولى
	ʾulā / hāʾulā	اولا \ هولا \ هاولا
	ʾulāʾi / hāʾulāʾi	أولاء \هولاء \ هاولاء
Far Pl. Dem.	*ʾulāka*	أولاك \ ألاك
	ʾulāʾika	ألانك \ أولانك
	ʾulālika	ألالك \ أولالك

Table 1. Plural proximal and distal demonstratives from ClAr and the Qurʾān.

Several aspects of the orthography of these forms in the Arabic script deserve attention. First, the *u* vowel is realised short in virtually all traditions behind ClAr, despite being written long in the majority of cases. Second, in all the forms in ClAr, and indeed virtually all modern dialects as well, the liquid [l] lacks gemination (Wright 2005, I, §339; Fischer and Jastrow 1980, 82; Magidow 2016). This is contrary to most cognate forms attested in the other Semitic languages (Hasselbach 2007; on which, see further below). Third, in some cases, long *ā* is spelled ى, that is, with what became in classical orthography the *alif maqṣūra*. It is clear, however, that instances in which ClAr *ā* are spelled with ى correspond to etymological diphthongs **ay*, and were still realised as either *ay* or *ē* in other pre- and early Islamic dialects such as, *e.g.*, southern Levantine Arabic and the Qurʾān (Rabin 1951, 115ff.; Al-Jallad 2017, §5.1; Van Putten 2017, §6). Thus the use of ى in some early ClAr texts suggests an underlying /ay/ or /ē/ in the final syllable of these forms. The majority of cases, in which the demonstrative is spelled with ا, are read /ā/. With Rabin (1951, 153ff.), it is probable that the original form ended in a diphthong /ay/. More common forms with *ā* in ClAr and modern Arabic should be understood in the context of an apparent shift word-final *ay* > *ā*, also witnessed in, *e.g.*, prepositions *ʿalā* (< **ʿalay*) "against" and *ʾilā* (< **ʾilā*) "to, toward." Against Rabin (*ibid.*), however, the proto-Arabic form can be reconstructed with only a single [l]: **ʾulay* (see below for further discussion).

While the foregoing argument is, I argue, the most parsimonious interpretation of the ClAr and Qurʾānic evidence, without the direct evidence afforded in these Safaitic inscriptions, the argument would be, at best, speculative, resting on a few instances in which ClAr *ʾulā* is spelled with ى. The Safaitic evidence now offers proof of the realization of the final sequence as a diphthong. The spelling of the plural demonstratives with [y], which in Safaitic must represent a consonant and never represents long *ā*, entails that the demonstrative was either realised as a diphthong /ay/ or /āy/, or possibly a triphthong /ayv/ or /āyv/. This further corroborates the phonetic reality underlying the use of ى to write final **ay* sequences in, *e.g.*, the Qurʾān and other early Arabic texts and inscriptions.

I have argued elsewhere (Stokes 2018) that the longer forms, *e.g.* *(hā)ʾulāʾi* and *ʾulāʾika*, are the result of a lengthening of **ay* > *āy*, possibly based on the clearer connection between **āy* and plurals, as in plural patterns like, *e.g.*, **ʾafʿilāyv* and **fuʿalāyv* (on the reconstruction of these patterns, see Van Putten 2018, 211). When pre-nominal, the heavy syllable *āy* + *l* (the definite article) led speakers to insert an epenthetic vowel *i*: *āy* + *l* > *āyi* + *l*. The ClAr shift of *āY/Wi* > *āʾi*, as in, *e.g.*, *qāyim* > *qāʾim* (Brockelmann 1908, 138; Al-Jallad 2014, 11-12) operated here as well, and led to the surface forms *ʾulāʾi(ka)* and *allāʾi*. Whether the Safaitic orthography represents the proto-Arabic form **ʾulay* unchanged, or rather a ClAr-like lengthened form **ʾulāy*, is unclear due to the limitations of the script and its lack of writing any vowels, long or short. It should be further noted that, because the Safaitic script does not indicate pure vowels, nor gemination, we cannot be sure of the quality of the vowel, nor whether the Safaitic *ʾly* represents an underlying single *l* /ʾulay/ or geminate *ll* /ʾullay/.

The Arabic plural demonstrative in the context of Semitic evidence

The Arabic data discussed above, and the reconstruction of **ʾulay* as the proto-Arabic plural demonstrative base, also has bearing on the discussion of the comparative Semitic data:

	Proximal Dem	Distal Dem
Old Babylonian	NA	msg. *ullûm* / fsg. *ullītum* mpl. *ullûtum* / fpl. *ulliātum*
Biblical Hebrew	*ʾēlle* (rarely *hāʾēl*)	
Biblical Aramaic	*ʾellē*	*ʾillēk*
Gəʿəz	*ʾəllu* / *ʾəllontu* *ʾəllā* / *ʾəllāntu*	*ʾəlləku*

Table 2. Sample demonstratives based on **ʾVl(lV)* in Semitic languages.

Scholars have largely agreed on several parts of the reconstruction of the proto-Semitic form. Regarding the initial vowel, scholars have typically reconstructed it either as short *u* on the basis of Arabic and Akkadian (cf. Barth 1913, 118-119), or as a short high vowel, as some languages, *e.g.* Akkadian and Arabic, attest short *u*, but others, *e.g.* Hebrew and Aramaic, attest short *i* (Hasselbach 2007, §3; Huehnergard and Pat-El 2018, 198).

Further, most languages attest long forms, in which the liquid *l* is geminate and followed by a vowel. The rare example of a shortened form, as in Hebrew *hāʾēl*, does not end in a vowel. On this basis, Hasselbach (2007, 23) reconstructs proto-Semitic **ʾVl(li)*, such that the long forms represent the proto-form **ʾVl* and a suffixed demonstrative element **li*, which is found in, *e.g.*, Arabic *ḏālika*. As I have noted elsewhere, however, Hebrew *ʾellē* and Aramaic *ʾillēk* and *ʾillēn*, cannot have come from **ʾVlli*, because, in both languages, short final vowels were lost (Stokes 2018). The Arabic evidence, confirmed by the Safaitic attestations published here, confirm that, at least for Central Semitic, these demonstratives ended in **ay*. In the Arabic data, further, forms with a single *l* end in a vowel – the diphthong *ay* or, more typically, the long vowel *ā* – and there is no equivalent *ʾVllay* / *ʾVllā* demonstrative. Thus the attested forms in Arabic are unique, combining the single *l* of the short form and a final vowel. This mismatch has, so far, not been widely acknowledged, nor satisfactorily explained.

Regarding the status of the liquid *l*, the ubiquity of the single *l* makes a reconstruction of the proto-Arabic plural demonstrative base as **ʾulay* secure. However, given the comparative evidence, Rabin was justified in expecting proto-Arabic **ʾullay*. Rabin's explanation (1951, 153ff.) relies the grammarians' reports of variation in the realization of the demonstratives, which could be read with either a short *u* or long *ū*. He thus concludes that there was originally a **ʾull* / *ʾūl* distinction, with the latter apparently being the progenitor of the single *l* forms. The problem with this, however, is that no such contrast is attested. Rather, if we accept the grammarians' reports, the contrast is still between *ʾul* /*ʾūl*, and *not ʾull* / *ʾūl*.

I suggest that there is only one plausible way to derive *ʾulay* from an earlier *ʾullay*, namely via analogy with the plural relative pronouns. While what became the default ClAr relative pronouns were based on the combination of the article *ʾal* + asseverative *la* + *ḏī*, and the plural series was created analogically by adding nominal plural morphology *allaḏīna*, other plural forms are attested:

Plural relative pronouns	*allāyi* / *allāʾi*	ألى
	ʾallāʾī	ألاءي
	al-ʾulā	الألى \ الأولى
	allāʾūna / *allāʾīna*	الأؤون \ الأئين

Table 3. Plural relatives attested in Classic Arabic grammatical tradition.

Huehnergard and Pat-El (2018, 196) have recently argued that **ʾVl*-based relatives are secondary, not original as previously assumed, one of several developments that constitute the *Sprachbund* of Arabian and Ethio-Semitic languages. I have argued that these relative forms are all based on the combination of the definite article *al* + the demonstrative *ʾulā* (< **ʾulay*). Except for the form *al-ʾulā*, the others attest the loss of unaccented short *u* in open syllables (Fischer 2002, §49 d). This resulted in forms like *allāʾi*, *etc.* It is possible that ClAr *ʾulā* and *al-ʾulā* were in fact originally **ʾullā* and **al-ʾullā*. Then, following the loss of open unaccented *u*, the relative forms became *allā*, *allāʾi*, *etc.* Since speakers recognised the first component as the definite article *al*, they backformed demonstratives, which lacked the definite article, as forms with a single *l*: *allāʾi* but *ʾulāʾi*.

Conclusion

This paper publishes for the first time two Safaitic inscriptions from Marabb Aš-Šurafāt in north-eastern Jordan, which provide evidence for the first attestations of the plural proximal demonstrative in the corpus. Based on the form of the two instances of the demonstrative, I argued that we can now securely reconstruct **ʾulay* as the proto-Arabic demonstrative. Further, the Arabic data was contextualised with other Semitic evidence, and several unsolved issues in the comparative Semitic data were addressed. It is hoped that this contribution, in addition to filling in the grammar of the Arabic varieties written in the Safaitic script, provides further demonstration of the potential value of Safaitic for illuminating the history of Arabic, and its relationship to other Semitic languages.

Sigla

JaS	Safaitic inscriptions published on OCIANA from ʿArʿar, Saudi Arabia, in the collection of I.A. al-Awshan
AbaNS	Ababneh, M.I. 2005. *Neue safaitische Inschriften und deren bildliche Darstellungen.* SSHB 6. Aachen: Shaker Verlag.

References

Al-Jallad, A. 2014. On the genetic background of the Rbbl bn Hfʿm grave inscription at Qaryat al-Fāw. *Bulletin of the School of Oriental and African Studies* 77, 1-21.

Al-Jallad, A. 2015. *An outline of the grammar of the Safaitic inscriptions*. Leiden: Brill.

Al-Jallad, A. 2018. The earliest stages of Arabic and its linguistic classification. In: Benmamoun, E. and Bassiouney, R. (eds.). *The Routledge handbook of Arabic linguistics*. London: Routledge, 315-331.

Barth, J. 1913. *Die Pronominalbildung in den semitischen Sprachen*. Leipzig: Hinrichs.

Brockelmann, C. 1908. *Grundriss der vergleichenden Grammatik der semitischen Sprachen. I. Band: Laut- und Formenlehre*. Berlin: Von Reuther & Reichard.

Brustad, K. 2000. *The syntax of spoken Arabic*. Georgetown: Georgetown University Press.

Fischer, W. 2002. *A grammar of Classical Arabic.* Third revised edition, translated from the German by Jonathan Rodgers. New Haven, CT: Yale University Press.

Fischer, W. and Jastrow, O. (eds.). 1980. *Handbuch der arabischen Dialekte*. Wiesbaden: Harrasowitz Verlag.

Hasselbach, R. 2007. Demonstratives in Semitic. *Journal of the American Oriental Society* 127, 1-27.

Huehnergard, J. and Pat-El, N. 2018. The origin of the Semitic relative marker. *Bulletin of the School of Oriental and African Studies* 81, 191-204.

Lane, E.W. 1863-1893. *An Arabic-English lexicon*. London: Williams & Norgate.

Macdonald, M. 2000. Reflections on the linguistic map of pre-Islamic Arabia. *Arabian Archaeology and Epigraphy* 11, 28-79.

Magidow, A. 2016. Diachronic dialect classification with demonstratives. *Al-ʿArabiyya: Journal of the American Association of Teachers of Arabic* 49, 91-115.

Rabin, Ch. 1951. *Ancient West-Arabian*. London: Taylor's Press.

Stokes, P. 2018. The plural demonstratives and relatives based on *ʾVl in Arabic and the origin of dialectal *illī*. In: Birnstiel, D. and Pat-El, N. (eds.). *Re-engaging comparative Semitic and Arabic studies*. Wiesbaden: Harrassowitz Verlag, 131-154.

Van Putten, M. 2017. The development of the triphthongs in Quranic and Classical Arabic. *Arabian Epigraphic Notes* 3, 47-74.

Van Putten, M. 2018. The feminine endings *-ay and *-āy in Semitic and Berber. *Bulletin of the School of Oriental and African Studies* 81, 205-225.

Wright, W. 2005 [reprint of 1896]. *Arabic grammar, revised by W. Robertson Smith and M. De Goeje*. Mineola, NY: Dover Publications.